Entropy Applications in Environmental and Water Engineering

Entropy Applications in Environmental and Water Engineering

Special Issue Editors

Huijuan Cui
Bellie Sivakumar
Vijay P. Singh

MDPI • Basel • Beijing • Wuhan • Barcelona • Belgrade

MDPI

Special Issue Editors

Huijuan Cui
Chinese Academy of Sciences
China

Bellie Sivakumar
The University of New South Wales, Australia
Indian Institute of Technology Bombay, India
Tsinghua University, China

Vijay P. Singh
Texas A&M University
USA

Editorial Office
MDPI
St. Alban-Anlage 66
Basel, Switzerland

This is a reprint of articles from the Special Issue published online in the open access journal *Entropy* (ISSN 1099-4300) from 2017 to 2018 (available at: http://www.mdpi.com/journal/entropy/special_issues/water_engineering)

For citation purposes, cite each article independently as indicated on the article page online and as indicated below:

LastName, A.A.; LastName, B.B.; LastName, C.C. Article Title. *Journal Name* **Year**, *Article Number, Page Range.*

ISBN 978-3-03897-222-8 (Pbk)
ISBN 978-3-03897-223-5 (PDF)

Contents

About the Special Issue Editors

Huijuan Cui, Associate Professor, was born in Shenyang, China, in 1986. She graduated with a B.S. degree in Hydraulic Engineering from Tsinghua University in 2009, and obtained her M.S. and Ph.D. in Water Management and Hydrological Sciences from Texas A&M University in 2011 and 2015, respectively, supervised by Prof. Vijay P. Singh. She is now an Associate Professor at the Institute of Geographical Sciences and Natural Resources Research, Chinese Academy of Sciences. Her research interests are stochastic hydrology, hydrological time series analysis, streamflow forecasting, climate change, and entropy-based hydrological modeling. She has published more than 30 scientific papers in international journals.

Bellie Sivakumar, Associate Professor, was born in Tamil Nadu, India in 1970. He graduated with a B.E. in Civil from Bharathiar University, India in 1992. He obtained an M.E. degree in Hydrology and Water Resources from Anna University, India in 1994, and a Ph.D. degree from the National University of Singapore in 1999. His research interests are in the fields of hydrology, hydroclimatology, and water resources. He has published extensively, including two books, over 150 articles in peer-reviewed international journals, and many book chapters. He is the Editor-in-Chief of *Hydrology*, an Associate Editor of several other international journals in hydrology and water resources, and also the sole/lead editor of seven journal Special Issues. He is a member of several professional societies/technical committees, and has also served as the President (currently Vice-President) of the Hydrological Sciences (HS) section of the Asia-Oceania Geosciences Society (AOGS).

Vijay P. Singh is a University Distinguished Professor, a Regents Professor, and Caroline and William N. Lehrer Distinguished Chair in Water Engineering at Texas A&M University. He received his B.S., M.S., Ph.D., and D.Sc. degrees in engineering. He is a registered professional engineer, a registered professional hydrologist, and an Honorary diplomate of ASCE-AAWRE. He has published more than 1075 journal articles, 28 textbooks, 67 edited reference books, 105 book chapters, and 315 conference papers in the areas of hydrology and water resources. He has received more than 90 national and international awards. He is a member of 10 international science/engineering academies. He has served as President of the American Institute of Hydrology (AIH) and is currently President-Elect of the American Academy of Water Resources Engineering. . He is editor-in-chief of three journals and two book series and serves on the editorial boards of more than 25 journals and three book series.

entropy

MDPI

Editorial

Entropy Applications in Environmental and Water Engineering

Huijuan Cui [1],*, Bellie Sivakumar [2,3,4] and Vijay P. Singh [5]

[1] Key Laboratory of Land Surface Pattern and Simulation, Institute of Geographic Sciences and Natural Resources Research, Chinese Academy of Sciences, Beijing 100101, China
[2] UNSW Water Research Centre, School of Civil and Environmental Engineering, The University of New South Wales, Sydney, NSW 2052, Australia; s.bellie@unsw.edu.au
[3] Department of Civil Engineering, Indian Institute of Technology Bombay, Powai, Mumbai, Maharashtra 400076, India
[4] State Key Laboratory of Hydroscience and Engineering, Tsinghua University, Beijing 100084, China
[5] Department of Biological and Agricultural Engineering & Zachry Department of Civil Engineering, Texas A and M University, College Station, TX 77843-2117, USA; vsiingh@tamu.edu
* Correspondence: cuihj@igsnrr.ac.cn

Received: 9 August 2018; Accepted: 9 August 2018; Published: 10 August 2018

Keywords: entropy theory; complex systems; hydraulics; hydrology; water engineering;environmental engineering

---

Entropy theory originated from the second law of thermodynamics, and its extension to information theory became a versatile tool for modeling complex systems and associated problems. Entropy has found applications in a wide range of problems in earth, environmental, and geographical sciences. This special issue focuses on the applications of entropy theory in environmental and water engineering.

Entropy is considered as a measure of uncertainty or the amount of information gained through measurements of a random variable. Baran et al. [1] defined entropy as an invariant measure function and extended the assessment of uncertainty. They stated that entropy did not mean an absolute measure of information, but as a measure of the variation of information, which intends to help solve information-related problems in hydrologic monitoring.

Based on the review of entropy modeling in water engineering by Singh [2], applications of entropy theory can be classified into three groups: (1) statistical or empirical, (2) physical, and (3) mixed. The first group focuses on probability determination and requires entropy maximization, including frequency analysis, parameter estimation, network evaluation and design, spatial and inverse spatial analysis, flow forecasting, and complexity analysis, and clustering. The second group involves deriving physical relations either in time or in space, such as rainfall-runoff modeling, infiltration, soil moisture, velocity distribution, and flow duration curve, among others. For instance, Zhang et al. [3] showed how the entropy parameter derived from the entropy-based flow duration curve is linked to the drainage area, impacted by reservoir operation, and possibly climate change. The third group is a mixture of the above two and includes applications such as the reliability of water distribution systems. Numerous examples of these can be found in the review article on the Tsallis entropy, by Singh et al. [4].

Applications in the first category are common in hydrology, and many studies in this special issue belong to this category. Using the principle of maximum entropy (POME), the four-parameter exponential gamma distribution, generalized gamma distribution, and generalized beta distribution were derived for flood frequency analysis in [5–7]. Chen et al. [8] showed that entropy-based generalized distributions can further be used for the analysis of extreme rainfall with Bayesian technique.

Keum et al. [9] reviewed applications of entropy in water monitoring network design, including precipitation, streamflow and water level, water quality, soil moisture, and groundwater network. The network designed by Yeh et al. [10] showed how to optimize the rainfall network with both radar and entropy. Santonastaso et al. [11] introduced flow entropy as a measure of network redundancy and a proxy of reliability in optimal network design procedures, which can identify the tradeoff between network cost and robustness. Besides water distribution networks, entropy was used to develop an integrated optimization model for the spatial optimization of agricultural land use based on crop suitability, spatial distribution of population density, and agricultural land use data [12]. Similar to optimization, entropy was also used to determine weights of evaluating indicators in a fuzzy system [13,14] and can be applied in combined forecasting of rainfall [15].

Applications in this special issue have used several different entropy formulations, such as the Shannon, Tsallis, Rényi, Burg, Kolmogorov, Kapur, configurational, and relative entropies, which can be derived in time, space or frequency domain. The sample entropy was used to investigate streamflow and water level complexity of the Poyang Lake over multiple time-scales [16]. The connection entropy was applied to establish a water resources vulnerability framework [17]. The generalized space q-entropy was employed for spatial scaling and complexity properties of Amazonian radar rainfall fields [18]. The Kolmogorov complexity and the Shannon entropy were combined to evaluate the randomness of turbulence [19]. Cheng et al. [20] employed several entropy measures, such as intensity entropy, apportionment entropy, and marginal entropy to investigate spatial and temporal precipitation variability. Defined in frequency or spectral power domain, entropy can be used for spectral analysis. In this way, entropy can be used in time series analysis and forecasting and, hence, for characterizing stochastic and periodic patterns [21].

Mutual information is a measure of mutual dependence between two variables and can be determined from marginal and joint entropies. It is an efficient tool to investigate linear or non-linear interactions, such as the relationship between vegetation pattern and hydro-meteorological elements [22], the relationship between annual streamflow, extreme precipitation and ENSO (El Niño–Southern Oscillation) [23], and the relationship between soil water content and its influencing factors [24].

More recently, entropy-based concepts have been coupled with other theories, including copula, wavelets, and ensemble filter, to study various issues associated with environmental and water resources systems. Guo et al. [25] developed a coupled maximum entropy-copula method for hydrologic risk analysis through deriving bivariate return periods, risk, reliability, and bivariate design events, which shows that the maximum entropy theory is beneficial for improving the performance of copulas. As a result, the distribution derived by the maximum entropy-copula model outperforms the conventional distributions for the probabilistic modeling of floods and extreme precipitation events. Foroozand et al. [26] combined entropy with ensemble filter method to evaluate model performance, which can mitigate the computational cost of the bootstrap aggregating method. The entropy concept was linked to the notion of elasticity to assess catchment resilience, which determined the changes in mean annual runoff [27].

Other than the above probabilistic entropy, the classical thermodynamic entropy concept was also visited in the special issue, by Koutsoyiannis [28], in the entropy production. Entropy production was explored within stochastics in logarithmic time, related to model identification and empirical fitting, which was applied to an extraordinarily long time series of turbulent velocity and showed how a parsimonious stochastic model can be identified and fitted.

The above contributions to this special issue show the enormous scope and potential of entropy theory in advancing research in the field of environmental and water engineering, including establishing and explaining physical connections between theory and reality.

Acknowledgments: We express our thanks to the authors of the contributions of this special issue, and to the journal *Entropy* and MDPI for their support during this work.

Conflicts of Interest: The authors declare no conflict of interest.

References

1. Baran, T.; Harmancioglu, N.B.; Cetinkaya, C.P.; Barbaros, F. An Extension to the Revised Approach in the Assessment of Informational Entropy. *Entropy* **2017**, *19*, 634. [CrossRef]
2. Singh, V.P. *Entropy Theory and Its application in Environmental and Water Engineering*; Wiley-Blackwell: Hoboken, NJ, USA, 2013; p. 640.
3. Zhang, Y.; Singh, V.P.; Byrd, A.R. Entropy Parameter *M* in Modeling a Flow Duration Curve. *Entropy* **2017**, *19*, 654. [CrossRef]
4. Singh, V.P.; Sivakumar, B.; Cui, H. Tsallis Entropy Theory for Modeling in Water Engineering: A Review. *Entropy* **2017**, *19*, 641. [CrossRef]
5. Song, S.; Song, X.; Kang, Y. Entropy-Based Parameter Estimation for the Four-Parameter Exponential Gamma Distribution. *Entropy* **2017**, *19*, 189. [CrossRef]
6. Chen, L.; Singh, V.P.; Xiong, F. An Entropy-Based Generalized Gamma Distribution for Flood Frequency Analysis. *Entropy* **2017**, *19*, 239. [CrossRef]
7. Chen, L.; Singh, V.P. Generalized Beta Distribution of the Second Kind for Flood Frequency Analysis. *Entropy* **2017**, *19*, 254. [CrossRef]
8. Chen, L.; Singh, V.P.; Huang, K. Bayesian Technique for the Selection of Probability Distributions for Frequency Analyses of Hydrometeorological Extremes. *Entropy* **2018**, *20*, 117. [CrossRef]
9. Keum, J.; Kornelsen, K.C.; Leach, J.M.; Coulibaly, P. Entropy Applications to Water Monitoring Network Design: A Review. *Entropy* **2017**, *19*, 613. [CrossRef]
10. Yeh, H.-C.; Chen, Y.-C.; Chang, C.-H.; Ho, C.-H.; Wei, C. Rainfall Network Optimization Using Radar and Entropy. *Entropy* **2017**, *19*, 553. [CrossRef]
11. Santonastaso, G.F.; Di Nardo, A.; Di Natale, M.; Giudicianni, C.; Greco, R. Scaling-Laws of Flow Entropy with Topological Metrics of Water Distribution Networks. *Entropy* **2018**, *20*, 95. [CrossRef]
12. Hao, L.; Su, X.; Singh, V.P.; Ayantobo, O.O. Spatial Optimization of Agricultural Land Use Based on Cross-Entropy Method. *Entropy* **2017**, *19*, 592. [CrossRef]
13. Chen, L.; Sun, C.; Wang, G.; Xie, H.; Shen, Z. Modeling Multi-Event Non-Point Source Pollution in a Data-Scarce Catchment Using ANN and Entropy Analysis. *Entropy* **2017**, *19*, 265. [CrossRef]
14. Zhou, R.; Pan, Z.; Jin, J.; Li, C.; Ning, S. Forewarning Model of Regional Water Resources Carrying Capacity Based on Combination Weights and Entropy Principles. *Entropy* **2017**, *19*, 574. [CrossRef]
15. Men, B.; Long, R.; Li, Y.; Liu, H.; Tian, W.; Wu, Z. Combined Forecasting of Rainfall Based on Fuzzy Clustering and Cross Entropy. *Entropy* **2017**, *19*, 694. [CrossRef]
16. Huang, F.; Chunyu, X.; Wang, Y.; Wu, Y.; Qian, B.; Guo, L.; Zhao, D.; Xia, Z. Investigation into Multi-Temporal Scale Complexity of Streamflows and Water Levels in the Poyang Lake Basin, China. *Entropy* **2017**, *19*, 67. [CrossRef]
17. Pan, Z.; Jin, J.; Li, C.; Ning, S.; Zhou, R. A Connection Entropy Approach to Water Resources Vulnerability Analysis in a Changing Environment. *Entropy* **2017**, *19*, 591. [CrossRef]
18. Salas, H.D.; Poveda, G.; Mesa, O.J. Testing the Beta-Lognormal Model in Amazonian Rainfall Fields Using the Generalized Space *q*-Entropy. *Entropy* **2017**, *19*, 685. [CrossRef]
19. Mihailović, D.; Mimić, G.; Gualtieri, P.; Arsenić, I.; Gualtieri, C. Randomness Representation of Turbulence in Canopy Flows Using Kolmogorov Complexity Measures. *Entropy* **2017**, *19*, 519. [CrossRef]
20. Cheng, L.; Niu, J.; Liao, D. Entropy-Based Investigation on the Precipitation Variability over the Hexi Corridor in China. *Entropy* **2017**, *19*, 660. [CrossRef]
21. Zhou, Z.; Ju, J.; Su, X.; Singh, V.P.; Zhang, G. Comparison of Two Entropy Spectral Analysis Methods for Streamflow Forecasting in Northwest China. *Entropy* **2017**, *19*, 597. [CrossRef]
22. Zhang, G.; Su, X.; Singh, V.P.; Ayantobo, O.O. Modeling NDVI Using Joint Entropy Method Considering Hydro-Meteorological Driving Factors in the Middle Reaches of Hei River Basin. *Entropy* **2017**, *19*, 502. [CrossRef]
23. Vu, T.M.; Mishra, A.K.; Konapala, G. Information Entropy Suggests Stronger Nonlinear Associations between Hydro-Meteorological Variables and ENSO. *Entropy* **2018**, *20*, 38. [CrossRef]
24. Wang, S.; Singh, V.P. Spatio-Temporal Variability of Soil Water Content under Different Crop Covers in Irrigation Districts of Northwest China. *Entropy* **2017**, *19*, 410. [CrossRef]

3

Entropy **2018**, *20*, 598

25. Guo, A.; Chang, J.; Wang, Y.; Huang, Q.; Guo, Z. Maximum Entropy-Copula Method for Hydrological Risk Analysis under Uncertainty: A Case Study on the Loess Plateau, China. *Entropy* **2017**, *19*, 609. [CrossRef]
26. Foroozand, H.; Radić, V.; Weijs, S.V. Application of Entropy Ensemble Filter in Neural Network Forecasts of Tropical Pacific Sea Surface Temperatures. *Entropy* **2018**, *20*, 207. [CrossRef]
27. Ilunga, M. Cross Mean Annual Runoff Pseudo-Elasticity of Entropy for Quaternary Catchments of the Upper Vaal Catchment in South Africa. *Entropy* **2018**, *20*, 281. [CrossRef]
28. Koutsoyiannis, D. Entropy Production in Stochastics. *Entropy* **2017**, *19*, 581. [CrossRef]

entropy

MDPI

Article

An Extension to the Revised Approach in the Assessment of Informational Entropy

Turkay Baran, Nilgun B. Harmancioglu *, Cem Polat Cetinkaya and Filiz Barbaros

Faculty of Engineering, Civil Engineering Department, Dokuz Eylul University, Tinaztepe Campus, Buca, 35160 Izmir, Turkey; turkay.baran@deu.edu.tr (T.B.); cem.cetinkaya@deu.edu.tr (C.P.C.); filiz.barbaros@deu.edu.tr (F.B.)
* Correspondence: nilgun.harmancioglu@deu.edu.tr; Tel.: +90-542-413-9300

Received: 29 September 2017; Accepted: 20 November 2017; Published: 29 November 2017

Abstract: This study attempts to extend the prevailing definition of informational entropy, where entropy relates to the amount of reduction of uncertainty or, indirectly, to the amount of information gained through measurements of a random variable. The approach adopted herein describes informational entropy not as an absolute measure of information, but as a measure of the variation of information. This makes it possible to obtain a single value for informational entropy, instead of several values that vary with the selection of the discretizing interval, when discrete probabilities of hydrological events are estimated through relative class frequencies and discretizing intervals. Furthermore, the present work introduces confidence limits for the informational entropy function, which facilitates a comparison between the uncertainties of various hydrological processes with different scales of magnitude and different probability structures. The work addresses hydrologists and environmental engineers more than it does mathematicians and statisticians. In particular, it is intended to help solve information-related problems in hydrological monitoring design and assessment. This paper first considers the selection of probability distributions of best fit to hydrological data, using generated synthetic time series. Next, it attempts to assess hydrometric monitoring duration in a netwrok, this time using observed runoff data series. In both applications, it focuses, basically, on the theoretical background for the extended definition of informational entropy. The methodology is shown to give valid results in each case.

Keywords: uncertainty; information; informational entropy; variation of information; continuous probability distribution functions; confidence intervals

---

1. Introduction

The concept of entropy has its origins in classical thermodynamics and is commonly known as "thermodynamic entropy" in relation to the second law of thermodynamics. Such a non-probabilistic definition of entropy has been used widely in physical sciences, including hydrology and water resources. Typical examples on the use of "thermodynamic entropy" in water resources involve problems associated with river morphology and river hydraulics [1,2].

Boltzmann's definition of entropy as a measure of disorder in a system was given in probabilistic terms and constituted the basis for statistical thermodynamics [3–5]. Later, Shannon [6] followed up on Boltzmann's definition, claiming that the entropy concept could be used to measure disorder in systems other than thermodynamic ones. Shannon's entropy is what is known as "informational entropy", which measures uncertainty (or, indirectly, information) about random processes. As uncertainty and information are the two most significant yet the least clarified problems in hydrology and water resources, researchers were intrigued by the concept of informational entropy. Thus, it has found a large number of diverse applications in water resources engineering.

Within a general context, the entropy principle is used to assess uncertainties in hydrological variables, models, model parameters, and water-resources systems. In particular, versatile uses of the concept range from specific problems, such as the derivation of frequency distributions and parameter estimation, to broader cases such as hydrometric data network design. The most distinctive feature of entropy in these applications is that it provides a measure of uncertainty or information in quantitative terms [7–19].

On the other hand, researchers have also noted some mathematical difficulties encountered in the computation of various informational entropy measures. The major problem is the controversy associated with the mathematical definition of entropy for continuous probability distribution functions. In this case, the lack of a precise definition of informational entropy leads to further mathematical difficulties and, thus, hinders the applicability the concept in hydrology. This problem needs to be resolved so that the informational entropy concept can be set on an objective and reliable theoretical basis and thereby achieve widespread use in the solution of water-resources problems based on information and/or uncertainty.

Some researchers [20,21] attempted to revise the prevailing definition of informational entropy, where entropy relates to the amount of reduction of uncertainty, or indirectly to the amount of information gained through measurements of a random variable. The study presented extends on the revised definition of Jaynes [20] and Guiasu [21] to describe informational entropy, not as an absolute measure of information, but as a measure of the variation of information. The mathematical formulation developed herein does not depend on the use of discretizing intervals when discrete probabilities of hydrological events are estimated through relative class frequencies and discretizing intervals. This makes it possible to obtain a single value for the variation of information instead of several values that vary with the selection of the discretizing interval. Furthermore, the extended definition introduces confidence limits for the entropy function, which facilitates a comparison between the uncertainties of various hydrological processes with different scales of magnitude and different probability structures.

It must be noted that the present work is intended for hydrologists and environmental engineers more than for mathematicians and statisticians. In particular, entropy measures have been used to help solve information-related problems in hydrological monitoring design and assessment. These problems are manifold, ranging from the assessment of sampling frequencies (both temporal and spatial) and station discontiuance to statistical analyses of observed data. For the latter, this paper considers the selection of probability distributions of best fit to hydrological data. Hence, the informational entropy concept is used here only in the temporal domain. To test another feature of entropy measures, the present work also attempts to assess hydrometric monitoring duration in a gauging network, this time using observed runoff data series. In both applications, the paper focuses, basically, on the theoretical background for the extended definition of informational entropy, and the results are shown to give valid results.

2. Mathematical Difficulties Associated with Informational Entropy Measures

Entropy is a measure of the degree of uncertainty of random hydrological processes. It is also a quantitative measure of information contained in a series of data since the reduction of uncertainty equals the same amount of gain in information [7,22]. Within the scope of Mathematical Communication Theory, later known as Information Theory, Shannon [6] and later Jaynes [23] defined informational entropy as the expectation of information or, conversely, as a measure of uncertainty. If S is a system of events, $E_1, E_2, \ldots, E_n$, and $p(E_k) = p_k$ the probability of the k-th event recurring, then the entropy of the system is:

$$H(S) = -\sum_{k=1}^{n} p_k \ln p_k \tag{1}$$

With,

$$\sum_{k=1}^{n} p_k = 1$$

Shannon's entropy as given in Equation (1) is originally formulated for discrete variables and always assumes positive values. Shannon extended this expression to the continuous case by simply replacing the summation with an integral equation as:

$$H(X) = -\int_{-\infty}^{+\infty} f(x) \cdot \ln f(x) \cdot dx \qquad (2)$$

With,

$$\int_{-\infty}^{+\infty} f(x) \cdot dx = 1$$

For the random variable $X \in (-\infty, +\infty)$, and where $H(X)$ is denoted as the marginal entropy of X, i.e., the entropy of a univariate process. Equation (2) is not mathematically justified, as it is not valid under the assumptions initially made in defining entropy for the discrete case. What researchers proposed for solving this problem has been to approximate the discrete probabilities p_k by $f(x)\Delta x$, where $f(x)$ is the relative class frequency and Δx, the size of class intervals. Under these conditions, the selection of Δx becomes a crucial problem, such that each specified class interval size gives a different reference level of zero uncertainty with respect to which the computed entropies are measured. In this case, various entropy measures become relative to the discretizing interval Δx and change in value as Δx changes. The unfavorable result here is that the uncertainty of a random process may assume different values at different selected values of Δx for the same variable and the same probability distribution function. In certain cases, the entropy of a random variable even becomes negative [16,17,22,24–27], a situation which contradicts Shannon's definition of entropy as the selection of particular Δx values produces entropy measures varying within the interval $(-\infty, +\infty)$. On the contrary, the theoretical background for the random variable X, $H(X)$ defines the condition:

$$0 \leq H(X) \leq \ln N \qquad (3)$$

where N is the number of events X assumes. The condition above indicates that the entropy function has upper (ln N) and lower (0 when X is deterministic) bounds, assuming positive values in between [6,8,10–13,16,17,22,24–28]. The discrepancies encountered in practical applications of the concept essentially result from the above errors in the definition of entropy for continuous variables.

Another significant problem is the selection of the probability distribution function to be used in the definition of entropy, as in Equation (2). The current expression for continuous entropy produces different values when different distribution functions are assumed for the same variable. In this case, there is the need for a proper selection of the distribution function which best fits the process analyzed. One may consider here a valid criterion in the form of confidence limits to assess the suitability of the selected distribution function for entropy computations.

Further problems are encountered when the objective is to compare the uncertainties of two or more random variables with widely varying means and thus with different scales of magnitude. For instance, if entropy values are computed, using the same discretizing interval Δx, for two variables with means of 100 units and 1 unit, respectively, the results become incomparable due to the improper selection of the reference level of zero uncertainty for each variable. Such a problem again stems from the inclusion of the discretizing interval Δx in the definition of entropy for continuous variables. Comparison of uncertainties of different variables is an important aspect of entropy-based hydrometric network design procedures, where the aforementioned problem leads to subjective evaluations of information provided by the network [7,19].

It follows from the above discussion that the main difficulty associated with the applicability of the informational entropy concept in hydrology is the lack of a precise definition for the case of the continuous variables. It is intended in this study to resolve this problem by extending the revised approach proposed by Guiasu [21] so that the informational entropy can be set on an objective and reliable theoretical basis in order to discard subjective assessments of information conveyed by hydrological data or of the uncertainty of hydrological processes.

3. The Revised Definition of Informational Entropy for Continuous Variables

To solve the difficulties associated with the informational entropy measure in the continuous case, some researchers have proposed the use of a function $m(x)$ such that the marginal entropy of a continuous variable X is expressed as:

$$H(X) = - \int_{-\infty}^{+\infty} f(x) \cdot \ln\left[\frac{f(x)}{m(x)}\right] \cdot dx \qquad (4)$$

"where $m(x)$ is an 'invariant measure' function, proportional to the limiting density of discrete points" [20]. The approach seemed to be statistically justified; however, it still remained uncertain what the $m(x)$ function might represent in reality. Jaynes [20] also discussed that it could be an a priori probability distribution function, but there were then controversies over the choice of a priori distribution such that the problem was unresolved [8].

In another study, Guiasu [21] referred to Shannon's definition of the informational entropy for the continuous case. He considered that the entropy {H_S} for the continuous variable X within an interval $[a, b]$ is:

$$H_S = - \int f(x) \cdot \ln f(x) \cdot dx \qquad (5)$$

When the random variable assumes a uniform probability distribution function as:

$$f(x) = \frac{1}{(b-a)} \quad x \in [a, b] \qquad (6)$$

Then the informational entropy H_S for the continuous case within this interval can be expressed as:

$$H_S = \ln(b - a) \qquad (7)$$

If the interval $[a, b]$ is discretized into N equal intervals, the variable follows a discrete uniform distribution and its entropy {H_N} can be expressed as:

$$x \in [a, b] \qquad (8)$$

When N goes to infinity, H_N will also approach infinity. In this case, Guiasu [21] claims that, although H_S and H_N are similarly defined, H_S will not approach H_N when $N{\to}\infty$. Accordingly, Guiasu [21] proposed an expression similar to that of Jaynes [20] for informational entropy in the continuous case as:

$$H(X/X^*) = - \int f(x) \cdot \ln\left[\frac{f(x)}{m(x)}\right] \cdot dx \qquad (9)$$

which he called as the variation of information. In Equation (9), X^* represents a priori information (i.e., information available before making observations on the variable X) and X is the a posteriori information (i.e., information obtained by making observations). Similarly, $m(x)$ is the a priori and $f(x)$ the a posteriori probability density function for the random variable X.

In previous studies by the authors [8,10–13], informational entropy has been defined as the variation of information, which indirectly equals the amount of uncertainty reduced by making observations. To develop such a definition, two measures of probability, p and q with (p and $q \in$ K), are considered in the probability space (Ω, K). Here, q represents a priori probabilities (i.e., probabilities prior to making observations). When a process is defined in such a probability space, the information conveyed when the process assumes a finite value A $\{A \in K\}$ in the same probability space is:

$$I = -\ln\left(\frac{p(A)}{q(A)}\right) \tag{10}$$

The process defined in Ω can assume one of the finite and discrete events $(A_1, \ldots, A_n) \in K$; thus, the entropy expression for any value A_n can be written as:

$$H(p/q) = -\ln\left(\frac{p(A_n)}{q(A_n)}\right)(n = 1, \ldots, N) \tag{11}$$

The total information content of the probability space (Ω, K) can be defined as the expected value of the information content of its elementary events:

$$H(p/q) = -\sum p(A_n) \cdot \ln\left(\frac{p(A_n)}{q(A_n)}\right) \tag{12}$$

Similarly, the entropy $H(X/X^*)$ of a random process X defined in the same probability space can be defined as:

$$H(X/X^*) = -\sum p(x_n) \cdot \ln\left(\frac{p(x_n)}{q(x_n)}\right) \tag{13}$$

where, $H(X/X^*)$ is in the form of conditional entropy, i.e., the entropy of X conditioned on X^*. Here, the condition is represented by an a priori probability distribution function, which can be described as the reference level against which the variation of information in the process can be measured.

Let us assume that the a priori $\{q(x)\}$ and a posteriori $\{p(x)\}$ probability distribution functions of the random variable X are known. If the ranges of possible values of the continuous variable X are divided into N discrete and infinitesimally small intervals of width Δx, the entropy expression for this continuous case can be given as:

$$H(X/X^*) = -\int p(x) \cdot \ln\left(\frac{p(x)}{q(x)}\right) \cdot dx \tag{14}$$

The above expression describes the variation of information (or, indirectly, the uncertainty reduced by making observations) to replace the absolute measure of information content given in Equation (2). This definition is essentially in conformity with those given by Jaynes [20] and Guiasu [21] for continuous variables. When the same infinitesimally small class interval Δx is used for the a priori and a posteriori distribution functions, the term Δx drops out in the mathematical expression of marginal entropy in the continuous case. Thus, this approach eliminates the problems pertaining to the use of Δx discretizing class intervals involved in the previous definitions of informational entropy [8,10–13].

At this point, the most important issue is the selection of a priori distribution. In case the process X is not observed at all, no information is available about it so that it is completely uncertain. In probability terms, this implies the selection of the uniform distribution. In other words, when no information exists about the variable X, the alternative events it may assume may be represented by equal probabilities or simply by the uniform probability distribution function.

If the a priori $\{q(x)\}$ is assumed to be uniform, and a posteriori $\{p(x)\}$ distribution of X is assumed to be normal, the informational entropy $H(X/X^*)$ can be expressed as:

$$H(X/X^*) = \ln\sqrt{2\pi} + \ln\sigma + \frac{1}{2} - \ln(b - a) \tag{15}$$

By integrating Equation (14). The first three terms in this equation represent the marginal entropy of X and the last term stands for the maximum entropy. Accordingly, the variation of information can be expressed simply as:

$$H(X/X^*) = H(X) - H_{\max} \tag{16}$$

If the a posteriori distribution of X is assumed to be lognormal, the informational entropy $H(X/X^*)$ becomes:

$$H(X/X^*) = \ln\sqrt{2\pi} + \ln\sigma_y + \mu_y + \frac{1}{2} - \ln(b-a) \tag{17}$$

with and μ_y and σ_y being the mean and standard deviation of $y = \ln x$.

If the a posteriori distribution of X is assumed to be 2-parameter gamma distribution with parameters α and β,

$$f_{(x)} = \frac{1}{\beta^\alpha\,\Gamma_{(\alpha)}}\, x^{\alpha-1}\, e^{\frac{-x}{\beta}}\ x \geq 0 \tag{18}$$

The informational entropy $H(X/X^*)$ becomes:

$$H(X/X^*) = \ln[\beta\cdot\Gamma(\alpha)] + \mu_x/\beta - (\alpha-1)\cdot\Phi(\alpha) - \ln(b-a) \tag{19}$$

where, μ_x is the mean of the series, α the shape parameter, and β the scale parameter.

In the above, entropy as the variation of information measures the amount of uncertainty reduced by making observations when the a posteriori distribution is estimated.

The maximum amount of information gained about the process X defined within the interval $[a, b]$ is $H_{\max}$. Thus, the expression in Equation (16) will assume negative values. However, since $H(X/X^*)$ describes entropy as the variation of information, it is possible to consider the absolute value of this measure.

When the a posteriori probability distribution function is correctly estimated, the information gained about the random variable will increase as the number of observations increases. Thus, when this number goes to infinity, the entropy $H(X/X^*)$ will approach zero. In practice, it is not possible to obtain an infinite number of observations; rather, the availability of sufficient data is important. By using the entropy measure $H(X/X^*)$, it possible to evaluate the fitness of a particular distribution function to the random variable and to assess whether the available data convey sufficient information about the process.

4. Mathematical Interpretation of the Revised Definition of Informational Entropy

4.1. The Distance between Two Continuous Distribution Functions as Defined by the Euclidian Metric

The approach used to obtain Equation (16) is essentially a means of measuring the distance between the points in probability space, described by the a priori $\{q(x)\}$ and a posteriori $\{p(x)\}$ distribution functions. The distance between these two functions can be determined by different measures like the metric concept, which enables one to see whether the two functions coincide.

According to the Euclidian metric, the distance between $p(x)$ and $q(x)$ functions defined in the same probability space (Ω, K) is:

$$I = \int [p(x) - q(x)]^2 dx \tag{20}$$

If $p(x)$ is the standard normal, and $q(x)$, the standard uniform distribution function, one obtains:

$$\Phi(x) = \frac{1}{\sqrt{2\pi}}\left\{\sqrt{1 + \frac{3\sqrt{\pi}}{x}\left[1 - \frac{4}{3}F(x)\right]}\right\} \tag{21}$$

By integrating Equation (20) to obtain the difference function $\Phi(x)$. The $F(x)$ function in Equation (21) represents the cumulative probabilities for the standard normal distribution. When the above difference function is equal to zero, $p(x)$ and $q(x)$, which are described as two points in the (Ω, K) probability space, will coincide at the same point. When the difference function assumes a minimum value, this will indicate a point of transition between $p(x)$ and $q(x)$, where the two functions can be expressed in terms of each other. The same point also refers to a minimum number of observations required to produce information about the process X. When the difference function is described as in Figure 1, the presence of such a minimum value can be observed. The difference function $\Phi(x)$ decreases until $x = x_0$, where it passes through a minimum value. At point x_0, the two functions $p(x)$ and $q(x)$ approach each other until the distance between them is approximated by a constant C. After this point, when x approaches infinity, the difference function gradually increases; and finally, the difference between $p(x)$ and $q(x)$ approaches zero at infinity. One may define x_0 as the point where the two probability functions can be used interchangeably with an optimum number of observations.

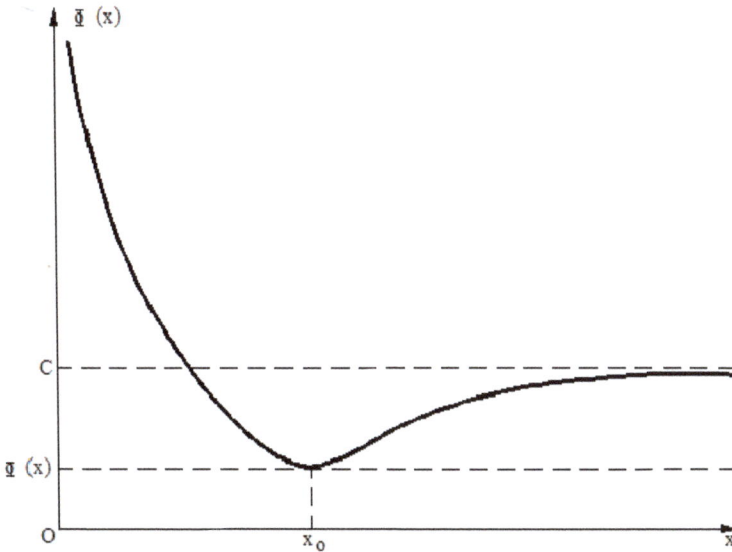

Figure 1. The difference function as defined by the Euclidian metric.

The purpose of observing the random variable X is to obtain a realistic estimate of its population parameters and to achieve reliable information about the process to allow for correct decisions in planning. On the other hand, each observation entails a cost factor; therefore, planners are interested in delineating how long the variable X has to be observed. Equation (16) is significant from this point of view. By defining the variation of information as the reduction of uncertainty via sampling, the point where no more increase or change in variation of information is obtained actually specifies the time point when sampling can be stopped. This is a significant issue which may be employed in considerations of gauging station discontinuance.

4.2. The Distance between Two Continuous Distribution Functions as Defined by Max-Norm

The max-norm can also be used to measure the distance between two functions defined in the probability space and to assess whether these two functions approach each other. According to the max-norm, the distance between two functions $p(x)$ and $q(x)$ is defined as:

$$\Delta(p, m) = \sup_{-\infty < x < +\infty} |p(x) - q(x)| \tag{22}$$

When $p(x)$ is used to represent the standardized normal and $q(x)$, the standardized uniform distribution functions, the difference function $\{h(x)\}$ will be:

$$h(x) = p(x) - q(x) \tag{23}$$

It may be observed in Figure 2 that, the critical points of the difference function are at h_0, h_1, and h_2 so that the difference between the two functions $\{\Delta(p,q)\}$ can be expressed as:

$$\Delta(p,q) = \max\{h_0, h_1, h_2\} \tag{24}$$

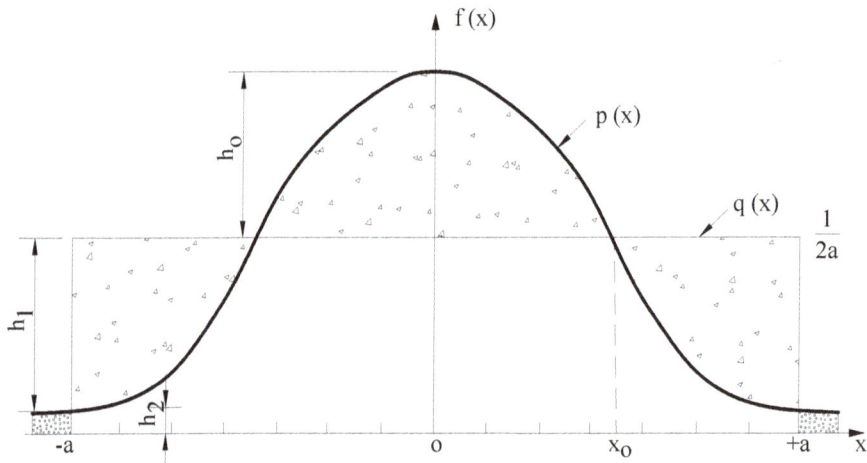

Figure 2. Critical values of the difference function as defined by the max-norm.

Based on the half-range value "a" in Figure 2, the critical points h_0, h_1, and h_2 can be obtained as:

$$h_0 = \frac{1}{\sqrt{2\pi}} \frac{1}{2a} \tag{25}$$

$$h_1 = \frac{1}{\sqrt{2\pi}} \frac{1}{2a} e^{(-a^2/2)} \tag{26}$$

$$h_2 = \frac{1}{\sqrt{2\pi}} e^{(-a^2/2)} \tag{27}$$

The problem here is then to find the distance between the two functions as the half-range value "a" which minimizes $\Delta(p,q)$ of Equation (24). The critical half-range value "a" that satisfies this supremum is:

$$a = \frac{3}{4}\sqrt{2\pi} \tag{28}$$

At the above critical half-range value "a", which is obtained by the max-norm, it is possible to use the two functions $p(x)$ and $q(x)$ interchangeably with an optimum number of observations.

When two points represented by the a posteriori and a priori distribution functions, $p(x)$ and $q(x)$, respectively, in the same probability space approach each other, this indicates, in information terms, an information increase about the random process analyzed. The case when the two points coincide represents total information availability about the process. Likewise, when $H(X/X^*)$ of Equation (16) approaches zero in absolute terms, this indicates a gain of total information about the process X defined within the interval $[a, b]$. One obtains sufficient information about the process when the variation information, as described by the Euclidian metric, approaches a constant value.

4.3. Asymptotic Properties of Shannon's Entropy

Vapnik [29] analyzed and provided proofs for some asymptotic properties of Shannon's entropy of the set of events on the sample size N. He used these properties to prove the necessary and sufficient conditions of uniform convergence of the frequencies of events to their probabilities.

In the work of Vapnik [23], it is shown that the sequence:

$$\frac{H(S)}{N}, \ N = 1, 2, \ldots, \tag{29}$$

has a limit c, when N goes to infinity. The lemma:

$$\lim_{N \to \infty} \frac{H(S)}{N} = c \ 0 \le c \le 1, \tag{30}$$

was proved by Vapnik [29] and was claimed to "repeat the proof of the analogous lemma in information theory for Shannon's entropy". Vapnik [29] also proved that, for any N, the sequence of Equation (29) is an upper bound for limit of Equation (30).

Vapnik [29] proved the above lemmas for Shannon's entropy, based on the discrete case of Equation (1). However, they are also valid for the continuous case as described by the Euclidian metric. Thus, it is possible to restate, using Vapnik's proofs, that the upper bound H_{max} of Shannon's entropy will be reached as the number of observations increases to approach the range of the population ($N \to \infty$) and that the variation of information of Equation (16) approaches a constant value "c".

In the next section, the derivation of the constant "c" is demonstrated for the case when the a priori distribution function is assumed to be uniform and the a posteriori function to be normal. These assumptions comply with the limits ($0 \le c \le 1$) defined for the discrete case as in Equation (30).

5. Further Development of the Revised Definition of the Variation of Information

If the observed range $[a, b]$ of the variable X is considered also as the population value of the range, R, of the variable, the maximum information content of the variable may be described as:

$$H_{max} = \ln R \tag{31}$$

With;

$$R = b - a \ a < x < b \tag{32}$$

When the a posteriori distribution of the variable is assumed to be normal, the marginal entropy of X becomes:

$$H(X) = \ln \sqrt{2\pi} + \ln \sigma + 1/2 \tag{33}$$

If the variable is actually normally distributed and if a sufficient number of observations are obtained, the entropy of Equation (16) will approach a value which can be considered to be within an acceptable region. This is the case where one may infer that sufficient information has been gained about the process.

When sufficient information is made available about X, it will be possible to make the best estimates for the mean (μ), variance (σ), and the range (R) of X. For this purpose, the variable has to be analyzed as an open series in the historic order. According to the approach used, the information gained about the process will continuously increase as the number of observations increase. Similarly, H_{max} and $H(X)$ will also increase, while $H(X/X^*)$ will decrease. When the critical point is reached, where the variable can be described by its population parameters, H_{max} will approach a constant value; $H(X)$ will also get closer to this value with $H(X/X^*)$ approaching a constant value of "c" as in Figure 3.

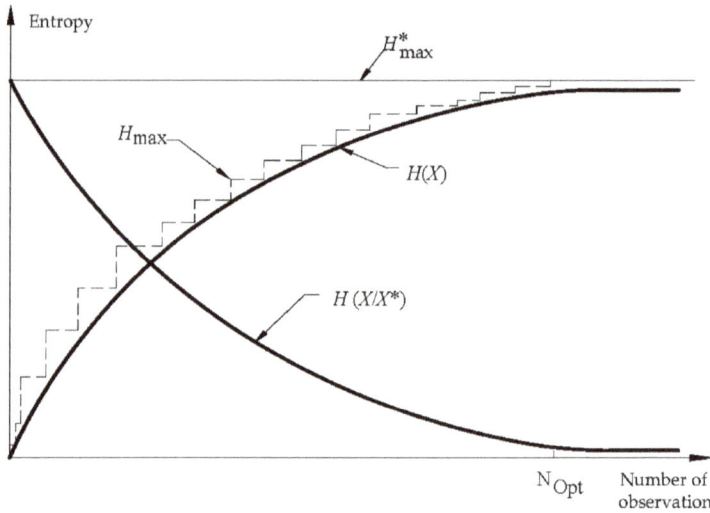

Figure 3. Maximum entropy $\{H_{max}\}$, marginal entropy $\{H(X)\}$ and entropy as the variation of information $\{H(X/X^*)\}$ versus the number of observations.

Determination on Confidence Limits for Entropy Defined by Variation of Information

The confidence limits (*acceptable region*) of entropy can be determined by using the a posteriori probability distribution functions. If the normal $\{N(0,1)\}$ probability density function is selected, the maximum entropy for the standard normal variable z is;

$$H_{max}(z) = \ln R_z, \tag{34}$$

with the range of z being,

$$R_z = 2a \tag{35}$$

Here, the value a describes the half-range of the variable. Then, the maximum entropy for variation x with $N(\mu,\sigma)$ is;

$$H_{max}(x) = \ln(R_z\sigma) \tag{36}$$

If the critical half-range value is foreseen as:

$$a = 4\sigma, \tag{37}$$

then the area under the normal curve may be approximated to be 1.

For the half-range value, replacing the appropriate values in Equation (16), one obtains the acceptable entropy value for the normal probability density function as:

$$H(X/X^*)_{cr} = 0.6605, \tag{38}$$

using natural logarithms. When the entropy $H(X/X^*)$ of the variable which is assumed to be normal remains below the above value, one may decide that the normal probability density function is acceptable and that a sufficient amount of information has been collected about the process.

If the a posteriori distribution function is selected as lognormal $LN(\mu_y, \sigma_y)$, the variation of information for the variable x can be determined as:

$$H(X/X^*) = \ln\left[2Sinh(a\sigma_y)\right] - \ln\sigma_y - 1.4189 \tag{39}$$

Here, since lognormal values will be positive, one may consider $0 \leq x \leq \infty$. Then the acceptable value of $H(X/X^*)$ for the lognormal distribution function will be;

$$H(X/X^*) = a\sigma_y - \ln \sigma_y - 1.4189 \qquad (40)$$

According to Equation (40), no single constant value exists to describe the confidence limit for lognormal distribution. Even if the critical half-range is determined, the confidence limits will vary according to the variance of the variable. However, if the variance of x is known, the confidence limits can be computed.

6. Application

6.1. Application to Synthetic Series to Test the Fit of Probability-Distribution Functions

It is often difficult in practice to find long series of complete hydrological data. Thus, it is preferred here to test the above methodology on synthetically generated data for the purposes of evaluating the fit of different probability distribution functions. For this purpose, normal $\{N\,(\mu, \sigma)\}$ and lognormal $\{LN\,(\mu_y, \sigma_y)\}$ distributed time series are produced, using uniformly distributed series derived by the Monte Carlo method. Ten-year time series are obtained with normal $\{N\,(\mu, \sigma)\}$ and lognormal $\{LN\,(\mu_y, \sigma_y)\}$ distributions, respectively. Each series covered a period of (10×365) days with cumulative data for each year as $(i \times 365;$ where $i = 1, ..., N)$.

To test the methodology, $N(8, 10)$ distributed 3650 synthetic data are divided into subgroups with 365 data in each. First, maximum informational entropy (H_{max}) is determined, using Equation (31) and the whole time series. Assuming that the a posteriori distribution is normal, marginal entropies $(H(X))$ and, finally, the informational entropy values $(H(X/X^*)$ are computed for the normal distribution using Equation (15). Consecutive values of these entropy measures are computed first for 365 generated data, next for 2×365 data, and for the last year 10×365 data. The confidence limits for the case of a posteriori normal distribution is determined by Equation (38). Figure 4 shows the results of this application. If a lognormal posteriori distribution is assumed for this series, which is actually normally distributed, this assumption is rejected on the basis of the computed confidence limits for normal distribution. Otherwise, the assumption is accepted. In Figure 4, the $H(X/X^*)$ values fall below the confidence level determined for normal distribution so that the assumption of a posteriori lognormal distribution is rejected.

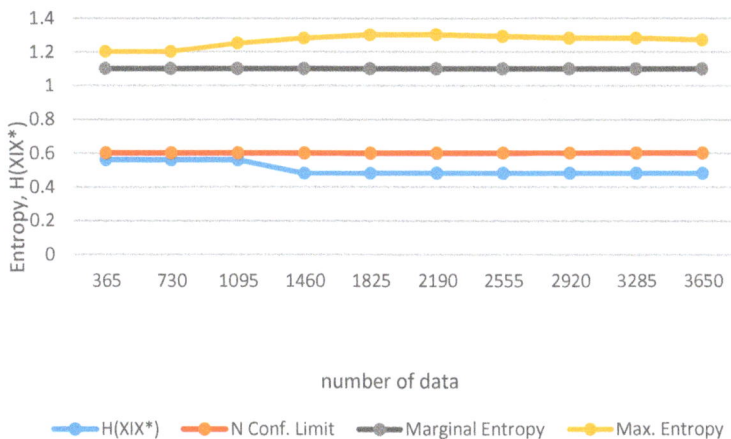

Figure 4. Normal distributed synthetic series, by the assumption of a posteriori normal distributed probability function (where; N Conf. Limit is the confidence limit for normal distribution).

If the same application is repeated by using the confidence limit for lognormal distribution, as in Figure 5, the assumption of a posterior lognormal distribution is rejected as the $H(X/X^*)$ values stay above the confidence level determined for lognormal distribution.

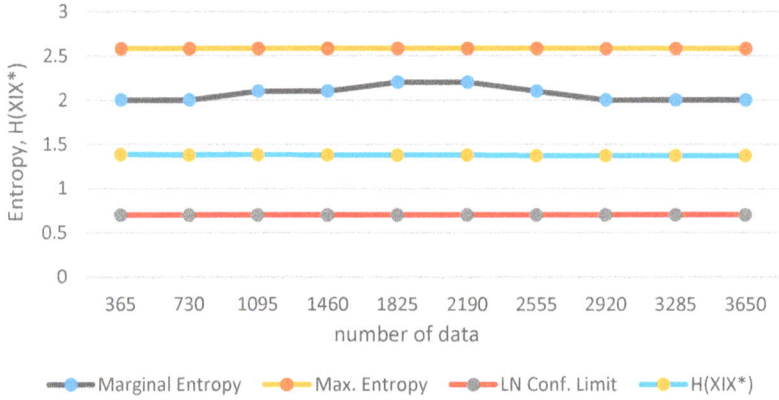

Figure 5. Normal distributed synthetic series, by the assumption of a posteriori lognormal distributed probability function (where; LN Conf. Limit is the limit of confidence for lognormal distribution).

Similar exercises may be run by generating lognormal distributed synthetic series and assuming the posteriori distribution first as lognormal (Figure 6) and then as normal distribution (Figure 7).

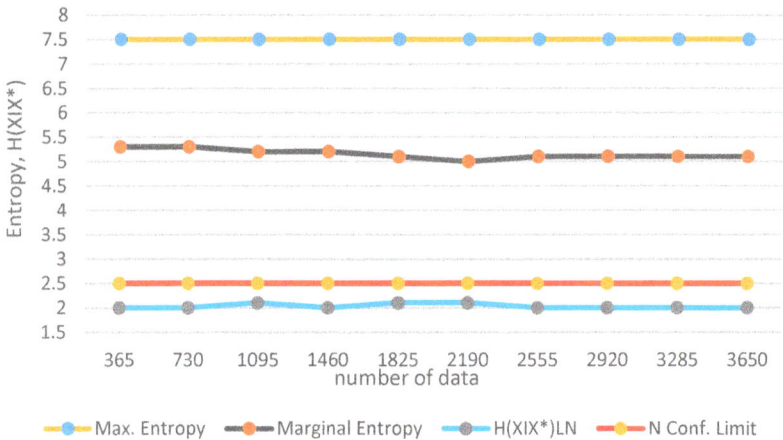

Figure 6. Lognormal distributed synthetic series, by the assumption of a posteriori lognormal distributed probability function (where; N Conf. Limit is the confidence limit for normal distribution).

The above exercises show that comparisons between assumptions of a posteriori normal and lognormal distributions on the basis of entropy-based confidence limits for each distribution give valid results by checking how the variation of information values behave with respect to the confidence limits.

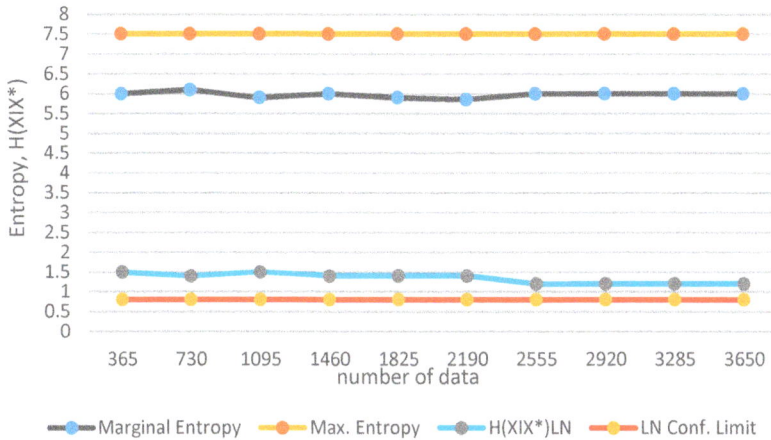

Figure 7. Lognormal distributed synthetic series, by the assumption of a posteriori normal distributed probability function (where; LN Conf. Limit is the limit of confidence for lognormal distribution).

6.2. Application to Runoff Data for Assessment of Sampling Duration

An important question regarding hydrometric data-monitoring networks is how long the observations should be continued. Considering the "data rich, information poor" data networks of our times, researchers and decisionmakers have wondered whether monitoring could be discontinued at certain sites, as data observation is a cost and labor-consuming activity [30,31]. To date, none of the approaches proposed for the problem of station discontinuance have found universal acceptance. Entropy measures as described in this work may as well be employed when a monitoring activity reaches an optimal point in time after which any new data does not produce new information. This feature of entropy measures is shown in Figure 3, where the marginal entropy of the process $H(X)$ approaches the total uncertainty Hmax as the number of observations (N) increase. Finally, a point is reached where H_{max} and $H(X)$ coincide after a certain number of observations, which can be defined as N_{opt}. After this point on, observed data do not produce new information, and thus monitoring can be discontinued. Certainly, the probability distribution of best fit to observed series must be selected first to evaluate this condition. This is an important feature of entropy measures as they can be used to infer about station discontinuance, based also on the selection of the appropriate distribution functions.

To test the above aspect of the entropy concept, observed runoff data at two monitoring stations (Kuskayasi and Akcil) in the Ceyhan river basin in Turkey are employed (Figure 8). The Ceyhan basin has been subject to several investigations and projects for the development of water schemes; thus, it is intended here to evaluate the monitoring activities in the basin in terms of entropy measures. Although there are other gauging stations along the river, their data are not homogeneous due to already-built hydraulic structures. Kuskayasi and Akcil are the two stations where natural flows are observed, although their common observation periods cover only 8 years.

The observations at Kuskayasi were discontinued after 1980 and Akcil after 1989. Thus, for the purposes of this application their common period between 1973 and 1980 is selected. Daily data for the observation period of 8 years are used, where the mean daily runoff at Kuskayasi is 10.8 m^3/s and that at Akcil is 27.18 m^3/s. The standard deviations are 11.77 m^3/s and 22.48 m^3/s, respectively.

Next, the fits of normal and lognormal distributions are tested at both stations again with the entropy concept. This analysis is followed by the computation of marginal entropies ($H(X)$, H_{max} and the variation of information $H(X)/H^*$) for these two distribution functions. The computations are carried out in a successive manner, using the first year's 365 data, the second year's 720, and so on until the total number of 2920 data are reached. Certainly, H_{max} changes with the total of data observed from the beginning of the observation period, assuming a ladder-like increase as in Figure 3, where H^* is used to represent H_{max} for the total observation period of 2920 daily data.

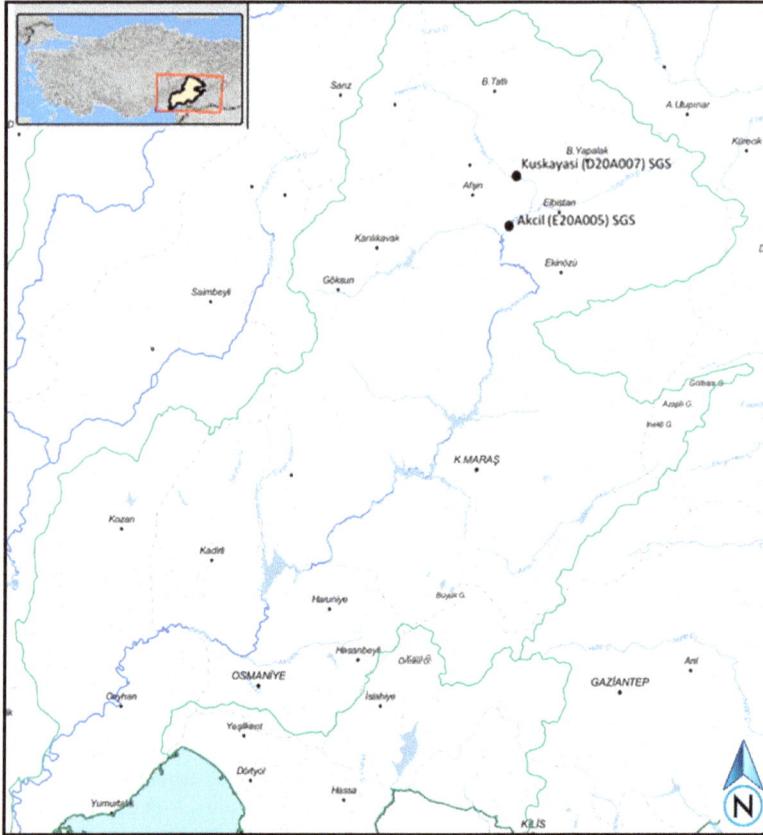

Figure 8. Ceyhan river basin in the south of Turkey and selected monitoring sites (Kuskayasi and Akcil).

Figures 9 and 10 show figures similar to Figure 3 under the assumption of normal and lognormal distributions fit to daily data for 8 years. Although both distributions seem to be sufficient, normal distribution shows more distinctively how $H(X)$ approaches the total entropy H^*. It may seem unusual for an upstream station with daily observations to reflect a normal distribution; yet this is physically due to karstic contributions to runoff, which stabilize the flows.

Whether the normal or lognormal distributions are selected, it can be observed in Figures 9 and 10 that 2920 observations are not sufficient to reach H^*. Although $H(X)$ approaches H^*, the optimal number of observations is not yet reached with only 8 years of observations.

Results for the downstream Akcil station are shown in Figures 11 and 12. Here, again, normal distribution appears to give a better fit to observed data. As can be observed especially in Figure 11, $H(X)$ closely approaches H^* for 8 years of data. If observations could be continued after 8 years of 2920 data, most probably the optimum number of observations would be reached.

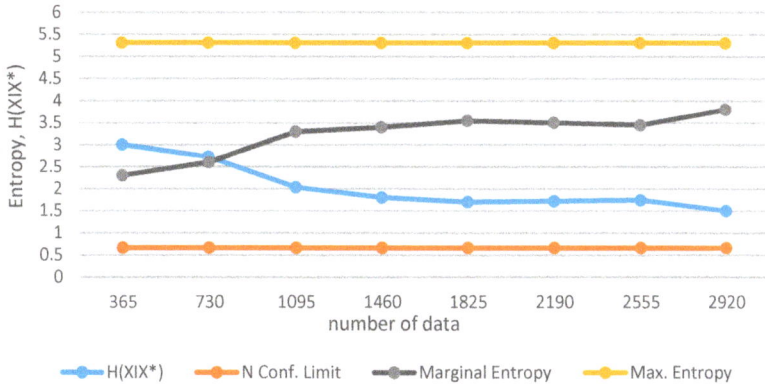

Figure 9. Kuskayasi (1973–1980), by the assumption of a posteriori normal distribution function (where; N Conf. Limit is the limit for normal distribution).

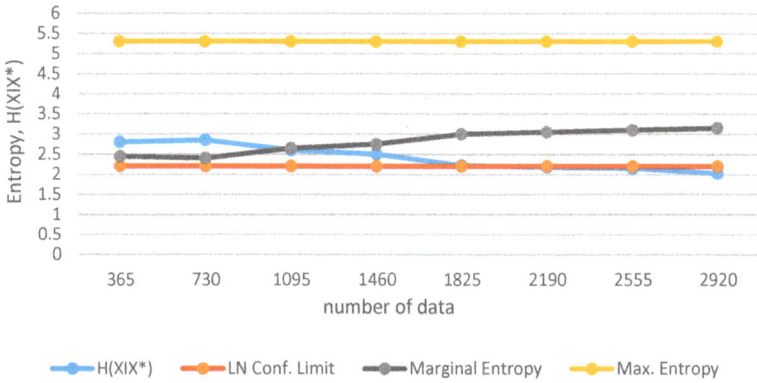

Figure 10. Kuskayasi (1973–1980), by the assumption of a posteriori lognormal distribution function (where; LN Conf. Limit is the limit of confidence for lognormal distribution).

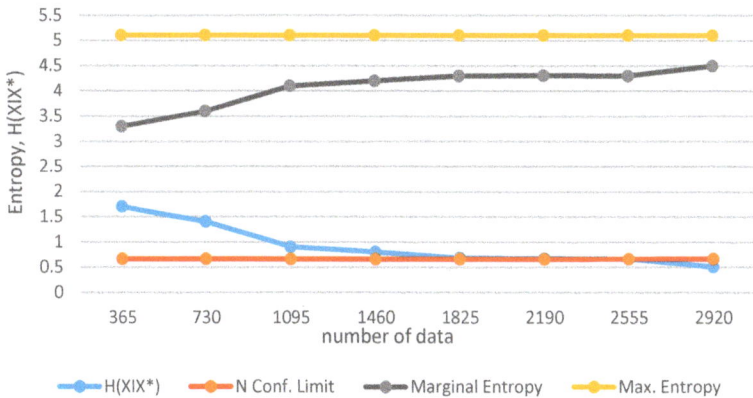

Figure 11. Akcil (1973–1980), by the assumption of a posteriori normal distribution function (where; N Conf. Limit is the limit of confidence for normal distribution).

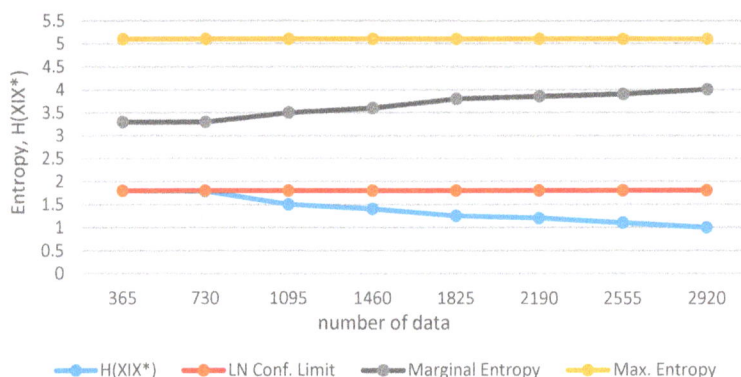

Figure 12. Akcil (1973–1980), by the assumption of a posteriori lognormal distribution function (where; LN Conf. Limit is the limit of confidence for lognormal distribution).

It is concluded on the basis of results obtained through the above application that, if sufficiently long observed time series are available, the entropy principle can be effectively used to infer on an important feature of hydrometric monitoring, i.e., sampling duration or station discontinuance.

7. Conclusions

The extension to the revised definition of informational entropy developed in this paper resolves further major mathematical difficulties associated with the assessment of uncertainty, and indirectly of information, contained in random variables. The description of informational entropy, not as an absolute measure of information but as a measure of the variation of information, has the following advantages:

- It eliminates the controversy associated with the mathematical definition of entropy for continuous probability distribution functions. This makes it possible to obtain a single value for the variation of information instead of several entropy values that vary with the selection of the discretizing interval when, in the former definitions of entropy for continuous distribution functions, discrete probabilities of hydrological events are estimated through relative class frequencies and discretizing intervals.
- The extension to the revised definition introduces confidence limits for the entropy function, which facilitates a comparison between the uncertainties of various hydrological processes with different scales of magnitude and different probability structures.
- Following from the above two advantages, it is further possible through the use of the concept of the variation of information to:

  o determine the contribution of each observation to information conveyed by data;
  o determine the probability distribution function which best fits the variable;
  o make decisions on station discontinuance.

The present work focuses basically on the theoretical background for the extended definition of informational entropy. The methodology is then tested via applications to synthetically generated data and observed runoff data and is shown to give valid results. For real-case observed data, long duration series with sufficient length and quality are needed. Currently, studies are being continued by the authors on long series of runoff, precipitation and temperature data.

It follows from the above discussions that the use of the concept of variation of information and of confidence limits makes it possible to:

- *determine the contribution of each observation to information conveyed by data;*
- *calculate the cost factors per information gained;*
- *determine the probability distribution function which best fits the variable;*
- *select the model which best describes the behavior of a random process;*
- *compare the uncertainties of variables with different probability density functions;*
- *make decisions on station discontinuance.*

The above points are different problems to be solved by the concept of entropy, and further extensions of the methodology are required to address each of them.

Acknowledgments: We gratefully acknowledge the support received from the authors' EU Horizon2020 Project entitled FATIMA (FArming Tools for external nutrient Inputs and water Management, Grant No. 633945) for providing the required funds to cover the costs towards publishing in open access.

Author Contributions: Turkay Baran and Nilgun B. Harmancioglu conceived and designed the experiments; Turkay Baran and Filiz Barbaros performed the experiments; Cem P. Cetinkaya and Filiz Barbaros analyzed the data; Turkay Baran, Nilgun B. Harmancioglu and Cem P. Cetinkaya contributed reagents/materials/analysis tools; Nilgun B. Harmancioglu wrote the paper.

Conflicts of Interest: The authors declare no conflict of interest.

References

1. Singh, V.P.; Fiorentino, M. (Eds.) A Historical Perspective of Entropy Applications in Water Resources. In *Entropy and Energy Dissipation in Water Resources*; Water Science and Technology Library; Kluwer Academic Publishers: Dordrecht, The Netherlands, 1992; Volume 9, pp. 155–173.
2. Fiorentino, M.; Claps, P.; Singh, V.P. An Entropy-Based Morphological Analysis of River Basin Networks. *Water Resour. Res.* **1993**, *29*, 1215–1224. [CrossRef]
3. Wehrl, A. General Properties of Entropy. *Rev. Mod. Phys.* **1978**, *50*, 221–260. [CrossRef]
4. Templeman, A.B. Entropy and Civil Engineering Optimization. In *Optimization and Artificial Intelligence in Civil and Structural Engineering*; NATO ASI Series (Series E: Applied Sciences); Topping, B.H.V., Ed.; NATO: Washington, DC, USA, 1989; Volume 221, pp. 87–105, ISBN 978-94-017-2490-6.
5. Schrader, R. On a Quantum Version of Shannon's Conditional Entropy. *Fortschr. Phys.* **2000**, *48*, 747–762. [CrossRef]
6. Shannon, C.E. A Mathematical Theory of Information. In *The Mathematical Theory of Information*; The University of Illinois Press: Urbana, IL, USA, 1948; Volume 27, pp. 170–180.
7. Harmancioglu, N.; Alpaslan, N. Water Quality Monitoring Network Design: A Problem of Multi-Objective Decision Making. *JAWRA* **1992**, *28*, 179–192. [CrossRef]
8. Harmancioglu, N.; Singh, V.P.; Alpaslan, N. Versatile Uses of The Entropy Concept in Water Resources. In *Entropy and Energy Dissipation in Water Resources*; Singh, V.P., Fiorentino, M., Eds.; Kluwer Academic Publishers: Dordrecht, The Netherlands, 1992; Volume 9, pp. 91–117, ISBN 978-94-011-2430-0.
9. Harmancioglu, N.; Alpaslan, N.; Singh, V.P. Assessment of Entropy Principle as Applied to Water Quality Monitoring Network Design. In *Stochastic and Statistical Methods in Hydrology and Environmental Engineering*; Water Science and Technology Library; Hipel, K.W., McLeod, A.I., Panu, U.S., Singh, V.P., Eds.; Springer: Dordrecht, The Netherlands, 1994; Volume 3, pp. 135–148, ISBN 978-94-017-3083-9.
10. Harmancioglu, N.B.; Yevjevich, V.; Obeysekera, J.T.B. Measures of information transfer between variables. In Proceedings of the Fourth International Hydrology Symposium—Multivariate Analysis of Hydrologic Processes, Fort Collins, CO, USA, 15–17 July 1985; Colorado State University: Fort Collins, CO, USA, 1986; pp. 481–499.
11. Harmancioglu, N.B.; Yevjevich, V. Transfer of hydrologic information among river points. *J. Hydrol.* **1987**, *91*, 103–118. [CrossRef]
12. Harmancioglu, N.; Cetinkaya, C.P.; Geerders, P. Transfer of Information among Water Quality Monitoring Sites: Assessment by an Optimization Method. In Proceedings of the EnviroInfo Conference 2004, 18th International Conference Informatics for Environmental Protection, Geneva, Switzerland, 21–23 October 2004; pp. 40–51.

13. Baran, T.; Bacanli, Ü.G. An Entropy Approach for Diagnostic Checking in Time Series Analysis. *SA Water* **2007**, *33*, 487–496.
14. Singh, V.P. The Use of Entropy in Hydrology and Water Resources. *Hydrol. Process.* **1997**, *11*, 587–626. [CrossRef]
15. Singh, V.P. The Entropy Theory as a Decision Making Tool in Environmental and Water Resources. In *Entropy Measures, Maximum Entropy Principle and Emerging Applications. Studies in Fuzziness and Soft Computing*; Karmeshu, Ed.; Springer: Berlin, Germany, 2003; Volume 119, pp. 261–297, ISBN 978-3-540-36212-8.
16. Harmancioglu, N.; Singh, V.P. Entropy in Environmental and Water Resources. In *Encyclopedia of Hydrology and Water Resources*; Herschy, R.W., Fairbridge, R.W., Eds.; Kluwer Academic Publishers: Dordrecht, The Netherlands, 1998; Volume 5, pp. 225–241, ISBN 978-1-4020-4497-7.
17. Harmancioglu, N.; Singh, V.P. Data Accuracy and Data Validation. In *Encyclopedia of Life Support Systems (EOLSS)*; Knowledge for Sustainable Development, Theme 11 on Environmental and Ecological Sciences and Resources, Chapter 11.5 on Environmental Systems; Sydow, A., Ed.; UNESCO Publishing-Eolss Publishers: Oxford, UK, 2002; Volume 2, pp. 781–798, ISBN 0 9542989-0-X.
18. Harmancioglu, N.B.; Ozkul, S.D. Entropy-based Design Considerations for Water Quality Monitoring Networks. In *Technologies for Environmental Monitoring and Information Production*; Nato Science Series (Series IV: Earth and Environmental Sciences); Harmancioglu, N.B., Ozkul, S.D., Fistikoglu, O., Geerders, P., Eds.; Springer: Dordrecht, The Netherlands, 2003; Volume 23, pp. 119–138, ISBN 978-94-010-0231-8.
19. Ozkul, S.; Harmancioglu, N.B.; Singh, V.P. Entropy-Based Assessment of Water Quality Monitoring Networks. *J. Hydrol. Eng.* **2000**, *5*, 90–100. [CrossRef]
20. Jaynes, E.T. *E. T. Jaynes: Papers on Probability, Statistics and Statistical Physics*; Rosenkrantz, R.D., Ed.; Springer: Dordrecht, The Netherlands, 1983; ISBN 978-94-009-6581-2.
21. Guiasu, S. *Information Theory with Applications*; Mc Graw-Hill: New York, NY, USA, 1977; 439p, ISBN 978-0070251090.
22. Harmancioglu, N.B. Measuring the Information Content of Hydrological Processes by the Entropy Concept. *J. Civ. Eng. Fac. Ege Univ.* **1981**, 13–88.
23. Jaynes, E.T. Information Theory and Statistical Mechanics. *Phys. Rev.* **1957**, *106*, 620–630. [CrossRef]
24. Harmancioglu, N. Entropy concept as used in determination of optimal sampling intervals. In Proceedings of the Hydrosoft 1984, International Conference on Hydraulic Engineering Software. Interaction of Computational and Experimental Methods, Portorož, Yugoslavia, 10–14 September 1984; Brebbia, C.A., Maksimovic, C., Radojkovic, M., Eds.; Editions du Tricorne: Geneva, Switzerland; pp. 99–110.
25. Harmancioglu, N.B. An Entropy-based approach to station discontinuance. In *Stochastic and Statistical Methods in Hydrology and Environmental Engineering*; Time Series Analysis and Forecasting; Hipel, K.W., McLeod, I., Eds.; Kluwer Academic Publishers: Dordrecht, The Netherlands, 1994; Volume 3, pp. 163–176.
26. Harmancioglu, N.B.; Alpaslan, N. *Basic Approaches to Design of Water Quality Monitoring Networks*; Water Science and Technology; Elsevier: Amsterdam, The Netherlands, 1994; Volume 30, pp. 49–56.
27. Harmancioglu, N.B.; Cetinkaya, C.P.; Barbaros, F. Environmental Data, Information and Indicators for Natural Resources Management. In *Practical Environmental Statistics and Data Analysis*; Rong, Y., Ed.; ILM Publications: Buchanan, NY, USA, 2011; Chapter 1, pp. 1–66.
28. Schultze, E. Einführung in die Mathematischen Grundlagen der Informationstheorie. In *Lecture Notes in Operations Research and Mathematical Economics*; Springer: Berlin, Germany, 1969; 116p, ISBN 978-3-642-86515-2.
29. Vapnik, V.N. *Statistical Learning Theory*; Wiley Interscience: New York, NY, USA, 1998; 736p, ISBN 978-0-47-03003-4.
30. Harmancioglu, N.B.; Singh, V.P.; Alpaslan, N. *Environmental Data Management*; Water Science and Technology Library; Kluwer Academic Publishers: Dordrecht, The Netherlands, 1998; 298p, ISBN 0792348575.
31. Harmancioglu, N.B.; Fistikoglu, O.; Ozkul, S.D.; Singh, V.P.; Alpaslan, N. *Water Quality Monitoring Network Design*; Water Science and Technology Library; Kluwer Academic Publishers: Dordrecht, The Netherlands, 1999; 290p, ISBN 978-94-015-9155-3.

entropy

MDPI

Article

Entropy Parameter *M* in Modeling a Flow Duration Curve

Yu Zhang [1], Vijay P. Singh [1,2,*] and Aaron R. Byrd [3]

[1]   Department of Biological and Agricultural Engineering, Texas A&M University,
      College Station, TX 77840, USA; zhangyu199002@tamu.edu
[2]   Zachry Department of Civil Engineering, Texas A&M University, College Station, TX 77843-2117, USA
[3]   Hydrologic Systems Branch, Coastal and Hydraulics Laboratory, Engineer Research Development Center,
      U.S. Army Corps of Engineers, Vicksburg, MS 39181, USA; Aaron.R.Byrd@erdc.dren.mil
*   Correspondence: vsingh@tamu.edu; Tel.: +1-979-845-7028

Received: 20 September 2017; Accepted: 30 November 2017; Published: 1 December 2017

Abstract: A flow duration curve (FDC) is widely used for predicting water supply, hydropower, environmental flow, sediment load, and pollutant load. Among different methods of constructing an FDC, the entropy-based method, developed recently, is appealing because of its several desirable characteristics, such as simplicity, flexibility, and statistical basis. This method contains a parameter, called entropy parameter *M*, which constitutes the basis for constructing the FDC. Since *M* is related to the ratio of the average streamflow to the maximum streamflow which, in turn, is related to the drainage area, it may be possible to determine *M* a priori and construct an FDC for ungauged basins. This paper, therefore, analyzed the characteristics of *M* in both space and time using streamflow data from 73 gauging stations in the Brazos River basin, Texas, USA. Results showed that the *M* values were impacted by reservoir operation and possibly climate change. The values were fluctuating, but relatively stable, after the operation of the reservoirs. Parameter *M* was found to change inversely with the ratio of average streamflow to the maximum streamflow. When there was an extreme event, there occurred a jump in the *M* value. Further, spatially, *M* had a larger value if the drainage area was small.

Keywords: flow duration curve; Shannon entropy; entropy parameter; modeling; spatial and dynamics characteristic

---

1. Introduction

A flow duration curve (FDC) is usually constructed empirically by plotting discharge against the percentage of time the discharge is equaled or exceeded during the year. Discharge from a gauge station can be daily, weekly, or monthly. The timescale of discharge depends on the use of FDC. For example, weekly discharge may be adequate for water supply, daily discharge for hydropower, and monthly discharge for sediment load and pollutant load [1,2]. Nonparametric methods use the record of discharge for the whole period for constructing an FDC and make no probabilistic statements about a given calendar or water year, because all the years of record are combined together into a whole period, so a return period cannot be assigned.

The methods for predicting an FDC are either deterministic or stochastic. For a given year of streamflow record at a station, an annual flow duration curve (AFDC) can be constructed [3,4]. With AFDCs of all the years at a given station, at each exceedance probability discharge percentiles can be determined given a return period. This leads to a final FDC with probabilistic statements by assigning return periods to individual AFDCs.

Singh [5] related dimensionless discharge with drainage area and constructed an exponential form of FDC using a deterministic model. Vogel and Fennessey [4] used an AFDC to define recurrence intervals for FDCs. Cigizoglu and Bayazit [6] modeled FDCs by introducing stream flow as a product of two variables, which represented the periodic and stochastic components. Castellarin et al. [7] developed a five-parameter stochastic model which combined annual flow distribution and standardized the daily flow distribution of the basin to simulate FDC and AFDC percentiles for the whole period of record. All of these studies made a series of assumptions because of statistical components, such as variables that are independent and identically distributed. Singh et al. [8] introduced Shannon entropy theory for modeling FDC, where the entropy of discharge or the probability density function (PDF) was used to express the uncertainty of flow. This method needs no fitting for the whole period of discharge record and no assumption about daily flow. This method contains an entropy parameter M which plays a fundamental role in the derivation of FDC. The objective of this paper was, therefore, to further study the temporal and spatial characteristics of the entropy parameter M in the entropy-based method and apply the method to 73 sites in Brazos River basin, Texas, USA.

2. Materials and Methods

The derivation of the FDC and the study area are described in this section. For the dataset, codes, and software information used in this paper, please see the Supplementary Materials. For the derivation of the FDC, first, the entropy of discharge is introduced, then the constraints for the probability density function (PDF) are determined. Second, entropy maximizing is conducted by using the method of Lagrange multipliers and solved numerically. Third, the cumulative probability distribution function (CDF) is embedded in the process and a relationship between the discharge and exceedance period is derived.

2.1. Derivation of FDC

The derivation of FDC using Shannon entropy is detailed in Singh et al. [8]. For the sake of completeness, a brief synopsis is given here. The Shannon entropy of discharge (Q) or $f(Q)$ [$H(Q)$] can be expressed as:

$$H = -\int_{Q_{min}}^{Q_{max}} f(Q) \ln[f(Q)]dQ \tag{1}$$

where Q_{min} and Q_{max} are the minimum and maximum discharges, respectively, and $f(Q)$ is the PDF of Q. The objective is to derive $f(Q)$ by maximizing H for which two constraints are defined as:

$$C1 = \int_{Q_{min}}^{Q_{max}} f(Q)dQ \tag{2}$$

$$C2 = \int_{Q_{min}}^{Q_{max}} Qf(Q)dQ = \overline{Q} = Q_m \tag{3}$$

where Q_m is the mean discharge. Entropy maximizing is done using the method of Lagrange multipliers:

$$L = -\int_{Q_{min}}^{Q_{max}} f(Q)lnf(Q)dQ - (\lambda_0 - 1)\left(\int_{Q_{min}}^{Q_{max}} f(Q)dQ - C1\right) - \lambda_1\left(\int_{Q_{min}}^{Q_{max}} Qf(Q)dQ - C2\right) \tag{4}$$

where L is the Lagrangian function, and λ_0 and λ_1 are the unknown Lagrangian multipliers. Differentiating Equation (4) with respect to $f(Q)$ and equating the derivative to zero yield the PDF of Q as:

$$f(Q) = exp(-\lambda_0 - \lambda_1 Q) \tag{5}$$

Substitution of Equation (5) in Equations (2) and (3) yields the solution for λ_0 and λ_1:

$$\lambda_0 = -ln\lambda_1 + ln[exp(-\lambda_1 Q_{min}) - exp(-\lambda_1 Q_{max})] \tag{6}$$

$$-\frac{1}{\lambda_1} - \frac{Q_{min}exp(-\lambda_1 Q_{min}) - Q_{max}exp(-\lambda_1 Q_{max})}{exp(-\lambda_1 Q_{min}) - exp(-\lambda_1 Q_{max})} = -\overline{Q} \tag{7}$$

The entropy parameter M is defined as $\lambda_1 Q_{max}$.

In order to construct an FDC, a relation between the CDF of Q and time needs to be hypothesized. A possible form of CDF can be expressed as:

$$F(Q) = 1 - a\left(\frac{t}{T}\right)^b \tag{8}$$

where a and b are coefficients, t is the number of days that discharge is being equaled or exceeded, and T is the total number of days for a year. Parameters a and b can be estimated by empirical fitting and it is hoped that they will be relatively stable.

Differentiating Equation (8) we obtain:

$$dF(Q) = f(Q)dQ = -ab\left(\frac{1}{T}\right)^b t^{b-1}dt \tag{9}$$

Substituting Equation (5) into Equation (9), integrating from Q to Q_{max}, replacing the term $exp(\lambda_0)$ from Equation (6) and replacing $\lambda_1 Q_{max}$ with M, the final FDC is obtained as:

$$\frac{Q}{Q_{max}} = -\frac{1}{M}ln\left\{exp(-M) - \left[exp(-M) - exp\left(-\frac{MQ_{min}}{Q_{max}}\right)\right]a\left(\frac{t}{T}\right)^b\right\} \tag{10}$$

Equation (10) contains Q_{max}, and Q_{min} which are known from observations, and M which can be calculated using Equation (7).

2.2. Study Area

The entropy parameter M was determined from observations and its space-time characteristics were then investigated. It was also related to the drainage area. Then, FDC was constructed and its reliability was assessed.

The study area was Brazos River basin (Figure 1) which extends from Eastern New Mexico to Southeastern Texas, up to the Gulf of Mexico. The basin has a length of approximately 1219 km and a width varying from about 133 km in the High Plains in the upper basin to a maximum of 210 km in the vicinity of the city of Waco, to about 19 km near the city of Richmond in the lower basin. The basin drainage area is approximately 116,550 square kilometers, with about 111,370 square kilometers in Texas and the remainder in New Mexico [9]. There are 73 gauging stations with discharge records 50 years long that were analyzed in this paper. Daily maximum, minimum and mean discharges; and reservoir and gauge station information were collected from the USGS website (https://waterdata.usgs.gov/nwis).

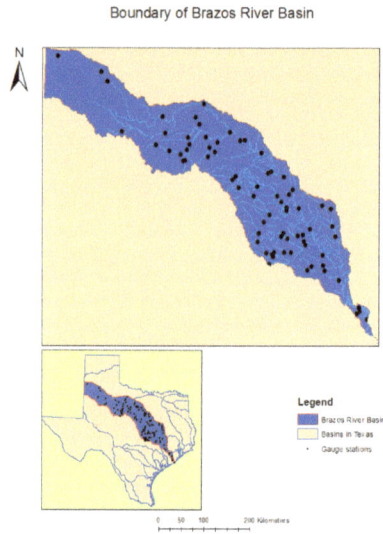

Figure 1. Brazos River basin.

3. Results and Discussion

3.1. Flow Duration Curve Estimation

The entropy parameter M is defined as $\lambda_1 Q_{max}$, where λ_1 can be obtained from Equation (7) by numerical solution with the observed Q_{max}, Q_{min} and $\overline{Q}$.

Using Equation (10) and the observed data, an FDC was constructed using the entropy parameter M, as shown in Figure 2.

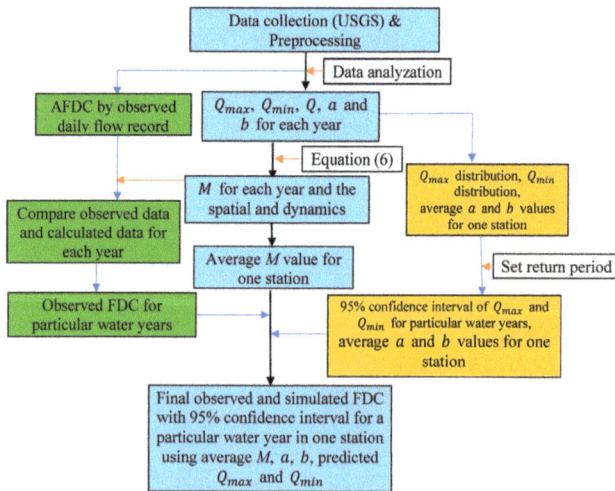

Figure 2. Schematic of the FDC construction.

First, the FDC of a specific year for a station was analyzed. Taking station 08093100 as an example, for 2009, M, calculated from Equation (7), equaled 10.47. After constructing the FDC for observations, parameters a and b were calculated using Equation (8) as $a = 1.021$ and $b = 0.778$. Substituting M, Q_{max}, Q_{min}, a, and b in Equation (10), we estimated the FDC. The correlation coefficient (R^2) between the observed and estimated FDCs was 0.969, which showed a good agreement, as shown in Figures 3 and 4.

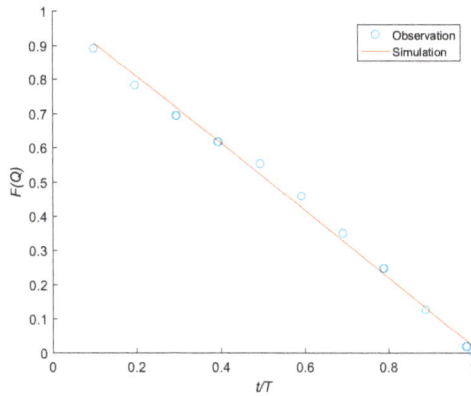

Figure 3. Relationship between F(Q) and t/T of station 08093100 in 2009.

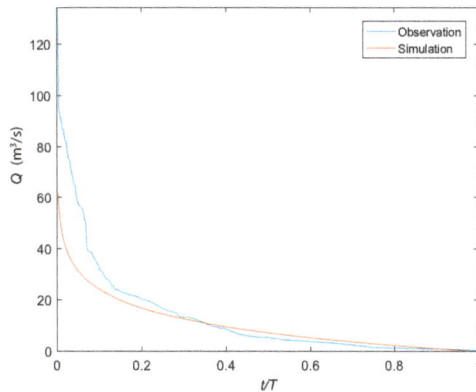

Figure 4. FDC of station 08093100 in 2009.

Second, the FDC was predicted for a particular hydrologic year using average values of M, a, and b for one station. For station 08093100, a, b, and M were calculated for each year and their histograms were constructed, as shown in Figures 5 and 6, and then their average values were estimated for the station. For the prediction of FDC, we needed to estimate Q_{max}, Q_{min}, and $\overline{Q}$ first by fitting the gamma distribution to each data set, as shown in Figures 7–9. For return periods of 1.3-year, 1.4-year, and 1.8-year, the estimated Q_{max}, Q_{min}, and $\overline{Q}$ with 95% confidence intervals were calculated, as shown in Table 1. The observed hydrologic years of 1.3-year, 1.4-year, and 1.8-year return periods were 2003, 2009, and 1994. The reason why we chose these years is that we wanted to focus on simulation for the recent years using parameters for a station. In addition, it showed that not all the stations followed good fitting, which is explained at the end of this section. Then, FDCs were predicted and compared with observed FDCs. The R^2 values of the predicted and observed FDCs were 0.979, 0.969, and 0.960, respectively. Figures 10–12 show that 95% intervals covered most of the observed data. The same was done for other stations in the basin.

Table 1. Q_{max}, Q_{min}, a, b, and M for different water years for station 08093100.

Water Year	Year	Q_{max}	Q_{min}	LI Q_{max}	LI Q_{min}	UI Q_{max}	UI Q_{min}	a	b	M	R^2
1.3	2003	121.26	0.21	44.89	0.08	302.58	0.54				0.979
1.4	2009	169.09	0.3	68.16	0.12	395.75	0.7	1.02	0.89	9.88	0.969
1.8	1994	257.14	0.46	114.31	0.2	558.27	0.99				0.96

Note: LI means lower interval, UI means upper interval, discharge unit is m^3/s.

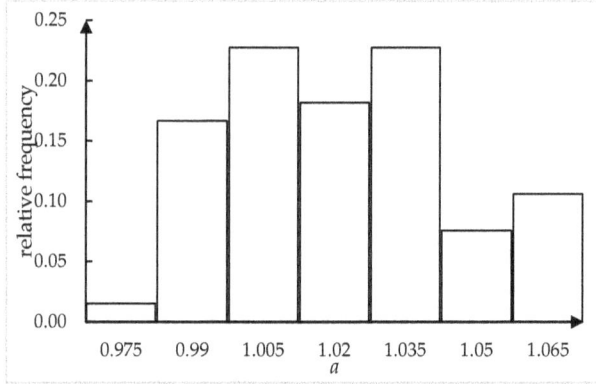

Figure 5. Relative frequency of parameter a at station 08093100.

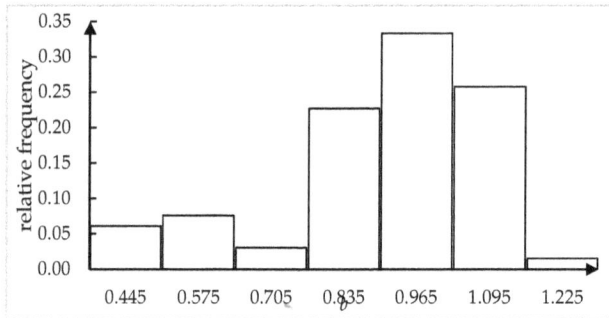

Figure 6. Relative frequency of parameter b at station 08093100.

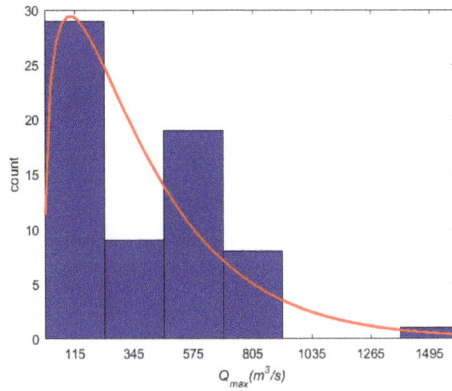

Figure 7. Gamma distribution fitting of the maximum discharge for station 08093100.

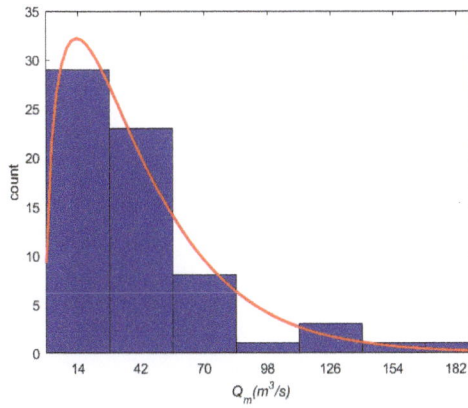

Figure 8. Gamma distribution fitting of the mean discharge for station 08093100.

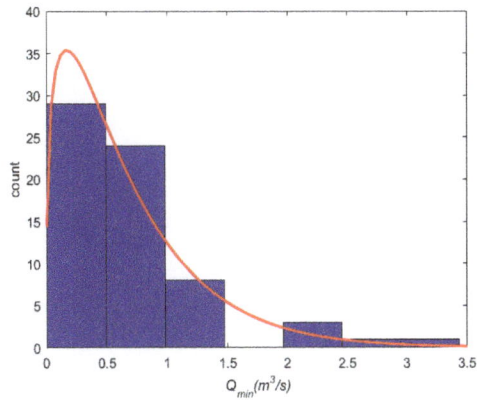

Figure 9. Gamma distribution fitting of the minimum discharge for station 08093100.

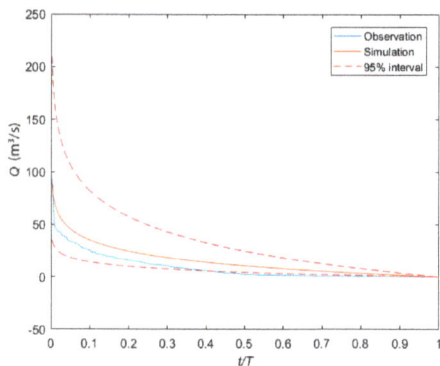

Figure 10. Estimation of FDC for 2003 using average *M*, *a*, and *b* values of station 08093100.

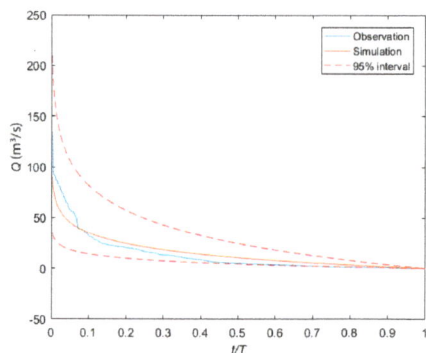

Figure 11. Estimation of FDC for 2009 using average *M*, *a*, and *b* values of station 08093100.

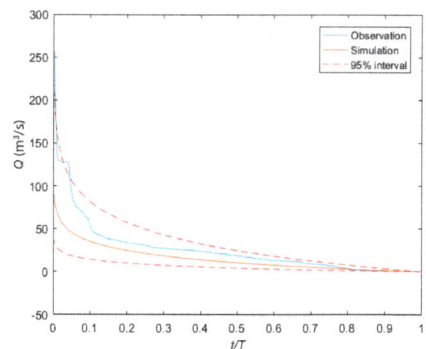

Figure 12. Estimation of the FDC for year 1994 using average *M*, *a*, and *b* values of station 08093100.

It was observed that the predicted FDCs fit well at most of the stations when discharges were relatively small, but were slightly poorer in the parts having large discharge values. Prediction for each year showed that R^2 was not always good. Figure 13 showed a good fit for the relationship with the ratio of $\overline{Q}$ and Q_{max}. When $\overline{Q}/Q_{max} \geq 0.10$, $R^2 \geq 0.90$. Further investigation could focus on making adjustments for better FDC prediction.

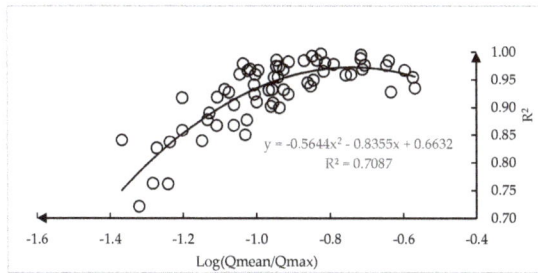

Figure 13. Relationship of R^2 and $\overline{Q}/Q_{max}$ for station 08093100.

3.2. Time Variability of M

The stream flow changes because of natural and anthropogenic factors, such as reservoir operation and climate change. First, we mapped the locations of reservoirs in the basin and analyzed the impact of reservoir on the time variability of M values. Reservoir locations, in part, are shown in Figure 14. As an example, we picked three stations, 08093100, 08099500, and 08093360, which were downstream of Whitney reservoir, Proctor reservoir, and Aquilla reservoir, respectively. The M values of these stations are shown in Figure 15a–c. For station 08093100, before 1951, the M value fluctuated, while after 1951 it was relatively stable because of the impact of the Whitney reservoir operation. The mean M value was 11.15 for the whole period, while the mean M value after 1951 was 9.88. It can be seen that the reservoir operation had a 12.85% influence on the M values for this station. However, our interest was in the period after 1951. Stations 08099500 and 08093360 had the same situation as did station 08093100, that is, the M values were fluctuating before the reservoir operation, but were stable thereafter. These stations were affected by the reservoirs by 189.15% and 43.82%, respectively. Similarly, there were other reservoirs in the basin which had an impact on the stations downstream of the reservoirs. For further analysis, we just chose record periods after the reservoir impact. After removing the impact of reservoirs, it was observed that the M values were relatively stable with time. At some stations, however, the M values jumped or fluctuated in some particular years.

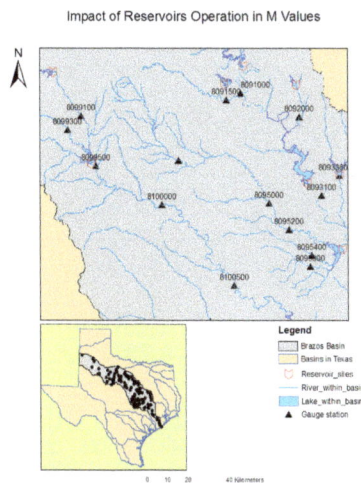

Figure 14. Locations of reservoirs and stations (middle part of the basin, the scalar applies to the basin panel).

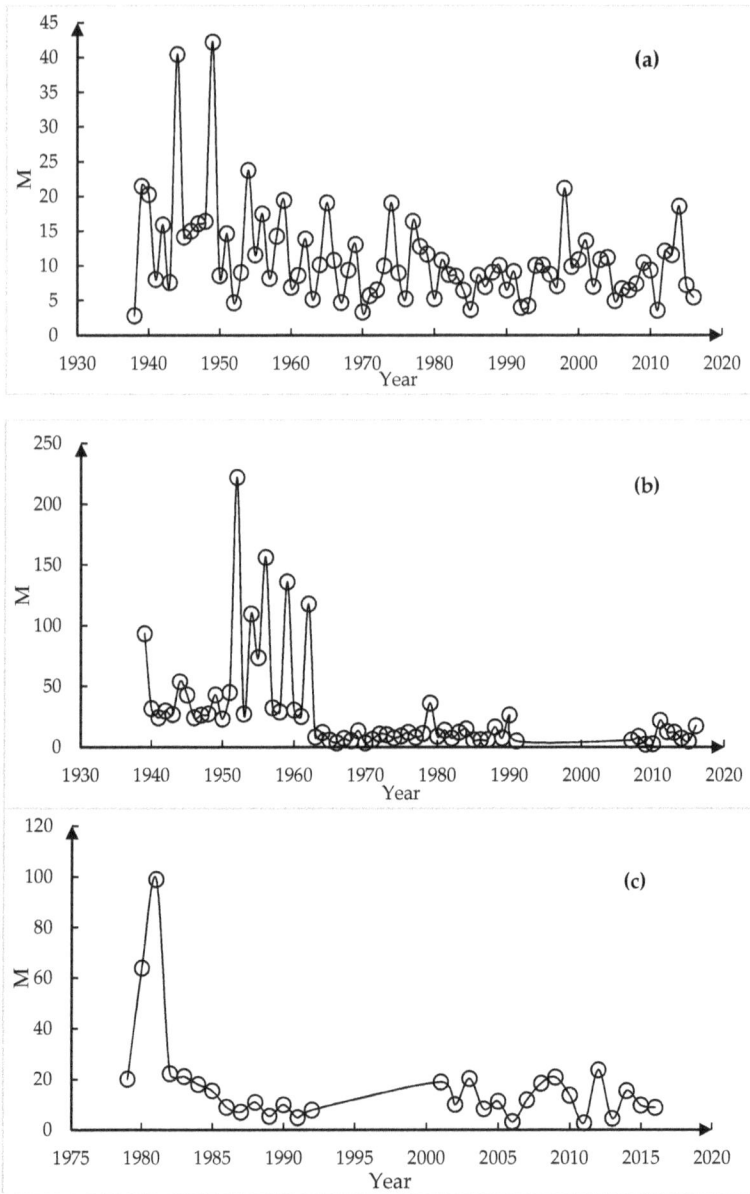

Figure 15. (a) M dynamics at station 08093100; (b) M dynamics at station 08099500; and (c) M dynamics at station 08093360.

Second, we determined the effect of climate change on the M values. M was defined as the Lagrange multiplier λ_1 times Q_{max}, as expressed by Equation (6), which relates it to Q_{max}, Q_{min}, and $\overline{Q}$. Though Equation (6) is slightly complicated, it can be simplified by setting Q_{min} equal to zero, which can usually be assumed to be near zero (it is true at most of the stations in the Brazos River basin). Then we found that M had an inverse relation with the ratio of $\overline{Q}$ and Q_{max}, as shown in figures plotting M and the ratio (Figures 16 and 17)

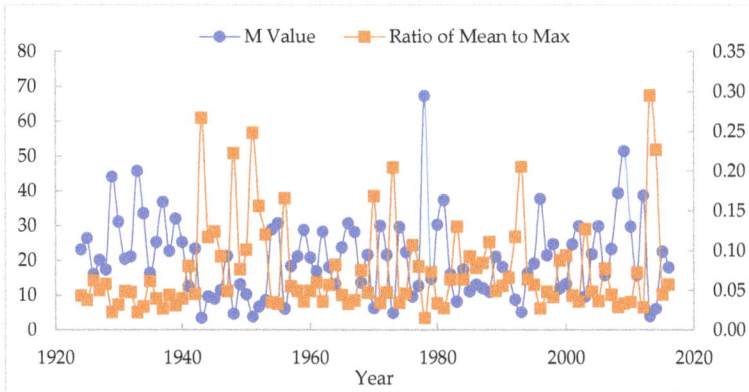

Figure 16. Correlation between M values and the ratio of $\overline{Q}$ and Q_{max} at station 08089000.

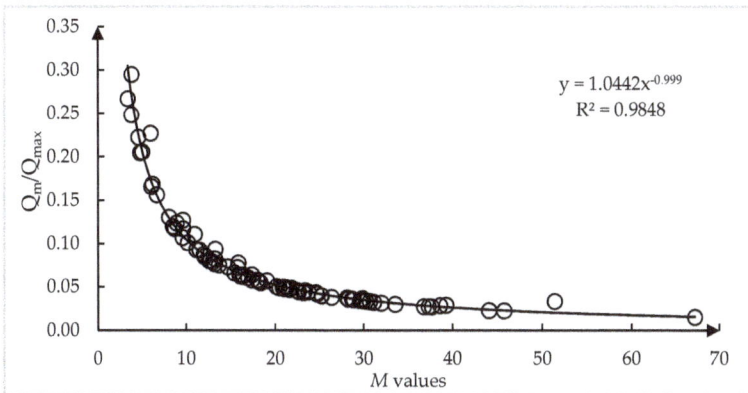

Figure 17. Powered relationship between M value and ratio of $\overline{Q}$ and Q_{max} at station 08089000.

Upon calculating M, the effect of climate change was determined. Studies on the impact of climate change on river discharge show that different parts of the basin have different impacts [10,11]. Discharge in a river can increase or decrease due to the impact of climate change and so can the ratio of $\overline{Q}$ and Q_{max}. Taking station 08089000 as an example, it can be seen from Figure 3 that the relation between M and the ratio had a correlation coefficient of -0.74, indicating that a high ratio is usually related to a low M value. At the same time, it was noticed that the M value had a dramatic jump in 1978 when Tropical Storm Amelia happened and caused a large storm in Texas [12]. It can be seen from Figure 18 that, in 1978, where there was an impact of the storm, there was a jump in the M value. This showed how M values reflected the change in flow characteristics related to the weather.

M values in 1978 compared to the mean M value

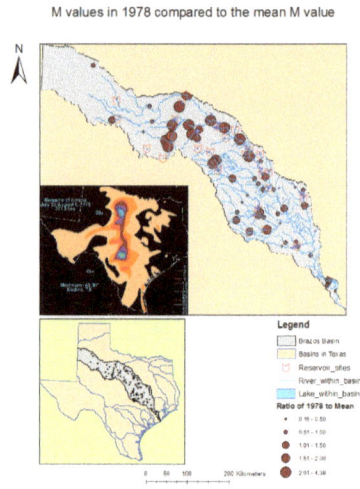

Figure 18. *M* values in 1978 compared to the mean *M* value.

The next step was to determine what other characteristics could be related to the M values, because the final goal was to apply this method to ungauged basins.

3.3. Spatial Variability

After calculating the *M* values for 73 stations and considering the impact of reservoirs, the mean *M* value was computed for each station. It was found that the *M* values ranged from 8.14 to 123.72. The lowest value occurred at gauge 08116650, which is located in the downstream part of the basin, and the highest value occurred at gauge 08086290, which is located in the middle-upper part of the basin. It can be seen from the map that most of the area in the upstream part had higher *M* values, higher than 55, the middle part had a range from 45 to 55, and the downstream areas had *M* less than 45. This showed a trend of decreasing *M* values from the upstream to the downstream part. It seems that the *M* values changed spatially because the drainage area changed, as shown in Figure 19, where if there was a small drainage area, then there was a large *M* value contour.

Spacial Characters of M Values in Brazos River Basin

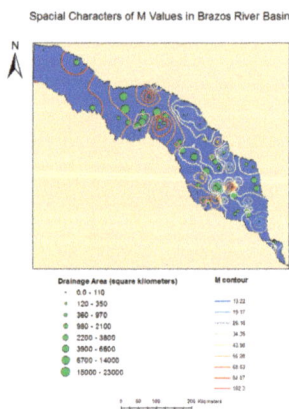

Figure 19. Spatial features of drainage areas and *M* values.

Fuller [13] developed a relation between Q_{max} and $\overline{Q}$ as:

$$\frac{Q_{max} - \overline{Q}}{\overline{Q}} = 1.5A^{-0.3} \tag{11}$$

where A is the drainage area (square kilometers). This relationship indicates that the ratio of $\overline{Q}$ and Q_{max} would increase with an increase in the drainage basin size. Since M has an inverse relation with the ratio of $\overline{Q}$ and Q_{max}, M also has an inverse relation with basin size, which can be reflected by the correlation coefficient -0.536 and the plot of M versus the drainage size (drainage area) in Figure 20.

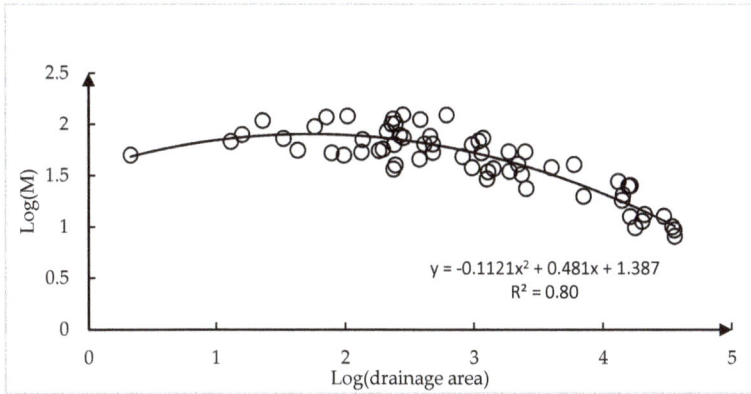

Figure 20. Relationship between the drainage area and M values.

3.4. Test for Ungauged Stations

We used station 08098290, assuming it as an ungauged station to test for the reliability of applying the function. First, following the schematic in Figure 2 and using the records from the station, we obtained the M value as the true value. Second, we estimated the M value using Equation (12):

$$\log(M) = -0.112[\log(A)]^2 + 0.481\log(A) + 1.387 \tag{12}$$

where A is the drainage area in square kilometers. The M value derived from records of observed data was 13.26. The M value simulated from the function was 14.23, which had a 7.31% difference. Third, we used both M values to form an FDC, compared to the empirical FDC, respectively, and calculated R^2 for both sides. Using the calculated M value led to a mean $R^2 = 0.91$, which ranged from 0.70 to 0.99, and simulated M led to mean $R^2 = 0.89$ which ranged from 0.68 to 0.95 which had a 2.20% difference with the calculated one. At last, we applied Equation (12) to all the stations in the basin and got simulated M for all the stations. The mean $R^2 = 0.86$ for the basin ranged from 0.58 to 0.93, while the calculated M from the records led to an $R^2 = 0.88$ and ranged from 0.61 to 0.95, which showed a mean difference between the results from the calculated and simulated M of 2.32%. Those test results indicated that the function can be applied to other ungauged stations.

4. Conclusions

This study analyzed in time and space the entropy parameter M which is basic to the entropy-based method for constructing the flow duration curve. Upon analysis of 73 stations in the basin, M ranged from 8.14 to 123.72, and was apparently impacted by anthropogenic and natural factors. Temporal patterns changed because of reservoir operation and flow characteristics. At the same time, M changed spatially with the drainage area. By analyzing the spatial and temporal

Entropy **2017**, *19*, 654

characteristics of M, a relation between M and drainage area was developed, a log-based function was fitted as y = $-0.112x^2$ + 0.388x + 1.567, which can be used in other basins. For most of the years, the average M yielded a good agreement between predicted and observed FDCs, where the mean R^2 was 0.92. Some years did not have good fit, especially in large discharge parts of the FDC; the reason why this occurred should be studied further. The procedure of applying the entropy parameter M for modeling the FDC can be extended to other basins. Further studies such as the adaptation to other basins, and improvement for the goodness of fit should be investigated.

Supplementary Materials: The following are available online at www.mdpi.com/1099-4300/19/12/654/s1, Section 1: Data Availability, Section 2: Code and Software.

Acknowledgments: This study was in part supported by project "Quantifying Uncertainty of Probable Maximum Flood (PMF)," project no. W912HZ-16-C-0027, funded by the U.S. Army Corps of Engineers, Engineering Research Development Center, Vicksburg, Mississippi, USA.

Author Contributions: Vijay P. Singh, Yu Zhang, and Aaron R. Byrd, conceived the idea and designed the experiments; Yu Zhang performed the experiments; Yu Zhang analyzed and Vijay P. Singh supervised the data; Aaron R. Byrd contributed to discussion and analysis; Yu Zhang and Vijay P. Singh wrote the paper; and Aaron R. Byrd helped with revision. All three authors, Yu Zhang, Vijay P. Singh, and Aaron R. Byrd, contributed to the paper throughout its preparation.

Conflicts of Interest: The authors declare no conflict of interest.

References

1. Atieh, M.; Taylor, G.; Sattar, A.M.; Gharabaghi, B. Prediction of Flow Duration Curves for Ungauged Basins. *J. Hydrol.* **2017**, *545*, 383–394. [CrossRef]
2. Atieh, M.; Gharabaghi, B.; Rudra, R. Entropy-Based Neural Networks Model for Flow Duration Curves at Ungauged Sites. *J. Hydrol.* **2015**, *529*, 1007–1020. [CrossRef]
3. LeBoutillier, D.V.; Waylen, P.R. A stochastic model of flow duration curves. *Water Resour. Res.* **1993**, *29*, 3535–3541. [CrossRef]
4. Vogel, R.M.; Fennessey, N.M. Flow-duration curves. I: New interpretation and confidence intervals. *J. Water Resour. Plan. Manag.* **1994**, *120*, 485–504.
5. Singh, K.P. Model flow duration and streamflow variability. *Water Resour. Res.* **1971**, *7*, 1031–1036. [CrossRef]
6. Cigizoglu, H.K.; Bayazit, M. A generalized seasonal model for flow duration curve. *Hydrol. Process.* **2000**, *14*, 1053–1067. [CrossRef]
7. Castellarin, A.; Vogel, R.M.; Brath, A. A stochastic index flow model of flow duration curves. *Water Resour. Res.* **2004**, *40*. [CrossRef]
8. Singh, V.P.; Byrd, A.; Cui, H. Flow duration curve using entropy theory. *J. Hydrol. Eng.* **2013**, *19*, 1340–1348. [CrossRef]
9. Wurbs, R.A.; Bergman, C.E.; Carriere, P.E.; Walls, W.B. *Hydrologic and Institutional Water Availability in the Brazos River Basin*; Technical and Special Reports; Texas Water Resources Institute: College Station, TX, USA, 1988.
10. Christensen, N.S.; Wood, A.W.; Voisin, N.; Lettenmaier, D.P.; Palmer, R.N. The effects of climate change on the hydrology and water resources of the Colorado River basin. *Clim. Chang.* **2004**, *62*, 337–363. [CrossRef]
11. Tao, B.; Tian, H.; Ren, W.; Yang, J.; Yang, Q.; He, R.; Lohrenz, S.E. Increasing Mississippi river discharge throughout the twenty-first century influenced by changes in climate, land use and atmospheric CO_2. In Proceedings of the AGU Fall Meeting Abstracts, San Francisco, CA, USA, 15–19 December 2014.
12. Roth, D. *Texas Hurricane History*; National Weather Service: Camp Springs, MD, USA, 2010.
13. Fuller, W.E. *Flood Flows*; Transactions of the American Society of Civil Engineers: New York, NY, USA, 1914; Volume 77, pp. 564–617.

entropy

MDPI

Review

Tsallis Entropy Theory for Modeling in Water Engineering: A Review

Vijay P. Singh [1,2,*], Bellie Sivakumar [3,4,*] and Huijuan Cui [5]

[1] Department of Biological and Agricultural Engineering, Texas A&M University, College Station, TX 77843-2117, USA
[2] Zachry Department of Civil Engineering, Texas A&M University, College Station, TX 77843-2117, USA
[3] School of Civil and Environmental Engineering, The University of New South Wales, Sydney, NSW 2052, Australia
[4] Department of Land, Air and Water Resources, University of California, Davis, CA 95616, USA
[5] Key Laboratory of Land Surface Pattern and Simulation, Institute of Geographic Sciences and Natural Resources Research, Chinese Academy of Sciences, Beijing 100101, China; cuihj@igsnrr.ac.cn
* Correspondence: vsingh@tamu.edu (V.P.S.); s.bellie@unsw.edu.au (B.S.); Tel.: +1-979-845-7028 (V.P.S.); +61-2-93855072 (B.S.)

Received: 15 September 2017; Accepted: 23 November 2017; Published: 27 November 2017

Abstract: Water engineering is an amalgam of engineering (e.g., hydraulics, hydrology, irrigation, ecosystems, environment, water resources) and non-engineering (e.g., social, economic, political) aspects that are needed for planning, designing and managing water systems. These aspects and the associated issues have been dealt with in the literature using different techniques that are based on different concepts and assumptions. A fundamental question that still remains is: Can we develop a unifying theory for addressing these? The second law of thermodynamics permits us to develop a theory that helps address these in a unified manner. This theory can be referred to as the *entropy theory*. The thermodynamic entropy theory is analogous to the Shannon entropy or the information theory. Perhaps, the most popular generalization of the Shannon entropy is the Tsallis entropy. The Tsallis entropy has been applied to a wide spectrum of problems in water engineering. This paper provides an overview of Tsallis entropy theory in water engineering. After some basic description of entropy and Tsallis entropy, a review of its applications in water engineering is presented, based on three types of problems: (1) problems requiring entropy maximization; (2) problems requiring coupling Tsallis entropy theory with another theory; and (3) problems involving physical relations.

Keywords: entropy; water engineering; Tsallis entropy; principle of maximum entropy; Lagrangian function; probability distribution function; flux concentration relation

1. Introduction

Water resources systems serve a multitude of human needs. They are needed for water supply, water transfer, water diversion, irrigation, land reclamation, drainage, flood control, hydropower generation, river training, navigation, coastal protection, pollution abatement, transportation and recreation, among others. Many of the systems (e.g., channels, culverts, impoundments) have been with us since the birth of human civilization. Some (e.g., spillways, small dams, levees) are several centuries old, while some others (e.g., large dams, long-distance water transfer structures) are of more recent origin. In the beginning, systems were designed more or less empirically. Then, engineering and economics constituted the sole foundation of design. About fifty years ago, planning and design of hydraulic structures went through a dramatic metamorphosis. These days, they are based on both engineering and non-engineering aspects. Engineering aspects encompass planning, development, design, operation and management, while non-engineering aspects include environmental impact assessment, socio-economic analysis, policy making and impact on society.

For designing water engineering systems (e.g., channels, levees, bridge piers, drainage structures, dams, reservoirs, spillways), some key questions that need to be addressed relate to the following, among others: peak discharge; velocity distribution; sediment yield, concentration and discharge; pollutant load, concentration and transport; river bed profile, meandering and braiding; downstream and at-a-point hydraulic geometry; flow depth, discharge and velocity routing; and seepage through a dam. For water supply systems, the key questions are concerned with, among others: reliability; loss of energy in the distribution system; and pipe sizes. In addition, pollutant concentration and transport as well as pollution abatement are also now considered as essential components of water resources system design. In a similar manner, because of increased public awareness, primarily triggered by environmental movement, non-engineering aspects of hydraulic design, dubbed under "socio-economic analyses," play a critical role. Besides engineering feasibility and economic viability, issues related to public health, political support, legal and judicial restrictions and social acceptability determine, to a large extent, if the water resources project will go off the ground.

A survey of the water engineering literature shows that there are myriad techniques for addressing questions pertaining to the design of water resources systems. The techniques range from empirical to semi-empirical to physically-based ones. Empirical techniques are data-based; examples are regression, time series analysis and other statistical methods. Semi-empirical methods, also sometimes referred to as conceptual or systems-based techniques, employ mass conservation and some empirical relationship or hypothesis; a good example is the unit hydrograph theory. Physically-based methods employ the laws of conservation of mass, momentum and energy; an example is the use of St. Venant's equations for flow routing or Richards' equation for computing infiltration. Strictly speaking, even physically-based methods also employ empirical parameters and, thus, are not entirely "physically-based." Indeed, all these three types of techniques employ some physics through data or hypotheses or laws. Extensive details of these methods are already well-documented in the literature [1–6]. A more recent and comprehensive account of these methods and applications is also presented in [7].

Because of the large diversity of these techniques, based on different hypotheses and assumptions, it is difficult to present the developments in any subject or field of interest in a unified and coherent manner. This becomes particularly challenging when undertaking water resources system engineering design. There are, of course, some theories that do apply to a wide variety of problems, such as kinematic wave theory and diffusion wave theory [8,9]. These theories can be applied to solve a wide variety of problems where the movement of water, sediment and/or pollutant is involved. However, many problems in water engineering design require a statistical treatment. For addressing such problems, entropy theory can serve as a unifying theory. During the past three decades, entropy theory has been applied to address a wide spectrum of problems in water engineering, including rainfall-runoff [1,3,8], infiltration [2], soil moisture [2], network design [10], velocity distributions [11–14], sediment concentration and discharge [15–17], hydraulic geometry [18–22] and reliability [23], among others. For recent comprehensive accounts of entropy theory applications in water and environmental engineering, see [24–26].

The origin of entropy theory is in the second law of thermodynamics. Koutsoyiannis [27] has presented a nice account of the historical background of entropy. The most commonly used measure of entropy is the Boltzmann-Gibbs-Shannon (BGS) entropy [28], which is often referred simply as the Shannon entropy [29]. Tsallis [30] introduced a more general entropy function for complex systems, which is now referred to as the Tsallis entropy. Tsallis entropy specializes in the Shannon entropy. During the last two decades or so, Tsallis entropy theory has found many applications in water and environmental engineering and there is certainly a great potential to extend the applications to a much wider spectrum of water systems and associated problems.

This paper aims to provide a review of the applications of Tsallis entropy theory in water engineering. It revisits the Tsallis entropy theory, presents a general methodology for application of the theory, shows how entropy theory couples statistical information with physical laws and how it can be employed to derive useful physical constructs in time and/or space and provides a review of physical applications of the Tsallis entropy theory in water engineering.

The rest of the paper is organized as follows. Section 2 reviews the Tsallis entropy theory. Sections 3–6 review the applications of the Tsallis entropy theory in water engineering. Section 3 presents an overview of three types of problems in water engineering. Section 4 reviews problems requiring entropy maximization. Section 5 reviews coupling entropy theory with another theory. Section 6 reviews physical relations. Section 7 draws some conclusions.

2. Tsallis Entropy Theory

2.1. Definition of Entropy

The concept of entropy is closely linked with the concept of uncertainty, information, chaos, disorder, surprise or complexity. Indeed, there are often different interpretations of entropy in different fields [24–26,31–33]. For instance, in statistics, entropy is regarded as a measure of randomness, objectivity or unbiasedness, dependence, or departure from the uniform distribution. In ecology, it is a measure of diversity of species or lack of concentration. In water engineering, it is a measure of information of uncertainty. In industrial engineering, it is a measure of complexity. In manufacturing, it is a measure of interdependence. In management, it is a measure of similarity. In social sciences, it is measure of equality. This may be illustrated as follows.

Consider a random variable X that takes on values x_i, $i = 1, 2, \ldots, N$, that occur with probabilities p_i, $i = 1, 2, \ldots, N$. If $p_i = 1$ (the event is certain to occur), $p_j = 0$ (the event is certain not to occur), $i \neq j$, then it can be said that there is no surprise about the occurrence or non-occurrence of event $X = x_i$ and the occurrence or non-occurrence of this event provides no information. The system that produces such an event has no complexity and is not chaotic or disorderly. On the other hand, if an event x_i occurs with very small probability p_i, say 0.01, then our anticipation of event x_i is highly uncertain and if x_i does indeed occur, then there will be a great deal of surprise about its occurrence. The occurrence of this event provides a great deal of information and the system producing such an event is complex, chaotic and disorderly. Intuitively, the information content of event x_i or the anticipatory uncertainty of x_i prior to the observation decreases as the value of probability $p(x_i)$ increases [24,26]. It may be noted that information is gained only if the variable takes on different values. The value that occurs with a higher probability conveys less information and vice versa.

It is logical to deduce that if a system has more uncertainty, then more information will be needed to characterize it and vice versa. That is, information reduces uncertainty, meaning that, for a system, more information means less uncertainty. Uncertainty increases the need for more information; that is, more uncertainty means more information is needed. Shannon [29] formulated entropy as the expected value of the probabilities of values that a variable or event may take on. The information gained is indirectly measured as the amount of reduction of uncertainty. Thus, entropy is defined as a measure of disorder, chaos, uncertainty, surprise, or information.

From an informational perspective, the information gain from the occurrence of any event x_i, $\Delta H(x_i)$, can be expressed as:

$$\Delta H(x_i) = -\log p_i \tag{1}$$

Equation (1) says that the information gained is minus the logarithm of the inverse of the probability of occurrence. For N events, the average or expected information gain, H_s, is the weighted average of Equation (1):

$$H_s = \sum_{i=1}^{N} H(x_i) = -\sum_{i=1}^{N} p_i \log p_i \tag{2}$$

Equation (2) is the Shannon entropy [29], also called informational entropy. Equation (1) shows that the information gain is directly a function of probability and is, hence, called gain function. Equation (1) can be generalized by expressing it as a power function, given by:

$$\Delta I(x_i) = \frac{1}{m-1}(1 - p_i^{m-1}), \sum_{i=1}^{N} p_i = 1 \tag{3}$$

where $\Delta I(x_i)$ is the gain in information from an event i which occurs with probability p_i and m is any real number. The gain function computed from Equation (3) for $m = -1, 0, 1$ and 2, as shown in Figure 1, decreases with an increase in the probability value regardless of the value of m. For increasing value of m, the gain function diminishes for the same probability value. The gain function has a much longer tail, showing very low values of gain as the probability increases.

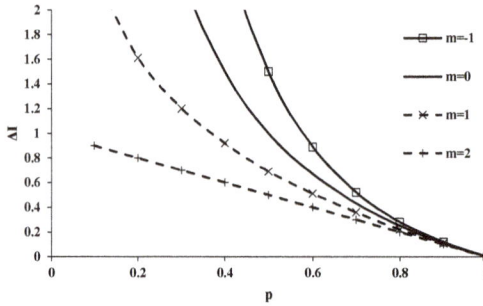

Figure 1. Gain function for $m = -1, 0, 1$ and 2.

Analogous to Shannon entropy, the average or expected gain function for N events, H_m, is the weighted average of Equation (3), given by:

$$H_m = E[\Delta I_i] = \sum_{i=1}^{N} p_i [\frac{1}{m-1}(1 - p_i^{m-1})] = \frac{1}{m-1}\sum_{i=1}^{N} p_i[1 - p_i^{m-1}] \tag{4}$$

where H_m is designated as the Tsallis entropy or m-entropy [30], which is often referred to as the non-extensive statistic, m-statistic, or Tsallis statistic. It can be shown that as $m \to 1$, Tsallis entropy converges to Shannon entropy. The quantity m is often referred to as the non-extensivity index, Tsallis entropy index, or simply entropy index and its value can be positive or negative. Entropy index m reflects the microscopic dynamics and the degree of nonlinearity of the system. Since almost all real systems (e.g., water systems) are inherently nonlinear in nature, the Tsallis entropy has a clear advantage when compared to the Shannon entropy. Tsallis [34] noted that super-extensivity, extensivity and sub-extensivity occur when $m < 1$, $m = 1$ and $m > 1$, respectively. Interestingly, if $m \geq 0$, $m < 1$ corresponds to the rare events ($0 \leq m < 1$) and $m > 1$ corresponds to frequent events [35,36], pointing to the stretching or compressing of the entropy curve to lower or higher maximum entropy positions.

If random variable X is non-negative continuous with a probability density function (PDF), $f(x)$, then the Shannon entropy can be written as:

$$H_s(X) = H_s(f) = -\int_0^\infty f(x) \log f(x) dx \tag{5}$$

Likewise, Tsallis entropy can be expressed as [37–39]:

$$H_m(X) = H_m(f) = \frac{1}{m-1} \int_0^\infty \{f(x) - [f(x)]^m\} dx = \frac{1}{m-1} \{1 - \int_0^\infty [f(x)]^m\} dx \tag{6}$$

From now onwards, subscript m will be deleted and H_m will be simply denoted by H.

A plot of H versus p for $m = -1, -0.5, 0, 0.5, 1$ and 2 is given in Figure 2. For $m < 0$, the Tsallis entropy is concave; and for $m > 0$, it becomes convex. For $m = 0$, $H = (N - 1)$ for all p_i's. For $m = 1$, it converges to Shannon entropy. For all cases, the Tsallis entropy decreases as m increases.

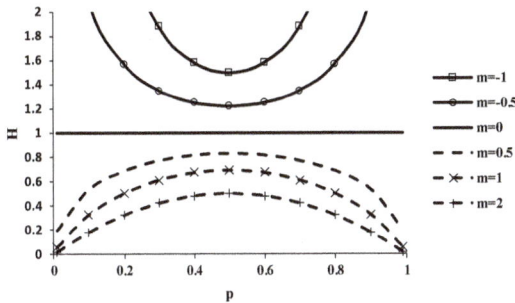

Figure 2. Plot of H for $m = -1, -0.5, 0, 0.5, 1$ and 2.

2.2. Properties of Tsallis Entropy

As mentioned earlier, the concept of entropy is closely linked with the concept of uncertainty, information, chaos, disorder, surprise or complexity. Tsallis entropy has some interesting properties [30,40] that are briefly summarized here.

(1) *m*-entropy: The *m*-surprise or *m*-unexpectedness is defined as $\log_m(1/p_i)$, where the logarithm is the base *m*. Hence, the *m*-entropy can be defined as:

$$H = E[\log_m \frac{1}{p_i}] \tag{7}$$

which coincides with the Tsallis entropy:

$$H = E[\frac{1 - p_i^{m-1}}{m-1}] \tag{8}$$

in which E is the expectation.

(2) Maximum value: Equation (4) attains an extreme value for all values of m when all p_i's are equal, i.e., $p_i = 1/N$. For $m > 0$, it attains a maximum value and for $m < 0$ it attains a minimum value. The extremum of H becomes

$$H = \frac{N^{m-1} - 1}{1 - m} \tag{9}$$

(3) Concavity: For two probability distributions $P = \{p_i, i = 1, 2, \ldots, N\}$ and $Q = \{q_i, i = 1, 2, \ldots, N\}$ corresponding to a unique set of N possibilities, an intermediate probability distribution $G = \{g_i, i = 1, 2, \ldots, N\}$ can be defined for a real a, such that $0 < a < 1$, as:

$$g_i = ap_i + (1-a)q_i \tag{10}$$

for all i's. It can be shown that for $m > 0$,

$$H[G] \geq aH[P] + (1-a)H[Q] \tag{11}$$

and for $m < 0$,

$$H[G] \leq aH[P] + (1-a)H[Q] \tag{12}$$

Functional $H(G) \geq 0$ if $m > 0$ and is, hence, concave; $H(G) = 0$ if $m = 0$; and $H(G) \leq 0$ if $m < 0$ and is, therefore, convex.

(4) Additivity: For two independent systems A and B with ensembles of configurational possibilities $E^A = \{1, 2, \ldots, N\}$ with probability distribution $P^A = \{p_i{}^A, i = 1, 2, \ldots, N\}$ and configurational possibilities $E^B = \{1, 2, \ldots, M\}$ with probability distribution $P^B = \{p_j{}^B, j = 1, 2, \ldots, M\}$, one can express the union of two systems $A \cup B$ and their corresponding ensembles of possibilities $E^{A \cup B} = \{(1, 1), (1, 2), \ldots, (i, j), \ldots, (N, M)\}$. If $p_{ij}{}^{A \cup B}$ represents the corresponding probabilities, then, by virtue of independence, the joint probability will be equal to the product of individual probabilities, i.e., $p_{ij}{}^{A \cup B} = p_i{}^A p_j{}^B$ or $p_{ij}(A + B) = p_i(A) p_j(B)$, for all i and j. One can write the entropy of the union of two systems A and B as:

$$H(A + B) = H(A) + H(B) + (1-m)H(A)H(B) \tag{13}$$

Equation (13) describes the additivity property, which can be extended to any number of systems. In all cases, $H \geq 0$ (non-negativity property). If systems A and B are correlated, then

$$p_{ij}{}^{A \cup B} \neq \left[\sum_{i=1}^{N} p_{ij}{}^{A \cup B}\right]\left[\sum_{j=1}^{M} p_{ij}{}^{A \cup B}\right] \tag{14}$$

for all (i, j). One may define mutual information of the two systems or transinformation S as:

$$T[(p_{ij}{}^{A \cup B})] = H^{A \cup B}[(p_{ij}{}^{A \cup B})] - H^A\left[\left(\sum_{i=1}^{N} p_{ij}\right)\right] - H^B\left[\left(\sum_{j=1}^{M} p_{ij}\right)\right] \tag{15}$$

Considering Equation (15), $T(p_{ij}) = 0$ for all m, if A and B are independent and Equation (15) will reduce to Equation (13). For correlated A and B, $T(p_{ij}) < 0$ for $m = 1$ and $T(p_{ij}) = 0$ for $m = 0$. For arbitrary values of m, it will be sensitive to p_{ij}; it can take on negative or positive values for both $m < 1$ and $m > 1$ with no particular regularity and can exhibit more than one extremum.

(5) Composability: The entropy $H(A + B)$ of a system comprised of two sub-systems A and B can be computed from the entropies of sub-systems, $H(A)$ and $H(B)$ and the entropy index m.

(6) Interacting sub-systems: Consider a set of N possibilities arbitrarily separated into two sub-systems with W_1 and W_2 possibilities, where $W = W_1 + W_2$. Defining $P_{W_1} = \sum_{i=1}^{W_1} p_i$ and $P_{W_2} = \sum_{j=1}^{W_2} p_j$, $P_{W_1} + P_{W_2} = 1 = \sum_{k=1}^{W} p_k$, it can be shown that

$$H(P_W) = H(P_{W_1}, P_{W_2}) + P_{W_1}{}^m H(p_i | P_{W_1}) + P_{N_2}{}^m H(p_j | P_{W_2}) \tag{16}$$

where $p_i | P_{W_1}$ and $p_j | P_{W_2}$ are the conditional probabilities. Note that $p_i{}^m > p_i$ for $m < 1$ and $p_i{}^m < p_i$ for $m > 1$. Hence, $m < 1$ corresponds to rare events and $m > 1$ to frequent events [41]. This property can be extended to any number R of interacting sub-systems: $W = \sum\limits_{j=1}^{R} W_j$. Then, defining $w_j = \sum\limits_{i=1}^{N_j} p_i, j = 1, 2, \ldots, N_j$, $\sum\limits_{j=1}^{N} w_j = 1$, Equation (16) can be generalized as:

$$H(\{p_i\}) = H(\{w_j\}) + \sum_{j=1}^{R} w_j{}^m H(\{p_i|w_j\}) \tag{17}$$

Here, $p_j = w_j$.

2.3. Principle of Maximum Entropy

Quite often, some information may be known about the random variable X. Then, it seems logical to choose a PDF of X that is consistent with the known information. Since there can be more than one PDF that may satisfy this condition, Jaynes [42,43] formulated the principle of maximum entropy (POME), which states that one should choose the distribution that has the highest entropy, subject to the given information. This distribution will be the least-biased distribution. Furthermore, it is equivalent to the minimum relative (cross) entropy condition when no prior is given, which is also called as the Kullback-Leibler principle [44]. The implication here is that POME takes into account all of the given information and, at the same time, avoids consideration of any information that is not given. This is consistent with Laplace's principle of insufficient reason (or principle of indifference), according to which all outcomes of an experiment should be considered equally likely unless there is information to the contrary. Therefore, POME enables entropy theory to achieve the probability distribution of a given random variable.

The procedures for applications of *POME* entails the following steps [24,26]: (1) definition of Tsallis entropy; (2) specification of constraints; (3) maximization of entropy, in accordance with POME, using of the method of Lagrange multipliers; (4) derivation of the probability distribution in terms of constraints; (5) determination of Lagrange multipliers; and (6) determination of the maximum entropy. Since the definition of entropy has already been reviewed earlier, it will not be repeated here. The remaining steps, (2)–(6), are briefly discussed here.

2.3.1. Specification of Constraints

For deriving the PDF of a random variable X using POME, appropriate constraints need to be defined. Papalexiou and Koutsoyiannis [28] suggested three considerations for defining constraints: (1) Simplicity and physical meaningfulness; (2) little variability in the future; and (3) definition in terms of laws of physics—mass, momentum and energy conservation and constitutive laws—as far as possible. For simplicity, constraints are often defined in terms of statistical moments and, fortunately, the moments are related to the laws of physics.

Let C_i denote the i-th constraint, $i = 0, 1, 2, \ldots, n$, where n is the number of constraints. Let $g_r(x)$ be an arbitrary function of X. Then, constraints $C_r, r = 0, 1, 2, \ldots, n$, can be defined as:

$$C_0 = \int_a^b f(x)dx = 1 \tag{18}$$

$$C_r = \int_a^b g_r(x) f(x) \, dx = E[g_r(x)] = \overline{g_r(x)}, r = 1, 2, \ldots, n \tag{19}$$

where $g_0(x) = 0$; $g_r(x), r = 1, 2, \ldots, n$, represents some function of x and $\overline{g_r(x)}$ is the expectation of $g_r(x)$. Equation (18) states the total probability theorem that the PDF must satisfy. If $r = 1$ (first moment) and $g_1(x) = x$, then Equation (19) represents the mean $\overline{x}$; likewise, for $r = 2$ (second moment) and

$g_2(x) = (x - \bar{x})^2$, it denotes the variance (σ^2) of X. The first moment corresponds to the conservation of mass and the second moment to the conservation of momentum. Similarly, $r = 3$ (third moment), measuring skewness, corresponds to the conservation of energy. It is noticed that the order of moments higher than 3 may be unreliable [45]. In water engineering, two or three constraints usually suffice.

2.3.2. Entropy Maximizing Using Lagrange Multipliers

Tsallis entropy, given by Equation (6), can be maximized, subject to constraints defined by Equations (18) and (19), using the method of Lagrange multipliers [42,43]. Therefore, the Lagrangian function L can be expressed as:

$$L = \int_0^\infty f(x) \frac{1 - [f(x)]^{m-1}}{m-1} dx + (\lambda_0 - \frac{1}{m-1})[\int_0^\infty f(x) dx - 1] + \sum_{i=1}^M \lambda_i [\int_0^\infty g_i(x) f(x) dx - \overline{g_i(x)}] \quad (20)$$

where λ_i, $i = 0, 1, 2, \ldots, M$, are the Lagrange multipliers. Note that $-1/(m-1)$ is added to the zeroth Lagrange multiplier for algebraic simplification. Using the Euler-Lagrange calculus of variation, differentiating Equation (20) with respect to $f(x)$ and equating the derivative to zero, we obtain:

$$\frac{dL}{df(x)} = 0 = \frac{1}{m-1} - [f(x)]^{m-1} + \lambda_0 - \frac{1}{m-1} + \sum_{i=1}^M \lambda_i g_i(x) \quad (21)$$

2.3.3. Determination of Probability Distribution

Equation (21) yields the least-biased probability distribution of X:

$$f(x) = [\lambda_0 + \sum_{i=1}^M \lambda_i g_i(x)]^{\frac{1}{m-1}} \quad (22)$$

Integrating Equation (22), the cumulative distribution function $F(x)$ is obtained as:

$$F(x) = \int_0^x [\lambda_0 + \sum_{i=1}^M \lambda_i g_i(x)]^{\frac{1}{m-1}} dx \quad (23)$$

The properties of the probability distribution will depend on the value of m, Lagrange multipliers, functions $g_i(x)$ and M.

2.3.4. Determination of the Lagrange Multipliers

Equation (22) contains M unknown Lagrange multipliers that can be determined with the use of Equations (18) and (19). Substituting Equation (22) in Equations (18) and (19), the result is, respectively:

$$\int_0^\infty [\lambda_0 + \sum_{i=1}^M \lambda_i g_i(x)]^{\frac{1}{m-1}} dx = 1 \quad (24)$$

$$\int_0^\infty g_i(x) [\lambda_0 + \sum_{i=1}^M \lambda_i g_i(x)]^{\frac{1}{m-1}} dx = \overline{g_i(x)}, \; i = 1, 2, \ldots, M \quad (25)$$

Solution of Equations (24) and (25) yields the unknown Lagrange multipliers λ_i, $i = 1, 2, \ldots, M$. It may be noted from Equation (24) that λ_0 can be expressed as a function of other Lagrange multipliers and, therefore, the unknown multipliers are λ_i, $i = 1, 2, \ldots, M$. Except for simple cases, Equations (24) and (25) do not have an analytical solution but can be solved numerically.

2.3.5. Determination of Maximum Entropy

Substitution of Equation (24) in Equation (6) leads to the maximum Tsallis entropy:

$$H = \frac{1}{m-1}\{1 - \int_0^\infty [\lambda_0 + \sum_{i=1}^M \lambda_i g_i(x)]^{\frac{m}{m-1}} dx\} \tag{26}$$

Equation (26) shows that the entropy of the probability distribution of X depends only on the specified constraints, since the Lagrange multipliers themselves can be expressed in terms of the specified constraints.

3. Applications in Water Engineering: Overview

The problems that can be addressed using the Tsallis entropy theory can be classified into three groups. The first group consists of problems that only require the maximization of the Tsallis entropy, which can be accomplished using POME. Examples of such problems include frequency analysis, parameter estimation, network evaluation and design, spatial and inverse spatial analysis, geomorphologic analysis, grain size distribution analysis, flow forecasting, complexity analysis and clustering [46–52].

The types of problems included in the second group require coupling with another theory, such as the theory of stream power or theory of minimum energy dissipation rate. Examples of problems in this group include hydraulic geometry [19,20] and evaporation [53]. Also included in this group are problems wherein first relations between entropy and design variables are derived and then relations between design variables and system characteristics are established. Examples include geomorphologic relations for elevation, slope and fall; and evaluation of water distribution systems [54,55].

The third group includes problems that require deriving a physical relation either in time or in space. This means, the domain of analysis requires a shift from the probability domain to a real domain (space or time), which is accomplished by invoking a relation between the cumulative probability distribution function and the design variable. Examples of such problems are infiltration capacity, soil moisture movement in vadose zone, groundwater head distribution, velocity distribution, rainfall-runoff relation, channel geometry, rating curve, flow-duration curve, erosion and sediment yield, sediment concentration and discharge, debris flow, longitudinal river profile, hydraulic geometry, channel cross-section shape and rating curve [19–22,56–61]. The objective in this class of problems is to derive a relation of the design variable as a function of time or space.

These three kinds of problems are further detailed in the following sections, with examples from water engineering.

4. Problems Requiring Entropy Maximization

Fundamental to solving problems that only require entropy maximization is the derivation of probability distribution. A multitude of problems in water engineering involve essentially data analysis for deriving either a probability distribution or computing entropy.

4.1. Frequency Distributions

The procedure for deriving a maximum entropy-based frequency distribution has been discussed above. The procedure is illustrated here using only two constraints: mean (first moment) and second moment. For any probability density function, $f(x)$, of a random variable X, the total probability must equal one, i.e.,

$$\int_0^\infty f(x)dx = 1 \tag{27}$$

The first and second moments can be defined, respectively, as:

$$\int_0^\infty xf(x)dx = E(x) = \mu_1 \tag{28}$$

and

$$\int_0^\infty x^2 f(x)dx = E(x^2) = \mu_2 \tag{29}$$

In order to obtain the least-biased $f(x)$, subject to Equations (27)–(29), Equation (6) can be maximized for $m > 0$ using the method of Lagrange multipliers. The Lagrange function L can be written as:

$$L = \int_0^\infty f(x)\frac{1 - [f(x)]^{m-1}}{m - 1}dx - \lambda_0[\int_0^\infty f(x)dx - 1] - \lambda_1[\int_0^\infty xf(x)dx - \mu_1] - \lambda_2[\int_0^\infty x^2 f(x)dx - \mu_2] \tag{30}$$

Differentiating L with respect to $f(x)$ and equating the derivative to zero, one obtains:

$$f(x) = \{\frac{1}{m} + \frac{(1 - m)}{m}[\lambda_0 + \lambda_1 x + \lambda_2 x^2]\}^{1/(m-1)} \tag{31}$$

Defining $k = (1-m)/m$ [37] and $\alpha_i = m\lambda_i$, $i = 0, 1, 2$, Equation (31) can be written as:

$$f(x) = (1 + k)^{-1-1/k}[1 + k(\alpha_0 + \alpha_1 x + \alpha_2 x^2)]^{-1-1/k} \tag{32}$$

Equation (32) is the entropy-based probability density function of power type.

The Lagrange multipliers, λ_i, $i = 0, 1, 2$ and, consequently, $\alpha_i = m\lambda_i$, $i = 0, 1, 2$, can be estimated using Equations (27) to (29). For simplification, let λ_2 be assumed as 0. Then, Equation (32) becomes:

$$f(x) = (1 + k)^{-1-1/k}[1 + k(\alpha_0 + \alpha_1 x)]^{-1-1/k} \tag{33}$$

whose parameters $\alpha_i = m\lambda_i$, $i = 0, 1$, are estimated using Equations (27) and (28). Substituting Equation (33) in Equation (27), one obtains:

$$\alpha_1 = (1 + k)^{-1-1/k}(1 + k\alpha_0)^{-1/k} \tag{34}$$

or

$$\alpha_0 = -1 + \frac{1}{k}[\alpha_1(1 + k)^{1+1/k}]^k \tag{35}$$

The other Lagrange multiplier can be determined by inserting Equation (33) in Equation (28) and is given as:

$$-\frac{(1 + k)^{-1-1/k}}{k(1 - \frac{1}{k})}(1 + k\alpha_0)^{1-1/k} = \alpha_1^2 \bar{x} \tag{36}$$

By solving Equations (34) or (35) with (36), Lagrange multipliers α_1 and α_0 can be determined. Inserting Equation (34) in Equation (33), one obtains

$$f(x) = (1 + k)^{-1-1/k}(1 + k\alpha_0)^{-1-1/k}[1 + k(1 + k)^{-1-1/k}k(1 + k\alpha_0)^{-1-1/k}x] \tag{37}$$

Let

$$\beta = [(1 + k)(1 + k\alpha_0)]^{-1-1/k} \tag{38}$$

Using Equation (38), Equation (37) becomes

$$f(x) = \frac{1}{\beta}[1 + \frac{kx}{\beta}]^{-1-1/k} \tag{39}$$

Equation (39) is a two-parameter generalized Pareto distribution.
Let

$$y = x + \frac{\beta}{k} \quad \Rightarrow \quad x = y - \frac{\beta}{k} \tag{40}$$

Equation (39) then becomes

$$f(y) = \frac{1}{\beta}[\frac{ky}{\beta}]^{-1-1/k} \tag{41}$$

Equation (41) is a two-parameter Pareto distribution. If $k \to 0$, then Equation (41) leads to an exponential distribution:

$$f(y) = \frac{1}{\beta}\exp[\frac{y}{\beta}] \tag{42}$$

Koutsoyiannis [37,38] proposed a generalization, following which Equation (32) can be expressed as:

$$f(x) = (1+k)^{-1-1/k}[1 + k(\alpha_0 + \alpha_1 x + \alpha_2 x^{c_1})]^{-1-1/k}x^{c_2-1} \tag{43}$$

where c_1 and c_2 are shape parameters. Equation (43) has four parameters: scale parameter α_1 and shape parameters k and c_1 and c_2. Note that α_0 is not a parameter, because it is a constant based on the satisfaction of Equation (27). It is, however, not clear as to what led to the generalized form of Equation (43). Koutsoyiannis [37] suggested that random variable X^{c_2} would have Beta Prime (also referred to as Beta of the second kind) distribution [62]. Then, the distribution of X would be referred to as the power-transformed Beta Prime (PBP). Koutsoyiannis [37] showed that Equation (43) can specialize into several exponential and power-type probability distributions, such as PBP-L1 $(k \to 0)$, gamma $(k \to 0, c_1 \to 1)$, Weibull $(k \to 0, c_2 = c_1)$, Pareto $(c_2 = c_1 = 1)$, Beta Prime $(c_1 = 1)$, PBP-L2 $(k \to \infty, k\alpha_0 \to k_0, k\alpha_1 \to \alpha_1)$ and others.

4.2. Network Evaluation and Design

Hydrometric data are required for an efficient planning, design, development, operation and management of water engineering systems. Many studies have applied the Shannon entropy theory to assess and optimize data collection networks (e.g., water quality, rainfall, streamflow, hydrometric, elevation, landscape, etc.) [63,64] but not the Tsallis entropy theory. The basic idea for developing a methodology for data collection network design is that it must take into account the information of each gaging station or potential gaging station in the network. A station with a higher information content is given a higher priority over other stations that have lower information content but the information content of a station must be tempered with the degree of use. That is, a station that is used by one user might be given a lower priority than a station that has diverse uses.

A framework for network design or evaluation considers a range of factors, such as: (a) objectives of sampling; (b) variables to be sampled; (c) locations of measurement stations; (d) frequency of sampling; (e) duration of sampling; (f) uses and users of data; and (g) socio-economic considerations. A network design has two modes: (1) number of gages and their locations (space evaluation); and (2) time interval for measurement (time evaluation).

Let there be two stations A and B with ensembles of configurational possibilities $E^A = \{1, 2, \ldots, N\}$ with probability distribution $P^A = \{p_i^A, i = 1, 2, \ldots, N\}$ and configurational possibilities $E^B = \{1, 2, \ldots, M\}$ with probability distribution $P^B = \{p_j^B, j = 1, 2, \ldots, M\}$. Then, the union of the two stations $A \cup B$ and their corresponding ensembles of possibilities are $E^{A \cup B} = \{(1, 1), (1, 2), \ldots, (i, j), \ldots, (N, M)\}$. If $p_{ij}{}^{A \cup B}$ represents the corresponding probabilities, then the mutual information or transinformation S can be expressed as:

$$T[(p_{ij}{}^{A \cup B})] = H^{A \cup B}[(p_{ij}{}^{A \cup B})] - H^A[(\sum_{i=1}^{N} p_{ij})] - H^B[(\sum_{j=1}^{M} p_{ij})] \tag{44}$$

From Equation (44), $T(p_{ij}) = 0$ for all m, if X and Y are independent. For correlated X and Y, $T(p_{ij}) < 0$ for $m = 1$ and $T(p_{ij}) = 0$ for $m = 0$. For arbitrary values of m, it will be sensitive to p_{ij}; it can take on negative or positive values for both $m < 1$ and $m > 1$ with no particular regularity and can exhibit more than one extremum.

4.3. Directional Information Transfer Index

By dividing by the marginal entropy, the mutual information T can be normalized [65] as:

$$\frac{T}{H} = DIT = \frac{(H - H_{Lost})}{H} = 1 - \frac{H_{Lost}}{H} \tag{45}$$

where H_{Lost} is the amount of information lost. The ratio of T by H is called the Directional Information Transfer (DIT) index. Mogheir and Singh [66] called it as the Information Transfer Index (ITI). The DIT varies from zero to unity and denotes the fraction of information transferred from one station to another. A zero value of DIT corresponds to the case where sites are independent and, therefore, no information is transmitted. A value of unity for DIT corresponds to the case where sites are fully dependent and no information is lost. Since $DIT_{XY} = T/H(X)$ is not the same as $DIT_{YX} = T/H(Y)$, DIT is not symmetrical. The term DIT_{XY} describes the fractional information inferred by station X about station Y, whereas DIT_{YX} describes the fractional information inferred by station Y about station X. Between two stations, the station with the higher DIT should be given higher priority because of its greater capability in inferring (predicting) the information at other sites. The DIT can be applied for regionalization of the network or watersheds.

4.4. Reliability of Water Distribution Networks

A water distribution system can be designed by minimizing head losses, costs, risks and departures from specified values of water quantity, pressure and quality and also maximizing reliability [67]. Thus, it becomes a multi-objective optimization problem. However, it is not uncommon to formulate the design problem as a single-objective optimization problem, where the system capital and operational costs are minimized and, at the same time, the laws of hydraulics are satisfied and the targets of water quantity and pressure at demand nodes are met. Fundamental to either type of optimization is reliability [68–70].

To develop a Tsallis entropy-based redundancy measure of the network with N nodes, where the nodes may be considered to constitute sub-systems, the Tsallis entropy of a node j can now be expressed in terms of W_{ij} as:

$$S_j = \frac{1}{m-1} [\sum_{i=1}^{n(j)} W_{ij} - W_{ij}{}^m)] = \frac{1}{m-1} \{\sum_{i=1}^{n(j)} [\frac{q_{ij}}{Q_j} - (\frac{q_{ij}}{Q_j})^m]\} \tag{46}$$

where m is the entropy index and is a real number and S_j is an entropic measure of redundancy at node j and is local redundancy. Maximizing S_j would maximize redundancy of node j and is equivalent to maximizing entropy at node j. The maximum value of S_j is achieved when all W_{ij}'s or q_{ij}/Q_j's are equal. This occurs when all q_{ij}'s are equal. For the entire water distribution network, redundancy is a function of redundancies S_j's of individual nodes in the network.

The overall network redundancy can be assessed in two ways. First, the network redundancy can be assessed by the relative importance of a link to its node and its importance recognized by q_{ij}/Q_j.

In this case, the redundancy is maximized at each node. It may, however, be noted that the network redundancy is not a sum of nodal redundancies. Second, the network redundancy can be assessed by the relative importance of a link to the total flow and its importance recognized by q_{ij}/Q_0. Here, the proposition is that the importance of a link relative to the local flow is not as important as it is to the total flow. In this case too, the network redundancy is not a sum of nodal redundancies. In order to acknowledge the relative importance of a link to the entire network, the nodal redundancy S_{j*} can be expressed as:

$$S_{j*} = \frac{1}{m-1} \sum_{i=1}^{n(j)} \left[\frac{q_{ij}}{Q_0} - \left(\frac{q_{ij}}{Q_0} \right)^m \right] = \frac{1}{m-1} \left[\frac{Q_j}{Q_0} - \sum_{i=1}^{n(j)} \left(\frac{q_{ij}}{Q_0} \right)^m \right] \tag{47}$$

It may be noted that S_{j*}, given by Equation (47), is similar to the S_j given by Equation (46). In this case too, the maximum value of S_{j*} will occur when the q_{ij} values are equal at the j-th node. It can also be shown that the maximum network redundancy will be achieved when all the q_{ij} values are equal. It may, however, be noted that

$$S_{j*} = \frac{1}{m-1} \sum_{i=1}^{n(j)} \left[\frac{q_{ij}}{Q_0} - \left(\frac{q_{ij}}{Q_0} \right)^m \right] \neq \frac{1}{m-1} \left[1 - \sum_{i=1}^{n(j)} \left(\frac{q_{ij}}{Q_0} \right)^{m-1} \right] \tag{48}$$

This is because,

$$\sum_{i=1}^{n(j)} q_{ij} = Q_j \neq Q_0 \text{ and } \sum_{i=1}^{n(j)} \frac{q_{ij}}{Q_0} \neq 1 \tag{49}$$

Therefore, in the second approach, Equation (47) can be used in the spirit of Tsallis entropy or considering it via partial Tsallis entropy [36].

The network redundancy for N nodes is a function of redundancies of individual nodes, S_j's, in the network. However, it will not be a simple summation of these nodal redundancies, because of the non-extensive property of the Tsallis entropy. For the first approach, it can be shown that the network redundancy (with N nodes) can be expressed as:

$$S_{1 \cup 2 \cup ... \cup N} = \sum_{1 \leq j \leq N} S_j + (1-m) \sum_{1 \leq j1 < j2 \leq N} S_{j1} S_{j2} + (1-m)^2 \sum_{1 \leq j1 < j2 < j3 \leq N} S_{j1} S_{j2} S_{j3} + \cdots$$
$$+ (1-m)^{N-1} \sum_{1 \leq j1 < j2 ... < jN \leq N} S_{j1} S_{j2} \cdots S_{jN} \tag{50}$$

In order to develop an appreciation for Equation (47), it will be instructive to expand Equation (47) in terms of flow quantities. Equation (47) is just the sum of nodal redundancies.

For the second approach, the network redundancy (with N nodes: $1, 2, \ldots, N$) can be expressed as:

$$S_{1 \cup 2 \cup 3 \cup 4 ... \cup N*} = (m-1)^{N-1} \prod_{j=1}^{N} S_{j*} - \frac{(m-1)^{N-2}}{Q_0} [Q_1 S_{2*} \ldots S_{N*} + Q_2 S_{1*} S_{3*} \ldots S_{N*} + \ldots Q_N S_{1*} S_{2*} \ldots S_{N-1*}]$$
$$+ \ldots + \frac{(m-1)}{Q_0^{N-1}} [Q_1 Q_2 \ldots Q_{N-1} S_{N*} + Q_1 Q_2 \ldots Q_{N-2} S_{N-1*} + \ldots + Q_2 Q_2 \ldots Q_N Q_1] \tag{51}$$

It can be seen that Equation (51) for network redundancy with the second approach is significantly different from Equation (50) with the first approach.

5. Problems Requiring Coupling with another Theory

There are some problems where the entropy theory can only be part of the solution methodology and needs to be coupled with another theory. Consider, for example, hydraulic geometry of a river or channel, which is defined by the relations between discharge and each of hydraulic variables (e.g., flow width, depth, velocity under bankfull conditions) and each of geometric variables (e.g., bed roughness, slope). Hydraulic geometry is of two types: (1) downstream; and (2) at-a-point. For either type, the values of hydraulic and geometric variables are average annual values corresponding to the equilibrium or stable condition of the river. At this condition, the river will try to spend the least amount of energy for transporting water and sediment. In response to the influx of water and sediment coming from the watershed, the river will adjust these variables or characteristics in order to attain the stable condition. This means, the river will spread the adjustment as equitably as the environment will allow and will follow the theory of minimum energy dissipation rate. The equal rate of adjustment can be described by the principle of maximum entropy (POME) that the river will follow. In this manner, the theory of minimum energy dissipation rate and entropy theory are combined for determining the hydraulic geometry of a river or designing a stable canal. The advantage of the entropy theory is that it explicitly allows to incorporate the constraints imposed by the watershed and the design. For example, if a river is leveed, then it cannot adjust its width and the adjustment will be shared by depth and velocity. Likewise, if a canal is lined, it cannot adjust its width, slope and roughness and will have to adjust its depth and velocity. In this manner, a whole hierarchy of hydraulic geometry relations can be obtained, depending on the constraints imposed. The other existing theories do not allow this flexibility. In what follows, some specific examples of problems requiring coupling entropy theory with other theories are presented.

5.1. Hydraulic Geometry

Hydraulic geometry is defined by relations between discharge (Q) and each of channel width (B), flow depth (d), flow velocity (V), slope (S) and roughness factor (say Manning's n). Hydraulic geometry is either downstream hydraulic geometry or at-a-station hydraulic geometry [71]. Langbein [72] and Yang et al. [73] reasoned that hydraulic geometry relations correspond to the equilibrium state of the channel. In order to attain this state, the channel adjusts its hydraulic variables and the adjustment is shared equally among these variables. In practice, the channel is seldom in equilibrium, meaning that the adjustment among hydraulic variables will be unequal. This, then, suggests that there will be a family of hydraulic geometry relations, depending on the adjustment of hydraulic variables and the adjustment should explain the variability in the parameters of hydraulic geometry relations. For downstream hydraulic geometry, Singh et al. [19,20] hypothesized that, for a given influx of discharge from the watershed, the channel would minimize its stream power by adjusting three controlling variables: depth, width and friction.

Coupling the theory of minimum energy dissipation rate with the principle of maximum entropy, three possibilities can occur corresponding to the spatial rate of adjustment of friction, the spatial rate of adjustment of width and the spatial rate of adjustment of flow depth. These possibilities then lead to the formulation of, respectively, the proportion of the adjustment of stream power (SP) by friction, the proportion of the adjustment of SP by channel width and the proportion of the adjustment of SP by flow depth and, hence, to four sets of hydraulic geometry relations, as follows: (1) the spatial change in SP is accomplished by an equal spatial adjustment between flow width B and resistance expressed by Manning's n; (2) the spatial variation in SP is accomplished by an equal spatial adjustment between flow depth and flow width; (3) the spatial variation in SP is accomplished by an equal spatial adjustment between flow depth and resistance; and (4) the spatial variation in SP is accomplished by an equal spatial adjustment between flow depth, flow width and resistance. These four possibilities can occur in the same river in different reaches or in the same reach at different times, or in different rivers at the same time or at different times.

The hydraulic geometry relations are expressed as:

$$B = aQ^b, \, d = cQ^f, \, V = kQ^m, \, n = NQ^p, \, S = sQ^y \qquad (52)$$

where a, c, k, N and s are parameters; and b, f, m, p and y are exponents. Values of these exponents and parameters depend upon the possibility under consideration and the entropy theory permits explicit expressions for the exponents and parameters. Singh et al. [19,20] showed that most of the downstream hydraulic geometry relations reported in the literature can be derived as special cases of the entropy-based equations.

For at-a-station hydraulic geometry, when discharge changes, a river cross-section can adjust its width, depth, velocity, roughness and slope or a combination thereof. Singh and Zhang [21,22] reasoned that the channel cross-section will adjust or minimize its SP by adjusting these four variables: (1) P_B can be interpreted as the proportion of the temporal change of *SP* due to the temporal rate of adjustment of width; (2) P_h as the proportion of the temporal change of *SP* due to the temporal rate of adjustment of depth; (3) P_α as the proportion of the temporal change of *SP* due to the temporal rate of adjustment of friction; and (4) P_S as the proportion of the temporal change of *SP* due to the temporal rate of adjustment of slope. These cases involve probabilities of four variables, meaning that any adjustment in hydraulic variables in combinations of two, three or four may occur. These give rise to different configurations of adjustment that do indeed occur in nature [74]. Thus, the equality among four probabilities yields 11 possibilities and, hence, leads to 11 sets of equations: (1) $P_B = P_h$; (2) $P_B = P_\alpha$; (3) $P_B = P_S$; (4) $P_h = P_\alpha$; (5) $P_h = P_S$; (6) $P_\alpha = P_S$; (7) $P_B = P_h = P_\alpha$; (8) $P_B = P_\alpha = P_S$; (9) $P_B = P_h = P_S$; (10) $P_h = P_\alpha = P_S$; and (11) $P_B = P_h = P_\alpha = P_S$. It should be noted that all eleven possibilities can occur in the same river cross-section at different times, or in different river cross-sections at the same time or at different times. Williams [75] explored 11 cases, which are similar to the above 11 possibilities. The resulting hydraulic geometry relations are of the same form as Equation (52) but expression for exponents and parameters therein are different.

5.2. Evaporation

Evaporation is a process by which liquid water is converted into water vapor. The process of evaporation entails four elements [76]: (1) supply of energy; (2) supply of water; (3) tendency of liquid water molecules to escape; and (4) turbulent transport. The source of energy that is needed for evaporation to occur from land surfaces is solar radiation, which can be defined by radiative flux. The supply of water can be from precipitation or irrigation, determining soil wetness, which is characterized by soil moisture. Fugacity refers to the tendency of water molecules to escape and is expressed by the saturated vapor pressure at the liquid-vapor interface. The turbulent transport of water vapor and heat is determined by wind speed and thermal instability of the surface layer and is defined by turbulent sensible heat flux into the atmosphere. Wang et al. [76] argued that, under thermodynamic equilibrium, thermal and hydrologic states of the land surface resulting from the interaction between land and atmospheric processes tend to maximize evaporation.

For a given radiative energy flux, the rate of evaporation depends on the combination of surface soil moisture, surface soil temperature and sensible heat flux, as well as the dynamic feedbacks among them at the land surface. There can be many combinations of ground and sensible heat fluxes, evaporation rate and net radiation that can satisfy the energy balance. Wang et al. [76] hypothesized that the preferred combination is the one that maximizes evaporation. Denoting the rate of evaporation by E, surface soil moisture by w, surface soil temperature by T, sensible heat flux into the atmosphere by H, ground heat flux by G and net radiative by R_n at the surface, maximizing evaporation, subject to the energy balance, the result is:

$$E = \max[E(w, T, H; R)|E + H + G = R_n] \tag{53}$$

for all combinations of independent variables w, T and H. Wang et al. [76] investigated three cases: (1) $R = R_n - G$, representing the turbulent energy budget, as described by the Bowen ratio; (2) $R = R_n$, corresponding to the partitioning of the net radiation into latent, sensible and ground heat fluxes; and (3) $R = R_i$, representing the budget of all surface energy fluxes. These cases express land-atmosphere interactions.

6. Problems Involving Physical Relations

In water engineering, we often need to determine physical relationships, such as infiltration rate as a function of time, runoff or discharge as a function of time, soil moisture as a function of depth from the soil surface, velocity as a function of flow depth measured from the bottom, sediment concentration as a function of flow depth and sediment discharge as a function of time. For deriving physical relationships, the probability domain and the physical domain need to be concatenated and this can be done by hypothesizing the cumulative distribution function (CDF) of a design (dependent) variable (e.g., flux, say discharge) in terms of independent (concentration) variable (e.g., stage of flow).

6.1. Hypotheses on Cumulative Probability Distribution Function

Different types of hypotheses have been formulated when applying entropy theory to derive relationships for design variables. Examples of a linear hypothesis include velocity distribution as a function of flow depth, wind velocity as a function of height, sediment concentration profile along the flow depth, rating curve and groundwater discharge along the horizontal direction of flow. It is noted that the CDF should have a one-to-one relation with the design variable (i.e., random variable) of interest and its value should only be between 0 and 1; 0 for the minimum value of the random variable and 1 for the maximum value. For deriving a two-dimensional velocity (*u*) distribution, Cui and Singh [77] hypothesized a general form of the CDF as:

$$F(u) = [1 - (\frac{x}{B})^2]^b (\frac{y}{D})^a \text{ for all } (x, y) \text{ on } I(u) \tag{54}$$

in which *y* is the vertical dimension, *x* is the transverse direction, *a* and *b* are shape parameters and *B* and *D* act as scale parameters or normalizing quantities. The CDF given by Equation (54) has a one-to-one relationship with the velocity value *u*; in other words, the CDF is unique on each isovel $I(u)$ and has a value of 0 at $I(0)$ and 1 at $I(u_{max})$. Also, CDF is 0 at $x = B$ or $y = 0$ and is 1 at $x = 0$ and $y = D$ (Figure 3). It is continuous and differentiable in both *x* and *y*.

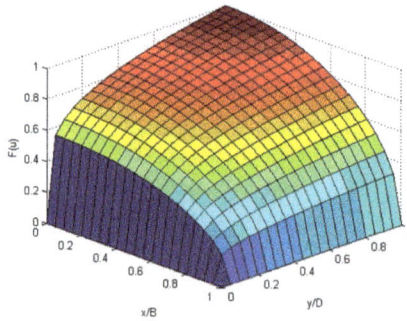

Figure 3. CDF of 2-D velocity distribution with $a = 0.5$, $b = 0.2$ in the idealized cross-section.

The two-dimensional (2-D) velocity distribution can also be expressed by transforming the Cartesian coordinates [vertical (*y*) and transverse (*x*)] into curvilinear *s-r* coordinate system in which *r* has a unique, one-to-one relation with a value of velocity and *s* (coordinate) curves are their orthogonal trajectories [56]. Then, the CDF of velocity in a channel cross section can be expressed as:

$$F(u) = \frac{r - r_0}{r_{max} - r_0} \tag{55}$$

where *r* is a coordinate between r_0 and r_{max} and corresponds to *y*.

For the one-dimensional (1-D) form of Equation (54) with $b = 0$, it is seen from Equation (54) that the CDF will remain between 0 and 1 when $0 \leq x \leq B$ and $0 \leq y \leq D$. Thus, the CDF for 1-D velocity distribution can be simplified as:

$$F(u) = \left(\frac{y}{D}\right)^{a} \text{ for all } (y) \text{ on } I(u) \tag{56}$$

This equation implies that the shape parameter a is critical. If $a = 1$, then

$$F(u) = \left(\frac{y}{D}\right) \text{ for all } (y) \text{ on } I(u) \tag{57}$$

In cases where the design variable is maximum at $y = 0$ and minimum at maximum y (i.e., $y = D$), as is the case for suspended sediment concentration, it is distributed from a maximum value at the channel bed and decreases towards the water surface. Hence, the CDF should be assumed in a way so that it is 1 at the channel bed ($y = 0$) and 0 at the water surface ($y = D$). The CDF of sediment concentration can then be expressed as:

$$F(c) = 1 - \left(\frac{y}{D}\right)^{a} \tag{58}$$

It is seen from Equation (58) that $F(c)$ will be between 0 and 1 if $0 \leq y \leq D$.

6.2. One-Dimensional Velocity Distribution

Velocity distribution is needed for determining flow discharge, scour around bridge piers, erosion and sediment transport, pollutant transport, energy and momentum distribution coefficients, hydraulic geometry, watershed runoff and river behavior. Velocity distributions have been derived using experimental, hydrodynamic, or entropy methods. Shannon entropy has been applied to derive one- and two-dimensional velocity distributions [26]. Singh and Luo [11,78] and Luo and Singh [12] employed Tsallis entropy to derive the 1-D velocity distribution. The Tsallis entropy-based velocity distribution has been shown to have an advantage over the Shannon entropy-based distribution. However, in these entropy-based velocity distributions, the CDF has been assumed to be linear, meaning the velocity is equally likely along the vertical from the channel bed to the water surface. This assumption is fundamental to the derivation of velocity distributions but has not been adequately scrutinized. Further, this assumption is weak and may partly explain the reason why these velocity distributions do not accurately describe the velocity near the channel bed.

For deriving 1-D velocity distribution, the constraint equation is derived using the continuity equation such that $g_1(u) = u$ and limits of integration are 0 and u_D (maximum velocity at the surface or flow depth D). The entropy theory then yields the probability density function of velocity u:

$$f(u) = \left[\frac{m-1}{m}(\lambda_* + \lambda_1 u)\right]^{\frac{1}{m-1}} \tag{59}$$

where $\lambda_* = \frac{1}{m-1} + \lambda_0$.

Following Cui and Singh [77], the CDF of u can be hypothesized as Equation (56). Then, a general velocity distribution based on the Tsallis entropy theory [14] is obtained as:

$$u = -\frac{\lambda_*}{\lambda_1} + \frac{1}{\lambda_1}\frac{m}{m-1}\left[\lambda_1\left(\frac{y}{D}\right)^{a} + \left(\frac{m-1}{m}\lambda_*\right)^{\frac{m}{m-1}}\right]^{\frac{m-1}{m}} \tag{60}$$

Equation (60) can be simplified by defining a dimensionless parameter G as:

$$G = \frac{\lambda_1 u_{max}}{\lambda_1 u_{max} + \lambda_*} \tag{61}$$

Parameter G can be regarded as an index of the velocity distribution uniformity. Equation (60) can be cast as:

$$\frac{u}{u_{max}} = 1 - \frac{1}{G}\left(1 - ((1-G)^{\frac{m}{m-1}} + \left(1 - (1-G)^{\frac{m}{m-1}}\right)\left(\frac{y}{D}\right)^a)^{\frac{m-1}{m}}\right) \tag{62}$$

Equation (62) shows that for a given m value, the velocity distribution can be obtained with only two parameters: a and G. A bigger G value tends to slow the growth of the velocity from the channel bed to the water surface, while the parameter a has an opposite effect. A lower value of G tends to linearize the velocity distribution and a higher value non-linearize. The opposite is the case for exponent a. The velocity distribution is more sensitive to a than to G.

6.3. Two-Dimensional Velocity Distribution

For two-dimensional (2-D) velocity distribution, Chiu [56] proposed a transformation that converts the Cartesian coordinates [vertical (y) and transverse (z)] into curvilinear s-r coordinate system, in which r has a unique, one-to-one relation with a value of velocity and s (coordinate) curves are orthogonal trajectories to r. Following the coordinate system of Chiu [56], Luo and Singh [12] employed the Tsallis entropy. Then, expressing the CDF of velocity in a channel cross section by Equation (57), the velocity distribution becomes

$$u = (\frac{m}{m-1})\frac{1}{\lambda_1}\left\{\left(\lambda_1\frac{r-r_0}{r_{max}-r_0}\right) + \left[(\frac{m-1}{m})\lambda_*\right]^{\frac{m}{m-1}}\right\}^{\frac{m-1}{m}} - \frac{\lambda_*}{\lambda_1} \tag{63}$$

Defining entropy parameter $M = \lambda_1 u^2_{max}$, the dimensionless velocity distribution can be expressed as:

$$\frac{u}{u_{max}} = \frac{2}{M}\left[M(\frac{r-r_0}{r_{max}-r_0}) + \frac{(4-M)^2}{16}\right]^{\frac{1}{2}} - \frac{4-M}{2M} \tag{64}$$

The Tsallis entropy-based approach of Luo and Singh [12] was either superior or comparable to Chiu's distribution for the data sets used therein for testing. However, due to the complexity of the coordinate system and a large number of parameters used, the application of these methods may be limited. Later, using normal x-y coordinate system, Cui and Singh [13] obtained the 2-D velocity distribution equation as:

$$u = -\frac{\lambda_*}{\lambda_1} + \frac{1}{\lambda_1}\frac{m}{m-1}\left[\lambda_1 F(u) + (\frac{m-1}{m}\lambda_*)^{\frac{m}{m-1}}\right]^{\frac{m-1}{m}} \tag{65}$$

where the CDF, $F(u)$, is defined as Equation (54). As defined in the case of one-dimensional velocity distribution, the dimensionless entropy parameter G is also used here as:

$$G = \frac{\lambda_1 u_{max}}{\lambda_1 u_{max} + \lambda_*} \tag{66}$$

Parameter G is found to be related to the ratio of mean and maximum velocity and a quadratic relation is obtained by computing from observed mean and maximum velocity values as [13]:

$$\frac{\bar{u}}{u_{max}} = \Psi(G) = 0.554G^2 - 0.077G + 0.568 \tag{67}$$

Now, with the use of Equation (67), the general velocity distribution Equation (65) can be cast as:

$$\frac{u}{u_{max}} = 1 - \frac{1}{G}\left(1 - \left[(1-G)^{\frac{m}{m-1}} + \left(1 - (1-G)^{\frac{m}{m-1}}\right)F(u)\right]^{\frac{m-1}{m}}\right) \tag{68}$$

Equation (68) is the general 2-D velocity distribution equation in terms of parameter G and maximum velocity.

6.4. Suspended Sediment Concentration

The sediment concentration distribution can also be derived using Tsallis entropy. For deriving the one-dimensional sediment concentration distribution, the constraint equation is derived using the mean concentration, such that $g_1(c) = c$ and limits of integration are $a = c_0$ (maximum concentration at the bed surface or flow depth $D = 0$) and $b = 0$ at the water surface, as [16]:

$$f(c) = \left[\frac{m-1}{m}(\lambda_* + \lambda_1 c)\right]^{\frac{1}{m-1}} \tag{69}$$

where $\lambda_* = \frac{1}{m-1} + \lambda_0$. Equation (69) is the least-biased entropy-based probability density function of sediment concentration, which is fundamental to determining the sediment concentration distribution. The dimensionless sediment concentration can then be obtained as:

$$\frac{c}{c_0} = 1 - \frac{1}{N}\left(1 - \left\{(1-N)^{\frac{m}{m-1}} + \left[1 - (1-N)^{\frac{m}{m-1}}\right]F(c)\right\}^{\frac{m-1}{m}}\right) \tag{70}$$

where N is a dimensionless parameter as a function of maximum concentration and the Lagrange multipliers expressed as:

$$N = \frac{\lambda_1 c_0}{\lambda_1 c_0 + \lambda_*} \tag{71}$$

By plotting the empirical observations of the mean over the maximum concentration values and corresponding N values, the implicit function can be regressed as a quadratic polynomial as:

$$\frac{\bar{c}}{c_0} = 0.176 + 0.5083N - 0.1561N^2 \tag{72}$$

with a coefficient of determination as 0.99. Thus, N can be used for deriving sediment concentration distribution instead of solving nonlinear equations for λ_1 and λ_*.

6.5. Sediment Discharge

The suspended sediment discharge can be computed simultaneously by integrating sediment concentration and velocity over the cross-section where the velocity distribution and sediment concentration can be obtained empirically as well as using the Tsallis entropy theory. That is, suspended sediment discharge can be derived using different combinations of entropy-based and empirical methods for velocity and sediment concentration distributions. Thus, the suspended sediment discharge can be computed with the use of entropy in three ways: (1) velocity distribution by a standard formula and concentration distribution by an entropy-based equation; (2) velocity distribution by an entropy-based equation and concentration distribution by a standard formula; and (3) velocity distribution and concentration distribution both by entropy-based equations. In the discussion that follows, only the combination where both are derived using Tsallis entropy is considered.

The Tsallis entropy-based velocity distribution [13], given by Equation (68) and sediment concentration [16], given by Equation (70), are integrated from bottom of the channel to the water surface to obtain the suspended sediment discharge, as:

$$q_s = u_{max}c_0 \int_0^D \left[1 - \frac{1}{N}\left(1 - ((1 + 0.5\ln N)F(c) - 0.5\ln N)^{2/3}\right)\right]$$
$$\left[1 - \frac{1}{G}(1 - ((1 + 0.5\ln G)F(u) - 0.5\ln G)^{2/3}\right]dy \tag{73}$$

To get an explicit solution of Equation (73) is difficult; however, it can be simplified with the use of mean values. The first term in the integration can be replaced by mean sediment concentration and the second term can be replaced by mean velocity, such that Equation (73) reduces to

$$q_s = Du_{max}c_0(0.554G^2 - 0.777G + 0.568)(0.554N^2 - 0.777N + 0.568) \tag{74}$$

which provides a simple method to compute sediment discharge. Entropy parameters G and N are fixed for each channel cross-section [17]. Thus, once the entropy parameters have been obtained for some known cross-section, with observed maximum velocity and sediment concentration, the sediment discharge can be obtained with ease.

6.6. Flow-Duration Curve

The flow-duration curve (FDC) is used for predicting the distribution of future flows, forecasting of future recurrence frequencies, determining low-flow thresholds for defining droughts, generating hydropower, constructing load-duration curves, determining power-duration curves and comparing watersheds. For deriving an FDC, it is assumed that temporally-averaged discharge Q is a random variable, varying from a minimum value Q_{min} to a maximum value Q_{max}, with a probability density function (PDF) denoted as $f(Q)$. The time interval for which the discharge is averaged depends on the purpose of constructing an FDC but it is often taken as one day. Considering $g_1(x)$ = mean $Q = Q_{mean}$, the PDF of Q can be derived as [79]:

$$f(Q) = \left[\frac{m-1}{m}\left(\frac{1}{m-1} - \lambda_0 - \lambda_1 Q\right)\right]^{\frac{1}{m-1}} \tag{75}$$

It is interesting to note that at $Q = 0$, $f(Q)$ becomes $\{[(m-1)/m][(1/(m-1)) - \lambda_0]\}^{1/(m-1)}$. Similar to velocity distribution, a dimensionless parameter M can be defined as:

$$M = \frac{\lambda_1 Q_{max}}{\lambda_1 Q_{max} - \lambda_*} \tag{76}$$

It was found [79] that M is linearly related to the ratio between the mean flow and the maximum flow, which, using regression, can be written as:

$$M = 2.246 - 4.891\frac{\overline{Q}}{Q_{max}} \tag{77}$$

with the squared correlation coefficient of 0.9972. The FDC can be expressed as:

$$\frac{Q}{Q_{max}} = 1 - \frac{1}{M} - \left(1 - \frac{1}{M}\right)\left\{-\left(\frac{1}{\lambda_*}\right)^{\frac{m}{m-1}}\lambda_1\left(\frac{m}{m-1}\right)^{\frac{1}{m-1}}F(Q) + 1\right\}^{\frac{m-1}{m}}$$
$$= 1 - \frac{1}{M}\left(1 - \left\{\left(\frac{m}{m-1}\right)^{\frac{m}{m-1}}\frac{M}{Q_{max}}[F(Q) - 1] + 1\right\}^{\frac{m-1}{m}}\right) \tag{78}$$

7. Conclusions

A survey of the literature shows that the Tsallis entropy theory has great potential to address a wide range of problems in water engineering and in many other fields, as reviewed in this paper; see also [80–85] for some more recent studies. As generalization of the Shannon entropy, the Tsallis entropy can be applied in generalized equilibrium and statics in physics. The advantage of using the theory is that it can combine statistical information with physical laws, permits deriving physical relations as functions of time or space and derives probability distributions in terms of the specified constraints. However, analysis using the Tsallis entropy theory becomes complicated when one or two constraints are involved and it is more difficult to obtain the analytical expressions. Besides, the *m*-statistic must be carefully chosen when applying the Tsallis entropy, since it sometimes involves complex operations. Although entropy is a thermodynamic quantity, development of the thermodynamic basis of entropy-based relations has not been accomplished yet. Until now, the Tsallis entropy has been applied to determine frequency distributions, network evaluation, hydraulic geometry, evaporation, velocity distribution, sediment concentration distribution, flow duration curves and several other problems, as reviewed above. The outcomes are certainly encouraging. Until now, as we have reviewed in this paper, the Tsallis entropy theory has been applied in water engineering particularly with the principle of maximum entropy (POME). There is, however, also potential to combine the Tsallis entropy with the minimum relative (cross) entropy, for the condition that prior assumptions can be made. There is no question that the Tsallis entropy theory has a much greater potential to study a wide spectrum of problems in water engineering. It is hoped that this review will stimulate further interest in this fascinating field.

Acknowledgments: Bellie Sivakumar acknowledges the financial support from the Australian Research Council (ARC) through the Future Fellowship grant (FT110100328). Huijuan Cui acknowledges the financial support from Key Research Program of the Chinese Academy of Sciences (ZDRW-ZS-2016-6-4).

Author Contributions: All three authors substantially contributed to the article. Vijay P. Singh conceived the ideas for this review article; Vijay P. Singh, Bellie Sivakumar and Huijuan Cui had detailed discussions about the contents, structure and organization of the manuscript; Vijay P. Singh prepared the initial draft; Bellie Sivakumar and Huijuan Cui contributed to modifications and improvements; Vijay P. Singh, Bellie Sivakumar and Huijuan Cui contributed to the responses to the review comments and manuscript revision.

Conflicts of Interest: The authors declare no conflict of interest.

Notation

B	channel width
c	sediment concentration
C_r	constraint
D	water depth
E	evaporation
$f(x)$	probability density function
G	ground heat flux
H	entropy
$\Delta I(x_i)$	gain in information
$k = (1 - m)/m$	
m	Tsallis entropy index
n	Manning's n
N	natural number representative
p_i	probability
Q	flow
q_s	sediment discharge
R	net radiative
S_j	entropic measure of redundancy
u	velocity

x	transverse direction
y	vertical dimension
σ^2	variance
λ_i	Langrange multiplier
μ_1	first moment
μ_2	second moment
w	surface soil moisture

References

1. Singh, V.P. *Hydrologic Systems: Vol. 1. Rainfall-Runoff Modeling*; Prentice Hall: Englewood Cliffs, NJ, USA, 1988.
2. Singh, V.P. *Hydrologic Systems: Vol. 2. Watershed Modeling*; Prentice Hall: Englewood Cliffs, NJ, USA, 1989.
3. Singh, V.P. (Ed.) *Computer Models of Watershed Hydrology*; Water Resources Publications: Highland Ranch, CO, USA, 1995.
4. Salas, J.D.; Delleur, J.W.; Yevjevich, V.; Lane, W.L. *Applied Modeling of Hydrologic Time Series*; Water Resources Publications: Littleton, CO, USA, 1995.
5. Beven, K.J. *Rainfall-Runff Modelling: The Primer*; Wiley: Chichester, UK, 2001.
6. Sivakumar, B.; Berndtsson, R. *Advances in Data-Based Approaches for Hydrologic Modeling and Forecasting*; World Scientific Publishing Company: Singapore, 2010.
7. Singh, V.P. *Handbook of Applied Hydrology*, 2nd ed.; McGraw-Hill Education: New York, NY, USA, 2017.
8. Singh, V.P. *Kinematic Wave Modeling in Water Resources: Surface Water Hydrology*; John Wiley & Sons: New York, NY, USA, 1995.
9. Singh, V.P. *Kinematic Wave Modeling in Water Resources: Environmental Hydrology*; John Wiley & Sons: New York, NY, USA, 1996.
10. Harmancioglu, N.B.; Fistikoglu, O.; Ozkul, S.D.; Singh, V.P.; Alpaslan, M.N. *Water Quality Monitoring Network Design*; Kluwer Academic Publishers: Boston, MA, USA, 1999; p. 299.
11. Singh, V.P.; Luo, H. Entropy theory for distribution of one-dimensional velocity in open channels. *J. Hydrol. Eng.* **2011**, *16*, 725–735. [CrossRef]
12. Luo, H.; Singh, V.P. Entropy theory for two-dimensional velocity distribution. *J. Hydrol. Eng.* **2011**, *16*, 303–315. [CrossRef]
13. Cui, H.; Singh, V.P. Two-dimensional velocity distribution in open channels using the Tsallis entropy. *J. Hydrol. Eng.* **2013**, *18*, 331–339. [CrossRef]
14. Cui, H.; Singh, V.P. One-dimensional velocity distribution in open channels using Tsallis entropy. *J. Hydrol. Eng.* **2014**, *19*, 290–298. [CrossRef]
15. Simons, D.B.; Senturk, F. *Sediment Transport Technology*; Water Resources Publications: Highland Ranch, CO, USA, 1976.
16. Cui, H.; Singh, V.P. Suspended sediment concentration in open channels using Tsallis entropy. *J. Hydrol. Eng.* **2014**, *19*, 966–977. [CrossRef]
17. Cui, H.; Singh, V.P. Computation of suspended sediment discharge in open channels by combining Tsallis entropy-based methods and empirical formulas. *J. Hydrol. Eng.* **2014**, *19*, 18–25. [CrossRef]
18. Singh, V.P. On the theories of hydraulic geometry. *Int. J. Sediment Res.* **2003**, *18*, 196–218.
19. Singh, V.P.; Yang, C.T.; Deng, Z.Q. Downstream hydraulic geometry relations: 1. Theoretical development. *Water Resour. Res.* **2003**, *39*, 1337. [CrossRef]
20. Singh, V.P.; Yang, C.T.; Deng, Z.Q. Downstream hydraulic geometry relations: 2. Calibration and testing. *Water Resour. Res.* **2003**, *39*, 1338. [CrossRef]
21. Singh, V.P.; Zhang, L. At-a-station hydraulic geometry: I. Theoretical development. *Hydrol. Process.* **2008**, *22*, 189–215. [CrossRef]
22. Singh, V.P.; Zhang, L. At-a-station hydraulic geometry: II. Calibration and testing. *Hydrol. Process.* **2008**, *22*, 216–228. [CrossRef]
23. Singh, V.P.; Oh, J. A Tsallis entropy-based redundancy measure for water distribution network. *Physica A* **2014**, *421*, 360–376. [CrossRef]
24. Singh, V.P. *Entropy Theory and Its Application in Environmental and Water Engineering*; John Wiley & Sons: Sussex, UK, 2013.

25. Singh, V.P. *Entropy Theory in Hydraulic Engineering: An Introduction*; ASCE Press: Reston, VA, USA, 2014.

26. Singh, V.P. *Entropy Theory in Hydrologic Science and Engineering*; McGraw-Hill Education: New York, NY, USA, 2015.

27. Koutsoyiannis, D. Physics of uncertainty, the Gibbs paradox and indistinguishable particles. *Stud. Hist. Philos. Mod. Phys.* **2013**, *44*, 480–489. [CrossRef]

28. Papalexiou, S.M.; Koutsoyiannis, D. Entropy based derivation of probability distributions: A case study to daily rainfall. *Adv. Water Resour.* **2012**, *45*, 51–57. [CrossRef]

29. Shannon, C.E. The mathematical theory of communications, I and II. *Bell Syst. Tech. J.* **1948**, *27*, 379–423. [CrossRef]

30. Tsallis, C. Possible generalization of Boltzmann-Gibbs statistics. *J. Stat. Phys.* **1988**, *52*, 479–487. [CrossRef]

31. Singh, V.P. The use of entropy in hydrology and water resources. *Hydrol. Process.* **1997**, *11*, 587–626. [CrossRef]

32. Singh, V.P. Hydrologic synthesis using entropy theory: Review. *J. Hydrol. Eng.* **2011**, *16*, 421–433. [CrossRef]

33. Singh, V.P. *Introduction to Tsallis Entropy Theory in Water Engineering*; CRC Press/Taylor and Francis: Boca Raton, FL, USA, 2016.

34. Tsallis, C. Entropic nonextensivity: A possible measure of complexity. *Chaos Solitons Fractals* **2002**, *12*, 371–391. [CrossRef]

35. Tsallis, C. On the fractal dimension of orbits compatible with Tsallis statistics. *Phys. Rev. E* **1998**, *58*, 1442–1445. [CrossRef]

36. Niven, R.K. The constrained entropy and cross-entropy functions. *Physica A* **2004**, *334*, 444–458. [CrossRef]

37. Koutsoyiannis, D. Uncertainty, entropy, scaling and hydrological stochastics. 1. Marginal distributional properties of hydrological processes and state scaling. *Hydrol. Sci. J.* **2005**, *50*, 381–404. [CrossRef]

38. Koutsoyiannis, D. Uncertainty, entropy, scaling and hydrological stochastics. 2. Time dependence of hydrological processes and time scaling. *Hydrol. Sci. J.* **2005**, *50*, 405–426. [CrossRef]

39. Koutsoyiannis, D. A toy model of climatic variability with scaling behavior. *J. Hydrol.* **2005**, *322*, 25–48. [CrossRef]

40. Tsallis, C. Nonextensive statistical mechanics: Construction and physical interpretation. In *Nonextensive Entropy-Interdisciplinary Applications*; Gell-Mann, M., Tsallis, C., Eds.; Oxford University Press: New York, NY, USA, 2004; pp. 1–52.

41. Tsallis, C. Nonextensive statistical mechanics and thermodynamics: Historical background and present status. In *Nonextensive Statistical Mechanics and Its Applications*; Abe, S., Okamoto, Y., Eds.; Series Lecture Notes in Physics; Springer: Berlin, Germany, 2001.

42. Jaynes, E.T. Information theory and statistical mechanics, I. *Phys. Rev.* **1957**, *106*, 620–630. [CrossRef]

43. Jaynes, E.T. Information theory and statistical mechanics, II. *Phys. Rev.* **1957**, *108*, 171–190. [CrossRef]

44. Kullback, S.; Leibler, R.A. On information and sufficiency. *Ann. Math. Stat.* **1951**, *22*, 79–86. [CrossRef]

45. Lombardo, F.; Volpi, E.; Koutsoyiannis, D.; Papalexiou, S.M. Just two moments! A cautionary note against use of high-order moments in multifractal models in hydrology. *Hydrol. Earth Syst. Sci.* **2014**, *18*, 243–255. [CrossRef]

46. Krstanovic, P.F.; Singh, V.P. A univariate model for longterm streamflow forecasting: I. Development. *Stoch. Hydrol. Hydraul.* **1991**, *5*, 173–188. [CrossRef]

47. Krstanovic, P.F.; Singh, V.P. A univariate model for longterm streamflow forecasting: II. Application. *Stoch. Hydrol. Hydraul.* **1991**, *5*, 189–205. [CrossRef]

48. Krstanovic, P.F.; Singh, V.P. A real-time flood forecasting model based on maximum entropy spectral analysis: I. Development. *Water Resour. Manag.* **1993**, *7*, 109–129. [CrossRef]

49. Krstanovic, P.F.; Singh, V.P. A real-time flood forecasting model based on maximum entropy spectral analysis: II. Application. *Water Resour. Manag.* **1993**, *7*, 131–151. [CrossRef]

50. Krasovskaia, I. Entropy-based grouping of river flow regimes. *J. Hydrol.* **1997**, *202*, 173–191. [CrossRef]

51. Krasovskaia, I.; Gottschalk, L. Stability of river flow regimes. *Nordic Hydrol.* **1992**, *23*, 137–154.

52. Singh, V.P. *Entropy-Based Parameter Estimation in Hydrology*; Kluwer Academic Publishers: Boston, MA, USA, 1998; p. 365.

53. Wang, J.; Bras, R.L. A model of evapotranspiration based on the theory of maximum entropy production. *Water Resour. Res.* **2011**, *47*, W03521. [CrossRef]

54. Yang, C.T. Unit stream power and sediment transport. *J. Hydraul. Div. ASCE* **1972**, *98*, 1805–1826.

55. Fiorentino, M.; Claps, P.; Singh, V.P. An entropy-based morphological analysis of river-basin networks. *Water Resour. Res.* **1993**, *29*, 1215–1224. [CrossRef]

56. Chiu, C.L. Entropy and 2-D velocity distribution in open channels. *J. Hydraul. Eng. ASCE* **1988**, *114*, 738–756. [CrossRef]

57. Cao, S.; Knight, D.W. Design of threshold channels. Hydra 2000. In Proceedings of the 26th IAHR Congress, London, UK, 11–15 September 1995.

58. Cao, S.; Knight, D.W. New Concept of hydraulic geometry of threshold channels. In Proceedings of the 2nd Symposium on the Basic Theory of Sedimentation, Beijing, China, 30 October–2 November 1995.

59. Cao, S.; Knight, D.W. Entropy-based approach of threshold alluvial channels. *J. Hydraul. Res.* **1997**, *35*, 505–524. [CrossRef]

60. Singh, V.P. Entropy theory for movement of moisture in soils. *Water Resour. Res.* **2010**, *46*, 1–12. [CrossRef]

61. Singh, V.P. Tsallis entropy theory for derivation of infiltration equations. *Trans. ASABE* **2010**, *53*, 447–463. [CrossRef]

62. Evans, M.; Hastings, N.; Peacock, B. *Statistical Distributions*; John Wiley & Sons: New York, NY, USA, 2000.

63. Krstanovic, P.F.; Singh, V.P. Evaluation of rainfall networks using entropy: 1. Theoretical development. *Water Resour. Manag.* **1992**, *6*, 279–293. [CrossRef]

64. Krstanovic, P.F.; Singh, V.P. Evaluation of rainfall networks using entropy: 2. Application. *Water Resour. Manag.* **1992**, *6*, 295–314. [CrossRef]

65. Yang, Y.; Burn, D.H. An entropy approach to data collection network design. *J. Hydrol.* **1994**, *157*, 307–324. [CrossRef]

66. Mogheir, Y.; Singh, V.P. Application of information theory to groundwater quality monitoring networks. *Water Resour. Manag.* **2002**, *16*, 37–49. [CrossRef]

67. Perelman, L.; Ostfeld, A.; Salomons, E. Cross entropy multiobjective optimization for water distribution systems design. *Water Resour. Res.* **2008**, *44*, W09413. [CrossRef]

68. Goulter, I.C. Current and future use of systems analysis in water distribution network design. *Civ. Eng. Syst.* **1987**, *4*, 175–184. [CrossRef]

69. Goulter, I.C. Assessing the reliability of water distribution networks using entropy based measures of network redundancy. In *Entropy and Energy Dissipation in Water Resources*; Singh, V.P., Fiorentino, M., Eds.; Kluwer Academic Publishers: Dordrecht, The Netherlands, 1992; pp. 217–238.

70. Walters, G. Optimal design of pipe networks: A review. In Proceedings of the 1st International Conference Computational Water Resources, Rabat, Morocco, 14–18 March 1988; Computational Mechanics Publications: Southampton, UK, 1988; pp. 21–31.

71. Leopold, L.B.; Maddock, T.J. *Hydraulic Geometry of Stream Channels and Some Physiographic Implications*; US Government Printing Office: Washington, DC, USA, 1953; p. 55.

72. Langbein, W.B. Geometry of river channels. *J. Hydraul. Div. ASCE* **1964**, *90*, 301–311.

73. Yang, C.T.; Song, C.C.S.; Woldenberg, M.J. Hydraulic geometry and minimum rate of energy dissipation. *Water Resour. Res.* **1981**, *17*, 1014–1018. [CrossRef]

74. Wolman, M.G. *The Natural Channel of Brandywine Creek, Pennsylvania*; US Government Printing Office: Washington, DC, USA, 1955.

75. Williams, G.P. *Hydraulic Geometry of River Cross-Sections-Theory of Minimum Variance*; US Government Printing Office: Washington, DC, USA, 1978.

76. Wang, J.; Salvucci, G.D.; Bras, R.L. A extremum principle of evaporation. *Water Resour. Res.* **2004**, *40*, W09303. [CrossRef]

77. Cui, H.; Singh, V.P. ON the cumulative distribution function for entropy-based hydrologic modeling. *Trans. ASABE* **2012**, *55*, 429–438. [CrossRef]

78. Singh, V.P.; Luo, H. Derivation of velocity distribution using entropy. In Proceedings of the IAHR Congress, Vancouver, BC, Canada, 9–14 August 2009; pp. 31–38.

79. Singh, V.P.; Byrd, A.; Cui, H. Flow duration curve using entropy theory. *J. Hydrol. Eng.* **2014**, *19*, 1340–1348. [CrossRef]

80. Weijs, S.V.; van de Giesen, N.; Parlange, M.B. HydroZIP: How hydrological knowledge can be used to improve compression of hydrological data. *Entropy* **2013**, *15*, 1289–1310. [CrossRef]

81. Chen, J.; Li, G. Tsallis wavelet entropy and its application in power signal analysis. *Entropy* **2014**, *16*, 3009–3025. [CrossRef]

82. Furuichi, S.; Mitroi-Symeonidis, F.-C.; Symeonidis, E. On some properties of Tsallis hypoentropies and hypodivergences. *Entropy* **2014**, *16*, 5377–5399. [CrossRef]

83. Lenzi, E.K.; da Silva, L.R.; Lenzi, M.K.; dos Santos, M.A.F.; Ribeiro, H.V.; Evangelista, L.R. Intermittent motion, nonlinear diffusion equation and Tsallis formalism. *Entropy* **2017**, *19*, 42. [CrossRef]

84. Evren, A.; Ustaoğlu, E. Measures of qualitative variation in the case of maximum entropy. *Entropy* **2017**, *19*, 204. [CrossRef]

85. Kalogeropoulos, N. The Legendre transform in non-additive thermodynamics and complexity. *Entropy* **2017**, *19*, 298. [CrossRef]

entropy

MDPI

Article

Entropy-Based Parameter Estimation for the Four-Parameter Exponential Gamma Distribution

Songbai Song *, Xiaoyan Song * and Yan Kang *

College of Water Resources and Architectural Engineering, Northwest A & F University, Yangling 712100, China
* Correspondence: ssb6533@nwsuaf.edu.cn (S.S.); songxiaoyan107@mails.ucas.ac.cn (X.S.);
  kangyan@nwsuaf.edu.cn (Y.K.); Tel.: +86-29-8708-2902 (S.S); Fax: +86-29-8708-2901 (S.S.)

Academic Editors: Huijuan Cui, Bellie Sivakumar, Vijay P. Singh and Kevin H. Knuth
Received: 6 March 2017; Accepted: 21 April 2017; Published: 26 April 2017

Abstract: Two methods based on the principle of maximum entropy (POME), the ordinary entropy method (ENT) and the parameter space expansion method (PSEM), are developed for estimating the parameters of a four-parameter exponential gamma distribution. Using six data sets for annual precipitation at the Weihe River basin in China, the PSEM was applied for estimating parameters for the four-parameter exponential gamma distribution and was compared to the methods of moments (MOM) and of maximum likelihood estimation (MLE). It is shown that PSEM enables the four-parameter exponential distribution to fit the data well, and can further improve the estimation.

Keywords: four-parameter exponential gamma distribution; principle of maximum entropy; precipitation frequency analysis; methods of moments; maximum likelihood estimation

1. Introduction

Hydrological frequency analysis is a statistical prediction method that consists of studying past events that are characteristic of a particular hydrological process in order to determine the probabilities of the occurrence of these events in the future [1,2]. It is widely used for planning, design, and management of water resource systems. The probability distributions containing four or more parameters may exhibit some useful properties [3]: (1) versatility and (2) ability to represent data from mixed populations. Among these distributions, some popular distributions are Wakeby, two-component lognormal, two-component extreme value distributions, and the four-parameter kappa distribution. Since the pioneering stream flow records frequency analysis of Herschel and Freeman during the period from 1880 to 1890, hydrological frequency analysis has undergone extensive further development. There are a multitude of methods for estimating parameters of hydrologic frequency distributions. Some of the popular methods include [3,4]: (1) the method of moments; (2) the method of probability weighted moments; (3) the method of mixed moments; (4) L-moments; (5) the maximum likelihood estimation; (6) the least square method; and (7) the entropy-based parameter estimation method.

Among the above parameter estimation methods, entropy, which is a measure of uncertainty of random variables, has attracted much attention and has been used for a variety of applications in hydrology [5–23]. For example, an entropy-based derivation of daily rainfall probability distribution [24], the Burr XII-Singh-Maddala (BSM) distribution function derived from the maximum entropy principle using the Boltzmann-Shannon entropy with some constraints [25]. "*Entropy-Based Parameter Estimation in Hydrology*" is the first book focusing on parameter estimation using entropy for a number of distributions frequently used in hydrology [3], including the uniform distribution, exponential distribution, normal distribution, two-parameter lognormal distribution, three-parameter lognormal distribution, extreme value type I distribution, log-extreme value type I distribution, extreme value type III distribution, generalized extreme value distribution, Weibull distribution, gamma distribution, Pearson type III distribution, log-Pearson type III distribution, beta distribution, two-parameter log-logistic distribution, three-parameter log-logistic distribution, two-parameter Pareto distribution, two-parameter generalized Pareto distribution, three-parameter generalized Pareto distribution and two-component extreme value distribution. Recently, two entropy-based methods, called the ordinary entropy method (ENT) and the parameter space expansion method (PSEM), that are both based on the principle of maximum entropy (POME) have been applied for estimating the parameters of the extended Burr XII distribution and the four-parameter kappa distribution [5,16]. The results of the estimation show that the entropy method enables these two distributions to fit the data better than the other estimation methods. In the above method of entropy-based parameter estimation of a distribution, the distribution parameters are expressed in terms of the given constraints, and then the method can provide a way to derive the distribution from the specified constraints. The general procedure for the ENT for a hydrologic frequency distribution involves the following steps [3]: (1) define the given information in terms of the constraints; (2) maximize the entropy subject to the given information; and (3) relate the parameters to the given information. The PSEM employs an enlarged parameter space and maximizes the entropy subject to the parameters and the Lagrange multipliers [3]. The parameters of the distribution can be estimated by the maximization of the entropy function.

The Pearson III distribution is recommended as a standard distribution to fit hydrological data in China. In addition, generalized Pareto distribution (GPD), generalized extreme value (GEV) and three-parameter Burr type XII distribution also have been applied flood frequency analysis [26].

Inspired in large part by the two-parameter gamma distribution, a four-parameter exponential gamma distribution has been developed to apply in many areas, such as wind and flood frequency in Yellow River basin, Yangtse River basin, Aumer Basin and Liaohe River basin of China [4]. Depending on the parameter values, the four-parameter exponential gamma distribution can be turned into a Pearson type III distribution, Weibull distribution, Maxwell distribution, Kritsky and Menkel distribution, Chi-square distribution, Poisson distribution, half-normal distribution and half-Laplace distribution. The properties of the four-parameter exponential gamma and relations between this distribution and other distributions have been investigated [4]. These investigations suggest that the four-parameter exponential gamma distribution may have a potential in hydrology. Despite the advances mentioned above, the entropy-based parameter estimation for the four-parameter exponential gamma distribution has received comparatively little attention from the hydrologic community.

The objective of this paper is to apply two entropy-based methods that both use the POME for the estimation of the parameters of the four-parameter exponential gamma distribution; compute the annual precipitation quantiles using this distribution for different return periods; and compare these parameters with those estimated when the methods of moments (MOM) and maximum likelihood estimation (MLE) were employed for parameter estimation.

2. Four-Parameter Exponential Gamma Distribution

2.1. Probability Density Function and Cumulative Distribution Function

The probability density function (PDF) of the four-parameter exponential gamma distribution can be expressed as [4]:

$$f(x) = \frac{\beta^{\alpha}}{b\Gamma(\alpha)}(x-\delta)^{\frac{\alpha}{b}-1}e^{-\beta(x-\delta)^{\frac{1}{b}}} ; \delta \leq x < \infty \tag{1}$$

where α β, δ and b are, respectively, the shape, scale, location and transformation parameter.

Depending on the values of the four parameters α β, δ and b, Equation (1) turns into the following special cases:

(1) If $b = 1$, then Equation (1) becomes the Pearson type III distribution:

$$f(x) = \frac{\beta^{\alpha}}{\Gamma(\alpha)}(x - \delta)^{\alpha-1}e^{-\beta(x-\delta)} \tag{2}$$

If $\delta = 0$, Equation (2) becomes a gamma distribution:

$$f(x) = \frac{\beta^{\alpha}}{\Gamma(\alpha)}x^{\alpha-1}e^{-\beta x} \tag{3}$$

(2) If $\delta = 0$, $b = \frac{1}{m}$, $\alpha = \frac{\alpha}{m}$ and $\beta = \frac{1}{d}$, then Equation (1) reduces to the Weibull distribution:

$$f(x) = \frac{m}{d^{\alpha}}(x - \delta)^{m-1}e^{-\frac{1}{d}(x-\delta)^m} \tag{4}$$

(3) If $\delta = 0$, $b = \frac{1}{m}$, $\alpha = \frac{\alpha}{m}$ and $\beta = \frac{1}{d}$, then Equation (1) becomes the three-parameter Weibull distribution:

$$f(x) = \frac{m}{d^{\frac{\alpha}{m}}\Gamma\left(\frac{\alpha}{m}\right)}x^{\alpha-1}e^{-\frac{x^m}{d}} \tag{5}$$

(4) If $\delta = 0$ and $\beta = \frac{\alpha}{\alpha b}$, then Equation (1) reduces to the Kritsky and Menkel distribution:

$$f(x) = \frac{\alpha^{\alpha}}{\alpha^{\frac{\alpha}{b}}b\Gamma(\alpha)}x^{\frac{\alpha}{b}-1}e^{-\alpha\left(\frac{x}{\alpha}\right)^{\frac{1}{b}}} \tag{6}$$

(5) If $\delta = 0$, $b = 1$, $\alpha = \frac{n}{2}$ and $\beta = \frac{1}{2}$, then Equation (1) becomes the Chi-square distribution:

$$f(x) = \frac{1}{2^{\frac{n}{2}}\Gamma\left(\frac{n}{2}\right)}x^{\frac{n}{2}-1}e^{-\frac{x}{2}} \tag{7}$$

(6) If $\delta = 0$, $b = 1$, $\alpha = k+1$ and $\beta = 1$, then Equation (1) reduces to the Poisson distribution

$$f(x) = \frac{x^k}{k!}e^{-x} \tag{8}$$

(7) If $\delta = a$, $b = \alpha = \frac{1}{2}$ and $\beta = \frac{1}{2\sigma^2}$, then Equation (1) becomes the half-normal distribution:

$$f(x) = \frac{2}{\sigma\sqrt{2\pi}}e^{-\frac{1}{2\sigma^2}(x-a)^2} \tag{9}$$

(8) If $\delta = 0$, $b = \frac{1}{2}$, $\alpha = \frac{3}{2}$ and $\beta = \frac{1}{a^2}$, then Equation (1) becomes the half-normal distribution:

$$f(x) = \frac{4}{a^3\sqrt{\pi}}x^2e^{-\frac{x^2}{a^2}} \tag{10}$$

(9) If $b = 1$, $\alpha = 1$ and $\beta = \frac{1}{\lambda}$, then Equation (1) reduces to the half-Laplace distribution:

$$f(x) = \frac{1}{2\lambda}e^{-\frac{x-\delta}{\lambda}} \tag{11}$$

The cumulative distribution function (CDF) of the four-parameter exponential gamma distribution can be expressed as:

$$P = F(X \leq x_p) = \int_{\delta}^{x_p} f(x)dx = \int_{\delta}^{x_p} \frac{\beta^{\alpha}}{b\Gamma(\alpha)}(x - \delta)^{\frac{\alpha}{b}-1}e^{-\beta(x-\delta)^{\frac{1}{b}}}dx \tag{12}$$

Let $t = \beta(x - \delta)^{\frac{1}{b}}$, then $x = \delta + \frac{1}{\beta^b} t^b$, $x - \delta = \frac{1}{\beta^b} t^b$ and $dx = \frac{b}{\beta^b} t^{b-1} dt$. Substitution of the above quantities in Equation (12) yields [4]:

$$P = \int_0^{t_p} \frac{1}{\Gamma(\alpha)} t^{\alpha-1} e^{-t} dt \tag{13}$$

where $t_p = \beta(x_p - \delta)^{\frac{1}{b}}$ and can be determined by the incomplete gamma function.

2.2. Quantile Corresponding to the Probability of Exceedance

The quantile corresponding to the probability of exceedance p, x_p, is obtained by Equation (14) or Equation (15):

$$x_p = \delta + \frac{1}{\beta^b} t_p^b \tag{14}$$

$$x_p = \bar{x}(1 + \Phi_p C_v) \tag{15}$$

here, $\bar{x}$ and C_v are the mean and coefficient of the variation of a sample, respectively, and Φ_p is the frequency factor corresponding to x_p. Given the expectation and variance of the population, the frequency factorr Φ_p is given by [4]:

$$\Phi_p = \frac{x_p - \mu'_1}{\sqrt{\mu_2}} \Phi_p = \frac{t_p^b \Gamma(\alpha) - \Gamma(\alpha + b)}{\sqrt{\Gamma(\alpha)\Gamma(\alpha + 2b) - \Gamma^2(\alpha + b)}} \tag{16}$$

where μ'_1 and μ_2 are the expectation and variance of the population.

If $b = 10$, the frequency factors of the four-parameter exponential gamma distribution, Φ_p, are very close to that of the log-normal distribution (Table 1). If $C_s = 1.1395$, the Φ_p values are very close to that of the Gumbel distribution (Table 2).

Table 1. Frequency factors of the four-parameter exponential gamma distribution and log-normal distribution ($b = 10$).

C_s	Distribution	P (%)							
		0.01	0.1	1	5	20	50	90	99
0.2	Four-parameter exponential gamma	4.17	3.39	2.47	1.70	0.830	−0.033	−1.26	−2.18
	log-normal	4.17	3.39	2.47	1.70	0.830	−0.033	−1.26	−2.18
2.0	Four-parameter exponential gamma	9.44	6.23	3.53	1.90	0.614	−0.240	−0.963	−1.26
	log-normal	9.51	6.24	3.52	1.89	0.614	−0.240	−0.967	−1.28

Table 2. Frequency factors of the four-parameter exponential gamma distribution and Gumbel distribution under $C_s = 1.1395$.

Distribution	P (%)							
	0.01	0.1	1	5	20	50	90	99
Four-parameter exponential gamma	6.80	4.92	3.12	1.87	0.728	−0.166	−1.10	−1.61
log-normal	6.80	4.94	3.14	1.87	0.728	−0.164	−1.10	−1.64

2.3. Cumulants and Moments

The first three cumulants of the four-parameter exponential gamma distribution are expressed as [4]:

$$k_1 = \delta + \frac{\Gamma(b + \alpha)}{\beta^b \Gamma(\alpha)} \tag{17}$$

$$k_2 = \frac{\Gamma(\alpha)\Gamma(\alpha + 2b) - \Gamma^2(\alpha + b)}{\beta^{2b}\Gamma^2(\alpha)} \tag{18}$$

$$k_3 = \frac{\Gamma^2(\alpha)\Gamma(\alpha + 3b) - 3\Gamma(\alpha)\Gamma(\alpha + b)\Gamma(\alpha + 2b) + 2\Gamma^3(\alpha + b)}{\beta^{3b}\Gamma^3(\alpha)} \tag{19}$$

Using the relations between moments and cumulants and Equations (17)–(19), the expression for the first four moments of the four-parameter exponential gamma distribution are given below:

$$v_1 = E(X) = \delta + \frac{\Gamma(\alpha + b)}{\beta^b \Gamma(\alpha)} \tag{20}$$

$$v_2 = \frac{\Gamma(\alpha + 2b)}{\beta^{2b}\Gamma(\alpha)} + 2\delta\frac{\Gamma(\alpha + b)}{\beta^b \Gamma(\alpha)} + \delta^2 \tag{21}$$

$$v_3 = \frac{\Gamma(\alpha + 3b)}{\beta^{3b}\Gamma(\alpha)} + 3\delta\frac{\Gamma(2b + \alpha)}{\beta^{2b}\Gamma(\alpha)} + 3\delta^2\frac{\Gamma(b + \alpha)}{\beta^b \Gamma(\alpha)} + \delta^3 \tag{22}$$

$$v_4 = \frac{\Gamma(\alpha + 4b)}{\beta^{4b}\Gamma(\alpha)} + 4\delta\frac{\Gamma(\alpha + 3b)}{\beta^{3b}\Gamma(\alpha)} + 6\delta^2\frac{\Gamma(\alpha + 2b)}{\beta^{2b}\Gamma(\alpha)} + 4\delta^3\frac{\Gamma(\alpha + b)}{\beta^b \Gamma(\alpha)} + \delta^4 \tag{23}$$

In next sections, we use two methods of parameter estimation, ENT and PSEM, to derive the parameters estimation expression of the four-parameter exponential gamma distribution.

3. Ordinary Entropy Method

For ENT, three steps are involved in the estimation of the parameters of a probability distribution: (1) specification of appropriate constraints, (2) derivation of the entropy function of the distribution, and (3) derivation of the relations between parameters and constraints [3,16].

3.1. Specification of Constraints

Taking the natural logarithm of Equation (1), we obtain:

$$\ln f(x) = \alpha \ln \beta - \ln b - \ln \Gamma(\alpha) + \left(\frac{\alpha}{b} - 1\right)\ln(x - \delta) - \beta(x - \delta)^{\frac{1}{b}} \tag{24}$$

Multiplying Equation (24) by [−f(x)] and integrating from δ to ∞, we obtain the entropy function:

$$S = -\int_\delta^\infty f(x)\ln f(x)dx = -\alpha\ln\beta + \ln b + \ln\Gamma(\alpha) - \left(\frac{\alpha}{b} - 1\right)E[\ln(x - \delta)] + \beta \cdot E\left[(x - \delta)^{\frac{1}{b}}\right] \tag{25}$$

To maximize S in Equation (25), the following constraints for Equation (25) should be satisfied

$$\int_\delta^\infty f(x)dx = 1 \tag{26}$$

$$\int_\delta^\infty \ln(x - \delta)f(x)dx = E[\ln(x - \delta)] \tag{27}$$

$$\int_\delta^\infty \ln(x - \delta)^{\frac{1}{b}} f(x)dx = E\left[\ln(x - \delta)^{\frac{1}{b}}\right] \tag{28}$$

$$\int_\delta^\infty (x - \delta)^{\frac{1}{b}} f(x)dx = E\left[(x - \delta)^{\frac{1}{b}}\right] \tag{29}$$

3.2. Construction of Partition Function and Zeroth Lagrange Multiplier

The least-biased pdf, $f(x)$, consistent with Equations (26) to (29) and corresponding to the POME takes the form:

$$f(x) = \exp\left[-\lambda_0 - \lambda_1 \ln(x - \delta) - \lambda_2 \ln(x - \delta)^{\frac{1}{b}} - \lambda_3(x - \delta)^{\frac{1}{b}}\right] \tag{30}$$

where $\lambda_0, \lambda_1, \lambda_2$ and λ_3 are Lagrange multipliers. Substitution of Equation (30) in Equation (26) yields:

$$\int_{\delta}^{\infty} \exp\left[-\lambda_0 - \lambda_1 \ln(x - \delta) - \lambda_2 \ln(x - \delta)^{\frac{1}{b}} - \lambda_3(x - \delta)^{\frac{1}{b}}\right] dx = 1 \tag{31}$$

The argument of the exponential function on the left side of Equation (31) has two parts: zeroth Lagrange multiplier without the random variable and four Lagrange multipliers with the random variable. The zeroth Lagrange multiplier part is separated out and is expressed as:

$$\exp(\lambda_0) = \int_{\delta}^{\infty} \exp\left[-\lambda_1 \ln(x - \delta) - \lambda_2 \ln(x - \delta)^{\frac{1}{b}} - \lambda_3(x - \delta)^{\frac{1}{b}}\right] dx \tag{32}$$

To calculate the above integral, let $y = \lambda_3(x - \delta)^{\frac{1}{b}}$, then $x = \left(\frac{y}{\lambda_3}\right)^b + \delta$, $dx = \frac{b}{\lambda_3^b} y^{b-1} dy$. Substituting the above quantities in Equation (32), we obtain:

$$\exp(\lambda_0) = \int_0^{\infty} \left(\frac{y}{\lambda_3}\right)^{-\left(\lambda_1 + \frac{\lambda_2}{b}\right)} \cdot \exp(-y) \frac{b}{\lambda_3^b} y^{b-1} dy$$

$$= \frac{b}{\lambda_3^{b-(b\lambda_1+\lambda_2)}} \int_0^{\infty} y^{b-(b\lambda_1+\lambda_2)-1} \cdot e^{-y} dy = \frac{b}{\lambda_3^{b-(b\lambda_1+\lambda_2)}} \Gamma[b - (b\lambda_1 + \lambda_2)] \tag{33}$$

Taking the logarithm of Equation (33) results in the zeroth Lagrange λ_0 multiplier as a function of Lagrange multipliers λ_1, λ_2 and λ_3, with the expression given as:

$$\lambda_0 = \ln b - [b - (b\lambda_1 + \lambda_2)] \ln \lambda_3 + \ln \Gamma[b - (b\lambda_1 + \lambda_2)] \tag{34}$$

$$\lambda_0 = \ln \int_{\delta}^{\infty} \exp\left[-\lambda_1 \ln(x - \delta) - \lambda_2 \ln(x - \delta)^{\frac{1}{b}} - \lambda_3(x - \delta)^{\frac{1}{b}}\right] dx \tag{35}$$

3.3. Relation between Lagrange Multiplier and Constraints

Differentiating Equation (35) with λ_1, λ_2 and λ_3, we obtain the derivatives of λ_0 with respect to λ_1, λ_2 and λ_3, the detailed derivations are given in Appendix B:

$$\frac{\partial \lambda_0}{\partial \lambda_1} = -E[\ln(x - \delta)] \tag{36}$$

$$\frac{\partial \lambda_0}{\partial \lambda_2} = -E\left[\ln(x - \delta)^{\frac{1}{b}}\right] \tag{37}$$

$$\frac{\partial \lambda_0}{\partial \lambda_3} = -E\left[(x - \delta)^{\frac{1}{b}}\right] \tag{38}$$

Furthermore, we can write:

$$\frac{\partial^2 \lambda_0}{\partial \lambda_3^2} = E\left[(x - \delta)^{\frac{1}{b}}\right]^2 - \left\{E\left[(x - \delta)^{\frac{1}{b}}\right]\right\}^2 \tag{39}$$

Additionally, differentiating Equation (34) with λ_1, λ_2 and λ_3, we obtain:

$$\frac{\partial \lambda_0}{\partial \lambda_1} = b \ln \lambda_3 - b\psi[b - (b\lambda_1 + \lambda_2)] \tag{40}$$

$$\frac{\partial \lambda_0}{\partial \lambda_2} = \ln \lambda_3 - \psi[b - (b\lambda_1 + \lambda_2)] \tag{41}$$

$$\frac{\partial \lambda_0}{\partial \lambda_3} = -\frac{b - (b\lambda_1 + \lambda_2)}{\lambda_3} \tag{42}$$

$$\frac{\partial^2 \lambda_0}{\partial \lambda_3^2} = \frac{b - (b\lambda_1 + \lambda_2)}{\lambda_3^2} \tag{43}$$

Equating Equations (36) and (40), we obtain:

$$- E[\ln(x - \delta)] = b \ln \lambda_3 - b\psi[b - (b\lambda_1 + \lambda_2)] \tag{44}$$

Equating Equations (37) and (41), we obtain:

$$- E\left[\ln(x - \delta)^{\frac{1}{b}}\right] = \ln \lambda_3 - \psi[b - (b\lambda_1 + \lambda_2)] \tag{45}$$

Equating Equations (38) and (42), we obtain:

$$- E\left[(x - \delta)^{\frac{1}{b}}\right] = -\frac{b - (b\lambda_1 + \lambda_2)}{\lambda_3} \tag{46}$$

Equating Equations (39) and (43), we obtain:

$$E\left[(x - \delta)^{\frac{1}{b}}\right]^2 - \left\{E\left[(x - \delta)^{\frac{1}{b}}\right]\right\}^2 = \frac{b - (b\lambda_1 + \lambda_2)}{\lambda_3^2} \tag{47}$$

3.4. Relation between Lagrange Multiplier and Parameters

Introduction of Equation (34) in Equation (30) produces:

$$f(x) = \exp\left\{-\ln b + [b - (b\lambda_1 + \lambda_2)]\ln \lambda_3 - \ln \Gamma[b - (b\lambda_1 + \lambda_2)] - \lambda_1 \ln(x - \delta) - \lambda_2 \ln(x - \delta)^{\frac{1}{b}} - \lambda_3(x - \delta)^{\frac{1}{b}}\right\}$$

$$= \exp(\ln b^{-1}) \cdot \exp\left[\ln \lambda_3^{b - (b\lambda_1 + \lambda_2)}\right] \cdot \exp\left\{\ln \frac{1}{\Gamma[b - (b\lambda_1 + \lambda_2)]}\right\}$$

$$\cdot \exp\left[\ln(x - \delta)^{-\lambda_1}\right] \cdot \exp\left[\ln(x - \delta)^{-\frac{\lambda_2}{b}}\right] \cdot \exp\left[-\lambda_3(x - \delta)^{\frac{1}{b}}\right] \tag{48}$$

$$= \frac{1}{b} \cdot \lambda_3^{b - (b\lambda_1 + \lambda_2)} \cdot \frac{1}{\Gamma[b - (b\lambda_1 + \lambda_2)]} \cdot (x - \delta)^{-\lambda_1} \cdot (x - \delta)^{-\frac{\lambda_2}{b}} \cdot \exp\left[-\lambda_3(x - \delta)^{\frac{1}{b}}\right]$$

$$= \frac{\lambda_3^{b - (b\lambda_1 + \lambda_2)}}{b \cdot \Gamma[b - (b\lambda_1 + \lambda_2)]} \cdot (x - \delta)^{-(\lambda_1 + \frac{\lambda_2}{b})} \cdot e^{-\lambda_3(x - \delta)^{\frac{1}{b}}}$$

a comparison of Equation (48) with Equation (1) shows that:

$$\lambda_3 = \beta \tag{49}$$

$$b\lambda_1 + \lambda_2 = b - \alpha \tag{50}$$

3.5. Relation between Parameters and Constraints

The four-parameter exponential gamma distribution has four parameters α, β, δ and b that are related to the Lagrange multipliers by Equations (49) and (50). In turn, these parameters are related to the known constrains by Equations (44)–(47). Eliminating the Lagrange multipliers among these four sets of Equations, we can obtain the following Equations:

$$\begin{cases} b \ln \beta - b\psi(\alpha) + E[\ln(x - \delta)] = 0 \\ \ln \beta - \psi(\alpha) + E\left[\ln(x - \delta)^{\frac{1}{b}}\right] = 0 \\ E\left[(x - \delta)^{\frac{1}{b}}\right] - \frac{\alpha}{\beta} = 0 \\ E\left[(x - \delta)^{\frac{1}{b}}\right]^2 - \left\{E\left[(x - \delta)^{\frac{1}{b}}\right]\right\}^2 - \frac{\alpha}{\beta^2} = 0 \end{cases} \tag{51}$$

4. Parameter Space Expansion Method

4.1. Specification of Constraints

Following reference [3], the constraints consistent with the POME method and appropriate for the four-parameter exponential gamma distribution are specified by Equations (26), (27) and (29).

4.2. Construction of Zeroth Lagrange Multiplier

The least-biased pdf corresponding to POME and consistent with Equations (26), (27) and (29) takes the form:

$$f(x) = \exp\left[-\lambda_0 - \lambda_1 \ln(x - \delta) - \lambda_2(x - \delta)^{\frac{1}{b}}\right] \tag{52}$$

where λ_0, λ_1 and λ_2 are Lagrange multipliers. Substitution of Equation (52) into Equation (26) yields

$$\int_\delta^\infty \exp\left[-\lambda_0 - \lambda_1 \ln(x - \delta) - \lambda_2(x - \delta)^{\frac{1}{b}}\right] dx = 1 \tag{53}$$

$$\exp(\lambda_0) = \int_\delta^\infty (x - \delta)^{-\lambda_1} \exp\left[-\lambda_2(x - \delta)^{\frac{1}{b}}\right] dx \tag{54}$$

Let $y = \lambda_2(x - \delta)^{\frac{1}{b}}$, then $x = \left(\frac{y}{\lambda_2}\right)^b + \delta$, $dx = \frac{b}{\lambda_2^b} y^{b-1} dy$. Substituting the above quantities in Equation (54) and changing the limits of integration, we obtain:

$$\exp(\lambda_0) = \int_0^\infty \left(\frac{y}{\lambda_2}\right)^{-b\lambda_1} \cdot \exp(-y) \frac{b}{\lambda_2^b} y^{b-1} dy$$
$$= \frac{b}{\lambda_2^{b-b\lambda_1}} \int_0^\infty y^{b-b\lambda_1-1} \cdot e^{-y} dy = \frac{b}{\lambda_2^{b-b\lambda_1}} \Gamma(b - b\lambda_1) \tag{55}$$

This yields the zeroth Lagrange multiplier:

$$\lambda_0 = \ln b - (b - b\lambda_1) \ln \lambda_2 + \ln \Gamma(b - b\lambda_1) \tag{56}$$

4.3. Derivation of Entropy Function

Introduction of Equation (56) into Equation (52) yields:

$$f(x) = \exp\left[-\ln b + (b - b\lambda_1) \ln \lambda_2 - \ln \Gamma(b - b\lambda_1) - \lambda_1 \ln(x - \delta) - \lambda_2(x - \delta)^{\frac{1}{b}}\right]$$
$$= \frac{\lambda_2^{b-b\lambda_1}}{b\Gamma(b-b\lambda_1)} (x - \delta)^{-\lambda_1} \exp\left[-\lambda_2(x - \delta)^{\frac{1}{b}}\right] \tag{57}$$

a comparison of Equation (57) with Equation (1) shows that:

$$\lambda_1 = 1 - \frac{\alpha}{b} \tag{58}$$

$$\lambda_2 = \beta \tag{59}$$

taking the logarithm of Equation (57) yields:

$$\ln f(x) = (b - b\lambda_1) \ln \lambda_2 - \ln b - \ln \Gamma(b - b\lambda_1) - \lambda_1 \ln(x - \delta) - \lambda_2(x - \delta)^{\frac{1}{b}} \tag{60}$$

then, making use of Equation (60), the entropy function can be written as:

$$S = -\int_\delta^\infty f(x) \ln f(x) dx$$

$$= \int_\delta^\infty \left[-(b - b\lambda_1) \ln \lambda_2 + \ln b + \ln \Gamma(b - b\lambda_1) + \lambda_1 \ln(x - \delta) + \lambda_2 (x - \delta)^{\frac{1}{b}} \right] f(x) dx \qquad (61)$$

$$= -(b - b\lambda_1) \ln \lambda_2 + \ln b + \ln \Gamma(b - b\lambda_1) + \lambda_1 E[\ln(x - \delta)] + \lambda_2 E\left[(x - \delta)^{\frac{1}{b}} \right]$$

4.4. Relation between Parameters and Constraints

Taking partial derivatives of (61) with respect to b, δ, λ_1, and λ_2, and equating each derivative to zero yields:

$$\frac{\partial S}{\partial \lambda_1} = 0 = b \ln \lambda_2 - b\psi(b - b\lambda_1) + E[\ln(x - \delta)] \qquad (62)$$

$$\frac{\partial S}{\partial \lambda_2} = 0 = -\frac{b - b\lambda_1}{\lambda_2} + E\left[(x - \delta)^{\frac{1}{b}} \right] \qquad (63)$$

$$\frac{\partial S}{\partial b} = 0 = -(1 - \lambda_1) \ln \lambda_2 + \frac{1}{b} + (1 - \lambda_1) \cdot \psi(b - b\lambda_1) - \frac{\lambda_2}{b^2} E\left[(x - \delta)^{\frac{1}{b}} \ln(x - \delta) \right] \qquad (64)$$

$$\frac{\partial S}{\partial \delta} = 0 = -\lambda_1 E\left(\frac{1}{x - \delta} \right) - \frac{\lambda_2}{b} E\left[(x - \delta)^{\frac{1}{b} - 1} \right] \qquad (65)$$

Introduction of Equations (58)–(59) into Equations (62)–(65) and recalling Equations (62)–(65) yields, respectively:

$$(66) \quad \begin{cases} b \ln \beta - b\psi(\alpha) + E[\ln(x - \delta)] = 0 \\ \frac{\alpha}{\beta} - E\left[(x - \delta)^{\frac{1}{b}} \right] = 0 \\ -\frac{\alpha}{b} \ln \beta + \frac{1}{b} + \frac{\alpha}{b} \cdot \psi(\alpha) - \frac{\beta}{b^2} E\left[(x - \delta)^{\frac{1}{b}} \ln(x - \delta) \right] = 0 \\ \left(\frac{\alpha}{b} - 1 \right) E\left(\frac{1}{x - \delta} \right) - \frac{\beta}{b} E\left[(x - \delta)^{\frac{1}{b} - 1} \right] = 0 \end{cases}$$

The expectations of Equation (66) are replaced by their sample estimates, and the simplification of Equation (66) leads to:

$$(67) \quad \begin{cases} -\frac{\alpha}{b} \sum_{i=1}^n \left(\frac{1}{x_i - \delta} \right) + \sum_{i=1}^n \left(\frac{1}{x_i - \delta} \right) + \frac{\beta}{b} \sum_{i=1}^n (x_i - \delta)^{\frac{1}{b} - 1} = 0 \\ n \ln \beta - n\psi(\alpha) + \frac{1}{b} \sum_{i=1}^n \ln(x_i - \delta) = 0 \\ \frac{n\alpha}{\beta} - \sum_{i=1}^n (x_i - \delta)^{\frac{1}{b}} = 0 \\ -\frac{n}{b} - \frac{\alpha}{b^2} \sum_{i=1}^n \ln(x_i - \delta) + \frac{\beta}{b^2} \sum_{i=1}^n \left[(x_i - \delta)^{\frac{1}{b}} \ln(x_i - \delta) \right] = 0 \end{cases}$$

Equations (51) has the second moments and results in some biases. Therefore, Equation (67) should be used for the estimation of the parameters.

5. Two Other Parameter Estimation Methods

Two other methods of parameter estimation frequently used in hydrology are the method of moments (MOM) and the MLE method.

5.1. Method of Moments

The four-parameter exponential gamma distribution has four parameters α, β, δ and b. Therefore, four moments are needed for the parameters estimation. The detailed derivation of the four moments is presented in Appendix A:

$$\begin{cases} \mu_2 = \frac{\Gamma(\alpha)\Gamma(\alpha+2b)-\Gamma^2(\alpha+b)}{\beta^{2b}\Gamma^2(\alpha)} \\ E\left[\frac{1}{x-\delta}\right] = \frac{\beta^b\Gamma(\alpha-b)}{\Gamma(\alpha)} \\ \mu'_1 = \frac{\Gamma(\alpha+b)}{\beta^b\Gamma(\alpha)} + \delta \\ E\left[(x-\delta)^{\frac{1}{b}}\right] = \frac{\alpha}{\beta} \end{cases} \quad (68)$$

For a sample, $x = \{x_1, x_2, \cdots, x_n\}$, the estimation equations become:

$$\begin{cases} \frac{1}{n}\sum_{i=1}^{n}(x_i-\overline{x})^2 = \frac{\Gamma(\alpha)\Gamma(\alpha+2b)-\Gamma^2(\alpha+b)}{\beta^{2b}\Gamma^2(\alpha)} \\ \frac{1}{n}\sum_{i=1}^{n}\left(\frac{1}{x_i-\delta}\right) = \frac{\beta^b\Gamma(\alpha-b)}{\Gamma(\alpha)} \\ \overline{x} = \frac{\Gamma(\alpha+b)}{\beta^b\Gamma(\alpha)} + \delta \\ \frac{1}{n}\sum_{i=1}^{n}(x_i-\delta)^{\frac{1}{b}} = \frac{\alpha}{\beta} \end{cases} \quad (69)$$

where n is the sample size; $\overline{x} = \frac{1}{n}\sum_{i=1}^{n} x_i$.

5.2. Method of Maximum Likelihood Estimation

For the MLE method, the log-likelihood function L for a sample $x = \{x_1, x_2, \cdots, x_n\}$ is given by:

$$\ln L = n\alpha\ln\beta - n\ln b - n\ln\Gamma(\alpha) + \frac{\alpha}{b}\sum_{i=1}^{n}\ln(x_i-\delta) - \sum_{i=1}^{n}\ln(x_i-\delta) - \beta\sum_{i=1}^{n}(x_i-\delta)^{\frac{1}{b}} \quad (70)$$

The MLE's of parameters $\hat{\alpha}$, $\hat{\beta}$, $\hat{\delta}$ and $\hat{b}$ are taken to be the values that yield the maximum of $\ln L$. Differentiating Equation (70) partially with respect to each parameter and equating each partial derivative to zero produces:

$$\begin{cases} -\frac{\alpha}{b}\sum_{i=1}^{n}\frac{1}{x_i-\delta} + \sum_{i=1}^{n}\frac{1}{x_i-\delta} + \frac{\beta}{b}\sum_{i=1}^{n}(x_i-\delta)^{\frac{1}{b}-1} = 0 \\ n\ln\beta - n\psi(\alpha) + \frac{1}{b}\sum_{i=1}^{n}\ln(x_i-\delta) = 0 \\ \frac{n\alpha}{\beta} - \sum_{i=1}^{n}(x_i-\delta)^{\frac{1}{b}} = 0 \\ -\frac{n}{b} - \frac{\alpha}{b^2}\sum_{i=1}^{n}\ln(x_i-\delta) + \frac{\beta}{b^2}\sum_{i=1}^{n}\left[(x_i-\delta)^{\frac{1}{b}}\ln(x_i-\delta)\right] = 0 \end{cases} \quad (71)$$

These are the parameter estimation Equations, and the obtained results are the same as those of the PSEM method.

6. Evaluation and Comparison of Parameter Estimation Methods

The PSEM as presented in this paper is used for six annual precipitation data sets observed from 1959 to 2008 without any missing records at the Weihe River basin of China. All data are obtained from the National Climate of China Meteorological Administration and are complete. The characteristics of these data are summarized in Table 3. Obviously, all annual precipitation records have very low first-order serial correlation coefficients, ρ. Using Anderson's test of independence, the results have shown that these gauge data have an independent structure at 90% confidence levels. Hence, they are suitable for the application of meteorological frequency analysis.

Table 3. Characteristics of data used for parameter estimation.

Site Name	Mean	Standard Deviation	Coefficient of Variation	Skewness	Kurtosis	First-Order Serial Correlation Coefficient
Xi'an	571.9	126.9575	0.2220	0.2938	3.1935	−0.11399
Zhouzhi	635.2	158.7627	0.2499	0.6613	3.6873	0.16198
Lantian	713.7	150.0908	0.2103	0.2787	3.1394	0.02126
Huxian	633.8	147.5611	0.2328	0.3582	3.1987	0.05029
Lintong	579.5	129.2021	0.2230	0.6108	3.5747	0.04745
Wugong	606.7	158.2829	0.2609	0.5710	3.0826	0.09791

None of the above-discussed three methods yielded explicit solutions for the estimation of parameters of the four-parameter exponential gamma distribution. The parameter estimation Equations were therefore solved for α, β, δ and b by the four-dimensional Levenberg–Marquardt method.

Equations (67)–(68) and (71) can be simplified as the form of $\begin{cases} F_1 = 0 \\ F_2 = 0 \\ F_3 = 0 \\ F_4 = 0 \end{cases}$. Then, according to the above procedures the Matlab (Version R2007b) computer codes were developed and used to calculate the parameters. To verify the validities of parameters, the left side functions F_1, F_2, F_3 and F_4 in Equations (67)–(68) and (71) are listed Table 4. It is seen that these compute quantities are close to zero, indicating satisfactory performance of the four dimensional Levenberg–Marquardt algorithm.

Table 4. The left side functions F_1, F_2, F_3 and F_4 in Equations (67)–(68) and (71).

Site Name	Methods	F_1	F_2	F_3	F_4
	PSEM	0.00581	0.00350	0.00000	−0.08445
Xi'an	MLE	0.00581	0.00350	0.00000	−0.08445
	MOM	0.00842	−0.00000	0.00000	0.07950
	PSEM	0.00144	0.00174	0.00000	−0.01937
Zhouzhi	MLE	0.00144	0.00174	0.00000	−0.01937
	MOM	0.01241	−0.00007	0.00000	0.02139
	PSEM	0.00244	0.00206	0.00000	−0.07525
Lantian	MLE	0.00244	0.00206	0.00000	−0.07525
	MOM	0.00493	−0.00000	0.00000	0.01965
	PSEM	0.00242	0.00199	0.00000	−0.05926
Huxian	MLE	0.00242	0.00199	0.00000	−0.05926
	MOM	0.00533	−0.00006	0.00000	0.08050
	PSEM	0.00086	0.00107	0.00000	−0.08474
Lintong	MLE	0.00086	0.00107	0.00000	−0.08474
	MOM	0.03094	−0.00002	0.00000	0.02216
	PSEM	−0.00047	−0.00132	0.00000	0.00726
Wugong	MLE	−0.00047	−0.00132	0.00000	0.00726
	MOM	0.00285	−0.00001	0.00000	0.00850

The values of the distribution parameters are given in Table 5. The results of PSEM and MLE are the same. To evaluate and compare the performance of the three methods, the relative error (RERR) was employed that can be defined as:

$$REER = \frac{1}{n} \sum_{i=1}^{n} \left(\frac{x_{0i} - x_{pi}}{x_{0i}} \right)^2 \tag{72}$$

where x_{0i} and x_{pi} are the observed and predicted values of a given (i-th) quantile, respectively, and n is the sample size. The RERR values are summarized in Table 5.

Table 5. Parameter values estimated by the three methods.

Site Name	Methods	$\hat{\delta}$	$\hat{\alpha}$	$\hat{\beta}$	$\hat{b}$	RERR
	PSEM	0.01000	88.34381	4.42388	2.11605	0.00127
Xi'an	MLE	0.01000	88.34381	4.42388	2.11605	0.00127
	MOM	7.20138	62.88730	1.79612	1.77889	0.00131
	PSEM	0.01000	79.70604	4.26549	2.19876	0.00189
Zhouzhi	MLE	0.01000	79.70604	4.26549	2.19876	0.00189
	MOM	210.63187	27.69157	1.27634	1.95521	0.00274
	PSEM	0.01000	94.39458	3.88381	2.05573	0.00128
Lantian	MLE	0.01000	94.39458	3.88381	2.05573	0.00128
	MOM	68.28209	60.56026	1.69273	1.80523	0.00136
	PSEM	0.01000	84.61866	4.25523	2.15294	0.00119
Huxian	MLE	0.01000	84.61866	4.25523	2.15294	0.00119
	MOM	212.47184	30.39572	1.33086	1.92191	0.00213
	PSEM	0.01000	90.81538	4.35225	2.08997	0.00129
Lintong	MLE	0.01000	90.81538	4.35225	2.08997	0.00129
	MOM	162.17872	36.92153	1.48868	1.87230	0.00161
	PSEM	0.01000	75.75303	4.39321	2.24406	0.00158
Wugong	MLE	0.01000	75.75303	4.39321	2.24406	0.00158
	MOM	114.20629	35.23074	1.35950	1.89735	0.00183

Examination of the data in Table 5 shows that the parameters estimated using PSEM and MLE are comparable to MOM in terms of RERR and it is thus difficult to distinguish them from one another. However, PSEM and MLE yield the best parameter estimates. Thus, the parameters estimated by PSEM should be employed as the ones of four-parameter exponential gamma distribution in case study sites.

To measure the agreement between a theoretical probability distribution and an empirical distribution for the samples, Kolmogorov–Smirnov (K–S) test D_n was used to assess the goodness-of-fit. Let $x_1 < x_2 < \cdots < x_n$ be order statistics for a sample size n whose population is defined by a continuous cumulative distribution function $F(x)$ and $F_0(x_i)$ be a specified distribution that contains a set of parameters θ ($\hat{\theta}$ is estimated value from a sample size n). For an annual precipitation series, the null hypothesis H_0 that the true distribution was F_0 with parameters θ was tested. K–S test D_n can be expressed as:

$$D_n = \max_{1 \leq i \leq n} \left(\hat{\delta}_i \right) \tag{73}$$

$$\hat{\delta}_i = \max \left[\frac{i}{n} - F_0\left(x_i; \hat{\theta}\right), F_0\left(x_i; \hat{\theta}\right) - \frac{i-1}{n} \right] \tag{74}$$

The sample values of K–S test statistic D_n, are shown in Table 6. The critical value D_n^* of the four-parameter exponential gamma distribution (at the significance level $a = 0.05$, for sample size n) is 0.18654. From Table 6 it can be seen that the statistics of observed annual precipitation are all less than their corresponding critical values, respectively. Therefore, it is concluded that annual precipitation series are all accepted by the K–S test.

Table 6. Sample values of K–S test statistic D_n of case study sites.

Site Name	D_n	Site Name	D_n
Xi'an	0.07764	Huxian	0.07268
Zhouzhi	0.07290	Lintong	0.07755
Lantian	0.04485	Wugong	0.10037

7. Conclusions

Hydrologic frequency analysis, in spite of having developed a great number of distribution models and parameter estimation methods for reliable parameters and quantiles estimates, comes up against practical difficulties imposed by the short sample ranges. The Pearson Type III distribution is recommended as a standard distribution in hydrological frequency analysis in China. A large number of studies have shown that fitting small and large return period segments of Pearson Type III distribution is affected by its skewness value. Different studies employing the same parameter estimation methods may obtain different results. The use of four-parameter exponential gamma distribution has emerged as an attempt to reduce the estimate errors of small and large return period segments. The advantage of the proposed entropy method is that the first moments are made about the calculation of the distribution parameters, instead of variance, skewness and kurtosis. The results of the case estimates show that the entropy method enables the four-parameter exponential gamma distribution to fit the data well. The entropy-based parameter estimation also provides a new way to estimate parameters of the four-parameter exponential gamma distribution. The disadvantage of the method is that it will be computationally cumbersome because four parameters are involved. However, this should not be an insurmountable difficultly, given the currently available numerical tools and computer progress. Also, there are significant differences between among the MOM, PSEM and MLE estimates. Such large differences may be caused by the system of non-linear equations of parameter estimation involving the second central moment of the variable for the MOM, first moments for PSEM and MLE. In addition, the confidence intervals of quantiles for the four-parameter exponential gamma distribution deserve thorough investigation.

In summary, the following conclusions can be drawn from the present study: (1) for parameter estimation, PSEM yields the same results as MLE, whereas MOM performs with the highest bias; (2) PSEM is comparable to the MOM; (3) the four-parameter exponential gamma distribution fits the observed annual precipitation data well; (4) the quantile discharge values estimated by the three methods are close to each other; (5) the four-parameter exponential gamma distribution is a versatile distribution and results in nine different distributions, depending on its parameter values.

Acknowledgments: The authors acknowledge the financial support of National Natural Science Foundation of China (Grant Nos. 51479171, 41501022 and 51409222). The authors also wish to express their cordial gratitude to the editors, and anonymous reviewers for their illuminating comments which have greatly helped to improve the quality of this manuscript.

Author Contributions: Songbai Song and Xiaoyan Song designed the computations; Xiaoyan Song and Yan Kang made Anderson's test of the data; Songbai Song and Xiaoyan Song wrote the paper. All authors have read and approved the final manuscript.

Conflicts of Interest: The authors declare no conflict of interest.

Appendix A. First Four Original Moments and Central Moments

Consider that the first original moments of the four-parameter exponential gamma distribution are μ'_1, μ'_2, μ'_3 and μ'_4, respectively; μ_2, μ_3, μ_4 are the second, third and fourth central moments, respectively; C_v is coefficient of variation; C_s is coefficient of skewness; C_e is coefficient of kurtosis. A detailed derivation of the above moments is given below:

$$\mu'_1 = \int_\delta^\infty x f(x)dx = \int_\delta^\infty \frac{\beta^\alpha}{b\Gamma(\alpha)} x(x-\delta)^{\frac{\alpha}{b}-1} e^{-\beta(x-\delta)^{\frac{1}{b}}} dx = \int_0^\infty \frac{\beta^\alpha}{\Gamma(\alpha)} \left(\delta + \frac{1}{\beta^b}t^b\right)\left(\frac{1}{\beta^b}t^b\right)^{\frac{\alpha}{b}-1} e^{-t}\frac{b}{\beta^b}t^{b-1}dt$$

$$= \int_0^\infty \frac{1}{\Gamma(\alpha)}\left(\delta + \frac{1}{\beta^b}t^b\right)t^{\alpha-1}e^{-t}dt = \frac{1}{\Gamma(\alpha)}\left[\delta\int_0^\infty t^{\alpha-1}e^{-t}dt + \frac{1}{\beta^b}\int_0^\infty t^{\alpha+b-1}e^{-t}dt\right] \tag{A1}$$

$$= \frac{1}{\Gamma(\alpha)}\left[\delta\cdot\Gamma(\alpha) + \frac{1}{\beta^b}\Gamma(\alpha+b)\right] = \frac{\Gamma(\alpha+b)}{\beta^b\Gamma(\alpha)} + \delta$$

$$\mu'_2 = \int_\delta^\infty x^2 f(x)dx = \int_\delta^\infty \frac{\beta^\alpha}{b\Gamma(\alpha)} x^2(x-\delta)^{\frac{\alpha}{b}-1} e^{-\beta(x-\delta)^{\frac{1}{b}}} dx$$

$$= \int_0^\infty \frac{\beta^\alpha}{b\Gamma(\alpha)}\left(\delta + \frac{1}{\beta^b}t^b\right)^2\left(\frac{1}{\beta^b}t^b\right)^{\frac{\alpha}{b}-1} e^{-t}\frac{b}{\beta^b}t^{b-1}dt = \int_0^\infty \frac{1}{\Gamma(\alpha)}\left(\delta + \frac{1}{\beta^b}t^b\right)^2 t^{\alpha-1}e^{-t}dt$$

$$= \frac{1}{\Gamma(\alpha)}\left[\delta^2\int_0^\infty t^{\alpha-1}e^{-t}dt + \frac{2\delta}{\beta^b}\int_0^\infty t^{\alpha+b-1}e^{-t}dt + \frac{1}{\beta^{2b}}\int_0^\infty t^{\alpha+2b-1}e^{-t}dt\right] \tag{A2}$$

$$= \frac{1}{\Gamma(\alpha)}\left[\delta^2\Gamma(\alpha) + \frac{2\delta}{\beta^b}\Gamma(\alpha+b) + \frac{1}{\beta^{2b}}\Gamma(\alpha+2b)\right]$$

$$= \delta^2 + \frac{2\delta}{\beta^b\Gamma(\alpha)}\Gamma(\alpha+b) + \frac{1}{\beta^{2b}\Gamma(\alpha)}\Gamma(\alpha+2b)$$

$$\mu'_3 = \int_\delta^\infty x^3 f(x)dx = \int_\delta^\infty \frac{\beta^\alpha}{b\Gamma(\alpha)} x^3(x-\delta)^{\frac{\alpha}{b}-1} e^{-\beta(x-\delta)^{\frac{1}{b}}} dx$$

$$= \int_0^\infty \frac{\beta^\alpha}{b\Gamma(\alpha)}\left(\delta + \frac{1}{\beta^b}t^b\right)^3\left(\frac{1}{\beta^b}t^b\right)^{\frac{\alpha}{b}-1} e^{-t}\frac{b}{\beta^b}t^{b-1}dt = \int_0^\infty \frac{1}{\Gamma(\alpha)}\left(\delta + \frac{1}{\beta^b}t^b\right)^3 t^{\alpha-1}e^{-t}dt$$

$$= \frac{1}{\Gamma(\alpha)}\left[\delta^3\int_0^\infty t^{\alpha-1}e^{-t}dt + \frac{3\delta^2}{\beta^b}\int_0^\infty t^{\alpha+b-1}e^{-t}dt + \frac{3\delta}{\beta^{2b}}\int_0^\infty t^{\alpha+2b-1}e^{-t}dt + \frac{1}{\beta^{3b}}\int_0^\infty t^{\alpha+3b-1}e^{-t}dt\right] \tag{A3}$$

$$= \frac{1}{\Gamma(\alpha)}\left[\delta^3\Gamma(\alpha) + \frac{3\delta^2}{\beta^b}\Gamma(\alpha+b) + \frac{3\delta}{\beta^{2b}}\Gamma(\alpha+2b) + \frac{1}{\beta^{3b}}\Gamma(\alpha+3b)\right]$$

$$\mu'_4 = \int_\delta^\infty x^4 f(x)dx = \int_\delta^\infty \frac{\beta^\alpha}{b\Gamma(\alpha)} x^4(x-\delta)^{\frac{\alpha}{b}-1} e^{-\beta(x-\delta)^{\frac{1}{b}}} dx$$

$$= \int_0^\infty \frac{\beta^\alpha}{b\Gamma(\alpha)}\left(\delta + \frac{1}{\beta^b}t^b\right)^4\left(\frac{1}{\beta^b}t^b\right)^{\frac{\alpha}{b}-1} e^{-t}\frac{b}{\beta^b}t^{b-1}dt = \int_0^\infty \frac{1}{\Gamma(\alpha)}\left(\delta + \frac{1}{\beta^b}t^b\right)^4 t^{\alpha-1}e^{-t}dt \tag{A4}$$

$$= \frac{1}{\Gamma(\alpha)}\left[\delta^4\int_0^\infty t^{\alpha-1}e^{-t}dt + \frac{4\delta^3}{\beta^b}\int_0^\infty t^{\alpha+b-1}e^{-t}dt + \frac{6\delta^2}{\beta^{2b}}\int_0^\infty t^{\alpha+2b-1}e^{-t}dt + \frac{4\delta}{\beta^{3b}}\int_0^\infty t^{\alpha+3b-1}e^{-t}dt + \frac{1}{\beta^{4b}}\int_0^\infty t^{\alpha+4b-1}e^{-t}dt\right]$$

$$= \delta^4 + \frac{4\delta^3\Gamma(\alpha+b)}{\beta^b\Gamma(\alpha)} + \frac{6\delta^2\Gamma(\alpha+2b)}{\beta^{2b}\Gamma(\alpha)} + \frac{4\delta\cdot\Gamma(\alpha+3b)}{\beta^{3b}\Gamma(\alpha)} + \frac{\Gamma(\alpha+4b)}{\beta^{4b}\Gamma(\alpha)}$$

$$\mu_2 = Var(x) = \mu'_2 - (\mu'_1)^1 = \delta^2 + \frac{2\delta}{\beta^b\Gamma(\alpha)}\Gamma(\alpha+b) + \frac{1}{\beta^{2b}\Gamma(\alpha)}\Gamma(\alpha+2b) - \left[\frac{\Gamma(\alpha+b)}{\beta^b\Gamma(\alpha)} + \delta\right]^2$$

$$= \delta^2 + \frac{2\delta}{\beta^b\Gamma(\alpha)}\Gamma(\alpha+b) + \frac{1}{\beta^{2b}\Gamma(\alpha)}\Gamma(\alpha+2b) - \frac{\Gamma^2(\alpha+b)}{\beta^{2b}\Gamma^2(\alpha)} - \frac{2\delta}{\beta^b\Gamma(\alpha)}\Gamma(\alpha+b) - \delta^2 \tag{A5}$$

$$= \frac{1}{\beta^{2b}\Gamma(\alpha)}\Gamma(\alpha+2b) - \frac{\Gamma^2(\alpha+b)}{\beta^{2b}\Gamma^2(\alpha)} = \frac{\Gamma(\alpha+2b)\Gamma(\alpha) - \Gamma^2(\alpha+b)}{\beta^{2b}\Gamma^2(\alpha)}$$

$$\mu_3 = E[X - E(X)]^3 = E\left[X^3 - 3X^2 E(X) + 3XE^2(X) - E^3(X)\right]$$

$$= E(X^3) - 3E(X^2)E(X) + 2E^3(X) = \mu'_3 - 3\mu'_2\mu'_1 + 2(\mu'_1)^3$$

$$= \tfrac{1}{\Gamma(\alpha)}\left[\delta^3\Gamma(\alpha) + \tfrac{3\delta^2}{\beta^b}\Gamma(\alpha+b) + \tfrac{3\delta}{\beta^{2b}}\Gamma(\alpha+2b) + \tfrac{1}{\beta^{3b}}\Gamma(\alpha+3b)\right]$$

$$-3\left[\delta^2 + \tfrac{2\delta\cdot\Gamma(\alpha+b)}{\beta^b\Gamma(\alpha)} + \tfrac{\Gamma(\alpha+2b)}{\beta^{2b}\Gamma(\alpha)}\right]\cdot\left[\tfrac{\Gamma(\alpha+b)}{\beta^b\Gamma(\alpha)} + \delta\right] + 2\left[\tfrac{\Gamma(\alpha+b)}{\beta^b\Gamma(\alpha)} + \delta\right]^3 \qquad\text{(A6)}$$

$$= (\delta^3 + 2\delta^3 - 3\delta^3) + \left[\tfrac{3\delta^2\Gamma(\alpha+b)}{\beta^b\Gamma(\alpha)} - \tfrac{3\delta^2\Gamma(\alpha+b)}{\beta^b\Gamma(\alpha)} - \tfrac{6\delta^2\cdot\Gamma(\alpha+b)}{\beta^b\Gamma(\alpha)} + \tfrac{6\delta^2\Gamma(\alpha+b)}{\beta^b\Gamma(\alpha)}\right]$$

$$+\left[\tfrac{3\delta\cdot\Gamma(\alpha+2b)}{\beta^{2b}\Gamma(\alpha)} - \tfrac{3\delta\cdot\Gamma(\alpha+2b)}{\beta^{2b}\Gamma(\alpha)}\right] + \left[\tfrac{6\delta\cdot\Gamma^2(\alpha+b)}{\beta^{2b}\Gamma^2(\alpha)} - \tfrac{6\delta\cdot\Gamma^2(\alpha+b)}{\beta^{2b}\Gamma^2(\alpha)}\right] + \tfrac{\Gamma(\alpha+3b)}{\beta^{3b}\Gamma(\alpha)} + \tfrac{2\Gamma^3(\alpha+b)}{\beta^{3b}\Gamma^3(\alpha)} - \tfrac{3\Gamma(\alpha+b)\Gamma(\alpha+2b)}{\beta^{3b}\Gamma^2(\alpha)}$$

$$= \tfrac{\Gamma^2(\alpha)\Gamma(\alpha+3b) - 3\Gamma(\alpha)\Gamma(\alpha+b)\Gamma(\alpha+2b) + 2\Gamma^3(\alpha+b)}{\beta^{3b}\Gamma^3(\alpha)}$$

$$\mu_4 = E[X - E(X)]^4 = E\left[X^4 - 4X^3 E(X) + 6X^2 E^2(X) - 4XE^3(X) + E^4(X)\right]$$

$$= E(X^4) - 4E(X^3)E(X) + 6E(X^2)E^2(X) - 3E^4(X)$$

$$= \mu'_4 - 4\mu'_3\mu'_1 + 6\mu'_2(\mu'_1)^2 - 3(\mu'_1)^4$$

$$= \delta^4 + \tfrac{4\delta^3\Gamma(\alpha+b)}{\beta^b\Gamma(\alpha)} + \tfrac{6\delta^2\Gamma(\alpha+2b)}{\beta^{2b}\Gamma(\alpha)} + \tfrac{4\delta\cdot\Gamma(\alpha+3b)}{\beta^{3b}\Gamma(\alpha)} + \tfrac{\Gamma(\alpha+4b)}{\beta^{4b}\Gamma(\alpha)}$$

$$-4\tfrac{1}{\Gamma(\alpha)}\left[\delta^3\Gamma(\alpha) + \tfrac{3\delta^2}{\beta^b}\Gamma(\alpha+b) + \tfrac{3\delta}{\beta^{2b}}\Gamma(\alpha+2b) + \tfrac{1}{\beta^{3b}}\Gamma(\alpha+3b)\right]\cdot\left[\tfrac{\Gamma(\alpha+b)}{\beta^b\Gamma(\alpha)} + \delta\right]$$

$$+6\left[\delta^2 + \tfrac{2\delta\cdot\Gamma(\alpha+b)}{\beta^b\Gamma(\alpha)} + \tfrac{\Gamma(\alpha+2b)}{\beta^{2b}\Gamma(\alpha)}\right]\cdot\left[\tfrac{\Gamma(\alpha+b)}{\beta^b\Gamma(\alpha)} + \delta\right]^2 - 3\left[\tfrac{\Gamma(\alpha+b)}{\beta^b\Gamma(\alpha)} + \delta\right]^4$$

$$= (\delta^4 - 4\delta^4 + 6\delta^4 - 3\delta^4) + \left[\tfrac{4\delta^3\Gamma(\alpha+b)}{\beta^b\Gamma(\alpha)} - \tfrac{4\delta^3\Gamma(\alpha+b)}{\beta^b\Gamma(\alpha)} - \tfrac{12\delta^3\Gamma(\alpha+b)}{\beta^b\Gamma(\alpha)} + \tfrac{12\delta^3\cdot\Gamma(\alpha+b)}{\beta^b\Gamma(\alpha)}\right] \qquad\text{(A7)}$$

$$+\left[\tfrac{12\delta^3\cdot\Gamma(\alpha+b)}{\beta^b\Gamma(\alpha)} - \tfrac{6\delta^3\cdot\Gamma(\alpha+b)}{\beta^b\Gamma(\alpha)} - \tfrac{6\delta^3\cdot\Gamma(\alpha+b)}{\beta^b\Gamma(\alpha)}\right] + \left[\tfrac{6\delta^2\Gamma(\alpha+2b)}{\beta^{2b}\Gamma(\alpha)} + \tfrac{6\delta^2\Gamma(\alpha+2b)}{\beta^{2b}\Gamma(\alpha)} - \tfrac{12\delta^2\Gamma(\alpha+2b)}{\beta^{2b}\Gamma(\alpha)}\right]$$

$$+\left[\tfrac{24\delta^2\cdot\Gamma^2(\alpha+b)}{\beta^{2b}\Gamma^2(\alpha)} - \tfrac{12\delta^2\Gamma^2(\alpha+b)}{\beta^{2b}\Gamma^2(\alpha)} - \tfrac{12\delta^2\cdot\Gamma^2(\alpha+b)}{\beta^{2b}\Gamma^2(\alpha)}\right] + \left[\tfrac{6\delta^2\Gamma^2(\alpha+b)}{\beta^{2b}\Gamma^2(\alpha)} - \tfrac{3\delta^2\Gamma^2(\alpha+b)}{\beta^{2b}\Gamma^2(\alpha)} - \tfrac{3\delta^2\Gamma^2(\alpha+b)}{\beta^{2b}\Gamma^2(\alpha)}\right]$$

$$+\left[\tfrac{4\delta\cdot\Gamma(\alpha+3b)}{\beta^{3b}\Gamma(\alpha)} - \tfrac{4\delta\cdot\Gamma(\alpha+3b)}{\beta^{3b}\Gamma(\alpha)}\right] + \left[\tfrac{12\delta\cdot\Gamma(\alpha+b)\Gamma(\alpha+2b)}{\beta^{3b}\Gamma^2(\alpha)} - \tfrac{12\delta\cdot\Gamma(\alpha+b)\Gamma(\alpha+2b)}{\beta^{3b}\Gamma^2(\alpha)}\right]$$

$$+\left[\tfrac{12\delta\cdot\Gamma^3(\alpha+b)}{\beta^{3b}\Gamma^3(\alpha)} - \tfrac{6\delta\cdot\Gamma^3(\alpha+b)}{\beta^{3b}\Gamma^3(\alpha)} - \tfrac{6\delta\cdot\Gamma^3(\alpha+b)}{\beta^{3b}\Gamma^3(\alpha)}\right] + \tfrac{\Gamma(\alpha+4b)}{\beta^{4b}\Gamma(\alpha)} - \tfrac{4\Gamma(\alpha+b)\Gamma(\alpha+3b)}{\beta^{4b}\Gamma^2(\alpha)} + \tfrac{6\Gamma^2(\alpha+b)\Gamma(\alpha+2b)}{\beta^{4b}\Gamma^3(\alpha)} - \tfrac{3\Gamma^4(\alpha+b)}{\beta^{4b}\Gamma^4(\alpha)}$$

$$= \tfrac{\Gamma(\alpha+4b)}{\beta^{4b}\Gamma(\alpha)} - \tfrac{4\Gamma(\alpha+b)\Gamma(\alpha+3b)}{\beta^{4b}\Gamma^2(\alpha)} + \tfrac{6\Gamma^2(\alpha+b)\Gamma(\alpha+2b)}{\beta^{4b}\Gamma^3(\alpha)} - \tfrac{3\Gamma^4(\alpha+b)}{\beta^{4b}\Gamma^4(\alpha)}$$

$$= \tfrac{\Gamma^3(\alpha)\Gamma(\alpha+4b) - 4\Gamma^2(\alpha)\Gamma(\alpha+b)\Gamma(\alpha+3b) + 6\Gamma(\alpha)\Gamma^2(\alpha+b)\Gamma(\alpha+2b) - 3\Gamma^4(\alpha+b)}{\beta^{4b}\Gamma^4(\alpha)}$$

$$C_v = \frac{(\mu_2)^{1/2}}{\mu'_1} = \frac{\left[\frac{\Gamma(\alpha+2b)\Gamma(\alpha) - \Gamma^2(\alpha+b)}{\beta^{2b}\Gamma^2(\alpha)}\right]^{1/2}}{\frac{\Gamma(\alpha+b)}{\beta^b\Gamma(\alpha)} + \delta} = \frac{\sqrt{\Gamma(\alpha+2b)\Gamma(\alpha) - \Gamma^2(\alpha+b)}}{\Gamma(\alpha+b) + \delta\cdot\beta^b\Gamma(\alpha)} \qquad\text{(A8)}$$

$$C_s = \frac{\mu_3}{(\mu_2)^{3/2}} = \frac{\frac{\Gamma^2(\alpha)\Gamma(\alpha+3b) - 3\Gamma(\alpha)\Gamma(\alpha+b)\Gamma(\alpha+2b) + 2\Gamma^3(\alpha+b)}{\beta^{3b}\Gamma^3(\alpha)}}{\left[\frac{\Gamma(\alpha+2b)\Gamma(\alpha) - \Gamma^2(\alpha+b)}{\beta^{2b}\Gamma^2(\alpha)}\right]^{3/2}}$$

$$= \frac{\Gamma^2(\alpha)\Gamma(\alpha+3b) - 3\Gamma(\alpha)\Gamma(\alpha+b)\Gamma(\alpha+2b) + 2\Gamma^3(\alpha+b)}{\beta^{3b}\Gamma^3(\alpha)}\frac{\beta^{3b}\Gamma^3(\alpha)}{\left[\Gamma(\alpha)\Gamma(\alpha+2b) - \Gamma^2(\alpha+b)\right]^{3/2}} \qquad\text{(A9)}$$

$$= \frac{\Gamma^2(\alpha)\Gamma(\alpha+3b) - 3\Gamma(\alpha)\Gamma(\alpha+b)\Gamma(\alpha+2b) + 2\Gamma^3(\alpha+b)}{\left[\Gamma(\alpha)\Gamma(\alpha+2b) - \Gamma^2(\alpha+b)\right]^{3/2}}$$

$$C_e = \frac{\mu_4}{(\mu_2)^2} - 3 = \frac{\Gamma^3(\alpha)\Gamma(\alpha+4b)-4\Gamma^2(\alpha)\Gamma(\alpha+b)\Gamma(\alpha+3b)+6\Gamma(\alpha)\Gamma^2(\alpha+b)\Gamma(\alpha+2b)-3\Gamma^4(\alpha+b)}{\beta^{4b}\Gamma^4(\alpha)}$$

$$\cdot \frac{\beta^{4b}\Gamma^4(\alpha)}{\left[\Gamma(\alpha)\Gamma(\alpha+2b)-\Gamma^2(\alpha+b)\right]^2} - 3$$

$$= \frac{\Gamma^3(\alpha)\Gamma(\alpha+4b)-4\Gamma^2(\alpha)\Gamma(\alpha+b)\Gamma(\alpha+3b)+6\Gamma(\alpha)\Gamma^2(\alpha+b)\Gamma(\alpha+2b)-3\Gamma^4(\alpha+b)}{\left[\Gamma(\alpha)\Gamma(\alpha+2b)-\Gamma^2(\alpha+b)\right]^2}$$

$$= \frac{\Gamma^3(\alpha)\Gamma(\alpha+4b)-4\Gamma^2(\alpha)\Gamma(\alpha+b)\Gamma(\alpha+3b)+6\Gamma(\alpha)\Gamma^2(\alpha+b)\Gamma(\alpha+2b)-3\Gamma^4(\alpha+b)}{\left[\Gamma(\alpha)\Gamma(\alpha+2b)-\Gamma^2(\alpha+b)\right]^2}$$

$$\qquad (A10)$$

$$+ \frac{-3\Gamma^2(\alpha)\Gamma^2(\alpha+2b)+6\Gamma(\alpha)\Gamma(\alpha+2b)\Gamma^2(\alpha+b)-3\Gamma^4(\alpha+b)}{\left[\Gamma(\alpha)\Gamma(\alpha+2b)-\Gamma^2(\alpha+b)\right]^2}$$

$$= \frac{\Gamma^3(\alpha)\Gamma(\alpha+4b)-4\Gamma^2(\alpha)\Gamma(\alpha+b)\Gamma(\alpha+3b)+12\Gamma(\alpha)\Gamma^2(\alpha+b)\Gamma(\alpha+2b)-3\Gamma^2(\alpha)\Gamma^2(\alpha+2b)-6\Gamma^4(\alpha+b)}{\left[\Gamma(\alpha)\Gamma(\alpha+2b)-\Gamma^2(\alpha+b)\right]^2}$$

Appendix B. Derivatives of λ_0 with Respect to λ_1, λ_2 and λ_3

$$\frac{\partial \lambda_0}{\partial \lambda_1} = \frac{-\int_\delta^\infty \ln(x-\delta)\cdot\exp\left[-\lambda_1\ln(x-\delta)-\lambda_2\ln(x-\delta)^{\frac{1}{b}}-\lambda_3(x-\delta)^{\frac{1}{b}}\right]dx}{\int_\delta^\infty \exp\left[-\lambda_1\ln(x-\delta)-\lambda_2\ln(x-\delta)^{\frac{1}{b}}-\lambda_3(x-\delta)^{\frac{1}{b}}\right]dx}$$

$$= \frac{-\int_\delta^\infty \ln(x-\delta)\cdot\exp\left[-\lambda_0-\lambda_1\ln(x-\delta)-\lambda_2\ln(x-\delta)^{\frac{1}{b}}-\lambda_3(x-\delta)^{\frac{1}{b}}\right]dx}{\int_\delta^\infty \exp\left[-\lambda_0-\lambda_1\ln(x-\delta)-\lambda_2\ln(x-\delta)^{\frac{1}{b}}-\lambda_3(x-\delta)^{\frac{1}{b}}\right]dx} \qquad (A11)$$

$$= \frac{-\int_\delta^\infty \ln(x-\delta)\cdot f(x)dx}{\int_\delta^\infty \cdot f(x)dx} = -E[\ln(x-\delta)]$$

$$\frac{\partial \lambda_0}{\partial \lambda_2} = \frac{-\int_\delta^\infty \ln(x-\delta)^{\frac{1}{b}}\cdot\exp\left[-\lambda_1\ln(x-\delta)-\lambda_2\ln(x-\delta)^{\frac{1}{b}}-\lambda_3(x-\delta)^{\frac{1}{b}}\right]dx}{\int_\delta^\infty \exp\left[-\lambda_1\ln(x-\delta)-\lambda_2\ln(x-\delta)^{\frac{1}{b}}-\lambda_3(x-\delta)^{\frac{1}{b}}\right]dx}$$

$$= \frac{-\int_\delta^\infty \ln(x-\delta)^{\frac{1}{b}}\cdot\exp\left[-\lambda_0-\lambda_1\ln(x-\delta)-\lambda_2\ln(x-\delta)^{\frac{1}{b}}-\lambda_3(x-\delta)^{\frac{1}{b}}\right]dx}{\int_\delta^\infty \exp\left[-\lambda_0-\lambda_1\ln(x-\delta)-\lambda_2\ln(x-\delta)^{\frac{1}{b}}-\lambda_3(x-\delta)^{\frac{1}{b}}\right]dx} \qquad (A12)$$

$$= \frac{-\int_\delta^\infty \ln(x-\delta)^{\frac{1}{b}}\cdot f(x)dx}{\int_\delta^\infty f(x)dx} = -E\left[\ln(x-\delta)^{\frac{1}{b}}\right]$$

$$\frac{\partial \lambda_0}{\partial \lambda_3} = \frac{-\int_\delta^\infty (x-\delta)^{\frac{1}{b}}\cdot\exp\left[-\lambda_1\ln(x-\delta)-\lambda_2\ln(x-\delta)^{\frac{1}{b}}-\lambda_3(x-\delta)^{\frac{1}{b}}\right]dx}{\int_\delta^\infty \exp\left[-\lambda_1\ln(x-\delta)-\lambda_2\ln(x-\delta)^{\frac{1}{b}}-\lambda_3(x-\delta)^{\frac{1}{b}}\right]dx}$$

$$= \frac{-\int_\delta^\infty (x-\delta)^{\frac{1}{b}}\cdot\exp\left[-\lambda_0-\lambda_1\ln(x-\delta)-\lambda_2\ln(x-\delta)^{\frac{1}{b}}-\lambda_3(x-\delta)^{\frac{1}{b}}\right]dx}{\int_\delta^\infty \exp\left[-\lambda_0-\lambda_1\ln(x-\delta)-\lambda_2\ln(x-\delta)^{\frac{1}{b}}-\lambda_3(x-\delta)^{\frac{1}{b}}\right]dx} \qquad (A13)$$

$$= \frac{-\int_\delta^\infty (x-\delta)^{\frac{1}{b}}\cdot f(x)dx}{\int_\delta^\infty f(x)dx} = -E\left[(x-\delta)^{\frac{1}{b}}\right]$$

Entropy **2017**, *19*, 189

Furthermore, we can write:

$$
\frac{\partial^2 \lambda_0}{\partial \lambda_3^2} = \frac{\int_\delta^\infty \left[(x-\delta)^{\frac{1}{b}} \right]^2 \cdot \exp\left[-\lambda_1 \ln(x-\delta) - \lambda_2 \ln(x-\delta)^{\frac{1}{b}} - \lambda_3(x-\delta)^{\frac{1}{b}} \right] dx}{\left\{ \int_\delta^\infty \exp\left[-\lambda_1 \ln(x-\delta) - \lambda_2 \ln(x-\delta)^{\frac{1}{b}} - \lambda_3(x-\delta)^{\frac{1}{b}} \right] dx \right\}^2}
$$

$$
\cdot \int_\delta^\infty \exp\left[-\lambda_1 \ln(x-\delta) - \lambda_2 \ln(x-\delta)^{\frac{1}{b}} - \lambda_3(x-\delta)^{\frac{1}{b}} \right] dx
$$

$$
+ \frac{-\int_\delta^\infty (x-\delta)^{\frac{1}{b}} \cdot \exp\left[-\lambda_1 \ln(x-\delta) - \lambda_2 \ln(x-\delta)^{\frac{1}{b}} - \lambda_3(x-\delta)^{\frac{1}{b}} \right] dx}{\left\{ \int_\delta^\infty \exp\left[-\lambda_1 \ln(x-\delta) - \lambda_2 \ln(x-\delta)^{\frac{1}{b}} - \lambda_3(x-\delta)^{\frac{1}{b}} \right] dx \right\}^2}
$$

$$
\cdot \int_\delta^\infty (x-\delta)^{\frac{1}{b}} \exp\left[-\lambda_1 \ln(x-\delta) - \lambda_2 \ln(x-\delta)^{\frac{1}{b}} - \lambda_3(x-\delta)^{\frac{1}{b}} \right] dx
$$

$$
= \frac{\int_\delta^\infty \left[(x-\delta)^{\frac{1}{b}} \right]^2 \cdot \exp\left[-\lambda_0 - \lambda_1 \ln(x-\delta) - \lambda_2 \ln(x-\delta)^{\frac{1}{b}} - \lambda_3(x-\delta)^{\frac{1}{b}} \right] dx}{\left\{ \int_\delta^\infty \exp\left[-\lambda_0 - \lambda_1 \ln(x-\delta) - \lambda_2 \ln(x-\delta)^{\frac{1}{b}} - \lambda_3(x-\delta)^{\frac{1}{b}} \right] dx \right\}^2} \qquad \frac{\partial \lambda_0}{\partial \lambda_3} = -\frac{b - (b\lambda_1 + \lambda_2)}{\lambda_3} \qquad \text{(A14)}
$$

$$
+ \frac{-\int_\delta^\infty (x-\delta)^{\frac{1}{b}} \cdot \exp\left[-\lambda_0 - \lambda_1 \ln(x-\delta) - \lambda_2 \ln(x-\delta)^{\frac{1}{b}} - \lambda_3(x-\delta)^{\frac{1}{b}} \right] dx}{\left\{ \int_\delta^\infty \exp\left[-\lambda_0 - \lambda_1 \ln(x-\delta) - \lambda_2 \ln(x-\delta)^{\frac{1}{b}} - \lambda_3(x-\delta)^{\frac{1}{b}} \right] dx \right\}^2}
$$

$$
\cdot \int_\delta^\infty (x-\delta)^{\frac{1}{b}} \exp\left[-\lambda_0 - \lambda_1 \ln(x-\delta) - \lambda_2 \ln(x-\delta)^{\frac{1}{b}} - \lambda_3(x-\delta)^{\frac{1}{b}} \right] dx
$$

$$
= \frac{\int_\delta^\infty \left[(x-\delta)^{\frac{1}{b}} \right]^2 \cdot f(x) dx}{\left\{ \int_\delta^\infty f(x) dx \right\}^2} \cdot \int_\delta^\infty f(x) dx - \frac{\int_\delta^\infty (x-\delta)^{\frac{1}{b}} \cdot f(x) dx}{\left\{ \int_\delta^\infty f(x) dx \right\}^2} \cdot \int_\delta^\infty (x-\delta)^{\frac{1}{b}} e f(x) dx
$$

$$
= \frac{\int_\delta^\infty \left[(x-\delta)^{\frac{1}{b}} \right]^2 \cdot f(x) dx}{\left\{ \int_\delta^\infty f(x) dx \right\}^2} \cdot \int_\delta^\infty f(x) dx - \frac{\int_\delta^\infty (x-\delta)^{\frac{1}{b}} \cdot f(x) dx}{\left\{ \int_\delta^\infty f(x) dx \right\}^2} \cdot \int_\delta^\infty (x-\delta)^{\frac{1}{b}} e f(x) dx
$$

$$
= E\left[(x-\delta)^{\frac{1}{b}} \right]^2 - \left\{ E\left[(x-\delta)^{\frac{1}{b}} \right] \right\}^2
$$

References

1. Rao, A.R.; Hamed, K.H. *Flood Frequency Analysis*; CRC Press: Boca Raton, FL, USA, 1999.
2. Meylan, P.; Favre, A.C.; Musy, A. *Predictive Hydrology: A Frequency analysis Approach*; CRC Press; Taylor & Francis Group, Sciences Publishers: Enfield, NH, USA, 2012.
3. Singh, V.P. *Entropy-Based Parameter Estimation in Hydrology*; Kluwer Academic Publishers: Boston, MA, USA, 1998.
4. Sun, J.; Qin, D.; Sun, H. *Generalized Probability Distribution in Hydrometeorology*; China Water and Power Press: Beijing, China, 2001.
5. Singh, V.P.; Deng, Z.Q. Entropy-Based Parameter Estimation for Kappa Distribution. *J. Hydrol. Eng.* **2003**, *8*, 81–92. [CrossRef]
6. Singh, V.P. *Entropy Theory and Its Application in Environmental and Water Engineering*; John Wiley & Sons: New York, NY, USA, 2013.
7. Singh, V.P. The use of entropy in hydrology and water resources. *Hydrol. Process.* **1997**, *11*, 587–626. [CrossRef]
8. Singh, V.P. *Entropy Theory in Hydrologic Science and Engineering*; McGraw-Hill Education: New York, NY, USA, 2015.
9. Singh, V.P. *Entropy Theory in Hydraulic Engineering: An Introduction*; ASCE Press: Reston, VA, USA, 2014.
10. Hao, Z.; Singh, V.P. Entropy-based method for bivariate drought analysis. *J. Hydrol. Eng.* **2013**, *18*, 780–786. [CrossRef]

11. Hao, Z.; Singh, V.P. Entropy-copula method for single-site monthly streamflow simulation. *Water Resour. Res.* **2012**, *48*, W06604. [CrossRef]
12. Hao, Z. Application of Entropy Theory in Hydrologic Analysis and Simulation. Ph.D. Thesis, Texas A&M University, College Station, TX, USA, May 2012.
13. Hao, Z.; Singh, V.P. Modeling multi-site streamflow dependence with maximum entropy copula. *Water Resour. Res.* **2013**, *49*, 7139–7143. [CrossRef]
14. Hao, Z.; Singh, V.P. Single-site monthly streamflow simulation using entropy theory. *Water Resour. Res.* **2011**, *47*. [CrossRef]
15. Hao, Z.; Singh, V.P. Entropy-based method for extreme rainfall analysis in Texas. *J. Geophys. Res.* **2013**, *118*, 263–273. [CrossRef]
16. Hao, Z.; Singh, V.P. Entropy-based parameter estimation for extended Burr XII distribution. *Stoch. Environ. Res. Risk Assess.* **2009**, *23*, 1113–1122. [CrossRef]
17. Mishra, A.K.; Özger, M.; Singh, V.P. An entropy-based investigation into the variability of precipitation. *J. Hydrol.* **2009**, *370*, 139–154. [CrossRef]
18. Cui, H.; Singh, V.P. On the Cumulative Distribution Function for Entropy-Based Hydrologic Modeling. *Trans. ASABE* **2012**, *55*, 429–438. [CrossRef]
19. Cui, H.; Singh, V.P. Two-Dimensional Velocity Distribution in Open Channels Using the Tsallis Entropy. *J. Hydrol. Eng.* **2013**, *18*, 331–339.
20. Cui, H.; Singh, V.P. Computation of Suspended Sediment Discharge in Open Channels Using Tsallis Entropy. *J. Hydrol. Eng.* **2013**, *19*, 18–25. [CrossRef]
21. Cui, H.; Singh, V.P. One Dimensional Velocity Distribution in Open Channels Using Tsallis Entropy. *J. Hydrol. Eng.* **2013**, *19*, 290–298. [CrossRef]
22. Cui, H.; Singh, V.P. Configurational Entropy Theory for Streamflow Forecasting. *J. Hydrol.* **2015**, *521*, 1–7. [CrossRef]
23. Cui, H.; Singh, V.P. Minimum Relative Entropy Theory for Streamflow Forecasting with Frequency as a Random Variable. *Stoch. Environ. Res. Risk Assess.* **2016**, *30*, 1545–1563. [CrossRef]
24. Papalexiou, S.M.; Koutsoyiannis, D. Entropy based derivation of probability distributions: A case study to daily rainfall. *Adv. Water Resour.* **2012**, *45*, 51–57. [CrossRef]
25. Brouers, F. The Burr XII Distribution Family and the Maximum Entropy Principle: Power-Law Phenomena Are Not Necessarily Nonextensive. *Open J. Stat.* **2015**, *5*, 730–741. [CrossRef]
26. Shao, Q.; Wong, H.; Xia, J.; Ip, W.C. Models for extremes using the extended three-parameter Burr XII system with application to flood frequency analysis. *Hydrol. Sci. J.* **2004**, *49*, 685–702. [CrossRef]

![entropy logo] *entropy*

MDPI

Article

An Entropy-Based Generalized Gamma Distribution for Flood Frequency Analysis

Lu Chen [1], Vijay P. Singh [2,3] and Feng Xiong [4,*]

[1]   College of Hydropower and Information Engineering, Huazhong University of Science and Technology, Wuhan 430074, China; chen_lu@hust.edu.cn
[2]   Department of Biological and Agricultural Engineering, Texas A&M University, College Station, TX 77843, USA; vsingh@tamu.edu
[3]   Zachry Department of Civil Engineering, Texas A&M University, College Station, TX 77843, USA
[4]   State Key Laboratory of Water Resources and Hydropower Engineering Science, Wuhan University, Wuhan 430072, China
*   Correspondence: fxiong07@whu.edu.cn

Academic Editor: Kevin Knuth
Received: 17 April 2017; Accepted: 19 May 2017; Published: 2 June 2017

Abstract: Flood frequency analysis (FFA) is needed for the design of water engineering and hydraulic structures. The choice of an appropriate frequency distribution is one of the most important issues in FFA. A key problem in FFA is that no single distribution has been accepted as a global standard. The common practice is to try some candidate distributions and select the one best fitting the data, based on a goodness of fit criterion. However, this practice entails much calculation. Sometimes generalized distributions, which can specialize into several simpler distributions, are fitted, for they may provide a better fit to data. Therefore, the generalized gamma (GG) distribution was employed for FFA in this study. The principle of maximum entropy (POME) was used to estimate GG parameters. Monte Carlo simulation was carried out to evaluate the performance of the GG distribution and to compare with widely used distributions. Finally, the *T*-year design flood was calculated using the GG and compared with that with other distributions. Results show that the GG distribution is either superior or comparable to other distributions.

Keywords: flood frequency analysis; generalized gamma (GG) distribution; principle of maximum entropy (POME)

---

1. Introduction

Flood frequency analysis (FFA) is needed for the design of water engineering and hydraulic structures. The sizing of bridges, culverts and other facilities; the design capacities of levees, spillways and other control structures; and reservoir operation or management all depend upon the estimated magnitude of various design flood values [1–3]. In FFA, flow data, such as the annual maximum data, are fitted using a theoretical frequency distribution, which is usually selected from a set of candidate distributions [4]. For example, the Pearson type three distribution (P3) has been recommended in China [5]. In the US, since 1967 the Log-Pearson type 3 distribution (LP3) has been the official distribution for all catchments which are fitted for planning and insurance purposes [6]. The UK has endorsed the GEV distribution [7,8] for FFA.

The choice of the appropriate model is one of the most important issues for FFA. The method commonly practiced is to try different distributions for the data at hand and choose the best fitted distribution using some particular goodness-of-fit measure [9]. One of the disadvantages of this method is that too many different distributions need to be tried and the selected distribution may be the best based on one goodness of fit criterion, but not based on another criterion. In order to overcome this disadvantage, some generalized frequency distributions have been recently used for FFA. The generalized gamma (GG) distribution is discussed in this study. It is a generalization of the two-parameter gamma distribution. The GG distribution includes as special cases the exponential distribution, the two-parameter gamma distribution, and the Weibull distribution, which provide sufficient flexibility to fit a large variety of data sets.

After deciding the distribution, the second issues is to estimate the parameters associated with the GG distribution. The popular techniques for parameter estimation include the methods of maximum likelihood (ML) [7], moments (MM) [10] and L-moments [11]. In addition, entropy theory can be used to derive more generalized distributions using different constraints [12]. The theory involves entropy maximizing in accord with the principle of maximum entropy (POME), in which the distribution parameter are determined, given the observed data and a set of constraints. Singh [12] indicated that the entropy method was reasonable and efficient for parameter estimation.

The objective of this study was therefore to propose an entropy based generalized gamma distribution for flood frequency analysis. The GG distribution parameters were estimated using POME. The GG distribution was tested using observed data sets. Also, Monte Carlo simulation was carried out to evaluate the predictive ability of the GG distribution and it was compared with some widely accepted distributions. Finally, the T-year design flood values were calculated and compared based on different FFA distributions.

2. Methodology

2.1. Generalized Gamma Distribution

Let X be a random variable and x be its specific value. The probability density function (PDF) of the generalized gamma (GG) distribution can be expressed as:

$$f(x) = \frac{r_2}{\beta \Gamma(\frac{r_1}{r_2})} (\frac{x}{\beta})^{(r_1-1)} \exp(-(\frac{x}{\beta})^{r_2}) \tag{1}$$

where $\Gamma(\bullet)$ is the gamma function, and r_1, r_2 are the shape parameters, and β is the scale parameter.

2.2. Estimation of Parameters of GB2 Distribution by POME

The GG distribution parameters were determined using the principle of maximum entropy (POME). The POME method involves the following steps: (1) specification of constraints; (2) maximization of entropy using the method of Lagrange multipliers; (3) derivation of the relation between Lagrange multipliers and constraints; (4) derivation of the relation between Lagrange multipliers and distribution parameters; and (5) derivation of the relation between distribution parameters and constraints. A flow chart showing the estimation procedure is shown in Figure 1.

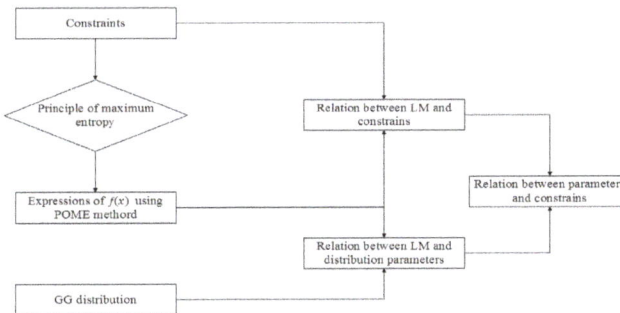

Figure 1. Flow chart of POME method for parameter estimation.

2.2.1. Specification of Constraints

Flood discharge is considered as a random variable X, which ranges from 0 to infinity. Its probability distribution function (PDF) and cumulative distribution function (CDF) are denoted as $f(x)$ and $F(x)$, respectively, where x is a specific value of X. Since constraints encode the information that can be given for the random variable, following Singh [12], the constraints for the GG distribution can be expressed as:

$$\int_0^\infty f(x)dx = 1 \tag{2a}$$

$$\int_0^\infty f(x)\ln x dx = E(\ln x) \tag{2b}$$

$$\int_0^\infty f(x)x^q dx = E(x^q) \tag{2c}$$

The first constraint is the total probability law, the second constraint is the mean of log values or the geometric mean, and the third constraint is the mean of values raised to a power q or log of scaled values raised to a power and then shifted by unity.

2.2.2. Maximization of Entropy Using the Method of Lagrange Multipliers

The Shannon entropy of X, $H(X)$, can be expressed as [13]:

$$H(X) = -\int_0^\infty f(x)\log f(x)dx \tag{3}$$

The $f(x)$ can be obtained by maximizing the Shannon entropy subject to given constraints in accord with the principle of maximum entropy (POME). Following Singh [14,15], maximization of Equation (3), subject to Equation (2a) to (2c), using the method of Lagrange multipliers leads to:

$$f(x) = \exp(-\lambda_0 - \lambda_1 \ln(x) - \lambda_2 x^q) \tag{4}$$

where $\lambda_0, \lambda_1, \lambda_2$ are the Lagrange multipliers that are not known.

2.2.3. Relation between Lagrange multipliers and parameters

Substitution of Equation (4) in Equation (2a) yields:

$$\int_0^\infty f(x)dx = \int_0^\infty \exp(-\lambda_0 - \lambda_1 \ln(x) - \lambda_2 x^q)dx = 1 \tag{5}$$

Equation (5) can be expressed as:

$$\exp(\lambda_0) = \int_0^\infty x^{(-\lambda_1)} \exp(-\lambda_2 x^q)dx \tag{6}$$

Let $t = \lambda_2 x^q$. Then $x = (\frac{t}{\lambda_2})^{\frac{1}{q}}$, and $dx = \frac{1}{q\lambda_2}(\frac{t}{\lambda_2})^{\frac{1}{q}-1}dt$. Then Equation (6) can be expressed as:

$$\exp(\lambda_0) = \int_0^\infty \frac{1}{q}\frac{1}{\lambda_2}(\frac{1}{\lambda_2})^{\frac{-\lambda_1+1}{q}} t^{\frac{-\lambda_1+1}{q}-1} \exp(-t)dt = \frac{1}{q}(\frac{1}{\lambda_2})^{\frac{-\lambda_1+1}{q}} \Gamma(\frac{-\lambda_1+1}{q}) \tag{7}$$

Substitution of Equation (7) in Equation (4) yields:

$$f(x) = \exp(-\lambda_0 - \lambda_1 \ln(x) - \lambda_2 x^q) = \frac{q\lambda_2^{\frac{-\lambda_1+1}{q}}}{\Gamma(\frac{-\lambda_1+1}{q})} x^{(-\lambda_1)} \exp(-\lambda_2 x^q) \tag{8}$$

Let $\frac{-\lambda_1+1}{q} = \frac{r_1}{r_2}$, and $\lambda_2 = (\frac{1}{\beta})^{r_2}$. Then, $r_2 = q$, and $r_1 = -\lambda_1 + 1$. Equation (8) can now be written as:

$$f(x) = \frac{r_2(\frac{1}{\beta})^{r_2 \frac{r_1}{r_2}}}{\Gamma(\frac{r_1}{r_2})} x^{(r_1-1)} \exp(-(\frac{1}{\beta})^{r_2} x^{r_2}) = \frac{r_2}{\beta\Gamma(\frac{r_1}{r_2})} (\frac{x}{\beta})^{(r_1-1)} \exp(-(\frac{x}{\beta})^{r_2}) \tag{9}$$

Equation (9) is the same as the generalized gamma distribution given by Equation (1). Hence, the relation between Lagrange multipliers and distribution parameters are given by:

$$\begin{cases} q = r_2 \\ \lambda_1 = 1 - r_1 \\ \lambda_2 = \beta^{-r_2} \end{cases} \tag{10}$$

2.2.4. Relation between Lagrange Multipliers and Constraints

Since the Lagrange multiplier λ_0 can be expressed by Equations (6) and (7), the set of equations can be used to obtain λ_0:

$$\lambda_0 = \ln(\int_0^\infty \exp(-\lambda_1 \ln(x) - \lambda_2 x^q) dx) \tag{11a}$$

$$\lambda_0 = -\ln q + \frac{\lambda_1 - 1}{q} \ln(\lambda_2) + \ln \Gamma(\frac{1 - \lambda_1}{q}) \tag{11b}$$

Differentiation of Equation (11a) with respect to λ_1 and λ_2 yields:

$$\begin{cases} \frac{\partial \lambda_0}{\partial \lambda_1} = \dfrac{\int_0^\infty \ln x \exp(-\lambda_1 \ln(x) - \lambda_2 x^q) dx}{\int_0^\infty \exp(-\lambda_1 \ln(x) - \lambda_2 x^q) dx} = -E(\ln x) \\[4mm] \frac{\partial \lambda_0}{\partial \lambda_2} = \dfrac{\int_0^\infty x^q \exp(-\lambda_1 \ln(x) - \lambda_2 x^q) dx}{\int_0^\infty \exp(-\lambda_1 \ln(x) - \lambda_2 x^q) dx} = -E(x^q) \end{cases} \tag{12}$$

Defining $b = \frac{1-\lambda_1}{q}$, and differentiating Equation (11b) with respect to λ_1 and λ_2, we obtain:

$$\begin{cases} \frac{\partial \lambda_0}{\partial \lambda_1} = \frac{1}{q} \ln \lambda_2 - \frac{1}{q} \partial \frac{\ln \Gamma(b)}{\partial b} = \frac{1}{q} \ln \lambda_2 - \frac{1}{q} \varphi(b) \\ \frac{\partial \lambda_0}{\partial \lambda_2} = \frac{1}{\lambda_2} \frac{\lambda_1 - 1}{q} = \frac{-b}{\lambda_2} \end{cases} \tag{13}$$

where $\varphi(\bullet)$ is a digamma function.

Based on Equations (12) and (13), the relation between Lagrange multipliers and constraints can be expressed as:

$$\begin{cases} \frac{\partial \lambda_0}{\partial \lambda_1} = \frac{1}{q} \ln \lambda_2 - \frac{1}{q} \varphi(b) = -E(\ln x) \\ \frac{\partial \lambda_0}{\partial \lambda_2} = \frac{-b}{\lambda_2} = -E(x^q) \end{cases} \tag{14}$$

Since there are three parameters, Equations (13) and (14) are not sufficient for calculating all the parameters, and one additional equation is therefore needed which is given as:

$$\frac{\partial^2 \lambda_0}{\partial^2 \lambda_1} = \frac{1}{q^2} \varphi'(b) = \text{var}(\ln x) \tag{15}$$

2.2.5. Relation between Parameters and Constraints

Based on the relation between parameters and constraints and between parameters and Lagrange multipliers, the relation between parameters and constraints can be expressed as:

$$\begin{cases} \frac{1}{r_2} \ln(\beta^{-r_2}) - \frac{1}{r_2} \varphi(\frac{r_1}{r_2}) = -E(\ln x) \\ \beta^{r_2} \frac{r_1}{r_2} = E(x^{r_2}) \\ \frac{1}{r_2^2} \varphi'(\frac{r_1}{r_2}) = \text{var}(\ln x) \end{cases} \tag{16}$$

where $\varphi(\bullet)$ is the digamma function; $\varphi'(\bullet)$ is the tri-gamma function. For a given data set X, the $E(\ln x)$ and var$(\ln x)$ can be calculated directly. There are three parameters and three equations in Equation (16). Therefore, this set of nonlinear functions can be solved by the widely used Newton iteration method (Deuflhard, [16]) for parameter estimation. The initial value of the three parameters are set to $(1, 1, 1)$. After multiple iterations, the optimal parameters can be obtained.

3. The Descriptive Ability of GG Distribution

Annual maximum (AM) flood peak data from 10 gauging stations, namely sites 1 to 10, were selected (Table 1). These ten stations are selected due to their diversity of statistical properties and climate types (arid, semi-arid and humid).

Table 1. Statistics of annual maximum flood data series for 10 sites.

Site No.	Gauging Station	Period	C_v	C_s	C_k
1	Rogue River at Raygold near Central Point, US	1906–2001	0.67	1.94	5.66
2	Quinault River at Quinault Lake, US	1912–2001	0.40	0.52	−0.6
3	Eel R A Scotla, US	1911–2001	0.51	0.61	−0.44
4	White River Near Meeker, US	1910–2001	0.34	0.65	0.77
5	Yellowstone River at Corwin Springs, US	1890–2001	0.30	0.66	0.59
6	Genesee River at Portageville, US	1909–2001	0.48	2.93	15.67
7	White River Near Meeker, US	1910–2001	0.34	0.65	0.77
8	Brokenstraw Creek at Younsville, US	1910–2001	0.33	0.75	0.65
9	Danjiangkou reservior at Danjiangkou, China	1929–2014	0.56	0.95	1.57
10	Geheyan reservior at Changyang, China	1951–2005	0.42	1.34	3.34

Besides AM series, partial-duration series can be also employed for the POME method. In this study, the AM series was considered since it is more widely used. The GG distribution was employed to fit the AM series of the 10 sites. The distribution parameters were estimated using Equations (16). The fitted GG distribution and the empirical frequency distribution of the AM series from sites 1, 5, 6 and 8 are shown in Figures 2–5. These four sites are selected because sites 5 and 8 have low skews, site 1 has moderate skew and site 6 has high skew, the cumulative distributions and histograms of AM series fitted by GG distribution for these sites can be representative. The line represents the fitted distribution and point represents the empirical frequencies of observations. Results show that the GG distribution fitted the empirical data well. Histograms of the AM flood peak series fitted by the GG distribution for the four sites are also shown in Figures 2–5 which also show that the GG distribution fitted the empirical histograms well. The skewness coefficient of AM series of sites 1, 5, 6 and 8 was 1.94, 0.66, 2.93 and 0.75, respectively, which showed that the GG distribution described both low and high skewed data well.

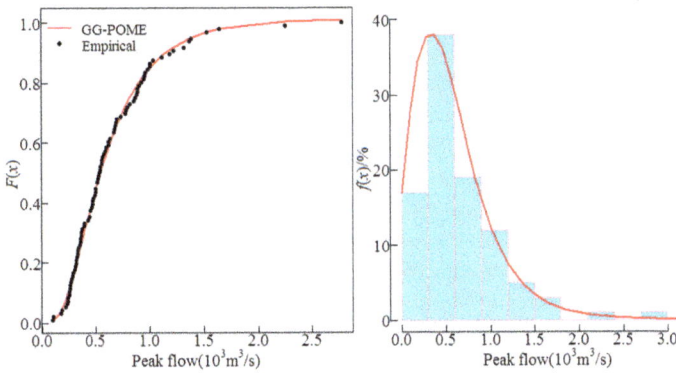

Figure 2. Cumulative distribution and histogram of AM flood peak series fitted by the GG distribution for site 1.

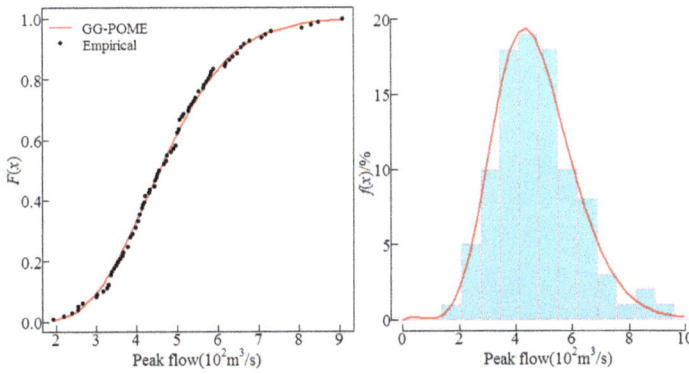

Figure 3. Cumulative distribution and histogram of AM flood peak series fitted by the GG distribution for site 5.

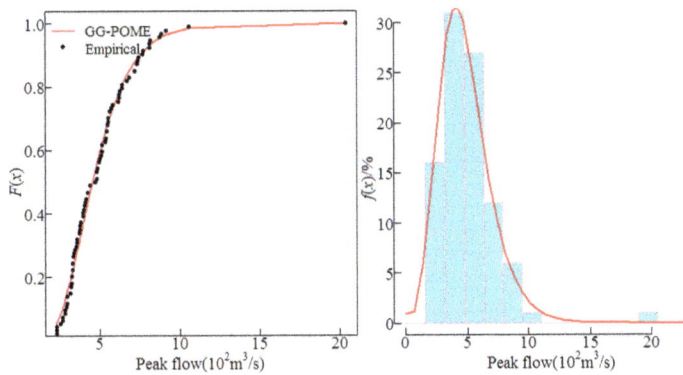

Figure 4. Cumulative distribution and histogram of AM flood peak series fitted by GG-POME model for site 6.

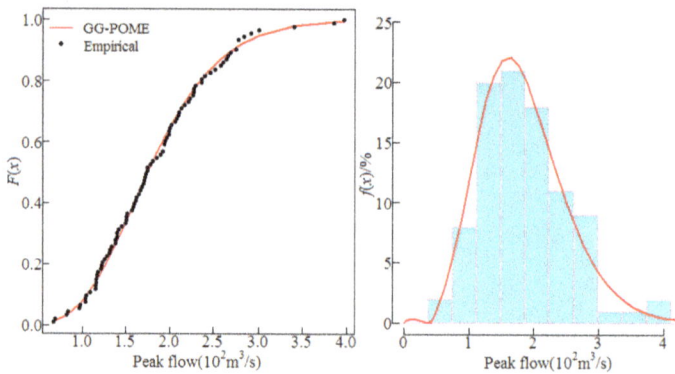

Figure 5. Cumulative distribution and histogram of AM flood peak series fitted by the GG distribution for site 8.

Several distributions, including normal (NM), exponential (EXP), generalized logistic (GLO), gamma (GM), generalized Pareto (GPA), Gumbel (GB), Weibull (WB), P3, GEV and LP3 distributions, in which the parameters of EXP, GLO, GM, GPA, GB, WB, P3, GEV distributions were estimated by the L-moment method (LM) [11,17], while the parameters of NM and LP3 distributions were estimated by MM [18,19]. These FFA models were also fitted to the AM series for the 10 sites and the values of *RMSE* and *AIC* were computed for each model using Equations (17) and (18) and listed in Table 2.

$$RMSE = \sqrt{\frac{\sum\limits_{i=1}^{n}\left(\hat{P}(i) - P(i)\right)^2}{n}} \tag{17}$$

$$AIC = n\left(\ln\left(\frac{1}{n}\sum\limits_{i=1}^{n}\left(\hat{P}(i) - P(i)\right)^2\right)\right) + 2K \tag{18}$$

where n denotes the sample size, K is the number of parameters of the distribution, $\hat{P}$ is the theoretical non-exceedance probability calculated by the distribution, and P is the empirical non-exceedance probability. Root mean square error (*RMSE*) is a frequently used measure of the differences between values (sample and population values) predicted by a model or an estimator and the values actually observed. The smaller *RMSE* values represent the better performance of the model. The Akaike information criterion (*AIC*) is a measure of the relative quality of statistical models for a given set of data. It also includes a penalty that is an increasing function of the number of estimated parameters. Given a set of candidate models for the data, the preferred model is the one with the minimum *AIC* value.

Table 2 illustrates that for sites 1, 2, 4, 5, 8, 9 and 10, the GG distribution had the smallest *RMSE* values, which means the GG distribution fitted the observed AM data best. In addition, the GG distribution had the smallest *AIC* values for sites 2, 5, 8, and 10. Table 2 also indicates that the average *RMSE* and *AIC* values of GG distribution are the smallest among all the compared distributions. Thus, the GG distribution performs better than other distributions.

Table 2. The RMSE and AIC values of different flood frequency analysis distributions.

Model	Site 1		Site 2		Site 3		Site 4		Site 5		Mean Value
	RMSE	AIC	RMSE	AIC	RMSE	AIC	RMSE	AIC	RMSE	AIC	RMSE
GG-POME	0.016	−508.12	0.028	−417.67	0.023	−430.63	0.021	−462.6	0.013	−578.63	0.020
NM-MM	0.078	−258.88	0.047	−306.32	0.048	−316.43	0.037	−394.91	0.038	−407.87	0.053
EXP-LM	0.021	−468.96	0.056	−273.3	0.054	−218.65	0.069	−244.39	0.058	−286.44	0.054
GLO-LM	0.025	−446.68	0.046	−303.81	0.035	−362.47	0.021	−474.16	0.015	−560.19	0.027
GM-LM	0.029	−447.1	0.032	−362.45	0.021	−447.48	0.021	−467.06	0.017	−550.61	0.027
GPA-LM	0.019	−504.1	0.029	−413.67	0.021	−444.15	0.042	−346.33	0.036	−389.21	0.030
GB-LM	0.039	−386.64	0.036	−343.98	0.025	−421.83	0.024	−430.95	0.015	−530.05	0.031
WB-LM	0.017	−522.29	0.029	−372.34	0.019	−459.48	0.025	−441.45	0.019	−515.48	0.024
P3-LM	0.016	−533.04	0.033	−351.64	0.023	−434.97	0.021	−468.95	0.015	−561.44	0.024
GEV-LM	0.019	−495.44	0.034	−347.79	0.025	−419.06	0.021	−472.43	0.014	−572.29	0.022
LP3-MM	0.016	−528.15	0.028	−376.18	0.021	−450.45	0.024	−451.6	0.016	−559.41	0.021

Model	Site 6		Site 7		Site 8		Site 9		Site 10		Mean Value
	RMSE	AIC	RMSE	AIC	RMSE	AIC	RMSE	AIC	RMSE	AIC	AIC
GG-POME	0.024	−427.53	0.021	−447.25	0.013	−562.19	0.021	−421.51	0.023	−269.84	−452.60
NM-MM	0.038	−407.87	0.113	−198.3	0.035	−400.26	0.046	−313.47	0.051	−210.33	−321.46
EXP-LM	0.058	−286.44	0.04	−373.78	0.059	−271.25	0.064	−233.3	0.061	−150.65	−280.72
GLO-LM	0.033	−377.21	0.014	−520.86	0.023	−449.16	0.031	−368.25	0.023	−267.35	−413.01
GM-LM	0.038	−393.42	0.046	−374.75	0.014	−541.17	0.023	−404.7	0.027	−264.86	−425.36
GPA-LM	0.024	−444.17	0.032	−359.81	0.025	−464.21	0.027	−384.59	0.044	−194.88	−394.51
GB-LM	0.033	−404.93	0.072	−282.68	0.019	−475.75	0.024	−406.35	0.024	−262.96	−394.61
WB-LM	0.022	−454.57	0.042	−312.87	0.013	−559.08	0.021	−429.65	0.030	−237.83	−430.50
P3-LM	0.022	−452.78	0.051	−275.86	0.014	−538.15	0.023	−419.04	0.027	−249.79	−428.57
GEV-LM	0.027	−411.21	0.021	−438.69	0.015	−525.05	0.024	−409.68	0.024	−260.26	−435.19
LP3-MM	0.026	−418.17	0.022	−450.08	0.014	−536.12	0.021	−426.08	0.026	−262.16	−445.84

Table 2 also shows that the GG, P3, GEV, LP3 distributions gave quite similar performances for most of the selected sites. However, it was observed that the GG distribution performed better at several sites. For site 2, the *RMSE* values for the GG, P3, and GEV distributions were 0.028, 0.033 and 0.034, respectively. The *AIC* values were −417.67, −351.64 and −347.79, respectively. Thus, the GG distribution performed much better than the P3 and GEV distributions for site 2. Compared with the LP3 distribution, the GG distribution was more appropriate for sites 5 and 7. For site 5, the *RMSE* and *AIC* values for the LP3 distribution (GG distribution) were 0.016(0.013) and −559.41 (−578.63), respectively. For site 7, the *RMSE* and *AIC* values for the LP3 distribution (GG distribution) were 0.019(0.016) and −545.85 (−571.71), respectively. Thus, the GG distribution outperformed the LP3 distribution for those two sites. The above discussions shows that the GG distribution is either superior or comparable to the commonly used distributions.

The maximum likelihood (ML) method was also employed for GG distribution and compared with the proposed GG-POME model for site 5 (low skew) and site 6 (high skew). Figure 6 gives comparisons of their probability density functions and indicates that the GG-POME model gives a better performance. The *RMSE* and *AIC* values of GG-ML model for sites 5 and 6 were also calculated. The *RMSE* and *AIC* values for the GG-ML (GG-POME) model are 0.023 (0.013) and −497.54 (−578.63), respectively for site 5. And the *RMSE* and *AIC* values for the GG-ML (GG-POME) model are 0.032 (0.024) and −379.24 (−427.53), respectively for site 6. Therefore it may imply that GG-POME model outperforms GG-ML model.

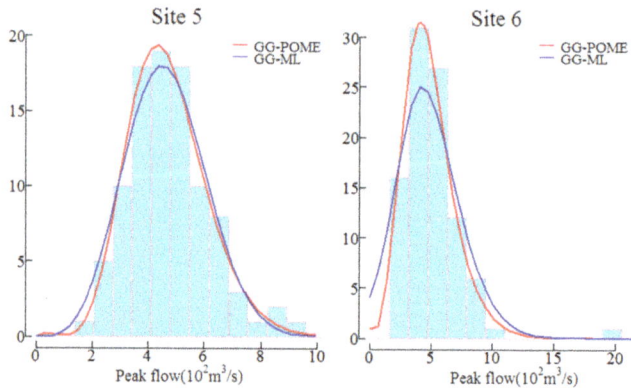

Figure 6. Comparisons of probability density functions of GG-POME and GG-ML models for sites 5 and 6.

4. Monte Carlo Simulation

The predictive ability of the GG distribution was evaluated using Monte Carlo simulation and compared with that of the P3, GEV, and LP3 distributions. To test how well a candidate distribution estimated the magnitude-return period relationship, a parent distribution which was not identical to any of the candidate distributions was chosen. Cunnane [20] recommended that such a parent distribution should be a Wakeby distribution with certain parameters. In this study, three kinds of data sets were generated from the Wakeby distribution with parameters as shown in Table 3. The Wakeby distribution has quantile function given as [21]:

$$x(F) = \xi + \frac{\alpha}{\beta}(1 - (1 - F)^{\beta}) - \frac{\gamma}{\delta}(1 - (1 - F)^{-\delta}) \qquad (19)$$

where F is the uniform (0, 1) variate; and $\xi, \alpha, \beta, \gamma, \delta$ are the parameters.

Table 3. Monte Carlo simulation data sets generated from the Wakeby distribution.

	ζ	α	β	γ	δ	C_v	C_s
Case 1	30.4	114.2	11.3	19.2	−0.5	0.2	0.16
Case 2	15.4	308.8	10.25	38.5	−0.3	0.36	0.48
Case 3	23.5	198.6	3.7	109.2	−0.2	0.55	0.95

Then, the real quantile value Q_T was computed. $S = 1000$ samples with size n ($n = 20, 50, 100$) were generated from each Wakeby distribution and fitted by the four distributions to estimate the events of $T = 10, 100$ and 1000-year return periods. Table 4 lists the *RB* and *RRMSE* values computed by each distribution using Equations (20) and (21):

$$RB = \frac{1}{S}\sum_{i=1}^{S}\frac{(\hat{Q}_T)_i - Q_T}{Q_T} \tag{20}$$

$$RRMSE = \sqrt{\frac{1}{S-1}\sum_{i=1}^{S}(\frac{(\hat{Q}_T)_i - Q_T}{Q_T})^2} \tag{21}$$

where Q_T is a given parent quantile, $(\hat{Q}_T)_1 \ldots (\hat{Q}_T)_S$ are the estimators for the samples generated from the Wakeby distribution, and S is the number of Monte Carlo trials. The relative bias (*RB*) and the relative root mean square errors (*RRMSE*) were used to evaluate the accuracy and efficiency of a candidate model, respectively.

Table 4. Calculated *RB* and *RRMSE* values for different FFA distributions.

			GG-POME		P3-LM		GEV-LM		LP3-MM	
			RB	RRMSE	RB	RRMSE	RB	RRMSE	RB	RRMSE
		$T = 10$	−1.71	4.54	−1.21	3.64	−0.87	3.78	−0.89	4.07
	$n = 20$	$T = 100$	6.11	7.95	4.6	9.26	4.89	9.31	2.56	8.84
		$T = 1000$	2.02	6.46	2.29	9.08	5.11	10.84	4.15	10.03
		$T = 10$	−1.32	2.73	−0.87	2.45	−0.75	2.43	−0.81	2.48
Case 1	$n = 50$	$T = 100$	5.53	6.5	4.38	6.28	6.12	8.02	4.25	6.77
		$T = 1000$	1.85	4.9	0.02	6.73	2.87	8.1	0.65	7.31
		$T = 10$	−1.06	1.82	−1.05	1.86	−0.72	1.84	−0.71	1.82
	$n = 100$	$T = 100$	6.02	6.55	4.97	5.72	5.47	6.34	4.74	5.82
		$T = 1000$	1.46	3.42	0.21	4.29	2.93	6.03	0.55	5.46
		$T = 10$	−2.65	9.78	−0.82	9.78	−0.58	9.34	−0.35	14.78
	$n = 20$	$T = 100$	1.93	10.23	2.25	16.85	3.64	19.23	5.97	39.86
		$T = 1000$	3.59	14.18	8.69	24.29	16.48	34.45	11.62	44.85
		$T = 10$	−1.93	8.44	−1.57	6.76	−1.76	6.21	−1.26	6.63
Case 2	$n = 50$	$T = 100$	1.66	7.55	1.76	10.27	1.37	10.74	−1.67	12.54
		$T = 1000$	3.27	10.33	8.64	16.46	6.67	18.96	4.32	21.43
		$T = 10$	−1.21	5.86	−1.18	4.73	−0.58	4.34	−0.86	4.52
	$n = 100$	$T = 100$	1.94	5.87	1.37	6.24	0.87	7.52	−1.24	7.88
		$T = 1000$	4.22	8.4	9.49	12.78	8.72	16.35	0.29	13.51
		$T = 10$	−0.42	11.46	−0.58	13.43	−2.34	12.57	2.64	14.79
	$n = 20$	$T = 100$	5.1	14	7.86	21.95	8.54	23.35	4.96	31.38
		$T = 1000$	17.67	28.16	13.57	29.87	21.82	42.72	26.58	57.28
		$T = 10$	-0.41	8.78	0.86	7.96	−1.25	7.38	1.78	8.67
Case 3	$n = 50$	$T = 100$	6.34	12.38	4.62	13.94	6.44	15.86	4.41	16.34
		$T = 1000$	16.86	22.85	14.53	24.63	21.02	33.68	8.29	27.32
		$T = 10$	−0.42	5.72	−0.87	5.87	−1.15	5.46	0.26	5.74
	$n = 100$	$T = 100$	6.37	8.54	4.75	8.98	6.65	11.87	3.54	12.64
		$T = 1000$	17.52	19.32	15.67	19.78	17.98	23.92	7.66	20.67

From Table 4, generally for all distributions and for all cases, it was observed that the *RB* and *RRMSE* values increased with the return period *T*. For a small return period (*T* = 10), the selected four distributions exhibited very similar behaviors regardless of the sample size. For moderate and large return periods (*T* = 100 and 1000), notable differences of *RB* and *RRMSE* values were observed. Thus, in the latter discussion, we would mainly focus on moderate and high return period quantile estimators.

For case 1 (C_v = 0.2, C_s = 0.16), it was observed that the GG and P3 distributions were superior to the GEV and LP3 distributions. When the sample size equaled 100 or 50, the P3 distribution quantile estimators had the smallest *RB* values for both moderate and large return periods (*T* = 100 and 1000). But the GG distribution quantile estimators had smaller *RRMSE* values for *T* = 1000 than other distributions. For a small sample size (*n* = 20), the GG distribution had the smallest *RB* and *RRMSE* values for both moderate and large return periods (*T* = 100 and 1000). For *T* = 1000, the *RRMSE* values of the GG, P3, GEV and LP3 distributions were 6.46, 9.08, 10.84 and 10.03, respectively. Apparently, the GG distribution performed much better when the sample size was small. This indicates that the GG distribution was more robust. Thus, for case 1, the P3 distribution was preferable when the sample size was large than 50, while the GG distribution was more appropriate when sample size did not exceed 50.

For case 2 (C_v = 0.36, C_s = 0.48), results indicated that for sample size *n* = 50 and *n* = 100, the GEV distribution quantile estimators had the smallest *RB* values for *T* = 100 and the LP3 distribution quantile estimators had the smallest *RB* values for *T* = 1000. However, their *RRMSE* values were quite large and increased significantly when the sample sizes decreased. For *T* = 1000, when the sample size decreased from 100 to 20, the *RRMSE* values of the GEV distribution rose from 16.35 to 34.45, and the *RRMSE* values of the LP3 distribution rose from 13.51 to 44.85. While the *RRMSE* values of the GG distribution rose slightly from 4.8 to 14.18. This was due to the poor accuracy of the GEV and LP3 distributions parameter estimators which had high variance for small sample sizes. In this case, the GG distribution performed significantly better than the other three distributions. Its *RB* values were quite small, and its *RRMSE* values were the smallest for all sample sizes and return periods. This was a good indication of the robustness of the GG distribution for this case.

For case 3 (C_v = 0.55, C_s = 0.97), all distribution quantile estimators had quite large *RB* and *RRMSE* values. For *n* = 50 and *n* = 100, *RB* and *RRMSE* of the GEV distribution were the highest, which amounted to 21.02 and 33.68, respectively, for *n* = 50, *T* = 1000, while the GG distribution yielded 16.86 and 22.85, respectively. Also for *n* = 50 and *n* = 100, the LP3 distribution quantile estimators had the smallest *RB* values for both *T* = 100 and *T* = 1000, and the other three distributions had similar *RB* values. But the LP3 distribution gave the worst performance for small sample sizes (*n* = 20). Its *RB* and *RRMSE* values were 26.58 and 57.28, respectively, for *T* = 1000, whereas the GG distribution yielded 17.67 and 28.16, respectively. In this case, the *RB* values of the GG distribution were comparable to the P3 and GEV distributions, and were a little larger than the LP3 distribution for *n* = 50 and *n* = 100, the *RRMSE* values of the GG distribution were the smallest for both moderate and large return periods (*T* = 100 and 1000) regardless of the sample size. Also, when the sample size decreased from 100 to 20, the *RB* and *RRMSE* values of the GG distribution rose from 17.52 and 19.32 to 17.67 and 28.16, respectively. This might imply that the distribution was less affected by sample size. Thus, the GG distribution was superior to other distributions for this case. Therefore, the predictive ability of the GG distribution was found to be comparable or superior to that of the other distributions, and it was more robust since it was less affected by sample size, and therefore, estimated the magnitude-return period relationships better.

5. *T*-Year Design Flood Calculation

The Danjiangkou reservoir lies in the upper Hanjiang basin and is the source of water for the Middle Route Project under the South-to-North Water Transfer Scheme in China [22]. The Geheyan reservoir, with a volume of 3.12 billion m^3, plays an important role in management of Qingjiang River [23]. Flood frequency analysis for these two sites was therefore considered in this study. The

T-year design flood calculated by different FFA distributions at Danjiangkou Reservoir and Geheyan Reservoir are listed in Table 5. Figures 7 and 8 compare frequency curves of different distributions at these two reservoir sites.

Table 5 indicates that design flood for small return periods was similar for these four distributions. However, significant differences were observed for large return periods. The 1000-year design flood calculated by the GG and LP3 distributions at Danjiangkou Reservoir were 55,234 m³/s and 48,822 m³/s, respectively. And the 1000-year design flood calculated by the GEV and LP3 distributions at Geheyan Reservoir were 15,746 m³/s and 13,877 m³/s, respectively.

Figure 7 indicates that the GG, P3, and GEV distributions had quite similar flood quantile estimators for large return periods at Danjiangkou Reservoir. However, the 1000-year design flood calculated by the LP3 distribution was smaller than by the other three distributions. Figure 8 indicates that the 1000-year design flood calculated by the GEV distribution at Geheyan Reservoir was the largest, and was the smallest for the LP3 distribution.

Table 5. Comparison of T-year design floods calculated by different FFA distributions at Danjiangkou and Geheyan sites.

Site	Model	Return Period (Year)				
		1000	500	100	50	10
Danjiangkou (m³/s)	GG-POME	55,234	51,432	42,204	38,803	27,398
	P3-LM	53,838	50,202	41,407	37,411	27,311
	GEV-LM	55,369	51,490	42,054	37,785	27,217
	LP3-MM	48,822	46,561	40,261	36,999	27,692
Geheyan (m³/s)	GG-POME	13,992	13,745	11,186	10,277	7957
	P3-LM	14,896	13,941	11,648	10,616	8039
	GEV-LM	15,746	14,594	11,910	10,746	7991
	LP3-MM	13,877	13,099	11,171	10,276	7963

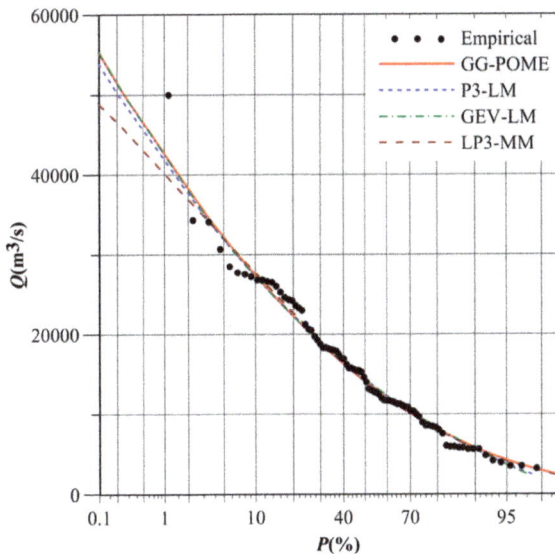

Figure 7. Frequency curves of different FFA distributions at Danjiangkou Reservoir.

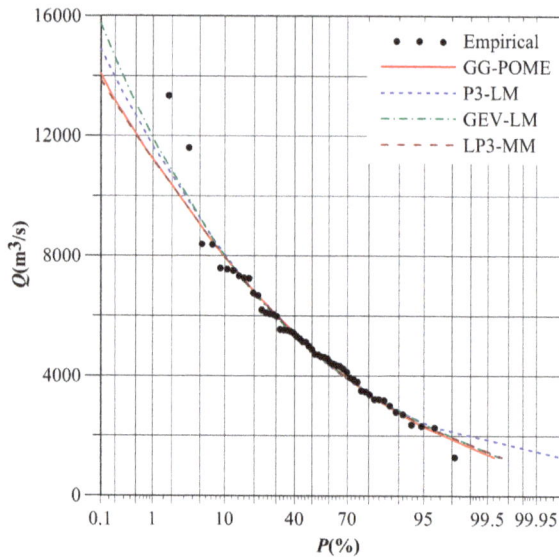

Figure 8. Frequency curves of different FFA distributions at Geheyan Reservoir.

The design flood calculated by the GG distribution was quite close to that by the LP3 distribution. Besides, the P3 distribution has been adopted in China as a uniform procedure for FFA [24,25]. Table 2 shows that *RMSE* and *AIC* values for the P3 distribution at Danjiangkou Reservoir were 0.023 and −419.04, respectively, and the GG distribution yielded 0.021 and −421.51, respectively. The *RMSE* and *AIC* values for the P3 distribution at Geheyan Reservoir were 0.027 and −249.79, respectively, and the GG distribution yielded 0.023 and −269.84, respectively. Thus, the performance of the GG distribution was better than that of the P3 distribution. Therefore, the design flood estimated by the GG distribution would be preferable in practice.

6. Conclusions

In this study, the GG distribution with parameters estimated by POME was applied for FFA. Ten gauging stations were selected as a case study to test the GG distribution. Frequency estimates from the GG distribution were also compared with those of commonly used distributions. A Monte Carlo simulation study was carried out to evaluate the predictive ability of the GG distribution and compare it with other distributions. In addition, some characteristics of frequency curves at Danjiangkou Reservoir and Geheyan Reservoir were evaluated. The following conclusions are drawn from this study:

(1) The GG distribution is appealing for FFA. The cumulative distributions and histograms show that the GG distribution can fit both low and high skewed data well.

(2) The parameters estimated by POME are found reasonable. Both the marginal distributions and histograms indicates that the GG distribution with so estimated parameters can successfully be fitted to empirical values.

(3) The performance of the GG distribution is comparable or superior to that of the other distributions. Results illustrate that for sites 1, 2, 4, 5, 8, 9 and 10, the GG distribution has the smallest *RMSE* values. In addition, the GG distribution has the smallest *AIC* values for sites 2, 5, 8, and 10. Thus, the GG distribution is preferred to other distributions for those sites. Furthermore, the GG, P3, GEV, and LP3 distributions give similar performance for most of the selected sites. However, the GG distribution fits better than them for a few sites.

(4) The predictive ability of the GG distribution is found to be comparable or superior to widely accepted distributions. The GG distribution performs significantly better than the other three distributions when sample sizes are small. Thus it is less effected by sample size and is more robust.

Acknowledgments: This study was supported by the National Natural Science Foundation of China (51679094; 51509273; 91547208) and Fundamental Research Funds for the Central Universities (2016YXZD048).

Author Contributions: All of the authors read and approved the final manuscript. Lu Chen and Feng Xiong conceived and designed the experiments; Vijay P. Singh analyzed the data and contributed materials; Lu Chen and Feng Xiong wrote the paper.

Conflicts of Interest: The authors declare no conflict of interest.

References

1. American Society of Civil Engineers. *Hydrology Handbook*, 2nd ed.; ACSE: New York, NY, USA, 1996.
2. Chen, L.; Guo, S.; Yan, B.; Liu, P.; Fang, B. A new seasonal design flood method based on bivariate joint distribution of flood magnitude and date of occurrence. *Hydrol. Sci. J.* **2010**, *55*, 1264–1280. [CrossRef]
3. Chen, L.; Singh, V.P.; Guo, S.; Hao, Z.; Li, T. Flood coincidence risk analysis using multivariate copula functions. *J. Hydrol. Eng.* **2012**, *17*, 742–755. [CrossRef]
4. Kasiviswanathan, K.S.; He, J.X.; Tay, J.H. Flood frequency analysis using multi-objective optimization based interval estimation approach. *J. Hydrol.* **2017**, *545*, 251–262. [CrossRef]
5. Chen, L.; Singh, V.P.; Guo, S.; Zhou, J.; Zhang, J. Copula-based method for multisite monthly and daily streamflow simulation. *J. Hydrol.* **2015**, *528*, 369–384. [CrossRef]
6. Benameur, S.; Benkhaled, A.; Meraghni, D.; Chebana, F.; Necir, A. Complete flood frequency analysis in Abiod watershed, Biskra (Algeria). *Nat. Hazards* **2017**, *86*, 519–534. [CrossRef]
7. Natural Environment Research Council. *Flood Studies Report*; Institute of Hydrology: London, UK, 1975.
8. Natural Environment Research Council. *Flood Studies Report*; Institute of Hydrology: Wallingford, UK, 1999.
9. Papalexiou, S.M.; Koutsoyiannis, D. Entropy based derivation of probability distributions: A case study to daily rainfall. *Adv. Water Resour.* **2012**, *45*, 51–57. [CrossRef]
10. Chebana, F.; Adlouni, S.; Bobée, B. Mixed estimation methods for Halphen distributions with applications in extreme hydrologic events. *Stoch. Environ. Res. Risk. Assess.* **2010**, *24*, 359–376. [CrossRef]
11. Hosking, J.R.M. L-moments: analysis and estimation of distributions using linear combinations of order statistics. *J. R. Stat. Soc.* **1990**, *52*, 105–124.
12. Singh, V.P. *Entropy-Based Parameter Estimation in Hydrology*; Springer: Dordrecht, The Netherlands, 1998.
13. Shannon, C.E. A mathematical theory of communication. *Bell Syst. Tech. J.* **1948**, *27*, 379–423. [CrossRef]
14. Singh, V.P. *Entropy Theory and Its Application in Environmental and Water Engineering*; John Wiley & Sons: Chichester, UK, 2013.
15. Singh, V.P. *Entropy Theory in Hydrologic Science and Engineering*; McGraw-Hill Education: New York, NY, USA, 2014.
16. Deuflhard, P. *Newton Methods for Nonlinear Problems: Affine Invariance and Adaptive Algorithms*; Springer: Berlin, Germany, 2011.
17. Liang, Z.; Hu, Y.; Li, B.; Yu, Z. A modified weighted function method for parameter estimation of Pearson type three distribution. *Water Resour. Res.* **2014**, *50*, 3216–3228. [CrossRef]
18. Rao, A.R.; Hamed, K.H. *Flood Frequency Analysis*; CRC Press: New York, NY, USA, 2000.
19. Griffis, V.W.; Stedinger, J.R. Log-Pearson type 3 distribution and its application in flood frequency analysis. II: Parameter estimation methods. *J. Hydrol. Eng.* **2007**, *12*, 482–491. [CrossRef]
20. Cunnane, C. Statistical distribution for flood frequency analysis. *Oper. Hydrol. Rep.* **1989**, *44*, 369.
21. Houghton, J.C. Birth of a parent: The Wakeby distribution for modeling flood flows. *Water Resour. Res.* **1978**, *14*, 1105–1109. [CrossRef]
22. Li, S.; Xu, Z.; Cheng, X.; Zhang, Q. Dissolved trace elements and heavy metals in the Danjiangkou Reservoir, China. *Environ. Geol.* **2008**, *55*, 977–983. [CrossRef]
23. Qi, S.; Yan, F.; Wang, S.; Xu, R. Characteristics, mechanism and development tendency of deformation of Maoping landslide after commission of Geheyan reservoir on the Qingjiang River, Hubei Province, China. *Eng. Geol.* **2006**, *86*, 37–51. [CrossRef]

24. Ji, X.; Ding, J.; Shen, H.; Salas, J.D. Plotting positions for Pearson type-III distribution. *J. Hydrol.* **1984**, *74*, 1–29.

25. Shao, Q.; Wong, H.; Xia, J.; Ip, W. Models for extremes using the extended three-parameter Burr XII system with application to flood frequency analysis. *Hydrol. Sci. J.* **2004**, *49*, 685–702. [CrossRef]

entropy

MDPI

Article

Generalized Beta Distribution of the Second Kind for Flood Frequency Analysis

Lu Chen [1],* and Vijay P. Singh [2]

[1] College of Hydropower & Information Engineering, Huazhong University of Science & Technology, Wuhan 430074, China
[2] Department of Biological and Agricultural Engineering & Zachry Department of Civil Engineering, Texas A&M University, College Station, TX 77843, USA; vsingh@tamu.edu
* Correspondence: chen_lu@hust.edu.cn; Tel.: +86-027-8754-3992

Academic Editor: Kevin H. Knuth
Received: 18 April 2017; Accepted: 26 May 2017; Published: 12 June 2017

Abstract: Estimation of flood magnitude for a given recurrence interval T (T-year flood) at a specific location is needed for design of hydraulic and civil infrastructure facilities. A key step in the estimation or flood frequency analysis (FFA) is the selection of a suitable distribution. More than one distribution is often found to be adequate for FFA on a given watershed and choosing the best one is often less than objective. In this study, the generalized beta distribution of the second kind (GB2) was introduced for FFA. The principle of maximum entropy (POME) method was proposed to estimate the GB2 parameters. The performance of GB2 distribution was evaluated using flood data from gauging stations on the Colorado River, USA. Frequency estimates from the GB2 distribution were also compared with those of commonly used distributions. Also, the evolution of frequency distribution along the stream from upstream to downstream was investigated. It concludes that the GB2 is appealing for FFA, since it has four parameters and includes some well-known distributions. Results of case study demonstrate that the parameters estimated by POME method are found reasonable. According to the RMSD and AIC values, the performance of the GB2 distribution is better than that of the widely used distributions in hydrology. When using different distributions for FFA, significant different design flood values are obtained. For a given return period, the design flood value of the downstream gauging stations is larger than that of the upstream gauging station. In addition, there is an evolution of distribution. Along the Yampa River, the distribution for FFA changes from the four-parameter GB2 distribution to the three-parameter Burr XII distribution.

Keywords: entropy theory; principle of maximum entropy (POME); GB2 distribution; flood frequency analysis

1. Introduction

Estimation of flood magnitude for a given recurrence interval T (T-year flood) at a given location is essential for the design of hydraulic and civil infrastructure facilities, such as dams, spillways, levees, urban drainage, culverts, road embankments, and parking lots. A key step in flood frequency estimation or analysis (FFA) is the selection of a suitable frequency distribution [1]. Commonly used distributions for flood frequency analysis include Gumbel, gamma, generalized extreme value (GEV), Pearson type III (P-III), log-Pearson type III (LP-III), Weibull, and log-normal (LN). Some of these distributions have been adopted in different countries. For example, the P-III distribution has been adopted in China and Australia as a standard method for hydrologic frequency analysis [2–4]. The LP-III distribution has been adopted in the United States and the GEV distribution in Europe.

Mielke and Johnson investigated the use of two special cases of the generalized beta distribution of the second kind, namely gamma and log normal distributions, for flood frequency analysis [5].

Wilks investigated the performance of eight three-parameter probability distributions for precipitation extremes using annual and partial duration data from stations in the northeastern and southeastern United States [6]. He found that the beta-κ distribution best described the extreme right tail of annual extreme series, and the beta-P distribution was best for the partial duration data.

Recently, some generalized frequency distributions have been used for hydrologic frequency analysis. For example, Perreault et al. presented a family of distributions, named Halphen distributions, for frequency analysis of hydrometeorological extremes [7]. Papalexiou and Koutsoyiannis used the generalized gamma distribution and generalized beta distribution of the second kind (GB2) for rainfall frequency analysis across the world and showed that these distributions were appropriate for worldwide rainfall data [8]. The greatest advantage of these generalized distributions is that they provide sufficient flexibility to fit a large variety of data sets, which facilitates the selection and comparison of different distributions. For instance, the GB2 distribution includes the exponential, Weibull, and gamma distributions as special cases. Since the GB2 distribution has four parameters, logically it should perform better than 3-parameter distributions, such as GEV, P-III, LP-III or LN-III. Papalexiou and Koutsoyiannis concluded that the GB2 distribution was a suitable model for rainfall frequency analysis because of its ability to describe both J-shaped and bell-shaped data [8]. The other advantages of the GB2 distribution can be summarized as: (1) the GB2 distribution can model positive or negative skewness which is an advantage over distributions, such as lognormal, with only positive skew; (2) it can jointly estimate both location and shape parameters, while many other distributions, such as exponential, logistic, normal, etc., usually focus on location only; and (3) it can better capture the long right or left tail. Because of these advantages, the GB2 distribution was employed in this study.

The second step in flood frequency analysis is to estimate parameters of the selected distribution. There are several standard parameter estimation methods, such as moments, maximum likelihood, L-moments, probability weighted moments, and least square. Among these methods, the maximum likelihood (ML) and L-moment methods are widely used in hydrology. In addition, the principle of maximum entropy (POME) has been applied to parameter estimation [9,10]. Singh and Guo indicated that POME method was comparable to ML and L-moment methods, and for certain situations, POME method was superior to these two methods [11]. Therefore, the POME method was considered in this study for parameter estimation.

Another aspect of FFA that is of interest is how the flood frequency distribution evolves from upstream to downstream along a river. The drainage area along the river increases from upstream to downstream. It is interesting to investigate if the same frequency distribution applies at all gauging stations along the stream.

The objective of this study therefore is to employ the GB2 distribution for flood frequency analysis (FFA). The specific objectives are to: (1) estimate the GB2 distribution parameters using the principle of maximum entropy; (2) evaluate the performance of the GB2 distribution and compare it with commonly used distributions in hydrology; (3) select the best distribution; and (4) discuss the evolution of frequency distribution and its parameters along the river.

2. GB2 Distribution

The generalized beta distribution of the second kind, denoted as GB2, is a four-parameter distribution and can be expressed as:

$$f(x) = \frac{r_3}{\beta B(r_1, r_2)} \left(\frac{x}{\beta}\right)^{r_1 r_3 - 1} \left(1 + \left(\frac{x}{\beta}\right)^{r_3}\right)^{-(r_2 + r_1)} \tag{1}$$

where $B(\cdot)$ is the beta function; β is the scale parameter, $\beta > 0$; and $r_1 > 0$, $r_2 > 0$, and $r_3 > 0$ are the shape parameters. Parameter r_3 represents the overall shape; parameter r_1 governs the left tail; parameter r_2 controls the right tail; and β is a scale parameter and depends on the unit of measurement. These parameters allow the distribution to be able to fit data having very different histogram shapes. It can simulate both the J-shaped and bell-shaped distributions. Parameters r_1 and

r_2 together determine the skewness of the distribution. The general shapes of GB2 probability density distribution were shown in Figure 1.

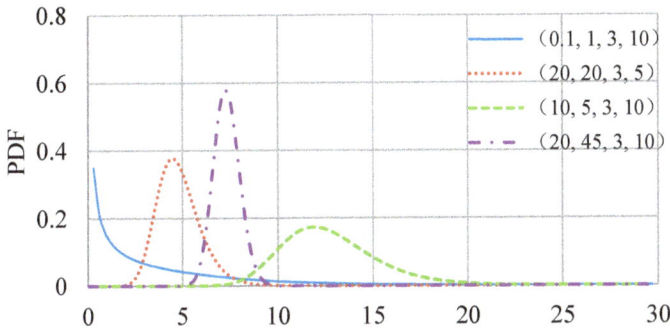

Figure 1. Shapes of PDF of GB2 distribution.

When analyzing extreme rainfall, Papalexiou and Koutsoyiannis showed that the GB2 distribution is a very flexible four-parameter distribution [8]. By fixing certain parameters, the GB2 distribution can yield some well-known distributions, such as the beta distribution of the second kind (B2), the Burr type XII, generalized gamma (GG), and so on. These distributions can be treated as special or limiting cases of the GB2 distribution, as shown in Figure 2. Some of these special cases have been applied in hydrological frequency analysis. For example, Shao et al. employed the Burr type XII distribution for flood frequency analysis [2].

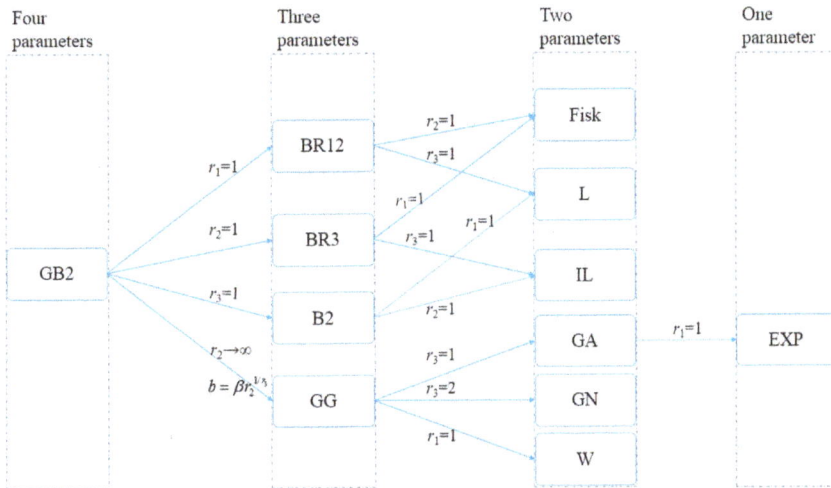

Figure 2. The GB2 distribution and its special cases (where BR12 means the Burr XII distribution; BR3 means the Burr III distribution; B2 means the beta distribution of second kind; Fisk means log-logistic distribution; L means the Lomax distribution; IL means inverse Lomax distribution; GA distribution means the gamma distribution; GN means the generalized normal distribution; W means the Weibull distribution and EXP means the exponential distribution).

3. Estimation of Parameters of GB2 Distribution by POME Method

The GB2 distribution parameters were determined using the principle of maximum entropy (POME). The POME method involves the following steps: (1) specification of constraints; (2) maximization of entropy using the method of Lagrange multipliers; (3) derivation of the relation between Lagrange multipliers and constraints; (4) derivation of the relation between Lagrange multipliers and distribution parameters; and (5) derivation of the relation between distribution parameters and constraints. These steps are discussed in Appendix A. Here only steps (1) and (5) are outlined.

Flood discharge is considered as a random variable X, which ranges from 0 to infinite. Its probability distribution function (PDF) and cumulative distribution function (CDF) are denoted as $f(x)$ and $F(x)$ respectively, where x is a specific value of X. Since constraints encode the information that can be given for the random variable, following Singh (1998), the constraints for the GB2 distribution can be expressed as:

$$\int_0^\infty f(x)dx = 1 \tag{2a}$$

$$\int_0^\infty f(x)\ln x\,dx = E(\ln x) \tag{2b}$$

$$\int_0^\infty f(x)\ln(1+(\tfrac{x}{\beta})^{r_3})dx = E(\ln(1+(\tfrac{x}{\beta})^{r_3})) \tag{2c}$$

The first constraint is the total probability law, the second constraint is the mean of log values or the geometric mean, and the third constraint is the mean of log of scaled values raised to a power and then shifted by unity.

Following the derivation in Appendix A, the relation between parameters and constraints can be expressed as:

$$
\begin{aligned}
-\ln\beta - \tfrac{1}{r_3}\varphi(r_1) + \tfrac{1}{r_3}\varphi(r_2) &= -E(\ln x) \\
\beta^{r_3}\varphi(r_2) - \beta^{r_3}\varphi(r_1+r_2) &= -E(\beta^{r_3}\ln(1+(\tfrac{x}{\beta})^{r_3})) \\
-\ln\beta + \tfrac{1}{r_3^2}\varphi'(r_1) + \tfrac{1}{r_3^2}\varphi'(r_2) &= \mathrm{var}(\ln x) \\
\varphi'(r_2) - \varphi'(r_1+r_2) &= \mathrm{var}(\ln(1+(\tfrac{x}{\beta})^{r_3}))
\end{aligned}
\tag{3}
$$

where $\phi(.)$ is the digamma function; and $\phi'(.)$ is the trigamma function. Detailed information for deriving these relationships can be found in Appendix A.

4. Flood Frequency Analysis

For FFA, three problems were addressed. First, the GB2 distribution was tested using observed flood data, and was compared with commonly used distributions in hydrology. Second, a method for selecting the best distribution was discussed. Third, flood frequency analysis was carried out at several gauging stations from upstream to downstream, and the evolution of frequency distribution along the stream was investigated.

4.1. Flood Data

Flood data from eight gauging stations on the Colorado River and its tributaries, as shown in Figure 3, were considered to test the performance of the GB2 distribution and discuss the evolution of frequency distribution along the river. The Colorado River is the principal river of the Southwestern United States and northwest Mexico. It rises in the central Rocky Mountains, flows generally southwest across the Colorado Plateau and through the Grand Canyon. The basin boundary consists of mountains that are 13,000 to 14,000 feet (3962.4 m to 4267.2 m) high in Wyoming, Colorado, and Utah; and the boundary drops to elevations of less than 1000 feet (304.8 m) at Hoover Dam. The northern part of the river basin in Colorado and Wyoming is a mountainous plateau that ranges from 5000 to 8000 feet (1524 m to 2438 m) in elevation, which encompasses deep canyons, rolling valleys, and intersecting mountain ranges. The central and southern portions of the basin in eastern Utah, northwestern New Mexico, and northern Arizona consist of rugged mountain ranges interspersed with rolling plateaus and broad valleys. In general, the mountains in the southern part of the basin are much lower than those in the northern part. Of the eight gauging stations considered in this study, gauging stations or sites 1, 2 and 3 are on the Yampa River which is a secondary tributary of the Colorado River. Sites 4, 5, 6, 7 and 8 are on the mainstream of the Colorado River. Site 8 is near the location of the Hoover Dam. The data of these gauging stations is directly downloaded from USGS (United States Geological Survey) website. The characteristics of flow data of these gauging stations, including length of the data, mean, standard deviation, skewness, and kurtosis, were calculated, as shown in Table 1. Since there is a dam, named Glenn Canyon, regulating the river flow past Lees Ferry (shown in Figure 3), the characteristics of the flow at the Hoover dam (site 8) are quite different from those at sites 4, 5, 6 and 7 upstream. It can be seen from Table 1 that for sites 1 to 7 the mean values increase from upstream to downstream, as more rainfall or water flows into the river. Since the standard deviation is related to the flood magnitude, it also increases with the mean value. For site 8, considering the impact of reservoir operation, some streamflow was stored in the reservoir, which leads that the streamflow at site 8 is reduced. The skewness is positive for all gauging stations, indicating that the right tail is longer or fatter than the left side and the mass of distribution is concentrated on the left side. Kurtosis is a measure of the peakedness of the probability distribution. The skewness and kurtosis values in the mainstream are generally lower than those in the tributaries.

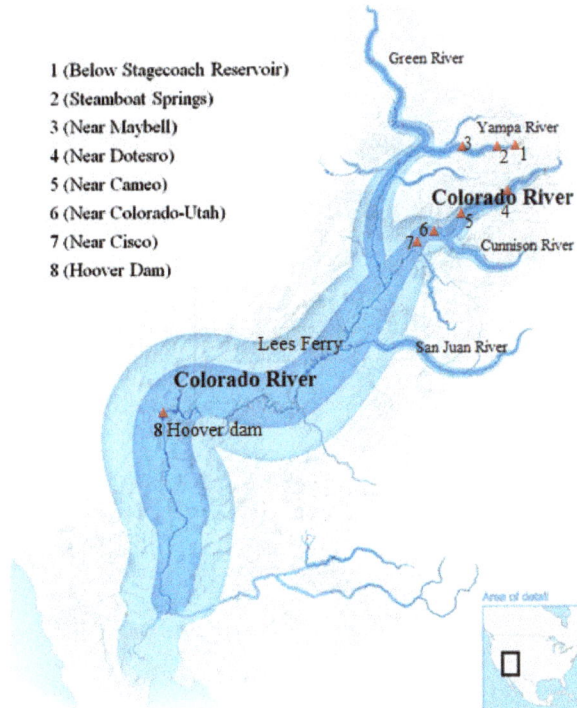

Figure 3. Locations of gauging stations on the Colorado River.

Table 1. Characteristics of the gauging stations used in the study.

River	No.	Gaging Station	Drainage Area (Square Miles)	Length of Data	Mean Value (ft³/s)	Standard Deviation	Skewness	Kurtosis
Yampa River	1	Below Stagecoach Reservoir	228	1957–2014	315	189	0.91	3.49
	2	Steamboat Springs	567	1904–2013	3630	1115	0.26	2.94
	3	Near Maybell	3383	1904–2013	10,419	3657	0.90	4.88
Colorado River	4	Near Dotsero	4390	1941–2013	9870	4450	0.39	2.59
	5	Near Cameo	7986	1934–2013	19,049	7687	0.26	2.68
	6	Near Colorado-Utah	17,847	1951–2013	26,714	13,936	0.84	3.53
	7	Near Cisco	24,100	1884–2013	34,329	16,520	0.36	2.31
	8	Hoover Dam	171,700	1934–2013	26,131	6831	1.37	5.83

4.2. Performance Measures

For evaluating the performance of the GB2 distribution, two measures were employed: (1) the root mean square deviation (RMSD); and (2) the Akaike information criterion (AIC). These methods assess the fitted distribution at a site by summarizing the deviations between observed discharges and computed discharges.

A frequently used method for assessing the goodness-of-fit of a function is the RMSD [12]. This method was used by NERC (1975) for ranking candidate distributions [13]. RMSD can be expressed as:

$$RMSD = \sqrt{\frac{1}{n}\sum_{i=1}^{n}\left(\frac{Q_{the}(i) - Q_{emp}(i)}{Q_{emp}(i)}\right)^2} \tag{4}$$

where n is the sample size; Q_{the} is the computed discharge at the ith plotting position. Q_{emp} denotes the observed i^{th} smallest discharge. The value of RMSD is from 0 to 1. The samller is, the better the distribution fits.

AIC is a measure of the relative quality of statistical models for a given set of data. It also includes a penalty that is an increasing function of the number of estimated parameters. The AIC value was calculated as [14]:

$$AIC = n(\ln(MSE)) + 2K \tag{5}$$

where K is the number of parameters of the distribution, and MSE was calculated by

$$MSE = \frac{1}{n}\sum_{i=1}^{n}\left(Q_{the}(i) - Q_{emp}(i)\right)^2 \tag{6}$$

Given a set of candidate models for the data, the preferred model is the one with the minimum AIC value.

4.3. Evaluation of GB2 Distribution

Annual maximum flood peak data from four gauging stations, namely sites 2, 6, 7 and 8 in Figure 3, were selected. The empirical frequencies were calculated first. The purpose of defining the empirical distribution is to compare it with selected theoretical distributions in order to verify whether they fit sample data.

Many plotting positions are proposed, most of which can be expressed in general form:

$$P_i = \frac{i - a}{n + 1 - 2a} \tag{7}$$

where a is a constant having values from 0 to 0.5 in different formula, 0.5 for Hazen's formula, 0.3 for Chegadayev's formula, zero for Weibull's formula, 3/8 for Blom's formula, 1/3 for Tukey's formula, and 0.44 for Gringorten's formula.

Among these formulars, Gringorten's formular is recoganized by lots of researchers, especially for GEV, gumbel, exponential, Generalized pareto distributions which have been widely used for flood frequency analysis [15–20]. The Gringorten formula is also used for GB2 distribution. For normal, generalized normal and Gamma distributions, the Blom's formula is recommended [21,22]. For Pearson type 3 and log Pearson type 3 distributions, Weibull's formula is recommended [18,21]. The GB2 distribution was employed to fit the annual maximum (AM) series of the four sites. The distribution parameters were estimated using Equation (3) and given in Table 2. The fitted GB2 distributions and empirical frequency of each AM series are shown in Figure 4. In the left of Figure 4, the line represents the fitted distribution and circle the empirical frequencies of observations. Results show that the marginal distributions fit the empirical data well. Histograms of AM flood peak series fitted by the GB2 distribution for the gauging stations on the Colorado River are shown in the right section of Figure 4. It also indicates that the GB2 distribution can successfully be fitted to empirical histograms.

Several distributions, including normal, exponential, gamma, Gumbel, generalized normal, pearson type III, log Pearson type III, generalized Pareto, and generalized extreme-value that are commonly used in hydrology, were fitted to the AM series at this site. The L-moment method was used to estimate the parameters of these distributions.

Table 2. Parameters of the GB2 distribution for the gauging stations along the Colorado River.

Number	Location	r_1	r_2	r_3	β
4	Near Dotsero	1.58	60.30	1.75	85.11
5	Near Cameo	1.12	77.57	2.53	112.93
6	Near Colorado-Utah	3.94	83.08	0.94	69.05
7	Near Cisco	2.73	76.82	1.07	80.90
8	Hoover Dam	10.59	434.72	1.31	43.62

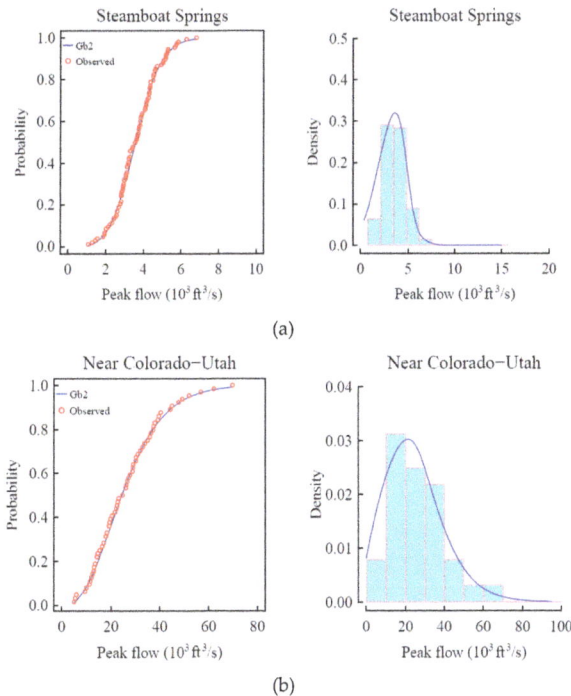

(a)

(b)

Figure 4. *Cont.*

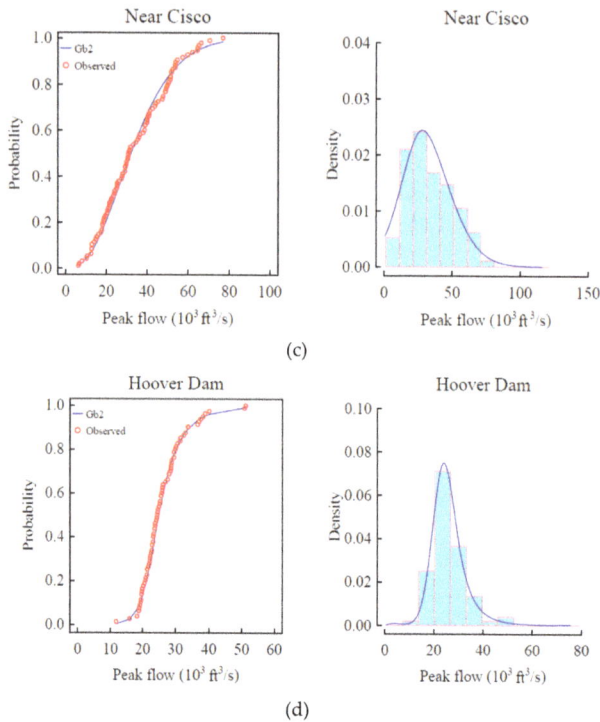

Figure 4. Marginal distributions and histograms of AM flood peak series fitted by the GB2 distribution for the gauging stations on the Colorado River. (**a**) Steamboat springs; (**b**) Near Colorado-Utah; (**c**) Near Cisco; (**d**) Hoover Dam.

Singh and Guo compared the POME method with the L-moment method, and indicated that the two methods are comparable [11,23,24]. Therefore no matter what method is used, it has little influence on the value of the T-year design discharge. The Kolmogorov-Smirnov test was used here to compare a sample with a reference probability distribution. The *p*-value was calculated and given in Table 3 as well. The higher or more close to 1 the *p*-value is the more similar the theoretical and empirical distributions are. It is indicated from Table 3 that the *p*-value of GB2 distribution is 1 or close to 1, which demonstrates that the GB2 distribution fit the data better. Table 3 also listed the RMSD and AIC values computed for the fitted GB2 distribution using Equations (4)–(7). The smaller the RMSD and AIC values are, the better the distribution fits. For the site streamboat springs, the GB2 and generalized normal distributions have the smallest RMSD values, which is equal to 0.025. For the site Near Cisco, the GB2 has the smallest RMSE values, which is equal to 0.061. For the site Near Colorado-Utah, the GB2 and gamma distributions have the smallest RMSE value. For the site Hoover dam, the GB2 distribution has the smallest RMSE value. Since the GB2 distribution have more parameters, the AIC values of GB2 distribution are larger than those of generalized normal, Gamma and GEV distributions. Thus, generally GB2 distribution gives a getter fit.

Table 3. RMSE and AIC values of different distributions.

Number	Distribution	Steamboat Springs			Near Cisco			Near Colorada-Utah			Hoover Dam		
		p-Value	RMSE	AIC	p-Value	RMSE	AIC	p-Value	RMSE	AIC	p-Value	RMSE	AIC
1	GB2	0.976	**0.025**	**924.1**	0.991	**0.061**	**1384.1**	1	**0.047**	**852.7**	1	**0.036**	**1098.8**
2	Normal	0.926	0.043	992.9	0.787	0.194	1502	0.839	0.221	1031	0.436	0.081	1236.2
3	Exponential	0.409	0.122	1306.8	0.336	0.171	1669.6	0.839	0.145	1036.4	0.919	0.055	1152.6
4	Gamma	1	0.045	1005.1	0.959	0.064	1512.8	1	**0.047**	**842.5**	0.692	0.057	1192.9
5	Gumbel	0.976	0.066	1143.5	0.959	0.088	1546.2	1	0.107	869.1	0.978	0.039	1122.3
6	Generalized normal	0.844	**0.025**	**922.9**	0.991	0.137	1455.5	1	0.083	852.8	0.978	0.039	1106.7
7	Pearson type III	0.976	0.035	953.2	0.991	0.1	1425	1	0.054	895.6	1	0.058	1146.8
8	Log Pearson type III	0.976	0.034	951.3	0.991	0.106	1431.5	1	0.054	893.1	1	0.052	1133
9	Generalized Pareto	0.976	0.078	1158	0.991	0.062	1386.1	1	0.09	960.2	1	0.054	1169
10	GEV	0.976	0.027	929.6	0.991	0.138	1450.1	1	0.128	865.5	1	0.036	1096.8

In order to compare the POME with the current used method, the maximum likelihood (ML) method was also employed for the parameter estimation of GB2 distribution. Taking the site Near Colorada-Utah for an example, the estimated parameters by POME and ML method are given in Table 4. The p-value, RMSE and AIC values are also given in Table 4. It is indicated that the parameters obtained by the two method are more or less the same. And the RMSE and AIC values based on the POME method are smaller.

Table 4. Parameters estimated by POME and ML methods for site Near Colorada-Utah.

Methods	r_1	r_2	r_3	β	p-Value	RMSE	AIC
POME	2.14	24.78	1.40	157.18	1	0.0169	−357.76
ML	2.26	30.85	1.35	158.55	1	0.0170	−357.80

4.4. Flood Frequency Analysis

The Hoover dam is a multi-purpose dam, serving the needs of flood control, irrigation, water supply, and hydropower generation. Therefore, it was desired to determine the most appropriate distribution for FFA at the dam site. The T-year design flood at Hoover dam was calculated using each distribution, as given in Table 5, and it can be seen that different distributions yielded significantly different values. For example, the 1000-year design flood values calculated by the GB2 and gamma distributions were 76,702 and 50,485 ft^3/s, respectively. The RMSD and AIC values for GB2 distribution (Gamma distribution) were 0.036 (0.057) and 1098.8 (1192.9), respectively, which indicates that the performance of GB2 distribution is much better than that of the gamma distribution. It concludes that if the gamma distribution were used, the design flood would be underestimated and potential flood risk would be higher.

Table 5. Comparison of T-year design flood discharges (10^3 ft^3/s) calculated by different distributions for the Hoover dam site.

Number	Return Period	1000	500	100	50	10
1	GB2	76.702	67.914	51.198	34.138	30.125
2	Normal	45.800	44.451	40.938	34.288	31.488
3	Exponential	68.561	63.583	52.024	35.486	30.508
4	Gamma	50.485	48.424	43.314	34.613	31.320
5	Gumbel	58.926	55.332	46.973	34.799	30.912
6	Generalized normal	50.513	49.271	45.325	35.732	31.485
7	Pearson type III	60.025	56.451	47.985	35.145	30.926
8	Log Pearson type III	69.568	64.494	52.713	35.639	31.858
9	Generalized Pareto	64.809	59.870	49.084	34.893	30.695
10	GEV	57.809	54.766	47.324	35.270	31.072

4.5. Change in Flood Frequency Distribution with Change in Drainage Area

The GB2 distribution was applied for FFA along the main stem of the Colorado River. Four gauging stations (sites 4, 5, 6 and 7) from upstream to downstream were used, as shown in Figure 3 and Table 6. These gauging stations were selected, because all these stations are on the mainstream and no dam has been built on this reach. The drainage area and statistical characteristics (including mean, skewness and kurtosis of the annual maximum data) of these stations were calculated, as given in Table 1. The T-year design flood of these gauging stations was calculated, as shown in Figure 5, in which the x-axis represents the return periods and the y-axis represents the design flood values. Figure 5 shows that for a given return period, the design flood value of the downstream gauging stations is larger than that of the upstream gauging stations. The increasing rates of drainage area and T-year design flood values between the adjacent gauging stations were computed, as given in Table 6, which indicates that the percentage increase of the drainage area was nearly the same as that of the design flood values. For instance, with the increase of drainage area up to 45% from the gauging station near Dotsero to that near Cameo, the flood value increased by 43% on average. It is also seen that from upstream to downstream, when the drainage area increased by 45%, 55% and 26%, the flood value increased by 43%, 42%, and 16%, respectively. It seems that in a mountainous watershed, the upstream the reach is, the greater the impact the drainage area has on flood. This may be because that the runoff coefficient is generally larger in the steep area.

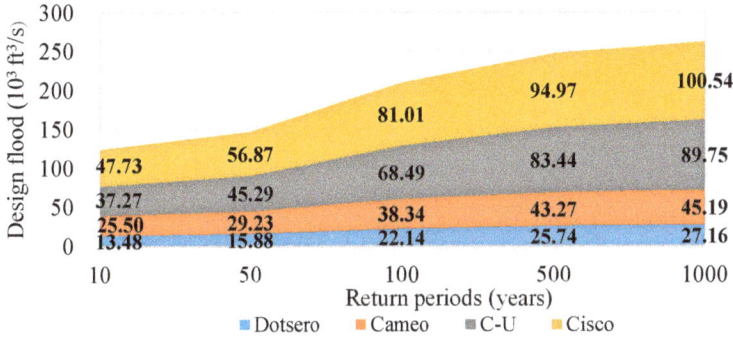

Figure 5. Flood values along the mainstream of the upper Colorado River.

Table 6. Statistical characteristics of the four gauging stations, the increasing rate of drainage area and flood discharge between adjacent gauging stations.

Number	Locations	Drainage Area (Square Miles)	Increase in Drainage Area (%)	Increase in in Flood Value (%)					
				1000	500	100	50	10	Mean
4	Near Dotsero	11370	45	40	41	42	47	46	43
5	Near Cameo	20683	55	50	48	44	32	35	42
6	Near Colorado-Utah	46228	26	11	12	15	22	20	16
7	Near Cisco	62419							

4.6. Evolution of Frequency Distribution along Stream

In order to determine the evolution of frequency distribution and its parameters along the river, data from the Yampa River were applied, because this river is taken as one of the west's last wild rivers and has only a few small dams and diversions. The Yampa River with a length of 402 km, located in northwestern Colorado, is a tributary of Green River and a secondary tributary of the Colorado River. Data from three gauging stations along this river, designated as sites 1, 2 and 3 in Figure 6, were used. The GB2 distribution was used to fit the AM series of each of the three gauging stations, as shown in Table 7. It can be seen that shape parameters r_1 and r_2 decreased along the river. The value of r_1 became close to be 1. When r_1 equals 1, the GB2 distribution becomes the Burr XII distribution [25]. This distribution has been shown to reasonably fit the income distribution data [20,26,27] and has recently been used in hydrology [2,28]. The PDF of Burr XII distribution can be written as:

$$f(x) = \frac{r_3}{bB(1, r_2)} \left(\frac{x}{b}\right)^{1 \times r_3 - 1} \left(1 + \left(\frac{x}{b}\right)^{r_3}\right)^{-(r_2+1)} = \frac{r_3 r_2}{b} \left(\frac{x}{b}\right)^{r_3 - 1} \left(1 + \left(\frac{x}{b}\right)^{r_3}\right)^{-(r_2+1)} \tag{8}$$

where b is the scale parameter. The Burr XII distribution was also used to fit the data at the gauging station near Maybell of Yampa River. The estimated parameters of Burr XII distribution were: $r_2 = 1.94$, $r_3 = 4.19$, and $b = 12.33$. The fitting results of the GB2 and Burr distributions for the gauging station near Maybell are shown in Figure 7. For the gauging station near Maybell, parameters of the GB2 distribution estimated by POME method are nearly as the same as the parameters of the Burr XII distribution estimated by MLE method. Thus, Burr XII distribution instead of GB2 distribution can be used for FFA at that station. In other words, the distribution for FFA changes from the four-parameter GB2 distribution to the three-parameter Burr XII distribution along the Yampa River. There is an evolution of distribution along this river. From Equation (1), the value of scale parameter β increases with the mean value, because more water flows into the stream. Parameters r_1 and r_2 govern the left and right tails, respectively. The smaller the value of r_1, the fatter the left tail is; and the smaller the value of r_2, the fatter the right tail is. It can be seen from Table 7 that both r_1 and r_2 decrease along the stream, which demonstrates that both the left and right tails become fatter, and the PDF values become larger in these areas and lower in the central area.

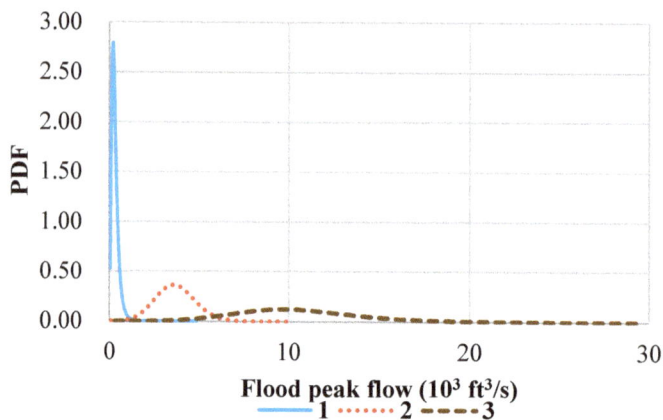

Figure 6. Evaluations of PDF of sites along the Yampa River.

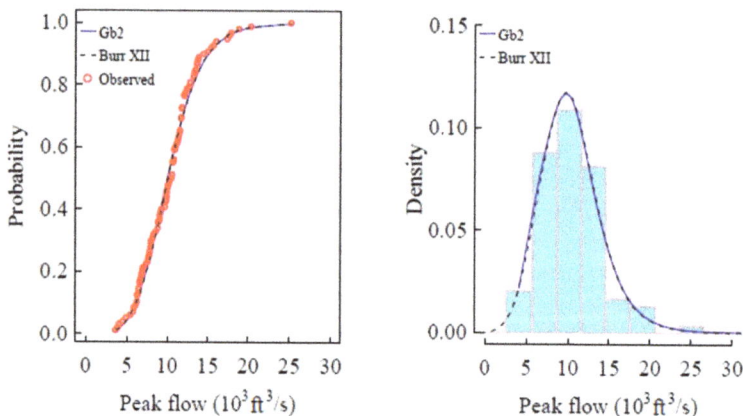

Figure 7. Marginal distribution and histograms of AM flood peak series fitted by the GB2 and Burr XII distributions for the gauging station near Maybell on the Yampa River.

Table 7. Parameters of the GB2 distribution for four gauging stations along the Yampa River.

Number	Location	r_1	r_2	r_3	β
1	Below stagecoach Reservoir	17.44	15.25	0.55	2.10
2	Steamboat springs	1.20	5.49	3.59	5.81
3	Near Maybell	1.14	2.07	3.92	12.11

5. Conclusions

The GB2 provides sufficient flexibility to fit a large variety of data sets. Papalexiou and Koutsoyiannis introduced this distribution in hydrology and used it for rainfall frequency analysis [8]. In this study, the generalized beta distribution of the second kind (GB2) is introduced for FFA for the first time. The POME method was proposed to estimate the parameters of GB2 distribution. Equations of POME method was deduced by ourselves and given in Appendix A. The Colorado River basin was selected as a case study to test the performance of GB2 distribution. Frequency estimates from the GB2 distribution were also compared with those of commonly used distributions in hydrology. In addition, some characteristics of FFA in mountainous areas are discussed. The conclusions can be summarized as follows:

(1) Results demonstrate that the GB2 is appealing for FFA, since it has four parameters which allows the distribution to be able to fit data having very different histogram shapes, such as the J-shaped and bell-shaped distributions. And by fixing certain parameters, the GB2 distribution can yield some well-known distributions, such as the beta distribution of the second kind (B2), the Burr type XII, generalized gamma (GG), and so on.

(2) The parameters estimated by POME method are found reasonable. Both the marginal distributions and histograms indicates that the GB2 distribution can successfully be fitted to empirical values using the POME method.

(3) The performance of the GB2 distribution is better than that of the widely used distributions in hydrology. For the site streamboat springs, the GB2 and generalized normal distributions have the smallest RMSD values. For the site Near Cisco, the GB2 has the smallest RMSE values. For the site Near Colorado-Utah, the GB2 and gamma distributions have the smallest RMSE value. For the site Hoover dam, the GB2 distribution has the smallest RMSE value. Since the GB2 distribution have more parameters, the AIC values of GB2 distribution are larger than those of generalized normal, Gamma and GEV distributions. Thus, generally GB2 distribution gives a getter fit.

(4) When using different distributions for FFA, significant different design flood values are obtained. It concludes that if the wrong distribution were used, the design flood would be underestimated and potential flood risk would be higher.

(5) The design flood value increase with the drainage area. For a given return period, the design flood value of the downstream gauging stations is larger than that of the upstream gauging stations. In this study, the percentage increase of the drainage area was nearly the same as that of the design flood values. It seems that in a mountainous watershed, the upstream the reach is, the greater the impact the drainage area has on flood. This may be because that the runoff coefficient is generally larger in the steep area.

(6) There is an evolution of distribution along this river. Along the Yampa River, the distribution for FFA changes from the four-parameter GB2 distribution to the three-parameter Burr XII distribution. And both r_1 and r_2 decrease along the stream, which demonstrates that both the left and right tails become fatter, and the PDF values become larger in these areas and lower in the central area, which means that when the drainage area become larger, the flood magnitudes has a more significant variation.

Acknowledgments: The project was financially supported by the National Natural Science Foundation of China (51679094, 51509273, 91547208 and 41401018), Fundamental Research Funds for the Central Universities (2017KFYXJJ194, 2016YXZD048).

Author Contributions: Vijay P. Singh conceived and designed the experiments; Lu Chen performed the experiments and analyzed the data; Lu Chen wrote the draft of the paper and Vijay P. Singh revised it. All authors have read and approved the final manuscript.

Conflicts of Interest: The authors declare no conflict of interest.

Appendix A. Estimation of Parameters of GB2 Distribution

The GB2 distribution parameters can be estimated by maximizing the Shannon entropy $H(X)$ which, for a random variable X, can be expressed as:

$$H(X) = -\int_0^\infty f(x) \log f(x) dx \tag{A1}$$

where $f(x)$ is the probability density function (PDF). The principle of maximum entropy (POME) indicates that the most appropriate PDF is the one that maximizes the value of entropy, given available data and a set of known constraints [29].

Specification of Constraints: Following Singh, the constraints for the GB2 distribution can be expressed as

$$\int_0^\infty f(x) dx = 1 \tag{A2a}$$

$$\int_0^\infty f(x) \ln x dx = E(\ln x) \tag{A2b}$$

$$\int_0^\infty f(x) \ln(1 + (\tfrac{x}{\beta})^{r_3}) dx = E(\ln(1 + (\tfrac{x}{\beta})^{r_3})) \tag{A2c}$$

Method of Lagrange Multipliers for Maximizing Entropy: In the search for an appropriate probability distribution for a given random variable, entropy should be maximized. In other words, the best fitted distribution is the one with the highest entropy. The method of Lagrange multipliers was used to obtain the appropriate probability distribution with the maximum entropy. Finally, the form of this distribution is given as:

$$f(x) = \exp(-\lambda_0 - \lambda_1 \ln(x) - \lambda_2' \ln(1 + (\tfrac{x}{\beta})^{r_3})) \tag{A3a}$$

in which λ_0, λ_1, and λ_2' are the Lagrange multipliers. Let $p = \beta^{-r_3}$. Then, Equation (A3a) can be written as

$$f(x) = \exp(-\lambda_0 - \lambda_1 \ln(x) - \lambda_2' \ln(1 + px^{r_3})) \tag{A3b}$$

Let $\lambda_2' = \tfrac{\lambda_2}{p}$ and $q = r_3$. Papalexiou and Koutsoyiannis defined the entropy-based PDF as:

$$f(x) = \exp(-\lambda_0 - \lambda_1 \ln(x) - \frac{\lambda_2}{p} \ln(1 + px^q)) \tag{A4}$$

Substitution of Equation (A4) in Equation (A2a) yields:

$$\int_0^\infty f(x) dx = \int_0^\infty \exp(-\lambda_0 - \lambda_1 \ln(x) - \frac{\lambda_2}{p} \ln(1 + px^q)) dx = 1 \tag{A5}$$

From Equation (A5):

$$
\begin{aligned}
\exp(\lambda_0) &= \int_0^\infty \exp(-\lambda_1 \ln x - \lambda_2 \ln(1 + px^q)/p) dx \\
&= \int_0^\infty \exp(-\lambda_1 \ln x) \exp(-\tfrac{\lambda_2}{p} \ln(1 + px^q)) dx \\
&= \int_0^\infty x^{(-\lambda_1)} (1 + px^q)^{(-\frac{\lambda_2}{p})} dx
\end{aligned}
\tag{A6}
$$

Let $t = px^q$. Then $x = (\frac{t}{p})^{\frac{1}{q}}$, and $dx = \frac{1}{pq}(\frac{t}{p})^{\frac{1}{q}-1}dt$. Thus, Equation (A6) can be expressed as:

$$
\begin{aligned}
\exp(\lambda_0) &= \int_0^\infty x^{(-\lambda_1)}(1 + px^q)^{(-\frac{\lambda_2}{p})}dx \\
&= \int_0^\infty (\tfrac{t}{p})^{\frac{-\lambda_1}{q}}(1+t)^{\frac{-\lambda_2}{p}}\tfrac{1}{pq}(\tfrac{t}{p})^{\frac{1}{q}-1}dt \\
&= \int_0^\infty \tfrac{1}{q}p^{\frac{\lambda_1-1}{q}}t^{\frac{-\lambda_1}{q}}(1+t)^{\frac{-\lambda_2}{p}}t^{\frac{1}{q}-1}dt
\end{aligned}
\tag{A7}
$$

Let $y = \frac{t}{1+t}$. Then $t = \frac{y}{1-y}$, and $dt = \frac{1}{(1-y)^2}dy$.
Since $y(0) = 0$ and $y(\infty) = 1$, $y \in [0,1]$.

$$
\begin{aligned}
\exp(\lambda_0) &= \int_0^1 \tfrac{1}{q}p^{\frac{\lambda_1-1}{q}}(\tfrac{y}{1-y})^{\frac{-\lambda_1}{q}}(1+\tfrac{y}{1-y})^{\frac{-\lambda_2}{p}}(\tfrac{y}{1-y})^{\frac{1}{q}-1}\tfrac{1}{(1-y)^2}dy \\
&= \int_0^1 \tfrac{1}{q}p^{\frac{\lambda_1-1}{q}}(\tfrac{y}{1-y})^{\frac{-\lambda_1+1}{q}-1}(1+\tfrac{y}{1-y})^{\frac{-\lambda_2}{p}}\tfrac{1}{(1-y)^2}dy \\
&= \int_0^1 \tfrac{1}{q}p^{\frac{\lambda_1-1}{q}}(\tfrac{y}{1-y})^{\frac{-\lambda_1+1}{q}-1}(\tfrac{1}{1-y})^{\frac{-\lambda_2}{p}+2}dy \\
&= \int_0^1 \tfrac{1}{q}p^{\frac{\lambda_1-1}{q}}(y)^{\frac{1-\lambda_1}{q}-1}(1-y)^{-\frac{1-\lambda_1}{q}+\frac{\lambda_2}{p}-1}dy \\
&= \tfrac{1}{q}p^{\frac{\lambda_1-1}{q}}B(\tfrac{1-\lambda_1}{q}, -\tfrac{1-\lambda_1}{q}+\tfrac{\lambda_2}{p})
\end{aligned}
\tag{A8}
$$

The Lagrange multiplier λ_0 can be calculated from Equation (A8) as:

$$
\lambda_0 = -\ln q + \frac{\lambda_1-1}{q}\ln(p) + \ln\Gamma(\frac{1-\lambda_1}{q}) + \ln\Gamma(-\frac{1-\lambda_1}{q}+\frac{\lambda_2}{p}) - \ln\Gamma(\frac{\lambda_2}{p})
\tag{A9}
$$

From Equation (A4), the other equation for calculating λ_0 can be defined as:

$$
\lambda_0 = \ln(\int_0^\infty \exp(-\lambda_1 \ln x - \frac{\lambda_2}{p}\ln(1 + px'^3))dx)
\tag{A10}
$$

Relation between Lagrange multipliers and constraints: Defining $a' = \frac{1-\lambda_1}{q}$ and $b' = -\frac{1-\lambda_1}{q}+\frac{\lambda_2}{p}$, differentiate Equation (A9) with respect to λ_1 and λ_2:

$$
\begin{aligned}
\frac{\partial\lambda_0}{\partial\lambda_1} &= \frac{\ln p}{q} + \frac{\partial\ln\Gamma(a')}{\partial a'}\frac{\partial a'}{\partial\lambda_1} + \frac{\partial\ln\Gamma(b')}{\partial(b')}\frac{\partial b'}{\partial\lambda_1} - \frac{\partial\ln\Gamma(a+b')}{\partial(a+b')}\frac{\partial(a+b')}{\partial\lambda_1} \\
&= \frac{\ln p}{q} - \tfrac{1}{q}\varphi(a') + \tfrac{1}{q}\varphi(b')
\end{aligned}
\tag{A11a}
$$

$$
\begin{aligned}
\frac{\partial\lambda_0}{\partial\lambda_2} &= \frac{\partial\ln\Gamma(b')}{\partial(b')}\frac{\partial b'}{\partial\lambda_2} - \frac{\partial\ln\Gamma(a'+b')}{\partial(a'+b')}\frac{\partial(a'+b')}{\partial\lambda_2} \\
&= \tfrac{1}{p}\varphi(b') - \tfrac{1}{p}\varphi(a'+b')
\end{aligned}
\tag{A11b}
$$

where $\varphi(.)$ is a digamma function. Differentiate Equation (A10) with respect to λ_1 and λ_2:

$$
\frac{\partial\lambda_0}{\partial\lambda_1} = \frac{\int_0^\infty \ln x \exp(-\lambda_1 \ln x - \frac{\lambda_2}{p}\ln(1+px^q))dx}{\int_0^\infty \exp(-\lambda_1 \ln x - \frac{\lambda_2}{p}\ln(1+px^q))dx} = -E(\ln x)
\tag{A12a}
$$

$$\frac{\partial \lambda_0}{\partial \lambda_2} = \frac{\int\limits_0^\infty x^q \exp(-\lambda_1 \ln x - \frac{\lambda_2}{p} \ln(1 + px^q))dx}{\int\limits_0^\infty \exp(-\lambda_1 \ln x - \frac{\lambda_2}{p} \ln(1 + px^q))dx} = -E(\frac{\ln(1 + px^q)}{p}) \tag{A12b}$$

Based on Equations (A11) and (A12), the relation between Lagrange multipliers and constraints can be expressed as:

$$\frac{\ln p}{q} - \frac{1}{q}\varphi(a) + \frac{1}{q}\varphi(b) = -E(\ln x) \tag{A13a}$$

$$\frac{1}{p}\varphi(b) - \frac{1}{p}\varphi(a+b) = -E(\frac{\ln(1 + px^q)}{p}) \tag{A13b}$$

Since there are four parameters, Equations (A13a) and (A13b) are not sufficient for calculating parameters, and two additional equations are needed that are given as:

$$\frac{\partial^2 \lambda_0}{\partial^2 \lambda_1} = \frac{1}{q^2}\varphi'(a') + \frac{1}{q^2}\varphi'(b') = var(\ln x) \tag{A14a}$$

$$\frac{\partial^2 \lambda_0}{\partial^2 \lambda_2} = \varphi'(r_2) - \varphi'(r_1 + r_2) = var(\ln(1 + (\frac{x}{\beta})^q)) \tag{A14b}$$

Relation between Lagrange multipliers and parameters: Substituting Equation (A8) in Equation (A4), it is known that:

$$f(x) = \frac{1}{\frac{1}{q}p^{\frac{\lambda_1 - 1}{q}} B(\frac{1-\lambda_1}{q}, -\frac{1-\lambda_1}{q} + \frac{\lambda_2}{p})} x^{-\lambda_1}(1 + px^q)^{-\frac{\lambda_2}{p}} \tag{A15}$$

Equation (A15) is the GB2 distribution. Comparing Equation (1) with Equation (A15), the following equations can be obtained:

$$\begin{aligned} \lambda_1 &= 1 - r_1 q \\ \lambda_2 &= p(r_2 + \frac{1-\lambda_1}{q}) \\ p &= (\frac{1}{\beta})^{r_3} \\ q &= r_3 \end{aligned} \tag{A16}$$

Relation between parameters and constraints: Based on the relation between parameters and constraints, and parameters and Lagrange multipliers, the relation between parameters and constraints can be expressed as:

$$\begin{aligned} -\ln \beta - \frac{1}{r_3}\varphi(r_1) + \frac{1}{r_3}\varphi(r_2) &= -E(\ln x) \\ \beta^{r_3}\varphi(r_2) - \beta^{r_3}\varphi(r_1 + r_2) &= -E(\beta^{r_3} \ln(1 + (\frac{x}{\beta})^{r_3})) \\ -\ln \beta + \frac{1}{r_3{}^2}\varphi'(r_1) + \frac{1}{r_3{}^2}\varphi'(r_2) &= var(\ln x) \\ \varphi'(r_2) - \varphi'(r_1 + r_2) &= var(\ln(1 + (\frac{x}{\beta})^{r_3})) \end{aligned} \tag{A17}$$

References

1. Beven, K.J.; Hornberger, G.M. Assessing the effect of spatial pattern of precipitation in modeling streamflow hydrographs. *J. Am. Water Resour. Assoc.* **1982**, *18*, 823–829. [CrossRef]
2. Shao, Q.; Wong, H.; Xia, J.; Ip, W. Models for extremes using the extended three-parameter Burr XII system with application to flood frequency analysis. *Hydrol. Sci. J.* **2004**, *49*, 685–702. [CrossRef]
3. Chen, L.; Guo, S.L.; Yan, B.W.; Liu, P.; Fang, B. A new seasonal design flood method based on bivariate joint distribution of flood magnitude and date of occurrence. *Hydrol. Sci. J.* **2010**, *55*, 1264–1280. [CrossRef]
4. Chen, L.; Singh, V.P.; Guo, S.; Hao, Z.; Li, T. Flood coincidence risk analysis using multivariate copula functions. *J. Hydrol. Eng.* **2012**, *17*, 742–755. [CrossRef]

5. Mielke, P.W., Jr.; Johnson, E.S. Some generalized beta distributions of the second kind having desirable application features in hydrology and meteorology. *Water Resour. Res.* **1974**, *10*, 223–226. [CrossRef]

6. Wilks, D.S. Comparison of three-parameter probability distributions for representing annual extreme and partial duration precipitation series. *Water Resour. Res.* **1993**, *29*, 3543–3549. [CrossRef]

7. Perreault, L.; Bobée, B.; Rasmussen, P. Halphen distribution system. I: Mathematical and statistical properties. *J. Hydrol. Eng.* **1999**, *4*, 189–199. [CrossRef]

8. Papalexiou, S.M.; Koutsoyiannis, D. Entropy based derivation of probability distributions: A case study to daily rainfall. *Adv. Water Resour.* **2012**, *45*, 51–57. [CrossRef]

9. Singh, V.P. *Entropy Based Parameter Estimation in Hydrology*; Kluwer Academic Publishers: Dordrecht, The Netherlands, 1998.

10. Singh, V.P. *Entropy-Based Parameter Estimation Hydrology*; Springer: Dordrecht, The Netherlands, 1998.

11. Singh, V.P.; Guo, H. Parameter estimation for 3-parameter generalized Pareto distribution by the principle of maximum entropy (POME). *Hydrol. Sci. J.* **1995**, *40*, 165–181. [CrossRef]

12. Karim, A.; Chowdhury, J.U. A comparison of four distributions used in flood frequency analysis in Bangladesh. *Hydrol. Sci. J.* **1995**, *40*, 55–66. [CrossRef]

13. Natural Environment Research Council. *Flood Studies Report*; Natural Environment Research Council: London, UK, 1975; Volumes 1–5.

14. Zhang, L.; Singh, V.P. Bivariate flood frequency analysis using the copula method. *J. Hydrol. Eng.* **2006**, *11*, 150–164. [CrossRef]

15. Ross, R. Graphical method for plotting and evaluating weibull distribution data. In Proceedings of the 4th International Conference on Properties and Application of Dielectric Materials, Brisbane, Austrialia, 3–8 July 1994; pp. 250–253.

16. Cunnane, C. Unbiased plotting positions—A review. *J. Hydrol.* **1978**, *37*, 205–222. [CrossRef]

17. Makkonen, L. Notes and correspondence plotting positions in extreme value analysis. *J. Appl. Meteorol. Clim.* **2006**, *45*, 334–340. [CrossRef]

18. Shabri, A. A Comparison of plotting formulas for the pearson type III distribution. *J. Technol.* **2002**, *36*, 61–74. [CrossRef]

19. Gringorten, I.I. A plotting rule for extreme probability paper. *J. Geophys. Res.* **1963**, *68*, 813–814. [CrossRef]

20. Dagum, C. A New Model of Personal Income Distribution: Specification and Estimation. In *Modeling Income Distributions and Lorenz Curves*; Springer: New York, NY, USA, 2008; pp. 3–25.

21. Mehdi, F.; Mehdi, J. Determination of plotting position formula for the normal, log-normal, pearson(III), log-pearson(III) and gumble distribution hypotheses using the probability plot correlation coefficient test. *World Appl. Sci. J.* **2011**, *15*, 1181–1185.

22. Kim, S.; Shin, H.; Kim, T.; Taesoon, K.; Heo, J. Derivation of the probability plot correlation coefficient test statistics for the generalized logistic distribution. In Proceedings of the International Workshop Advances in Statistical Hydrology, Taormina, Italy, 23–25 May 2010; pp. 1–8.

23. Singh, V.P.; Guo, H. Parameter estimation for 2-parameter log-logistic distribtuion distribution (LLD2) by maximum entropy. *Civ. Eng. Syst.* **1995**, *12*, 343–357. [CrossRef]

24. Singh, V.P.; Guo, H. Parameter estimations for 2-parameter Pareto distribution by pome. *Stoch. Hydrol. Hydraul.* **1980**, *9*, 81–93.

25. Burr, I.W. Cumulative Frequency Functions. *Ann. Math. Stat.* **1942**, *13*, 215–232. [CrossRef]

26. Kleiber, C.; Kotz, S. *Statistical Size Distributions in Economics and Actuarial Sciences*; John Wiley & Sons: Hoboken, NJ, USA, 2003.

27. Singh, S.K.; Maddala, G.S. A function for size distribution of incomes. *Econometrica* **1976**, *44*, 963–970. [CrossRef]

28. Hao, Z.; Singh, V.P. Entropy-based parameter estimation for extended Burr XII distribution. *Stoch. Environ. Res. Risk Assess.* **2008**, *23*, 1113–1122. [CrossRef]

29. Singh, V.P. Hydrologic synthesis using entropy theory: Review. *J. Hydrol. Eng.* **2011**, *16*, 421–433. [CrossRef]

![entropy logo] *entropy*

MDPI

Article

Bayesian Technique for the Selection of Probability Distributions for Frequency Analyses of Hydrometeorological Extremes

Lu Chen [1],*, Vijay P. Singh [2] and Kangdi Huang [1]

[1] College of Hydropower & Information Engineering, Huazhong University of Science & Technology, Wuhan 430074, China; hkd921110@163.com
[2] Department of Biological and Agricultural Engineering & Zachry Department of Civil Engineering, Texas A&M University, College Station, TX 77843-2117, USA; vsingh@tamu.edu
* Correspondence: chen_lu@hust.edu.cn; Tel.: +86-139-8605-1604

Received: 13 November 2017; Accepted: 16 January 2018; Published: 11 February 2018

Abstract: Frequency analysis of hydrometeorological extremes plays an important role in the design of hydraulic structures. A multitude of distributions have been employed for hydrological frequency analysis, and more than one distribution is often found to be adequate for frequency analysis. The current method for selecting the best fitted distributions are not so objective. Using different kinds of constraints, entropy theory was employed in this study to derive five generalized distributions for frequency analysis. These distributions are the generalized gamma (GG) distribution, generalized beta distribution of the second kind (GB2), Halphen type A distribution (Hal-A), Halphen type B distribution (Hal-B), and Halphen type inverse B (Hal-IB) distribution. The Bayesian technique was employed to objectively select the optimal distribution. The method of selection was tested using simulation as well as using extreme daily and hourly rainfall data from the Mississippi. The results showed that the Bayesian technique was able to select the best fitted distribution, thus providing a new way for model selection for frequency analysis of hydrometeorological extremes.

Keywords: entropy theory; frequency analysis; hydrometeorological extremes; Bayesian technique; rainfall

---

1. Introduction

Frequency analysis of hydrometeorological extremes plays an important role in the design of structures, such as dams, bridges, culverts, levees, highways, sewage disposal plants, waterworks, and industrial buildings [1–5]. From a frequency analysis, the probability of an extreme event can be estimated, and the value of a T-year design event (e.g., rainfall or flood) can be calculated. One of the objectives of frequency analysis of hydrometeorological extremes therefore is to establish a relationship between a flood or rainfall magnitude and its recurrence interval or return period.

A multitude of distributions have been employed for frequency analysis of hydrometeorological extremes. For example, the Pearson Type three (P-III) distribution is recommended in China; the Log-Pearson type three (LPT 3) is used in the U.S and Australia; and generalized extreme value (GEV) distribution is usually employed in Europe. Frequency analysis of hydrometeorological extremes at a given site or location is usually performed based on an appropriate probability distribution, which is selected on the basis of statistical tests for extreme hydrometeorological data [6]. However, no single distribution has gained global acceptance [7,8]. The traditional method is to try a variety of distributions and choose the best fitted distribution based on a particular mathematical norm, such as a least square error or a likelihood norm [9]. The disadvantages of this method of choosing are that it is laborious because too many different distributions need to be tried and empirical choices of candidate distributions make the results subjective [9–11]. In order to overcome these disadvantages, the generalized distributions have recently gained a lot of attention because they have been shown to be an effective tool for frequency analysis of hydrometeorological extremes. The greatest advantage of these generalized distributions is that they provide sufficient flexibility to fit a large variety of data sets, which facilitates the selection and comparison of different distributions. For example, Papalexiou and Koutsoyiannis [9] concluded that the generalized beta distribution of the second kind (GB2), which includes commonly used exponential, Weibull, and gamma distributions as special cases, was a suitable model for rainfall frequency analysis because of its ability to describe both J-shaped and bell-shaped data. Chen et al. [10] and Chen and Singh [11] also used the generalized gamma (GG) and GB2 distributions for hydrological frequency analysis, respectively. The results demonstrated that these two distributions could fit hydrometeorological data well. The generalized distributions can be derived using entropy theory by specifying appropriate constraints. The theory also provides a way for efficient parameter estimation [12].

Selection of the most appropriate distribution is of fundamental importance in hydrometeorological frequency analysis, since a wrong choice could lead to significant error and bias in design flood or rainfall estimates, particularly for higher return periods, leading to either under- or over-estimation, which may have serious implications in practice [13].

A distribution is often selected on the basis of statistical tests or by graphical methods [14]. Selection criteria based on the Akaike Information Criterion (AIC), the Bayesian Information Criterion (BIC), and the Anderson–Darling Criterion (ADC) are widely used in hydrology [4,15]. Laio et al. [16] presented an objective model selection criterion based on the AIC, the Bayesian Information Criterion (BIC), and the Anderson–Darling Criterion (ADC). Using a rigorous numerical framework, they found that the ability of these criteria to recognize the correct parent distribution from the available data varied from case to case, and these were more effective for two parameter distributions [13]. In this study, a more objective method based on a Bayesian technique is introduced to select the distribution with more parameters for frequency analysis of hydrometeorological extremes.

Bayesian method has been widely used for hydrological analysis, such as model selection and hydrological uncertainty analysis. Duan et al. [17] used Bayesian model averaging for multi-model ensemble hydrologic prediction. Hsu et al. [18] used a sequential Bayesian approach for hydrologic model selection and prediction. Najafi et al. [19] used Bayesian Model Averaging method to assess the uncertainties of hydrologic model selection. Robertson and Wang [20] introduced a predictor selection method for the Bayesian joint probability modeling approach to seasonal streamflow forecasting at multiple sites. In addition, Bayesian model method was also used for model uncertainty analysis [21,22].

The objective of this study is therefore to present a more objective method based on a Bayesian technique to select the most appropriate generalized distribution for frequency analysis of hydrometeorological extremes. The entropy theory was employed to derive generalized distributions for hydrometeorological extremes and estimate their parameters based on the principle of maximum entropy. A simulation test was carried out to evaluate the performance of the proposed Bayesian model selection technique. The proposed method was then tested using annual maximum hourly and daily precipitation data from Mississippi.

2. Entropy Theory

Since the entropy theory was used for the derivation of these generalized distributions and estimation of their parameters, in this section, the entropy theory combined with the principle of maximum entropy (POME) method is introduced.

The entropy, defined by Shannon in 1848, can be expressed by

$$H(X) = -\int_{0}^{\infty} f(x) \log f(x) \mathrm{d}x \tag{1}$$

where $f(x)$ is the probability density function (PDF) of X. $f(x)$ can be derived by maximizing the entropy subject to given constraints, which can be expressed by

$$\max H(X) \tag{2a}$$

$$\text{s.t.} \int_0^\infty f(x)\mathrm{d}x = 1; \tag{2b}$$

$$\int_0^\infty g_i(x)f(x)\mathrm{d}x = C_i \qquad (i = 1, \ldots, m) \tag{2c}$$

Employing the method of Lagrange multipliers, the PDF of X from Equations (1) and (2) can be derived as

$$f(x) = \exp(-\lambda_0 - \lambda_1 g_1(x) - \lambda_2 g_2(x) - \ldots - \lambda_m g_m(x)) \tag{3}$$

where m is the number of constraints; and $\lambda_0, \ldots, \lambda_m$ are the Lagrange multipliers. According to (2b), λ_0 can be defined as

$$\exp(\lambda_0) = \int_0^\infty \exp(-\lambda_1 g_1(x) - \lambda_2 g_2(x) - \ldots - \lambda_m g_m(x))\mathrm{d}x. \tag{4}$$

When different constraints are used, different PDFs can be obtained. According to the POME theory, all of the generalized distributions discussed in the following can be written in the form of Equation (3).

3. Generalized Distributions

Five generalized distributions, namely the GG distribution, the GB2 distribution, and three Halphen family distributions, were used in this study. The principle of maximum entropy (POME) method was used for parameter estimation, and it involves the following steps: (1) specification of constraints and maximization of entropy using the method of Lagrange multipliers; (2) derivation of the relation between Lagrange multipliers and constraints; (3) derivation of the relation between Lagrange multipliers and distribution parameters; and (4) derivation of the relation between distribution parameters and constraints. Detailed information on obtaining the equations for parameter estimation of those generalized distributions is given in [10,11,23]. In this paper, we mainly focus on model selection based on the Bayesian method.

3.1. Generalized Gamma Distribution

The probability density function of the GG distribution is given by

$$f(x) = \frac{r_2}{\beta \Gamma(\frac{r_1}{r_2})} \left(\frac{x}{\beta}\right)^{(r_1 - 1)} \exp\left(-\left(\frac{x}{\beta}\right)^{r_2}\right) \tag{5a}$$

where $\Gamma(\cdot)$ is the gamma function; r_1 and r_2 are the shape parameters, $r_1 > 0$, $r_2 > 0$; and beta is the scale parameter, $\beta > 0$.

For deriving Equation (5a) from the entropy theory, the following constraints are specified:

$$\int_0^\infty f(x)\mathrm{d}x = 1 \tag{5b}$$

$$\int_0^\infty f(x) \ln x dx = E(\ln X) \tag{5c}$$

$$\int_0^\infty f(x) x^q dx = E(X^q). \tag{5d}$$

The probability density function (PDF) of the GG distribution can then be expressed as [10]:

$$f(x) = \exp(-\lambda_0 - \lambda_1 \ln(x) - \lambda_2 x^q) \tag{6}$$

where λ_0, λ_1, and λ_2 are the Lagrange multipliers, and q is the parameter $q = r_2$ [10].
The relations between Lagrange multipliers and parameters can be summarized as

$$\begin{cases} q = r_2 \\ \lambda_1 = 1 - r_1 \\ \lambda_2 = \beta^{-r_2} \end{cases}. \tag{7}$$

The equations for parameter estimation based on the POME method can be given as [10]

$$\begin{cases} \frac{1}{r_2} \ln(\beta^{-r_2}) - \frac{1}{r_2} \varphi(\frac{r_1}{r_2}) = -E(\ln X) \\ \beta^{r_2} \frac{r_1}{r_2} = -E(\ln(X^{r_2})) \\ \frac{1}{r_2^2} \varphi'(\frac{r_1}{r_2}) = \text{var}(\ln X) \end{cases} \tag{8}$$

where $\varphi(\cdot)$ is the digamma function; and $\varphi'(\cdot)$ is the tri-gamma function.

As seen in Equation (8), there are three unknown parameters, r_1, r_2, and β, in the three equations, and the variable X represents the observed hydrometeorological extreme series, which have been known before. By solving this equation set, the parameter of the GG distribution can be determined. The estimation procedures for other distributions are the same as those for the GG distribution.

3.2. Generalized Beta Distribution of the Second Kind

The PDF of the GB2 distribution is given by

$$f(x) = \frac{r_3}{bB(r_1, r_2)} (\frac{x}{b})^{r_1 r_3 - 1} (1 + (\frac{x}{b})^{r_3})^{-(r_2 + r_1)} \tag{9a}$$

where $B(\cdot)$ is the beta function; and r_1, r_2, and r_3 are the shape parameters, $r_1 > 0$, $r_2 > 0$ and $r_3 > 0$; and b is the scale parameter, $b > 0$.

For deriving Equation (9a) from the entropy theory, the following constraints are specified:

$$\int_0^\infty f(x) dx = 1 \tag{9b}$$

$$\int_0^\infty f(x) \ln x dx = E(\ln X) \tag{9c}$$

$$\int_0^\infty f(x) \ln (1 + px^q)^{1/p} dx = E(\ln (1 + pX^q)^{1/p}). \tag{9d}$$

According the maximum entropy theory, the PDF of the GB2 distribution can be expressed as [11]

$$f(x) = \exp(-\lambda_0 - \lambda_1 \ln(x) - \lambda_2 \ln (1 + px^q)^{1/p}) \tag{10}$$

where p and q are two parameters, which are also related to the parameters of the GB2 distribution, $p = (\frac{1}{\beta})^{r_3}$, and $q = r_3$.

The relations between Lagrange multipliers and parameters can be summarized as

$$\begin{cases} \lambda_1 = 1 - r_1 q \\ \lambda_2 = p(r_2 + \frac{1-\lambda_1}{q}) \\ p = (\frac{1}{\beta})^{r_3} \\ q = r_3 \end{cases} \tag{11}$$

The equations for parameter estimation based on the POME method can be given as [11]

$$\begin{cases} -\ln \beta - \frac{1}{r_3} \varphi(r_1) + \frac{1}{r_3} \varphi(r_2) = -E(\ln X) \\ \varphi(r_2) - \varphi(r_1 + r_2) = -E(\ln(1 + (\frac{X}{\beta})^{r_3})) \\ \frac{1}{r_3^2} \varphi'(r_1) + \frac{1}{r_3^2} \varphi'(r_2) = \text{var}(\ln X) \\ \varphi'(r_2) - \varphi'(r_1 + r_2) = \text{var}(\ln(1 + (\frac{X}{\beta})^{r_3})) \end{cases} \tag{12}$$

3.3. Halphen Type A (Hal-A) Distribution

The PDF of the Hal-A distribution is given as

$$f(x) = \frac{1}{2m^v K_v(2\alpha)} x^{v-1} \exp[-\alpha(\frac{x}{m} + \frac{m}{x})] \qquad x > 0 \tag{13a}$$

where $K_0(\cdot)$ is the modified Bessel function of the second kind of order v, $v \in R$; and m and α are parameters, $m > 0$ and $\alpha > 0$.

For deriving Equation (13a) from the entropy theory, the following constraints are specified:

$$\int_0^\infty f(x)dx = 1 \tag{13b}$$

$$\int_0^\infty f(x) \ln x dx = E(\ln X) \tag{13c}$$

$$\int_0^\infty x f(x)dx = E(X) \tag{13d}$$

$$\int_0^\infty \frac{1}{x} f(x)dx = E(\frac{1}{X}). \tag{13e}$$

From the entropy theory, the PDF of the Halphen type A distribution can be expressed as [23]

$$f(x) = \exp(-\lambda_0 - \lambda_1 \ln x - \lambda_2 x - \lambda_3 \frac{1}{x}) \quad x > 0 \tag{14}$$

where λ_3 is also the Lagrange multiplier.

The relations between Lagrange multipliers and parameters can be summarized as

$$\begin{cases} \lambda_1 = 1 - v \\ \lambda_2 = \frac{\alpha}{m} \\ \lambda_3 = m\alpha \end{cases} \tag{15}$$

The equations for parameter estimation based on the POME method can be given as

$$\begin{cases} \ln m + \frac{1}{K_v(2\alpha)}\frac{\partial K_v(2\alpha)}{\partial v} = E(\ln X) \\ \frac{mK_{v+1}(2\alpha)}{K_v(2\alpha)} = E(X) \\ \frac{K_{v-1}(2\alpha)}{mK_v(2\alpha)} = E(\frac{1}{X}) \end{cases} . \tag{16}$$

3.4. Halphen Type B (Hal-B) Distribution

The PDF of the Hal-B distribution can be given as

$$f(x) = \frac{2}{m^{-2v}ef_v(\alpha)}x^{2v-1}\exp[-(\frac{x}{m})^2 + \alpha(\frac{m}{x})] \qquad x > 0 \tag{17a}$$

where $ef_v(\cdot)$ is the exponential factorial function, defined as $ef_v(\alpha) = 2\int\limits_0^\infty x^{2v-1}\exp[-x^2 + \alpha x]dx\ (x > 0)$, $m > 0$ are scale parameters, and $v > 0$ and $\alpha \in \Re$ are shape parameters.

For deriving Equation (17a) from the entropy theory, the following constraints are specified:

$$\int\limits_0^\infty f(x)dx = 1 \tag{17b}$$

$$\int\limits_0^\infty f(x)\ln xdx = E(\ln X) \tag{17c}$$

$$\int\limits_0^\infty x^2 f(x)dx = E(X^2) \tag{17d}$$

$$\int\limits_0^\infty xf(x)dx = E(X). \tag{17e}$$

From the entropy theory, the PDF of the Halphen type B distribution can be expressed as [23]

$$f(x) = \exp(-\lambda_0 - \lambda_1 \ln x - \lambda_2 x^2 - \lambda_3 x) \qquad x > 0. \tag{18}$$

The relations between Lagrange multipliers and parameters can be summarized as

$$\begin{cases} \lambda_1 = 1 - 2v \\ \lambda_2 = \frac{1}{m^2} \\ \lambda_3 = -\frac{\alpha}{m} \end{cases} . \tag{19}$$

The equations for parameter estimation based on the POME method can be given as

$$\begin{cases} \ln m + \frac{1}{2ef_v(\alpha)}\frac{\partial ef_v(\alpha)}{\partial v} = E(\ln X) \\ \frac{m^2 ef_{v+1}(\alpha)}{ef_v(\alpha)} = E(X^2) \\ \frac{m \cdot ef_{v+\frac{1}{2}}(\alpha)}{ef_v(\alpha)} = E(X) \end{cases} . \tag{20}$$

3.5. Halphen Type Inverse B (Hal-IB) Distribution

The PDF of the Hal-IB distribution can be given as

$$f(x) = \frac{2}{m^{-2v}ef_v(\alpha)}x^{-2v-1}\exp[-(\frac{m}{x})^2 + \alpha(\frac{m}{x})] \quad x > 0 \tag{21a}$$

where $m > 0$ is a scale parameter, and $\alpha \in \Re$ and $v > 0$ are shape parameters.

For deriving Equation (21a) from the entropy theory, the following constraints are specified:

$$\int_0^\infty f(x)dx = 1 \tag{21b}$$

$$\int_0^\infty f(x)\ln x dx = E(\ln X) \tag{21c}$$

$$\int_0^\infty \frac{1}{x^2}f(x)dx = E(\frac{1}{X^2}) \tag{21d}$$

$$\int_0^\infty \frac{1}{x}f(x)dx = E(\frac{1}{X}). \tag{21e}$$

From the entropy theory, the PDF of the Halphen type inverse B can be expressed as [23]

$$f(x) = \exp(-\lambda_0 - \lambda_1 \ln x - \lambda_2\frac{1}{x^2} - \lambda_3\frac{1}{x}) \quad x > 0. \tag{22}$$

The relations between Lagrange multipliers and parameters can be summarized as

$$\begin{cases} \lambda_1 = 2v + 1 \\ \lambda_2 = m^2 \\ \lambda_3 = -m\alpha \end{cases}. \tag{23}$$

The equations for parameter estimation based on the POME method can be given as

$$\begin{cases} \ln m - \frac{1}{2ef_v(\alpha)}\frac{\partial ef_v(\alpha)}{\partial v} = E(\ln X) \\ \frac{ef_{v+1}(\alpha)}{m^2ef_v(\alpha)} = E(\frac{1}{X^2}) \\ \frac{ef_{v+\frac{1}{2}}(\alpha)}{mef_v(\alpha)} = E(\frac{1}{X}) \end{cases}. \tag{24}$$

4. Model Selection Based on the Bayesian Technique

First, the five generalized distributions given above were used to fit a given data set **D**, and the equation sets derived by the POME method were applied for estimating their parameters. Second, the Bayesian technique introduced as follows was used to select the most appropriate distribution from the set of distributions for the data set **D**. In this study, the data **D** can be simulated data and observed data.

Let *I* be the background information. The posterior probabilities over a set of distributions can be expressed as

$$P(M_i|\mathbf{D}, I) = \frac{P(M_i|I) \cdot P(\mathbf{D}|M_i, I)}{P(\mathbf{D}|I)} \tag{25}$$

where $P(M_i|\mathbf{D}, I)$ is the posterior probability of distribution or model M_i and indicates the probability of this distribution to be true given the data series **D** and background information *I*. The largest approximate posterior probability among all of the distributions should be chosen as the most appropriate distribution. $P(M_i|I)$ is the prior model probability of distribution M_i; $P(\mathbf{D}|M_i, I)$ is the probabilistic evidence or integrated likelihood of data $\mathbf{D}$ conditional on model M_i. $P(\mathbf{D}|I)$ is a normalization constant and is calculated using the sum and product rules of probability theory as

$$P(\mathbf{D}|I) = \sum_{i=1}^{N} P(M_i|I)P(\mathbf{D}|M_i, I) \tag{26}$$

where N is the number of distributions that are used for the frequency analysis.

To obtain the posterior probability, one needs to calculate the probabilistic evidence $P(\mathbf{D}|M_i, I)$, which can be obtained by integrating a joint distribution $P(\boldsymbol{\lambda}, \mathbf{D}|M_i, I)$ with respect to vector $\boldsymbol{\lambda}$, and can be expressed as

$$P(\mathbf{D}|M_i, I) = \int_{-\infty}^{+\infty} P(\boldsymbol{\lambda}, \mathbf{D}|M_i, I)d\boldsymbol{\lambda} \tag{27}$$

since

$$P(\boldsymbol{\lambda}, \mathbf{D}|M_i, I) = P(\boldsymbol{\lambda}|M_i, I)P(\mathbf{D}|\boldsymbol{\lambda}, M_i, I) \tag{28}$$

where $P(\boldsymbol{\lambda}|M_i, I)$ is the prior PDF for the Lagrangian multipliers given distribution M_i and background information I. Equation (27) can be obtained as

$$P(\mathbf{D}|M_i, I) = \int_{-\infty}^{+\infty} P(\boldsymbol{\lambda}|M_i, I)P(\mathbf{D}|\boldsymbol{\lambda}, M_i, I)d\boldsymbol{\lambda} = E[P(\mathbf{D}|\boldsymbol{\lambda}, M_i, I)] \tag{29}$$

where $P(\mathbf{D}|\boldsymbol{\lambda}, M_i, I)$ is the likelihood function of the data in terms of the set of Lagrangian multipliers, and can be expressed by

$$P(\mathbf{D}|\boldsymbol{\lambda}, M_i, I) = \prod_{k=1}^{n} f(D_k|\boldsymbol{\lambda}, M_i, I) \tag{30}$$

where n is the sample size, and D_k denotes a specific value in data set $\mathbf{D}$. For a given sample size $\mathbf{D}$, model M_i and background information I, $P(\mathbf{D}|\boldsymbol{\lambda}, M_i, I)$ can be calculated by the multiplication of all PDF values of D_k.

The multivariate Gaussian distribution was selected as the prior distribution for the Lagrangian multiplier vector $\boldsymbol{\lambda}$. The mean value of Lagrangian multipliers was the estimated $\boldsymbol{\lambda}$. The covariance matrix Σ was calculated based on the Hessian matrix H, $\Sigma = \mathbf{H}^{-1}$. The equation for calculating the Hessian matrix can be expressed as

$$H = \begin{bmatrix} \frac{\partial^2 \lambda_0}{\partial \lambda_1^2} & \frac{\partial^2 \lambda_0}{\partial \lambda_1 \partial \lambda_2} & \cdot & \cdot & \frac{\partial^2 \lambda_0}{\partial \lambda_1 \partial \lambda_m} \\ \frac{\partial^2 \lambda_0}{\partial \lambda_2 \partial \lambda_1} & \frac{\partial^2 \lambda_0}{\partial \lambda_2^2} & \cdot & \cdot & \frac{\partial^2 \lambda_0}{\partial \lambda_2 \partial \lambda_m} \\ \cdot & \cdot & \cdot & \cdot & \cdot \\ \cdot & \cdot & \cdot & \cdot & \cdot \\ \frac{\partial^2 \lambda_0}{\partial \lambda_m \partial \lambda_1} & \frac{\partial^2 \lambda_0}{\partial \lambda_m \partial \lambda_2} & \cdot & \cdot & \frac{\partial^2 \lambda_0}{\partial \lambda_m^2} \end{bmatrix}. \tag{31}$$

From Equation (29), $P(D|M_i, I)$ can be obtained by integration. Since the integration in Equation (29) is often a complex and high-dimensional function in Bayesian statistics, the quantity $P(\mathbf{D}|M_i, I)$ was calculated based on the calculation of $E[P(\mathbf{D}|\boldsymbol{\lambda}, M_i, I)]$.

A Markov Chain Monte Carlo (MCMC) method was used in this study to calculate $P(\mathbf{D}|M_i, I)$ and the posterior distribution of each distribution. The idea of MCMC sampling was first introduced by [24]. Since the target distribution is very complex, we cannot sample from it directly. The indirect method for obtaining samples from the target distribution is to construct an Markov chain with state space E, and whose stationary (or invariant) distribution is $\pi(\cdot)$, as discussed in [25]. Then, if we run the chain for sufficiently long, simulated values from the chain can be treated as a dependent sample from the target distribution. Using the MCMC simulation, pairs of Lagrangian multipliers λ were drawn from the joint distribution $P(\lambda, \mathbf{D}|M_i, I)$. The quantity $P(D|M_i, I)$ was finally calculated based on the calculation of $E[P(D|\lambda, M_i, I)]$.

In the following, simulated data and real-world data were used for testing the proposed method. The flow chart can be found in Figure 1.

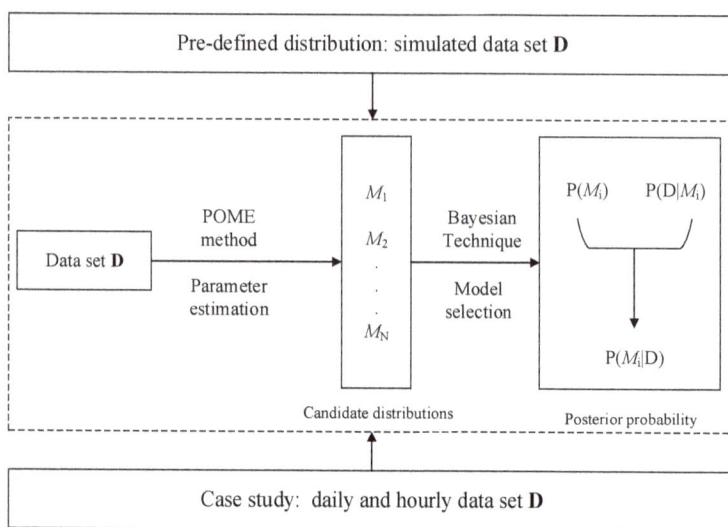

Figure 1. Flowchart of the whole paper. POME: principle of maximum entropy.

5. Performance Evaluation

Before using the proposed method in a practical application, a simulation test was carried out to evaluate the performance of the proposed Bayesian technique for model selection. The simulation test involves the following steps.

First, a distribution with given parameters was pre-defined.

Second, simulated datasets D were randomly drawn from the pre-defined distributions.

Third, the Gaussian, lognormal, Gamma, and Weibull distributions were used to fit the data set $\mathbf{D}$, and the POME method was applied for parameter estimation.

Fourth, the proposed Bayesian technique was applied for model selection, and the best fitted distributions with the highest posterior probabilities were determined. The results were compared with the pre-defined distributions.

Fifth, the Bayesian model selection technique was compared with commonly used methods in hydrology, such as the root mean square error of the empirical and theoretical probabilities and the AIC criterion.

According to the steps mentioned above, this test focuses on the evaluation of the reliability of the Bayesian model selection for different distributions and data sample sizes. In order to show the performance of the proposed method, some simple and widely used distributions were considered, including the Gaussian, lognormal, Gamma, and Weibull distributions, which involve the Gaussian and non-Gaussian cases. The parameters used for the simulation are given in Table 1. Simulated datasets were randomly drawn from the pre-defined distributions given in Table 1 with sample sizes of 40, 80, 120, 160, 200, and 240. The proposed Bayesian technique was then applied to determine the best fitted distributions for each dataset. The multivariate Gaussian distribution was used for the prior distribution, in which the mean values are the estimated Lagrangian multiplier, and the covariance matrix Σ was calculated based on the Hessian matrix H, $\Sigma = H^{-1}$. Usually, the estimated parameters were around the true values, so the Gaussian distribution was used. Additionally, the Hessian matrix was calculated to represent the covariance matrix. It is not straightforward to try other distributions, since it is a multivariate problem for which the multivariate Gaussian distribution is widely used.

Table 1. Parameters of different distributions for simulation test.

Number	Distribution	Probability Density Function (PDF)	Parameters
1	Gaussian	$f(x) = \frac{1}{\sqrt{2\pi}\sigma} \exp(-\frac{(x-\mu)^2}{2\sigma^2})$	$\mu = 10,$ $\sigma = 3.162$
2	Lognormal	$f(x; \mu, \sigma) = \begin{cases} \frac{1}{\sqrt{2\pi}\sigma x} \exp[-\frac{1}{2\sigma^2}(\ln x - \mu)^2], & x > 0 \\ 0, & x \le 0 \end{cases}$	$\mu = 2,$ $\sigma = 0.6$
3	Gamma	$f(x; \beta, \alpha) = \frac{\beta^{\alpha}}{\Gamma(\alpha)} x^{\alpha-1} e^{-\beta x}, x > 0$	$\alpha = 10,$ $\beta = 1$

The simulation results are shown in Figure 2, which indicate that when the data was sampled from the Gaussian distribution, for all of the sample size, the posterior probabilities of the Gaussian distribution were the highest. For the other tests, namely the lognormal distribution and the gamma distribution as the pre-defined distributions, respectively, the highest posterior probabilities for all of the sample size were the lognormal distribution and gamma distribution as well. Therefore, the proposed Bayesian technique can select the best fitted distribution even for a small sample size (sample size = 40).

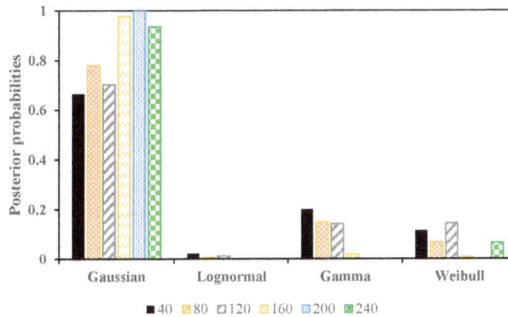

(a) Gaussian distribution

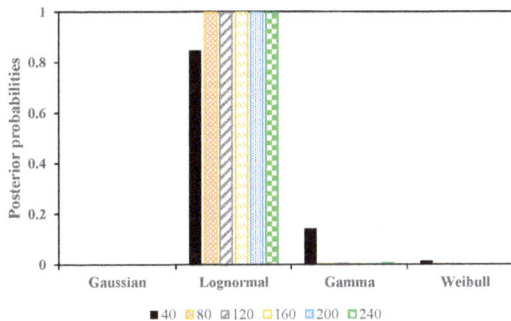

(b) Lognormal distribution

Figure 2. *Cont.*

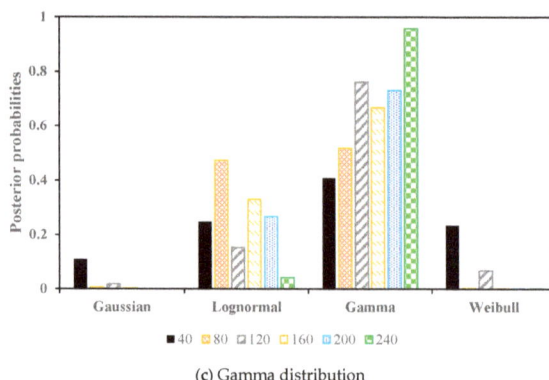

(c) Gamma distribution

Figure 2. The posterior probabilities of the simulation tests with the Gaussian distribution, Lognormal distribution, and Gamma distribution as the pre-defined distributions, respectively.

The proposed method was compared with the traditional root mean square error (RMSE) and AIC values, which are also used to select the most appropriate distribution. The results are given in Tables 2 and 3, in which the best fitted distributions with the smallest RMSE and AIC values are in bold. According to the smallest RMSE and AIC values, the correct distribution cannot always be selected. Take the Gaussian distribution as an example. When the sample size was 40, 80, 120, and 160, the best fitted distribution was, respectively, gamma, Weibull, Weibull, and Weibull. When the sample size became larger, greater than 160, the Gaussian distribution was detected as the correct distribution. The RMSE and AIC values of different distributions did not show significantly different results. In other words, the differences in the RMSE and AIC values among those distributions were not large. In Table 3, generally the AIC and RMSE values can show the best fitted distribution. However, in some cases the RMSE and AIC values of different distributions were nearly the same, such as the sample size equaling 160 and 200 in Table 3.

According to the performance test, the Bayesian technique can obtain the correct distribution at any time no matter what the sample size is. On the contrary, the traditional RMSE and AIC do not always work effectively. The RMSE and AIC for the data fitted using different distributions do not shown large differences. Therefore, the proposed method can provide an effective way for model selection in hydrological frequency analysis.

Table 2. The root mean square error (RMSE) and Akaike Information Criterion (AIC) values for the simulation test, in which the pre-defined distribution is the Gaussian distribution.

Distributions	Criteria	40	80	120	160	200	240
Gaussian	RMSE	0.03	0.0175	0.0247	0.0243	**0.0118**	**0.0115**
	AIC	−169.07	−443.78	−577.56	−792.76	**−1244.68**	**−1482.07**
lognormal	RMSE	0.0385	0.044	0.025	0.05	0.057	0.0437
	AIC	−155.12	−279.77	−558.39	−525.84	−683.75	−913.47
Gamma	RMSE	**0.0239**	0.0293	0.0161	0.0377	0.0367	0.026
	AIC	**−177.05**	−332.5793	−642.25	−607.66	−801.37	−1116.59
Weibull	RMSE	0.03	**0.0168**	**0.016**	**0.0168**	0.017	0.0118
	AIC	−167.34	**−441.3**	**−661.23**	**−858.42**	−1082.21	−1457.45

Table 3. The RMSE and AIC values for the simulation test, in which the pre-defined distribution is the Gamma distribution.

Distributions	Criteria	40	80	120	160	200	240
Gaussian	RMSE	0.0409	0.0317	0.33	0.0333	0.0318	0.033
	AIC	−152.98	−372.77	−509.91	−693.5	−879.39	−1030.29
lognormal	RMSE	0.0217	0.0161	0.0211	**0.0098**	0.0158	0.0179
	AIC	−185.92	−437.1	−582.86	−1018.23	**−1094.31**	−1311.04
Gamma	RMSE	**0.0229**	**0.0141**	**0.0169**	0.0108	**0.0124**	0.0126
	AIC	**−189.06**	**−486.4**	**−632.77**	−1026.04	−1191.77	**−1421.8**
Weibull	RMSE	0.042	0.0398	0.0314	0.036	0.0344	0.0332
	AIC	−150.55	−373.95	−523.74	−672.99	−853.67	−1053.41

6. Case Study

Rainfall data for many different timescales were investigated. The timescales of these rainfall dates in the Mississippi River basin ranged from hourly to yearly. The annual maximum daily and hourly series were extracted for frequency analysis, and detailed information of daily and hourly data is shown in Table 4, in which the length of data, the mean value, standard deviation, and the minimum and maximum values are shown. The daily and hourly rainfall histograms for each gauging station are given in Figure 3.

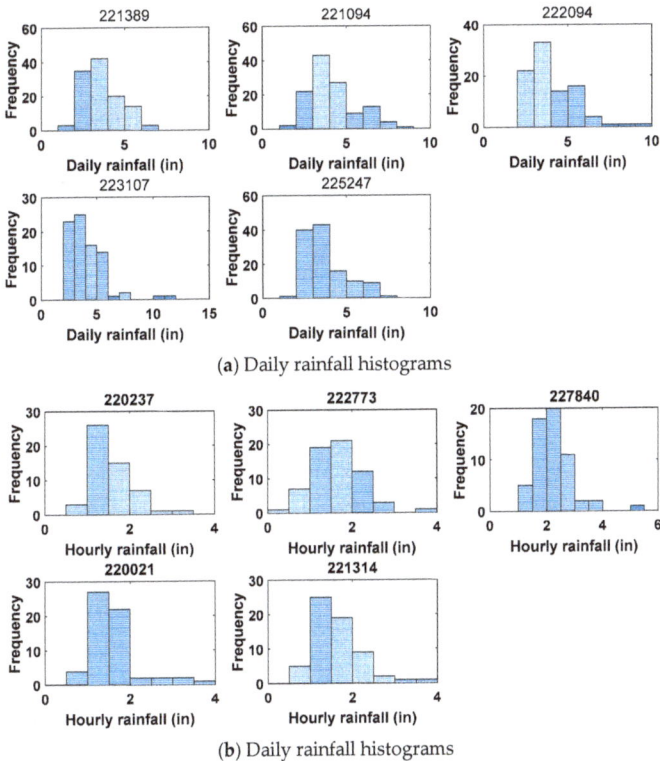

(a) Daily rainfall histograms

(b) Daily rainfall histograms

Figure 3. Daily and Hourly rainfall histograms for each gauging station.

Table 4. Detailed information of daily and hourly annual maximum rainfall series.

Times	Gauging Station	Number	Length of Data	Mean Values	SD	Max	Min
Daily	Canton gauging	221389	1893–2012	3.6	1.1	6.8	1.65
	Brookhaven City, MS	221094	1894–2014	4.14	1.44	8.08	1.85
	Crvstal Spgs Exp Stn, MS	222094	1893–1954, 1985–2014	4	1.4	9.04	2.02
	Forest, MS	223107	1930–2012	4.02	1.66	11.75	2
	Louisville, MS	225247	1895–2014	3.65	1.28	7.47	1.7
Hourly	Arkabutla dam	220237	1949–2001	1.57	0.46	3.12	0.88
	Enid dam MS	222773	1949–2012	1.62	0.59	4	0.2
	Saucier experimental forest MS	227840	1955–2013	2.23	0.7	5.13	1.2
	Aberdeen MS	220021	1952–2011	1.55	0.62	3.8	0.7
	Calhoun city MS	221314	1948–2009	1.6	0.57	3.88	0.8

The five generalized distributions were used to fit the data set, and the entropy method was used to estimate the parameters of these distributions, as given in Table 5 (for daily data) and Table 6 (for hourly data). A full Newton method was used to find the solution of the non-linear equation sets derived before. The "nleqslv" package in R language was used to solve the equation set. The initial value was set as 1 for all potential parameters. The proposed Bayesian technique was used to select the most appropriate distribution for rainfall frequency analysis. The multivariate Gaussian distribution was used for the prior distribution, in which the mean values are the estimated Lagrangian multiplier, and the covariance matrix Σ was calculated based on the Hessian matrix H, $\Sigma = H^{-1}$. The posterior probabilities are also in Table 5 (for daily data) and Table 6 (for hourly data). The RMSE, AIC, and BIC were also calculated as given in Tables 5 and 6. Both the AIC and BIC indexes are based on the likelihood values, and a penalty term was introduced for the number of parameters in the model. However, the differences between them are that the penalty term is larger in BIC than in AIC. In this study, it is seen from Tables 5 and 6 that the selected model by the two methods is the same. Therefore, only the results given by AIC are discussed hereafter. The results indicate that for some of the cases, the selected model based on the three criteria are the same, e.g., gauging stations 225247, 220237, 227840, 220021, and 221314. For some of the stations, the results given by the three methods were not coincident. However, for these cases, the distribution with the lowest AIC value usually had the second-highest posterior probability. Take the gauging station 221094 in Table 5 for example. The AIC and RMSE criteria suggested that the GB2 distribution was the best, for which the posterior probability was 0.34, smaller than the highest one 0.58 (Hal-A). According to the simulation test in Section 4, the performance of the proposed method was better than the traditional AIC and RMSE values. The Bayesian method amplified the differences among the generalized distributions. In order to further compare the performance of these models, the theoretical and empirical exceedance probabilities of the daily rainfall data for the gauging station 223107 are shown in Figure 4a.

According to the results given in Table 5, the best fitted distribution for the gauging station 223107 recommended by the RMSE, AIC, and Bayesian methods, was GB2, Hal-A, and Hal-IB, respectively. As shown in Figure 4a, if the Hal-A distribution was used, the design values for large return periods would be underestimated. The fitting curves of the GB2 and Hal-IB distributions were nearly the same. Thus, the distribution Hal-A recommended by the AIC is not appropriate, and compared with GB2, the Hal-IB with less parameters and higher posterior probability was chosen finally.

The theoretical and empirical exceedance probabilities of the hourly rainfall data for the gauging station 222773 are shown in Figure 4b. According to the results given in Table 6, the best fitted distribution for the gauging station 222773 recommended by the RMSE, AIC and Bayesian methods was GG, Hal-B, and GB2, respectively. As shown in Figure 4b, if the GG and Hal-B distributions were used, the design values for large return periods would be underestimated.

Table 5. Parameters, RMSE, AIC, and posterior probabilities for daily data calculated by five generalized distributions.

Number	Distribution	Para1	Para2	Para3	Para4	RMSE	AIC	BIC	Posterior Probabilities
	GG	13.251	0.8397	0.1335		0.0205	−613.05	−604.76	0.02
	Gb2	0.9955	1.5808	34.95	16.84	0.0195	−605.68	−597.39	0.08
221389	Hal-A	5.235	3.431	0.0327		**0.0183**	**−624.56**	**−616.27**	0.14
	Hal-B	−20.927	7.363	5.381		0.0237	−593.21	−584.93	0.02
	Hal-IB	−9.103	2.855	4.776		0.0233	−544.21	−535.92	**0.74**
	GG	13.308	0.684	0.0527		0.0362	−482.72	−477.32	0.08
	Gb2	1.678	2.203	11.994	4.894	**0.0203**	**−607.37**	**−601.98**	0.34
221094	Hal-A	3.507	7.812	−5.534		0.0229	−584.34	−578.95	**0.58**
	Hal-B	−20.918	8.0322	5.6835		0.0557	−409.26	−410.15	0.00
	Hal-IB	−10.755	2.319	3.699		0.0525	−539.94	−531.55	0.00
	GG	14.855	0.628	0.0255		0.0307	−412.5	−404.94	0.04
	Gb2	1.3803	1.38	24.977	6.573	0.0213	−462.61	−455.04	0.20
222094	Hal-A	2.5669	12.3193	−7.9616		**0.021**	**−466.21**	**−458.64**	0.32
	Hal-B	−12.2589	7.581	3.5971		0.04	−385.6	−378.04	0.01
	Hal-IB	−9.546	2.2708	4.2153		0.0218	−464.18	−456.61	**0.44**
	GG	13.5999	0.5606	0.0131		0.0294	−409.36	−402.11	0.00
	Gb2	2.199	1.837	9.931	2.4389	**0.0164**	−466.07	−458.81	0.08
223107	Hal-A	0.977	30.104	−8.194		0.0168	**−479.18**	**−471.93**	0.21
	Hal-B	−27.5605	14.5778	3.8889		0.0406	−362.58	−355.33	0.00
	Hal-IB	−3.8918	3.7076	3.2608		0.0165	−457.91	−450.65	**0.71**
	GG	13.596	0.6657	0.0384		0.0437	−433.23	−424.86	0.00
	Gb2	1.822	1.156	23.62	3.641	0.02456	−565.73	−557.36	0.00
225247	Hal-A	2.0554	14.775	−8.756		0.0287	−533.57	−525.21	0.04
	Hal-B	−46.515	14.128	6.0834		0.0653	−363.04	−354.68	0.00
	Hal-IB	−3.361	3.938	3.602		**0.0232**	**−575.37**	**−567.01**	**0.96**

Table 6. Parameters, RMSE, AIC, and posterior probabilities for hourly data calculated by five generalized distributions.

Number	Distribution	Para1	Para2	Para3	Para4	RMSE	AIC	BIC	Posterior Probabilities
	GG	−19.179	0.6794	0.0284		0.048	−187.74	−181.83	0.20
	GB2	2.963	2.5391	6.8892	2.4497	**0.0308**	**−233.74**	**−227.83**	0.57
220237	Hal-A	3.0259	15.972	−12.521		0.0344	−215.46	−209.55	0.08
	Hal-B	−27.458	8.384	6.6655		0.0523	−178.6	−172.69	0.10
	Hal-IB	−7.185	4.2898	5.7116		0.0312	−222.62	−216.72	0.06
	GG	4.169	1.858	1.656		**0.0318**	−291.31	−284.83	0.13
	GB2	4.5	3.221	0.775	2.153	0.0337	−291.92	−285.44	**0.70**
222773	Hal-A	5.36×10^{-2}	1.90×10^{-2}	6.8204		0.0333	−274.18	−267.70	0.05
	Hal-B	1.3105	1.6132	1.5702		0.0338	**−292.69**	**−286.21**	0.12
	Hal-IB	−50.5719	0.1565	3.0044		0.0419	−231.43	−224.95	0.00
	GG	17.0415	0.6977	0.0225		0.0473	−266.6	−260.37	0.05
	GB2	2.013	1.2966	11.7277	4.6042	**0.0266**	**−275.55**	−266.32	0.84
227840	Hal-A	2.7995	8.6585	−11.0919		0.0271	−274.68	**−268.45**	0.00
	Hal-B	−23.6777	4.7218	5.8388		0.0422	−249.55	−243.32	0.11
	Hal-IB	−10.601	1.865	5.7014		0.0316	−264.11	−257.88	0.00
	GG	11.875	0.65	0.0173		0.0582	−189.32	−183.04	0.10
	GB2	4.1081	0.9324	3.7284	0.9621	0.0468	−221.54	−212.26	0.33
220021	Hal-A	1.1704	10.6076	−8.8785		0.0532	−217.75	−211.47	0.15
	Hal-B	−30.3037	5.6947	4.203		0.0761	−181.74	−175.42	0.00
	Hal-IB	0.7828	2.248	2.3074		**0.0455**	**−228.4**	**−222.12**	**0.42**
	GG	14.1465	0.6738	0.0171		0.0405	−281.77	−275.39	0.00
	GB2	1.5734	0.3406	44.9999	4.516	0.0318	−293.19	−286.81	0.05
221314	Hal-A	1.7786	7.979	−9.475		0.0266	**−316.06**	**−309.68**	0.11
	Hal-B	−27.792	4.9387	4.622		0.0432	−263.62	−257.24	0.00
	Hal-IB	−6.217	1.438	4.185		**0.0249**	−309.73	−303.35	**0.84**

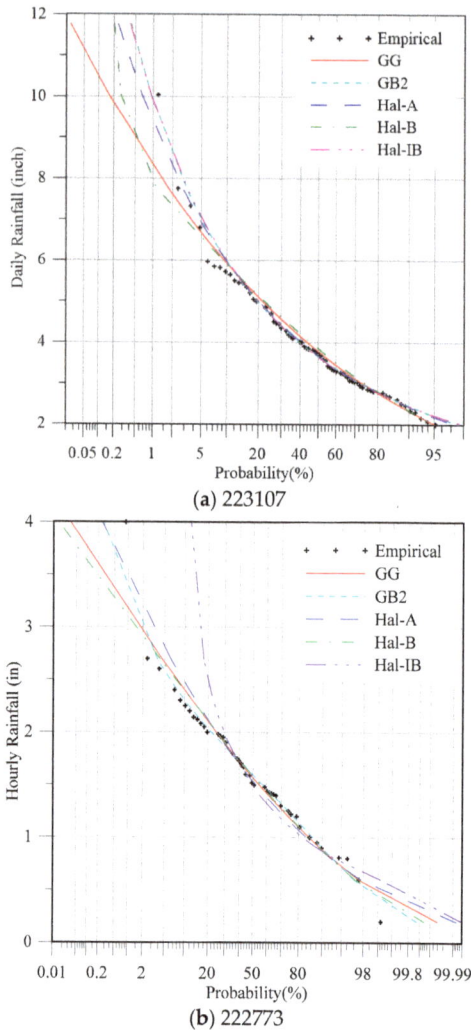

Figure 4. Theoretical and empirical exceedance probabilities of the annual maximum rainfall data at the stations 223107 and 222773.

In order to compare the fitting results more comprehensively, the Q-Q plot, P-P plot, and S-P plot were represented for the daily rainfall data from the gauging station 223107 as shown in Figure 5. It can be seen from Figure 5a that the fitting results of GB2 and Hal-IB are nearly the same. When the GG, Hal-A, and Hal-B distributions were used, the design rainfall for a large quantile would be underestimated, since the theoretical rainfall values calculated by the GG, Hal-A, and Hal-B distributions are significantly lower than the observed ones. For the P-P and S-P plots, the differences for large probability are not so obvious, and the plots in Figure 5b,c are well-distributed compared with the Q-Q plot. In Figure 5b, it is easily observed that the Hal-B distribution fits the worst, and the empirical probabilities in the middle part are significantly larger than the theoretical ones. S-P plots remove the impact of variance on the plot, and it is seen that the plots in the S-P figure are much more concentrated than those in the P-P figure.

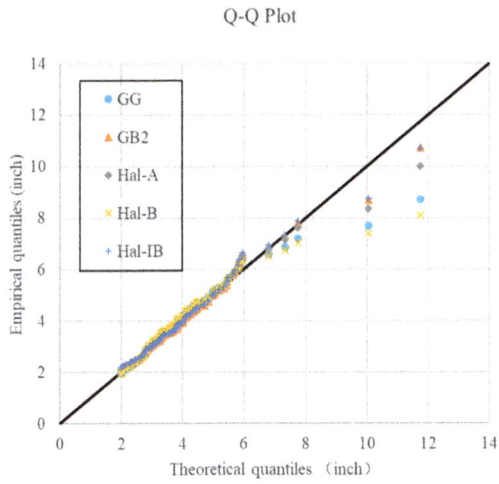

(a) Q-Q plot

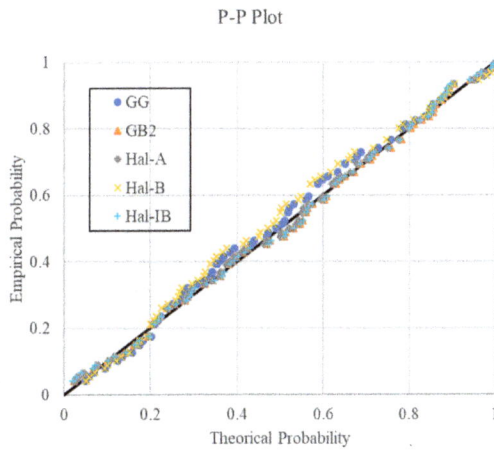

(b) P-P plot

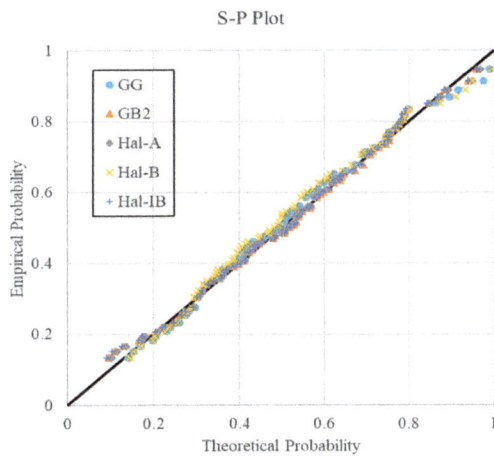

(c) S-P plot

Figure 5. Q-Q, P-P, and S-P plots for the daily rainfall data from the gauging station 223107.

Furthermore, in the U.S., the Log-Pearson three (LP3) distribution has been recommended for hydrological frequency analysis [26,27]. In order to compare the five generalized distributions with the commonly used LP3 distribution, the six distributions were considered and the proposed Bayesian method was used to select the best fitted one. The results are given in Table 7.

Table 7. Parameters, RMSE, AIC, and poster probabilities for 223107 daily data calculated by five generalized distributions and the Log-Pearson three (LP3) distribution.

Distributions	Para1	Para2	Para3	Para4	RMSE	AIC	BIC	Posterior Probabilities
GG	13.60	0.56	0.013		0.0294	−406.36	−402.11	0.002
GB2	2.20	1.84	9.93	2.44	**0.0164**	−463.07	−460.81	0.071
Hal-A	0.98	30.10	−8.19		0.0168	**−476.18**	**−471.93**	0.197
Hal-B	−27.56	14.58	3.89		0.0406	−359.58	−355.33	0.0003
Hal-IB	−3.89	3.70	3.26		0.0165	−454.91	−450.77	**0.674**
LP3	14.29	0.09	−0.03		0.0167	−448.65	−444.39	0.056

7. Conclusions and Discussion

The paper proposes a model selection approach based on a Bayesian technique to choose the best fitted distribution for hydrological frequency analysis. Five generalized distributions, including GG, GB2, Hal-A, Hal-B, and Hal-IB, which are also widely used in hydrology, were considered. The entropy-based method was used to express these distributions and the POME method was applied for parameter estimation. A simulation test was carried out to evaluate the performance of the proposed Bayesian method. Daily rainfall data from five stations and hourly rainfall data from another five stations from the Mississippi basin were selected as case studies. The main conclusions are summarized as follows.

(1) The entropy-based five generalized distributions are given, and their corresponding equation sets for parameter estimation are introduced. The results of simulation test and case study show that the POME method can provide an effective way for parameter estimation.
(2) Results of the simulation test demonstrate that the Bayesian technique can choose the most suitable distribution. Compared with the commonly used RMSE and AIC values, the proposed method gives a better performance.
(3) Results of the case study indicate that when using different criteria for model selection, the results are not always the same. For some of the cases, the three criteria choose the same distribution. For others, the results are slightly different. Since choosing the probable distribution for hydraulic design is very significant, especially for extreme magnitudes, the distribution should be selected carefully. According to the posterior probabilities calculated by the proposed method for daily and hourly data from 10 gauging stations, generally the Hal-IB distributions give better fits for daily data and GB2 distributions give better fits for hourly data.
(4) According to the results of the simulation test and case studies, the Bayesian model selection technique can give a more reliable result than the traditional RMSE and AIC values. Thus, the proposed method provides an effective way for model selection for hydrological frequency analysis.
(5) The significant contribution of this paper is that compared with the traditional method, the proposed method is based on entropy theory, and the posterior probabilities were calculated based on the generation of Lagrange multipliers. In addition, the five generalized distributions were involved in this paper, since previous research mainly focus on the commonly used distribution or standard distributions.

This contribution of this paper mainly concentrates on univariate hydrometeorological frequency analysis. Recently, multivariate hydrological analysis has also surged up, such as [2,4,28–31]. However, univariate frequency analysis is the basis of multivariate frequency analysis, which can provide the marginal distributions for joint distribution. Thus, before establishing the multivariate distributions, the univariate distribution should be built rationally and appropriately first.

In addition, in the common hydrological frequency analysis, the hydrological data set is assumed to be independent and identically distributed [1]. Since there are influences of climate change and human activities on streamflow, it is possible that the mean value or the variation of the whole series would be changed. In other words, the data set is non-stationary. Non-stationary hydrological frequency analysis is also another hot and difficult topic in hydrology recently. In this paper, we mainly focus on the stationary frequency analyses of hydrometeorological extremes. Non-stationary hydrological frequency analysis will be discussed in future research.

Although this paper discussed the model selection method based on the five generalized distributions, the traditional commonly used distribution, the LP3 distribution, is still an effective tool for frequency analysis and can be used for design rainfall or flood calculation.

Acknowledgments: This study was supported by the National Natural Science Foundation of China (51679094), the National Key R&D Program of China (2017YFC0405900), National Natural Science Foundation of China (51509273; 91547208), and Fundamental Research Funds for the Central Universities (2017KFYXJJ194, 2016YXZD048).

Author Contributions: Vijay P. Singh conceived and designed the experiments; Lu Chen and Kangdi Huang performed the experiments and analyzed the data; Lu Chen and Vijay P. Singh wrote the paper.

Conflicts of Interest: The authors declare no conflict of interest.

References

1. Rosbjerg, D. *Partial Duration Series in Water Resources*; Technical University of Denmark: Kongens Lyngby, Denmark, 1993.
2. Lin-Ye, J.; García-León, M.; Gracia, V.; Sánchez-Arcilla, A. A multivariate statistical model of extreme events: An application to the Catalan coast. *Coast. Eng.* **2016**, *117*, 138–156. [CrossRef]
3. Chen, L.; Zhang, Y.; Zhou, J.; Singh, V.P.; Guo, S.L.; Zhang, J. Real-time error correction method combined with combination flood forecasting technique for improving the accuracy of flood forecasting. *J. Hydrol.* **2015**, *521*, 157–169. [CrossRef]
4. Chen, L.; Singh, V.P.; Guo, S.; Zhou, J. Copula-based method for Multisite Monthly and Daily Streamflow Simulation. *J. Hydrol.* **2015**, *528*, 369–384. [CrossRef]
5. Chen, L.; Singh, V.P.; Guo, S.L.; Zhou, J.Z.; Zhang, J.H.; Liu, P. An objective method for partitioning the entire flood season into multiple sub-seasons. *J. Hydrol.* **2015**, *528*, 621–630. [CrossRef]
6. Yoon, P.; Kim, T.M.; Yoo, C. Rainfall frequency analysis using a mixed GEV distribution: A case study for annual maximum rainfalls in South Korea. *Stoch. Environ. Res. Risk Assess.* **2013**, *27*, 1143–1153. [CrossRef]
7. Perreault, L.; Bobée, B.; Rasmussen, P. Halphen distribution system. I: Mathematical and statistical properties. *J. Hydrol. Eng.* **1999**, *4*, 189–199. [CrossRef]
8. Zhang, J.; Chen, L.; Singh, V.P.; Cao, W.; Wang, D. Determination of the distribution of flood forecasting error. *Nat. Hazards* **2015**, *1*, 1389–1402. [CrossRef]
9. Papalexiou, S.M.; Koutsoyiannis, D. Entropy based derivation of probability distribution: A case study to daily rainfall. *Adv. Water Resour.* **2012**, *45*, 51–57. [CrossRef]
10. Chen, L.; Singh, V.P.; Xiong, F. An Entropy-Based Generalized Gamma Distribution for Flood Frequency Analysis. *Entropy* **2017**, *19*, 239. [CrossRef]
11. Chen, L.; Singh, V.P. Generalized Beta Distribution of the Second Kind for Flood Frequency Analysis. *Entropy* **2017**, *19*, 254. [CrossRef]
12. Singh, V.P. *Entropy Based Parameter Estimation in Hydrology*; Kluwer Academic Publishers: Dordrecht, The Netherlands, 1998.
13. Rahman, A.S.; Rahman, A.; Zaman, M.A.; Haddad, K.; Ahsan, A.; Imteaz, M. A study on selection of probability distributions for at-site flood frequency analysis in Australia. *Nat. Hazards* **2013**, *69*, 1803–1813. [CrossRef]
14. Bobee, B.; Perreault, L.; Ashkar, F. Two kinds of moment ratio diagrams and their applications in Hydrology. *Stoch. Hydrol. Hydraul.* **1993**, *7*, 41–65. [CrossRef]

15. Chen, L.; Singh, V.P.; Guo, S.L.; Zhou, J.; Ye, L. Copula entropy coupled with artificial neural network for rainfall-runoff simulation. *Stoch. Environ. Res. Risk Assess.* **2014**, *28*, 1755–1767. [CrossRef]

16. Laio, F.; Baldassarre, G.D.; Montanari, A. Model selection techniques for the frequency analysis of hydrological extremes. *Water Resour. Res.* **2009**, *45*, 162–174. [CrossRef]

17. Duan, Q.; Ajami, N.K.; Gao, X.; Sorooshian, S. Multi-model ensemble hydrologic prediction using Bayesian model averaging. *Adv. Water Resour.* **2007**, *30*, 1371–1386. [CrossRef]

18. Hsu, K.L.; Moradkhani, H.; Sorooshian, S. A sequential Bayesian approach for hydrologic model selection and prediction. *Water Resour. Res.* **2009**, *45*, 1079. [CrossRef]

19. Najafi, M.R.; Moradkhani, H.; Jung, I.W. Assessing the uncertainties of hydrologic model selection in climate change impact studies. *Hydrol. Process.* **2011**, *25*, 2814–2826. [CrossRef]

20. Robertson, D.E.; Wang, Q.J. A Bayesian Approach to Predictor Selection for Seasonal Streamflow Forecasting. *J. Hydrometeorol.* **2012**, *13*, 155–171. [CrossRef]

21. Vrugt, J.A.; Robinson, B.A. Treatment of uncertainty using ensemble methods: Comparison of sequential data assimilation and Bayesian model averaging. *Water Resour. Res.* **2007**, *43*, 223–228. [CrossRef]

22. Parrish, M.A.; Moradkhani, H.; Dechant, C.M. Toward reduction of model uncertainty: Integration of Bayesian model averaging and data assimilation. *Water Resour. Res.* **2012**, *48*, 3519. [CrossRef]

23. Chen, L.; Singh, V.P. Entropy-based derivation of generalized distributions for hydrometeorological frequency analysis. *J. Hydrol.* **2018**, *557*, 699–712. [CrossRef]

24. Metropolis, N.; Rosenbluth, A.W.; Rosenbluth, M.N.; Teller, A.H.; Teller, E. Equations of state calculations by fast computing machines. *J. Chem. Phys.* **1953**, *21*, 1087–1091. [CrossRef]

25. Smith, A.E.M.; Roberts, G. Bayesian computation via the Gibbs sampler and related Markov chain Monte Carlo methods. *J. R. Stat. Soc. B* **1993**, *55*, 3–23.

26. Griffis, V.W.; Stedinger, J.R. Log-Pearson Type 3 Distribution and Its Application in Flood Frequency Analysis. I: Distribution Characteristics. *J. Hydrol. Eng.* **2007**, *12*, 482–491. [CrossRef]

27. Lamontagne, J.R.; Stedinger, J.R.; Yu, X.; Whealton, C.A.; Xu, Z. Robust flood frequency analysis: Performance of EMA with multiple Grubbs-Beck outlier tests. *Water Resour. Res.* **2016**, *52*, 3068–3084. [CrossRef]

28. Lin-Ye, J.; Garcia-Leon, M.; Gracia, V.; Ortego, M.I.; Lionello, P.; Sanchez-Arcilla, A. Multivariate statistical modelling of future marine storms. *Appl. Ocean Res.* **2017**, *65*, 192–205. [CrossRef]

29. Chen, L.; Singh, V.P.; Guo, S.L. Measure of correlations for rivers flows based on copula entropy method. *J. Hydrol. Eng.* **2013**, *18*, 1591–1606. [CrossRef]

30. Chen, L.; Ye, L.; Singh, V.P.; Zhou, J.; Guo, S.L. Determination of input for artificial neural networks for flood forecasting using the copula entropy method. *J. Hydrol. Eng.* **2014**, *19*, 217–226. [CrossRef]

31. Chen, L.; Singh, V.P.; Lu, W.; Zhang, J.; Zhou, J.; Guo, S. Streamflow forecast uncertainty evolution and its effect on real-time reservoir operation. *J. Hydrol.* **2016**, *540*, 712–726. [CrossRef]

entropy

MDPI

Review

Entropy Applications to Water Monitoring Network Design: A Review

Jongho Keum [1,*], Kurt C. Kornelsen [2], James M. Leach [1] and Paulin Coulibaly [1,2]

[1] Department of Civil Engineering, McMaster University, Hamilton, ON L8S4L8, Canada;
   leachjm@mcmaster.ca (J.M.L.); couliba@mcmaster.ca (P.C.)
[2] School of Geography and Earth Sciences, NSERC Canadian FloodNet, McMaster University, Hamilton,
   ON L8S4L8, Canada; kornelkc@mcmaster.ca
* Correspondence: jkeum@mcmaster.ca; Tel.: +1-905-525-9140 (ext. 27875)

Received: 16 October 2017; Accepted: 10 November 2017; Published: 15 November 2017

Abstract: Having reliable water monitoring networks is an essential component of water resources and environmental management. A standardized process for the design of water monitoring networks does not exist with the exception of the World Meteorological Organization (WMO) general guidelines about the minimum network density. While one of the major challenges in the design of optimal hydrometric networks has been establishing design objectives, information theory has been successfully adopted to network design problems by providing measures of the information content that can be deliverable from a station or a network. This review firstly summarizes the common entropy terms that have been used in water monitoring network designs. Then, this paper deals with the recent applications of the entropy concept for water monitoring network designs, which are categorized into (1) precipitation; (2) streamflow and water level; (3) water quality; and (4) soil moisture and groundwater networks. The integrated design method for multivariate monitoring networks is also covered. Despite several issues, entropy theory has been well suited to water monitoring network design. However, further work is still required to provide design standards and guidelines for operational use.

Keywords: entropy; water monitoring; network design; hydrometric network; information theory; entropy applications

1. Introduction

Water monitoring networks account for all aspects of the water-related measurement system including precipitation, streamflow, water quality, groundwater, soil moisture, etc. [1–3]. Adequate water monitoring networks and quality data from them comprise one of the first and primary steps towards efficient water resource management. The basic principles of water monitoring network design have simply been a number of monitoring stations, locations of the stations and data period or sampling frequency [4,5]. Recent technological advances have allowed gradual transitions from manual sampling to the automated observations, while some water quality parameters still require field and/or lab analyses of water or other environmental samples. One may expect that the more data we collect, the more water resource problems are solved efficiently. However, this is not always true because irrelevant, inadequate or inefficient data in the wrong location or at the wrong time can inhibit the quality of a dataset [1,6,7]. More seriously, the decline of water monitoring networks has been a general trend due to financial limitations and changes of monitoring priority [8–10]. Therefore, determining the adequate number of monitoring stations and their locations has become critical to network design. However, a standardized methodology for a proper water monitoring network design process has not been drawn yet due to the practical and socioeconomic complexity in diverse design cases [1,11].

The existing reviews have investigated the broad range of the water monitoring network design methodologies, such as statistical analysis, spatial interpolation, application of information theory, optimization techniques, physiographic analysis, user survey or expert recommendations and combinations of multiple methods [4,6,10–15]. A prior comprehensive review by Mishra and Coulibaly [10] reviewed evidence of declining hydrometric network density, highlighted the importance of quality data from well-designed networks and considered a range of approaches by which networks were designed. They also compared statistical, spatial interpolation, physiographic, sampling-driven and entropy-based approaches to hydrometric network design. Mishra and Coulibaly [10] were able to draw several conclusions about the importance of high quality hydrometric data for water resource management that remain valid. They also concluded that one of the most promising approaches for network design was the application of entropy methods highlighting early studies using the principle of maximum entropy and information transfer. Therefore, this review focuses on the recent studies that have applied information theory. Information theory was initially developed by Shannon in 1948 [16] to measure the information content in a dataset and has been applied to solve water resource problems. Recently, its applications extended to water monitoring network design by adopting the concept that the entropy would be able to explain the inherent information content in a monitoring station or a monitoring network. The basic objective has naturally been to have the maximum amount of information. In other entropy approaches, stations in a monitoring network would have the least sharable or common information, which is called transinformation. To achieve this, the stations should be as independent from each other as possible.

The scope of this paper includes water monitoring network design and evaluation studies that (1) applied entropy theory in the design process and (2) were published after the existing comprehensive review by Mishra and Coulibaly in 2009 [10]. To the best of our knowledge, no review exists that has focused on entropy applications to water monitoring network design previous to the 2009 study by Mishra and Coulibaly [10]. However, there has been considerable progress in the application of entropy theory to monitoring network design following the previous review, including new entropy-based measures, optimization techniques and approaches to estimating information content at ungauged stations. Therefore, a need was identified to consolidate knowledge and recent advances on the subject. For publications prior to 2010, the reader is referred to Mishra and Coulibaly [10]. This review firstly describes entropy concepts and various terms that are typically used in network design. The previous studies are then summarized by categorizing the type of networks; i.e., precipitation, streamflow and water level, soil moisture and groundwater and water quality monitoring networks. The integrated design method for multiple types of networks is also reviewed. We define some terminology hereafter to ensure a common understanding for the readers.

The term network evaluation is used when the network quality is assessed without changing any station, while network design is a general term that suggests some changes in stations. Specifically, network design includes network reduction, network expansion and network redesign. Network reduction is applied where some monitoring stations should be removed from the network. On the other hand, if financial flexibility meets the monitoring needs, further stations can be added to the existing network, called network expansion. Network redesign refers to rearranging stations without changing the number of stations. The term optimal network is to be used only if the network consists of optimal locations of stations that are identified by the actual use of an optimization technique.

2. Definitions of Entropy Terms as Applied to Water Monitoring Networks

2.1. Entropy Concept

In thermodynamics, entropy has been understood as a measure of disorder or randomness of a system. Shannon [16] extended the entropy concept to information theory by recognizing that uncertainty in a system will be decreased when information is added to the system. Therefore, the term entropy in information theory introduced by Shannon [16] in 1948 describes the amount of information content in a random variable. The likelihood of an event is typically given by probability p. If a probability of an event is very high, such as 0.9999 or one, one will not be surprised, but can certainly anticipate the outcome. On the other hand, any low probability event is highly uncertain, so that a considerable amount of information can be given if this happens. Hence, the information from an event that occurred is inversely related to its probability, $1/p$ [17]. Suppose that there are two independent events A and B with their probabilities p_A and p_B, respectively. The probability of the joint occurrence of the events A and B can be $p_A\,p_B$, and the information gained by the joint event is then $1/(p_A\,p_B)$. However, the sum of information from each individual event is not equal to the information from the joint event, that is:

$$\frac{1}{p_A\,p_B} \neq \frac{1}{p_A} + \frac{1}{p_B} \tag{1}$$

The only transition that will make both sides of Equation (1) be equal is the logarithm [17–19], which can be written as:

$$\log\frac{1}{p_A\,p_B} = \log\frac{1}{p_A} + \log\frac{1}{p_B} = -\log p_A - \log p_B \tag{2}$$

Likewise, Tribus [20] showed that the uncertainty of an event with probability p is $-\log p$, which became a basis of the Shannon entropy, which is further described hereafter.

2.2. Marginal Entropy

When information is provided in a system, one can expect that the uncertainty of the system would be reduced; therefore, the amount of information that was given to a system by knowing a variable is called marginal entropy. If a random variable X is expected to have N outcomes with a probability distribution $P = \{p_1, p_2, \cdots, p_N\}$, the (weighted) average information provided by the N joint events is given by:

$$H(X) = -p_1\log p_1 - p_2\log p_2 - \cdots - p_N\log p_N = -\sum_{i=1}^{N} p_i\log p_i, \quad \sum_{i=1}^{N} p_i = 1,\ p_i \geq 0 \tag{3}$$

where $H(X)$ is the marginal entropy of a random variable X. Any base of the logarithm can be used in Equation (3), the choice depending on the problem given. In binary questions (i.e., yes or no questions), the base of two should be used, and the corresponding unit of entropy is bit. Similarly, unit trit for base 3, unit nat for base e and unit decibels or decit for base 10 are some example units of information. Recall that this review covers entropy applications for hydrometric network design, and the expected answer of the design process can be either "use/include/install the station" and "do not use/include/install the station" for the network to be optimal. Therefore, the logarithm in Shannon entropy calculation for hydrometric network design is most appropriate with a base of two. Then, the $H(X)$ value from Equation (3) will be understood as the information contents of a station X that can be delivered if installed.

If a variable K has a known value, the probability of an event will be one, while all the other alternative probabilities are zero. The information content in the variable K, $H(K)$, will be zero from Equation (3) representing that there is no uncertainty or a certain outcome. On the other hand, if a variable U has a uniform distribution (i.e., probability of each event is equal, 1/N), the entropy of the variable U will be:

$$H(U) = \log N \tag{4}$$

The value of Equation (4) is often called as maximum entropy or saturated entropy. These two entropies, $H(K)$ and $H(U)$, define the minimum and maximum boundaries of entropy values, that is:

$$0 \leq H(X) \leq \log N \tag{5}$$

2.3. Multivariate Joint Entropy

While the marginal entropy described in Section 2.2 explains a univariate entropy, one can imagine how to calculate entropy values in a bivariate or a multivariate case. The total information contents from N variables can be calculated by using joint probability instead of univariate probability in Equation (3), given by:

$$H(X_1, X_2, \cdots, X_N) = -\sum_{i_1=1}^{n_1}\sum_{i_2=1}^{n_2}\cdots\sum_{i_N=1}^{n_N} p(x_{1,i_1}, x_{2,i_2}, \cdots, x_{N,i_N}) \log_2 p(x_{1,i_1}, x_{2,i_2}, \cdots, x_{N,i_N}) \quad (6)$$

where $H(X_1, X_2, \cdots, X_N)$ is the joint entropy of N variables, $p(x_{1,i_1}, x_{2,i_2}, \cdots, x_{N,i_N})$ is the joint probability of N variables and $n_1, n_2, \cdots, n_N$ are the numbers of class intervals of corresponding variable distributions [21]. If all variables are stochastically independent, the joint entropy from Equation (6) will be equal to the sum of marginal entropies, which becomes the maximum value of joint entropy. Therefore, the joint entropy is bounded by [21]:

$$0 \leq H(X_1, X_2, \cdots, X_N) \leq \sum_{i=1}^{N} H(X_i) \leq N \log_2 N \quad (7)$$

2.4. Conditional Entropy

Conditional entropy explains a measure of information content of one variable that is not deliverable by other variables. If two random variables, A and B, are correlated, providing information from one variable may clear some uncertainty that the other variable has. In the case of no correlation between variables, the conditional entropy is equal to marginal entropy. That is:

$$H(A|B) = H(A, B) - H(B) \leq H(A) \quad (8)$$

where $H(A|B)$ is conditional entropy of the variable A when the information contents of the variable B is given. One can rewrite Equation (8) as:

$$H(A, B) = H(A|B) + H(B) = H(B|A) + H(A) \quad (9)$$

Furthermore, conditional entropy can be also presented mathematically using joint and conditional probabilities and Bayes theorem as:

$$H(A|B) = -\sum_{i=1}^{N_A}\sum_{j=1}^{N_B} p(a_i, b_j) \log p(a_i|b_j) = -\sum_{i=1}^{N_A}\sum_{j=1}^{N_B} p(a_i, b_j) \log \frac{p(a_i, b_j)}{p(b_j)} \quad (10)$$

2.5. Transinformation

The two variables, A and B, described in Section 2.4 will have some common or shared information, which is called transinformation or mutual information, because they are correlated.

$$T(A, B) = H(A) - H(A|B) = H(B) - H(B|A) = T(B, A) \quad (11)$$

where $T(A, B)$ is transinformation between the variables A and B. The larger the transinformation is, the higher those variables depend on each other. In other words, the transinformation indicates how much information content is transferrable from other variables. Similar to Equation (10), transinformation shall be written as [19]:

$$T(A, B) = \sum_{i=1}^{N_A}\sum_{j=1}^{N_B} p(a_i, b_j) \log \frac{p(a_i|b_j)}{p(a_i)} = \sum_{i=1}^{N_A}\sum_{j=1}^{N_B} p(a_i, b_j) \log \frac{p(a_i, b_j)}{p(a_i)p(b_j)} \quad (12)$$

Transinformation is typically used for measuring mutual information between two variables or two groups of variables as the generalized form for multivariate transinformation is given as:

$$T[(X_1, X_2, \cdots, X_k); \quad (X_{k+1}, X_{k+2}, \cdots, X_N)] = H(X_1, X_2, \cdots, X_k) - H[(X_1, X_2, \cdots, X_k)|(X_{k+1}, X_{k+2}, \cdots, X_N)] \quad (13)$$

2.6. Total Correlation

While transinformation and mutual information have the same definition, total correlation is not equivalent to them as the total correlation is a simple estimate that defines the amount of shared information typically of multiple variables. Simply, the total correlation is defined by the difference between the sum of marginal entropy of N variables and their joint entropy [22,23], which is given as:

$$C(X_1, X_2, \cdots, X_N) = \sum_{i=1}^{N} H(X_i) - H(X_1, X_2, \cdots, X_N) \quad (14)$$

If $N = 2$ in Equation (14), the total correlation will be equal to the transinformation or mutual information. However, the transinformation is only meaningful to two random variables as shown in Equations (11) to (13); therefore, the total correlation and the transinformation values would be different if $N > 2$.

2.7. Other Entropy Terms

The entropy terms described above (i.e., marginal entropy, joint entropy, conditional entropy, transinformation and total correlation) are the basic measures that have been typically used in entropy applications to water monitoring network design. While many studies developed specific approaches and applied for case studies using the basic entropy terms, some have extended the terms beyond them by deriving from or combining the basic measures. The detailed descriptions of the extended entropy terms are not included in this review, but briefly explained when needed in Section 3. Interested readers may refer to the original references.

3. Applications of Entropy to Water Monitoring Network Design

This section summarizes the recent applications of entropy theory to design water monitoring networks. The review was categorized by the types of networks, such as precipitation, streamflow or water level, soil moisture or groundwater and water quality networks. Then later, a hybrid design method for multivariate water monitoring networks was discussed. Table 1 presents brief summaries including network types, methods and key findings of the selected research articles that applied entropy theory for designing the water monitoring network and were published in 2010 or after to cover the most recent contributions since the existing review [10].

Entropy **2017**, *19*, 613

Table 1. Summary of significant contributions to water monitoring network design using entropy (author alphabetical order).

Authors/Year	Network Types	Study Areas	Methods/Entropy Measures	Key Findings
Alameddine et al., 2013 [24]	Water quality	Neuse River Estuary, NC, USA	-Total system entropy -Standard violation entropy -Multiple attribute decision making process -Analytical hierarchical process	-Networks designed using total system entropy and violation entropy of dissolved oxygen were similar -When measured water quality parameters have a low probability of violating water quality standards, their violation entropy is less informative
Alfonso et al., 2010 [25]	Water level	Pijnacker Region, The Netherlands	-Directional information transfer (DIT)	-Introduced total correlation for determining multivariate dependence in water monitoring network design -Information content and redundancy is dependent on the DIT between monitoring stations (DIT$_{XY}$ or DIT$_{YX}$)
Alfonso et al., 2010 [26]	Water level	Pijnacker Region, The Netherlands	-Max(Joint Entropy) min(Total Correlation) -Non-dominated Sorting Genetic Algorithm II (NSGA-II)	-Total correlation should be combined with joint entropy to get most information out of monitoring network
Alfonso et al., 2013 [27]	Streamflow	Magdalena River, Colombia	-Max(Joint Entropy) min(Total Correlation) -Rank-based iterative approach	-Rank method is useful in finding extremes on Pareto front -When iteratively selecting stations, the information content of the network is not guaranteed to be maximum if the network contains the station with the most information
Alfonso et al., 2014 [28]	Water level	North Sea, The Netherlands	-Max(Joint Entropy) min(Total Correlation) -Ensemble entropy -NSGA-II	-By creating an ensemble of solutions through varying the bin size of the initial Pareto optimal solution set, the authors highlight the uncertainty related to choosing bin size
Boroumand and Rajaee, 2017 [29]	Water quality	San Francisco Bay, CA, USA	-Transinformation-distance (T-D) curve	-Using T-D curve they were able to reduce the network from 37 to 21 monitoring stations. -New network covered entire study area without having redundant data
Brunsell, 2010 [30]	Precipitation	Continental United States	-Relative entropy -Wavelet multi-resolution analysis	-The temporal scaling regions identified (1) synoptic, (2) monthly to annual, (3) interannual patterns -Little correlation between relative entropy and annual precipitation except for breakpoint at 95° W Lat
Fahle et al., 2015 [31]	Water level/Groundwater level	Spreewald region, Germany	-MIMR, max(Joint Entropy + Transinformation − Total Correlation) -Subsets of time series data	-Found using subsets of the available time series data could better identify important stations -Showed water levels across network react similarly during high precipitation and are more unique during dry periods -Consequently method can allow for design of network which focuses on floods or droughts

Table 1. *Cont.*

Authors/Year	Network Types	Study Areas	Methods/Entropy Measures	Key Findings
Hosseini and Kerachian, 2017 [32]	Groundwater level	Dehgolan plain, Iran	-Marginal entropy -Data fusion of spatiotemporal kriging and ANN model -Value of information (VOI)	-Network reduction from 52 to 42 (35 high priority and 7 low priority) stations while standard deviation of average estimation error variance stayed the same -Found sampling frequency of high priority stations should be every 20 days and low priority should be every 32, based on analysis of stations selected using VOI
Hosseini and Kerachian, 2017 [33]	Groundwater level	Dehgolan plain, Iran	-Bayesian maximum entropy (BME) -Multi-criteria decision making based on ordered weighted averaging	-Network reduction from 52 to 33 stations while standard deviation of average estimation error variance stayed the same -Sampling frequency increased from 4 weeks to 5 weeks
Keum and Coulibaly, 2017 [34]	Precipitation/Streamflow	Columbia River basin, BC, Canada. Southern Ontario, Canada	-Dual Entropy and Multiobjective Optimization (DEMO) to max(Joint Entropy) and min(Total Correlation)	-Found that networks obtain significant amount of information from 5 to 10 years of data periods, and total correlation tends to be stabilized within 5 years by applying daily time series -Recommended minimum 10 years data periods for designing precipitation or streamflow networks using daily time series
Keum and Coulibaly, 2017 [35]	Integrated	Southern Ontario, Canada	-DEMO to max(Joint Entropy), min(Total Correlation), and max(Conditional Entropy) -Sturge, Scott and rounding binning methods	-Precipitation and streamflow networks were designed simultaneously. -Binning methods were compared and concluded that the optimal networks can be altered due to the binning methods
Kornelsen and Coulibaly, 2015 [36]	Soil Moisture	Great Lakes Basin, Canada-USA	-DEMO to Max(Joint Entropy) min(Total Correlation) -SMOS satellite data	-Optimum networks were different for ascending and descending overpasses -Combining overpass data resulted in complimentary spatial distribution of stations
Leach et al., 2015 [37]	Streamflow	Columbia River basin, BC, Canada. Southern Ontario, Canada	-DEMO to Max(Joint Entropy) min(Total Correlation) -Streamflow signatures -Indicators of hydrologic alteration (IHA)	-Found that including streamflow signatures as design objective increases network coverage in headwater areas. -Found including IHAs increases network coverage in downstream and urban areas.
Leach et al., 2016 [38]	Groundwater level	Southern Ontario, Canada	-DEMO to Max(Joint Entropy) min(Total Correlation) -Annual recharge	-Found that considering spatial distribution of annual recharge can improve network coverage

Table 1. *Cont.*

Authors/Year	Network Types	Study Areas	Methods/Entropy Measures	Key Findings
Lee, 2013 [39]	Water quality	Hagye Basin, South Korea	-Marginal entropy analogous cost function -Genetic algorithm	-Developed computationally efficient way to design a monitoring network in an ungauged basin
Lee et al., 2014 [40]	Water quality	Sanganmi Basin, South Korea	-Multivariate transinformation -Genetic algorithm	-Developed method based on maximizing information content to design a water quality monitoring network in a sewer system
Li et al., 2012 [41]	Streamflow/Water level	Brazos River basin, USA. Pijnacker, The Netherlands	-MIMR, max[Joint Entropy + Transinformation − Total Correlation)	-Developed maximum information minimum redundancy method (MIMR) -Found it to better at locating high information content stations for a monitoring network
Mahjouri and Kerachian, 2011 [42]	Water quality	Jajrood River, Iran	-Information transfer index (ITI) distance and time curves -Micro genetic algorithm (MGA)	-The MGA was used to find the optimal combination of monitoring stations which minimize the temporal and spatial ITI -Found that the sampling frequency and number of stations could be increased in the monitoring network
Mahmoudi-Meimand et al., 2016 [43]	Precipitation	Karkheh, Iran	-Transinformation entropy -Kriging error variance -Weighted cost function to select from Monte Carlo generated networks	-Consideration of spatial analysis error and transinformation entropy improved network design
Masoumi and Kerachian, 2010 [44]	Groundwater quality	Tehran, Iran	-Transinformation-distance (T-D) curve -Transinformation-time (T-T) curve -C-mean clustering -Hybrid genetic algorithm (HGA)	-Developed different T-D curves based on homogeneous clusters of existing monitoring stations -Used HGA to find optimal network with maximum spatial coverage and minimum transinformation -Showed that sampling frequency could be optimized in the same way
Memarzadeh et al., 2013 [45]	Water quality	Karoon River, Iran	-Information transfer index (ITI) distance curve -Homogenous zone clustering -Dynamic factor analysis (DFA)	-Increased monitoring network without increasing redundant information

Table 1. Cont.

Authors/Year	Network Types	Study Areas	Methods/Entropy Measures	Key Findings
Mishra and Coulibaly, 2010 [46]	Streamflow	Selected basins across Canada	-Transinformation index -Marginal, joint, and transinformation	-Used information theory to highlight critical areas across Canada in need of monitoring -Found that several watersheds are information deficient and would benefit from increased monitoring
Mishra and Coulibaly, 2014 [47]	Streamflow	Selected basins across Canada	-Transinformation index -Seasonal streamflow information (SSI)	-Evaluated and highlighted the effects of seasonal climate on streamflow network design
Mondal and Singh, 2012 [48]	Groundwater level	Kodaganar River basin, India	-Marginal entropy, joint entropy, transinformation -Information transfer index (ITI)	-Identified high priority monitoring stations using marginal entropy -ITI was used to evaluate monitoring network, showed that it could be reduced
Samuel et al., 2013 [49]	Streamflow	St. John and St. Lawrence River basins, Canada	-Combined Regionalization-DEMO -Max(Joint Entropy) min(Total Correlation)	-Proposed combined regionalization dual entropy multi-objective optimization approach to design of minimum optimal network that meets World Meteorological Organization (WMO) guidelines -Found that the location of new monitoring stations added to a network depends on the current network density
Santos et al., 2013 [50]	Precipitation	Portugal	-ANN sensitivity analysis -Mutual Information criteria -K-means with Euclidean distance	-Compared three clustering methods to reduce station density -Best method was case dependent -All subset networks reproduced spatial precipitation pattern
Stosic et al., 2017 [51]	Streamflow	Brazos River, TX, USA	-Joint permutation entropy	-Used joint permutation entropy to account for ordering of time series data to better account for station information -Found that the most efficient measurement window was seven days when compared to daily and monthly
Su and You, 2014 [52]	Precipitation	Shihmen Reservoir Taiwan	-Developed 2D transinformation-distance (T-D) model -T-D model used to interpolate network information	-Network designed by maximizing additional information provided by station given regionalized transinformation -Temporal scale has significant influence on information delivery

Table 1. *Cont.*

Authors/Year	Network Types	Study Areas	Methods/Entropy Measures	Key Findings
Uddameri and Andruss, 2014 [53]	Groundwater level	Victoria County Groundwater Conservation District, TX, USA	-Marginal entropy -Monitoring priority index (MPI)	-Compared MPI found using kriging to MPI found using marginal entropy -Showed entropy derived MPI to be more conservative measure
Wei et al., 2014 [54]	Precipitation	Taiwan University Experimental Forest, Taiwan	-Joint Entropy of hourly, monthly, dry/wet months and annual rainfall at 1, 3, 5 km grids	-Station priority changes at different spatiotemporal scales -Temporal scales have more significant changes on joint entropy values than spatial scales -Long time and short spatial scales require fewer stations for stable joint entropy
Werstuck and Coulibaly, 2016 [55]	Streamflow	Ottawa River Basin, Canada	-Transinformation index -DEMO to Max(Joint Entropy) min(Total Correlation)-Streamflow signatures -Indicators of hydrologic alteration (IHA)	-Compared regionalized data from McMaster University-Hydrologiska Byråns Vattenbalansavdelnin (MAC-HBV) and Inverse Distance Weighting—Drainage Area Ratio (IDW-DAR) and found IDW-DAR to be more adequate for generating synthetic time series for potential monitoring stations -Critical areas highlighted by TI index method were the same areas where additional stations were added using DEMO method
Werstuck and Coulibaly, 2017 [56]	Streamflow	Ottawa River Basin, Canada	-Transinformation index -DEMO to max(Joint Entropy) min(Total Correlation)	-Transinformation index analysis is not significantly affected by scaling -Scaling effects are noticeable when DEMO method was applied
Xu et al., 2015 [57]	Precipitation	Xiangjiang River Basin, China	-Mutual Information (MI) of rain gauges -Designed network by min(Σ[MI]), min(bias), max(NSE) -Resampled rainfall used in Xinanjiang and SWAT models	-Lumped model performance was stable with different Pareto optimal networks -Distributed model performance improves with number of stations
Yakirevich et al., 2013 [58]	Groundwater quality	OPE3 research site, Maryland, USA	-Principle of minimum cross entropy (POMCE) -Hydrus-3D	-Using POMCE with two variants of Hydrus-3D, additional monitoring stations were added where the difference between the models was greatest

3.1. Precipitation Networks

The design of a representative precipitation monitoring network is an important and still challenging task for which an entropy approach is well suited. High quality precipitation information is necessary for streamflow and flood forecasting, surface water management, agricultural management, climate process understanding and many other applications. However, precipitation is well known to be highly variable in both space and time [59] and often statistically represented by highly skewed distributions [60] making the application of parametric analysis methods difficult. These challenges also extend to entropy-based approaches for precipitation monitoring. For example, the marginal entropy has been found to be well correlated with total precipitation in northern Brazil because the probability distribution in regions with higher rainfall tended to be more uniform and less skewed [61]. In contrast, Mishra et al. [59] found that the marginal disorder index (MDI), which is the ratio of observed entropy to the maximum possible entropy at a given site, was inversely related to mean annual rainfall in the U.S. state of Texas, where MDI was found to vary seasonally. Brunsell [30] studied the entropy from monitoring stations across the United States where little correlation was found between precipitation and marginal entropy with the exception of a breakpoint in entropy at $-95°$ longitude corresponding to high temporal variability precipitation patterns. It has also been noted by several studies that the temporal sampling of precipitation is an important consideration for calculating entropy and for designing precipitation networks [54,59]. At finer timescales (hourly to daily), precipitation is highly variable resulting in higher overall entropy, whereas longer time periods (monthly to annual) have less variability resulting in lower marginal entropy [30,52,59]. The dependence on spatial and temporal scales has also been identified in a network design application. Wei et al. [54] prioritized potential stations in Central Taiwan to maximize the joint entropy of the network at hourly, monthly and annual temporal scales, as well as 1-, 3- and 5-km spatial scales. They found that priority stations changed with both spatial and temporal scales, where changes in temporal scales resulted in more significant changes in station priority than spatial-rescaling. The decrease in entropy at longer timescales also had an impact on station density where fewer stations were required to reach a stable joint entropy value for longer time scales [54]. These findings demonstrate the important first consideration of network objectives when determining the spatial and temporal sampling used to calculate entropy. However, the research on this topic is still limited, and more work is needed to provide robust guidance on sampling strategies.

Several approaches have been proposed to design or redesign a precipitation monitoring network using one or more entropic measures. Many of these approaches are initialized by building a network around a central station usually selected as the station with the highest marginal entropy [43,62–64]. In urban Rome, Ridolfi et al. [62] selected stations for the precipitation network by sequentially finding the next station that minimized the conditional entropy of the network and adding that station to the network. A similar approach was taken by Yeh et al. [63] to expand a precipitation network in Taiwan. Hourly rainfall data were normalized with a Box-Cox transform and kriging used to interpolate rainfall to candidate grid cells. The joint entropy of the network was calculated using an analytic equation for joint entropy valid for normal data [65], and stations were added sequentially that had the lowest conditional entropy with the rest of the network. The final number of stations needed by the network was accepted when 95% of the network information was captured [63]. Awadallah [64] applied multiple entropy measures sequentially to add stations to a precipitation network. The first new stations were selected as those with the highest entropy. The second station was chosen to minimize the mutual information and the third as the station that maximized conditional entropy.

The aforementioned approaches all sequentially add single stations to a monitoring network based on a single criterion. Mahmoudi-Meimand et al. [43] presented a methodology to add stations to a network based on a multi-variate cost function. Precipitation data were spatially interpolated from existing stations using the kriging approach where the kriging error associated with the rainfall estimation is calculated as the kriging error variance. Their method selected the station that maximized transinformation entropy and minimized error variance using a weighted average of both measures as an objective [43]. This approach balanced the information content in the network with the errors in the interpolation method. Xu et al. [57] used a multi-objective approach to simultaneously select a subset of stations that minimized the sum of pairwise mutual information, minimized bias and maximized Nash–Sutcliffe efficiency. Solutions were generated via Monte Carlo sampling, and network solutions falling along the Pareto front were found as compromise solutions. Coulibaly and Keum [66] and Samuel and Coulibaly [67] also used a multi-objective approach to add stations to snow monitoring networks in Canada. Their approach used a genetic algorithm to find networks that maximized the joint entropy and minimized the total correlation of the network to form a Pareto front of optimal network designs, some of which also included network cost in the optimization [35,67].

A challenge to an entropy-based approach to adding stations to a precipitation network is the requirement to have data available for candidate points. For precipitation, this can be challenging because data at shorter time scales in particular are well known to be non-normal. Most studies use the kriging approach for interpolation [43,63,64] and address the need for normally distributed data using a Box-Cox transform. Samuel and Coulibaly [67] addressed the interpolation problem by using the external data from the Snow Data Assimilation System (SNODAS) for candidate stations. Su and You [52] presented a unique approach to adding stations that maximized the information content of the network. In most literature cases, entropic measures at ungauged sites are determined by interpolating observations of precipitation across a watershed. Su and You [52] calculated the transinformation between neighbouring stations to develop a 2D transinformation-distance relationship. In contrast to transferring data to ungauged stations, this approach transferred transinformation to ungauged stations and selected a site with the maximum transinformation. This approach should be further tested and contrasted with the data transfer approach.

As previously stated, precipitation data are of critical importance for a variety of applications. Despite this, few studies have explored the impact of precipitation networks designed with an entropy approach for actual water resource applications. Applications found in the literature have taken the reasonable approach of using entropy to reduce network density for comparison to a network that included all stations. In Portugal, Santos et al. [50] compared artificial neural networks, K-means clustering and mutual information (MI) criteria for reducing the density of a precipitation network for drought monitoring at different time scales. They found the best performing reduction method was case dependent depending on the region and time scale applied, but noted that all methods performed well. They also found that all subset networks could reliably reproduce the spatial precipitation pattern. Xu et al. [57] used the multi-objective approach previously described to select a subset of precipitation stations from a dense network in the Xiangjiang River Basin in China. Rainfall from the subset networks was used to force the lumped Xinanjiang hydrological model [68] and the distributed SWAT hydrological model [69]. The author's found that lumped model performance became stable with a subset of 20 to 25 stations, whereas the distributed model's performance continued to increase as more stations were added to the network [57]. These analyses are important to demonstrate the utility of precipitation networks and the advantages of entropy-based approaches in designing precipitation networks.

3.2. Streamflow and Water Level Networks

Water quantity monitoring, such as streamflow rates and water level, is one of the essential tasks for water management to prevent damage to nature and human beings from flooding. A successful floodplain management or flood forecasting and warning system can be feasible through expert forecasters who implement well-calibrated models and reliable tools using quality data [70]. The design of water quantity monitoring network has been well implemented because of not only the good performance of entropy-based methods, but also the unaffectedness by the zero effect, which is caused by discontinuity of probability density function due to zero values in data, except for the ephemeral or intermittent streams. To deal with the zero effect in entropy calculations, Chapman [71] and Gong et al. [60] separated the marginal entropy Equation (3) to nonzero terms and zero values, which are certain. While Gong et al. [60] summarized the possible issues in entropy calculations from hydrologic data as effects due to zero values, histogram binning including skewness consideration and measurement errors, some studies noticed that the length and the location of time window also affect entropy calculations and the corresponding network design. Fahle et al. [31] observed the temporal variability of station rankings by shifting the time window for the design of water level network of a ditch system in Germany. Mishra and Coulibaly [47] also found the dependency of the seasonality on the efficiency of hydrometric networks. Stosic et al. [51] found an inverse relationship between the network density and sampling time interval as the larger number of monitoring stations is required if the time interval is shorter and vice versa. Keum and Coulibaly [34] analyzed the temporal changes of entropy measures and optimal networks by applying daily time series for streamflow network design. They found that the information gain of a monitoring network is not significant when the length of time series is longer than 10 years, and the total correlation tends to stabilize within five years of data. The optimal networks using the data lengths of 5, 10, 15 and 20 years also show that there are no significant differences in the results from 10 years or longer while the optimal network using five years of data was evidently different from others. Werstuck and Coulibaly [56] analyzed scaling effects by considering two study areas. Specifically, one study area is a small watershed, which is a part of another study area. After applying the transinformation analysis and the multi-objective optimization, they concluded that the optimal networks tend to be affected by scaling while transinformation index does not.

Mishra and Coulibaly [46] evaluated the effects of the class intervals and the infilling missing data by applying the linear regression method to daily time series and concluded that the station rankings based on the transinformation values were not significantly changed. Li et al. [41] also investigated the changes of station rankings based on the maximum information minimum redundancy (MIMR) approach and obtained the similar conclusion. However, Fahle et al. [31] and Keum and Coulibaly [35] drew the opposite opinion that station rankings can be affected by the binning method that defines the class intervals. The conflict comes from the selection of the binning methods compared. The former group applied different parameters to a single binning method, the mathematical floor function. However, the latter group compared other binning methods with the floor function. Considering that Alfonso et al. [25] found that the design solutions were not common in some cases from the sensitivity analysis of the parameter of the mathematical floor function, it is not recommended to use a specific binning method without any consideration.

As discussed in the review of precipitation networks in Section 3.1, network redesign and network expansion require data at candidate locations, which are ungauged. Alfonso et al. [27] applied a one-dimensional hydrodynamic model to generate the discharge time series. The model estimated discharge at each segment, which divides rivers with approximately 200 m increments longitudinally. The use of hydrodynamic model enabled to determine the critical monitoring locations in the main stream and its tributaries. On the other hand, Samuel et al. [49] combined regionalization techniques with entropy calculation in order to estimate the discharge at candidate locations. They compared the performance of various regionalization methods including not only a conceptual hydrologic model, but also spatial proximity, physical similarity and their combinations with drainage area ratio. Based on the performance statistics by applying multiple basins, inverse distance weighting coupled with drainage area ratio performed the best, and this conclusion has been adopted in several studies [34,35,37,55].

Some studies have extended the entropy applications for the streamflow monitoring network design. Stosic et al. [51] proposed the concept of permutation entropy, which is able to differentiate based on the order of sequential observations, as well as the histogram frequency in basic Shannon entropy measures. Even though histograms from two different observations are the same, the permutation entropy value tends to be higher if there are more variations between time steps. However, the network design studies using the permutation entropy are still limited. On the other hand, Leach et al. [37] applied additional features to the network design. While the common objectives in water monitoring network design using an optimization technique are to maximize the information and to minimize the redundancy in the network, they additionally considered the physical properties of watersheds, such as the streamflow signatures [72,73] and the indicators of hydrologic alterations (IHAs) [74,75]. After the comparison of the optimal streamflow monitoring networks with and without considerations of the streamflow signatures and IHAs, it was concluded that inclusion of basin physical characteristics yielded a better coverage of the selected locations of the optimal networks.

3.3. Soil Moisture and Groundwater Networks

Soil moisture is a critical water variable as the interface between the atmosphere and subsurface. Unfortunately, the monitoring of soil moisture is very sparse compared to its spatial variability. To design an optimum network for monitoring soil moisture in the Great Lakes Basin, Kornelsen and Coulibaly [36] proposed using data from the Soil Moisture and Ocean Salinity (SMOS) satellite [76] to design a soil moisture monitoring network using the DEMO algorithm of Samuel et al. [49]. Grid cells were selected to add monitoring stations that optimally maximized joint entropy while minimizing total correlation using only the satellite data. The ascending and descending overpasses were found to contain different information, and the spatial distribution of a network designed with both overpasses was found to contain complimentary features from both datasets [36].

Groundwater monitoring allows for a better understanding of the hydrogeology in an area. This is achieved through groundwater quality and quantity monitoring. Groundwater quality monitoring is used to detect contaminant plumes or for long-term monitoring (LTM) of post remediation effects, and groundwater quantity monitoring is used to determine available water for drinking, irrigation and industry. However, monitoring groundwater is inherently difficult due to physical barriers between observers and the water. Through the understanding of subsurface flow physics and with flow and contaminant transport models such as MODFLOW, MODPATH and MT3D [77–79], we can simulate the behaviour of groundwater. Unfortunately, our simulations are not always accurate, and the models require real-world observations to be calibrated and validated. Due to constraints such as accessibility and financial cost, it is not feasible to monitor at every possible location in an area of interest. It is instead ideal for an optimal monitoring network to be designed to allow for the best placement of monitoring stations and to determine the ideal measurement frequencies. The merit of using information theory entropy has been shown in several cases of groundwater network design [31–33,38,44,48,53,58].

Various methods that utilize information theory entropy have been developed for use in designing optimal groundwater monitoring networks. These include the use of entropy measures in both single and multi-objective optimization problems and are used in network reduction [32,33], expansion [38,58] and redesign [44], as well as have been used to highlight vulnerable areas in an area that should be monitored [53]. In identifying vulnerable areas in the Victoria County Groundwater Conservation District (VCGCD) in Texas, USA, Uddameri and Andruss [53] developed a monitoring priority index (MPI) based on a weighted stakeholder preference to highlight the areas of interest. They compared kriging standard deviation and marginal entropy as metrics to characterize groundwater variability and found entropy to be the more conservative metric.

In areas where there is excessive monitoring, Mondal and Singh [48] showed the information transfer index (ITI), the quotient of joint entropy and transinformation, could be used to evaluate the existing monitoring network. Through this evaluation, redundant monitoring stations (wells) could be identified and removed from the groundwater monitoring network. It may also be the case that the existing groundwater monitoring network is not adequate and additional monitoring stations are needed. Yakirevich et al. [58] developed a method that utilizes minimum cross entropy (MCE) to sequentially add monitoring stations to a network. MCE was used as a metric to quantify the difference between two variants of a Hydrus-3D model [80], and the monitoring stations were added to the network where the difference between models was largest. A multi-objective approach for adding monitoring stations to a groundwater monitoring network was applied by Leach et al. [38], which utilized two entropy measures, total correlation and joint entropy and a metric used to quantify the spatial distribution of annual recharge; the results of which were used to develop maps that highlight areas in which additional monitoring stations should be added. The majority of network design experiments look at the entire available time series when calculating entropy measures; however, Fahle et al. [31] showed that using a combination of MIMR and subsets of the data series could be more ideal. The subsets were used to represent the intra-annual variability of groundwater levels. This method identified locations which were consistently important through each subset and found that monitoring stations showed similarities during wet periods and uniqueness during dry periods. Fahle et al. [31] also suggest that a consequence of using subsets of data allows for the design of a network, which can be focused on floods or droughts.

One issue that can arise with entropy-based methods is the need for lengthy data series to produce accurate measures of entropy. Unfortunately, the area of interest for new monitoring stations will not have available data for all possible locations. To work around this limitation transinformation-distance (T-D) curves have been applied in the design of optimal groundwater monitoring networks [44,81]. In these studies, T-D curves were developed for sub areas within the desired study area based in different clustering methods. Additionally, Masoumi and Kerachian [44] showed that this method could be applied temporally as transinformation-time curves which could then be used to optimize the temporal sampling frequency of the stations. It should be noted that both previously mentioned studies were applied in the same study area using slightly different methods for clustering monitoring stations, and both produced different groundwater monitoring networks that could be considered optimal. This highlights an issue with optimal monitoring network design in that it can be subjective and does not have a singular solution. A comparison of Hosseini and Kerachian [32,33] also illustrates this issue, where through the use of different entropy measures, marginal entropy and Bayesian maximum entropy and optimization techniques, one experiment found the optimal monitoring network included 42 monitoring stations while the other only included 33 stations.

3.4. Water Quality Networks

The importance of water quality monitoring networks is their ability to assist in identifying those parameters that exceed water quality standards. Several water quality monitoring strategies, including two methods that utilized entropy measures [42,45], were recently reviewed by Behmel et al. [15]. This review found that identifying a single approach to water quality monitoring network design would be virtually impossible. Despite this, various applications of the transinformation-distance curve methods have shown promise in the optimal redesign and reduction of water quality monitoring networks [29,42,45]. Lee [39] found that by maximizing the multivariate transinformation between chosen and unchosen stations, using the storm water management model to simulate the total suspended solids and a GA for optimization, an optimal water quality network could be designed for a sewer system. Banik et al. [82] compared information theory, detection time and reliability measures for the design of a sewer system monitoring network through both single and multi-objective optimization approaches. It was shown that for a small monitoring network, the methods had similar performances, while the single objective detection time-based method had slightly better performance when the number of monitoring station is larger. Alameddine et al. [24] used exceedance probabilities to determine violation entropy of dissolved oxygen and chlorophyll-a in the Neuse River estuary. Along with violation entropy, the total system entropy was used as a measure to identify areas of importance of monitoring. A multi-objective optimization scheme based on expert assigned weights was used to develop a compromise solution from the three entropy measures. Ultimately, the method allowed for the identification of high uncertainty areas, which would benefit from future water quality monitoring. Data availability is an issue when using entropy methods, particularly when attempting to use them in the design of a monitoring network in an ungauged basin. To address this, Lee et al. [40] developed a method that uses a measure analogous to marginal entropy. This method uses characteristics of the basin such as the length and number of reaches in the river network as part of the cost function, which is then optimized using a combined GA and filtering algorithm. This was shown to be a computationally-efficient method for use in optimal network design of an ungauged river basin.

3.5. Integrated Network Design

To the best of our knowledge, almost all of the previous studies about water monitoring network design have focused on a specific network type (i.e., considering a single hydrologic variable in each study) as reviewed in the previous sections. However, considering that hydrologic processes are interconnected in a water cycle, there are causes and effects between hydrologic variables. For instance, if a noticeable amount of precipitation occurs, streamflow or groundwater level is likely increased; hence, the information content of a variable may affect that of other variables. Keum and Coulibaly [35] developed a multivariate network design method by taking conditional entropy as the measure of information that is independent to a given variable. In their study, the method designed precipitation and streamflow monitoring networks simultaneously. Specifically, the method followed the traditional multi-objective approach that maximizes joint entropy and minimizes total correlation, but added another objective that maximizes conditional entropy of streamflow network given precipitation network to mimic the direction of the water cycle as streamflow may fluctuate due to precipitation. After comparing the integrated design with the single-variable design, their results showed that the effectiveness of network integration mostly came from reducing the number of additional precipitation stations. It was also found that the integrated network design approach allows adding a precipitation station at a location that will benefit the stream gauge network.

4. Conclusions and Recommendations

It is evident that successful water management cannot be achieved without proper water monitoring networks. Although there has been much progress in network design methods and applications, a standardized design methodology has not yet emerged. After the pioneering invention of information theory in the 1940s, entropy concepts have been applied in various applications with recent efforts on network design problems. The unique benefit of this approach is that a water monitoring network can be evaluated or designed based on the information the network monitors, which is in contrast to the set station densities proposed by WMO guidelines; the advantage of the former being that a network could be better tailored to specific applications or optimized to provide the most gain at densities lower than those suggested in WMO guidelines. In addition, when combined with multi-objective optimization techniques, users' specific criteria can be included in the optimal network design process.

This manuscript provides a comprehensive review of the recent research attainments and their applications in entropy-based water monitoring network design. The literature has demonstrated the use of various information theory measures and adaptations thereof for use in network design with an emerging consensus that the goal of these network design methods is to select the stations that provide the most information to the monitoring network while simultaneously being independent of each other. Through rigorous testing, information theory has proven to be a robust tool to use when evaluating and designing an optimal water monitoring network. However, when it comes to evaluating the optimal design, there are still issues that need to be addressed.

The first is that an optimal monitoring network design can be found based on specified design criteria; however, the practical application of the new optimal monitoring network is rarely evaluated in a hydrologic or other model [11,57]. This type of numerical experiment is an important requirement to evaluate the utility of a network rather than just identifying its optimality or information content. Further, it is an important exercise to identify the benefits of entropy-based network designs in order to convince decision makers of the importance of adopting entropy approaches.

Another issue with the optimal network is that it can be subjective, based on choices made in the calculation of entropy and the design method chosen, especially when additional objective functions are considered in the design. This extends to the method selected for finding the optimal monitoring network, whether it is found using an iterative method where one station is added at a time or a collection of stations is added all at once. Research has also shown that data length, catchment scale and ordering can influence the design of an optimal network [31,34,51]. Finally, when using discrete entropy, the binning method has been shown to influence the final network design [35]. The influence of binning on entropy calculation has received greater attention in other geophysical network design applications [83–85], and similar consideration should be given in the field of water resources, particularly owing to the unique and difficult nature of water variables (e.g., streamflow, precipitation) spatial and temporal distribution [30,60]. Thus, explicit consideration is needed when choosing the bins based on the intended application of the monitoring network and further research to provide guidance specific to water monitoring networks. Therefore, despite the possibility of finding an optimal network design in a formal sense, the subjectivity induced by the designer's choices, and the lack of standardized design methods, must be recognized. Future research should focus on comparative studies among multiple entropy design methods, discretization approaches and data characteristics. The current literature provides many novel entropy design approaches and the evolution of concepts, but rigorous comparisons are critical to provide generic guidelines for network design. Despite the potential sources of subjectivity identified, entropy methods remain one of the most objective approaches for network design.

In particular, more work is needed on spatial and temporal scaling of data for entropy calculation to provide robust guidance to decision makers. Many new methods and optimization techniques have been reviewed herein, but few examples were found in the literature that explored the data characteristics used in those techniques. Further research is required to provide guidance on the proper length of data in water monitoring network design [34], the sampling frequency of the data [54] and the spatial scale at which information should be measured for various monitoring network applications.

The aforementioned issues are considered crucial gaps that need to be filled to enable practical recommendations or guidelines for a widespread adoption of entropy approaches for designing optimal water monitoring networks. In addition, the comparative studies of entropy-based methods reviewed herein should be robustly compared to network design methods from other disciplines, such as geostatistics, to identify areas of equivalence and disparity [10]. Considerable advances have occurred over the past decade as reviewed herein, and measures derived from Shannon's base equation [16] have reached a high level of maturity for the task of network design. We challenge the research community to put a similar creativity into the joint consideration of the nexus of data characteristics, network design and applications, all of which are intricately linked.

Acknowledgments: This research was supported by the Natural Science and Engineering Research Council (NSERC) of Canada, Grant NSERC Canadian FloodNet (NETGP-451456).

Author Contributions: The structure of this review paper was built and the contents were written by all authors. Specifically, Jongho Keum contributed Sections 1, 2, 3.2 and 3.5. Kurt C. Kornelsen contributed Sections 3.1 and 3.3. James M. Leach contributed Sections 3.3, 3.4 and 4. Paulin Coulibaly defined the scope of the review and supervised and reviewed the entire manuscript. Although the primary contributing sections are as indicated above, all authors cross-reviewed other co-authors' sections. All authors have read and approved the final manuscript.

Conflicts of Interest: The authors declare no conflict of interest.

References

1. Langbein, W.B. Overview of Conference on Hydrologic Data Networks. *Water Resour. Res.* **1979**, *15*, 1867–1871. [CrossRef]
2. Herschy, R.W. *Hydrometry: Principles and Practice*, 2nd ed.; John Wiley and Sons Ltd.: Chichester, UK, 1999.
3. Boiten, W. *Hydrometry*; A.A. Balkema Publishers: Lisse, The Netherlands, 2003.
4. Nemec, J.; Askew, A.J. Mean and variance in network-design philosophies. In *Integrated Design of Hydrological Networks (Proceedings of the Budapest Symposium)*; Moss, M.E., Ed.; International Association of Hydrological Sciences Publication: Washington, DC, USA, 1986; pp. 123–131.

5. Rodda, J.C.; Langbein, W.B. *Hydrological Network Design—Needs, Problems and Approaches*; World Meteorological Organization: Geneva, Switzerland, 1969.

6. World Meteorological Organization. *Casebook on Hydrological Network Design Practice*; Langbein, W.B., Ed.; World Meteorological Organization: Geneva, Switzerland, 1972.

7. Davis, D.R.; Duckstein, L.; Krzysztofowicz, R. The Worth of Hydrologic Data for Nonoptimal Decision Making. *Water Resour. Res.* **1979**, *15*, 1733–1742. [CrossRef]

8. Pilon, P.J.; Yuzyk, T.R.; Hale, R.A.; Day, T.J. Challenges Facing Surface Water Monitoring in Canada. *Can. Water Resour. J.* **1996**, *21*, 157–164. [CrossRef]

9. U.S. Geological Survey. *Streamflow Information for the Next Century—A Plan for the National Streamflow Information Program of the U.S. Geological Survey*; U.S. Geological Survey: Denver, CO, USA, 1999.

10. Mishra, A.K.; Coulibaly, P. Developments in Hydrometric Network Design: A Review. *Rev. Geophys.* **2009**, *47*. [CrossRef]

11. Chacon-hurtado, J.C.; Alfonso, L.; Solomatine, D.P. Rainfall and streamflow sensor network design: A review of applications, classification, and a proposed framework. *Hydrol. Earth Syst. Sci.* **2017**, *21*, 3071–3091. [CrossRef]

12. Moss, M.E. *Concepts and Techniques in Hydrological Network Design*; World Meteorological Organization: Geneva, Switzerland, 1982.

13. Van der Made, J.W.; Schilperoort, T.; van der Schaaf, S.; Buishand, T.A.; Brouwer, G.K.; van Duyvenbooden, W.; Becinsky, P. *Design Aspects of Hydrological Networks*; Commissie voor Hydrologisch Onderzoek TNO: The Hague, The Netherlands, 1986.

14. Pyrce, R.S. *Review and Analysis of Stream Gauge Networks for the Ontario Stream Gauge Rehabilitation Project*, 2nd ed.; Watershed Science Centre: Peterborough, ON, Canada, 2004.

15. Behmel, S.; Damour, M.; Ludwig, R.; Rodriguez, M.J. Water quality monitoring strategies—A review and future perspectives. *Sci. Total Environ.* **2016**, *571*, 1312–1329. [CrossRef] [PubMed]

16. Shannon, C.E. A Mathematical Theory of Communication. *Bell Syst. Tech. J.* **1948**, *27*, 379–423. [CrossRef]

17. Batty, M. Space, scale, and scaling in entropy maximizing. *Geogr. Anal.* **2010**, *42*, 395–421. [CrossRef]

18. Singh, V.P. *Entropy Theory in Hydrologic Science and Engineering*; McGraw-Hill Education: New York, NY, USA, 2015.

19. Lathi, B.P. *An Introduction to Random Signals and Communication Theory*; International Textbook Company: Scranton, PA, USA, 1968.

20. Tribus, M. *Rational Descriptions, Decisions and Designs*; Irvine, T.F., Hartnett, J.P., Eds.; Pergamon Press: Oxford, UK, 1969.

21. Krstanovic, P.F.; Singh, V.P. Evaluation of rainfall networks using entropy: I. Theoretical development. *Water Resour. Manag.* **1992**, *6*, 279–293. [CrossRef]

22. McGill, W.J. Multivariate information transmission. *Psychometrika* **1954**, *19*, 97–116. [CrossRef]

23. Watanabe, S. Information Theoretical Analysis of Multivariate Correlation. *IBM J. Res. Dev.* **1960**, *4*, 66–82. [CrossRef]

24. Alameddine, I.; Karmakar, S.; Qian, S.S.; Paerl, H.W.; Reckhow, K.H. Optimizing an estuarine water quality monitoring program through an entropy-based hierarchical spatiotemporal Bayesian framework. *Water Resour. Res.* **2013**, *49*, 6933–6945. [CrossRef]

25. Alfonso, L.; Lobbrecht, A.; Price, R. Information theory-based approach for location of monitoring water level gauges in polders. *Water Resour. Res.* **2010**, *46*. [CrossRef]

26. Alfonso, L.; Lobbrecht, A.; Price, R. Optimization of water level monitoring network in polder systems using information theory. *Water Resour. Res.* **2010**, *46*. [CrossRef]

27. Alfonso, L.; He, L.; Lobbrecht, A.; Price, R. Information theory applied to evaluate the discharge monitoring network of the Magdalena River. *J. Hydroinform.* **2013**, *15*, 211–228. [CrossRef]

28. Alfonso, L.; Ridolfi, E.; Gaytan-Aguilar, S.; Napolitano, F.; Russo, F. Ensemble Entropy for Monitoring Network Design. *Entropy* **2014**, *16*, 1365–1375. [CrossRef]

29. Boroumand, A.; Rajaee, T. Discrete entropy theory for optimal redesigning of salinity monitoring network in San Francisco bay. *Water Sci. Technol. Water Supply* **2017**, *17*, 606–612. [CrossRef]

30. Brunsell, N.A. A multiscale information theory approach to assess spatial-temporal variability of daily precipitation. *J. Hydrol.* **2010**, *385*, 165–172. [CrossRef]

31. Fahle, M.; Hohenbrink, T.L.; Dietrich, O.; Lischeid, G. Temporal variability of the optimal monitoring setup assessed using information theory. *Water Resour. Res.* **2015**, *51*, 7723–7743. [CrossRef]
32. Hosseini, M.; Kerachian, R. A data fusion-based methodology for optimal redesign of groundwater monitoring networks. *J. Hydrol.* **2017**, *552*, 267–282. [CrossRef]
33. Hosseini, M.; Kerachian, R. A Bayesian maximum entropy-based methodology for optimal spatiotemporal design of groundwater monitoring networks. *Environ. Monit. Assess.* **2017**, *189*, 433. [CrossRef] [PubMed]
34. Keum, J.; Coulibaly, P. Sensitivity of Entropy Method to Time Series Length in Hydrometric Network Design. *J. Hydrol. Eng.* **2017**, *22*. [CrossRef]
35. Keum, J.; Coulibaly, P. Information theory-based decision support system for integrated design of multi-variable hydrometric networks. *Water Resour. Res.* **2017**, *53*, 6239–6259. [CrossRef]
36. Kornelsen, K.C.; Coulibaly, P. Design of an Optimal Soil Moisture Monitoring Network Using SMOS Retrieved Soil Moisture. *IEEE Trans. Geosci. Remote Sens.* **2015**, *53*, 3950–3959. [CrossRef]
37. Leach, J.M.; Kornelsen, K.C.; Samuel, J.; Coulibaly, P. Hydrometric network design using streamflow signatures and indicators of hydrologic alteration. *J. Hydrol.* **2015**, *529*, 1350–1359. [CrossRef]
38. Leach, J.M.; Coulibaly, P.; Guo, Y. Entropy based groundwater monitoring network design considering spatial distribution of annual recharge. *Adv. Water Resour.* **2016**, *96*, 108–119. [CrossRef]
39. Lee, J.H. Determination of optimal water quality monitoring points in sewer systems using entropy theory. *Entropy* **2013**, *15*, 3419–3434. [CrossRef]
40. Lee, C.; Paik, K.; Yoo, D.G.; Kim, J.H. Efficient method for optimal placing of water quality monitoring stations for an ungauged basin. *J. Environ. Manag.* **2014**, *132*, 24–31. [CrossRef] [PubMed]
41. Li, C.; Singh, V.P.; Mishra, A.K. Entropy theory-based criterion for hydrometric network evaluation and design: Maximum information minimum redundancy. *Water Resour. Res.* **2012**, *48*. [CrossRef]
42. Mahjouri, N.; Kerachian, R. Revising river water quality monitoring networks using discrete entropy theory: The Jajrood River experience. *Environ. Monit. Assess.* **2011**, *175*, 291–302. [CrossRef] [PubMed]
43. Mahmoudi-Meimand, H.; Nazif, S.; Abbaspour, R.A.; Sabokbar, H.F. An algorithm for optimisation of a rain gauge network based on geostatistics and entropy concepts using GIS. *J. Spat. Sci.* **2016**, *61*, 233–252. [CrossRef]
44. Masoumi, F.; Kerachian, R. Optimal redesign of groundwater quality monitoring networks: A case study. *Environ. Monit. Assess.* **2010**, *161*, 247–257. [CrossRef] [PubMed]
45. Memarzadeh, M.; Mahjouri, N.; Kerachian, R. Evaluating sampling locations in river water quality monitoring networks: Application of dynamic factor analysis and discrete entropy theory. *Environ. Earth Sci.* **2013**, *70*, 2577–2585. [CrossRef]
46. Mishra, A.K.; Coulibaly, P. Hydrometric Network Evaluation for Canadian Watersheds. *J. Hydrol.* **2010**, *380*, 420–437. [CrossRef]
47. Mishra, A.K.; Coulibaly, P. Variability in Canadian Seasonal Streamflow Information and Its Implication for Hydrometric Network Design. *J. Hydrol. Eng.* **2014**, *19*. [CrossRef]
48. Mondal, N.C.; Singh, V.P. Evaluation of groundwater monitoring network of Kodaganar River basin from Southern India using entropy. *Environ. Earth Sci.* **2011**, *66*, 1183–1193. [CrossRef]
49. Samuel, J.; Coulibaly, P.; Kollat, J.B. CRDEMO: Combined Regionalization and Dual Entropy-Multiobjective Optimization for Hydrometric Network Design. *Water Resour. Res.* **2013**, *49*, 8070–8089. [CrossRef]
50. Santos, J.F.; Portela, M.M.; Pulido-Calvo, I. Dimensionality reduction in drought modelling. *Hydrol. Process.* **2013**, *27*, 1399–1410. [CrossRef]
51. Stosic, T.; Stosic, B.; Singh, V.P. Optimizing streamflow monitoring networks using joint permutation entropy. *J. Hydrol.* **2017**, *552*, 306–312. [CrossRef]
52. Su, H.T.; You, G.J.Y. Developing an entropy-based model of spatial information estimation and its application in the design of precipitation gauge networks. *J. Hydrol.* **2014**, *519*, 3316–3327. [CrossRef]
53. Uddameri, V.; Andruss, T. A GIS-based multi-criteria decision-making approach for establishing a regional-scale groundwater monitoring. *Environ. Earth Sci.* **2014**, *71*, 2617–2628. [CrossRef]
54. Wei, C.; Yeh, H.C.; Chen, Y.C. Spatiotemporal scaling effect on rainfall network design using entropy. *Entropy* **2014**, *16*, 4626–4647. [CrossRef]
55. Werstuck, C.; Coulibaly, P. Hydrometric network design using dual entropy multi-objective optimization in the Ottawa River Basin. *Hydrol. Res.* **2016**, *48*, 1–13. [CrossRef]

56. Werstuck, C.; Coulibaly, P. Assessing Spatial Scale Effects on Hydrometric Network Design Using Entropy and Multi-Objective Methods. *JAWRA J. Am. Water Resour. Assoc.* **2017**, in press.

57. Xu, H.; Xu, C.-Y.; Sælthun, N.R.; Xu, Y.; Zhou, B.; Chen, H. Entropy theory based multi-criteria resampling of rain gauge networks for hydrological modelling—A case study of humid area in southern China. *J. Hydrol.* **2015**, *525*, 138–151. [CrossRef]

58. Yakirevich, A.; Pachepsky, Y.A.; Gish, T.J.; Guber, A.K.; Kuznetsov, M.Y.; Cady, R.E.; Nicholson, T.J. Augmentation of groundwater monitoring networks using information theory and ensemble modeling with pedotransfer functions. *J. Hydrol.* **2013**, *501*, 13–24. [CrossRef]

59. Mishra, A.K.; Özger, M.; Singh, V.P. An entropy-based investigation into the variability of precipitation. *J. Hydrol.* **2009**, *370*, 139–154. [CrossRef]

60. Gong, W.; Yang, D.; Gupta, H.V.; Nearing, G. Estimating information entropy for hydrological data: One dimensional case. *Water Resour. Res.* **2014**, *50*, 5003–5018. [CrossRef]

61. Silva, V.; da Silva, V.d.P.R.; Belo Filho, A.F.; Singh, V.P.; Almeida, R.S.R.; da Silva, B.B.; de Sousa, I.F.; de Holanda, R.M. Entropy theory for analysing water resources in northeastern region of Brazil. *Hydrol. Sci. J.* **2017**, *62*, 1029–1038.

62. Ridolfi, E.; Montesarchio, V.; Russo, F.; Napolitano, F. An entropy approach for evaluating the maximum information content achievable by an urban rainfall network. *Nat. Hazards Earth Syst. Sci.* **2011**, *11*, 2075–2083. [CrossRef]

63. Yeh, H.C.; Chen, Y.C.; Wei, C.; Chen, R.H. Entropy and kriging approach to rainfall network design. *Paddy Water Environ.* **2011**, *9*, 343–355. [CrossRef]

64. Awadallah, A.G. Selecting optimum locations of rainfall stations using kriging and entropy. *Int. J. Civ. Environ. Eng* **2012**, *12*, 36–41.

65. Ahmed, N.A.; Gokhale, D.V. Entropy expressions and their estimators for multivariate distributions. *IEEE Trans. Inf. Theory* **1989**, *35*, 688–692. [CrossRef]

66. Coulibaly, P.; Keum, J. *Snow Network Design and Evaluation for La Grande River Basin*; Hydro-Quebec: Hamilton, ON, Canada, 2016.

67. Samuel, J.; Coulibaly, P. *Design of Optimum Snow Monitoring Networks Using Dual Entropy Multiobjective Optimization (DEMO) with Remote Sensing Data: Case Study for Columbia River Basin*; BC-Hydro: Hamilton, ON, Canada, 2013.

68. Zhao, R.J. The Xinanjiang model applied in China. *J. Hydrol.* **1992**, *135*, 371–381.

69. Arnold, J.G.; Srinivasan, R.; Muttiah, R.S.; Williams, J.R. Large Area Hydrologic Modeling and Assessment Part I: Model Development. *J. Am. Water Resour. Assoc.* **1998**, *34*, 73–89. [CrossRef]

70. United Nations. *Guidelines for Reducing Flood Losses*; Pilon, P.J., Ed.; United Nations: Geneva, Switzerland, 2004.

71. Chapman, T.G. Entropy as a measure of hydrologic data uncertainty and model performance. *J. Hydrol.* **1986**, *85*, 111–126. [CrossRef]

72. Yadav, M.; Wagener, T.; Gupta, H. Regionalization of constraints on expected watershed response behavior for improved predictions in ungauged basins. *Adv. Water Resour.* **2007**, *30*, 1756–1774. [CrossRef]

73. Sawicz, K.; Wagener, T.; Sivapalan, M.; Troch, P.A.; Carrillo, G. Catchment classification: Empirical analysis of hydrologic similarity based on catchment function in the eastern USA. *Hydrol. Earth Syst. Sci.* **2011**, *15*, 2895–2911. [CrossRef]

74. Richter, B.D.; Baumgartner, J.V.; Powell, J.; Braun, D.P. A Method for Assessing Hydrologic Alteration within Ecosystems. *Conserv. Biol.* **1996**, *10*, 1163–1174. [CrossRef]

75. Monk, W.A.; Peters, D.L.; Allen Curry, R.; Baird, D.J. Quantifying trends in indicator hydroecological variables for regime-based groups of Canadian rivers. *Hydrol. Process.* **2011**, *25*, 3086–3100. [CrossRef]

76. Kerr, Y.H.; Waldteufel, P.; Wigneron, J.P.; Delwart, S.; Cabot, F.; Boutin, J.; Escorihuela, M.J.; Font, J.; Reul, N.; Gruhier, C.; et al. The SMOS Mission: New Tool for Monitoring Key Elements of the Global Water Cycle. *Proc. IEEE* **2010**, *98*, 666–687. [CrossRef]

77. Harbaugh, A.W. *MODFLOW-2005, The U.S. Geological Survey Modular Ground-Water Model—The Ground-Water Flow Process*; U.S. Geological Survey: Reston, VA, USA, 2005.

78. Pollock, D.W. *User Guide for MODPATH Version 7—A Particle-Tracking Model for MODFLOW*; U.S. Geological Survey: Reston, VA, USA, 2016.

79. Bedekar, V.; Morway, E.D.; Langevin, C.D.; Tonkin, M.J. *MT3D-USGS Version 1: A U.S. Geological Survey Release of MT3DMS Updated with New and Expanded Transport Capabilities for Use with MODFLOW*; U.S. Geological Survey: Reston, VA, USA, 2016.

80. Šimůnek, J.; van Genuchten, M.T.; Šejna, M. *The HYDRUS Software Package for Simulating Two- and Three-Dimensional Movement of Water, Heat, and Multiple Solutes in Variably-Saturated Porus Media*; PC Progress: Prague, Czech Republic, 2012.

81. Owlia, R.R.; Abrishamchi, A.; Tajrishy, M. Spatial-temporal assessment and redesign of groundwater quality monitoring network: A case study. *Environ. Monit. Assess.* **2011**, *172*, 263–273. [CrossRef] [PubMed]

82. Banik, B.K.; Alfonso, L.; di Cristo, C.; Leopardi, A.; Mynett, A. Evaluation of Different Formulations to Optimally Locate Sensors in Sewer Systems. *J. Water Resour. Plan. Manag.* **2017**, *143*. [CrossRef]

83. Ruddell, B.L.; Kumar, P. Ecohydrologic process networks: 1. Identification. *Water Resour. Res.* **2009**, *45*. [CrossRef]

84. Ruddell, B.L.; Kumar, P. Ecohydrologic process networks: 2. Analysis and characterization. *Water Resour. Res.* **2009**, *45*, 1–14. [CrossRef]

85. Kang, M.; Ruddell, B.L.; Cho, C.; Chun, J.; Kim, J. Agricultural and Forest Meteorology Identifying CO_2 advection on a hill slope using information flow. *Agric. For. Meteorol.* **2017**, *232*, 265–278. [CrossRef]

entropy

MDPI

Article

Rainfall Network Optimization Using Radar and Entropy

Hui-Chung Yeh [1], Yen-Chang Chen [2], Che-Hao Chang [2], Cheng-Hsuan Ho [2] and Chiang Wei [3,*]

[1] Department of Natural Resources, Chinese Culture University, Taipei 11114, Taiwan;
    xhz@faculty.pccu.edu.tw
[2] Department of Civil Engineering, National Taipei University of Technology, Taipei 10608, Taiwan;
    yenchen@ntut.edu.tw (Y.-C.C.); chchang@ntut.edu.tw (C.-H.C.); mazdaabuse@gmail.com (C.-H.H.)
[3] Experimental Forest, National Taiwan University, NanTou 55750, Taiwan
*   Correspondence: d87622005@ntu.edu.tw; Tel.: +886-49-2658412; Fax: +886-49-2631943

Received: 9 September 2017; Accepted: 16 October 2017; Published: 19 October 2017

Abstract: In this study, a method combining radar and entropy was proposed to design a rainfall network. Owing to the shortage of rain gauges in mountain areas, weather radars are used to measure rainfall over catchments. The major advantage of radar is that it is possible to observe rainfall widely in a short time. However, the rainfall data obtained by radar do not necessarily correspond to that observed by ground-based rain gauges. The in-situ rainfall data from telemetering rain gauges were used to calibrate a radar system. Therefore, the rainfall intensity; as well as its distribution over the catchment can be obtained using radar. Once the rainfall data of past years at the desired locations over the catchment were generated, the entropy based on probability was applied to optimize the rainfall network. This method is applicable in remote and mountain areas. Its most important utility is to construct an optimal rainfall network in an ungauged catchment. The design of a rainfall network in the catchment of the Feitsui Reservoir was used to illustrate the various steps as well as the reliability of the method.

Keywords: entropy; information transfer; optimization; radar; rainfall network

1. Introduction

Rainfall data form the fundamental basis for hydraulic and hydrological engineering. Adequate and long-term rainfall data are essential in planning and management of water resources. The definition of rainfall is any product of atmospheric water that reaches the surface of Earth in the form of droplets of water [1]. Thus, rain gauges are the gold standard of precipitation measurement [2]. They are the principal source of rainfall data for rainfall network design. However, accurate and reliable rainfall data of catchments depend on well-designed rainfall networks. Ideally, a higher number of rainfall gauges in a catchment provides a clearer picture of the aerial distribution of the rainfall. Usually, network density and rainfall gauge distribution depend on the particular application. Many factors may affect the number and locations of rain gauges. However, there is no definite rule for constructing a rainfall network. The actual density of a rainfall network is significantly poorer than the values recommended by the World Meteorological Organization (WMO) [3]. Therefore, various methods have been used in the past to investigate the density of rainfall networks and the optimization of these networks. The WMO recommends certain densities of rain gauge stations for different types of catchments. In flat regions of temperate zones, 500 km^2 per station is recommended. For small mountainous islands with irregular precipitation, 25 km^2 per station is recommended [3]. Langbein [4] suggested that the densities of rain gauge stations are usually proportional to population density. The use of statistical characteristics is generally desirable in the design of rainfall networks; Rodriguez-Iturbe and Mejia [5] used a random process technique to develop design curves for estimating the mean of a rainfall event. Shih [6] introduced various steps based on a covariance factor among rain gauge stations to design a rainfall network. Patra [1] applied the coefficient of variance and allowable percentage of error to estimate the optimal number of rain gauge stations. Basalirwa et al. [7] attempted to design a minimum rainfall network by using principal component analysis. Likewise, geostatistics is frequently used in the design of rainfall networks. Kassim and Kottegoda [8] prioritized rain gauges with respect to their contribution in error reduction in the network through comparative Kriging methods. Chen et al. [9] developed a method by using Kriging and entropy that can determine the optimum number and spatial distribution of rain gauge stations in a catchment. Chebbi et al. [10] proposed an algorithm composed of a geostatistical variance-reduction method and simulated annealing to expand the existing rainfall network. Ridolfi et al. [11] introduce an entropy approach for evaluating the maximum information content achievable by an urban rainfall network. Shaghaghian and Abedini [12] selected an optimal subset of stations in the network by using Kriging, factor analysis, and clustering techniques to achieve the optimum rainfall network. Chebbi et al. [13] identified the optimal network using an intensity-duration-frequency curve and a variance-reduction method. Related works based on entropy since Krstanovic and Singh [14] is widen and undergoing [15–18]. Wei et al. [19] introduced entropy to evaluate the effect of spatiotemporal scaling on rainfall network design.

Two major scientific problems need to be addressed in studies on rainfall networks: the first problem is the number of rain gauge stations required to provide adequate representation of a catchment's rainfall characteristics, and the second problem is that the positioning of these rain gauge stations. These two issues are essential to the optimal network design and addressed by previous studies in river points [20], water level networks in polders [21,22], cross-section spacing for river modeling [23], groundwater quality monitoring [24,25], monitoring network design [26,27], and the homogeneity of the study region, also discussed by new clustering method based on entropy [28]. As these are interrelated problems, they need to be considered in conjunction during the design of rainfall networks. In this work, the data of Project Quantitative Precipitation Estimation and Segregation Using Multiple Sensors (QPESUMS) were used to estimate the spatial distribution of rainfall, whereas entropy was used to evaluate the uncertainty of each rain gauge station and to determine how the uncertainty and spatial distribution of rainfall interact with each other. For this purpose, a comprehensive evaluation of the Feitsui Reservoir's rain gauge station locations and distribution were performed using a three-step procedure: the first step was to use the currently available rainfall data to calibrate the radar parameters of the QPESUMS. The second step was to apply the radar to estimate the rainfall data of the candidate rain gauge stations. The third step was to use information entropy to determine the priority of the candidate rain gauge stations and to estimate the minimum required number of rain gauge stations. By combining the use of the radar of the QPESUMS and information entropy, we can determine the locations where new stations should be set up and the number of rain gauge stations required for a rainfall network. Therefore, during the determination of rain gauge station locations using this method, new rain gauge stations will be suggested if a rainfall network has fewer stations than the saturation number determined by our method, to provide an adequate quantity of catchment rainfall data. Conversely, if a rainfall network has a higher number of rain gauge stations than the saturation number, the rainfall data provided by the excess rain gauge stations will be limited, thus the removal of rain gauge stations would be suggested to improve the cost efficiency of hydrological information systems.

In this study, the Feitsui Reservoir's catchment was used to demonstrate how the proposed method can be used to construct an optimal rainfall network. As global climate change may have altered the hydrological characteristics of the Feitsui Reservoir's catchment, long-term hydrological data was used to assess whether an adjustment was needed for the rain gauge stations in this area, so that rain gauge stations may be added to ensure the operational safety of the reservoir. For this purpose, the measured rainfall data of the rain gauge stations were used to calibrate the parameters of the weather radar in this area, which was subsequently used to estimate the spatial distribution of rainfall in the reservoir catchment and the historical rainfall data of the candidate stations. In addition, the information transfer theories of information entropy were used to derive the importance of each candidate station and to construct an optimal rainfall network for the catchment after information was accumulated up to a specified saturation level. Finally, suggestions were provided on the appropriateness of rain gauge station additions.

2. Methodology

2.1. Radar Estimation of Rainfall

Radar was recognized for the measurement of precipitation in the late 1940s. An equation that relates the intensity of rainfall with radar echo factors was constructed after the Second World War, which gave birth to studies on the application of radar in the observation of rainfall. Radar observation is based on the scattering and reflection of high-power electromagnetic waves (emitted by radar antennas towards the atmosphere) when they encounter droplets of water or ice in clouds or raindrops. The energy of the reflected electromagnetic waves received by the radar antennas may then be used to estimate the quantity of rainfall.

The operation of radars is based on the Doppler effect, and radar is mainly used in meteorology to measure rainfall and track storms. Radars operate by performing 360° scans at different elevation angles (from the highest to lowest elevation angles), and the observation of precipitation may then be performed using the Doppler principle. As the topography of Taiwan is highly complex and mountainous, radar scans at low elevation angles are often blocked by mountains. The current solution for circumventing this issue is to use radar echo data from higher elevations to replace the data of topographically obstructed regions, i.e., by selecting echo factors with the lowest (unobstructed) elevation angle for an obstructed region, as shown in Figure 1.

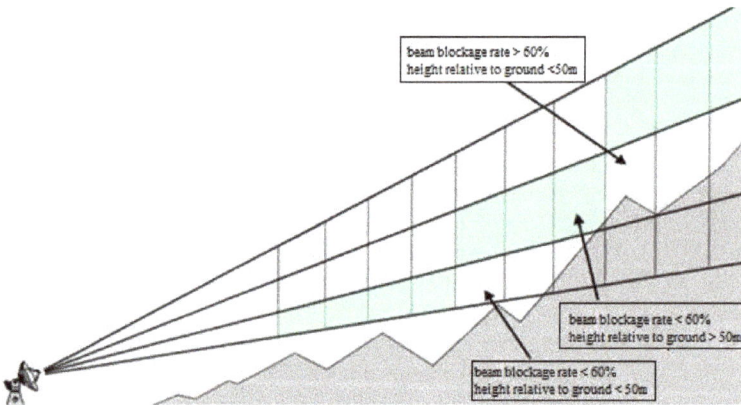

Figure 1. Schematic diagram of radar echoes with elevation (Zhang, 2006). When the beam blockage of a pulse volume's echo factor exceeds 60%, or if the central point of a pulse volume is less than 50 m above ground, its data is then replaced by the echo factor of the next (higher) elevation angle.

The basic meteorological radar equation for the quantification of rainfall is

$$P_r = \frac{\pi^3}{2^{10} \ln} \frac{P_t g^2 \theta h}{\lambda^2 r_0^2} |k_w|^2 Z \tag{1}$$

In this equation, P_r is the power of the echoes received by the radar antenna, P_t is the transmission power of the radar antenna, g is the radar antenna's gain, θ is the beam width, h is the spatial pulse length, k_w is the dielectric constant of the medium, r_0 is the distance from the radar to the area of precipitation, and Z is the radar reflectivity factor.

When the raindrops are very small in diameter and homogeneously distributed in space, the radar reflectivity factor and the raindrop diameter within a unit volume of the radar beam are proportionally related by a power of 6. Hence, the echo factor (the value of Z) may be expressed as

$$Z = \frac{1}{\Delta V}\sum_i D_i^6 = \int_0^\infty D^6 N(D)dD \tag{2}$$

In this equation, ΔV is the unit volume, D is the diameter of the raindrops, $N(D)$ is the raindrop diameter distribution function for a diameter, D (i.e., the raindrop density). The raindrop diameter distribution proposed by Marshall & Palmer [29] is

$$N(D) = N_0 e^{-\Lambda D} \tag{3}$$

Here, N_0 is a constant, and Λ is the rainfall rate function. Therefore, the function for the total number of raindrops may be expressed as

$$N_{total} = \int_0^\infty D^0 N(D)dD \tag{4}$$

The rainfall intensity (R) (mm/h) of the rainfall rate is the total rainfall on each unit of surface area per unit time. Hence, the relationship between rainfall intensity and raindrop diameter is

$$R = \frac{\pi}{6}\sigma_w \int_0^\infty D^3 N(D)(W_t - W)dD \tag{5}$$

In this equation, σ_w is the density of liquid water, W_t is the terminal velocity of the raindrops, and W is the flow rate of ascending airflows.

Equations (2) and (5) demonstrate that Z and R are both related to the raindrop density function; according to statistical analyses of the observed rainfall intensity and rainfall density data, the empirical function that relates rainfall intensity to the radar echo wave is

$$Z = aR^b \tag{6}$$

In this equation, a and b are parameters that may be derived from regression analysis with the rainfall data collected on the ground. In this study a and b are 32.5 and 1.65, respectively [30].

2.2. Information Transfer by Using Entropy

In 1948, Shannon proposed the probability-based concept of information entropy, which is quite different from thermodynamic entropy [31].

$$H(x) = -\sum_i p(x_i)\ln p(x_i) \tag{7}$$

In this equation, $H(x)$ is the entropy value. x represents an event, and $p(x)$ represents the probability of this event.

The data acquired by the rain gauge stations of a rainfall network may overlap with each other. If we treat the data of two rain gauge stations as two variables, x and y, the joint probability of x and y, p_{ij}, may then be expressed as

$$p_{ij} = p(x = x_i, y = y_j) \tag{8}$$

The total information content may be deduced from the joint entropy, which is

$$H(x,y) = -\sum_i \sum_j p_{ij} \ln\left(p_{ij}\right) \tag{9}$$

Equation (9) represents the uncertainty between two rain gauge stations. Like the characteristics of the joint probability distribution, the sum of the marginal entropies of x and y should be larger than, or equal to, the joint probability

$$H(x,y) \le H(x) + H(y) \tag{10}$$

Similarly, the joint probability of three rain gauge stations (x, y, and z) is

$$H(x,y,z) = -\sum_i \sum_j \sum_k p_{ijk} \ln p_{ijk} \tag{11}$$

In this equation, p_{ijk} is the joint probability between rain gauge stations x, y and z.

When a rainfall signal is measured by station x, the residual uncertainty of station y may be expressed by the conditional entropy. The conditional probability for an event occurring at x when an event has occurred at y may be expressed as

$$p(x|y) = p_{i|j} = \frac{p_{ij}}{p_j} \tag{12}$$

Hence,

$$
\begin{aligned}
H(x,y) &= -\sum_i \sum_j p_{ij} \ln\left(p_{ij}\right) \\
&= -\sum_i \sum_j p(x|y)p(y) \ln(p(x|y)p(y)) \\
&= -\sum_i \sum_j p(x|y)p(y)[\ln(p(x|y)) + \ln(p(y))] \\
&= -\sum_j p(y)\sum_i p(x|y) \ln(p(x|y)) - \sum_j p(y) \ln(p(y))\sum_i p(x|y)
\end{aligned}
\tag{13}
$$

The first term of Equation (13) is the conditional entropy, $H(x\,|\,y)$, while the second term is the conditional probability; therefore, $\sum_i \sum_j p(x|y) = 1$ becomes the entropy value of y. $\sum_i \sum_j p(x|y) = 1$. Equation (12) may then be written as

$$H(x,y) = H(x|y) + H(y) \tag{14}$$

and

$$p_{ij} = p(x|y)p(y) = p(y|x)p(x) \tag{15}$$

Therefore,

$$H(x,y) = H(y|x) + H(x) \tag{16}$$

It may be inferred from Equations (13) and (16) that

$$H(x|y) \le H(x) \tag{17}$$

Furthermore, the conditional entropy may be inferred from the equation below

$$H(x|y) = -\sum_j \sum_i p_{ij} \ln p_{i|j} \tag{18}$$

In the conditional entropy of a single rain gauge station, there will be no uncertainty. Hence, the conditional entropy of a single station is

$$H(x|x) = 0 \tag{19}$$

The calculation of transferable information may be used to determine whether two rain gauge stations will possess shared or redundant information, which would allow the rainfall at Station y to be deduced from the data of Station x. The equation for calculating the transferable information is

$$
\begin{aligned}
T(x,y) &= H(y) - H(y|x) \\
&= H(x) - H(x|y) \\
&= H(x) + H(y) - H(x,y)
\end{aligned}
\tag{20}
$$

or

$$T(x,y) = \sum_i \sum_j p_{ij} \ln \frac{p_{ij}}{p_i p_h} \tag{21}$$

The importance of each rain gauge station in a rainfall network is described by its entropy value, and the priority of the rain gauge stations may be displayed by sorting the stations by entropy. The rain gauge station that has the largest entropy value will have the highest uncertainties, and it should be the first station selected for entry into the rainfall network. After the first rain gauge station has been determined, the rain gauge stations with the lowest quantity of redundant information should be systematically added to the rainfall network one after another, to reduce the uncertainty of the system. Hence, the criterion for determining the second most important rain gauge station for addition to the rainfall network is

$$Min\{H(x_1) - H(x_1|x_2)\} \tag{22}$$

This selects for the station with the highest $H(x_1 | x_2)$ value. The selection criterion for the j-th most important rain gauge station is then

$$Min\{H(x_1, x_2, \cdots, x_{j-1}) - H[(x_1, x_2, \cdots, x_{j-1}|x_j)]\} \tag{23}$$

The stations with the highest value of $H[(x_1, x_2, \cdots, x_{j-1}|x_j)]$ may then be selected from the calculations. This yields a ranking of all of the rain gauge stations in a rainfall network by the redundancy of their data, and the station with the highest redundancy will be the last station to be added to the network.

The ranking of rain gauge stations by importance derived from their entropy values may be used as an ordering for the removal of stations, and the increase in uncertainty may be used as a criterion for determining the removal of a station. After a certain number of stations have been added to the network, the value of $H[(x_1, x_2, \cdots, x_{j-1}|x_j)]$ will no longer increase or change significantly, and the information provided by further additions will be limited. Hence, an exponential model may be defined using the number of rain gauge stations and the value of $H[(x_1, x_2, \cdots, x_{j-1}|x_j)]$ to find the critical quantity of information and the required number of rain gauge stations, as shown below

$$H(n) = \omega \left[1 - \exp\left(\frac{-n}{c}\right)\right] \tag{24}$$

where ω and c are to-be-determined parameters.

If the number of rain gauge stations in a rainfall network is larger than the required number of stations, then the stations that rank lower than the required number may then be removed. Conversely, if the number of rain gauge stations in a rainfall network is lower than the required number of stations, more stations then need to be added.

The ranking of rain gauge stations by importance based on their entropy values may be used as an ordering for the removal of stations. The maximization of entropy is the objective of each station selection stage. The addition of each station should increase the joint entropy, but after a certain number of stations have been added to the network, it may be observed that the entropy value, $H(n)$ no longer increases or changes significantly, and converges to a fixed value instead. This indicates that all further additions to the system will only be able to provide a limited quantity of information. The index model of this study was used to plot the relationship between $H\left[(x_1, x_2, \cdots, x_{j-1}|x_j)\right]$ and the number of rain gauge stations, to find the critical quantity of information and the number of required stations. Here, we define a k_m coefficient, which represents the ratio between the entropy value of the m-th station added to the system and the total entropy of the study area. Hence, k_m may be used to represent the quantity of information provided by the m-th station added to the network.

Suppose that the study area has n measurement stations; after the base station has been selected, each subsequent addition is performed with the objective of maximizing entropy. The definition of k_m is then

$$k_m = \frac{H(x_1, x_2, \ldots, x_m)}{H(x_1, x_2, \ldots, x_m, \ldots, x_{n-1}, x_n)}, m < n \tag{25}$$

and $k_1, k_2, \ldots, k_m, \ldots, k_{n-1}, k_n < 1$.

When determining the number of rain gauge stations for an area, a threshold value, k_m^* also needs to be determined. When $k_m > k_m^*$, the number of rain gauge stations for a study area may then be obtained. The determination of the threshold value may be determined by the increase in efficacy, as revealed by the increase in k_m. The threshold is usually defined as $k_m = 0.95$, i.e., 95% of the information content. If the number of rain gauge stations in a rainfall network is larger than the required number of stations, the stations that rank lower than the required number may then be removed. Conversely, if the number of rain gauge stations in a rainfall network is smaller than the required number of stations, more stations then need to be added.

3. Study Area and Data Description

The Feitsui Reservoir is located southeast of Taipei. The main river of the Feitsui Reservoir's catchment is approximately 50 km long, with a drainage area of about 303 km^2 (Figure 2). The dendritic drainage system of this catchment includes four tributaries (the Daiyuku Creek, the Jingualiao Creek, the Houkengzi Creek, and the Huoshaozhang Creek) joined together into the main river, the Beishi River. The source of the Beishi River lies in the western slope at the northern end of the Xueshan Range, and it flows into Pinglin before feeding into the Daiyuku Creek and Jingualiao Creek. After this point, the Beishi River broadens and slows before it flows westward into the Feitsui Reservoir.

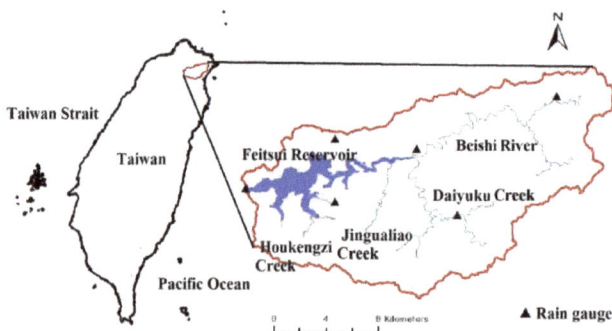

Figure 2. The Feitsui Reservoir's catchment and the locations of the rain gauge stations.

The Feitsui Reservoir catchment is within the subtropical climate zone. Cold and wet northeasterly monsoons prevail in winter, and the cold Arctic air will invade southwards from time to time. Cold snaps, low clouds and drizzles thus have a high probability of occurrence in winter. The southwesterly monsoons in summer have a minimal impact on this region, due to the obstruction of the Xueshan range. However, local showers often occur in the afternoon since solar radiation on the river valleys and hillsides has a significant impact on local convection. During the transition between summer and autumn, typhoons will bring about warm and highly humid airflows, and extremely heavy and intense rainfalls. The temporal and spatial distributions of rainfall were already taken into consideration during the initial design of the current rain gauge stations. For this reason, the Taipei Feitsui Reservoir Administration has established an integrated weather station for the Feitsui Reservoir in one location, and rain gauge stations in five locations, as shown in Figure 2.

The maximum, minimum and standard deviation of the monthly rainfall in the watershed are shown in Table 1. The rainy season is from August to November, and the variance in mean monthly rainfall often exceeds 100 mm. The rainfall tends to fluctuate from year to year; the annual rainfall ranges between 2520 and 5740 mm, and the mean annual rainfall is approximately 3760 mm. From August to October, this area is impacted by typhoons and storms that frequently bring about heavy rainfall, whereas the rainfall frequency peaks in the period between October and January, due to the impacts of the northeasterly monsoon. During 2006–2015, at least 30 typhoons invaded Taiwan. The rainfall data during 2006 and 2015, including the heavy rain coming with the typhoons, is used for this study.

Table 1. The maximum, minimum and standard deviation of the monthly rainfall of six rain gauges in Feitsui Reservoir's catchment.

Station	Taiping	Sirsangoo	Pingling	Feitsui	Geochungan	Beefu
Grid number	19	61	66	102	129	157
Maximum (mm/M)	2668.5	1723.0	2113.5	1902.0	1925.5	2330.0
Minimum (mm/M)	0.5	0.0	12.0	43.0	26.5	0.0
Mean (mm/M)	449.3	282.3	294.7	299.0	302.5	333.2
Std. Dev. (mm/M)	358.1	218.4	263.0	234.1	234.4	234.4

The weather radar station was installed by the Central Weather Bureau of Taiwan to provide real-time severe weather information. It is located around the most northeastern Taiwan. The radar was tested in July 1996, however it was destroyed by a typhoon in August 1996. The radar was repaired in 1998, and the data acquisition system was upgraded in 2006. In order to make the data consistent, only the data after 2006 is used in this study. The radar has a wavelength of 10 cm, a spatial resolution of 1.3×1.3 km^2, and makes rainfall observations once every 10 min. Therefore, the catchment of the Feitsui Reservoir may be divided into 217 grids according to the radar's spatial resolution, with the center of each grid being the location of a candidate rain gauge station, as shown in Figure 3. The 19th, 61st, 66th, 102nd, 129th, and 157th grids in Figure 3 correspond to the locations of the six currently existing rain gauge stations.

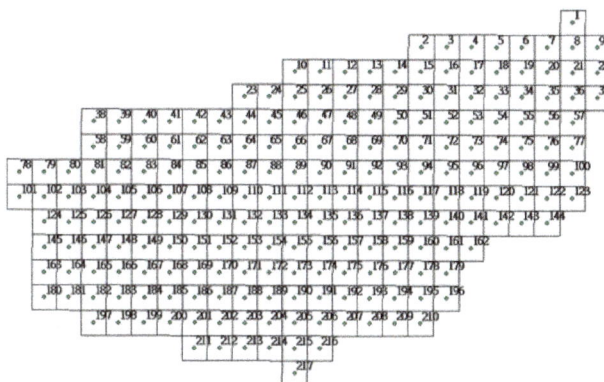

Figure 3. The positions and numbering of the candidate rain gauges in the Feitsui Reservoir's catchment.

Prior to this study, it has not been assessed whether increases in the number of rain gauge stations are necessary in light of alterations to the spatial and temporal distribution of rainfall induced by global climate changes, to ensure that the rainfall data being acquired is sufficient for the smooth operation of the reservoir. To address this shortfall, the mean monthly rainfall data from January 2006 up to December 2015 (120 months in total) was used to evaluate the efficacy of the current rainfall network. Figure 4 illustrates the monthly rainfall map of the Feitsui Reservoir's catchment from 2006 to 2015. Heavy rainfalls always occur in summer due to the arrival of typhoons, while the Meiyu front brings rain in spring, and the northeasterly monsoon also brings rain in winter, thus resulting in rainfall over long periods of the year. The minimum in rainfall usually occurs in autumn. The Feitsui Reservoir provides the water supply and usually operates a regulation line for 10-days. According to this, the rainfall temporal period for six-month/year seems too long for the reservoir operation and the analysis, hourly data may contribute to the flood control, while the month scale is reasonable for reservoir operation and also reflecting the variation during dry and wet seasons.

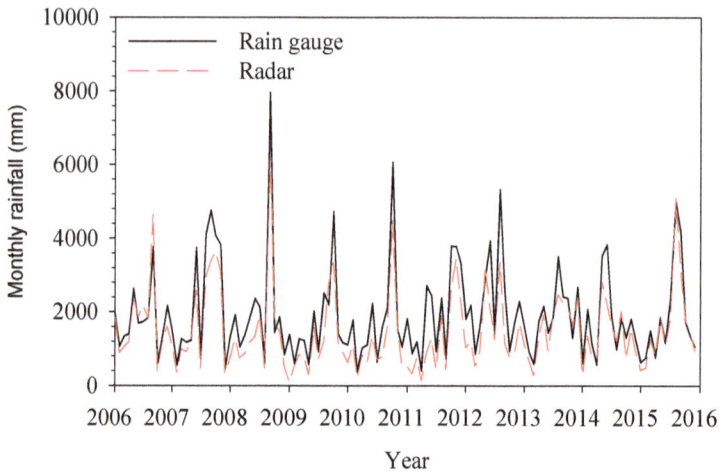

Figure 4. The monthly rainfall in the Feitsui Reservoir's catchment from 2006 to 2015.

4. Results and Discussion

As rainfall network evaluations will require considerations on the long-term temporal and spatial distributions of rainfall, the monthly rainfall data was selected for data analysis in this work. The arithmetic mean was used to estimate the mean rainfall of the catchment to minimize the impact of the spatial distribution of rainfall, and because the six currently existing rain gauge stations are spatially distributed in a uniform manner.

To estimate the rainfall data of the candidate rain gauge stations for the evaluation of the rainfall network, the observed mean monthly rainfall of the catchment was derived from the rainfall data of the six currently existing rain gauge stations, while the radar of QPESUMS was used to estimate the monthly rainfall at these six locations to obtain the radar-derived mean monthly rainfall. A linear regression was then used to probe the relationship between these quantities. Figure 5 displays the relationship between the rainfalls measured by the six currently existing rain gauge stations and the rainfalls estimated using the radar system. The horizontal axis (R_{QP}) indicates the monthly rainfall observed by the radar, while the vertical axis (R_G) represents the mean monthly rainfall of the catchment measured by the rain gauge stations. It was found that these quantities were related by $R_G = 1.1R_{QP} + 36.13$, with a correlation coefficient of 0.87. Since the correlation is quite strong, this equation and the radar's rainfall data may then be used to estimate the mean monthly rainfalls of all the candidate rain gauge stations in the 211 grids of the catchment, over the last 20 years.

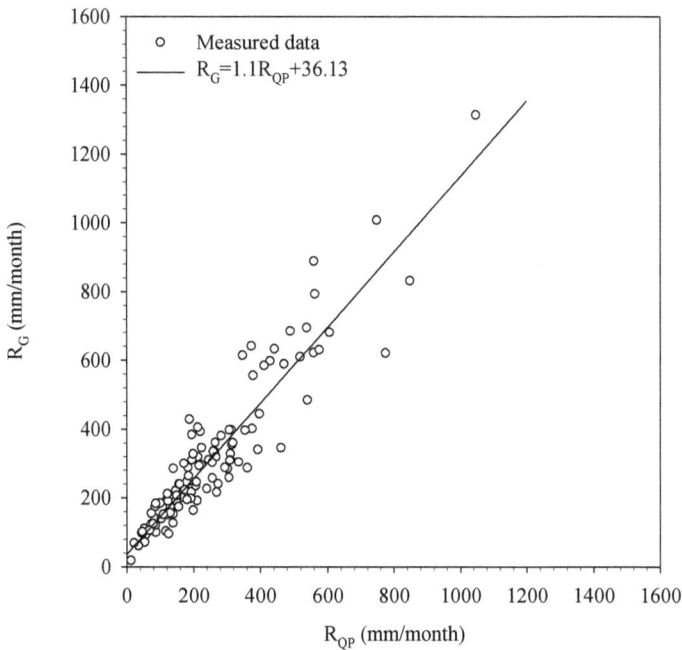

Figure 5. The relationship between the measured mean monthly rainfall and the estimated mean monthly rainfall.

In the study area, stations were selected according to transferable information calculations and joint entropy-based ordering. Hence, the entropy values of the selected stations in the study area were calculated using joint entropy and transferable information, and the stations were then ordered by sorting the calculated values. The ranking of rain gauge stations by importance based on their entropy values may be used as an ordering for the removal of stations. The maximization of entropy is the objective of each station selection stage. Each station addition should increase the joint entropy, but after the number of added stations has reached a certain value, it may be observed that the entropy value, $H(n)$, no longer increases or changes significantly, and converges to a fixed value; this indicates that all further additions to the system will only provide a limited quantity of information.

The study area has 217 grids. The grids corresponding to the six currently existing stations were selected as necessary locations for the reservoir's operation, while the remaining grids were added one after another, based on the principle of entropy maximization. A threshold value is usually defined to determine the number of stations required by some area. In this case, the threshold, k_m^*, was defined as 0.95, which corresponds to a threshold value of 95%. By doing so, almost all of the rainfall information of a region can be acquired using only a few rain gauge stations. Hence, when $k_m \geq k_m^*$, the number of stations that need to be added to the study area and the location of these stations may then be obtained. The relationship between the entropy value and the optimal number of stations is illustrated in Figure 6; an index function was used in this work (as shown in Equation (24)) to obtain an estimate for the optimal number of stations.

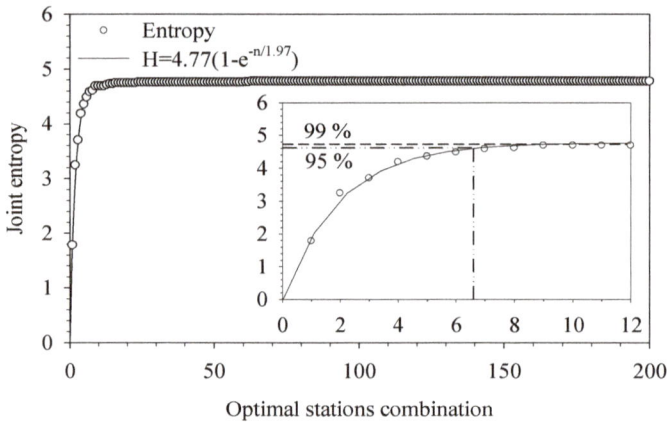

Figure 6. The relationship between joint entropy value and the optimal number of stations; all of the grids and the first 12 grids within the graph.

The grids labeled 1–6 in Figure 7 correspond to the locations of the six currently existing rain gauge stations in the catchment. The entropy value of these six stations is already ~94%, which indicates that these stations are sufficient for normal operation of the water reservoir. Nonetheless, if an improvement in the completeness of the data is desired, the addition of a station at Grid 7 will increase k_m from 0.938 to 0.958, which is larger than the $k_m^* = 0.95$ threshold. Therefore, we propose that one more rain gauge station should be added to the study area, in addition to the pre-existing rain gauge stations. The location of this station is indicated by Grid 7 in Figure 7, and it is located at the boundary of the catchment in the southeastern part of the study area.

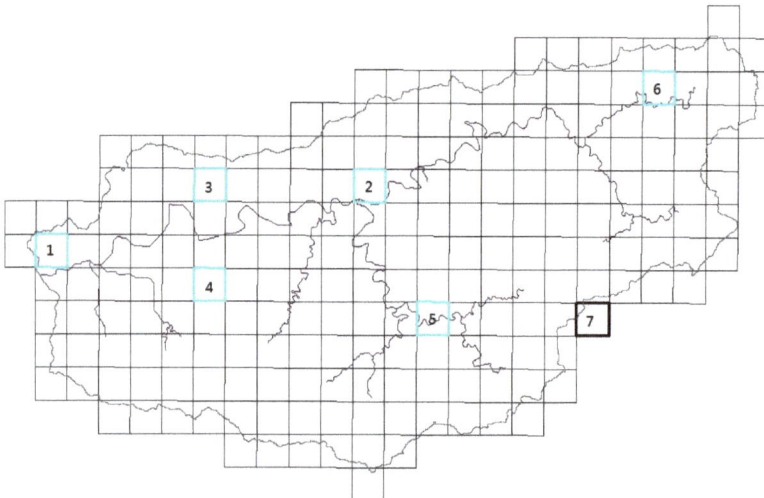

Figure 7. The position of the recommended rain gauge station (the bold grid) for addition to the rainfall network; Blue grids 1 to 6 are the existed rainfall stations.

5. Conclusions

In this study, surface and radar rainfall data were employed in unison to estimate the rainfall data of the candidate rain gauge sites, and information entropy was used to evaluate the information content and uncertainty of each rain gauge station. An optimal rainfall network may then be constructed by combining these methods. Unlike previous estimations of rainfall based on the Kriging method, the rainfall data obtained using radar is an actual measurement, whereas rainfalls estimated using statistical methods are at best, estimated values, which may not accurately reflect on the temporal and spatial distributions of rainfall in a catchment. Previously, it was impossible to obtain an exact answer for the location of rain gauge stations and the minimum required number of stations; with information entropy, it is now possible to simultaneously obtain an answer for both of these questions. This method will provide an important basis for the management of watersheds and the establishment of rain gauge stations. Furthermore, this method may also be used to assess the sufficiency of the rainfall data provided by currently existing rain gauge stations. More stations need to be added to a rainfall network if the data is insufficient, whereas stations need to be removed if the redundancy in data is too high. It is hoped that this method will be used to evaluate or adjust currently existing rainfall networks in the catchment.

To identify and test the robustness of this method of rainfall network design in catchments, the rainfall network at the Feitsui Reservoir's catchment was evaluated to highlight the applicability and reliability of our method. A grid was defined every 1.3 km in the study area, and actual radar data was used to reconstruct the historical rainfall data of these grids. Information entropy was then used to evaluate the spatial information content and the uncertainty of the information, and the estimated entropy values were used to add stations to the network one after another. It was shown that seven stations is the optimal number of rain gauge stations for the selected study area, as this is the number of stations required to obtain 95% of the study area's rainfall information, additional manpower and resources for initial establishing and maintaining more stations can be saved.

Acknowledgments: This paper is based on work partially supported by Ministry of Science and Technology, Taiwan (Grant No. MOST 106-2221-E-027-031-). The authors would like to thank the Taipei Feitsui Reservoir Administration for providing the fundamental, rainfall data and related assistance.

Author Contributions: Hui-Chung Yeh and Yen-Chang Chen conceived and designed the experiments; Che-Hao Chang analyzed the radar data; Cheng-Hsuan Ho performed the materials/analysis tools; Chiang Wei performed the analysis and wrote the manuscript. All authors have read and approved the final manuscript.

Conflicts of Interest: The authors declare no conflict of interest.

References

1. Patra, K.C. *Hydrology and Water Resources Engineering*; Alpha Science: Oxford, UK, 2010.
2. Strangeways, I. *Precipitation: Theory, Measurement and Distribution*; Cambridge University Press: Cambridge, UK, 2007.
3. WMO. *Guide to Hydrological Practices, WMO-164*; WMO: Geneva, Switzerland, 1994.
4. Langbein, W.B. Hydrologic data networks and methods of extrapolating or extending available hydrologic data. In *Hydrologic Networks and Method*; United Nations: Bangkok, Thailand, 1960.
5. Rodriguez-Iturbe, I.; Mejia, J.M. The design of rainfall networks in time and space. *Water Resour. Res.* **1974**, *23*, 181–190. [CrossRef]
6. Shih, S.F. Rainfall variation analysis and optimization of gaging systems. *Water Resour. Res.* **1982**, *18*, 1269–1277. [CrossRef]
7. Basalirwa, C.P.K.; Ogallo, L.J.; Mutua, F.M. The design of regional minimum rain gauge network. *Int. J. Water Resour. Dev.* **2007**, *9*, 411–424. [CrossRef]
8. Kassim, A.H.M.; Kottegoda, N.T. Rainfall network design through comparative kriging methods. *Hydrol. Sci. J.* **1991**, *36*, 223–240. [CrossRef]
9. Chen, Y.C.; Wei, C.; Yeh, H.C. Rainfall network design using kriging and entropy. *Hydrol. Process.* **2008**, *22*, 340–346. [CrossRef]

10. Chebbi, A.; Bargaoui, Z.K.; Cunha, M.D.C. Optimal extension of rain gauge monitoring network for rainfall intensity and erosivity index interpolation. *J. Hydrol. Eng. ASCE* **2011**, *16*, 665–676. [CrossRef]
11. Ridolfi, E.; Montesarchio, V.; Russo, F.; Napolitano, F. An entropy approach for evaluating the maximum information content achievable by an urban rainfall network. *Nat. Hazards Earth Syst. Sci.* **2011**, *11*, 2075–2083. [CrossRef]
12. Shaghaghian, M.R.; Abedini, M.J. Rain gauge network design using coupled geostatistical and multivariate techniques. *Sci. Iran.* **2013**, *20*, 259–269. [CrossRef]
13. Chebbi, A.; Bargaoui, Z.K.; Cunha, M.D.C. Development of a method of robust rain gauge network optimization based on intensity-duration-frequency results. *Hydrol. Earth Syst. Sci.* **2013**, *17*, 4259–4268. [CrossRef]
14. Krstanovic, P.F.; Singh, V.P. Evaluation of rainfall networks using entropy: I. Theoretical development. *Water Resour. Manag.* **1992**, *6*, 279–293. [CrossRef]
15. Yoo, C.; Jung, K.; Lee, J. Evaluation of Rain Gauge Network Using Entropy Theory: Comparison of Mixed and Continuous Distribution Function Applications. *J. Hydrol. Eng.* **2008**, *13*, 226–235. [CrossRef]
16. Leach, J.M.; Kornelsen, K.C.; Samuel, J.; Coulibaly, P. Hydrometric network design using streamflow signatures and indicators of hydrologic alteration. *J. Hydrol.* **2015**, *529*, 1350–1359. [CrossRef]
17. Chacon-Hurtado, J.C.; Alfonso, L.; Solomatine, D.P. Rainfall and streamflow sensor network design: A review of applications, classification, and a proposed framework. *Hydrol. Earth Syst. Sci. Discuss.* **2017**, *21*, 3071–3091. [CrossRef]
18. Stosic, T.; Stosic, B.; Singh, V.P. Optimizing streamflow monitoring networks using joint permutation entropy. *J. Hydrol.* **2017**, *552*, 306–312. [CrossRef]
19. Wei, C.; Yeh, H.C.; Chen, Y.C. Spatiotemporal scaling effect on rainfall network design using entropy. *Entropy* **2014**, *16*, 4626–4647. [CrossRef]
20. Harmancioglu, N.; Yevjevich, V. Transfer of hydrologic information among river points. *J. Hydrol.* **1987**, *91*, 103–118. [CrossRef]
21. Alfonso, L.; Lobbrecht, A.; Price, R. Information theory-based approach for location of monitoring water level gauges in polders. *Water Resour. Res.* **2010**, *46*. [CrossRef]
22. Alfonso, L.; Lobbrecht, A.; Price, R. Optimization of water level monitoring network in polder systems using information theory. *Water Resour. Res.* **2010**, *46*, W12553. [CrossRef]
23. Ridolfi, E.; Alfonso, L.; Baldassarre, G.D.; Dottori, F.; Russo, F.; Napolitano, F. An entropy approach for the optimization of cross-section spacing for river modelling. *Hydrol. Sci. J.* **2013**, *59*, 126–137. [CrossRef]
24. Mogheir, Y.; Singh, V.P. Application of information theory to groundwater quality monitoring networks. *Water Resour. Manag.* **2002**, *16*, 37–49. [CrossRef]
25. Mogheir, Y.; de Lima, J.L.M.P.; Singh, V.P. Assessment of spatial structure of groundwater quality variables based on the entropy theory. *Hydrol. Earth Syst. Sci.* **2003**, *7*, 707–721. [CrossRef]
26. Alfonso, L.; Ridolfi, E.; Gaytan-Aguilar, S.; Napolitano, F.; Russo, F. Ensemble entropy for monitoring network design. *Entropy* **2014**, *16*, 1365–1375. [CrossRef]
27. Ridolfi, E.; Yan, K.; Alfonso, L.; Baldassarre, G.D.; Napolitano, F.; Russo, F.; Bates, P.D. An entropy method for floodplain monitoring network design. *AIP Conf. Proc.* **2012**, *1479*, 1780–1783. [CrossRef]
28. Ridolfi, F.; Rianna, E.; Trani, G.; Alfonso, L.; Baldassarre, G.D.; Napolitano, G.; Russo, F. A new methodology to define homogeneous regions through an entropy based clustering method. *Adv. Water Resour.* **2016**, *96*, 237–250. [CrossRef]
29. Marshall, J.S.; Palmer, W.M. The distribution of raindrops with size. *J. Meteorol.* **1948**, *5*, 165–166. [CrossRef]
30. Xin, L.; Reuter, G.; Larochelle, B. Reflectivity-rain rate relationships for convective rainshowers in Edmonton: Research note. *Atmos. Ocean* **1997**, *35*, 513–521. [CrossRef]
31. Shannon, C.E. A mathematical theory of communication. *Bell Syst. Tech. J.* **1948**, *27*, 623–656. [CrossRef]

![entropy logo] *entropy*

MDPI

Article

Scaling-Laws of Flow Entropy with Topological Metrics of Water Distribution Networks

Giovanni Francesco Santonastaso [1,2], Armando Di Nardo [1,2], Michele Di Natale [1,2], Carlo Giudicianni [1] and Roberto Greco [1,2,*]

[1] Dipartimento di Ingegneria Civile, Design, Edilizia e Ambiente, Università degli Studi della Campania "Luigi Vanvitelli", via Roma 29, 81031 Aversa, Italy; giovannifrancesco.santonastaso@gmail.com (G.F.S.); armando.dinardo@unicampania.it (A.D.N.); michele.dinatale@unicampania.it (M.D.N.); carlo.giudicianni@gmail.com (C.G.)
[2] Action Group CTRL+SWAN of the European Innovation Partnership on Water, EU, B-1049 Brussels, Belgium
*   Correspondence: roberto.greco@unicampania.it; Tel.: +39-081-501-0207

Received: 28 December 2017; Accepted: 26 January 2018; Published: 30 January 2018

Abstract: Robustness of water distribution networks is related to their connectivity and topological structure, which also affect their reliability. Flow entropy, based on Shannon's informational entropy, has been proposed as a measure of network redundancy and adopted as a proxy of reliability in optimal network design procedures. In this paper, the scaling properties of flow entropy of water distribution networks with their size and other topological metrics are studied. To such aim, flow entropy, maximum flow entropy, link density and average path length have been evaluated for a set of 22 networks, both real and synthetic, with different size and topology. The obtained results led to identify suitable scaling laws of flow entropy and maximum flow entropy with water distribution network size, in the form of power–laws. The obtained relationships allow comparing the flow entropy of water distribution networks with different size, and provide an easy tool to define the maximum achievable entropy of a specific water distribution network. An example of application of the obtained relationships to the design of a water distribution network is provided, showing how, with a constrained multi-objective optimization procedure, a tradeoff between network cost and robustness is easily identified.

Keywords: scaling laws; power laws; water distribution networks; robustness; flow entropy

1. Introduction

The topology of water distribution networks (WDN) is being deeply studied with respect to its relationship with their robustness, i.e., their capability of effectively delivering the demanded flows to the users with the required pressure under unfavorable operating conditions [1]. In fact, evaluating the performance of a WDN requires the complex calibration of a hydraulic model of the network, and often a number of time-consuming simulations. Hence, establishing relationships, linking topological metrics of a WDN, easily achievable from the mere knowledge of the network layout, with its hydraulic behavior, would represent a powerful tool for the design, rehabilitation and management of WDN. In this respect, aiming at quantitative comparison of different network layouts, it is important to understand how topological metrics change with the size of the considered network.

In fact, the size variation of a system can cause changes in the order of predominance of physical phenomena; this is called scaling effect, and the laws that govern such an effect are called scaling laws. The scaling laws are relationships linking any parameter associated with an object (or system) with its length scale [2]. They constitute a very useful tool to predict the behavior and the properties of a large system by experimenting on a small-sized scale model, since the characteristics of a system can be expressed through various parameters in such a way that any change in size (i.e., scale) does not affect the magnitudes of these quantities. Scaling laws represent useful tools for understanding the interplay among various physical phenomena and geometric characteristics of complex systems, and often it happens that simple scaling laws can provide clues to some fundamental aspects of the system. In many fields, scaling laws have been identified. For example, scaling laws have been experimentally determined over a huge range of scales in probability distributions describing river basin morphology [3], whose geometrical description is of great importance for a deeper understanding of how some related natural events occur. The existence of a scaling law relating point precipitation depth records to duration has been known for at least 60 years through published tabulations of data and the associated graphs [4,5], even if there is no explanation of the mechanism underlying this remarkably robust relationship, making it even more tantalizing [6]. Scaling laws have been also identified in fluid mechanics, to describe turbulent energy distribution across scales [7,8], and in meteorology, to describe scaling of clouds [9], atmospheric variability [10], and fluctuations of Arctic sea ice [11]. In the field of network topology, it has been found that many real networks exhibit power–law shaped node degree distribution, where the degree is the number of connected links to each node. Such networks have been named scale-free networks [12], because power–laws have the property of retaining the same functional form at all scales. These networks result in the simultaneous presence of a few nodes (the hubs) linked to many other nodes, and a large number of poorly connected elements [13]. The World Wide Web (WWW) is one of the most famous scale-free networks. It is formed by the hyperlinks between different Web pages, and, with more than 10^8 nodes, it is the largest network ever studied.

Differently, water distribution networks (WDN) do not present hubs, as each node is connected only to a few nodes located in its immediate surroundings. The connections between nodes in a WDN ensure multiple possible flow paths, so to cope with abnormal working conditions, such as unexpected water requests by the users and failure of some elements [14]. In this respect, several topological metrics aimed at quantifying WDN connectivity have been proposed as proxies for network robustness and reliability [15].

Reliability, in a WDN, can be defined as the probability of the system being capable of supplying the water demands both under normal and abnormal conditions [16,17]. The assessment of reliability is influenced by many factors: spatial and temporal demand distribution, possible failure of one or more components, pressure-flow relationship, connectivity of the network, etc. Therefore, there is not an established measure of WDN reliability, and a review of different methods to evaluate it can be found in [18]. Reliability measures are categorized into three groups: topological, hydraulic and entropic. Topological reliability is based on the probability of node connectivity/reachability [19]; hydraulic reliability is focused on the probability of delivering design water demands, (e.g., [17]); and the last category adopts the informational entropy as a surrogate of the reliability [18].

The concept of informational entropy [20] has been widely applied in hydraulics and hydrology (i.e., to estimate velocity distribution in open channels, suspended sediment concentration profile, suspended sediment discharge, or precipitation variability, moisture profiles, etc.) [21]. In the field of WDN, Shannon's entropy has been proposed as a measure of connectivity and so as a proxy for reliability [22].

The adoption of entropy as a surrogate for network reliability was investigated by several authors [23–26]. The basic idea is that entropy is a measure of the uniformity of pipe flow rate [27], thus it is related to looped network redundancy, which makes it potentially more capable of facing unfavorable working conditions, such as concentrated peaks of demand or failure of pipes (e.g., [1]). Hence, redundancy increases network robustness, and so, indirectly, its reliability.

Hence, many studies [28–30] have proposed multi-objective optimization for water distribution network design or rehabilitation based on minimizing costs for construction, operation, and maintenance, coupled with the maximization of the entropy as a measure of robustness.

Traditionally, the robustness of water distribution networks was assured by means of densely looped layouts, so to provide alternative paths for each demand node [31]. More recently, Di Nardo et al. [32]

have studied the topological redundancy of a water supply network, with regard to pipe failures, applying the complex network theory [33,34]. In fact, many water supply systems consisting of up to tens of thousands nodes and hundreds of looped paths can be considered as complex networks [13]. Thus, it is possible to compute topological metrics [32,35–37] to analyze the robustness of a water distribution network.

Recently, comparisons between entropy and other indirect measures of robustness [1,26,38–40] such as resilience index [41], network resilience [42] and Surplus Power Factor [43] have been proposed, but the obtained results are contradictory. According to some authors [26,38,39], informational entropy is a good measure of network robustness. Conversely, other studies indicate that the resilience index estimates better the network hydraulic performance than entropy in the case of pipe failures [1] and for multi-objective design optimization [40].

The advantage of using informational entropy to evaluate network robustness is that only pipe flows and topology are required for its computation [39], while the main drawback is that there is not a reference value of entropy allowing for defining an acceptable level of robustness for a given WDN, nor to compare different WDN layouts. In this respect, the definition of scaling laws of flow entropy with the topological dimension of the network could be useful for WDN design and rehabilitation purposes.

This work investigates the possible relationship between topological metrics, borrowed from complex network theory, and flow entropy, through the analysis of the values that they assume for several WDNs, both real and synthetic. In particular, for each network, five of the coarsest topological characteristics of a network, the number of nodes n and links m, the average node degree k, the link density q and the average path length APL have been calculated. The results show that the flow entropy of a WDN is strongly linked to its size and topology, and that it can be expressed as a function of topological metrics. Furthermore, the maximum achievable flow entropy value has been calculated for each WDN. Scaling-laws of flow entropy with the size of the networks have been identified. Two examples of the application of the obtained results to the design of WDNs are finally provided.

2. Methods

The study of WDN using innovative topological metrics, borrowed from the theory of complex networks [13], already led to interesting results for the analysis of water network vulnerability [32,35,44], as well as for water network partitioning [45,46]. In the following sections, the topological and entropy metrics used in this paper are briefly described, and finally the deviation of actual entropy from maximum entropy is introduced as a possible measure of network robustness.

2.1. Topological Metrics

The average node degree, k, represents the mean number of links concurring in the nodes of the network, and is given by:

$$k = \frac{2 \cdot m}{n} \tag{1}$$

in which n is the number of nodes and m the number of links of the network.

The link density, q, expresses the ratio between the total number of network edges and the number of edges of a globally coupled network with the same number of nodes, thus providing a measure of network redundancy:

$$q = \frac{2 \cdot m}{n \cdot (n-1)} \tag{2}$$

The average path length, APL [33], is the average number of steps along the shortest paths between all possible pairs of nodes in the network:

$$APL = \frac{\sum_{\forall s \neq t} \sigma(s,t)}{\frac{1}{2} n \cdot (n-1)} \tag{3}$$

where $\sigma(s,t)$ is the number of edges along the shortest path connecting node s to node t (when there is no path between a pair of nodes, the path length is assumed to be infinite) [47]. A short average path length indicates a more interconnected network, while a long one indicates greater overall topological distances between nodes. Consequently, a network with a large *APL* value may be considered more fragmented [48].

2.2. Entropic Metrics

The Shannon's information entropy [49] is a statistical measure of the amount of uncertainty associated with the probability distribution of any discrete random variable, defined as follows:

$$E = -\sum_{k=1}^{l} p_k \ln p_k, \tag{4}$$

where E is the entropy, p_k is the probability, and l is the number of values that the variable can assume. Tanyimboh and Templeman [28], with the use of the conditional entropy formula of [50], considered all the possible flow paths from sources to demand nodes, and introduced the flow entropy S of a water distribution system by defining the probability of the water to flow along the k-th path as the ratio between the flow rate reaching the end node of the path and the total delivered flow rate [42]. The following recursive formula [24] allows the calculation of S, which is regarded as a measure of pipe flow rates uniformity:

$$S = -\sum_{i=1}^{NS} \frac{Q_i}{T} \ln\left(\frac{Q_i}{T}\right) - \frac{1}{T}\sum_{j=1}^{NN} T_j \left[\frac{Q_j}{T_j} \ln\left(\frac{Q_j}{T_j}\right) + \sum_{ji \in N_j} \frac{q_{ji}}{T_j} \ln\left(\frac{q_{ji}}{T_j}\right)\right] \tag{5}$$

On the right hand side of Equation (5), the first term is the entropy of supply nodes and the second is the entropy of demand nodes; NS is the number of supply nodes; T is the total supplied flow rate; NN is the number of demand nodes; Q_i represents the inflow at the i-th source node; T_j is the total flow rate reaching the j-th demand node; Q_j is the water demand at the j-th demand node; q_{ij} is the flow rate in the pipe connecting node j with surrounding node i; and N_j is the number of pipes carrying water from the j-th demand node towards other surrounding nodes.

The data required to assess the flow entropy are the topological layout, the water supply and the demand at all nodes, and the flow direction along each pipe. To this purpose, the hydraulic simulation of the network, carried out with the solver EPANET 2 [51], provides the flow rate and direction along each pipe.

2.3. Maximum Entropy and Network Robustness

The maximization of Equation (3) can be used to compute the maximum value of the flow entropy, MS, and in this case only the source flow rates, the water demands at nodes and the flow directions along the links are required. Specifically, MS is here computed here by means of a non-iterative procedure for multi-source networks, proposed in [52]. The entropy deficit, i.e., the deviation between the flow entropy S and the corresponding values of MS, given by Equation (5), is assumed to be representative of how much a network is robust, based on the idea that networks, designed to supply maximum entropy flows, would be the most robust for a given source pressure excess compared to the design pressure at nodes [23].

$$\Delta S = 1 - \frac{S}{MS} \tag{6}$$

3. Results and Discussions

Topological metrics and flow entropy metrics were computed for a set of 22 WDNs, both real and synthetic. The maximum entropy *MS* of each network was calculated adopting the same flow directions along the pipes as for the calculation of flow entropy *S* (i.e., the directions provided by the hydraulic simulation of the network for the actual set of pipe sizes). Therefore, the obtained *MS* cannot be considered as the maximum possible values of flow entropy, as a different choice of flow directions could lead to a higher value of *MS*. However, as the flow directions are mainly dictated by the position of sources and demand nodes, and by the assumed water demand at nodes, it is expected that flow directions would be only slightly (and locally) affected by changes in the size of some of the pipes. In Table 1, the computed values of the metrics are reported for all the considered networks.

Table 1. Topological metrics: number of nodes (*n*) and links (*m*), density (*q*), average path length (*APL*), flow entropy (*S*) and maximum flow entropy (*MS*), for all WDN (* denotes synthetic networks).

Network	*n*	*m*	*q*	*APL*	*S*	*MS*	Δ*S*
Two Loop * [53]	7	8	0.5333	1.90	2.063	2.296	0.101
Two Reservoirs * [54]	12	17	0.3778	2.59	2.829	3.008	0.059
Anytown * [55]	25	43	0.1861	2.94	4.172	5.048	0.174
GoYang * [56]	23	30	0.1299	3.75	3.113	3.658	0.149
Blacksburg * [57]	32	35	0.0805	4.37	3.358	3.473	0.033
Hanoi * [58]	32	34	0.0731	5.31	3.384	3.395	0.003
BakRyan * [59]	36	58	0.0975	4.30	3.243	3.709	0.126
Fossolo [60]	37	58	0.0921	3.67	3.677	4.441	0.172
Pescara [60]	72	99	0.0435	8.69	4.273	4.572	0.065
BWSN2008-1 * [61]	127	168	0.0213	10.15	3.939	5.567	0.292
Skiathos [62]	176	189	0.0124	11.52	5.551	6.196	0.104
Parete [1]	184	282	0.0171	8.80	6.561	9.331	0.297
Villaricca [1]	199	249	0.0130	11.29	5.206	5.497	0.053
Monteruscello [63]	206	231	0.0110	20.24	5.211	5.385	0.032
Modena [60]	272	317	0.0089	14.04	5.436	5.764	0.057
Celaya [64]	338	477	0.0086	11.81	6.8	7.734	0.121
Balerma Irrigation [65]	448	454	0.0046	23.89	6.091	6.489	0.061
Castellammare	1231	1290	0.0017	32.25	7.583	8.094	0.063
Matamoros [66]	1293	1651	0.0020	27.76	9.896	13.325	0.257
Wolf Cordera Ranch [67]	1786	1985	0.0013	25.94	7.905	9.865	0.199
Exnet * [68]	1893	2465	0.0014	20.60	10.466	12.882	0.188
San Luis Rio Colorado [66]	1908	2681	0.0015	28.86	8.097	9.443	0.143

The set of networks used as case study includes water distribution networks with very different characteristics, as indicated by the very different values assumed by the metrics:

- dimension: the smallest network has a number of nodes *n* = 6 (Two Loop), while the largest has *n* = 1890 (Exnet);
- layout: looped networks as well as branched ones are included, i.e., Balerma Irrigation can be considered a tree-network, while networks such as Parete and Sector Centro Real are very looped; compact and elongated networks are included, with low values of APL coupled with high values of density being representative of compact network layouts;
- robustness: the set of networks includes systems with very small deviation of actual entropy from maximum entropy, like Hanoi and Modena (the entropy deviation Δ*S* is equal to 0.0032 and 0.0616, respectively), and networks with high deviation of entropy, like Parete and BWSN2008-1(entropy deviations of 0.297 and 0.292, respectively).

These differences indicate that the adopted set is suitable to analyze the entropy metrics from a topological point of view in a general sense.

Figure 1 shows the scatter plots of the values of *S* vs. various topological metrics, and the best fitting power–law equations. The diagrams show that an increasing trend exists in the relationship between flow entropy and number of nodes (Figure 1a), and between flow entropy and number of links (Figure 1b), as well as a decreasing trend for the relationship between flow entropy and link density (Figure 1c). Although in Figure 1d a positive trend of flow entropy vs. average path length is also observable, it is less clearly defined than the previous ones.

The scatter plots of *MS* vs. the same topological metrics, reported in Figure 2, confirm similar trends as in Figure 1. Specifically, a clear increasing relation of *MS* with the number of links (Figure 2b) can be noted, while Figure 2d shows a weak relation between *MS* and *APL*. The determination coefficients, R^2, of *S* (Figure 1) are slightly greater than the ones computed for the *MS* (Figure 2). Anyway, both *S* and *MS* are clearly related to network topology.

It is worth to noting that, although *APL* is considered a proxy of the topological robustness of a network [32] and the entropy has been proposed as a surrogate of network reliability, the relationships between *S* and *APL*, as well as between *MS* and *APL*, are less consistent than expected.

The best fitting relationships are the power–laws linking *S* and *MS* with the number of pipes *m*, as indicated by $R^2 = 0.94$ for the flow entropy and $R^2 = 0.90$ for the maximum flow entropy (Figures 1b and 2b, respectively). The clear dependence of *MS* and *S* on *m*, indicating that flow entropy is related to the size of the network, suggested to investigate the ratios *S/m* and *MS/m* to characterize the redundancy of a network regardless of its dimension.

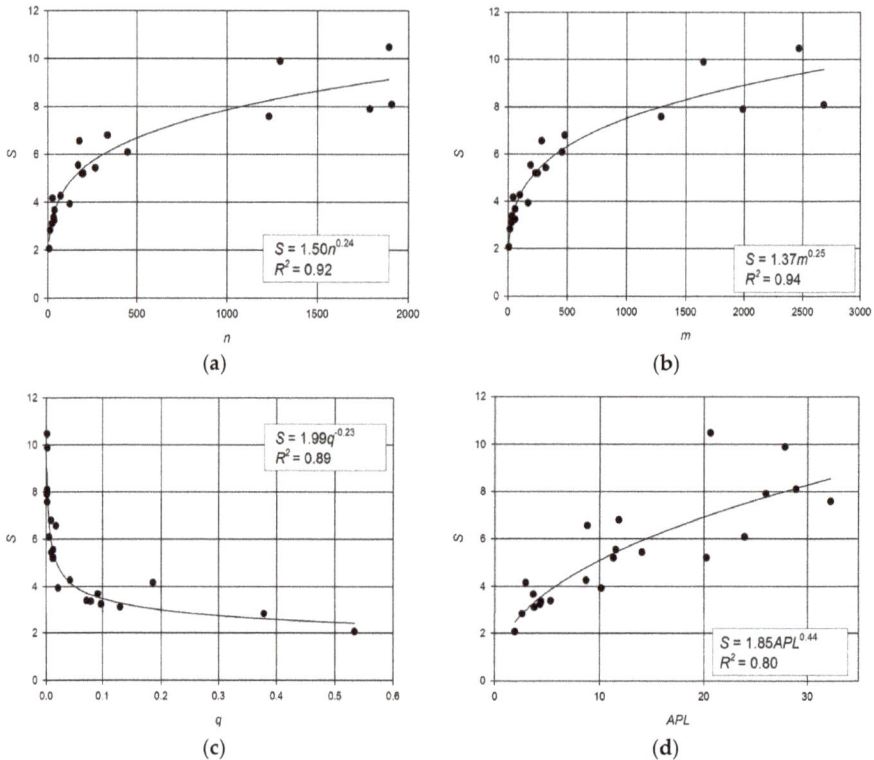

Figure 1. Scatter plots and best fitting power–laws of: (**a**) entropy vs. number of nodes; (**b**) entropy vs. number of pipes; (**c**) entropy vs. link density; (**d**) entropy vs. network average path length.

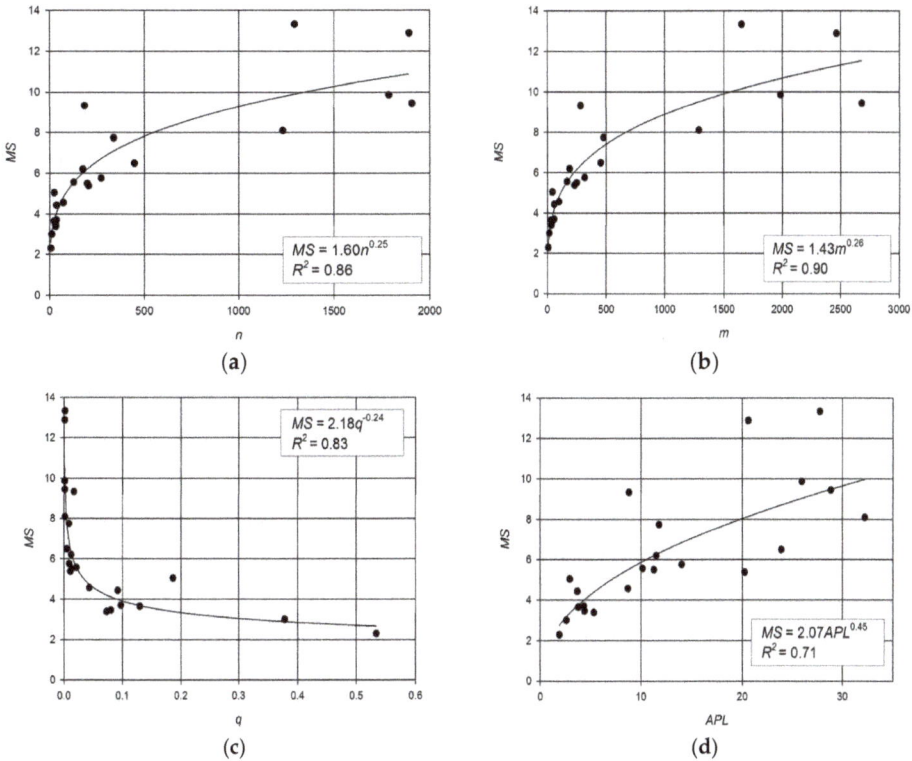

Figure 2. Scatter plots and best fitting power–laws of: (**a**) maximum entropy vs. number of nodes; (**b**) maximum entropy vs. number of pipes; (**c**) maximum entropy vs. link density; (**d**) maximum entropy vs. network average path length.

The scatter plots of S/m vs. n and MS/m vs. n, and the coefficients of determination of the relevant best fitting power–laws, are reported in Figure 3. A distinct trend is clearly visible for both the flow entropy measures, as indicated by $R^2 = 0.99$ for both the relationships. The obtained best fitting power–law equations are:

$$S = 1.05m \times n^{-0.74} \tag{7}$$

$$MS = 1.12m \times n^{-0.72} \tag{8}$$

Looking at Equation (4), and keeping in mind the adopted definition of the probability of the water flowing along a path from a source node to a demand node, it becomes clear that the maximum theoretical flow entropy (i.e., all flow paths sharing the same probability) should scale with $\ln n$. In fact, the number of possible paths in a network scales with the number of nodes (e.g., in a network with a single source, the total number of flow paths to all nodes equals the number of links $m = n + l - 1$, l being the number of loops). Therefore, it is expected that

$$\frac{MS}{m} \sim \frac{-\sum_n \frac{1}{n} \ln \frac{1}{n}}{n} = \frac{\ln n}{n} \tag{9}$$

The curve of Equation (9), also plotted in Figure 3, is not far from the scaling behavior exhibited by the maximum entropy of the considered WDNs. The observed difference can be ascribed to the fact that water flows must obey the flow balance equations at nodes, so that equal probabilities of all the flow paths are not physically possible.

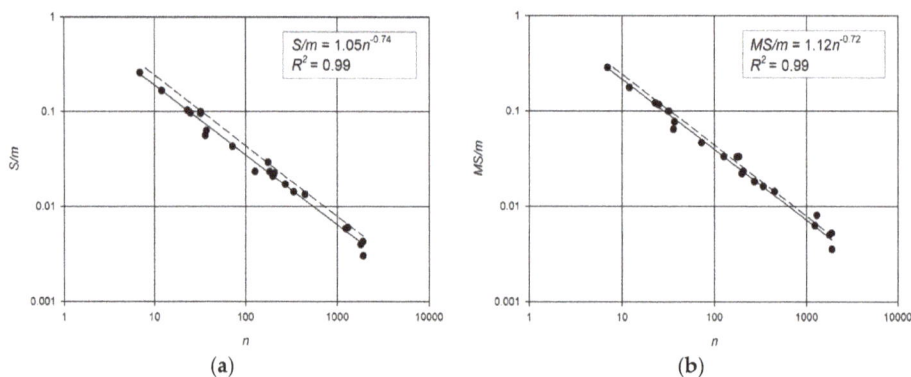

Figure 3. Scatter plots and best-fitting power law equations: (a) S/m vs. number of nodes; (b) MS/m vs. number of nodes. The dashed lines represent the expected scaling of flow entropy for a network with equiprobable flow paths.

It looks clear how both actual and maximum flow entropy strictly depend on network size and topology. The very good alignment of the values of S of WDNs designed with different criteria along a single power–law can be seen as an indirect confirmation of its suitability as a measure of network robustness. In fact, regardless of the criteria adopted for the design of pipe diameters, the smaller the hydraulic resistance of pipes (i.e., larger diameter and shorter length), the higher the flows that spontaneously tend to develop through them. The flow distribution along pipes, and so the flow entropy of the network, is thus determined by the hydraulic laws governing energy dissipation along pipes, which lead to the delivery of the demand at nodes with the minimum dissipated power [41] and, at the same time, set limits to the "disorder" of flow distribution.

The small scatter of the points from the curve of Equation (8), comparable to that of Equation (7), is likely due to the imperfect calculation of MS, as already discussed in the previous section, due to the a priori assumption of flow directions along pipes. However, as expected, the obtained trend seems not to be significantly affected by such an issue.

Equations (7) and (8) shed some light on the link between flow entropy and topology of a WDN. In fact, introducing the relationship $m = n + l - 1$, it is possible to compare the flow entropy of networks with different size and different number of loops. In example, Figure 4 shows the dependence of MS on n and l according to Equation (8). It looks clear that the more looped the network is, the higher is its entropy, thus confirming that flow entropy is a suitable measure of WDN redundancy. On the same graph, the curves representing the maximum flow entropy of WDNs with average node degree $k = 2$ and $k = 4$ are also plotted, delimiting the part of the plane to which WDNs belong. In fact, owing to the physical constraints of pipe connections at nodes, the average node degree of most WDNs falls between such values, as confirmed by the positions of the dots representing the 22 considered networks.

The obtained relationships indicate that, thanks to the high values of the coefficients of determination, it is possible to assess the maximum achievable flow entropy of a network starting from mere basic topological information such as the numbers of links and nodes.

In particular, Equation (8) provides a simple way to compute MS, without the need of a preliminary determination of flow pipe orientations, which can be easily implemented in the design of water supply networks aiming at taking into account the positive effect of redundancy on network robustness [39].

It is worth highlighting that the obtained relationships (7) and (8) have been derived for very different WDNs, both real and synthetic, from different countries, with quite different topological and hydraulic characteristics. Nonetheless, they show a clear scaling behavior in the form of power–laws, indicating that the values of the informational flow entropy are strongly related to some intrinsic and scale-invariant topological characteristic of WDNs, which likely reflects the spatial embedding of these networks, limiting their topological "disorder" (e.g., the degree connectivity of WDNs assumes a nearly constant value as the size of the network increases [32]).

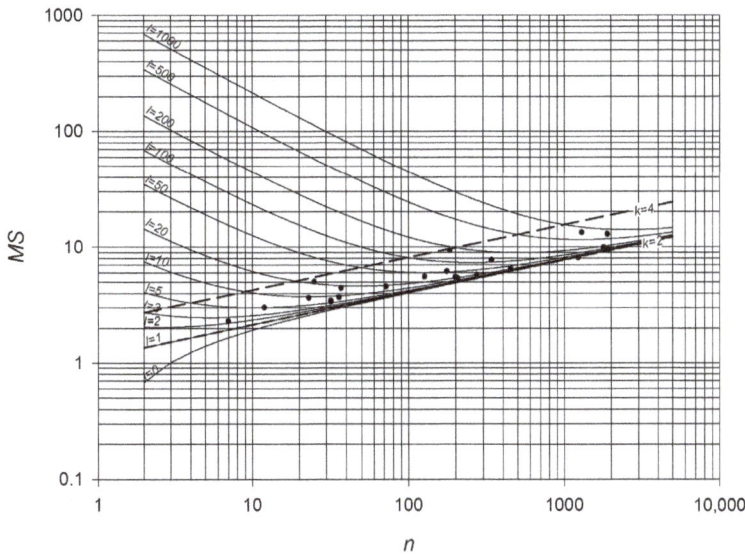

Figure 4. Scaling of maximum flow entropy with number of nodes, for networks with various numbers of loops *l*. The dashed lines represent maximum flow entropy of networks with fixed average node degree $k = 2$ and $k = 4$. The dots represent the considered set of 22 WDNs.

4. Examples of Application

In this section, practical examples are given of how the maximum flow entropy value *MS*, computed by Equation (8), can be used for WDN design or rehabilitation. Starting from the value of maximum entropy, estimated only by means of topological information, the design of the water supply network can be carried out by means of a multi-objective optimization procedure, based on the minimization of entropy deviation and pipe costs, in compliance with hydraulic constraints (i.e., the required minimum pressure at demand nodes). Specifically, the optimization problem consists of defining the optimal choice of the diameters of all pipes in the network, by minimizing the following multi-objective function (*MOF*):

$$
\begin{cases}
MOF = \left\{ \Delta S; \ C = C\prime \cdot \sum_{j=1}^{m} L_j D_j^{\beta} \right\} \\
\text{constraint}: h_i > \overline{h}_i \ i = 1,,n
\end{cases}
\tag{10}
$$

In Equation (10), the first component of *MOF*, ΔS, represents the deviation of flow entropy, calculated with Equation (5), from the maximum flow entropy, estimated by means of Equation (8) as a function of n and m; the second component C represents the total cost of the pipes of the network ($C\prime$ is the unit cost of pipes; L_j and D_j are the length and the diameter of the j-th pipe, respectively; β is a coefficient expressing the dependence of the cost of a pipe on its diameter, for which the value

$\beta = 1.5$ has been proposed [69]); h_i and $\overline{h}_i$ are, respectively, the actual and the design pressure heights at the i-th node of the network.

The application of the proposed WDN design optimization procedure, summarized by Equation (10) , has been carried out for the real water supply networks of Fossolo [60], a neighborhood of the city of Bologna (Italy), and of the town of Skiathos (Greece) [62]. The first network consists of 36 nodes and 58 polyethylene pipes, and the design pressure was assumed equal to $\overline{h} = 30$ m at all nodes. The second network, made with cast iron pipes, has $n = 175$ and $m = 189$, with $\overline{h} = 22$ m. In Figure 5, the sketches of the WDN of Fossolo and Skiathos are reported.

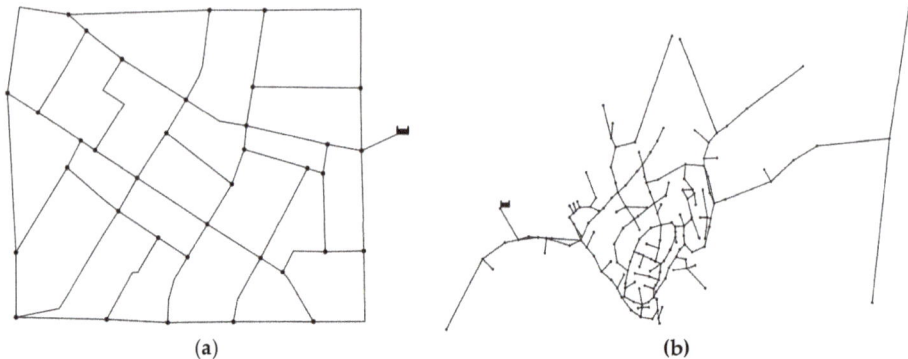

(a)   (b)

Figure 5. Layouts of the water distribution networks for which the multi-objective optimal design procedure based on maximum flow entropy has been applied: (**a**) Fossolo; (**b**) Skiathos.

The minimization of *MOF* was carried out by a heuristic optimization method based on a Genetic Algorithm (GA), a minimum search technique based on mimicking the process of natural selection in the evolution of species [70]. Such an evolutionary algorithm allows for easily introducing constraints on the unknown parameters, at the same time avoiding local minima by introducing random variations to parameter vectors. The GA parameters are the following: each individual of the population is a sequence of chromosomes corresponding to the diameters of all the pipes of the network, which can assume only the values of the existing commercial pipes reported in Table 2. The number of GA generations, the size of the population and the crossover percentage were set to 100, 100 individuals and 0.8, respectively.

The application of the proposed network design procedure led to the definition of the Pareto frontsreported in Figure 6, which represent, in the plane (C, S), the set of all the optimal solutions obtained by minimizing ΔS and the total pipe cost, in compliance with the hydraulic constraints of Equation (10). In addition, the red dots in Figure 6 represent the entropy deviation and the total pipe cost of the original network layouts. Without limiting the general validity of the obtained results, the unit cost of pipes has been assumed $C\prime = 1$. The obtained Pareto fronts show that the smallest values of ΔS correspond to the highest values of total pipe cost, as the more a network is robust, the more investment is needed for its realization (e.g., [71]). For the network of Fossolo, the minimum value of $\Delta S = 0.0004$, corresponding to a flow entropy $S = 4.66$, implies an increase of pipe cost, compared with the original layout, of about 58%. However, a flow entropy $S = 4.55$ can be obtained with an increase of cost smaller than 25%, which represents a good tradeoff between reliability improvement and cost increase. For the case of Skiathos, instead, it is worth noting that nearly the maximum flow entropy $S = 6.72$ can be achieved without any increment of overall pipe cost compared to the existing network.

Table 2. Pipe diameters of the networks of Fossolo (polyethylene pipes) and Skiathos (cast iron pipes).

FossoloDN (mm)		SkiathosDN (mm)	
16.00	73.60	40.00	125.00
20.40	90.00	50.00	140.00
26.00	102.20	63.00	150.00
32.60	147.20	75.00	160.00
40.80	184.00	80.00	225.00
51.40	204.6	90.00	
61.40	229.2	110.00	

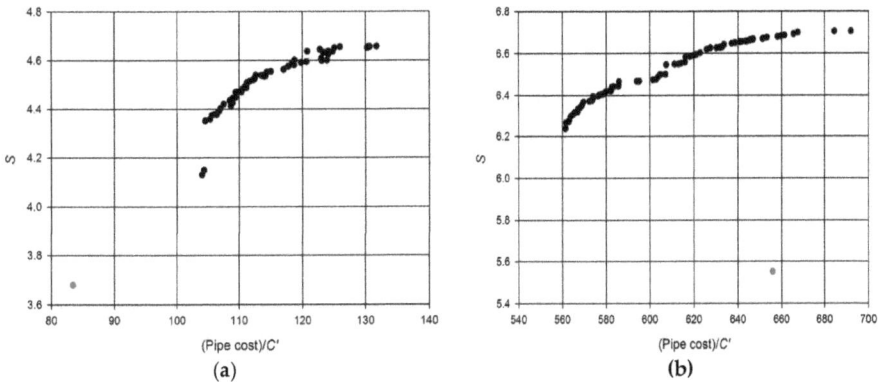

Figure 6. Pareto fronts of the proposed multi-objective optimal network design procedure (flow entropy and total cost of network pipes): (**a**) Fossolo; (**b**) Skiathos. The red dots correspond to network layouts before optimization.

5. Conclusions

The study investigates the scaling-law of informational flow entropy of water distribution networks, often assumed as a surrogate of network robustness, with their topological size. To such aim, the relationships between informational flow entropy, S, maximum informational flow entropy, MS, and some suitable topological metrics (namely, number of nodes, n; number of links, m; network link density, q; network average path length, APL) are investigated for a set of 22 networks, both real and synthetic, with different characteristics.

A clear dependence of flow entropy on topological metrics is observed, and, in particular, power–law relationships, strongly linking S/m and MS/m to the number of nodes of the network (i.e., $R^2 = 0.99$), are identified. The obtained scaling laws result in being close to the expected scaling of flow entropy in networks with equiprobable flow paths (i.e., the same flow carried to the end of any flow path connecting sources to demand nodes), although the actual flow paths cannot be equiprobable, as they must obey flow balance equations at nodes. Such a scale-invariant behavior, testified by the power–laws, probably reflects the peculiar topological feature of water distribution networks, in which each node is connected only to a few immediately surrounding nodes, thus limiting the topological "disorder" of the network, i.e., the number of possible flow paths from each node.

The obtained power–laws, providing an easy estimate of actual and maximum flow entropy of a network, allow to quantify the entropy deficit of a network, i.e., the distance of the flow entropy of a network of given topology from its maximum achievable flow entropy, which can be used in network design and rehabilitation as a measure of network robustness. In this respect, examples of application to multi-objective design of real water distribution networks show how optimal solutions in terms of pipe cost and overall network robustness are easily identified.

Acknowledgments: This research is part of the Ph.D. project "Water distribution network management optimization through Complex Network theory" within the Doctoral Course "A.D.I." granted by Università degli Studi della Campania "L. Vanvitelli".

Author Contributions: All the Authors contributed to the development of the research idea, to the conceivement of the paper, and to the discussion and comment of the results; G.F.S. and C.G. carried out the numerical experiments; G.F.S. and R.G. mostly wrote the paper, which was commented and edited by all the others.

Conflicts of Interest: The authors declare no conflict of interest.

References

1. Greco, R.; Di Nardo, A.; Santonastaso, G.F. Resilience and entropy as indices of robustness of water distribution networks. *J. Hydroinform.* **2012**, *14*, 761–771. [CrossRef]
2. Ghosh, A. Scaling Laws. In *Mechanics over Micro and Nano Scales*; Chakraborty, S., Ed.; Springer: New York, NY, USA, 2011.
3. Rodriguez-Iturbe, I.; Rinaldo, A. *Fractal River Basins: Chance and Self Organization*; Cambridge University Press: Cambridge, UK, 1996; ISBN 0521473985.
4. Hubert, P.; Tessier, Y.; Lovejoy, S.; Schertzer, D.; Schmitt, F.; Ladoy, P.; Carbonnel, J.P.; Violette, S. Multifractals and extreme rainfall events. *Geophys. Res. Lett.* **1993**, *20*, 931–934. [CrossRef]
5. Jennings, A.H. World's greatest observed point rainfalls. *Mon. Weather Rev.* **1950**, *78*, 4–5. [CrossRef]
6. Galmarini, S.; Steyn, D.G.; Ainslie, B. The scaling law relating world point-precipitation records to duration. *Int. J. Climatol.* **2004**, *24*, 533–546. [CrossRef]
7. Kolmogorov, A.N. The local structure of turbulence in incompressible viscous fluid for very large Reynolds numbers. *Proc. R. Soc. A* **1991**, *434*, 9–13. [CrossRef]
8. Frisch, U.; Sulem, P.; Nelkin, M. A simple dynamical model of intermittent fully developed turbulence. *J. Fluid Mech.* **1978**, *87*, 719–736. [CrossRef]
9. Arrault, J.; Arnéodo, A.; Davis, A.; Marshak, A. Wavelet based multifractal analysis of rough surfaces: Application to cloud models and satellite data. *Phys. Rev. Lett.* **1997**, *79*, 75–78. [CrossRef]
10. Badin, G.; Domeisen, D.I.V. Nonlinear stratospheric variability: Multifractal detrended fluctuation analysis and singularity spectra. *Proc. R. Soc. A* **2016**, *472*, 20150864. [CrossRef] [PubMed]
11. Agarwal, S.; Moon, W.; Wettlaufer, J. Trends, noise and re-entrant long-term persistence in Arctic sea ice. *Proc. R. Soc. A* **2012**, *468*, 2416–2432. [CrossRef]
12. Albert, R.; Barabási, A.-L. Statistical mechanics of complex networks. *Rev. Mod. Phys.* **2002**, *74*, 47. [CrossRef]
13. Boccaletti, S.; Latora, V.; Moreno, Y.; Chavez, M. Hwanga, D.U. Complex networks: Structure and dynamics. *Phys. Rep.* **2006**, *424*, 175–308. [CrossRef]
14. Maier, H.R.; Lence, B.J.; Tolson, B.A.; Foschi, R.O. First-order reliability method for estimating reliability, vulnerability, and resilience. *Water Resour. Res.* **2001**, *37*, 779–790. [CrossRef]
15. Yazdani, A.; Jeffrey, P. A complex network approach to robustness and vulnerability of spatially organized water distribution networks. *Phys. Soc.* **2010**, *15*, 1–18.
16. Bao, Y.; Mays, L.W. Model for water distribution system reliability. *J. Hydraul. Eng.* **1990**, *116*, 1119–1137. [CrossRef]
17. Gargano, R.; Pianese, D. Reliability as a tool for hydraulic network planning. *J. Hydraul. Eng. ASCE* **2000**, *126*, 354–364. [CrossRef]
18. Ostfeld, A. Reliability analysis of water distribution systems. *J. Hydroinform.* **2004**, *6*, 281–294. [CrossRef]
19. Wagner, J.M.; Shamir, U.; Marks, D.H. Water distribution reliability: Analytical methods. *J. Water Resour. Plan. Manag.* **1988**, *114*, 253–274. [CrossRef]
20. Shannon, C.E. A mathematical theory of communication. *Bell Syst. Tech. J.* **1948**, *27*, 379–423. [CrossRef]
21. Singh, V.P. *Entropy Theory and Its Application in Environmental and Water Engineering*, 1st ed.; John Wiley and Sons, Ltd.: Oxford, UK, 2013; ISBN 978-1-119-97656-1.
22. Awumah, K.; Goulter, I.; Bhatt, S.K. Assessment of reliability in water distribution networks using entropy based measures. *Stoch. Hydrol. Hydraul.* **1990**, *4*, 309–320. [CrossRef]
23. Tanyimboh, T.T.; Templeman, A.B. A quantified assessment of the relationship between the reliability and entropy of water distribution systems. *Eng. Optim.* **2000**, *33*, 179–199. [CrossRef]

24. Tanyimboh, T.T.; Sheahan, C. A maximum entropy based approach to the layout optimization of water distribution systems. *Civ. Eng. Environ. Syst.* **2002**, *19*, 223–253. [CrossRef]

25. Setiadi, Y.; Tanyimboh, T.T.; Templeman, A.B. Modelling errors, entropy and the hydraulic reliability of water distribution systems. *Adv. Eng. Softw.* **2005**, *36*, 780–788. [CrossRef]

26. Liu, H.; Savic, D.; Kapelan, Z.; Zhao, M.; Yuan, Y.; Zhao, H. A diameter-sensitive flow entropy method for reliability consideration in water distribution system design. *Water. Resour. Res.* **2014**, *50*, 5597–5610. [CrossRef]

27. Tanyimboh, T.T.; Templeman, A.B. Calculating maximum entropy flows in networks. *J. Oper. Res. Soc.* **1993**, *44*, 383–396. [CrossRef]

28. Tanyimboh, T.T.; Templeman, A.B. Using entropy in water distribution networks. In *Integrated Computer Applications in Water Supply*, 1st ed.; Coulbeck, B., Ed.; Research Studies Press Ltd.: Taunton, UK, 1993; Volume 1, pp. 77–90, ISBN 086380-154-4.

29. Perelman, L.; Housh, M.; Ostfeld, A. Robust optimization for water distribution systems least cost design. *Water Resour. Res.* **2013**, *49*, 6795–6809. [CrossRef]

30. Saleh, S.H.A.; Tanyimboh, T.T. Optimal design of water distribution systems based on entropy and topology. *Water Resour. Res.* **2014**, *28*, 3555–3575. [CrossRef]

31. Mays, L.W. *Water Distribution Systems Handbook*, 1st ed.; McGraw-Hill: New York, NY, USA, 2000; ISBN 780071342131.

32. Di Nardo, A.; Di Natale, M.; Giudicianni, C.; Greco, R.; Santonastaso, G.F. Complex network and fractal theory for the assessment of water distribution network resilience to pipe failures. *Water Sci. Technol. Water Supply* **2017**, *17*, ws2017124. [CrossRef]

33. Watts, D.; Strogatz, S. Collective dynamics of small world networks. *Nature* **1998**, *393*, 440–442. [CrossRef] [PubMed]

34. Barabasi, A.L.; Albert, R. Emergence of scaling in random networks. *Science* **1999**, *286*, 797–817. [CrossRef]

35. Yazdani, A.; Jeffrey, P. Robustness and Vulnerability Analysis of Water Distribution Networks Using Graph Theoretic and Complex Network Principles. In Proceedings of the 12th Annual Conference on Water Distribution Systems Analysis (WDSA), Tucson, AZ, USA, 12–15 September 2010.

36. Gutiérrez-Pérez, J.A.; Herrera, M.; Pérez-García, R.; Ramos-Martínez, E. Application of graph-spectral methods in the vulnerability assessment of water supply networks. *Math. Comput. Model.* **2013**, *57*, 1853–1859. [CrossRef]

37. Herrera, M.; Abraham, E.; Stoianov, I. A graph-theoretic framework for assessing the resilience of sectorised water distribution networks. *Water Resour. Res.* **2016**, *30*, 1685–1699. [CrossRef]

38. Gheisi, A.; Naser, G. Multistate Reliability of Water-Distribution Systems: Comparison of Surrogate Measures. *J. Water Res. Plan. Manag.* **2015**, *141*. [CrossRef]

39. Tanyimboh, T.T. Informational entropy: A failure tolerance and reliability surrogate for water distribution networks. *Water Resour. Res.* **2017**, *31*, 3189–3204. [CrossRef]

40. Creaco, E.; Fortunato, A.; Franchini, M.; Mazzola, M.R. Comparison between entropy and resilience as indirect measures of reliability in the framework of water distribution network design. *Proc. Eng.* **2014**, *70*, 379–388. [CrossRef]

41. Todini, E. Looped water distribution networks design using a resilience index based heuristic approach. *Urban Water* **2000**, *2*, 115–122. [CrossRef]

42. Prasad, T.D.; Park, N.S. Multi-Objective Genetic Algorithms for Design of Water Distribution Networks. *J. Water Res. Plan. Manag.* **2004**, *130*, 73–82. [CrossRef]

43. Vaabel, J.; Ainola, L.; Koppel, T. Hydraulic power analysis for determination of characteristics of a water. In Proceedings of the Eighth Annual Water Distribution Systems Analysis Symposium (WDSA), Cincinnati, OH, USA, 27–30 August 2006.

44. Yazdani, A.; Jeffrey, P. Complex network analysis of water distribution systems. *Chaos* **2011**, *21*, 016111. [CrossRef] [PubMed]

45. Di Nardo, A.; Di Natale, M.; Giudicianni, C.; Greco, R.; Santonastaso, G.F. Water supply network partitioning based on weighted spectral clustering. *Stud. Comp. Intell.* **2017**, *693*, 797–807. [CrossRef]

46. Herrera, M.; Izquierdo, J.; Pérez-Garcìa, R.; Montalvo, I. Multi-agent adaptive boosting on semi supervised water supply clusters. *Adv. Eng. Softw.* **2012**, *50*, 131–136. [CrossRef]

47. Guest, G.; Namey, E.E. *Public Health Research Methods*, 1st ed.; SAGE Publications, Inc.: Thousand Oaks, CA, USA, 2014.

48. Di Nardo, A.; Di Natale, M.; Giudicianni, C.; Musmarra, D.; Santonastaso, G.F.; Simone, A. Water Distribution System Clustering and Partitioning Based on Social Network Algorithms. *Proc. Eng.* **2015**, *119*, 196–205. [CrossRef]

49. Khinchin, A.I. *Mathematical Foundations of Statistical Mechanics*; Dover: New York, NY, USA, 1953.

50. Ang, W.K.; Jowitt, P.W. Path entropy method for multiple-source water distribution networks. *J. Eng. Optim.* **2005**, *37*, 705–715. [CrossRef]

51. Rossman, L.A. *EPANET2 Users Manual*; US EPA: Cincinnati, OH, USA, 2000.

52. Yassin-Kassab, A.; Templeman, A.B.; Tanyimboh, T.T. Calculating maximum entropy flows in multi-source, multi-demand networks. *Eng. Optim.* **1999**, *31*, 695–729. [CrossRef]

53. Alperovits, E.; Shamir, U. Design of optimal water distribution systems. *Water Resour. Res.* **1977**, *13*, 885–900. [CrossRef]

54. Gessler, J. Pipe network optimization by enumeration. Proceedings of Computer Applications in Water Resources, ASCE Specialty Conference, Buffalo, NY, USA, 10–12 June 1985.

55. Walski, T.; Brill, E.; Gessler, J.; Goulter, I.; Jeppson, R.; Lansey, K.; Lee, H.; Liebman, J.; Mays, L.; Morgan, D.; et al. Battle of the network models: Epilogue. *J. Water Res. Plan. Manag.* **1987**, *113*, 191–203. [CrossRef]

56. Kim, J.H.; Kim, T.G.; Kim, J.H.; Yoon, Y.N. A study on the pipe network system design using non-linear programming. *J. Korean Water Resour. Assoc.* **1994**, *27*, 59–67.

57. Sherali, H.D.; Subramanian, S.; Loganathan, G. Effective relaxations and partitioning schemes for solving water distribution network design problems to global optimality. *J. Global. Optim.* **2001**, *19*, 1–26. [CrossRef]

58. Fujiwara, O.; Khang, D.B. A two-phase decomposition method for optimal design of looped water distribution networks. *Water Resour. Res.* **1990**, *26*, 539–549. [CrossRef]

59. Lee, S.C.; Lee, S.I. Genetic algorithms for optimal augmentation of water distribution networks. *J. Korean Water Resour. Assoc.* **2001**, *34*, 567–575.

60. Bragalli, C.; D'Ambrosio, C.; Lee, J.; Lodi, A.; Toth, P. On the Optimal Design of Water Distribution Networks: A Practical MINLP Approach. *Optim. Eng.* **2012**, *13*, 219–246. [CrossRef]

61. Ostfeld, A.; Uber, J.G.; Salomons, E.; Berry, J.W.; Hart, W.E.; Phillips, C.A.; Watson, J.; Dorini, G.; Jonkergouw, P.; Kapelan, Z.; et al. The Battle of the Water Sensor Networks (BWSN): A Design Challenge for Engineers and Algorithms. *J. Water Resour. Plan. Manag.* **2008**, *134*, 556–568. [CrossRef]

62. Di Nardo, A.; Di Natale, M.; Giudicianni, C.; Laspidou, C.; Morlando, F.; Santonastaso, G.F.; Kofinas, D. Spectral analysis and topological and energy metrics for water network partitioning of Skiathos island. *Eur. Water* **2017**, *58*, 423–428.

63. Di Nardo, A.; Di Natale, M.; Gisonni, C.; Iervolino, M. A genetic algorithm for demand pattern and leakage estimation in a water distribution network. *J. Water Supply Res. Tech. Aqua* **2015**, *64*, 35–46. [CrossRef]

64. Herrera, M.; Canu, S.; Karatzoglou, A.; Perez-Garcıa, R.; Izquierdo, J. An approach to water supply clusters by semi-supervised learning. In Proceedings of theInternational Congress on Environmental Modelling and Software Modelling (iEMSs) 2010 for Environment's Sake, Fifth Biennial Meeting, Ottawa, ON, Canada, 5–8 July 2010.

65. Reca, J.; Martinez, J. Genetic algorithms for the design of looped irrigation water distribution networks. *Water Resour. Res.* **2006**, *42*, W05416. [CrossRef]

66. Tzatchkov, V.G.; Alcocer-Yamanaka, V.H.; Ortìz, V.B. Graph Theory Based Algorithms for Water Distribution Network Sectorization Projects. In Proceedings of the Eighth Annual Water Distribution Systems Analysis (WDSA) Symposium, Cincinnati, OH, USA, 27–30 August 2006.

67. Lippai, I. Colorado Springs Utilities Case Study: Water System Calibration Optimization. In Proceedings of the ASCE Pipeline Division Specialty Conference, Houston, TX, USA, 21–24 August 2005.

68. Farmani, R.; Savic, D.A.; Walters, G.A. Evolutionary multi-objective optimization in water distribution network design. *Eng. Optim.* **2005**, *37*, 167–183. [CrossRef]

69. Tanyimboh, T.T.; Saleh, S.H.A. Global maximum entropy minimum cost design of water distribution systems. In Proceedings of the ASCE/EWRI World Environmental and Water Resources Congress 2011, Palm Springs, CA, USA, 22–26 May 2011.

70. Goldberg, D.E. *Genetic Algorithms in Search, Optimization and Machine Learning*; Addison-Wesley Longman Publishing Co. Inc.: Boston, MA, USA, 1989.

71. Tricarico, C.; Gargano, R.; Kapelan, Z.; Savic, D.; de Marinis, G. Economic Level of Reliability for the Rehabilitation of Hydraulic Networks. *J. Civ. Eng. Environ. Syst.* **2006**, *23*, 191–207. [CrossRef]

entropy

MDPI

Article

Spatial Optimization of Agricultural Land Use Based on Cross-Entropy Method

Lina Hao [1], Xiaoling Su [1,*], Vijay P. Singh [2] and Olusola O. Ayantobo [1,3]

[1] College of Water Resources and Architectural Engineering, Northwest A & F University,
     Yangling 712100, China; haolina@nwafu.edu.cn (L.H.); Skollar24k@yahoo.com (O.O.A.)
[2] Department of Biological & Agricultural Engineering and Zachry Department of Civil Engineering,
     Texas A & M University, College Station, TX 77843-2117, USA; vsingh@tamu.edu
[3] Department of Water Resources Management and Agricultural-Meteorology,
     Federal University of Agriculture, PMB 2240 Abeokuta, Nigeria
*   Correspondence: xiaolingsu@nwsuaf.edu.cn; Tel.: +86-29-8708-2902; Fax: +86-29-8708-2901

Received: 4 September 2017; Accepted: 2 November 2017; Published: 7 November 2017

Abstract: An integrated optimization model was developed for the spatial distribution of agricultural crops in order to efficiently utilize agricultural water and land resources simultaneously. The model is based on the spatial distribution of crop suitability, spatial distribution of population density, and agricultural land use data. Multi-source remote sensing data are combined with constraints of optimal crop area, which are obtained from agricultural cropping pattern optimization model. Using the middle reaches of the Heihe River basin as an example, the spatial distribution of maize and wheat were optimized by minimizing cross-entropy between crop distribution probabilities and desired but unknown distribution probabilities. Results showed that the area of maize should increase and the area of wheat should decrease in the study area compared with the situation in 2013. The comprehensive suitable area distribution of maize is approximately in accordance with the distribution in the present situation; however, the comprehensive suitable area distribution of wheat is not consistent with the distribution in the present situation. Through optimization, the high proportion of maize and wheat area was more concentrated than before. The maize area with more than 80% allocation concentrates on the south of the study area, and the wheat area with more than 30% allocation concentrates on the central part of the study area. The outcome of this study provides a scientific basis for farmers to select crops that are suitable in a particular area.

Keywords: cross-entropy minimization; land suitability evaluation; spatial optimization

---

1. Introduction

Scarcity of agricultural water and land resources is becoming severe due to the growing population and continued economic development, and has become a critical issue in formulating sustainable developmental policies [1–3]. Therefore, appropriate and efficient allocation of agricultural water and land resources has become necessary in regional agricultural sustainable development [4]. Agricultural land use allocation is the core issue of agricultural land and water resource allocation optimization [5,6]. Not only can it guide decision makers in assessing land demand for different crop types, but it can also identify the optimum land spatial unit with characteristics that are related to their geographical locations for each crop type [7–9], as well as simultaneously seeking the best land use layout [10,11].

The process of agricultural land-use allocation would be undertaken in three main stages: (i) demand assessment; (ii) agricultural land suitability evaluation; and (iii) spatial distribution of crop types [7].

To assess the agricultural land suitability of crops, the required environmental and socio-economic conditions are considered [12]. The Food and Agricultural Organization (FAO) developed crop-specific maps of crop suitability classes using the spatial data on soil, topography, features, and crop characteristics [13]. Many studies have developed methods for land suitability allocation based on the FAO framework. In these methods, input attributes and suitability indices were classified into different classes, and weights were assigned to the attributes, depending on their relative importance [5,14–17]. However, the main shortcoming of FAO-based methods is that their crop suitability data are usually available at a 5 min (approximately 9 km × 9 km at the equator) grid and such resolution is too coarse to satisfy the research demand at the irrigation district scale. Furthermore, when evaluating land suitability, very few studies have taken into account the actual distribution of various crops. For example, You and Wood [18] improved the pre-allocation by providing the existing crop distribution maps. To a significant extent, land suitability are determined by biophysical and soil conditions, such as organic matter, total nitrogen, total phosphorus. When the observed data of these attributes cannot be completely covered on the entire region, the actual distribution of crops can provide the supplement of crop distribution information. Using the research findings of Peng [19], an integrated model combined with multi-source remote sensing data was developed to generate maps of spatial distribution with 1 km × 1 km resolution suitable for crop planting.

The main objective of agricultural land use allocation is the spatial allocation of crop types to different spatial units having characteristics related to their geographical locations, for the purpose of seeking the best land use layout [11]. There is extensive literature on various methods of optimization for agricultural cropping patterns, such as linear programming [20], non-linear programming [21], multi-objective programming [22], fuzzy programming [23,24], and stochastic optimization [25]. However, these methods have ignored the effective unification of quantity and space, and have merely focused on the quantity optimization. In other words, these models can provide an optimal cropping pattern but cannot analyze the optimal spatial distribution of crops, which has an important guiding significance in actual production work.

Since numerous variables are involved in spatial optimization, conventional mathematical models are deemed unfit to determine the optimal solution within an acceptable timeframe [12]. Various heuristic algorithms for land-use spatial optimization have also been developed, including particle swarm algorithms [26,27], colony algorithms [28], and genetic algorithms [11,29]. Although these algorithms have a significant global optimization capability, they involve complex patch coding, resulting in programming difficulties. Other methods adopted cellular automation models, based on land-use conversion rules for local areas, to generate land-use patterns under different conditions using a bottom-up approach [12,29,30]. However, cellular automaton is restricted by neighborhood rules, and cannot search across space. In this study, spatial crop optimization is defined in the framework of minimum cross entropy. The principle of minimum cross-entropy (POMCE) was formulated by Kullback and Leibler [31] and is detailed by Kullback [32]. The cross entropy can measure the variation between different information contents, which seems an ideal approach to resolve the spatial allocation problem [33]. However, the cross entropy method has usually been applied in determining spatial-temporal changes of land use and applied a meso-scale model for the spatial disaggregation of crop production [34–36], but has rarely been coupled with spatial optimization modeling [18]. For example, You and Wood [37] described an entropy-based approach to conduct a spatially disaggregated assessment of the distribution of crop production. Considering the difficulty in coupling spatial variables with non-spatial variables, a loosely coupled model, based on minimum cross entropy and nonlinear optimization, is constructed in this study. The crop spatial allocation can be performed by determining the minimum cross entropy between the prior distribution and the desired distribution. The prior distribution produces crop-related and environment-related information obtained from integrated multi-source analysis in agricultural land suitability.

This study develops an integrated optimization model for the spatial distribution of agricultural cropping. The model is based on the spatial distribution of crop suitability, spatial distribution of population density, and agricultural land use data. It combines multi-source remote sensing data with constraints of optimal crop areas, which are obtained from an agricultural cropping pattern optimization model. Minimization of cross-entropy is applied to build the model. The model determines the suitable planting region for a specified crop well.

2. Materials and Methods

2.1. Study Area

The study area is the middle reaches of the Heihe River basin (98°30′–101° E, 38°30′–40° N), lying in an arid region of Gansu Provence, northwest of China, and covers an area of 11,427 km². The Heihe River basin is the second largest inland river basin in China. In this region, crop production mainly depends on agricultural irrigation, because the mean annual precipitation and evaporation are about 117 mm and 1065 mm, respectively. Water consumption from agricultural irrigation accounts for approximately 90% of the total water consumption in this region [38]. Therefore, the optimal distribution of limited irrigation water and land resources is a key factor for agricultural development and sustainability [4,39]. The study area presents a higher terrain in the southeast and low in the northwest, and the elevation is between 1235 m and 3634 m. there is heterogeneity of biophysical and soil condition in this region. The content of organic matter and nitrogen is higher in the southern part than in the northern part [19]. The middle reaches of Heihe River basin is a commodity grain production base, which is made up of 17 irrigation districts (as shown in Figure 1). The main crops in this region are maize, potato, seed maize, cotton, oil crops, and vegetable. Therefore, planting crops in a suitable region, as well as integration between the quantity structure optimization and spatial allocation optimization, are of primary importance to agricultural production management [40].

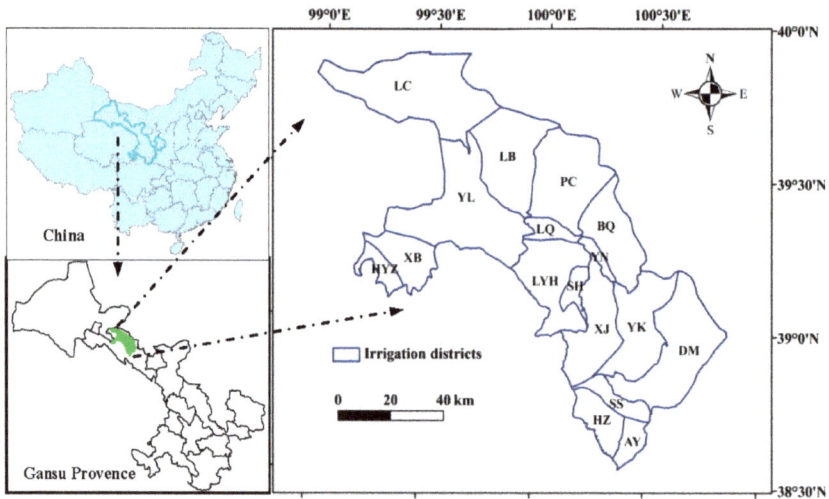

Figure 1. Location of study area in China.

The letters in the figure represent the names of irrigation districts. LC is for Luocheng irrigation district, LB is for Liuba irrigation district, YL is for Youlian irrigation district, XB is for Xinba irrigation district, HYZ is for Hongyazi irrigation district, PC is for Pingchuan irrigation district, LQ is for Liaoquan irrigation district, LYH is for Liyuanhe irrigation district, BQ is for Banqiao irrigation district, YN is for Yanuan irrigation district, SH is for Shahe irrigation district, XJ is for Xijun irrigation district, YK is for Yingke irrigation district, DM is for Daman irrigation district, SS is for Shangsan irrigation district, HZ is for Huazhai irrigation district, and AY is for Anyang irrigation district.

2.2. Methods and Data

Figure 2 shows an overview of the integrated agricultural cropping spatial distribution optimization model. This model is based on minimizing cross-entropy, re-aggregating administrative statistics data, and multi-source remote sensing information data, such as spatial distribution of crop suitability, spatial distribution of population density, and agricultural land use data in a logical framework.

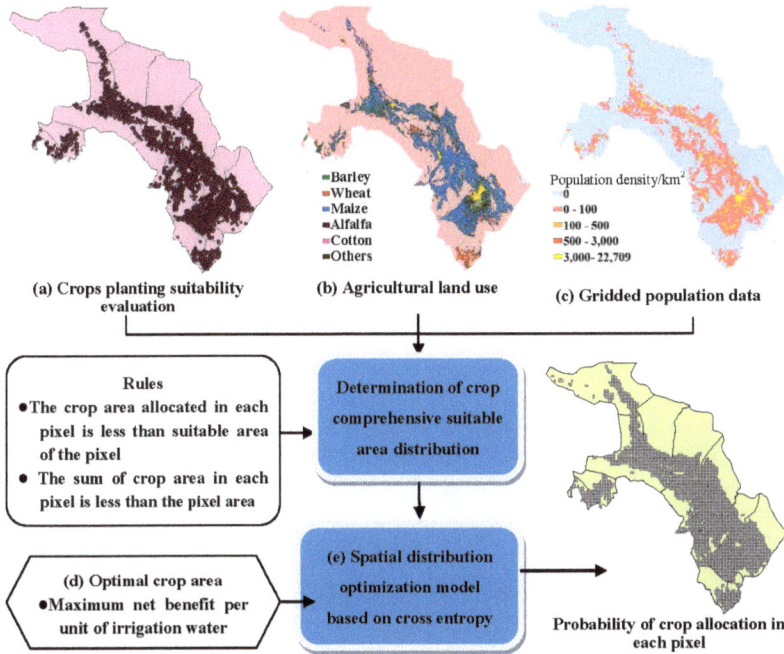

Figure 2. Overview of integrated agricultural cropping spatial distribution optimization model.

2.2.1. Evaluation of Crop Planting Suitability

The evaluation of crop planting suitability is important for agricultural land use allocation. It provides the essential data for the optimization of crop spatial framework in order to realize reasonable utilization of land resources as well as providing references for the scientific management and sustainable utilization of cultivated land resources.

The suitability of crop cultivation in this study is referenced from Peng [19]. Based on the ecological niche fitness theory, Peng [19] selected climatic and environment factors (rainfall, temperature, ET_0), soil characteristics (organic matter, total nitrogen, total phosphorus, total potassium, pH, bulk density), and geographical factors (terrain elevation, slope and aspect). These factors are closely related to crop growth and are used for evaluating the ecological niche and crop planting suitability. Figure 3 shows the spatial distribution of crop planting suitability index. Owing to the limitation of data collection, Peng's [19] study on the maize and wheat planting suitability was employed.

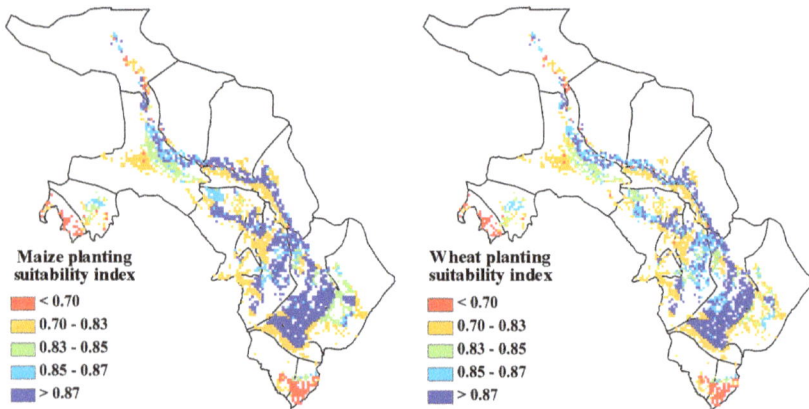

Figure 3. Spatial distribution of crop planting suitability index [19].

2.2.2. Agricultural Land Use

The Heihe River basin land use and land cover data set (HiWATER: Land cover map of Heihe River basin) were obtained from the Cold and Arid Regions Sciences Data Center (http://westdc. westgis.ac.cn/) which provided the 2011–2015 monthly data covering the type of surface. The data set is based on China's domestic satellite HJ/CCD data which has a high temporal and spatial resolution (30 M) [41,42]. This data set increases the classification of cultivated crops, including barley, wheat, maize, alfalfa, cotton and others, and can provide the current crop spatial distribution to determine the crop suitable distribution area.

2.2.3. Gridded Population Data

The spatial distribution of population density data is from the gridded population data of the Heihe River basin and provides the spatial distribution of population density in a 1 km × 1 km grid. The data set is given by the Cold and Arid Regions Sciences Data Center at Lanzhou (http://westdc.westgis.ac.cn) [40].

2.2.4. Optimization Model of Agricultural Cropping Pattern

To realize the appropriate allocation and efficient use of agricultural water and land resources and provide the primary input data for spatial allocation, the optimization model for agricultural cropping pattern is established with the objective of maximum agricultural net benefit per unit of irrigation water under certain agricultural water resources. The planting area of crops in an irrigation district is a decision variable for the model. Crops considered include maize, wheat, potato, maize seed, cotton, oil crop, and vegetable, while the basic data of 2013 used in the model are crop yield, crop prices, cost, agricultural irrigation quota and water availability. This model references the results of the project supported by National Natural Science Fund in China (91,425,302).

(1) Objective function

The agricultural cropping pattern optimization model can be written as follows, the variables x_{ij} is expected area of crop j in irrigation district i (ha); and the objective of optimization model is the maximum agricultural net benefit per unit of irrigation water:

$$\max f = \sum_{i=1}^{17} \sum_{j=1}^{7} \left((y_{ij} v_{ij} - c_{ij}) \cdot x_{ij} / ET_{ij} \right) / \sum_{i=1}^{17} \sum_{j=1}^{7} x_{ij} \tag{1}$$

where f is the net benefit per unit of irrigation water (RMB/m^3); i (i = 1, 2, . . . , n) is irrigation district identifier, of which there are 17 within the study area; j (j = 1, 2, . . . , 7) is the crop type identifier, of which 7 main types are considered in this model (maize, wheat, potato, maize seed, cotton, oil crops, and vegetable); v_{ij} is the price of crop j in the irrigation district i (RMB/kg); y_{ij} is the yield of crop j in the irrigation district i (kg/ha); c_{ij} is the cost of crop j in the irrigation district i (RMB/ha); ET_{ij} is the net irrigation quota of crop j in the irrigation district i (m^3/ha).

(2)   Constraints

The irrigation water of irrigation district i should be less than the available water supply:

$$\sum_{i=1}^{n}\sum_{j=1}^{7} m_{ij}x_{ij} \leq Q_i \tag{2}$$

where m_{ij} is the gross irrigation quota of crop j in the irrigation district i (m^3/ha); Q_i is the available water supply in the irrigation district i (m^3).

The irrigation area of irrigation district i would be less than the effective irrigation area X_i (ha):

$$\sum_{i=1}^{n}\sum_{j=1}^{7} x_{ij} \leq X_i \tag{3}$$

The agriculture product would be to meet the local demand:

$$\sum_{i=1}^{n}\sum_{j=1}^{4} x_{ij} \cdot y_{ij} \geq K \cdot P \cdot FN \tag{4}$$

where P is the population in the study area; FN is the per person grain demand, 135 kg/per; and VN is the per person vegetable demand, 140 kg/per; K is demand coefficient, when the agriculture product meet the local demand, K = 1.

$$\sum_{i=1}^{n}\sum_{j=9} x_{ij} \cdot y_{ij} \geq K \cdot P \cdot VN \tag{5}$$

non-negative constraint:

$$x_{ij} \geq 0 \tag{6}$$

2.2.5. Spatial Distribution Optimization Model Based on Cross Entropy

Shannon (1948) introduced information entropy to measure the uncertainty of the expected information. He defined entropy $H(p)$ as a weighted sum of the information [34]. The entropy of a random variable with probability distribution P ($p_1, p_2, \ldots, p_k$) can be expressed using Equation (7) [43]

$$H(P) = -\sum_{i=1}^{k} p_i \ln p_i \tag{7}$$

Jaynes (1957) proposed the maximum entropy principle in statistical inference: the least informative probability distribution P ($p_1, p_2, \ldots, p_k$) can be found by maximizing the entropy $H(p)$ [44]. In Equation (7), the solutions are: $p_i = 1/n$, $i = 1,2, \ldots, n$, $H(p) = \ln n$ [34,35].

The cross-entropy formulation is based on the Shannon Entropy theory [36]. Cross-entropy was formulated by Kullback and Leibler [31] and is detailed by Kullback [32]. It measures the divergence between the prior distribution and the desired distribution. The principle of minimum cross entropy (POMCE), also referred to as the principle of minimum discrimination information, is obtained by minimizing cross-entropy with respect to the given prior distribution, subject to given constraints [33]. POMCE can be expressed as

$$D(P,Q) = \sum_{i=1}^{k} p_i \ln\left(\frac{p_i}{q_i}\right) \tag{8}$$

where D is the cross-entropy or the discrimination information and the objective is to minimize D. $P\,(p_1, p_2, \dots, p_k)$ is the desired distribution, $Q = (q_1, q_2, \dots, q_k)$ is prior distribution chosen based on all the given information, but does not satisfy the prescribed constraints [45–47].

In this study, we considered a comprehensive crop suitable distribution based on the integrated multi-source data analysis as prior knowledge in the POMCE. Let q_{ij} represent the suitable cultivated land area shares of crop j on pixel i. Therefore,

$$q_{ij} = \frac{Suitable_{ij}}{\sum_i Suitable_{ij}} \tag{9}$$

Based on the spatial distribution of crop suitability, spatial distribution of population density, and agricultural land use data, the spatial distribution optimization model based on cross entropy, subject to the constraints of optimal crop area obtained from agricultural cropping pattern optimization model, is used to determine the optimal spatial distribution of crops.

The objective of crop spatial distribution optimization is to minimize cross-entropy of comprehensive crop distribution probability and desired distribution probability, subject to area constraint on the pixel scale and other related limitations.

The spatial distribution optimization model, based on cross entropy, can be written as follows, variables p_{ij} represent the desired area shares of crop j on pixel i:

$$\min_{p_{ij}} D\left(p_{ij}, q_{ij}\right) = \min\left(\sum_i p_{ij} \ln p_{ij} - \sum_i p_{ij} \ln q_{ij}\right) \tag{10}$$

subject to the following constraints:

$$\sum_i p_{ij} = 1 \quad 0 \le p_{ij} \le 1 \tag{11}$$

making sure the allocated area of crop j on pixel i would be less than suitable cultivated land area of crop j on pixel i:

$$Area_j \times p_{ij} \le Suitable_{ij} \tag{12}$$

making sure the allocated area on pixel i would be less than the arable land area on pixel i:

$$\sum_j Area_{ij} \times p_{ij} \le Available_i \tag{13}$$

where $i = 1, 2, 3, \dots$, represents the pixel identifier within the study area; $j = 1, 2$, represents the crop identifier within the study area; q_{ij} represent the suitable area shares of crop j on pixel i; $Available_i$ represent the arable land area on pixel i; $Suitable_{ij}$ represent the suitable cultivated land area of crop j on pixel i; and $Area_j$ represent the optimal crop areas, which are obtained from the agricultural cropping pattern optimization model.

3. Results and Discussion

3.1. Crop Demand Assessment

Based on limited data on crop planting suitability, this study focuses only on the spatial allocation of two crops, maize and wheat. Thus, in this section, the maize and wheat demands in the middle reaches of Heihe River basin are analyzed.

As shown in Table 1, seed maize controls the priority of water allocation, due to its higher unit net irrigation benefit and lower water requirement in these irrigation districts compared with other crops. Compared with the actual situation in 2013, the proportion of the total planting areas of corn and wheat through optimization all accounts for about 87%. However, the area of maize should increase, and the area of wheat should decrease in a region-wide range.

Table 1. Comparison of crop structure between present and optimal allocation (ha).

	Maize	Wheat	Potato	Seed Maize	Cotton	Oil Crops	Vegetable	Sum
Actual situation	19,680	14,241	794	75,252	3573	1116	11,258	125,915
Optimization	14,111	12,818	725	82,785	4174	990	10,375	125,977

Table 2 compares maize and wheat areas between present and optimal allocations in different irrigation districts. Comparing the actual crop structure, results revealed that except for the XJ, SS, PC, SH, YL, LC, XB irrigation districts, the maize area in the remaining irrigation districts increased, especially in DM and YK. The change was apparent with 2809 and 2057 ha increases, respectively, while the maize area in YL decreased by 2128 ha. However, DM, YK, PC, BQ, LYH and LC would be more favorable to wheat because of lower water requirement, while in other irrigation districts, the area of wheat decreased. For example, it would reduce to 504 ha and 436 ha in LQ and HYZ district, respectively.

Table 2. Comparison of maize and wheat area between present and optimal allocations in different irrigation district (ha).

Irrigation Districts	Maize		Wheat	
	Actual Situation	Optimization	Actual Situation	Optimization
DM	11,805	14,614	160	167
YK	15,049	17,106	1053	1067
XJ	17,598	17,567	579	400
SS	6474	6421	87	78
AY	13	18	561	545
HZ	0	30	353	199
PC	3013	3000	1180	1268
LQ	2367	2433	1253	749
BQ	3900	3931	167	204
YN	2113	2331	947	679
SH	2873	2210	1007	924
LYH	10,220	10,433	3707	3850
YL	14,540	12,412	867	800
LB	1880	1882	56	52
LC	1400	1303	13	73
XB	1593	1080	640	585
HYZ	93	125	1613	1177
Sum	94,932	96,896	14,241	12,817

3.2. Crop Comprehensive Suitable Area Distribution

Using the ArcGIS platform, the whole research area was divided into 9041 1 km × 1 km grids. According to crop planting suitability evaluation, different colors were selected to represent different crop suitability indices, and grids were filled to generate the spatial distribution map of crop planting suitability index.

In light of the spatial distribution of maize and wheat planting suitability obtained above, combined with the spatial distribution of population density and agricultural land use data, the crop comprehensive suitable area distribution can be determined according to the following principles. Figure 4 shows an overview of this process.

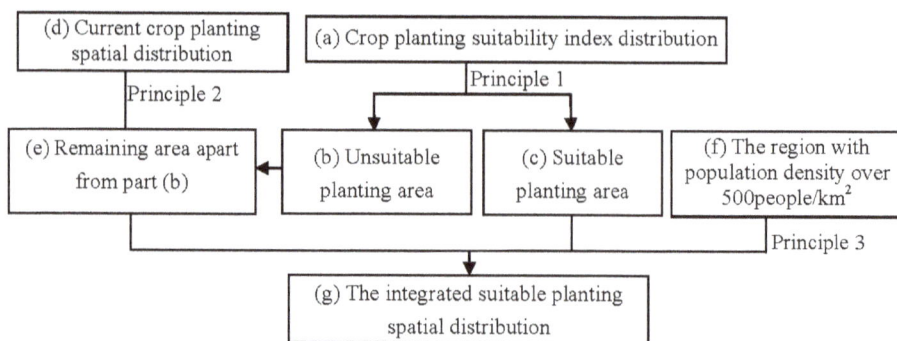

Figure 4. Overview of comprehensive suitable planting spatial distribution.

Principle 1: Preferentially assigning crops to the area with high suitability. First, rank crop planting suitability index (part a in Figure 4) in descending order, then contrast the area corresponding to the crop planting suitability index with the optimized area needed to be allocated, when the former is just larger than the latter, select the corresponding suitability index as a threshold of crops planting suitability index.

For example, Table 3 shows the maize planting suitability index and its corresponding area. When the index is greater than or equal to 0.83, its area is 103,337 ha, which is large enough to allocate the expected area of 96,896 ha obtained from the optimal model. Similarly, Table 4 shows that when the wheat planting suitability index is greater than or equal to 0.83, the area is 20,315 ha which can allocate the expected area of 12,817 ha. As a result, we can distinguish the unsuitable planting area (part b in Figure 4) and the suitable planting area (part c in Figure 4).

Table 3. Area of maize planting suitability index.

Maize Planting Suitability Index	0.83–0.84	0.84–0.85	0.85–0.86	0.86–0.87	0.87–0.88	0.88–0.89	0.89–0.9	0.9–0.92	Sum
Area (ha)	11,295	8242	7006	11,579	30,977	22,851	8899	2486	103,337

Table 4. Area of wheat planting suitability index.

Wheat Planting Suitability Index	0.88–0.89	0.89–0.90	Sum
Area (ha)	16,942	3373	20,315

Principle 2: The evaluation of crop planting suitability index is based on fixed-point sampling, and owing to the limitation of field investigation, there is inevitably some missing information on crop spatial distribution. In order to make up for this deficiency, the current crop planting spatial distribution (part d in Figure 4) is taken into account. Taking part b away from part d, the remaining area (part e in Figure 4) can be considered a supplement of crop distribution information.

Principle 3: According to [18], when the population density exceeds 500 people/km^2, it would be unsuitable for crop growing area. The land would be urban with little agriculture. Combined with the spatial distribution of population density, the region with population density over 500 people/km^2 (part f in Figure 4) is deducted from the suitable planting area.

Figures 5 and 6 show the formation process of maize and wheat comprehensive suitable planting spatial distribution.

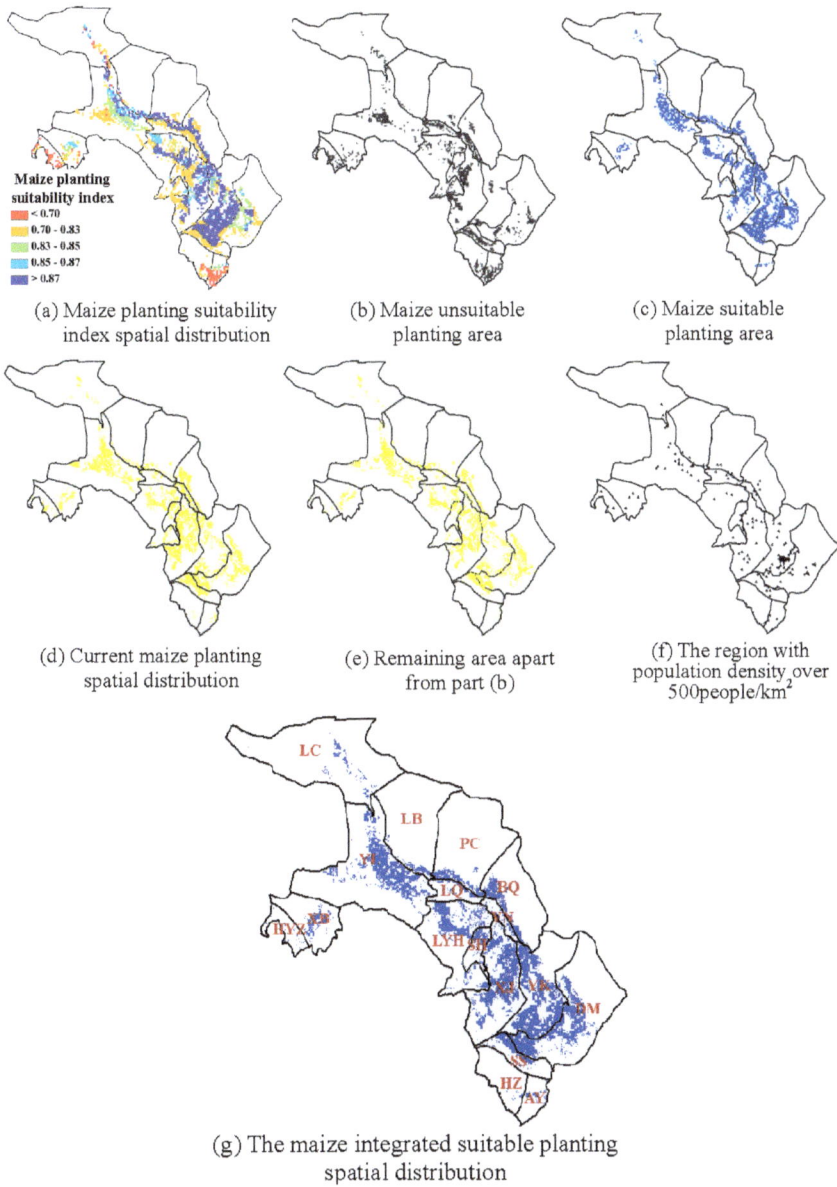

(a) Maize planting suitability index spatial distribution

(b) Maize unsuitable planting area

(c) Maize suitable planting area

(d) Current maize planting spatial distribution

(e) Remaining area apart from part (b)

(f) The region with population density over 500people/km^2

(g) The maize integrated suitable planting spatial distribution

Figure 5. Development of maize comprehensive suitable planting spatial distribution.

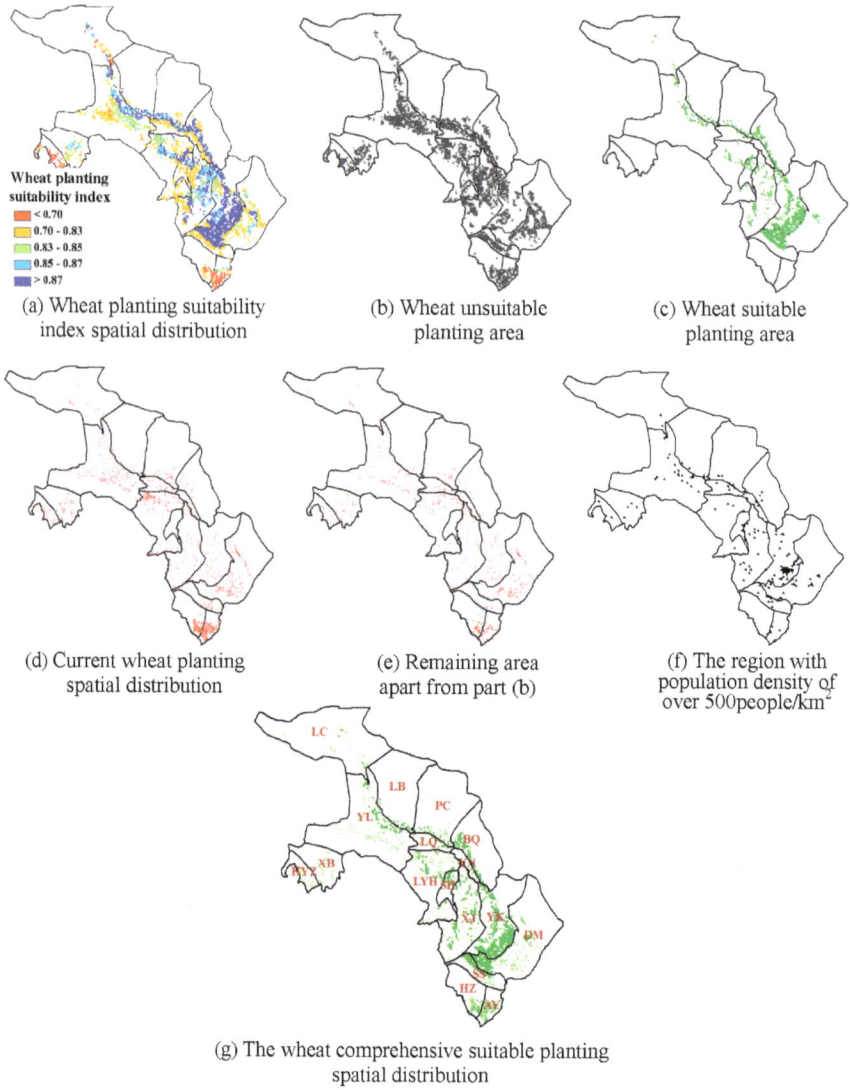

(a) Wheat planting suitability index spatial distribution

(b) Wheat unsuitable planting area

(c) Wheat suitable planting area

(d) Current wheat planting spatial distribution

(e) Remaining area apart from part (b)

(f) The region with population density of over 500people/km²

(g) The wheat comprehensive suitable planting spatial distribution

Figure 6. Development of wheat comprehensive suitable planting spatial distribution.

Results show that the maize comprehensive suitable planting spatial distribution is consistent with the current maize planting spatial distribution. The suitable area mainly focuses on the east of the YL irrigation district, the north of the LYH irrigation district, and most areas of the XJ, YK, DM irrigation districts. However, there are fewer areas suitable for growing maize in HYZ, HZ, and LQ.

In YK and DM irrigation district, the suitable area for planting maize is larger than the current area; therefore, an increase in the maize area is suggested in these districts. On the contrary, in YL, XJ and XB irrigation district, the maize area should be reduced because the suitable area is smaller than the current planting area.

Results show that the wheat comprehensive suitable planting spatial distribution has exhibited its obvious characteristics and the regular difference from the current wheat planting spatial distribution.

For example, the planting areas of wheat are relatively centralized in the HZ and AY irrigation districts and evenly distributed in other regions under the existing circumstances. However, the suitable areas for planting wheat are mainly concentrated in the YK, DM and SS irrigation district.

In the LQ, YN and HYZ irrigation districts, the suitable area for planting wheat is smaller than current area; therefore, a reduced area of wheat is suggested in these districts. However, in the BQ, YK, DM and SS irrigation districts, the maize area should be increased, for the suitable area is larger than current planting area.

3.3. Comparison of Crop Area Spatial Distribution between Present and Optimal Allocations

Comparison of maize area spatial distribution between present and optimal allocations is shown in Figure 7. Results show that after optimization, it tends to centralize grids, and the proportion of maize area in each grid is greater than 80% in the south of study area. This mainly focuses on the SS irrigation district, as the suitable area for planting maize in SS irrigation approaches the optimized area needed to be allocated.

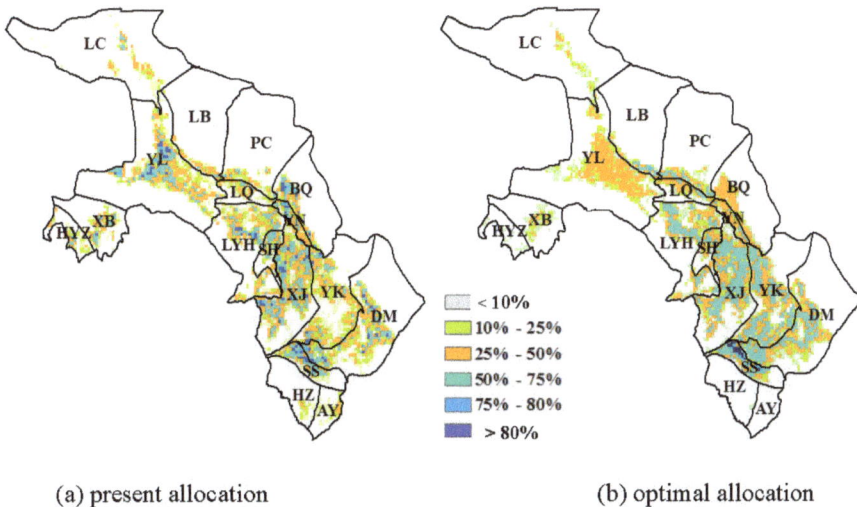

(a) present allocation   (b) optimal allocation

Figure 7. Comparison of maize area spatial distribution between present and optimal allocations.

The number of grids with higher proportion, including the proportion greater than 80% and between 75% and 80% is less than the actual situation, mainly owing to the deduction of unsuitable area for planting maize according to the principles in Section 3.1.

Meanwhile, the optimal spatial distribution shows more centralization than before optimization, because of the number of grids with a proportion of 50–75% increase, while the amount of grids with a proportion of 10–20% and less than a 10% decrease.

Figure 8 shows a comparison of wheat area spatial distribution between present and optimal allocations. Similarly, it indicates that grids with a high proportion of wheat area in each grid are more concentrated than in the actual situation, mainly distributed in LYH, LQ and SH irrigation districts.

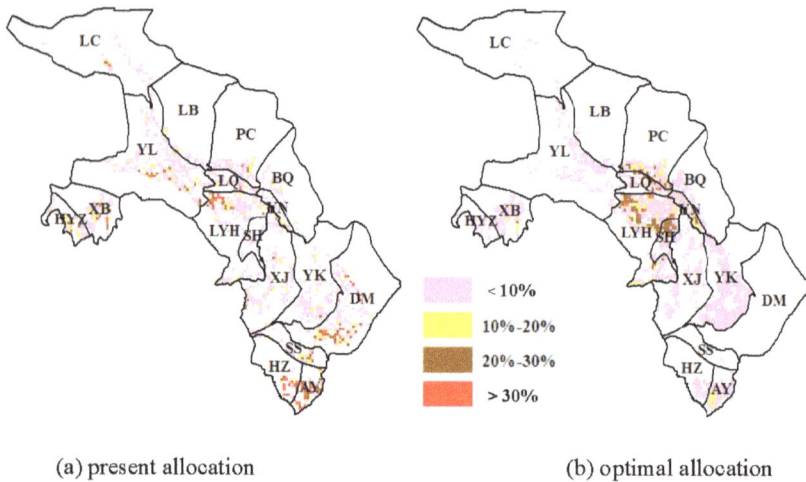

(a) present allocation   (b) optimal allocation

Figure 8. Comparison of wheat area spatial distribution between present and optimal allocations.

Compared with the maize proportion in each grid, the wheat proportion is lower, owing to less optimal wheat area and the decentralization of wheat comprehensive suitable planting spatial distribution. The number of grids with proportion greater than 30% decreases, while the amount of grids with proportion less than 10% and between 10% and 30% increases after spatial optimization.

In the south of the study area (AY and HZ irrigation district), based on the comparison of crop area spatial distribution between present and optimal allocations, the wheat area was more overloaded than its area suitable to plant in the actual situation. Therefore, it is suggested that the wheat area should be reduced in these two districts.

The optimization of crop spatial distribution, which is based on crop planting suitability evaluation and agricultural cropping pattern optimization, can improve the efficient utilization of agricultural water and land resources. The study ensures that crops are planted in suitable areas to provide the agricultural planting results of specialization and visualization.

4. Conclusions

In this study, an integrated agricultural cropping spatial distribution optimization is achieved at the irrigation district scale, based on the combination of multi-source remote sensing information data with optimal crop area. Minimizing cross-entropy was applied to build the model. This study considers the maize and wheat comprehensive suitable area distribution in the middle reaches of Heihe River basin and maize and wheat spatial distribution optimization is obtained. The high proportion of maize and wheat areas is more concentrated than before optimization. The integrated model contributed a new idea to cropping pattern spatial optimization. An optimizing approach based on cross-entropy minimization can lead to the efficient allocation of water resources and appropriate crop spatial distribution simultaneously. This study can ensure that crops are planted in a suitable region and provide a scientific basis for farmers to make crop selection decisions, which has an important guiding significance in the actual production work. However, due to the limitations of data on the planting suitability evaluation of other crops, only the main crops (maize and wheat) were studied. Further studies are therefore recommended using more crops.

Acknowledgments: We are grateful for the grant support from the National Natural Science Fund in China (91425302), National Key Research and Development Program during the 13th Five-year Plan in China (2016YFC0401306) and China Water Resources Industry Research Special Funds for Public Welfare Projects (No. 201301016).

Entropy **2017**, *19*, 592

Author Contributions: Lina Hao and Xiaoling Su designed the computations; Lina Hao and Xiaoling Su analyzed the data and wrote the paper; Vijay P. Singh and Olusola O. Ayantobo provides an English editing service and some improvement of style where necessary.

Conflicts of Interest: The authors declare no conflict of interest.

References

1. Li, M.; Guo, P. A coupled random fuzzy two-stage programming model for crop area optimization—A case study of the middle Heihe River basin, China. *Agric. Water Manag.* **2015**, *155*, 53–66. [CrossRef]
2. Dai, Z.Y.; Li, Y.P. A multistage irrigation water allocation model for agricultural land-use planning under uncertainty. *Agric. Water Manag.* **2013**, *129*, 69–79. [CrossRef]
3. Mianabadi, H.; Mostert, E.; Zarghami, M.; van de Giesen, N. A new bankruptcy method for conflict resolution in water resources allocation. *J. Environ. Manag.* **2014**, *144*, 152–159. [CrossRef] [PubMed]
4. Su, X.; Li, J.; Singh, V.P. Optimal Allocation of Agricultural Water Resources Based on Virtual Water Subdivision in Shiyang River Basin. *Water Resour. Manag.* **2014**, *28*, 2243–2257. [CrossRef]
5. Karimi, M.; Sharifi, M.A.; Mesgari, M.S. Modeling land use interaction using linguistic variables. *Int. J. Appl. Earth Obs. Geoinf.* **2012**, *16*, 42–53.
6. Ma, S.; He, J.; Liu, F.; Yu, Y. Model of urban land-use spatial optimization based on particle swarm optimization algorithm. *Trans. Chin. Soc. Agric. Eng.* **2010**, *26*, 321–326.
7. Pilehforooshha, P.; Karimi, M.; Taleai, M. A GIS-based agricultural land-use allocation model coupling increase and decrease in land demand. *Agric. Syst.* **2014**, *130*, 116–125. [CrossRef]
8. Cao, K.; Ye, X. Coarse-grained parallel genetic algorithm applied to a vector based land use allocation optimization problem: The case study of Tongzhou Newtown, Beijing, China. *Stoch. Environ. Res. Risk Assess.* **2012**, *27*, 1133–1142.
9. Cao, K.; Huang, B.; Wang, S.; Lin, H. Sustainable land use optimization using Boundary-based Fast Genetic Algorithm. *Comput. Environ. Urban Syst.* **2012**, *36*, 257–269. [CrossRef]
10. Baja, S.; Arif, S.; Neswati, R. Developing a User Friendly Decision Tool for Agricultural Land Use Allocation at a Regional Scale. *Mod. Appl. Sci.* **2017**, *11*, 11. [CrossRef]
11. Li, X.; Parrott, L. An improved Genetic Algorithm for spatial optimization of multi-objective and multi-site land use allocation. *Comput. Environ. Urban Syst.* **2016**, *59*, 184–194. [CrossRef]
12. Liu, Y.; Tang, W.; He, J.; Liu, Y.; Ai, T.; Liu, D. A land-use spatial optimization model based on genetic optimization and game theory. *Comput. Environ. Urban Syst.* **2015**, *49*, 1–14. [CrossRef]
13. Maleki, F.; Kazemi, H.; Siahmarguee, A.; Kamkar, B. Development of a land use suitability model for saffron (*Crocus sativus* L.) cultivation by multi-criteria evaluation and spatial analysis. *Ecol. Eng.* **2017**, *106*, 140–153. [CrossRef]
14. Reshmidevi, T.V.; Eldho, T.I.; Jana, R. A GIS-integrated fuzzy rule-based inference system for land suitability evaluation in agricultural watersheds. *Agric. Syst.* **2009**, *101*, 101–109. [CrossRef]
15. Sorel, L.; Viaud, V.; Durand, P.; Walter, C. Modeling spatio-temporal crop allocation patterns by a stochastic decision tree method, considering agronomic driving factors. *Agric. Syst.* **2010**, *103*, 647–655. [CrossRef]
16. Akıncı, H.; Özalp, A.Y.; Turgut, B. Agricultural land use suitability analysis using GIS and AHP technique. *Comput. Electron. Agric.* **2013**, *97*, 71–82. [CrossRef]
17. Nouri, H.; Mason, R.J.; Moradi, N. Land suitability evaluation for changing spatial organization in Urmia County towards conservation of Urmia Lake. *Appl. Geogr.* **2017**, *81*, 1–12. [CrossRef]
18. You, L.; Wood, S. Assessing the spatial distribution of crop areas using a cross-entropy method. *Int. J. Appl. Earth Obs. Geoinf.* **2005**, *7*, 310–323. [CrossRef]
19. Peng, C. *Research on the Optimization of Temporal and Spatial Distribution of Regional Crop Evapotranspiration and Its Application in Middle Reaches of Heihe River*; China Agricultural University: Beijing, China, 2016.
20. Sahoo, B.; Lohani, A.K.; Sahu, R.K. Fuzzy Multiobjective and Linear Programming Based Management Models for Optimal Land-Water-Crop System Planning. *Water Resour. Manag.* **2006**, *20*, 931–948. [CrossRef]
21. Henseler, M.; Wirsig, A.; Herrmann, S.; Krimly, T.; Dabbert, S. Modeling the impact of global change on regional agricultural land use through an activity-based non-linear programming approach. *Agric. Syst.* **2009**, *100*, 31–42. [CrossRef]

22. Sarker, R.; Ray, T. An improved evolutionary algorithm for solving multi-objective crop planning models. *Comput. Electron. Agric.* **2009**, *68*, 191–199. [CrossRef]

23. Niu, G.; Li, Y.P.; Huang, G.H.; Liu, J.; Fan, Y.R. Crop planning and water resource allocation for sustainable development of an irrigation region in China under multiple uncertainties. *Agric. Water Manag.* **2016**, *166*, 53–69. [CrossRef]

24. Biswas, A.; Pal, B.B. Application of fuzzy goal programming technique to land use planning in agricultural system. *Omega* **2005**, *33*, 391–398. [CrossRef]

25. Liu, J.; Li, Y.P.; Huang, G.H.; Zhuang, X.W.; Fu, H.Y. Assessment of uncertainty effects on crop planning and irrigation water supply using a Monte Carlo simulation based dual-interval stochastic programming method. *J. Clean. Prod.* **2017**, *149*, 945–967. [CrossRef]

26. Liu, X.; Ou, J.; Li, X.; Ai, B. Combining system dynamics and hybrid particle swarm optimization for land use allocation. *Ecol. Model.* **2013**, *257*, 11–24. [CrossRef]

27. Liu, Y.; Wang, H.; Ji, Y.; Liu, Z.; Zhao, X. Land use zoning at the county level based on a multi-objective particle swarm optimization algorithm: A case study from Yicheng, China. *Int. J. Environ. Res. Public Health* **2012**, *9*, 2801–2826. [CrossRef] [PubMed]

28. Yu, J.; Chen, Y.; Wu, J. Modeling and implementation of classification rule discovery by ant colony optimisation for spatial land-use suitability assessment. *Comput. Environ. Urban Syst.* **2011**, *35*, 308–319. [CrossRef]

29. Liu, Y.; Yuan, M.; He, J.; Liu, Y. Regional land-use allocation with a spatially explicit genetic algorithm. *Landsc. Ecol. Eng.* **2014**, *11*, 209–219. [CrossRef]

30. Li, X.; Yeh, A.G.-O. Neural-network-based cellular automata for simulating multiple land use changes using GIS. *Int. J. Geogr. Inf. Sci.* **2002**, *16*, 323–343. [CrossRef]

31. Kullback, S.; Leibler, R.A. On information and sufficiency. *Ann. Math. Stat.* **1951**, *22*, 79–86. [CrossRef]

32. Kullback, S. *Information Theory and Statistics*; John Wiley: Hoboken, NJ, USA, 1959.

33. Singh, V.P. *Entropy Theory and Its Application in Environmental and Water Engineering*; John Wiley & Sons: Hoboken, NJ, USA, 2013.

34. You, L.; Wood, S.; Wood-Sichra, U. Generating plausible crop distribution maps for Sub-Saharan Africa using a spatially disaggregated data fusion and optimization approach. *Agric. Syst.* **2009**, *99*, 126–140. [CrossRef]

35. You, L.; Wood, S.; Wood-Sichra, U.; Wu, W. Generating global crop distribution maps: From census to grid. *Agric. Syst.* **2014**, *127*, 53–60. [CrossRef]

36. Tan, J.; Yang, P.; Liu, Z.; Wu, W.; Zhang, L.; Li, Z.; You, L.; Tang, H.; Li, Z. Spatio-temporal dynamics of maize cropping system in Northeast China between 1980 and 2010 by using spatial production allocation model. *J. Geogr. Sci.* **2014**, *24*, 397–410. [CrossRef]

37. You, L.; Wood, S. An entropy approach to spatial disaggregation of agricultural production. *Agric. Syst.* **2006**, *90*, 329–347. [CrossRef]

38. Li, M.; Guo, P.; Singh, V.P.; Yang, G. An uncertainty-based framework for agricultural water-land resources allocation and risk evaluation. *Agric. Water Manag.* **2016**, *177*, 10–23. [CrossRef]

39. Hao, L.; Su, X. Determination for suitable scale of oasis and cultivated land in middle reaches of Heihe River basin. *Trans. Chin. Soc. Agric. Eng.* **2015**, *31*, 262–268.

40. Li, X.; Tong, L.; Niu, J.; Kang, S.; Du, T.; Li, S.; Ding, R. Spatio-temporal distribution of irrigation water productivity and its driving factors for cereal crops in Hexi Corridor, Northwest China. *Agric. Water Manag.* **2017**, *179*, 55–63. [CrossRef]

41. Zhong, B.; Ma, P.; Nie, A.; Yang, A.; Yao, Y.; Lü, W.; Zhang, H.; Liu, Q. Land cover mapping using time series HJ-1/CCD data. *Sci. China Earth Sci.* **2014**, *57*, 1790–1799. [CrossRef]

42. Zhong, B.; Yang, A.; Nie, A.; Yao, Y.; Zhang, H.; Wu, S.; Liu, Q. Finer resolution land-cover mapping using multiple classifiers and multisource remotely sensed data in the Heihe river basin. *IEEE J. Sel. Top. Appl. Earth Obs. Remote Sens.* **2015**, 1–20. [CrossRef]

43. Shannon, C.E. A mathematical theory of communication, Part I, Part II. *Bell Syst. Tech. J.* **1948**, *27*, 623–656. [CrossRef]

44. Jaynes, E.T. Information Theory and Statistical Mechanics. *Phys. Rev.* **1957**, *106*, 620–630. [CrossRef]

45. Tan, J.; Li, Z.; Yang, P.; Liu, Z.; Li, Z.; Zhang, L.; Wu, W.; You, L.; Tang, H. Spatiotemporal changes of maize sown area and yield in Northeast China between 1980 and 2010 using spatial production allocation model. *Acta Geogr. Sin.* **2014**, *69*, 353–364.

46. Tang, P.; Yang, P.; Chen, Z. Using cross-entropy method simulates spatial distribution of rice in Northeast China. *Trans. Chin. Soc. Agric. Eng.* **2013**, *29*, 96–104.
47. Liu, Z.; Li, Z.; Tang, P.; Li, Z.; Wu, W.; Yang, P.; You, L.; Tang, H. Change analysis of rice area and production in China during the past three decades. *J. Geogr. Sci.* **2013**, *23*, 1005–1018. [CrossRef]

entropy

MDPI

Article

Modeling Multi-Event Non-Point Source Pollution in a Data-Scarce Catchment Using ANN and Entropy Analysis

Lei Chen, Cheng Sun, Guobo Wang, Hui Xie and Zhenyao Shen *

State Key Laboratory of Water Environment Simulation, School of Environment, Beijing Normal University, Beijing 100875, China; chenlei1982bnu@bnu.edu.cn (L.C.); 13756892980@163.com (C.S.); wangguobo@yeah.net (G.W.); imbahui@163.com (H.X.)
* Correspondence: zyshen@bnu.edu.cn; Tel.: +86-10-5880-0398

Received: 5 May 2017; Accepted: 7 June 2017; Published: 19 June 2017

Abstract: Event-based runoff–pollutant relationships have been the key for water quality management, but the scarcity of measured data results in poor model performance, especially for multiple rainfall events. In this study, a new framework was proposed for event-based non-point source (NPS) prediction and evaluation. The artificial neural network (ANN) was used to extend the runoff–pollutant relationship from complete data events to other data-scarce events. The interpolation method was then used to solve the problem of tail deviation in the simulated pollutographs. In addition, the entropy method was utilized to train the ANN for comprehensive evaluations. A case study was performed in the Three Gorges Reservoir Region, China. Results showed that the ANN performed well in the NPS simulation, especially for light rainfall events, and the phosphorus predictions were always more accurate than the nitrogen predictions under scarce data conditions. In addition, peak pollutant data scarcity had a significant impact on the model performance. Furthermore, these traditional indicators would lead to certain information loss during the model evaluation, but the entropy weighting method could provide a more accurate model evaluation. These results would be valuable for monitoring schemes and the quantitation of event-based NPS pollution, especially in data-poor catchments.

Keywords: non-point source pollution; ANN; entropy weighting method; data-scarce; multi-events

1. Introduction

Non-point source (NPS) pollution has resulted in the deterioration of water bodies and has become a major environmental threat among most counties [1,2]. The quantification of the rainfall–runoff process and the resulting NPS pollutants is essential for developing mitigation strategies, which are the basis for watershed management [3]. The rainfall process is the major driving force for NPS, thus rainfall-runoff-pollutant (R-R-P) relationships have become the focus of watershed research [4,5]. Many studies have been conducted in the fields of rainfall–runoff relationships but have rarely involved the runoff–pollutant relationship, especially for the event-based estimation of NPS loads [6–8].

The NPS processes can be expressed from the event-step to long-term steps. Event-based NPS exports and the resulting change in water quality can provide detailed features of the NPS, which is more appropriate for the design of storm-based management practices [9]. Models are developed to construct the runoff–pollutant relationship, and the discrepancies of the collected measured data in different rainfall patterns would have a considerable influence on the model construction. Identifying the correlation among the series of rainfall, runoff and pollutant loads for multiple rainfall events is inevitable for NPS model construction. Although many models are well suited for offline water quality analyses, Soil and Water Assessment Tool (SWAT) is more representative than any other models [10].

However, owing to limited human resources, data scarcity has become one of the key barriers to establish the R-R-P relationship, especially for event-based process [11,12]. Thus, the application of watershed models such as SWAT for assessing NPS pollution is also limited by temporal resolution which ranges from annual to sub-hourly averages. The SWAT model usually operates continuously at a daily time step, which ensures that the long-term impacts of NPS can be quantified. Sub-daily calculations of runoff, erosion, and sediment transport are also available in new version of SWAT by sub-daily rainfall input and Green and Ampt method, though few attempts have reached to that higher temporal resolution. In the future, we would develop other more appropriate models to solve this problem. Currently, acceptable rainfall and streamflow data sets are more readily available, especially because of the recent development of data centers and satellite data observations. However, hourly or sub-hourly flow data for high-frequency time series are still limited, especially with respect to event-based hydrological studies for data-poor regions [7]. Water quality records, which are based on periodic monitoring by human resources, are thus even scarcer. Therefore, data scarcity for NPS predictions is unavoidable for multiple-rainfall event simulations. Typically, we collected samples during multiple rainfall events in the monitoring process but discarded some of events from further analysis, especially for light rainfall, for which only a few data points exist. This treatment of incomplete data would result in the loss of information, especially for multiple rainfall events among data-scarce regions.

Currently, statistical models have been widely used to estimate rainfall–runoff relationships for its ease of application without considering a large amount of delicate formulas and parameters [13]. For example, unit hydrographs (UH), as one of the most famous methods, is used to estimate a direct runoff hydrograph of a given rainfall duration. Meanwhile, statistical models are used to simulate pollutant loads based on the established runoff–pollutant relationship. For example, Park and Engel [14] developed Load Estimator (LOADEST) to predict pollutant concentration (or load) on days when flow data were measured, and the results showed that absolute values of errors in the annual sediment load estimation decreased from 39.7% to 10.8%. Meanwhile, most of the findings demonstrate that the LOADEST model could provide more accurate results and may be useful for simulating runoff–pollutant processes [15–17]. However, the LOADEST model has strict requirements on the number of data points, which should include continuous flow data and dispersed water quality data, and its calibration process is relatively complicated.

Owing to the limited measured data, the black-box model might be a substitution to construct the logical relationships between runoff and pollutant loads for multiple-events processes. The artificial neural network (ANN) with the characteristics of self-learning and adaptability has become the most commonly used tool in environmental prediction, and it is also available for poor-data regions. This method is applicable to simulate the imaginal thinking of the human brain, for which the most prominent characteristic is the parallel processing of information and distributed storage. As an example, Melesse et al. [18] used the ANNto estimate suspended sediment loads for three major rivers. The results showed that daily predictions were better than weekly predictions. Therefore, it can be seen that ANN models have flexible structures that allow multi-input and multi-output modeling. This is particularly important in streamflow forecasting where inflows at multiple locations are considered within a given catchment [19]. Though the application of ANN in the field of load production has proliferated in recent years, the impact of data scarcity on its prediction capabilities during different rainfall patterns still creates limitations [20].

Simulation evaluation is the most important step for the setup of statistical models [21]. In traditional applications, the model evaluation is usually performed using a single regression goodness-of-fit indicator, the most common of which is the point-to-point pairs (a series of single data pairs) of the predicted and measured data. However, this might lead to the loss of specific information, resulting in dubitable simulation results. In this case, a joint evaluation should be a substitute for the traditional single indicator. With the high precision and objectivity, the entropy regulates the uncertainty of different criteria from different perspectives [22]. Compared with the traditional single indicators, it can combine different indicators to evaluate the discrepancy between causes comprehensively. For instance, Khosravi et al. [23] sought to map the flooding susceptibility using different bivariate methods, including Shannon's entropy, the statistical index and the weighting factor. Yuan et al. [24] developed an entropy method to find the weight sum of the information entropy maximum to allocate the reduction of pollutants for the main seven valleys in China. The entropy weighting method may be an efficient way to evaluate the regulation of the simulation results and to balance the strengths and weaknesses of the results. However, these studies do not provide much attention to event-based NPS predictions, especially for data-scarce catchments.

This study surveys the motivation for a methodology of action, looks at the difficulties posed by data scarcity and outlines the need for the development of possibility methods to cope with data scarcity in multiple rainfall events. The objectives of this work are: (1) to identify the impacts of different rainfall patterns on the model construction using a complete data series; (2) to simulate the scarce pollutant data in other data-scarce rainfall events; and (3) to test the application of the entropy weighting method for the evaluation of ANN.

2. Materials and Methods

A prediction-evaluation framework is proposed for the NPS prediction for data-scarce catchments, the flow chart of methods is shown in Figure 1. The ANN is proposed to simulate the missing data points during multiple-events, and the entropy weighting method is used as a comprehensive indicator to construct the model. As a necessary supplement, the interpolation method is used for tail correction during multiple rainfall events. Data-scarce rainfall events denote the absence of data, especially for measured flow and water quality, in a given period of time due to human mistakes during high-resolution monitoring process. Instead, complete rainfall events are defined if there are no measured flow and water quality data scarcity. The demonstration of traditional indicators is shown in Section 2.2.

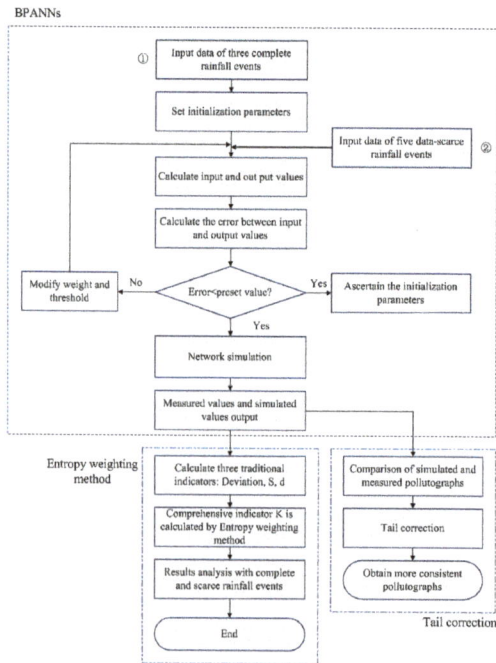

Figure 1. The methods presented in a flow chart.

2.1. The Description of the ANN

The back propagation algorithm is a supervised learning method based on the commonly used steepest descent method to minimize global errors [25], while it is also the multilayer feedforward network based on the error back propagation algorithm [2,26]. It accumulates an abundant mapping relation of the input-output pattern and does not need to reveal mathematical equations to describe the mapping relation before calculation. The ANN may be an efficient method to adjust the weights and thresholds through back propagation to minimize the sum of the squared errors. As shown in Figure 2, the topological structure of the ANN consists of an input layer, a hidden layer and an output layer [27].

The learning mechanism of the ANN is shown in Figure 2, where x_i is the input signal and w_i is the weight coefficient. The outside input samples $x_1, x_2, \dots, x_n$ are accepted into the input layer, and the network weight coefficients are adjusted during training. The discrete values, 0 and 1, are selected as the input sampling signals. By comparing the network output signals and the expected output signals to generate the error signals, the weight coefficients of the learning system can be rectified based through iterative adjustments to minimize the errors until reaching an acceptable range [28]. In this process, the expected output signals are regarded as the teacher signals, which are compared with the actual output, and the errors produced are applied to rectify the weight coefficients. At the point when the actual output values and expected values are nearly the same, the process is concluded [26]. Finally, the results are produced through and equation of *U* based on the weight coefficients and are exported by the output layers. In the ANN training process, three prime criteria can be summarized: the error surface gradient can converge rapidly, the mean squared error is below the error of the preset level, and the correlation coefficient of the training results is more than 0.9, indicating that training results are an improvement [29]. This section briefly surveys the measurement for methodology, while the ANN should be judged for whether each indicator can or cannot reach the given standards.

In this study, multiple rainfall events are used as the input conditions. Multiple rainfall events are divided into either the training process or the simulation process based on the data conditions. To establish the black-box model, data of three complete rainfall events are first input into the layer, including light, moderate, and heavy rainfall patterns. The training results also indicate that the ANN is applicable for various rainfall patterns. In the simulation process, the flow data for all the rainfall and water quality information for the data-complete rainfall events for the same rainfall pattern are regarded as the input layer. The hidden layer contains the water quality data for the data-scarce rainfall events which correspond to all the flow and water quality data in the input layer. To obtain the output layer, the training layer feedbacks the results into the prediction interval. Finally, the output layer is simulated using the input data of the input layer.

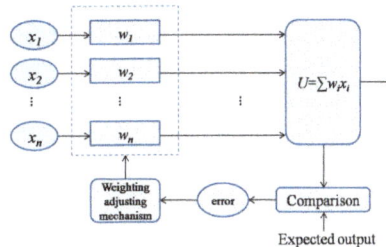

Figure 2. The learning mechanism of the artificial neural network (ANN).

2.2. The Description of the Entropy Weighting Method

Three commonly used indicators, the mean relative error ($\bar{d}$), the standard deviation of the relative error (*S*), and the load deviation percentage (*deviation*), are selected to evaluate the simulation results [30,31]. The formulas are shown as followed:

$$deviation = \frac{O_i^{origin} - O_i}{O_i^{origin}} \times 100\% \tag{1}$$

$$\bar{d} = \sum_{i=1}^{n}(O_i - P_i)/n \tag{2}$$

$$S = \sqrt{\frac{\sum_{i=1}^{n}\left(d_i - \bar{d}\right)^2}{n-1}} \tag{3}$$

where O_i is the set of measured data, P_i is the set of predicted data, and O_i^{origin} denotes the total loads of the original conditions, and is the mean value of the measured data.

Each of the three indicators represents the credibility of the measurements based on the discrepancy between the measured and simulated values. Lower indicator values indicate that the fitting between the simulated and measured data is improved, and the model is considered to have a satisfactory performance. However, single indicators have limitation on amount of information loss. Therefore, these indicators are handled with the entropy weighting method for a more comprehensive assessment of the ANN. Based on the fundamental principles of information theory, information is a measurement of the degree of order for a given system, and the entropy is a measurement of the degree of disorder [32]. The entropy weighting method serves as a mathematic method and considers the information provided by each factor [33]. Information entropy is negatively associated with the increase in information provided by different indicators, and a smaller information entropy result in higher weights for each single indicator. As an objective and comprehensive method, the entropy weighting method considers the advantages of every indicator and makes a synthetic evaluation. This principle is as follows:

Firstly, an $n \times m$ origin data matrix is established according to the selected evaluation indicators:

$$X = \begin{bmatrix} x_{11} & \cdots & x_{1n} \\ \vdots & \ddots & \vdots \\ x_{n1} & \cdots & x_{nm} \end{bmatrix}_{n \times m} \tag{4}$$

where m denotes the evaluation indicator, and individual rows represent different evaluation objects. Therefore, matrix X is known.

A second, positive matrix should be established with a transformation following same trend. The transformed matrix is

$$Y = \begin{bmatrix} y_{11} & \cdots & y_{1n} \\ \vdots & \ddots & \vdots \\ y_{n1} & \cdots & y_{nm} \end{bmatrix}_{n \times m} \tag{5}$$

Matrix Y is normalized, and the ratio of each column vector y_{ij} and the sum of all elements in this matrix should be normalized. The formulas for these calculations are:

$$Z_{ij} = \frac{y_{ij}}{\sum_{i=1}^{n} Y_{ij}}(j = 1, 2, \ldots, m) \tag{6}$$

where Z_{ij} are the elements of the normalized matrix.

The operational formula in the process of generating the entropy weights of the evaluation indicators is

$$H(x_j) = -k \sum_{i=1}^{n} z_{ij} \ln z_{ij}(j = 1, 2, \ldots, m) \tag{7}$$

where k is a normalizing constant, $k = 1/\ln n$, and Z_{ij} is the j-th the probability of the element of the i-th evaluation unit. Entropy values of the evaluation indicators should be transformed into the weighted values:

$$w_j = \frac{1 - H(x_j)}{m - \sum_{j=1}^{m} H(x_j)} \quad j = 1, 2, \ldots, m, \tag{8}$$

where $0 \leq w_i \leq 1$ and $\sum_{j=1}^{m} w_j = 1$ are the acquired weighted values. Finally, the comprehensive weighting values for each evaluation indicator should be ensured. The weighted values of each indicator are multiplied with the corresponding indicators and summed. The evaluation model is

$$U = \sum_{j=1}^{m} w_i z_{ij} \ (j = 1, 2, \ldots, n) \tag{9}$$

where U represents the comprehensive evaluation function of the entropy weights for each evaluation indicator. This function reflects the comprehensive characteristics of the evaluation objective, which avoids limiting these indicators [34].

The principle of the entropy weighting method is that information for each evaluation unit will be qualified and synthesized, while every factor is weighted to simplify the evaluation process [35]. Therefore, the weight values can be ascertained with the entropy weighting method, and we choose the *deviation*, $\bar{d}$, and S as the evaluation indicators.

2.3. Method for Tail Correction

Statistical models would result in tail deviation problems if data scarcity exists in this study. This problem addressed through data interpolation for the tail deviation. Therefore, linear interpolation, as a common-used method, is used to obtain the missing values of the other data points. Two values of the function $f(x)$ are used to reduce the errors in the tail of the pollutographs. This approach is relatively straightforward and is used widely in the field of mathematics or computer graphics. The error of the approximate method can be defined as follows:

$$R_T = f(x) - \rho(x) \tag{10}$$

where ρ represents the linear interpolated polynomial:

$$\rho(x) = f(x_0) + \frac{f(x_1) - f(x_0)}{x_1 - x_0}(x - x_0) \tag{11}$$

As a result of Rolle's theorem, if $f(x)$ has two continuous derivatives, the error range is

$$|R_T| \leq \frac{(x_1 - x_0)^2}{8} \max_{x_0 \leq x \leq x_1} |f'(x)| \tag{12}$$

As shown in Formula (12), the approximate error of the linear interpolation increases with the function curvature.

3. Case Study

3.1. Study Areas

As shown in Figure 3, the Zhangjiachong catchment, which is a representative area in the Three Gorges Reservoir Region (TGRR), is selected as a case study [36]. It covers a drainage area of 1.62 km², and the landscape is primarily mountainous, with an elevation between 148 m and 530 m above the Yellow Sea level. Agriculture and forests cover the majority of the total area. The main local crops are tea, corn, oil seed rape, and chestnuts [37]. The background values of nitrogen and phosphorus are higher because the fertilizer usage is relatively high, resulting in a high risk of nutrient loss into nearby streams [38].

The average annual temperature is approximately 18 °C, and the average annual precipitation is approximately 1439 m, 80% of which occurs from May to August. Thus, soil erosion frequently occurs during wet seasons, and results in an increase in the pollutant loads with increased runoff. We consider that the variation of rainfall might impact the model accuracy. Therefore, identifying the classification of rainfall patterns should be determined before any simulations. According to the investigation results of existing rainfall data, rainfall patterns are divided into light, moderate, and heavy events. Meanwhile, based on our monitoring data, a majority of rainfall events in the Zhangjiachong catchment are considered moderate events, while heavy events are rare.

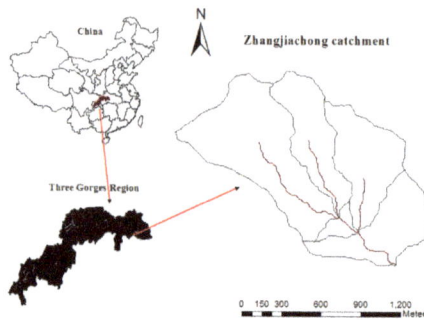

Figure 3. The location of the Zhangjiachong catchment.

3.2. Field Monitoring and Data Record

In this study, field monitoring data were collected from 1 January 2013 to 31 December 2014 and the rainfall, streamflow and pollutant data during eight rainfall events were recorded. The data used in this study represent three complete rainfall events, which include light, middle, and heavy rainfall (21 April 2014, 24 July 2014, and 5 August 2014), and five other data-scarce events (15 April 2014, 23 August 2014, 20 July 2014, 5 July 2013, and 28 August 2013). Data-scarce rainfall events denote the absence of data, especially for measured flow and water quality, in a given period of time due to human mistakes during high-resolution monitoring process. Instead, complete rainfall events are defined if there are no measured flow and water quality data scarcity. The equations with explicit parameters are constructed through a training process with complete data of the three complete rainfall events, and the constructed ANN is used to predict the missing NPS data in the other five data-scarce rainfall events. The output layer includes pollutant load data for five data-scarce rainfall events.

The weather station (Skye Lynx Standard) provided continuous records for climate data and a float-operator sensor (WGZ-1) was located at the catchment outlet, where high-frequency sampling was recorded in approximately 15 min steps. Base flows were measured before the runoff started, and water samples were collected every 15 min in the first hour after runoff began and every 30 min over the following two hours. After water levels had stabilized, water samples were collected once every hour until the end of the event. All water samples were placed in pre-cleaned glass jars with aluminumfoil liners along the lids and stored at −20 °C during transportation to the laboratory for processing and analysis. Specifically, the total nitrogen of NPS (NPS-TN) levels were measured via Alkaline persulfate oxidation-UV spectrophoto metric with the detection limitation from 0.05 mg/L to 4.0 mg/L, while the total phosphorus of NPS (NPS-TP) levels in the samples were measured via Potassium persulfate oxidation-molybdenum blue colorimetric methods. The main instrument is ultraviolet spectrophotometer. Finally, the recorded rainfall, flow and pollutant levels were used for the following analysis.

Table 1. Complete data for the three rainfall events.

	21 April 2014				24 July 2014				5 August 2014		
Time	Flow (m³/s)	NPS-TN (mg/L)	NPS-TP (mg/L)	Time	Flow (m³/s)	NPS-TN (mg/L)	NPS-TP (mg/L)	Time	Flow (m³/s)	NPS-TN (mg/L)	NPS-TP (mg/L)
2:45	0.002	4.75	0.063	22:45	0.003	0.84	0.11	21:30	0.008	2.66	0.30
3:00	0.009	8.89	0.193	23:00	0.380	6.26	0.76	21:45	0.678	7.29	0.84
3:15	0.015	15.29	0.300	23:15	0.647	6.96	1.04	22:00	1.107	11.30	1.12
3:30	0.016	13.31	0.301	23:30	0.726	7.06	0.87	22:15	1.400	11.90	1.26
3:45	0.031	25.28	0.369	23:45	0.336	6.77	0.92	22:30	2.227	9.85	0.94
4:00	0.037	14.14	0.297	0:00	0.971	8.77	1.12	23:00	1.647	15.00	1.20
4:30	0.065	22.64	0.652	0:30	0.570	8.08	0.70	23:30	0.585	9.67	0.62
5:00	0.071	24.48	0.441	1:00	0.294	5.53	0.48	0:00	0.945	9.56	0.68
6:00	0.086	26.49	0.469	2:00	0.266	5.05	0.57	1:00	0.410	7.01	0.35
7:00	0.126	23.89	0.443	3:00	0.172	5.11	0.36	2:00	0.237	2.03	0.28
8:00	0.146	16.97	0.286	4:00	0.191	5.79	0.34	3:00	0.183	7.60	0.37
9:00	0.264	11.60	0.171	5:00	0.041	5.85	0.20	4:00	0.166	6.81	0.27
10:00	0.278	11.00	0.117	6:00	0.090	5.59	0.18	5:00	0.115	7.01	0.20
11:00	0.288	10.73	0.104	7:00	0.064	7.78	0.15	6:00	0.126	6.71	0.19
12:00	0.296	10.94	0.109	8:00	0.048	5.29	0.14	7:00	0.102	6.08	0.12
13:00	0.411	9.64	0.113					8:00	0.073	7.03	0.24
14:00	0.593	9.48	0.089					9:00	0.063	5.68	0.11
15:00	0.593	9.65	0.087					10:00	0.086	5.75	0.09
16:00	0.602	9.26	0.072					11:00	0.075	6.44	0.18
17:00	0.770	9.93	0.068					12:00	0.045	6.33	0.21
								13:00	0.029	5.88	0.14

Table 2. Five rainfall events with data scarcity.

15 April 2014 (1.213 mm/h)

Time	Flow	TN	TP
6:00	0.0127	7.65	0.050
6:30	0.024	12.27	0.412
6:45	0.0375	11.49	0.366
7:00	0.0427	23.53	0.579
7:15	0.0401	16.13	0.363
7:30	0.0406	15.19	0.350
8:00	0.0327	16.92	0.373
8:30	0.0351	14.29	0.363
9:00	0.0327	11.26	0.291
10:00	-	-	-
10:30	0.0351	8.67	0.228
11:00	0.0127	7.65	0.050

28 August 2013 (2.027 mm/h)

Time	Flow	TN	TP
15:00	0.0375	1.13	0.13
15:30	-	-	-
16:00	0.0964	6.29	0.71
16:30	0.1020	5.57	0.3
17:00	0.0964	5.81	0.32
17:30	0.0547	5.2	0.2
18:00	0.0427	4.95	0.14
18:30	-	-	-
19:00	0.0375	5.14	0.13
19:30	-	-	-
20:00	0.0427	5.21	0.13

20 July 2014 (2.013 mm/h)

Time	Flow	TN	TP
19:00	0.0127	5.51	0.27
19:30	0.0163	7.02	0.56
20:00	0.0485	6.92	0.86
20:30	0.1190	5.56	1.15
21:00	0.0866	5.14	0.53
21:30	0.0327	7.37	0.4
22:00	-	-	-
22:30	0.0375	5.64	0.24
23:00	-	-	-
23:30	0.0182	6.49	0.19

5 July 2013 (2.380 mm/h)

Time	Flow	TN	TP
14:30	0.0127	5.03	0.06
15:00	-	-	-
15:30	-	-	-
16:00	0.0127	8.81	1.50
16:30	0.0127	6.37	2.56
17:00	0.0375	1.06	2.11
17:30	-	-	-
18:00	0.4140	-	-
18:30	0.4440	-	-
19:00	0.2220	3.20	0.41
19:30	-	-	-
20:00	0.1590	4.08	0.28
20:30	-	-	-
21:00	0.0964	4.26	0.11
21:30	-	-	-
22:00	0.0775	3.78	0.22
22:30	-	-	-
23:00	0.0616	3.80	0.13
23:30	-	-	-
0:30	0.0547	-	-
1:30	0.0427	-	-

28 August 2013 (2.647 mm/h)

Time	Flow	TN	TP
19:00	0.03	5.59	0.05
19:30	0.06	12.65	0.33
20:00	1.78	15.24	1.39
20:30	2.22	16.38	0.84
21:00	1.94	13.88	0.69
21:30	0.80	14.95	0.44
22:00	0.39	13.43	0.55
22:30	0.26	14.81	0.56
23:00	-	-	-
23:30	-	-	-
0:00	-	-	-
0:30	-	-	-
1:00	0.21	10.17	0.14
1:30	-	-	-
2:00	0.15	9.83	0.19
2:30	-	-	-
3:00	0.09	6.54	0.08

Note: the units of flow are in m^3/s; the units of NPS-TN and NPS-TP are mg/L; - denotes that data are missing at this time.

However, flow and water quality data were limited because of the use of flow instruments via manual collection. Rainfall levels were recorded to divide the rainfall into light, moderate, and heavy events. The rainfall levels for 21 April 2014, 24 July 2014, and 5 August 2014 are 1.308 mm/h, 3.000 mm/h, and 6.054 mm/h, respectively. The flow data were replenished with unit hydrographs as the basis for the ANN. In addition, this catchment is dominated by agriculture, so fertilizer use results in deteriorated water quality. Therefore, the NPS-TN and NPS-TP are selected as the evaluation indicators. All the data for the three complete rainfall events and five typical data-scarce rainfall events are shown in Tables 1 and 2, respectively, including the rainfall intensity, flow data, and the pollutant concentration of the NPS-TN and NPS-TP. As shown in Table 1, complete data are used as the input of the ANN and represent the impacts of the rainfall patterns on the model applicability. As shown in Table 2, data scarcity of the five random rainfall events is simulated, and the impacts of the data scarcity are quantified.

4. Results and Discussion

4.1. Training Results of the ANN Using the Complete Data

This section demonstrates the training process with data for the three complete rainfall events, illustrating that the applicability of ANN in different rainfall patterns. The training results for the ANN areas followed (the figure is shown in the Supplementary Materials): the error surface gradient rapidly converges to a flat surface for both the NPS-TN and NPS-TP. The mean squared error of the training results for the NPS-TP prediction reaches the 10^{-3}, 10^{-2}, and 10^{-1} orders of magnitude for the light, moderate, and heavy rainfall events, respectively. However, the mean squared error for the NPS-TN prediction reaches the 1.0, 10^{-3}, and 1.0 orders of magnitude during the light, moderate, and heavy rainfall events, respectively, indicating that all the results fall within the range of permissible errors or rapidly reach a flat surface. The correlation coefficients are more than 0.9, indicating that all the training results are good. In this respect, it can be said that the ANN is applicability to simulate the NPS for different rainfall patterns, and we extrapolated ANN for pollutant load simulations in the data-scarce rainfall events.

To better understand the simulation results, the entropy weighting method was used in the evaluation process. As shown in Table 3, K results are all higher than 0.9, indicating that there is no obvious deviation between the simulated and measured values. Meanwhile, the K values for the NPS-TP are higher than the NPS-TN for different rainfall patterns, and the K value for the light rainfall is higher than the other rainfall patterns. Therefore, it is apparent that the NPS-TP simulation is an improvement over the NPS-TN, and the simulation is better suited for the light rainfall events for both the NPS-TP and NPS-TN. It is obvious that the flow have different shear force in different rainfall patterns. The soil particles and pollutants act differently with different rainfall levels and intensities. It is possible that our monitoring scheme is more appropriate in light rainfall patterns in this experiment, and the peak data cannot be monitored during heavy rainfall patterns [39]. The NPS-TN concentration peak and flow peak appear to be consistent. When one of the flow or load peaks is missing, it is the same as both of them missing simultaneously, resulting in a poor simulation effect. However, the apparent time of the NPS-TP concentration peak and flow peak is inconsistent in different rainfall patterns. Xu et al. [40] introduced the support vector regression (SVR) model to develop a quantitative relationship between the environmental factors and the eutrophic indices compared with the ANN. The results show that the correlation coefficients of the NPS-TP are greater than those for the NPS-TN, indicating that the model effect of the NPS-TP is improved over the NPS-TN. This study verifies this conclusion with the ANN model.

Table 3. Evaluation of the simulation results of the pollutant loads for different rainfall patterns.

Rainfall Events	Comprehensive Indicators K	
	NPS-TP	NPS-TN
21 April 2014	0.986	0.953
24 July 2014	0.973	0.938
5 August 2014	0.958	0.921

4.2. Simulated Results of the ANN for Data-Scarce Events

Five typical data-scarce rainfall events were used to discuss the impact of different data-scarce patterns on the NPS predictions. As shown in Figure 4, the NPS-TP training results for the NPS-TN have a faster convergence rate for the grads and lower mean squared errors. The training values for the NPS-TN are represented by an R^2 value that is more than 0.9, and the mean squared errors are under the permissible values or reach the flat surface rapidly. However, only one event (5 July 2013) was observed to have lower grads beyond the preset value, and its training effect was the worst because this rainfall event has peak scarcity.

The entropy values in the five data-scarce rainfall patterns are shown in Table 4. Combined with the complete data events, it is apparent that the simulated effect for 5 July 2013 has a worse fit compared with the other rainfall events, which reflects the poor training effect when there is a scarcity of peak concentration data. The peak data are the key information, and reflect the overall process of the rainfall events. However, the peak scarcity is unintentional and due to system errors. In addition, the training effect of the NPS-TP is improved over the NPS-TN.

Table 4. Evaluation effect of the data scarcity on the models for the five data-scarce rainfall events.

Rainfall Events	Comprehensive Indicators K		Traditional Indicators (NPS-TP)		
	NPS-TN	NPS-TP	Deviation	$\bar{d}$	S
15 April 2014	0.546	0.971	0.989	0.956	0.966
23 August 2014	0.937	0.924	0.934	0.892	0.943
20 July 2014	0.959	0.930	0.964	0.882	0.941
5 July 2013	0.340	0.948	0.997	0.908	0.938
28 August 2013	0.948	0.982	0.996	0.980	0.970

We further compared the evaluation results between the traditional methods and the entropy weighting method, which are shown in Table 4. As shown in the results, the rank order of the effects of the simulation results with the traditional indicators (high to low) as the following: deviations: 5 July 2013, 23 August 2013, 15 April 2014, 20 July 2014, and 23 August 2014; $\bar{d}$: 23 August 2014, 5 July 2013, 20 July 2014, 23 August 2013, and 15 April 2014; and S: 23 August 2014, 5 July 2013, 23 August 2013, 15 April 2014, and 20 July 2014. The application of a single indicator is limited by the indicator selection so that we cannot sum them up simply or select one of them. For instance, the effect of 5 July 2013 showed the best *deviation* but the worst S, which represents the rainfall amount and the average rainfall, respectively. Therefore, choosing these traditional indicators would lead to information loss during the model evaluation. Conversely, the entropy weighting method considers the advantages and characteristics of each traditional indicators and assesses the simulation results comprehensively form different perspectives [34,35]. Thus, the K values are more accurate and easier to compare.

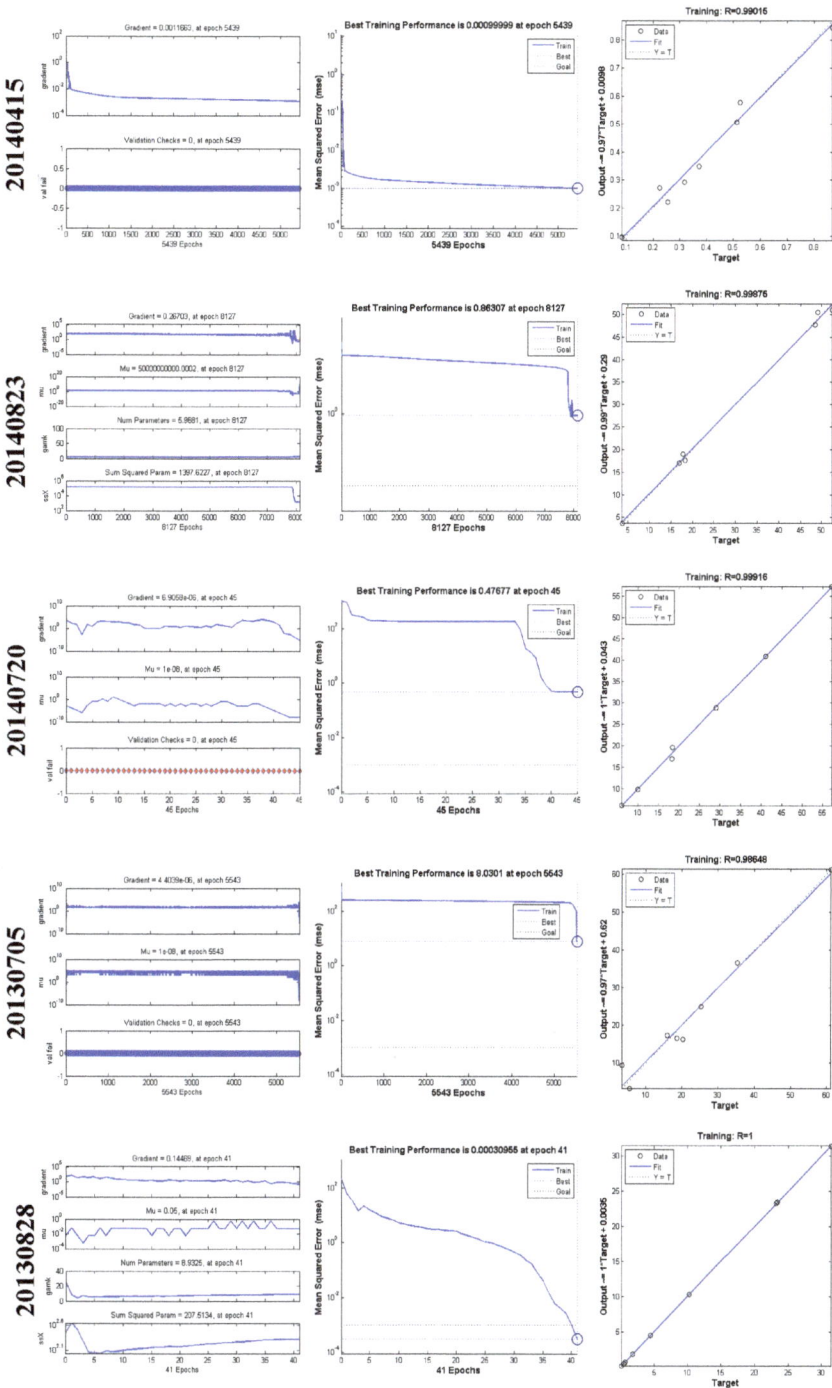

(a)

Figure 4. *Cont.*

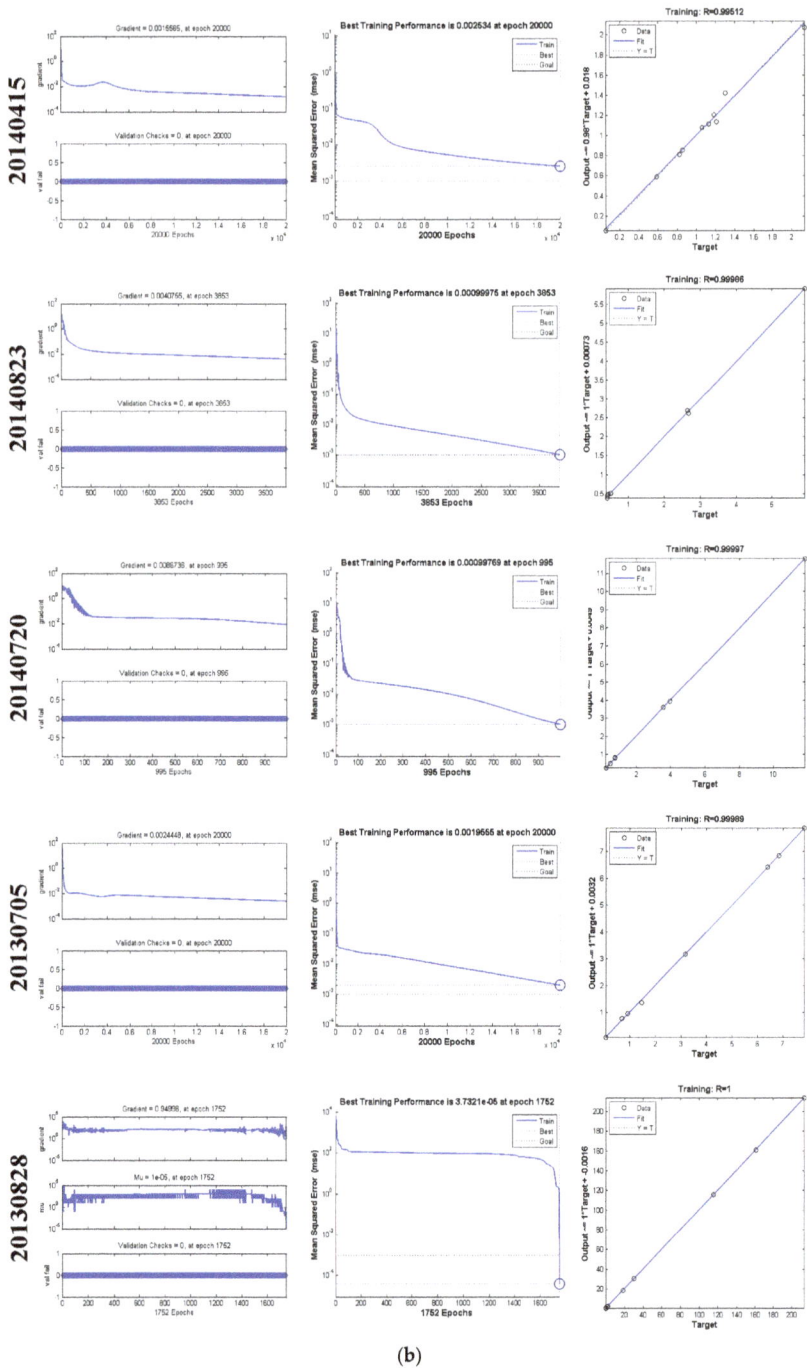

Figure 4. Training results for the loads in five data-scarce rainfall events: (**a**) the total nitrogen of non-point source (NPS-TN); and (**b**) the total phosphorus of non-point source (NPS-TP). Note: the pink line with dots represent the epochs are smaller.

Owing to the limited water quality data, we randomly selected 30% of the measured values as verification points, and the simulated data points were compared to the selected data to test the accuracy of the ANN during data-scarce conditions. The evaluated results are shown in Table 5, and the intuitionistic indicator is the mean percentage of the load deviation. As shown in Table 5, the effects of the training results for the runoff–pollutant load process are better in different rainfall events, and each of the load deviations is smaller. The mean percentage load deviations of the three events (15 April 2014, 23 August 2014, 28 August 2013) are higher than the other events. This is because these pollutant load data for the three individual rainfall events have peak loads nearby the flow peaks. It is apparent that the flow peak and these high values have major impacts on the training and predictive values of the ANN. In general, the simulation effects are improved, which shows that this method is feasible for estimating scarce pollutant load data.

Table 5. Evaluation of the predicted effects of the verification points.

Rainfall Events	Mean Load Deviation Percentage of Verification Points (%)	
	NPS-TP	NPS-TN
15 April 2014	0.150	0.0770
23 August 2014	0.153	0.029
20 July 2014	0.041	0.094
5 July 2013	0.002	0.038
28 August 2013	0.120	0.022

4.3. Implication for NPS Studies of Multi-Events

Figure 5 compares pollutographs of complete rainfall events with pollutographs simulated by ANN in the same rainfall pattern (data-scarce rainfall events in light and moderate rainfall patterns). Most of the pollutographs conform to the ordinary rules (pollutographs in complete data rainfall events), and the overall tendency is consistent with the hydrographs with complete data, indicating that the method is reliable. Moreover, the model performance is worse under conditions of missing peak data, which is consistent with the abovementioned conclusions. According to the comparisons, the pollutographs of the measured points are more consistent with the ordinary rules than when the tails have missing data. Meanwhile, tail scarcity often appears in actual monitoring to reduce manpower [41]. The tails of the pollutographs have stronger linear characteristics, so a linear interpolation is used to amend the incomplete tails [42]. The pollutographs amended by linear interpolation are shown in Figure 6, indicating that the hydrographs with the tail correction are more coincident.

During the monitoring process, emphasis is placed on the discrepancy in the monitoring mechanism under different rainfall conditions. Based on the abovementioned analysis, the NPS prediction performs the best during the light rainfall events and is the worst during heavy rainfall events. Therefore, the monitoring process for the NPS can be appropriately focused on heavy rainfall conditions. Researchers should pay more attention to monitoring time to avoid peak data scarcity, especially for the NPS-TN monitoring [43]. As already suggested, peak concentration appeared after nearly five hours of runoff during light rainfall events and after nearly three hours of runoff during moderate events. Therefore, we promote peak monitoring techniques, for example, anautomatic sampler with programming, we can appropriately shorten sampling intervals for the peak lag times. Meanwhile, based on the pollutographs improved by this study, we can design the sampling scheme and avoid the risk time in order to require complete water quality data. In addition, the entropy weighting method can be effectively used to evaluate the measured and simulated data [44], showing that it can be used to comprehensively assess the discrepancies more accurately and to easily compare the results, which can be generalized to other catchments.

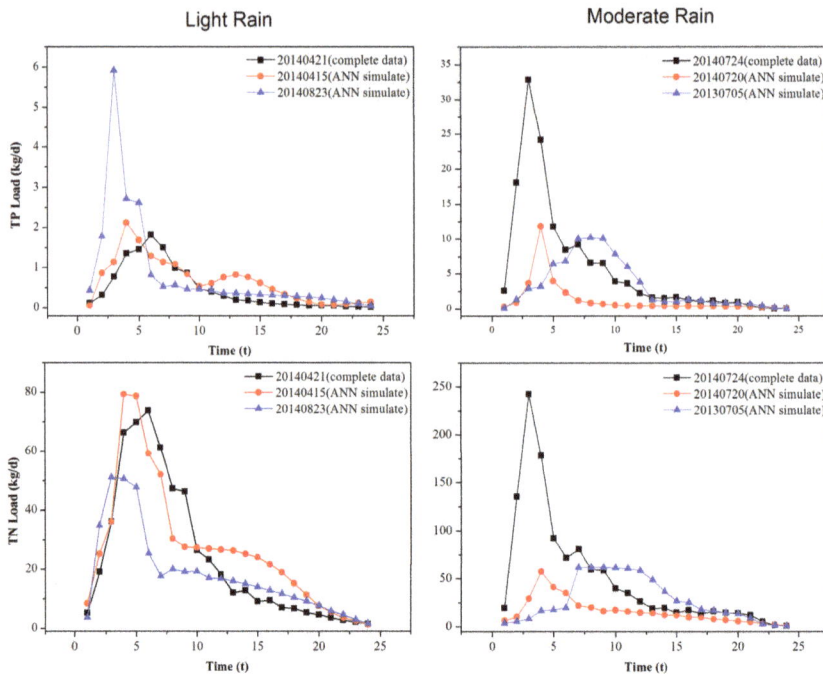

Figure 5. The pollutographs for different rainfall patterns.

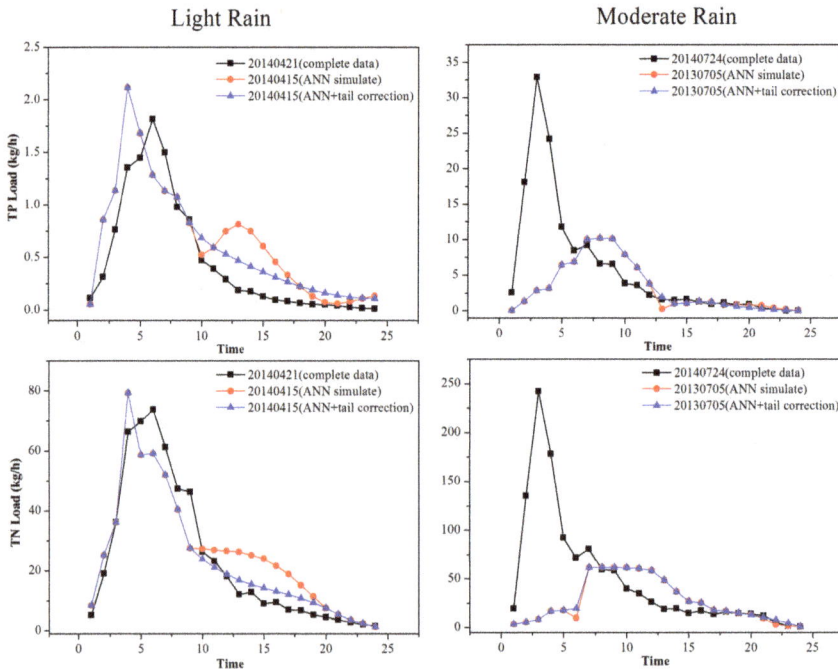

Figure 6. The tail amendment of the pollutographs for different rainfall patterns.

Entropy **2017**, *19*, 265

5. Conclusions

In this study, a new framework is proposed for the event-based NPS prediction and evaluation in data-scarce catchments. The results obtained from this study indicate that the proposed ANN had an improved performance over the NPS simulation of light rainfall events, and the NPS-TP model was always more accurate than the NPS-TN under scarce data conditions. In addition, the scarcity of the peak pollutant data has a significant impact on the model performance, so more attention should be given to the monitoring scheme of the event-based NPS studies, especially for the NPS-TN monitoring and the lag time of the peak data. Compared to the traditional indicators, the entropy weighting method can provide a more accurate ANN by considering all of the information during model evaluation. These tools could be extended to other catchments to quantify the event-based NPS pollution, especially data-poor catchments.

However, we should pay more attention to the mechanism of the NPS during multiple rainfall events because the NPS pollution was not the simple consequence of current rainfall events. Additionally, because of the computational burden, the errors and the related uncertainty of the model results were not explored, so more studies are suggested to test this new framework among more diverse regions. Meanwhile, data-driven black-box models are not good at long-term forecasting, nor are they good for examining the effect of BMPs.

Supplementary Materials: The following are available online at www.mdpi.com/1099-4300/19/6/265/s1, Figure S1: Training results of the loads in the different rainfall patterns: (a) NPS-TN; and (b) NPS-TP.

Acknowledgments: This research was funded by the National Natural Science Foundation of China (Nos. 51579011 and 51409003) and the Fund for the Innovative Research Group of the National Natural Science Foundation of China (No. 51421065).

Author Contributions: Lei Chen constructed the research framework and designed the study. Cheng Sun was responsible for the data analysis and code programming. Guobo Wang performed the experiments and conducted data analysis. Hui Xie and Zhenyao Shen reviewed and edited the manuscript. All of the authors have read and approved the final manuscript.

Conflicts of Interest: The authors declare no conflict of interest.

References

1. Ongley, E.D.; Zhang, X.L.; Yu, T. Current status of agricultural and rural non-point source Pollution assessment in China. *Environ. Pollut.* **2010**, *158*, 1159–1168. [CrossRef] [PubMed]
2. Li, X.F.; Xiang, S.Y.; Zhu, P.F.; Wu, M. Establishing a dynamic self-adaptation learning algorithm of the BP neural network and its applications. *Int. J. Bifurc. Chaos* **2015**, *25*, 1540030. [CrossRef]
3. Gong, Y.W.; Liang, X.Y.; Li, X.N.; Li, J.Q.; Fang, X.; Song, R.N. Influence of rainfall characteristics on total suspended solids in urban runoff: A case study in Beijing, China. *Water* **2016**, *8*, 278. [CrossRef]
4. Coulliette, A.D.; Noble, R.T. Impacts of rainfall on the water quality of the Newport River Estuary (Eastern North Carolina, USA). *J. Water Health* **2008**, *6*, 473–482. [CrossRef] [PubMed]
5. Chen, C.L.; Gao, M.; Xie, D.T.; Ni, J.P. Spatial and temporal variations in non-point source losses of nitrogen and phosphorus in a small agricultural catchment in the Three Gorges Region. *Environ. Monit. Assess.* **2016**, *188*, 257. [CrossRef] [PubMed]
6. Sajikumar, N.; Thandaveswara, B.S. A non-linear rainfall-runoff model using an artificial neural network. *J. Hydrol.* **1999**, *216*, 32–55. [CrossRef]
7. Bulygina, N.; McIntyre, N.; Wheater, H. Conditioning rainfall-runoff model parameters for ungauged catchments and land management impacts analysis. *Hydrol. Earth Syst. Sci.* **2009**, *13*, 893–904. [CrossRef]
8. Maniquiz, M.C.; Lee, S.; Kim, L.H. Multiple linear regression models of urban runoff pollutant load and event mean concentration considering rainfall variables. *J. Environ. Sci.* **2010**, *22*, 946–952. [CrossRef]
9. Chen, N.W.; Hong, H.S.; Cao, W.Z.; Zhang, Y.Z.; Zeng, Y.; Wang, W.P. Assessment of management practices in a small agricultural watershed in Southeast China. *J. Environ. Sci. Health Part A Toxic Hazard. Subst. Environ. Eng.* **2006**, *41*, 1257–1269. [CrossRef] [PubMed]
10. Sun, A.Y.; Miranda, R.M.; Xu, X. Development of multi-meta models to support surface water quality management and decision making. *Environ. Earth Sci.* **2015**, *73*, 423–434. [CrossRef]

11. Yuceil, K.; Baloch, M.A.; Gonenc, E.; Tanik, A. Development of a model support system for watershed modeling: A case study from Turkey. *CLEANSoil Air Water* **2007**, *35*, 638–644. [CrossRef]
12. Shope, C.L.; Maharjan, G.R.; Tenhunen, J.; Seo, B.; Kim, K.; Riley, J.; Arnhold, S.; Koellner, T.; Ok, Y.S.; Peiffer, S.; et al. Using the SWAT model to improve process descriptions and define hydrologic partitioning in South Korea. *Hydrol. Earth Syst. Sci.* **2014**, *18*, 539–557. [CrossRef]
13. Jeong, J.; Kannan, N.; Arnold, J.; Glick, R.; Gosselink, L.; Srinivasan, R. Development and integration of sub-hourly rainfall-runoff modeling capability within a watershed model. *Water Resour. Manag.* **2010**, *24*, 4505–4527. [CrossRef]
14. Park, Y.S.; Engel, B.A. Identifying the correlation between water quality data and LOADEST model behavior in annual sediment load estimations. *Water* **2016**, *8*, 368. [CrossRef]
15. Park, Y.S.; Engel, B.A.; Frankenberger, J.; Hwang, H. A web-based tool to estimate pollutant loading using LOADEST. *Water* **2015**, *7*, 4858–4868. [CrossRef]
16. Das, S.K.; Ng, A.W.M.; Perera, B.J.C.; Adhikary, S.K. Effects of climate and landuse activities on water quality in the Yarra River catchment. In Proceedings of the 20th International Congress on Modelling and Simulation (Modsim2013), Adelaide, Australia, 1–6 December 2013; pp. 2618–2624.
17. Chen, D.J.; Hu, M.P.; Guo, Y.; Dahlgren, R.A. Reconstructing historical changes in phosphorus inputs to rivers from point and nonpoint sources in a rapidly developing watershed in eastern China, 1980–2010. *Sci. Total Environ.* **2015**, *533*, 196–204. [CrossRef] [PubMed]
18. Melesse, A.M.; Ahmad, S.; Mcclain, M.E.; Wang, X.; Lim, Y.H. Suspended sediment load prediction of river systems: An artificial neural network approach. *Agric. Water Manag.* **2011**, *98*, 855–866. [CrossRef]
19. Tran, H.D.; Muttil, N.; Perera, B.J.C. Investigation of artificial neural network models for streamflow forecasting. In Proceedings of the 19th International Congress on Modelling and Simulation (Modsim 2011), Perth, Australia, 12–16 December 2011; pp. 1099–1105.
20. Hassan, M.; Shamim, M.A.; Sikandar, A.; Mehmood, I.; Ahmed, I.; Ashiq, S.Z.; Khitab, A. Development of sediment load estimation models by using artificial neural networking techniques. *Environ. Monit. Assess.* **2015**, *187*, 686. [CrossRef] [PubMed]
21. Dhiman, N.; Markandeya; Singh, A.; Verma, N.K.; Ajaria, N.; Patnaik, S. Statistical optimization and artificial neural network modeling for acridine orange dye degradation using in-situ synthesized polymer capped ZnO nanoparticles. *J. Colloid Interface Sci.* **2017**, *493*, 295–306. [CrossRef] [PubMed]
22. Wang, H.W.; Ai, Z.W.; Cao, Y. Information-entropy based load balancing in parallel adaptive volume rendring. In Proceedings of the International Conferences on Interfaces and Human Computer Interaction 2015, Game and Entertainment Technologies 2015, and Computer Graphics, Visualization, Computer Vision and Image Processing 2015, Las Palmas de Gran Canaria, Spain, 22–24 July 2015; pp. 163–169.
23. Khosravi, K.; Pourghasemi, H.R.; Chapi, K.; Bahri, M. Flash flood susceptibility analysis and its mapping using different bivariate models in Iran: A comparison between Shannon's entropy, statistical index, and weighting factor models. *Environ. Monit. Assess.* **2016**, *188*, 656. [CrossRef] [PubMed]
24. Yuan, Y.M.; Wei, G.A. Empirical studies of unblocked index for urban freeway traffic flow states. In Proceedings of the 2009 12th International IEEE Conference on Intelligent Transportation Systems, St. Louis, MO, USA, 4–7 Octorber 2009; pp. 1–6.
25. Kan, J.M.; Liu, J.H. Self-Tuning PID controller based on improved BP neural network. In Proceedings of the 2009 Second International Conference on Intelligent Computation Technology and Automation, Changsha, China, 10–11 Octorber 2009; pp. 95–98.
26. Chen, X.Y.; Chau, K.W. A hybrid double feedforward neural network for suspended sediment load estimation. *Water Resour. Manag.* **2016**, *30*, 2179–2194. [CrossRef]
27. Guo, Z.H.; Wu, J.; Lu, H.Y.; Wang, J.Z. A case study on a hybrid wind speed forecasting method using BP neural network. *Knowl. Based Syst.* **2011**, *24*, 1048–1056. [CrossRef]
28. Ju, Q.; Yu, Z.B.; Hao, Z.C.; Ou, G.X.; Zhao, J.; Liu, D.D. Division-based rainfall-runoff simulations with BP neural networks and Xinanjiang model. *Neurocomputing* **2009**, *72*, 2873–2883. [CrossRef]
29. Jing, J.T.; Feng, P.F.; Wei, S.L.; Zhao, H. Investigation on surface morphology model of Si3N4 ceramics for rotary ultrasonic grinding machining based on the neural network. *Appl. Surf. Sci.* **2017**, *396*, 85–94. [CrossRef]
30. Ullrich, A.; Volk, M. Influence of different nitrate-N monitoring strategies on load estimation as a base for model calibration and evaluation. *Environ. Monit. Assess.* **2010**, *171*, 513–527. [CrossRef] [PubMed]

31. Wilson, D.R.; Apreleva, M.V.; Eichler, M.J.; Harrold, F.R. Accuracy and repeatability of a pressure measurement system in the patellofemoral joint. *J. Biomech.* **2003**, *36*, 1909–1915. [CrossRef]

32. Ai, Y.T.; Guan, J.Y.; Fei, C.W.; Tian, J.; Zhang, F.L. Fusion information entropy method of rolling bearing fault diagnosis based on n-dimensional characteristic parameter distance. *Mech. Syst. Signal Process.* **2017**, *88*, 123–136. [CrossRef]

33. Liu, F.; Zhao, S.; Weng, M.; Liu, Y. Fire risk assessment for large-scale commercial buildings based on structure entropy weight method. *Saf. Sci.* **2017**, *94*, 26–40. [CrossRef]

34. Sun, L.Y.; Miao, C.L.; Yang, L. Ecological-economic efficiency evaluation of green technology innovation in strategic emerging industries based on entropy weighted TOPSIS method. *Ecol. Indic.* **2017**, *73*, 554–558. [CrossRef]

35. Huang, Z.Y. Evaluating intelligent residential communities using multi-strategic weighting method in China. *Energy Build.* **2014**, *69*, 144–153. [CrossRef]

36. Shen, Z.Y.; Gong, Y.W.; Li, Y.H.; Liu, R.M. Analysis and modeling of soil conservation measures in the Three Gorges Reservoir Area in China. *Catena* **2010**, *81*, 104–112. [CrossRef]

37. Shen, Z.Y.; Gong, Y.W.; Li, Y.H.; Hong, Q.; Xu, L.; Liu, R.M. A comparison of WEPP and SWAT for modeling soil erosion of the Zhangjiachong Watershed in the Three Gorges Reservoir Area. *Agric. Water Manag.* **2009**, *96*, 1435–1442. [CrossRef]

38. Shen, Z.; Qiu, J.; Hong, Q.; Chen, L. Simulation of spatial and temporal distributions of non-point source pollution load in the Three Gorges Reservoir Region. *Sci. Total Environ.* **2014**, *493*, 138–146. [CrossRef] [PubMed]

39. Gottschalk, L.; Weingartner, R. Distribution of peak flow derived from a distribution of rainfall volume and runoff coefficient, and a unit hydrograph. *J. Hydrol.* **1998**, *208*, 148–162. [CrossRef]

40. Xu, Y.F.; Ma, C.Z.; Liu, Q.; Xi, B.D.; Qian, G.R.; Zhang, D.Y.; Huo, S.L. Method to predict key factors affecting lake eutrophication—A new approach based on Support Vector Regression model. *Int. Biodeterior. Biodegrad.* **2015**, *102*, 308–315. [CrossRef]

41. Wagner, P.D.; Fiener, P.; Wilken, F.; Kumar, S.; Schneider, K. Comparison and evaluation of spatial interpolation schemes for daily rainfall in data scarce regions. *J. Hydrol.* **2012**, *464–465*, 388–400. [CrossRef]

42. Croke, B.; Islam, A.; Ghosh, J.; Khan, M.A. Evaluation of approaches for estimation of rainfall and the unit hydrograph. *Hydrol. Res.* **2011**, *42*, 372–385. [CrossRef]

43. Ryu, J.; Jang, W.S.; Kim, J.; Jung, Y.; Engel, B.A.; Lim, K.J. Development of field pollutant load estimation module and linkage of QUAL2E with watershed-scale L-THIA ACN model. *Water* **2016**, *8*, 292. [CrossRef]

44. Li, P.Y.; Qian, H.; Wu, J.H. Groundwater quality assessment based on improved water quality index in Pengyang County, Ningxia, Northwest China. *J. Chem.* **2010**, *7*, S209–S216.

entropy

MDPI

Article

Forewarning Model of Regional Water Resources Carrying Capacity Based on Combination Weights and Entropy Principles

Rongxing Zhou [1,2], Zhengwei Pan [1,2], Juliang Jin [3,*], Chunhui Li [4] and Shaowei Ning [3]

[1] School of Civil Engineering and Environmental Engineering, Anhui Xinhua University, Hefei 230088, China; zhourx11@163.com (R.Z.); pzhwei1023@163.com (Z.P.)
[2] Institute of Safety and Environmental Assessment, Anhui Xinhua University, Hefei 230088, China
[3] School of Civil Engineering, Hefei University of Technology, Hefei 230009, China; ning@hfut.edu.cn
[4] Key Laboratory for Water and Sediment Sciences of Ministry of Education, School of Environment, Beijing Normal University, Beijing 100875, China; chunhuili@bnu.edu.cn
* Correspondence:JINJL66@126.com; Tel.: +86-0551-62903357

Received: 5 September 2017; Accepted: 19 October 2017; Published: 25 October 2017

Abstract: As a new development form for evaluating the regional water resources carrying capacity, forewarning regional water resources of their carrying capacities is an important adjustment and control measure for regional water security management. Up to now, most research on this issue have been qualitative analyses, with a lack of quantitative research. For this reason, an index system for forewarning regional water resources of their carrying capacities and grade standards, has been established in Anhui Province, China, in this paper. Subjective weights of forewarning indices can be calculated using a fuzzy analytic hierarchy process, based on an accelerating genetic algorithm, while objective weights of forewarning indices can be calculated by using a projection pursuit method, based on an accelerating genetic algorithm. These two kinds of weights can be combined into combination weights of forewarning indices, by using the minimum relative information entropy principle. Furthermore, a forewarning model of regional water resources carrying capacity, based on entropy combination weight, is put forward. The model can fully integrate subjective and objective information in the process of forewarning. The results show that the calculation results of the model are reasonable and the method has high adaptability. Therefore, this model is worth studying and popularizing.

Keywords: water resource carrying capacity; forewarning model; entropy of information; fuzzy analytic hierarchy process; projection pursuit; accelerating genetic algorithm

1. Introduction

With the continuous development of the economy and the increase in population in China, the contradiction between the rapid and steady development of the economy, the health of the ecological environment and the sustainable development of water resources has become increasingly prominent [1]. The water resources carrying capacity is a comprehensive index, which measures whether the water resources system is sustainable or not. The water resources carrying capacity is the largest scale for the development of economy and society that can be supported by regional water resources in a particular area and the specific historical stage, which is based on the designed available water resources, the predictable technology and the development level of economic and social factors, and takes into account the principles of sustainable development and a good ecological environment. This index is a "bottle neck" that determines whether the water resources can support the population and the coordinated development of economy and environment in a water shortage area [2]. Therefore, the forewarning and control of the regional water resources carrying capacity are significant for ensuring the safety of water resources.

Based on a comprehensive analysis of the factors affecting the regional water resources carrying capacity, the forewarning index system and corresponding grade standard of regional water resources carrying capacity have been set up. The actual values of the forewarning indices in different periods of the study area are compared with the corresponding standard values at different levels, to judge and identify the warning status of the study area. Currently, there are many studies on the evaluation of the water resources carrying capacity. Xiang et al. simplified seven factors, which included irrigation rate, the utilization rate of water resources, the developmental degree of water resources, a water supply module, a water requirement module, the per capita water supply and the ecological environment water use rate, into two principal components, with a principal component analysis. In addition, the corresponding principal component evaluation criteria of the water resources carrying capacity were obtained, based on the classification criteria of the evaluation parameters. On this basis, the water resources carrying capacity of each district was evaluated [3]. Zhou et al. calculated the comprehensive score of the water resources carrying capacity in Guiyang City, according to the three principal components of the water resources carrying capacity that were selected by the principal component analysis and their weights were calculated by entropy [4]. Gong et al. evaluated the current situation of the water resource capacity in Lanzhou and its dynamic trends, by using the method of fuzzy comprehensive evaluation, based on historical data over a 40-year period in Lanzhou [5]. Based on the analysis of the main factors affecting the carrying capacity, Liang et al. used a factor analysis method to determine their weights. Then the comprehensive index value of each partition in the Shiyang river basin was calculated, to evaluate the water resources carrying capacity [6]. Zhao et al. analyzed the water resources carrying capacity in Ningxia, during 2004–2014, by using the principal component analysis and factor analysis. Then, based on these analyses, the water resources utilization index and classification criteria in different areas of Ningxia were obtained [7]. Chen et al. put forward a new fuzzy model, according to the Jaynes maximum entropy in information theory, and applied this model in assessing the water resources carrying capacity in Henan province [8]. Wang et al. constructed a set pair analysis model, based on entropy weight, and calculated the average connection degree of each region with this model, to rank their water resources carrying capacities [9]. Li et al. simulated changes in the carrying capacity of water resources under the conditions of future water management policies, by using a system dynamics model [10]. However, from a theoretical and practical point of view, the regional water resources carrying capacity warning is a new form of development for evaluating the regional water resources carrying capacity, and is still in its infancy.

Anhui Province was chosen as the study area in this paper. This province lies in the Eastern part of China. Floods and droughts occur frequently in this province. Water use in agriculture and industry, accounts for the majority of water consumption in this area. The water resources in Anhui account for only 2.7% of the total water resources in China and the water resources per capita are much lower than the world average. Thus, it is necessary to study the water resources carrying capacity in Anhui Province. The forewarning index system and grade standards of regional water resources carrying capacity were established according to the regional situation. The combination weights, which are a combination of subjective weights and objective weights, were calculated based on the principle of minimum relative information entropy. The forewarning model of the water resources carrying capacity was established according to the corresponding relationship between the "non-warning" membership degree, and the comprehensive forewarning index, of the standard grade sample value series.

2. Summary of the Study Area

2.1. Hydrological and Climatic Conditions

Anhui Province, located in the east of China's mainland (Figure 1), is a total area of 1.396×10^5 square kilometers, accounting for about 1.45% of the area of China. The main rivers in Anhui Province are the Yangtze river, Huaihe river and Xin'An river, as well as one of the five largest freshwater lakes, Chaohu lake. It belongs to the transition zone between the warm temperate zone and the subtropical zone, whose climate is warm and humid, and the monsoon climate is obvious. The annual average temperature is about 14–17 °C and the relative humidity is about 70–80%. The distribution of precipitation in time and space is uneven. The precipitation in the south is more than in the north and the precipitation in mountain areas is more than in plain and hilly areas. The variation of annual precipitation is great (Figure 2) and under the influence of the monsoon climate; the precipitation in summer is abundant and accounts for 40–60% of the total annual precipitation. In addition, the monthly precipitation also changes greatly, which often leads to drought and flood disasters. The Northern droughts often appear in spring and summer, while the droughts south of the Huaihe river often appear in summer and autumn. Floods occur mostly in the Huaihe river basin and secondarily, in the Yangtze river basin.

Figure 1. The location of Anhui Province in China.

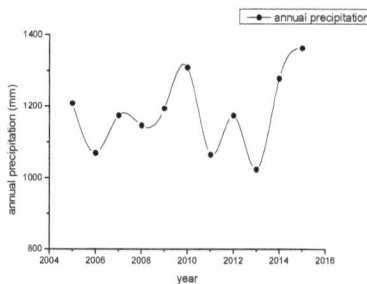

Figure 2. Annual variation curve of precipitation in Anhui Province (2005–2015).

2.2. Situation of Water Resources

The average annual water resources in Anhui was about 72 billion m³, accounting for only 2.7% of the total water resources in China. The Per capita water resources was about 1190 m³, which accounts for about 59% of the per capita water resources of China. Similar to the precipitation distribution, the water resources also showed a trend of decreasing from the south to the north in space. At the same time, the amount of water resources had an uneven distribution over time (Figure 3).

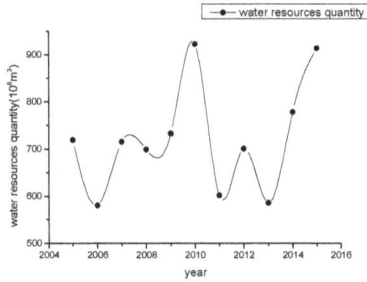

Figure 3. Annual variation curve of water resources in Anhui Province (2005–2015).

2.3. Situation of Water Consumption

The annual water consumption in Anhui Province varied from 20 to 30 billion cubic meters in 2005–2015, which included a rapid growth trend from 2005–2009, but tended to be stable after 2010 (Figure 4). The structure of water consumption is shown in Figure 5. Agricultural water accounted for about 56% of total water consumption and industrial water, domestic water and ecological water accounted for 33%, 10% and 1%, respectively.

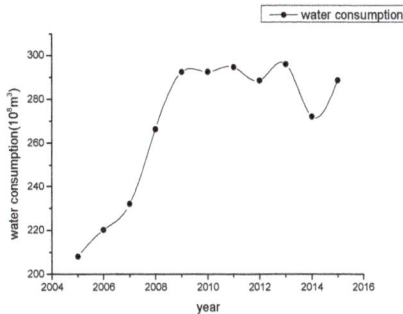

Figure 4. Annual variation curve of water consumption in Anhui Province (2005–2015).

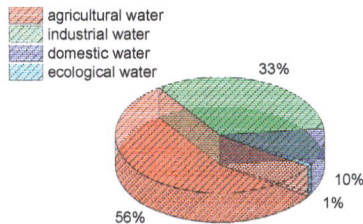

Figure 5. The structure of water consumption in Anhui Province.

3. Forewarning Model and Application

In order to illustrate the modeling steps more clearly, the forewarning model of the water resources carrying capacity in Anhui, China was used as an example to show how this model was established. The procedure of modeling is shown in the diagram of the technical route (Figure 6).

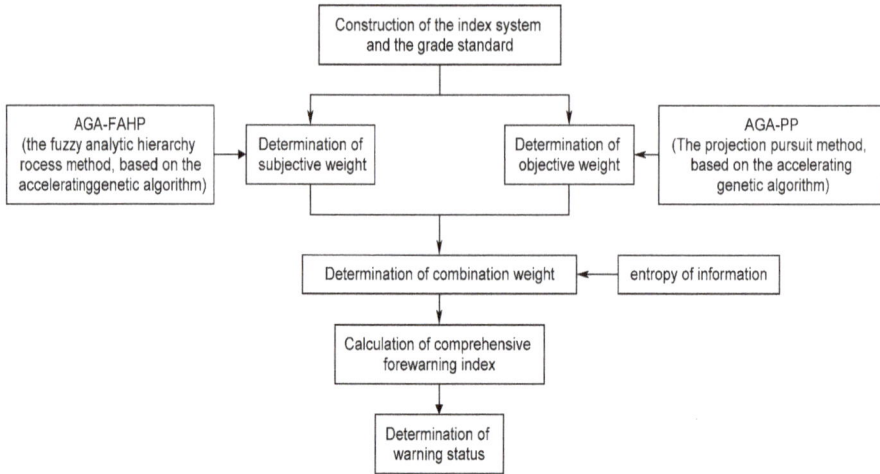

Figure 6. Diagram of the technical route.

3.1. Construction of the Index System and the Grade Standard

According to the specific situation of Anhui Province, and the construction experience and principles of the index system, the index system of the water resources carrying capacity for Anhui Province was established, as shown in Table 1.

Table 1. Forewarning Index System of the Water Resources Carrying Capacity for Anhui Province and its Grade Standard.

Forewarning Index System			Grade Standard		
Target Layer	Subsystem Layer	Index Layer	First Grade	Second Grade	Third Grade
	holding power subsystem	x_1 m^3	$[1670, +\infty)$	$[1000, 1670)$	$[0, 1000)$
		x_2 10^4 m^3/km^2	$[80, +\infty)$	$[50, 80)$	$[0, 50)$
		x_3 m^3/year	$[450, +\infty)$	$[350, 450)$	$[0, 350)$
		x_4%	$[40, +\infty)$	$[25, 40)$	$[0, 25)$
	regulate and control subsystem	x_5%	$[0, 40]$	$(40, 70]$	$[70, +\infty)$
Water Resources Carrying Capacity		x_6 yuan	$[24,840, +\infty)$	$[6624, 24,840)$	$[0, 6624)$
		x_7%	$[90, +\infty)$	$[70, 90)$	$[0, 70)$
		x_8%	$[95, +\infty)$	$[70, 95)$	$[0, 70)$
		x_9%	$[5, +\infty)$	$[1, 5)$	$[0, 1)$
	pressure subsystem	x_{10} L/d	$[0, 70]$	$(70, 180]$	$[180, +\infty)$
		x_{11} L/d	$[0, 100]$	$(100, 400]$	$[400, +\infty)$
		x_{12} m^3	$[0, 50]$	$(50, 200]$	$[200, +\infty)$
		x_{13} people/km^2	$[0, 200]$	$(200, 500]$	$[500, +\infty)$
		x_{14}%	$0, 50]$	$(50, 80]$	$[80, +\infty)$
		x_{15} m^3/mu	$[0, 250]$	$(250, 400]$	$[400, +\infty)$

x_1 is the per capita water resources. x_2 is the modulus of water production. x_3 is the water supply quantity. x_4 is the vegetation coverage rate. x_5 is the water resources development and utilization rate. x_6 is the per capita gross domestic product (GDP). x_7 is the wastewater discharge compliance rate.

x_8 is the compliance rate of the water function zone. x_9 is the ecological water use rate. x_{10} is the daily water consumption per capita. x_{11} is the water consumption per ten thousand yuan GDP. x_{12} is the water demand per ten thousand yuan of industrial added value. x_{13} is the population density. x_{14} is the urbanization rate. x_{15} is the irrigation quota of farmland.

3.2. Determination of Subjective Weight by AGA-FAHP

The fuzzy analytic hierarchy process method, based on the accelerating genetic algorithm (AGA-FAHP) [11–16], was used to determine the subjective weights of each forewarning index. The fuzzy complementary judgment matrix can be defined as:

$$A = (a_{ij})_{n \times n}$$
$$0 \le a_{ij} \le 1, \quad a_{ij} + a_{ji} = 1 \quad \forall i, j = 1, 2, \cdots, n \tag{1}$$

where a_{ij} is the relative importance between indices, i and j. If $a_{ij} > 0.5$, it means that index i is more important than index j. The bigger a_{ij} is, the more important index i is. If $a_{ij} = 0.5$, it means that index i is as important as index j. If $a_{ij} < 0.5$, it means that index j is more important than index i.

If A has complete consistency, then there is [17]:

$$\sum_{i=1}^{n} \sum_{j=1}^{n} |0.5(n-1)[w(i) - w(j)] + 0.5 - a_{ij}|/n^2 = 0 \tag{2}$$

where the bars | | denote the absolute value; $\{w(j), j = 1,2, \ldots, n\}$ is the subjective weight of each index.

In practical applications, due to the fuzziness and complexity of the evaluation system and the diversity and instability of a person's cognition, there is no uniform and exact yardstick to measure the importance of indices. Therefore, in practical applications, conditions that matrix A, given by the decision-maker, cannot satisfy with consistency, often occur. Thus,

$$minCIC(n) = \sum_{i=1}^{n} \sum_{j=1}^{n} |0.5(n-1)[w(i) - w(j)] + 0.5 - a_{ij}|/n^2$$
$$\text{s.t.} \sum_{j=1}^{n} w(j) = 1, w(j) > 0 \tag{3}$$

where the objective function $CIC(n)$ is the consistency index coefficient; n is the order of matrix A.

The weight of each index $w(j)$ can be calculated according to Equation (3). If $CIC(n) < 0.1$, A is considered to have satisfactory consistency and the corresponding $w(j)$ is the subjective weight of each index [17].

By comparing the indices one by one, the fuzzy complementary judgment matrixes of each subsystem are obtained:

$$A_1 = \begin{bmatrix} 0.5 & 0.5 & 0.6 & 0.9 \\ 0.5 & 0.5 & 0.6 & 0.9 \\ 0.4 & 0.4 & 0.5 & 0.8 \\ 0.1 & 0.1 & 0.2 & 0.5 \end{bmatrix} \quad A_2 = \begin{bmatrix} 0.5 & 0.6 & 0.8 & 0.6 & 0.9 \\ 0.4 & 0.5 & 0.7 & 0.55 & 0.6 \\ 0.2 & 0.3 & 0.5 & 0.4 & 0.6 \\ 0.4 & 0.45 & 0.6 & 0.5 & 0.8 \\ 0.1 & 0.4 & 0.4 & 0.2 & 0.5 \end{bmatrix}$$

$$A_3 = \begin{bmatrix} 0.5 & 0.2 & 0.4 & 0.2 & 0.4 & 0.2 \\ 0.8 & 0.5 & 0.6 & 0.5 & 0.6 & 0.5 \\ 0.6 & 0.4 & 0.5 & 0.4 & 0.4 & 0.4 \\ 0.8 & 0.5 & 0.6 & 0.5 & 0.6 & 0.5 \\ 0.6 & 0.4 & 0.6 & 0.4 & 0.5 & 0.4 \\ 0.8 & 0.5 & 0.6 & 0.5 & 0.6 & 0.5 \end{bmatrix}$$

If the nonlinear optimization problem is calculated with Equation (3) by AGA-FAHP, the subjective weights of the indices in each subsystem can be obtained.

The subjective weights of indices x_1 to x_4, in the holding power subsystem of the water resources carrying capacity, are $w_1 = (0.3335, 0.3332, 0.2665, 0.0668)$.

The corresponding *CIC* is $(n) = 2.60 \times 10^{-4} < 0.1$.

The subjective weights of indices x_5 to x_9 in the regulatory and control subsystems of the water resources carrying capacity are $w_2 = (0.2968, 0.2427, 0.1469, 0.2167, 0.0969)$.

The corresponding *CIC* is $(n) = 0.03 < 0.1$.

The subjective weights of indices x_{10} to x_{15} in the regulatory and control subsystems of the water resources carrying capacity are $w_3 = (0.0824, 0.1997, 0.1561, 0.2024, 0.1597, 0.1997)$.

The corresponding *CIC* is $(n) = 0.02 < 0.1$.

Then, the importance of three subsystems and the fuzzy complementary judgment matrix can be compared with the following:

$$A' = \begin{bmatrix} 0.5 & 0.7 & 0.5 \\ 0.3 & 0.5 & 0.3 \\ 0.5 & 0.7 & 0.5 \end{bmatrix}$$

Using AGA-FAHP, the weights of these three subsystems can be calculated: $w' = (0.4, 0.2, 0.4)$. The corresponding *CIC* is $(n) = 1.19 \times 10^{-8} < 0.1$.

The subjective weights of each index relative to the water resources carrying capacity (w_s) can be obtained by multiplying the weights of each index, relative to the subsystems (w_1, w_2, w_3) and by the weights of each subsystem (w'), as shown in Table 2. For example, $w_{s1} = 0.3335 \times 0.4 = 0.1334$.

Table 2. The Subject Weights, the Objective Weights and the Combination Weights of Indices.

Weight	x_1	x_2	x_3	x_4	x_5	x_6	x_7	x_8
w_s	0.1334	0.1333	0.1066	0.0267	0.0594	0.0485	0.0294	0.0433
w_o	0.0515	0.0589	0.0022	0.0523	0.0884	0.0309	0.0555	0.0137
w_c	0.0952	0.1018	0.0176	0.0429	0.0833	0.0445	0.0464	0.0280

Weight	x_9	x_{10}	x_{11}	x_{12}	x_{13}	x_{14}	x_{15}
w_s	0.0194	0.033	0.0799	0.0624	0.081	0.0639	0.0799
w_o	0.0715	0.1344	0.1302	0.0001	0.1464	0.0868	0.0771
w_c	0.0428	0.0765	0.1172	0.0029	0.1251	0.0856	0.0902

3.3. Determination of Objective Weight by AGA-PP

The projection pursuit method, based on the accelerating genetic algorithm (AGA-PP) [16,18,19], was used to determine the objective weight of each index: $\{w_o(j), j = 1,2, \ldots, n\}$. According to the water resources carrying capacity evaluation standard, the water resources carrying capacity was divided into three grades: 1 (available load state), 2 (critical state), and 3 (overload state). When the grade of the carrying capacity of regional water resources is greater than 1, the state of the water resources carrying capacity is changed from available load to overload. According to the principle of forewarning, the system will enter into the "warning" state. The "non-warning" state of the regional water resources carrying capacity is a fuzzy set, and its membership degree (f) can be defined as follows: When the water resources carrying capacity is in an available load state, the f value is 1; when the water resources carrying capacity is in an overload state, the f value is 0. The "non-warning" membership degree (f) is divided into five sub-intervals: [0.8, 1.0], [0.6, 0.8], [0.4, 0.6], [0.2, 0.4], and [0.0, 0.2], which are defined as the non-warning interval, light warning interval, middle warning interval, heavy warning interval, and great warning interval, respectively. In order to determine the objective weights of each index, nine standard grade sample values were interpolated linearly between the critical values of each index. At the same time, the "non-warning" membership (f) was also interpolated linearly in [0, 1]. Therefore, together with the two critical sample values, there were 11 standard rank sample values of each index obtained, which are shown as the sample numbers 1–11 in Table 3. In Table 3, the meanings of the warning indices x_1–x_{15} are the same as those in Table 1. The sample numbers 12–22 are the sample value series of the water resources carrying capacity of Anhui Province in 2005–2015.

Table 3. Sample Values of the Regional Water Resources Carrying Capacity Forewarning and its "Non-Warning" Membership Degrees (f) and Comprehensive Forewarning Indices (s).

Serial Number	Values of Forewarning Indices								
	x_1	x_2	x_3	x_4	x_5	x_6	x_7	x_8	x_9
1	1670.00	80.00	450.00	40.00	40.00	24,840.00	90.00	95.00	5.00
2	1603.00	77.00	440.00	38.50	43.00	23,018.40	88.00	92.50	4.60
3	1536.00	74.00	430.00	37.00	46.00	21,196.80	86.00	90.00	4.20
4	1469.00	71.00	420.00	35.50	49.00	19,375.20	84.00	87.50	3.80
5	1402.00	68.00	410.00	34.00	52.00	17,553.60	82.00	85.00	3.40
6	1335.00	65.00	400.00	32.50	55.00	15,732.00	80.00	82.50	3.00
7	1268.00	62.00	390.00	31.00	58.00	13,910.40	78.00	80.00	2.60
8	1201.00	59.00	380.00	29.50	61.00	12,088.80	76.00	77.50	2.20
9	1134.00	56.00	370.00	28.00	64.00	10,267.20	74.00	75.00	1.80
10	1067.00	53.00	360.00	26.50	67.00	8445.60	72.00	72.50	1.40
11	1000.00	50.00	350.00	25.00	70.00	6624.00	70.00	70.00	1.00
12	1135.70	51.57	319.26	26.06	28.92	7685.40	69.81	52.20	0.66
13	950.10	41.62	333.95	26.06	25.46	8820.18	56.41	64.30	0.59
14	1164.50	51.08	347.59	26.06	32.57	10,016.20	58.26	59.10	0.69
15	1139.80	50.13	395.13	26.06	38.09	11,261.22	71.02	72.20	0.61
16	1195.70	52.56	430.32	26.06	39.81	12,699.43	66.36	75.40	0.66
17	1578.20	67.33	428.45	27.53	31.15	15,092.34	67.03	78.50	0.79
18	1008.80	43.17	428.49	27.53	48.94	16,992.92	64.32	74.30	1.34
19	1170.60	50.53	418.08	27.53	41.17	17,072.43	69.42	71.40	1.31
20	974.54	41.99	427.22	27.53	50.55	17,262.21	71.92	70.50	1.37
21	1279.78	55.81	392.29	28.65	34.95	17,956.27	73.01	73.60	1.71
22	1495.31	65.54	415.40	28.65	31.58	18,665.40	73.68	78.90	1.70

Serial Number	Values of Forewarning Indices								Warning Status
	x_{10}	x_{11}	x_{12}	x_{13}	x_{14}	x_{15}	$s(i)$	$f(i)$	
1	70.00	100.00	50.00	200.00	50.00	250.00	0.9449	1.00	–
2	81.00	130.00	65.00	230.00	53.00	265.00	0.8573	0.90	–
3	92.00	160.00	80.00	260.00	56.00	280.00	0.7697	0.80	–
4	103.00	190.00	95.00	290.00	59.00	295.00	0.6821	0.70	–
5	114.00	220.00	110.00	320.00	62.00	310.00	0.5945	0.60	–
6	125.00	250.00	125.00	350.00	65.00	325.00	0.5069	0.50	–
7	136.00	280.00	140.00	380.00	68.00	340.00	0.4193	0.40	–
8	147.00	310.00	155.00	410.00	71.00	355.00	0.3317	0.30	–
9	158.00	340.00	170.00	440.00	74.00	370.00	0.2441	0.20	–
10	169.00	370.00	185.00	470.00	77.00	385.00	0.1565	0.10	–
11	180.00	400.00	200.00	500.00	80.00	400.00	0.0689	0.00	–
12	89.04	386.10	369.10	465.00	35.50	320.30	0.3694	–	heavy
13	85.69	393.80	361.90	470.00	37.10	337.60	0.3022	–	heavy
14	91.13	305.90	287.90	479.00	38.70	287.30	0.4019	–	heavy
15	96.10	288.80	244.90	483.00	40.05	349.20	0.3827	–	heavy
16	101.21	290.30	227.60	487.00	42.10	366.10	0.3780	–	heavy
17	107.62	238.50	166.30	490.00	43.20	358.60	0.5154	–	middle
18	109.53	195.00	123.10	493.00	44.80	341.50	0.3683	–	heavy
19	109.63	167.60	120.90	495.00	46.50	362.60	0.4212	–	middle
20	109.99	155.50	110.20	496.00	47.86	313.90	0.3911	–	heavy
21	110.35	131.00	97.00	497.00	49.15	259.00	0.5439	–	middle
22	110.94	131.20	96.80	498.00	50.50	282.40	0.5961	–	light

The sample value series are denoted as $\{x(i,j) \mid i = 1,2, \dots , m; j = 1,2, \dots , n\}$. For this example: $m = 22$ and $n = 15$. In order to eliminate the dimension effect of each index and make the modeling universal, it was necessary to standardize the $x(i,j)$. The standardized result was recorded as $y(i,j)$. For positive indices (the larger the index value is, the greater the water resources carrying capacity is)

$$y(i, j) = [x(i, j) - x_{\min}(j)] / [x_{\max}(j) - x_{\min}(j)] \tag{4}$$

For negative indices (The larger the index value, the smaller the carrying capacity of water resources)

$$y(i,j) = [x_{\max}(j) - x(i,j)]/[x_{\max}(j) - x_{\min}(j)] \tag{5}$$

where the $x_{\min}(j)$ is the minimum value of the j index in the sample value; $x_{\max}(j)$ is the maximum value of the j index in the sample value series. In addition, in this example, the indices, x_1–x_4 and x_6–x_8 are the positive indices. The negative indices include x_5 and x_{10}–x_{15}.

Projection pursuit (PP) [20,21] is a type of statistical technique which involves finding the optimal possible projections in multidimensional data. In this example, it was used to make fifteen-dimensional data $\{y(i,j)\}$ into a one-dimensional projection value date $p(i)$, whose unit length projection direction was $a = (a(1), a(2), \ldots, a(15))$:

$$p(i) = \sum_{j=1}^{15} a(j)y(i,j) \ (i = 1, 2, \cdots, 22) \tag{6}$$

$$a(j) > 0, \sum_{j=1}^{15} a^2(j) = 1 \tag{7}$$

In the projection process, the projection value, $p(i)$, should extract the change information in $\{y(i,j)\}$ as much as possible, and make the projection of the clusters as far as possible. This means the standard deviation (S) of $p(i)$ is as large as possible. The local projection points should be as dense as possible, and it is best to condense into a number of clusters. In addition, according to Jaynes maximum entropy principle [22], when only partial information is grasped, the distribution of $a^2(j)$ which makes the information entropy maximum and meets the condition of Equation (7), should be adopted. Based on this, the projection index function can be constructed as [16]:

$$Q(a) = S(a) + D(a) + E(a) \tag{8}$$

where $S(a)$ is the standard deviation of $p(i)$; $D(a)$ is the local density of $p(i)$; and $E(a)$ is the information entropy of $a^2(j)$.

$$S(a) = \left| \sum_{i=1}^{m} (p(i) - p)^2/(m-1) \right|^{0.5} \tag{9}$$

$$D(a) = \sum_{i=1}^{m}\sum_{j=1}^{n} (R - r_{ij})u(R - r_{ij}) \tag{10}$$

$$E(a) = -\sum_{j=1}^{n} a^2(j)\ln a^2(j) \tag{11}$$

where p is the meaning of the sequence $\{p(i)\}$. R is the window radius for calculating the local density, the selection of which should not only make the average number of projection points contained in the window not too little, but also make it not increase too fast with an increase in sample size. Normally, $R = 0.1S$. $r_{ij} = |p(i) - p(j)|$. $u(t)$ is a unit step function. When $t < 0$, the function value of $u(t)$ is 0. Otherwise the function value of $u(t)$ is 1.

When the sample value series are given, the projection index function ($Q(a)$) will only change with the projection direction a. In addition, the classification and sorting feature structure of high-dimensional index data can be exposed by the most likely optimal projection. Therefore, the optimal projection direction can be estimated by solving the maximization of the projection index function

$$\max Q(a) = S(a) + D(a) + E(a)$$
$$\text{s.t. } a(j) > 0, \sum_{j=1}^{n} a^2(j) = 1 \tag{12}$$

where $a^2(j)$ is the objective weight $w_o(j)$ of the index x_j.

This is a nonlinear optimization problem with $a(j)$ as a variable. It can be solved by AGA [13,23], which simulates biological population evolution and genetic mechanisms. For this example, the best projection direction is

$$a = (0.2270, 0.2426, 0.0470, 0.2288, 0.2973, 0.1759, 0.2356, 0.1171, 0.2674, 0.3666,$$
$$0.3609, 0.0100, 0.3826, 0.2946, 0.2777)$$

The objective weights of each index are $w_o(j) = a^2(j)$, as shown in Table 2.

3.4. Determination of Combination Weight by Entropy of Information

After combining the subjective weights, $w_s(j)$ and objective weight, $w_o(j)$, the combination weight $w_c(j)$ of each index can be obtained. Obviously, $w_c(j)$, $w_s(j)$ and $w_o(j)$ should be as close as possible. According to the principle of minimum relative entropy [24]

$$\min F = \sum_{j=1}^{n} w_c(j)[\ln w_c(j) - \ln w_s(j)] + \sum_{j=1}^{n} w(j)[\ln w_c(j) - \ln w_o(j)]$$

$$\text{s.t. } \sum_{j=1}^{n} w_c(j) = 1; \; w_c(j) > 0 \tag{13}$$

where F is the relative entropy.

The Lagrange multiplier method [25] can be used to solve the above optimization problems:

$$w_c(j) = \frac{[w_s(j)w_o(j)]^{0.5}}{\sum_{j=1}^{15} [w_s(j)w_o(j)]^{0.5}} \tag{14}$$

Equation (14) shows that in all combinations of weights which satisfy Equation (13), the geometric average requires the least amount of information. Taking other forms of combination weights adds tangible or intangible additional information that is not actually available. According to the index of Equation (14), the combination weights of indices 1–15 can be calculated, as shown in Table 2.

3.5. Calculation of Comprehensive Forewarning Index

$$s(i) = \sum_{j=1}^{n} wc(j)y(i,j) \quad (i = 1,2,\cdots,m) \tag{15}$$

where $s(i)$ is the comprehensive forewarning index of the sample (i). The greater the forewarning index $s(i)$ is, the greater the water resources carrying capacity of the sample (i) is. In addition, the corresponding "non-warning" membership degree $f(i)$ is greater. The calculation results of $s(i)$ are shown in Table 3.

According to the corresponding relationship between the "non-warning" membership degree $f(i)$ and the forewarning comprehensive index $s(i)$, the forewarning interval corresponding to the forewarning comprehensive index can be obtained. The corresponding relationship is a forewarning model of the regional water resources carrying capacity based on entropy combination weights.

4. Results and Discussions

4.1. Determination of the Warning Status of the Water Resources Carrying Capacity

According to the previous introduction, we know that when the "non-warning" membership degree is $f(i) \in [0.8, 0.1]$, it means that the forewarning state of the water resources carrying capacity in that year is non-warning. As shown in Table 3, the interval of the comprehensive forewarning index corresponding to the non-warning interval is [0.7697, 0.9449]. Similarly, when $s(i)$ belongs to [0.5945, 0.7697], [0.4193, 0.5945], [0.2441, 0.4193], [0.0689, 0.2441], the corresponding warning status of the water resources carrying capacity is light warning status, middle warning status, heavy warning status, and great warning status, respectively. Taking 2005 as an example, the comprehensive forewarning index of 2005 which is equal to 0.3694 belongs to [0.2441, 0.4193]. This means that the warning status of the water resources carrying capacity in 2005 is a heavy warning status. Similarly, the warning status of the water resources carrying capacity over the years 2006–2015 can be obtained (Table 3).

Figure 7 shows the yearly variation curve of the comprehensive warning indices and the range of each warning interval. As shown in Figure 7, the water resources carrying capacity in Anhui showed an overall trend of improvement. But most years had a heavy warning status. Although the state of the water resources carrying capacity did not enter the overload state, it still should arouse our attention and measures must be taken to prevent further reduction of the water resources carrying capacity.

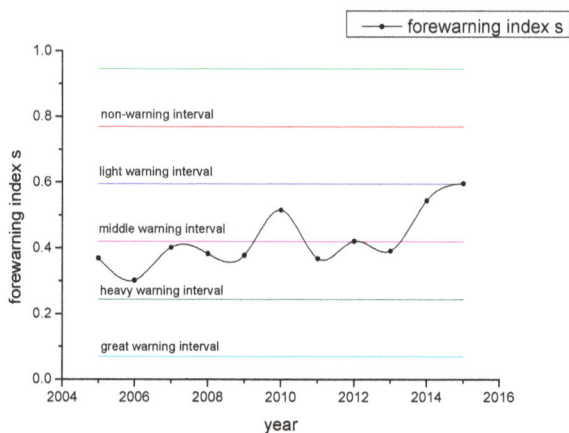

Figure 7. The change chart of the forewarning comprehensive indices.

4.2. Results Comparison

Because the research on the water resources carrying capacity forewarning is less established, the results of this study are compared with only two other studies on the water resources carrying capacity in Anhui.

Huang and Chen selected twelve indicators from aspects of water resources, environmental development, economic development and the ecological environment, and evaluated the water resources carrying capacity in Anhui Province in 2006–2015. The water resources carrying capacity index (CCI) was used to determine the status of water resources carrying capacity. If CCI < 1, it means that the carrying capacity of water resources is in the loaded state. If CCI > 1, it means that the carrying capacity of water resources is in the overload state. And if CCI = 1, it means that the carrying capacity of water resources is in the critical load state. This standard shows that there was hardly ever a critical state. The results show that the value of CCI was between 0.9–1.3 [26]. This is basically consistent with the results of this paper which show that most warning statuses of water resources are in the middle and heavy warning statuses.

Liu and Yu used entropy weight and an analytic hierarchy process to evaluate the water resources carrying capacity in Anhui Province in 2007–2010. The comprehensive score of the water resources carrying capacity was used to measure the size of the water resources carrying capacity. The closer the value is to 0, the higher the carrying capacity is, and the closer is to 1, the lower the carrying capacity is. The result showed that the water resources carrying capacity comprehensive evaluation values in 2007–2010 were between 0.46 and 0.49, which means that the state was a critical load state [27].

This implies that the water resources in Anhui have been developed to a certain extent, no longer in the status of higher carrying capacity, but they can still provide necessary water for life and production. This is also consistent with the result of this paper.

4.3. Results Analysis

It is obvious that the forewarning comprehensive index of Anhui Province in 2015 was the highest from the last 11 years, which means that the water resources carrying capacity water resources in 2015 was the highest in the last 11 years. With reference to Figure 3, we can see that the water resources quantity in Anhui Province in 2015 was close to that of 2010, both of which were much larger than other years. However, the water resources carrying capacity in 2015 was larger than in 2010, obviously. This was because the irrigation quota of farmland in Anhui province in 2010 (358.6 cubic meters per mu) was far more than that in 2015 (282.4 cubic meters per mu) and water consumption per 10,000 GDP only accounted for about 55% in 2010 (Table 3). In addition, the water resources quantities in 2006, 2011 and 2013 were less, but the water resources carrying capacity gradually increased. This is because the water consumption per 10,000 yuan GDP of these three years had gradually decreased (Table 3). This indicates that the water use efficiency has a great influence on the water resources carrying capacity when the water resources quantity is similar. Correspondingly, when the water efficiency gap is not large, the total amount of water resources has a significant impact on the size of the water resources carrying capacity, such as in the years of 2005 and 2006.

The above analysis results are consistent with the results from the weights of influence factors analysis. According to the combination weights, the five indices which have the greatest influence on water resources carrying capacity are population density, 10,000 yuan GDP water consumption, water production modulus, per capita water resources and irrigation quota of farmland. The population density changed slowly over the past 11 years. Thus, the water use efficiency and water resources quantity are the most important factors affecting the water resources carrying capacity in Anhui Province.

Therefore, in order to improve the regional water resources carrying capacity, it is necessary to improve the efficiency of water use. In addition, some methods can be used to increase the amount of available water resources, such as improving the sewage discharge rate and utilization rate of reclaimed water.

5. Conclusions

(1) In this model, for the determination of the weight of each index, the forewarning index system and the grade standards of water resources carrying capacity were established, according to the actual situation of the study area. AGA-FAHP was used to calculate the subjective weight of each index. AGA-PP was used to calculate the objective weight of each index. In addition, the minimum relative information entropy principle was used to calculate the combination weights. Then, the forewarning comprehensive index was calculated to find out the relationship between the forewarning comprehensive index and the membership degree of "non-warning". On this basis, the forewarning model of the regional water resources carrying capacity was established. This model combined expert experience and objective information in the research area. In addition, the calculation results were reasonable. This can be used in other forewarning systems by modifying the index system and grade standards for its universality.

(2) As a basic natural resources and strategic economic resources, water resources are an important guarantee for the sustainable development of economy and society. Therefore, the water resources carrying capacity forewarning, based on theory and technology, has an important significance for guiding the water resources development and management, scientifically and reasonably. With this forewarning model, we can evaluate the status of the water resources carrying capacity, study the variation trend of water resources carrying capacity and analyze the main influence factors of the water resources carrying capacity. Based on this we can forecast the status of the

Entropy **2017**, *19*, 574

water resources carrying capacity, guide the strategy of the development and utilization of water resources, according to the results of forewarning and the main influence factors, and find ways to improve the water resources carrying capacity.

(3) In order to improve the water resources carrying capacity, product consumption must be reduced and the water recycling rate must be improved through scientific management and technological innovation. The efficient use of water resources, the reasonable development and utilization of water resources must be ensured through the optimal allocation of water resources and the adjustment of economic structure. Overall, it is time to promote the construction of a water-saving society and building a harmonious environment between human and water.

Acknowledgments: The authors would like to thank the support of the National Key Research and Development Program of China under Grant No. 2016YFC0401305 and No. 2016YFC0401303, the National Natural Science Foundation of China under Grant No. 51579059 and Higher Education Supporting Project of China under Grant No. JZ2016YYPY0065.

Author Contributions: Rongxing Zhou was responsible for the data analysis and code programming; Rongxing Zhou wrote the paper; Zhengwei Pan contributed the data; Juliang Jin constructed the research framework and designed the study; Chunhui Li and Shaowei Ning reviewed and edited the manuscript. All of the authors have read and approved the final manuscript.

Conflicts of Interest: The authors declare no conflict of interest.

References

1. Mei, H. Advances in study on water resources carrying capacity in China. *Procedia Environ. Sci.* **2010**, *2*, 1894–1903. [CrossRef]
2. Xiong, H.; Fu, J.H.; Wang, K.-L. Evaluation of water resource carrying capacity of Qitai Oasis in Xinjiang by entropy method. *Chin. J. Eco-Agric.* **2012**, *20*, 1382–1387. [CrossRef]
3. Fu, X.; Ji, C.M. A comprehensive evaluation of the regional water resource carrying capacity—Application of Main Component Analysis Method. *Resour. Environ. Yangtza Basin* **1999**, *2*, 168–173.
4. Zhou, L.; Liang, H. A Study on the Evolution of Water Resource Carrying Capacity in Karst Area Based on Component Analysis and Entropy. *J. Nat. Resour.* **2006**, *21*, 827–833.
5. Gong, L.; Jin, C. Fuzzy comprehensive evaluation for carrying capacity of regional water resources. *Water Resour. Manag.* **2009**, *23*, 2505–2513. [CrossRef]
6. Liang, M.; Wang, Z.; Liu, J. Evaluation model of water resources carrying capacity based on factor analysis. *Yellow River* **2010**, *32*, 62–65.
7. Zhao, Z.; Li, W.; Wang, C. Study on Water Resources Carrying Capacity in Ningxia Based on Principal Component Analysis and Factor Analysis. *Hydrology* **2017**, *37*, 64–72.
8. Chen, N.X.; Ban, P.L.; Zhang, W.B. Fuzzy evaluation of the water resources carrying capacity based on the maximum entropy theory. *J. Irrig. Drain.* **2008**, *27*, 57–60.
9. Li, F.; Chen, Y.; Li, W. The Application of Set Pair Analysis Based on Entropy Weight to Evaluation of Sustainable Water Resources Utilization—A Case Study in the Three Sources of Tarim River. *J. Glaciol. Geocryol.* **2010**, *32*, 723–730.
10. Feng, L.H.; Zhang, X.C.; Luo, G.Y. Application of system dynamics in analyzing the carrying capacity of water resources in Yiwu City, China. *Math. Comput. Simul.* **2008**, *79*, 269–278. [CrossRef]
11. Guo, Q.; Tian, Z.; Zeng, L. Application of AGA-FAHP in Bids Evaluation of Construction Projects. *Water Sav. Irrig.* **2010**, *9*, 62–64.
12. Jin, J.; Yang, X.; Wei, Y. System Evaluation Method Based on Fuzzy Preferential Relation Matrix. *Syst. Eng. Theory Methodol. Appl.* **2005**, *14*, 364–368.
13. Jin, J.; Yang, X.; Ding, J. An Improved Simple Genetic Algorithm—Accelerating Genetic Algorithm. *Syst. Eng. Theory Methodol. Appl.* **2001**, *21*, 8–12.
14. Mon, D.L.; Cheng, C.H.; Lin, J.C. Evaluating weapon system using fuzzy analytic hierarchy process based on entropy weight. *Fuzzy Sets Syst.* **1994**, *62*, 127–134. [CrossRef]
15. Tesfamariam, S.; Sadiq, R. Risk-based environmental decision-making using fuzzy analytic hierarchy process (F-AHP). *Stoch. Environ. Res. Risk Assess.* **2006**, *21*, 35–50. [CrossRef]

16. Wen, J.; Li, J.; Jin, J. Forewarning Model of Regional Water Resources Sustainable Utilization Based on Combined Weights and Entropy Principles. *Water Resour. Power* **2006**, *24*, 6–10.

17. Xu, Z. Research on Compatibility and Consistency of Fuzzy Complementary Judgment Matrices. *PLA UST* **2002**, *3*, 94–96.

18. Wang, M.; Jin, J.; Li, L. Application of method based on raga to assessment of sand liquefaction potential. *Chin. J. Rock Mech. Eng.* **2004**, *23*, 631–634.

19. Jin, J.; Wang, M.; Wei, Y. Objective Combined Evaluation Model for Optimizing Water Resource Engineering Schemes. *Syst. Eng. Theory Methodol. Appl.* **2004**, *24*, 111–116.

20. Friedman, J.H.; Tukey, J.W. A projection pursuit algorithm for exploratory data analysis. *IEEE Trans. Comput.* **1974**, *100*, 881–890. [CrossRef]

21. Jones, M.C.; Sibson, R. What is projection pursuit. *J. R. Stat. Soc. Ser. A (Gen.)* **1987**, *150*, 1–37. [CrossRef]

22. Jaynes, E.T. On the rationale of maximum-entropy methods. *Proc. IEEE* **1982**, *70*, 939–952. [CrossRef]

23. Jin, J.; Yang, X. Real Coding Based Acceleration Genetic Algorithm. *J. Sichuan Univ. Eng. Sci. Ed.* **2000**, *32*, 20–24.

24. Woodbury, A.D.; Ulrych, T.J. Minimum relative entropy: Forward probabilistic modeling. *Water Resour. Res.* **1993**, *29*, 2847–2860. [CrossRef]

25. Bertsekas, D.P. *Nonlinear Programming*; Athena Scientific: Belmont, MA, USA, 1999.

26. Huang, T.; Chen, C. Evaluation of water resources carrying capacity in Anhui Province based on analytic hierarchy process. *J. Chifeng Univ. (Nat. Sci. Ed.)* **2017**, *33*, 19–21.

27. Liu, C.; Yu, W. Fuzzy comprehensive evaluation of water resources carrying capacity in Anhui based on entropy weight and AHP. *Harnessing Huaihe River* **2013**, *2*, 21–22.

entropy

MDPI

Article

Combined Forecasting of Rainfall Based on Fuzzy Clustering and Cross Entropy

Baohui Men [1,*], Rishang Long [2], Yangsong Li [1], Huanlong Liu [1], Wei Tian [3,4] and Zhijian Wu [1]

1   Beijing Key Laboratory of Energy Safety and Clean Utilization, North China Electric Power University, Renewable Energy Institute, Beijing 102206, China; lys18811551843@163.com (Y.L.); liuhuanlongHD@163.com (H.L.); wzjdalanqiu@163.com (Z.W.)
2   State Key Laboratory of New Energy Power System, North China Electric Power University, Beijing 102206, China; lrs18810667721@163.com
3   Key Laboratory of Water Cycle and Related Land Surface Process, Institute of Geographic Sciences and Natural Resources Research, Chinese Academy of Sciences, Beijing 100101, China; tianweiBT@163.com
4   University of Chinese Academy of Sciences, Beijing 100049, China
*   Correspondence: menbh@ncepu.edu.cn

Received: 31 August 2017; Accepted: 14 December 2017; Published: 19 December 2017

Abstract: Rainfall is an essential index to measure drought, and it is dependent upon various parameters including geographical environment, air temperature and pressure. The nonlinear nature of climatic variables leads to problems such as poor accuracy and instability in traditional forecasting methods. In this paper, the combined forecasting method based on data mining technology and cross entropy is proposed to forecast the rainfall with full consideration of the time-effectiveness of historical data. In view of the flaws of the fuzzy clustering method which is easy to fall into local optimal solution and low speed of operation, the ant colony algorithm is adopted to overcome these shortcomings and, as a result, refine the model. The method for determining weights is also improved by using the cross entropy. Besides, the forecast is conducted by analyzing the weighted average rainfall based on Thiessen polygon in the Beijing–Tianjin–Hebei region. Since the predictive errors are calculated, the results show that improved ant colony fuzzy clustering can effectively select historical data and enhance the accuracy of prediction so that the damage caused by extreme weather events like droughts and floods can be greatly lessened and even kept at bay.

Keywords: rainfall forecast; cross entropy; ant colony fuzzy clustering; combined forecast

1. Introduction

Rainfall forecasts play an important role in agricultural production, urban industry and life. The accurate prediction of rainfall has significant economic and social value. It can provide data support for the relevant departments and help detect droughts and floods and reduce the degree of harm. However, affected by complex factors such as geographical environment, ocean currents, air pressure, temperature, etc. [1], rainfall exhibits strong randomness and nonlinear characteristics that often hamper the forecast of rainfall.

The rainfall prediction methods discussed in this paper are based on mathematical models and algorithms, through the full mining of historical data to establish the forecasting model. At present, a variety of approaches have been applied to predict rainfall at home and abroad and basically they can be classified into five categories: (1) *Numerical prediction model*. This model is based on the physical model of process. Its advantage lies in the fast speed and easy procedure, but due to the impact of the memory required in can only reasonably be used for monthly forecasts and for longer periods (such as an annual forecast) is difficult to use [2,3]; (2) *Time series model exponential smoothing*. The moving average method and Autoregressive Integrated Moving Average Model (ARIMA) [4] belong to this type; they can better describe the linear change process yet sometimes they are not suitable for non-stationary random processes; (3) *Probabilistic simulation methods*. These include the grey model (GM) [5] and Monte Carlo method [6]. The prediction of exponential trends is more accurate with the grey model, but it is only suitable for short and medium term forecasting and the longer the forecast lasts, the larger the errors that may occur. The Monte Carlo method is characterized by describing the random process, but the support of data is also required; (4) *Artificial intelligence methods* such as radical basis function (RBF), genetic algorithm (GA), wavelet analysis (WA) [7–10], can better simulate nonlinear processes with higher prediction accuracy, but may fail due to issues like local optima, overlearning and weak generalization ability; (5) in addition there are some other methods, like Numerical Weather Prediction (NWP) [11,12], R/S analysis [13], trend analysis [14], etc. that can predict rainfall from different angles, but two problems should not be ignored: first, in the long-term forecast, rainfall is a random process hence any single forecasting method cannot ensure the stability during the process of prediction. Additionally, serious errors may occur at certain times resulting in the failure of prediction. Second, obtaining useful information with a single method is one-sided, and overlooks different factors from all perspectives.

For the prediction of rainfall, a large amount of historical data is necessary to ensure the accuracy of the forecast. Nonetheless, historical data inevitably contains some errors or abnormal information and this affects the accuracy of the forecast because rainfall is associated with many factors such as temperature, climate, and human activities. Therefore, the forecast method based on fuzzy clustering prediction method has been applied in recent years [15,16]. However, the traditional fuzzy clustering algorithm approach easily falls into local optimal solutions, and it is difficult to deal with a large number of high-dimensional data from a time performance point of view [17]. In this paper, the ant colony algorithm is proposed to improve the fuzzy clustering. With this method, the reliability and computational efficiency of data filtering and processing are greatly increased.

Bates and Granger established the combined forecasting method based on weights in 1969 [18]. The approach combines the different methods and the features of data to improve the accuracy of forecasting and reduce the risk of failure. The combined forecasting method has been widely used in various fields, including electric power load forecasting, economics, logistics, etc., and facts have proven how effective the method is [19–22]. Nonetheless, as a simple combination of several single methods, the previous method neglects the bias of the selection of a single method. Furthermore, there is no detailed analysis of the time characteristics of historical data. At present, some scholars are paying attention to combined forecasting in the field of rainfall forecasting [23–25]. In Cui's study [23], the wavelet analysis method aims to determine the weight reconstruction of the rainfall forecast, yet the time distribution of historical data is not considered. In Xiong's study [24] and Lu's study [25], real-time river flow or flood forecasting methods have been studied, but they are not suitable for medium and long-term prediction.

The concept of entropy propounded by the German physicist Clausius in 1877 is a function of the state of the system, but the reference value and the variation of entropy are often analyzed and compared. Cross Entropy (CE) is a kind of entropy that reflects the similarity between variables from the perspective of probability. The application of entropy theory in hydrology mainly include the derivation of the distributions and estimation of the corresponding parameters for hydrometeorological variables [26–28], dependence analysis [29] and runoff forecasting [30–34]. The cross entropy is introduced into the combination forecasting by Li et al. [24,25]. Their research put forward a new method of determining weight, which improves the stability of the prediction results. However, the probability density function [24] is not suitable for the prediction of radial flow. The wind power load forecasting method based on normal distribution is proposed by Chen et al. [25]. The time characteristic of historical data is not considered in this method and the solution is too complex to be implemented.

The key of the prediction method based on historical data is not only a prediction model, but also a validity of historical data. And that is the category of data mining. The choice of historical data is fundamentally a clustering process, so clustering method is very important. In terms of the weaknesses of the fuzzy clustering method, the ant colony algorithm attempts to improve the model. Meanwhile, the method for determining weights is also improved by using the cross entropy (CE).

2. Improved Fuzzy Clustering Model

2.1. Research Data

The Beijing–Tianjin–Hebei region is located on the east coast of Eurasia, mid-latitude coastal and inland transfer zone. Influenced by a temperate climate with alternating moist and dry seasons, the annual rainfall in this area ranges from 400 to 800 mm. The study analyses the rainfall from 1969 to 2010 in the Thiessen rainfall station and the forecast is conducted based on the data of rainfall from the Taisen Station in the Beijing–Tianjin–Hebei region. Based on the Taisen Station rainfall data, the forecast is conducted. The results show that improved ant colony fuzzy clustering can effectively select historical data and improve the accuracy of prediction.

On the basis of the data of rainfall from 26 stations, the weight is determined by the Thiessen polygon method, and the weighted average rainfall data sequence is obtained. The Beijing–Tianjin– Hebei administrative divisions and the change of monthly rainfall from 1960 to 2013 in this area are shown in Figure 1.

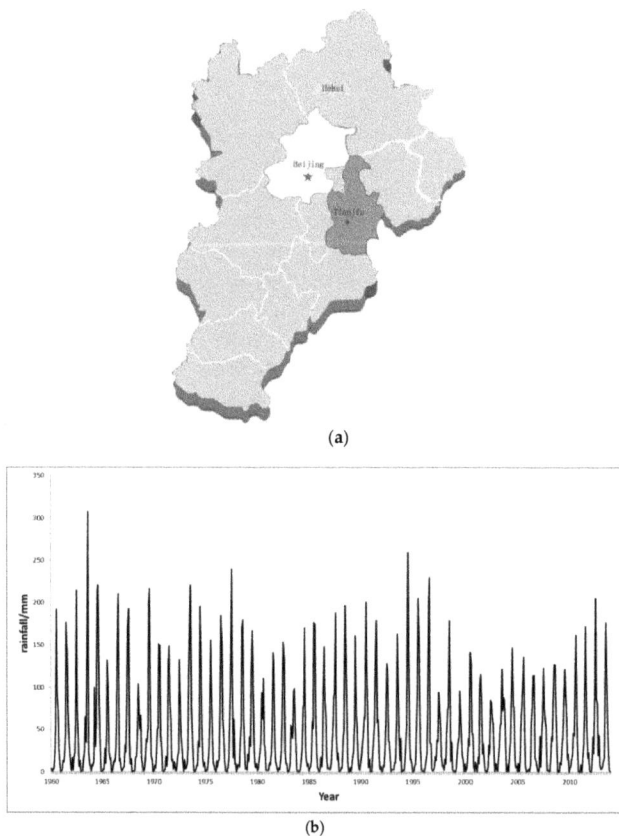

(a)

(b)

Figure 1. The Beijing–Tianjin–Hebei administrative divisions and change of monthly rainfall from 1960 to 2013. (**a**) Beijing–Tianjin–Hebei administrative divisions; (**b**) monthly rainfall from 1960 to 2013.

Figure 2 gives the results of the wavelet analysis of the data from 1960–2006. Wavelet analysis is a localized analysis of the time (space) frequency. It multiplies the signal (function) step by step through the telescopic translation operation, finally reaches the time subdivision at high frequency, subdivides the frequency at low frequency, and can automatically adapt to the data analysis.

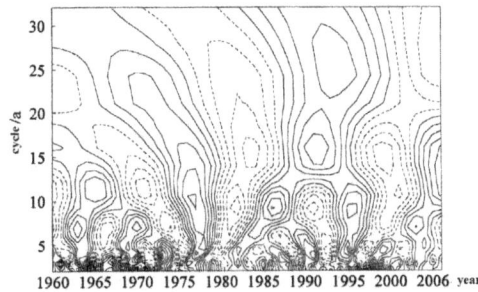

Figure 2. Wavelet analysis of the rainfall data from 1960–2006.

There are four obvious characteristic time scales, namely, 3a, 9a, 14a and 24a, respectively, where the characteristic time scale of 3a is always present from 1960–2006 and the period oscillation is stable. What is more, with 24a time scale cycle time oscillation throughout the study period, the performance is relatively stable. In the middle of the 1960s, Beijing–Tianjin–Hebei area had experienced four dry and wet alternations: from the mid-1960s to the late 1970s, the precipitation was abundant. In the 1980s, the precipitation was relatively low. In the 1990s, the precipitation again entered an abundant period. After the 21st century, the precipitation began to decrease. The characteristics of the two feature scales, 9a and 14a, are similar. Before the mid-1970s, the oscillation of the cycle time was more obvious, and after the rich period of 1970 to 1980, there was a slight increase in the feature scale, respectively, about 10a and about 15a.

The analysis reveals that the periodic variation of rainfall is obvious. Therefore, it is important to predict the future rainfall by grasping the key information of the rainfall in the historical year and using the data mining technology to classify the rainfall-related data reasonably.

2.2. An Introduction of Ant Colony Algorithm

We take the Travelling Salesman Problem (TSP) as an example to illustrate Ant Colony (AC) Algorithm. Suppose there are m cities, d_{ij} is the distance between city i and city j. $\tau_{ij}(t)$ is the amount of information between city i and city j at time t. We use it to simulate the actual ant anterin, set a total of m ants, the term $p_{ij}(t)$ represents the probably of the k-th ant being transferred between city i and city j at time t:

$$p_{ij}(t) = \frac{\tau_{ij}^a(t) \cdot \eta_{ij}^b}{\sum\limits_{(i,k) \in S, k \notin U} \tau_{ik}^a(t) \cdot \eta_{ik}^b} \tag{1}$$

where U is the part of the path that the ants have searched for, and S is the set of cities that the next step of the ant k allowed to pass, a indicates the amount of information on the path to the path chosen by the ants, η_{ij} indicates the degree of transfer expectation between city i and j. When $a = 0$, the algorithm is the traditional greedy algorithm; and when $b = 0$, it becomes a pure positive feedback heuristic algorithm. After n moments, the ants can finish all the cities and complete a cycle. In this case, the amount of information on each path is updated according to the following formula:

$$\tau_{ij}^{new} = (1 - \rho)\tau_{ij}^{old} + \Delta \tau_{ij} \tag{2}$$

where $\rho \in (0, 1)$ represents the amount of information that fades with time. The information increment is expressed as:

$$\Delta\tau_{ij} = \sum_{k=1}^{m} \Delta\tau_{ij}^{k} \tag{3}$$

where $\Delta\tau_{ij}^{k}$ is the amount of information left by ant k between city i and j. It can be expressed as:

$$\Delta\tau_{ij}^{k} = \begin{cases} Q/L_k, i, j \in S \\ 0, \text{otherwise} \end{cases} \tag{4}$$

where Q is a constant and L_k is the length of the path traveled by the ant k in this cycle. After several cycles, the calculation can be terminated according to the appropriate stop condition.

2.3. Basic Principles of Fuzzy Clustering

Among the many fuzzy clustering algorithms, the most widely used and successful is Fuzzy C-means (FCM). The FCM algorithm divides n vectors x_k ($k = 1, 2, \ldots, n$) into m fuzzy clusters and obtains the clustering center of each cluster so that the objective function is minimized. The objective function is defined as:

$$J = \sum_{k=1}^{n} \sum_{i=1}^{m} (\mu_{ik})^h d(x_k, c_i) \tag{5}$$

where μ_{ij} is the membership function, c_i is the i-th clustering center, h is the fuzzy weight index. $\mu_{ij} \in (0, 1)$ and:

$$\sum_{i=1}^{m} \mu_{ik} = 1 \tag{6}$$

$$d(x_k, c_i) = ||x_k - c_i|| \tag{7}$$

In order to minimize the objective function, the update of the cluster center and membership function is as follows:

$$c_i = \frac{\sum\limits_{k=1}^{n} \mu_{ik}^h x_k}{\sum\limits_{k=1}^{n} \mu_{ik}^h} \tag{8}$$

$$\mu_{ik} = \frac{1}{\sum\limits_{j}^{m} (d_{ik}/d_{jk})^{1/(h-1)}} \tag{9}$$

Since the solution of a multi-constrained optimization problem is complex, the commonly used method is to fix one of the parameters to optimize the other amounts, and the solution alternates until the difference between two consecutive functions is less than a very small value (the precision requirement).

The flaw of this algorithm is that it needs to be given multiple c values for repeated calculations, and the result is usually a local optimal solution, and the computation time is large because the time required for a matrix multiplication is $O(n^3)$, the time complexity of the first step of the algorithm is reached $O(n^4 \log n)$.

2.4. Improvement of Fuzzy Clustering by Ant Colony Algorithm

One of the keys to improve the speed of fuzzy clustering is to select the initial point of the membership function. If we can get the membership degree approximation result of each parameter point to each cluster, we will improve the speed of the fuzzy clustering algorithm and the ant colony algorithm can achieve this function.

The basic idea is to treat the data as an ant with different attributes. The clustering center is regarded as the "food source" that the ants are looking for, so the data clustering is seen as the process of the ant looking for food sources. The specific process can be described as follows: each ant travels from each cluster center, searches for the next sample point in the entire solution space, and then starts from the cluster center and searches for another sample point in the entire solution space. When the sample point reaches the total number of the original sample points of the cluster, the ant is considered to have completed a search for a path so that the ants do not repeat the same sample point in the search for the same path and set a taboo for each ant $tabu(N)$. If $tabu(N) = 1$, then the node j can choose the search sample point, when the ants selected the node j, $tabu(N)$ will be set 0, then the ants cannot choose the node.

Assume $X = \{X_i \mid X_j = (x_{i1}, x_{i2}, \dots, x_{im})\}$, $i = 1, 2, \dots, n$ is a collection of data to be clustered, $\tau_{ij}(t)$ is amount of information between X_i and X_j at time t. When all the ants have completed a path search, it is said the algorithm carried out a search cycle. In the t search period, the path selection probability can be expressed as:

$$
p_{ij}(t) = \begin{cases} \dfrac{\tau_{ij}^a(t) \cdot \eta_{ij}^b}{\sum\limits_{(i,s)\in S, k \notin tabu(j)} \tau_{is}^a(t) \cdot \eta_{is}^b}, & j = 1, 2, \dots, m \\ 0, & \text{otherwise} \end{cases} \tag{10}
$$

where $S = \{X_s \mid d_{sj} \le r_j, s = 1, 2, \dots, N\}$, and the other parameters are consistent with the above.

When the i value is determined, make j from 1 to m, search the maximum $p_{ij}(t)$, then X_i is merged into the X_j field. Make $C_j = \{X_i \mid d_{ij} < r_j, i = 1, 2, \dots, k\}$, C_j represents all the data sets that are merged into X_j, and we find the cluster center:

$$
\bar{c}_j = \frac{1}{k} \sum_{i=1}^{k} X_i \tag{11}
$$

When the ant colony completes a search period, the probability of each parameter point attributed to a cluster is obtained according to $p_{ij}(t)$, and the general initial value of the fuzzy clustering membership matrix is obtained, and $\bar{c}_j$ is used as the initial of fuzzy clustering center.

As the ant colony algorithm itself has a certain computational complexity, each fuzzy clustering cycle using multiple ant colony algorithm will produce over optimization phenomenon. We adopt the following strategy: in the first initial cycles, we use ant colony algorithm to determine the initial value $p_{ij}(t)$ (in this paper we set 4 cycles) and $\bar{c}_j$, then iterate according to Equations (8) and (9). When the optimization process slows down, we use the ant colony algorithm once or twice to optimize until the accuracy requirements are reached.

2.5. Method Validation

In terms of monthly rainfall, the relevant factors include historical monthly mean temperature, mean air pressure, mean humidity, season etc. Because it is the region's total rainfall forecast, we do not take into account the effects of terrain and ocean currents.

In this way, a five-dimensional vector is formed, namely monthly rainfall, monthly average temperature, monthly mean humidity, monthly mean pressure, seasonal type. Through the fuzzy clustering, the historical data will be grouped to form a database. In this way, when we predict future weather data more precisely, we can search from the historical database to find useful data for rainfall forecast. For seasonal data, we need to map the value, as shown in Table 1.

Table 1. Seasonal mapping values.

Month	12, 1, 2	3, 4, 5	6, 7, 8	9, 10, 11
value	0.1	0.3	0.5	0.7

In order to test the performance of the algorithm, on a computer with the processor i7, memory 4g, the membership degree matrix is simulated. The first is to generate (0, 1) random number, the second is to use ant colony optimization (that is $p_{ij}(t)$). When calculating $p_{ij}(t)$, set $\rho = 0.7$, $a = 1$, $b = 1$, $\eta = 1$, $\tau_{ij}(0) = 0$. The results are shown in Figure 3.

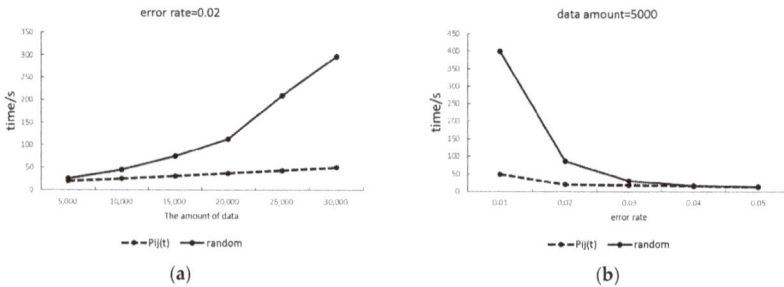

Figure 3. Clustering calculation time. (a) calculation time when error rate is constant; (b) calculation time when data amount is constant.

Figure 3a shows that the fuzzy clustering with the random number as the initial value of the membership matrix increases rapidly with the increase of the data volume, while the fuzzy clustering algorithm with $p_{ij}(t)$ improves the number of samples with the increase of the number. Figure 3b shows that when the error rate is large, the random number and fuzzy clustering calculation using $p_{ij}(t)$ are not very different, but when the error rate becomes smaller, the $p_{ij}(t)$ fuzzy clustering is not changed, and the fuzzy clustering using random number calculation time is rising rapidly. Therefore, the clustering method adopted in this paper is more scientific and effective.

3. Rainfall Forecasting Model Based on CE

The historical data is huge and contains useful data. There are some abnormal data, so we have to choose specific methods according to actual situations in order to ensure the accuracy of the forecast. It is important to predict the future rainfall by grasping the key information of the rainfall in the historical year and using the data mining technology to classify the rainfall-related data reasonably; So Fuzzy Clustering by Ant Colony Algorithm is used for data clustering and the historical data are classified.

3.1. Combined Forecasting Model

The combined forecasting model comprises m single forecasting models and the relative effectiveness of a single forecasting model determined by the historical data. If the combined forecast value at time t is y_t, ω_{ij} is the weight of the i-th model at time t, and $\hat{y}_{it}$ is the predicted value of the i-th model at time t, then the problem of combined forecasting is described as follows:

$$y_t = \sum_{i=1}^{m} \omega_{it} \hat{y}_{it} \tag{12}$$

Here two factors influence the final results of combined forecasting: a single model and the weight of a single forecasting model. In this study, we focus on the latter.

There are no uniform rules for selecting a single method, but instead we must consider the actual problem and the needs of the model. The factors considered in this study include: independence, diversity, and the accuracy of the algorithm. We use a single forecasting method to include the ARIMA time series model, GM, and the RBF.

3.2. The CE Model

According to the definition of entropy, a method for calculating the difference in information between two random vectors is defined as the CE. The CE model determines the extent of the mutual support degree by assessing the degree of intersection between different information sources. Moreover, the mutual support degree can be used to determine the weights of the information sources, where a greater weight represents higher mutual support. This is also called the *Kullback-Leibler* (K-L) distance. The CE of two probability distributions is expressed as $D(f \mid g)$ [26,27].

For the discrete case:

$$D(g||f) = \sum_{1}^{n} g_i \ln \frac{g_i}{f_i} \tag{13}$$

and for the continuous case:

$$\begin{aligned} D(g||f) &= \int g(x) \ln \frac{g(x)}{f(x)} dx \\ &= \int g(x) \ln g(x) dx - \int g(x) \ln f(x) dx \end{aligned} \tag{14}$$

where $D(g||f)$ represents the f to g distance, and f and g denote the probability vector in the discrete case and the probability density function in the continuous case, respectively.

The CE model quantifies the "distance" between the amounts of information. However, the K-L distance is not the real length distance, but instead it is the difference between two probability distributions. In this paper, g is the combined forecast function, f is the single method. CE value should be smallest when two pdf are identical. For the combined forecasting model based on CE, the CE model represents the support for combined forecasting. Therefore, the objective function is to assign weights between different single methods, so that there is the most similar case between the total predictive function and the true value.

To use the CE model, two major problems should be solved: establishing the probability density function and generating the CE objective function and solving the weight coefficient by iteration.

The rainfall is treated as a sequence of discrete random variables in the forecast period. For a certain point in the sequence, the value of the rainfall at a certain prediction time is continuous, so it can be regarded as a continuous random variable. Therefore, rainfall prediction can be treated as a sequence of discrete times but continuous values.

The probability density function for predicting rainfall $f(x)$ can be regarded as the probability density function $f_i(x)$ of the single forecasting method multiplied by the corresponding weight. According to the central limit theorem, if a variable is the sum of many independent random factors, we can treat the variable as following a normal distribution, and thus the rainfall value at a certain time can be considered as satisfying a normal distribution. The minimum CE is used to determine the probability distribution of the different forecasting methods, so the combined probability distribution of the rainfall is obtained.

The probability density function for method i is ($i = 1, 2, \ldots, m$):

$$f_i(x) = \frac{1}{\sqrt{2\pi}\sigma_i} e^{-(x-\mu_i)^2/(2\sigma_i^2)} \tag{15}$$

where μ_i is the mean value and σ_i is the variance.

Thus, the combined probability density function of the predicted rainfall can be obtained based on the probability density function of the single prediction method:

$$f(x) = \sum_{i=1}^{m} \omega_{it} f_i(x) \tag{16}$$

where ω_{ij} is the weight of the i-th single method.

Therefore:

$$f(x) \sim N\left(\sum_i^m \omega_{it} \mu_{it}, \sum_i^m (\omega_{it} \sigma_{it})^2\right) \qquad (17)$$

From (17), the objective function of the minimum CE optimization problem is set as:

$$\min F = \min D(f_i(x) \| f(x)) \qquad (18)$$

s.t.

$$\begin{cases} 0 \le \omega_{it} \le 1 \\ \sum_{i=1}^m \omega_{it} = 1 \end{cases}$$

Selecting the appropriate weight vector to obtain the minimum F involves determining the support for different algorithms.

The weight coefficient is derived based on the Lagrange function method. The K-L distance can be transformed into a sampling function $g^*(x)$ and $f(x;\omega_{ij})$ to ensure that $-\int g^*(x) \ln f(x;\omega_{it}) dx$ reaches the minimum value, which is equivalent to the maximum value problem:

$$\max \int g^*(x) \ln f(x;\omega_{it}) dx \qquad (19)$$

where:

$$g^*(x) = \frac{I_{\{S(x)>\gamma\}} f(x, \omega_0)}{l} \qquad (20)$$

and $I_{\{S(x)>\gamma\}}$ is called the indicator function:

$$I_{\{S(x)>\gamma\}} = \begin{cases} 1, S(x) > \gamma \\ 0, S(x) \le \gamma \end{cases} \qquad (21)$$

where $S(x)$ is also $f(x;\omega_{ij})$, ω_0 is the initial weight, γ is the target estimation parameter, and L represents the estimated target value of a low probability event.

Based on the idea of CE, a low probability sampling method is used to convert the optimization problem into the following CE problem:

$$\max \overline{D}(\omega_{it}) = \max \frac{1}{N} \sum_{k=1}^N I_{\{S(x)>\gamma\}} \ln f(x_k;\omega_{it}) \qquad (22)$$

where N is a random number of samples.

Note that $\sum_{i=1}^m \omega_{it} = 1$, and thus we can construct a Lagrange function:

$$H(x) = \frac{1}{N} \sum_{k=1}^N I_{\{S(x)>\gamma\}} \ln f(x_k;\omega_{it}) + \lambda \left(\sum_{i=1}^m \omega_{it} - 1\right) \qquad (23)$$

where λ is the Lagrange multiplier.

Note that:

$$\ln f(x_k;\omega_{it}) = -\sum_{i=1}^m \frac{(x_k - \mu_{it})^2}{2\sigma_{it}^2} \ln \frac{\omega_{it}}{\sqrt{2\pi}\sigma_{it}} \qquad (24)$$

By taking the partial derivative to ω_{ij} and λ to zero, we can obtain:

$$\omega_{it} = \sqrt{\frac{\pi}{2}} \frac{\sum_{k=1}^N I_{\{S(x)>\gamma\}} (x_k - \mu_{it})^2}{N\sigma_{it}\lambda} \qquad (25)$$

By substituting this into $\sum\limits_{i=1}^{m} \omega_{it} = 1$, we can obtain:

$$\lambda = \frac{1}{N}\sqrt{\frac{\pi}{2}} \sum_{i=1}^{m} \frac{\sum\limits_{k=1}^{N} I_{\{S(x)>\gamma\}}(x_k - \mu_{it})^2}{N\sigma_{it}} \tag{26}$$

The expression for the weight coefficient is obtained as follows:

$$\omega_{it} = \frac{\sum\limits_{k=1}^{N} I_{\{S(x)>\gamma\}} \frac{(x_k - \mu_{it})^2}{\sigma_{it}}}{\sum\limits_{i=1}^{m} \sum\limits_{k=1}^{N} I_{\{S(x)>\gamma\}} \frac{(x_k - \mu_{it})^2}{\sigma_{it}}} \tag{27}$$

Iterative process:

A. Set $t = 1$;

B. Set $w_{it} = w_0$, set iteration number $z = 1$;

C. Generate sample sequence $X = \{X_1, X_2, \ldots, X_N\}$ by $f(x; \omega_{it})$, and sort it from small to large, calculate $S(x_k) = f(X_k, \omega_{it})$, and thus the estimated value γ is:

$$\gamma(z) = f(X_{1-\rho}; \omega_{it}) \tag{28}$$

D. Calculate (27) and obtain the z-th iteration result $\omega_{it}(z)$. Set $z = z + 1$;

E. Return to Step B to obtain $\gamma(z)$, and calculate $|\gamma(z) - \gamma(z-1)|$. If the results is less than a certain error ε, return to F; otherwise, return to C;

F. Stop the iterations, where $\omega_{it}(z)$ is the optimal weight and the rainfall prediction value is

$$f_t = \sum_{i=1}^{m} \omega_{it}(z)f_{it} \tag{29}$$

G. Set $t = t + 1$. Assess whether t is less than or equal to T. If yes, return to step 2 to calculate some combined forecast values at other times; if not, finish the computation.

The overall forecasting process is shown in Figure 4.

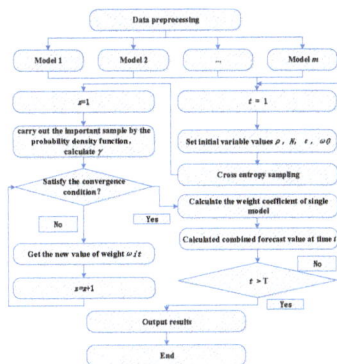

Figure 4. Flowchart illustrating the algorithm.

4. Results and Analysis

This study selects the monthly and annual rainfall data from 1960–2010 as training samples and chooses the rainfall data from 2011–2013 as test samples. Then a monthly rainfall forecast is carried on.

The selected single models include ARIMA, GM, RBF, CE. The performance of the single forecasting model and combined forecasting model is characterized by Root mean squared error (RMSE) and Maximum relative percentage error (MRPE).

4.1. Predictive Stability Comparison Results

After the single algorithm is determined, the rainfall from 2011 to 2013 is predicted and the absolute error is calculated. To compare the stability of various algorithms, true value, predictive value and the absolute error trend curve (a total of 36 prediction points) are shown as follows:

From Figure 5, it is clear that in a single algorithm, the absolute error of prediction is large or small, and stable results cannot be maintained over a long time scale. As a result, the reliability is not high; in combined prediction, the absolute error is relatively flat, which greatly improves the stability of prediction. The prediction results are more authentic. The error analysis of the results of the 2011–2013 is shown in Table 2.

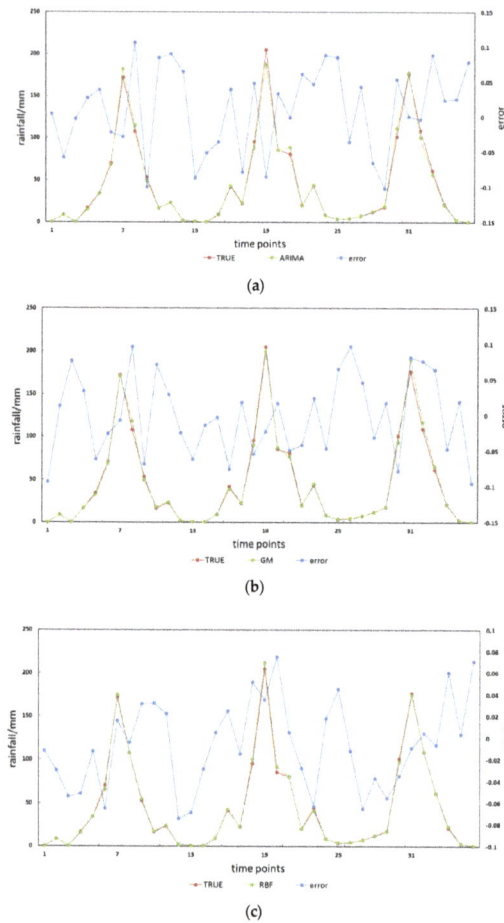

(a)

(b)

(c)

Figure 5. *Cont.*

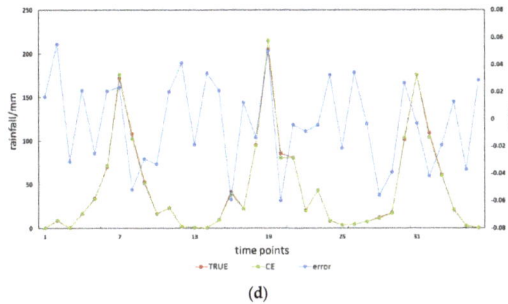

(d)

Figure 5. Prediction curve. (**a**) ARIMA; (**b**) GM; (**c**) RBF; (**d**) CE.

Table 2. Error analysis of the results.

	ARIMA		GM		RBF		CE	
	MRPE	**RMSE**	**MRPE**	**RMSE**	**MRPE**	**RMSE**	**MRPE**	**RMSE**
2011	12.13%	5.53%	11.15%	4.49%	10.99%	4.21%	10.02%	3.93%
2012	9.44%	4.99%	13.15%	4.97%	8.65%	5.01%	9.45%	4.32%
2013	11.10%	7.12%	7.22%	6.19%	7.01%	4.49%	6.88%	4.17%
Average	10.89%	5.88%	10.51%	5.22%	8.88%	4.57%	8.78%	4.14%

From Table 2 and Figure 5 it can be seen that: (1) compared to a single method of RMSE, the combination of forecasting methods in a certain prediction point may not be optimal, but the overall error is small compared with GM, ARIMA, RBF of RMSE, and the error reduces. The combined forecasting method has a higher accuracy: in the average MRPE index, compared to ARIMA, GM, RBF, CE reduces error of 2.11%, 1.73%, 0.1%; in the average RMSE index, compared to ARIMA, GM, RBF, CE were reduced by 1.74%, 1.08%, 0.43%; (2) the MRPE index in a single method can be very high (MRPE of GM in 2012), and there is a risk of failure of the model. The combined forecasting method greatly reduces the maximum error, and has better prediction stability (MRPE of CE in 2012); (3) in a single method, the prediction error sometimes is large, the error of the combined forecast will be also relatively large (for example in 2012), and the accuracy of the single forecasting model has a certain effect on the accuracy of the combined model.

4.2. The Influence of Clustering Method on Prediction Results

To illustrate the effect of clustering on the accuracy of prediction, two scenarios are created:

Scenario 1: Traditional c-means clustering method
Scenario 2: We do not cluster historical data.

The error results are shown in Table 3 and Figure 6.

Table 3. Influence of clustering method on prediction results.

	S1		S2	
	MRPE	**RMSE**	**MRPE**	**RMSE**
2011	10.04%	3.89%	13.01%	7.12%
2012	9.41%	4.34%	12.04%	8.14%
2013	6.92%	4.17%	10.29%	7.71%
Average	8.79%	4.19%	11.78%	7.66%

(a)

(b)

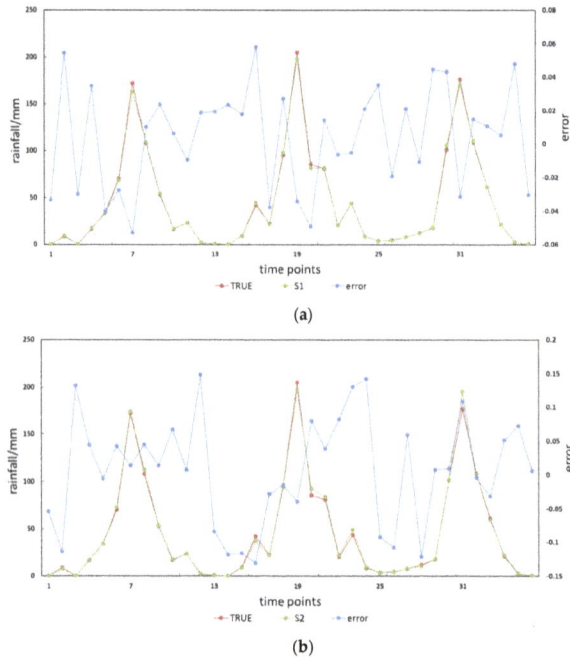

Figure 6. Prediction curve. (a) S1; (b) S2.

From the comparison between Tables 2 and 3, it is easy to know that the clustering method used in this paper is consistent with the accuracy of traditional clustering method to meet the forecasting requirements. In the average MRPE index, compared to S1, S2, the clustering method used in this paper reduces the error by 0.01%, 3%; in the average RMSE index, compared to ARIMA, GM and RBF, CE were reduced by 0.05%, 1.08%, 3.52%. However, from Figure 3, the method has great advantages in calculating the speed when the data volume is large. In addition, comparing with the case of S2, we can see that there is a large deviation in the result of prediction in the case of no clustering, which indicates that the selection of historical data has important influence on the precision of prediction and it is necessary to fully excavate and classify the historical information to reach reasonable effect.

5. Conclusions

Because the meteorological conditions are stochastic, the rainfall variation is a non-stationary time series, and the accuracy of prediction and reliability of traditional single forecasting methods cannot be ensured. The combined forecasting model based on CE is proposed to solve these problems. Besides, the forecasting results of several single methods are used as the input variables for training, and the output weights are predicted by analyzing the total rainfall. The simulation results show that the combined forecasting model enhances the accuracy and reliability of the rainfall forecasting model. Additionally, clustering method is modeled and analyzed so as to improve the accuracy of prediction. The results demonstrate that the accuracy of combined forecasting can be improved by using the fuzzy clustering. The chosen method characterizes the change laws of the rainfall well and to improve the stability of the prediction. The prediction results can help agriculture, water conservancy departments to improve the ability of drought and flood disaster prevention and control. In the future, more steps can be taken for improvement: (1) the single forecasting method with higher accuracy and more suitable single forecasting method should be chosen and the accuracy of combined model prediction should be further improved to explore the rule of selecting different single methods in combined forecasting method; (2) the data in similar year should be collected to predict and historical data should be utilized more effectively so that the result is more scientific and convincing.

Acknowledgments: This work was supported by the National Key R&D Program of China (Grant No. 2016YFC0401406), and the Famous Teachers Cultivation planning for Teaching of North China Electric Power University (the Fourth Period), the 2014 Education Reform Project of North China Electric Power University (Beijing Department) (Grant No. 2014JG57).

Author Contributions: The combined forecasting method based on SVM was developed to forecast the rainfall. The results demonstrate that this model can improve the accuracy of daily rainfall forecasting and it can be applied in practice. Baohui Men designed the study, Rishang Long and Yangsong Li performed the experiments and wrote the paper, Baohui Men reviewed and edited the manuscript. All authors read and approved the manuscript.

Conflicts of Interest: The authors declare no conflict of interest.

References

1. Serreze, M.C.; Etringer, A.J. Precipitation characteristics of the Eurasian Arctic drainage system. *Int. J. Climatol.* **2003**, *23*, 1267–1291. [CrossRef]
2. Bustamante, J. Evaluation of April 1999 Rainfall Fore-casts Over South American using the Eta Model Climanalise, Divulgacao Científica. *Cachoeira Paulista* **1999**, *6*, 3563–3569.
3. Black, T. The New NMC Mesoscale ETA Model: Descriptionand Forecast Examples. *Whether Forecast.* **1994**, *9*, 265–278. [CrossRef]
4. Mossad, A.; Alazba, A.A. Drought Forecasting Using Stochastic Models in a Hyper-Arid Climate. *Atmosphere* **2015**, *6*, 410–430. [CrossRef]
5. Bian, H.J.; Lei, H.J.; Wang, Y. Application of Grey Theory to Regional Rainfall Rorecast. *Anhui Agri. Sci.* **2009**, *37*, 6059–6060.
6. Yang, J.L.; Wu, Y.N.; Xie, M.; Li, H.Y. Application of Monte Carlo Method in Rainfall Forecast of Flood Season in Nenjiang Basin. *S. N. Water Transf. Water Sci. Technol.* **2011**, *9*, 28–32.
7. Ramirez, M.C.V.; de Campos Velho, H.F.; Ferreira, N.J. Artificial Neural Network Technique for Rainfall Forecasting Applied to the Sao Paulo Region. *J. Hydrol.* **2005**, *301*, 146–162. [CrossRef]
8. Manzato, A. Sounding-derived indices for neural network based short-term thunderstorm and rainfall forecasts. *Atmos. Res.* **2007**, *83*, 349–365. [CrossRef]
9. Yang, S.; Rui, J.; Feng, H. Application of Support Vector Machine (SVM) Method in Precipitation Classification Forecast. *J. Southwest. Agric. Univ.* **2006**, *28*, 252–257.
10. Cui, L.; Chi, D.; Qu, X. Based on Wavelet De-noising of Stationary Time Series Analysis Method in Rainfall Forecasting. *China Rural Water Hydropower* **2010**, *9*, 31–35.
11. Merlinde, K. The Application of TAPM for Site Specific Wind Energy Forecasting. *Atmosphere* **2016**, *7*, 23. [CrossRef]
12. Lauret, P.; Lorenz, E.; David, M. Solar Forecasting in a Challenging Insular Context. *Atmosphere* **2016**, *7*, 18. [CrossRef]
13. Men, B.; Liu, C.; Xia, J.; Liu, S.; Lin, Z. Application of R/S Analysis Method of Water Runoff Trend in West Route of South-to-North Water Transfer Project. *J. Glaciol. Geocryol.* **2005**, *27*, 568–573.
14. Soro, G.E.; Noufé, D.; Goula Bi, T.A.; Shorohou, B. Trend Analysis for Extreme Rainfall at Sub-Daily and Daily Timescales in Côte d'Ivoire. *Climate* **2016**, *4*, 37. [CrossRef]
15. Chen, C.-S.; Duan, S.; Cai, T. Short-term photovoltaic generation forecasting system based on fuzzy recognition. *Trans. China Electrotech. Soc.* **2011**, *26*, 83–88.
16. Liu, Y.; Fu, X.; Zhang, W. Short-term load forecasting method based on fuzzy pattern recognition and fuzzy cluster theory. *Trans. China Electrotech. Soc.* **2002**, *17*, 83–86.
17. Kolentini, E.; Sideratos, G.; Rikos, V. Developing a Matlab tool while exploiting neural networks for combined prediction of hour's ahead system load along with irradiation, to estimate the system loadcovered by PV integrated systems. In Proceedings of the IEEE Conferences on Clean Electrical Power, Capri, Italy, 9–11 June 2009; pp. 182–186.
18. Bates, J.; Granger, C. The combination of forecast. *Oper. Res. Q.* **1969**, *20*, 451–468. [CrossRef]

19. Wu, Q.; Peng, C. Wind Power Generation Forecasting Using Least Squares Support Vector Machine Combined with Ensemble Empirical Mode Decomposition, Principal Component Analysis and a Bat Algorithm. *Energies* **2016**, *9*, 261. [CrossRef]

20. Pedersen, J.W.; Lund, N.S.V.; Borup, M.; Löwe, R.; Poulsen, T.S.; Mikkelsen, P.S.; Grum, M. Evaluation of Maximum a Posteriori Estimation as Data Assimilation Method for Forecasting Infiltration-Inflow Affected Urban Runoff with Radar Rainfall Input. *Water* **2016**, *8*, 381. [CrossRef]

21. Box, G.E.P.; Jenkins, G.M. *Time Series Analysis, Forecasting and Control*; Holden-Day: San Francisco, CA, USA, 1976.

22. Men, B.; Long, R.; Zhang, J. Combined Forecasting of Stream flow Based on Cross Entropy. *Entropy* **2016**, *18* 336. [CrossRef]

23. Cui, D. Application of Combined Model in Rainfall Forecast. *Comput. Simul.* **2012**, *29*, 163–166.

24. Xiong, L.; O'Connor, K.M.; Kieran, M. Comparison of four updating models for real-time river flow forecasting. *Hydrol. Sci. J.* **2002**, *47*, 621–640. [CrossRef]

25. Lu, C.; Zhang, Y.; Zhou, J.; Vijay, P.; Guo, S.; Zhang, J. Real-time error correction method combined with combination flood forecasting technique for improving the accuracy of flood forecasting. *J. Hydrol.* **2015**, *521*, 157–169.

26. Singh, V.P. *Entropy Based Parameter Estimation in Hydrology*; Kluwer Academic Publishers: Dordrecht, The Netherlands, 1998.

27. Cui, H.; Singh, V.P. Entropy spectral analyses for groundwater forecasting. *J. Hydrol. Eng.* **2017**, *22*, 06017002. [CrossRef]

28. Chen, L.; Singh, V.P.; Xiong, F. An Entropy-Based Generalized Gamma Distribution for Flood Frequency Analysis. *Entropy* **2017**, *19*, 239. [CrossRef]

29. Chen, L.; Singh, V.P. Generalized Beta Distribution of the Second Kind for Flood Frequency Analysis. *Entropy* **2017**, *19*, 254. [CrossRef]

30. Chen, L.; Singh, V.P.; Guo, S.L.; Zhou, J.; Ye, L. Copula entropy coupled with artificial neural network for rainfall-runoff simulation. *Stoch. Environ. Res. Risk Assess.* **2014**, *28*, 1755–1767. [CrossRef]

31. Li, R.; Liu, H.L.; Lu, Y.; Han, B. A combination method for distribution transformer life prediction based on cross entropy theory. *Power Syst. Prot. Control* **2014**, *42*, 97–101.

32. Chen, N.; Sha, Q.; Tang, Y.; Zhu, L. A Combination Method for Wind Power Predication Based on Cross Entropy Theory. *Proc. CSEE* **2012**, *32*, 29–34.

33. Mehdi, N.; Hossein, N. Image denoising in the wavelet domain using a new adaptive thresholding function. *Neurocomputing* **2009**, *72*, 1012–1025.

34. Asgari, M.S.; Abbasi, A. Comparison of ANFIS and FAHP-FGP methods for supplier selection. *Kybernetes* **2016**, *45*, 474–489. [CrossRef]

entropy

MDPI

Article

Investigation into Multi-Temporal Scale Complexity of Streamflows and Water Levels in the Poyang Lake Basin, China

Feng Huang [1,2,*], **Xunzhou Chunyu** [1,2], **Yuankun Wang** [3], **Yao Wu** [2,4], **Bao Qian** [5], **Lidan Guo** [6], **Dayong Zhao** [1,2] **and Ziqiang Xia** [1,2]

1   State Key Laboratory of Hydrology-Water Resources and Hydraulic Engineering, Hohai University, Nanjing 210098, China; zodiacnix@163.com (X.C.); dyzhao@hhu.edu.cn (D.Z.); zqxia@hhu.edu.cn (Z.X.)
2   College of Hydrology and Water Resources, Hohai University, Nanjing 210098, China; wuyao@jxsl.gov.cn
3   School of Earth Sciences and Engineering, Nanjing University, Nanjing 210023, China; yuankunw@nju.edu.cn
4   Poyang Lake Hydro Project Construction Office of Jiangxi Province, Nanchang 330046, China
5   Bureau of Hydrology, Changjiang River Water Resources Commission, Wuhan 430012, China; jacber@163.com
6   International River Research Centre, Hohai University, Nanjing 210098, China; ldguohhu@163.com
*   Correspondence: hfeng0216@163.com; Tel./Fax: +86-25-8378-6845

Academic Editors: Huijuan Cui, Bellie Sivakumar and Vijay P. Singh
Received: 24 December 2016; Accepted: 9 February 2017; Published: 10 February 2017

Abstract: The streamflow and water level complexity of the Poyang Lake basin has been investigated over multiple time-scales using daily observations of the water level and streamflow spanning from 1954 through 2013. The composite multiscale sample entropy was applied to measure the complexity and the Mann-Kendall algorithm was applied to detect the temporal changes in the complexity. The results show that the streamflow and water level complexity increases as the time-scale increases. The sample entropy of the streamflow increases when the time-scale increases from a daily to a seasonal scale, also the sample entropy of the water level increases when the time-scale increases from a daily to a monthly scale. The water outflows of Poyang Lake, which is impacted mainly by the inflow processes, lake regulation, and the streamflow processes of the Yangtze River, is more complex than the water inflows. The streamflow and water level complexity over most of the time-scales, between the daily and monthly scales, is dominated by the increasing trend. This indicates the enhanced randomness, disorderliness, and irregularity of the streamflows and water levels. This investigation can help provide a better understanding to the hydrological features of large freshwater lakes. Ongoing research will be made to analyze and understand the mechanisms of the streamflow and water level complexity changes within the context of climate change and anthropogenic activities.

Keywords: complexity; streamflow; water level; composite multiscale sample entropy; trend; Poyang Lake basin

1. Introduction

The complexity of a hydrological time series, e.g., precipitation, streamflow, and water level means that there is a degree of uncertainty, randomness, or irregularity of the time series. Many researchers have studied the hydrological complexity [1–5]. Chou analyzed the complexity of the correlation between rainfall and runoff, using the multiscale entropy approach, and found that: (1) the entropy measures of rainfall are higher than those of runoff at all scale factors; (2) the entropy measures of the runoff coefficient series lie between the entropy measures of the rainfall and runoff at various scale factors; and (3) the entropy values of rainfall, runoff and runoff coefficient series increase as scale factors increase. Liu et al. studied the complexity features of regional groundwater depth and found that human activities are the main driving force, causing the complexity of regional groundwater depth [6]. However, there seems to have been few attempts made to investigate the hydrological complexities of large lakes. Therefore, this study looks to the Poyang Lake basin as the study area and investigates its streamflow and water level complexity.

Poyang Lake, which is located in the Yangtze River basin, is the largest freshwater lake in China. The lake is the major body of water in an important global ecoregion, and plays an important role in maintaining the water resources of the Yangtze River and a healthy aquatic ecosystem in the region [7]. The hydrological processes of Poyang Lake are essential to the lacustrine and wetland ecosystems; and the disturbance of hydrological processes may break the longstanding ecological balance and influence the distinctive biodiversity have in this region [8]. The hydrological changes of Poyang Lake have attracted worldwide attention [9–11]. The Poyang Lake basin has experienced six extreme droughts during the past 60 years, thus resulting in the reduction of water resources from the five tributaries flowing into Poyang Lake [12]. The Three Gorges Reservoir has changed the Yangtze River streamflow and has further impacted the interrelationship between the Yangtze River and Poyang Lake [13]. Since the start of operations of the Three Gorges Reservoir, the seasonal water level of Poyang Lake has had a great magnitude of fluctuation [14].

Based on these previous studies, the present study investigates the hydrological changes of Poyang Lake from another perspective: hydrological complexity. Using the streamflow and water level data in the Poyang Lake basin from 1954 to 2013, this study aims to accomplish the following objectives: to investigate the streamflow and water level complexity over various time-scales; and secondly to detect the temporal changes in the streamflow and water level complexity. The first objective will reveal the streamflow and water level complexity changes versus the time-scale changes. Will the complexity increase, keep stable or decrease when the time-scale increases from a daily scale to a seasonal scale? The second objective will reveal the streamflow and water level complexity changes in the recent 60 years. The 1954–2013 streamflows and water levels are divided into 51 subseries. The complexity of each subseries is analyzed for a specific time-scale, and a complexity series can be obtained. The trends in the complexity series are further investigated to reveal what the temporal changes are. The results will add an understanding to the hydrological features and the changes of Poyang Lake and will also provide scientific references for other similar large freshwater lakes.

2. Study Area and Data

Poyang Lake is located in the Jiangxi Province in the middle to lower reaches of the Yangtze River. It receives inflows from five main tributaries: the Ganjiang, Fuhe, Xinjiang, Raohe, and Xiushui rivers. The lake exchanges water with the Yangtze River through a narrow channel in the north (Figure 1). Six hydrometric stations (Table 1) monitor the inflow processes from the tributaries, and one hydrometric station (Hukou) monitors the water exchanges of the Poyang Lake and the Yangtze River. The streamflows of the Hukou station have both positive and negative values. Recorded positive values are when the lake discharges into the river. When the river water discharges into the lake, the streamflow data of the Hukou station are recorded as negative values. Because the streamflow data is dominated by the positive values, the streamflow processes at the Hukou station are described as the outflow processes of Poyang Lake. Four hydrometric stations monitor the water levels of Poyang Lake. The hydrological data are daily recorded and the period of all hydrological data spans from 1954 through 2013.

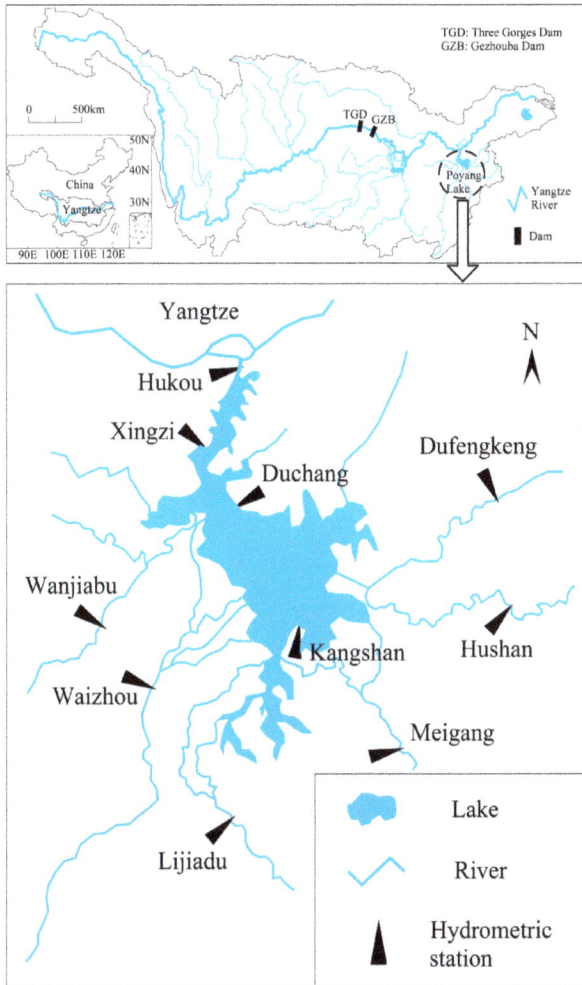

Figure 1. Locations of Poyang Lake and the hydrometric stations.

Table 1. Hydrometric stations in the Poyang Lake basin.

Hydrometric Station	Data Type	Location	Longitude and Latitude	Drainage Area (km^2)
Waizhou	Streamflow	Ganjiang River	[115°50′E, 28°38′N]	80,948
Lijiadu	Streamflow	Fuhe River	[116°10′E, 28°13′N]	15,811
Meigang	Streamflow	Xinjiang River	[116°49′E, 28°26′N]	15,535
Hushan	Streamflow	Raohe River	[117°16′E, 28°55′N]	6374
Dufengkeng	Streamflow	Raohe River	[117°12′E, 29°16′N]	5013
Wanjiabu	Streamflow	Xiushui River	[115°39′E, 28°51′N]	3548
Kangshan	Water level	Poyang Lake	[116°25′E, 28°53′N]	/
Duchang	Water level	Poyang Lake	[116°11′E, 29°15′N]	/
Xingzi	Water level	Poyang Lake	[116°2′E, 29°27′N]	/
Hukou	Streamflow, Water level	Poyang Lake	[116°13′E, 29°45′N]	162,225

3. Methodology

3.1. Analysis Procedure

The streamflow and water level complexity are measured by sample entropy. Composite multiscale sample entropy is applied to reveal the streamflow and water level complexity over various time-scales. Both static and dynamic sample entropies over various time-scales are calculated to investigate the multiscale complexity [15]. The static sample entropy reveals different complexities over various time-scales, and the dynamic sample entropy reveals temporal complexity trends over various time-scales. First, the static sample entropy is analyzed, which ignores the complexity changes over time. The composite multiscale sample entropy is calculated using the entire hydrological data records. Second, the dynamic sample entropy is analyzed, which reflects the complexity changes over time. To obtain how the streamflow and water level complexity changed over time, a ten year sliding window was selected for the hydrological data, which divides the original entire hydrological data records. Thus, the 1954–2013 series is divided into 51 subseries: 1954–1963, 1955–1964, 1956–1965, ..., 2002–2011, 2003–2012, and 2004–2013. The composite multiscale sample entropy is calculated using each subseries, respectively. A sample entropy series can be obtained for a time-scale, which reflects the complexity changes over time. Temporal trends in the sample entropy series are further analyzed applying the Mann-Kendall algorithm.

3.2. Sample Entropy

The hydrological complexity means the degree of uncertainty or the rate of information production of the hydrological series, e.g., the streamflow and the water level. Techniques for measuring the hydrological complexity typically involve the calculations of: the Lyapunov exponent, correlation dimension, fractal dimension, Kolmogorov-Sinai entropy, spectral entropy, approximate entropy, and sample entropy [5]. The approximate entropy calculation solves the problem of insufficient number of data points, and is applicable to noisy, medium-sized data sets. However, it lacks relative consistency and its results depend on data length. Improved on the basis of the approximate entropy calculation, the sample entropy calculation is an unbiased estimation of the conditional probability that two similar sequences of m consecutive data points (m is the embedded dimension) will remain similar when one more consecutive point is included [16]. It is largely independent of data length and keeps relative consistency without counting self-matches [17]. For a time series of N points $\{x(i), i = 1, 2, \ldots, N\}$, the sample entropy (*SampEn*) is calculated by the following steps:

(1) Constitute vectors of m dimensions:

$$X(i) = \{x(i), x(i+1), \ldots, x(i+m-1)\} \quad (i = 1, 2, \ldots, N-m+1) \tag{1}$$

(2) Define the Euclidean distance between $X(i)$ and $X(j)$, $d[X(i), X(j)]$ as the maximum absolute difference of their corresponding scalar components:

$$d[X(i), X(j)] = \underset{k=0 \to m-1}{\text{Max}} \{|X(i+k) - X(j+k)|\} \tag{2}$$

(3) Take n_i^m as the number of sequences in the time series that match (without self-matching) the template with the length m within the tolerance criterion r. Then, define $C_i^m(r)$ and $C^m(r)$ separately as the following equations:

$$C_i^m(r) = \frac{n_i^m}{N-m} \tag{3}$$

$$C^m(r) = \frac{1}{N-m+1} \sum_{i=1}^{N-m+1} C_i^m(r) \tag{4}$$

(4) Change the dimension of the vector $X(i)$ to $m+1$ and calculate $C^{m+1}(r)$ similarly.

(5) Finally, *SampEn* is defined as:

$$SampEn = -\ln \frac{C^{m+1}(r)}{C^m(r)}$$ (5)

When calculating the sample entropy, the template length m is set to be 2, and the tolerance criterion r is set to be 0.15σ, where σ denotes the standard deviation of the original time series.

3.3. Composite Multiscale Sample Entropy

To completely measure the underlying dynamics of complex systems under consideration, displaying their disorderliness over multiple time-scales, Costa et al. proposed the multiscale entropy algorithm. This algorithm calculates the sample entropy of a time series over various time-scales [18]. Although the multiscale entropy algorithm has been successfully applied in a number of different fields (including the analysis of the human gait dynamics, heart rate variability, rainfall, and river streamflow), it encounters a problem in that the statistical reliability of the sample entropy of the coarse-grained series is reduced as the time-scale is increased. To overcome this limitation, Wu et al. proposed the concept of composite multiscale sample entropy, which better presents data in both simulation and real world data analysis [19].

The principles and calculation procedures of the composite multiscale sample entropy can be found in Wu et al. [19]. When the sample entropy is calculated with the template length $m = 2$, there are two and three coarse-grained time series divided from the original time series for scale factors of 2 and 3, respectively. The kth coarse-grained time series for a scale factor of τ, $y_k^{(\tau)}$ is defined as:

$$y_{k,j}^{(\tau)} = \frac{1}{\tau} \sum_{i=(j-1)\tau+k}^{j\tau+k-1} x_i \quad (1 \le j \le \tfrac{N}{\tau}, 1 \le k \le \tau)$$ (6)

At a scale factor of τ, the sample entropy of all coarse-grained time series are calculated and the composite multiscale sample entropy (*CMSE*) value is defined as the means of τ entropy values:

$$CMSE(x, \tau, m, r) = \frac{1}{\tau} \sum_{k=1}^{\tau} SampEn(y_k^{(\tau)}, m, r)$$ (7)

3.4. Mann-Kendall Algorithm

The trend detection methods include Spearman's rho test, Mann-Kendall test, seasonal Kendall test, linear regression test, and so on [20]. One of the most favored trend detection methods is the Mann-Kendall algorithm, which has been employed to detect the trends in hydrological series, including precipitation [21], runoff [22], sample entropy [5], coefficient of variation, and concentration degree [23]. Detailed information on the Mann-Kendall algorithm can be found in the published papers [24]. The Mann-Kendall test is based on the test statistic S:

$$S = \sum_{i=1}^{n-1} \sum_{j=i+1}^{n} \text{sgn}(x_j - x_i)$$ (8)

where the x_j and x_i are the sequential data values, n is the length of the data set, and:

$$\text{sgn}(\theta) = \begin{cases} 1 & if \quad \theta > 0 \\ 0 & if \quad \theta = 0 \\ -1 & if \quad \theta < 0 \end{cases}$$ (9)

When $n \ge 8$, the statistic S is approximately normally distributed with the mean and the variance as follows:

$$E(S) = 0 \tag{10}$$

$$Var(S) = \frac{n(n-1)(2n+5) - \sum\limits_{i=1}^{n} t_i i(i-1)(2i+5)}{18} \tag{11}$$

where t_i is the number of ties of extent i. The standardized test statistic Z is computed by

$$Z = \begin{cases} \frac{S-1}{\sqrt{Var(S)}} & S > 0 \\ 0 & S = 0 \\ \frac{S+1}{\sqrt{Var(S)}} & S < 0 \end{cases} \tag{12}$$

Z follows the standard normal distribution with a mean of zero and variance of one. A positive or negative value of Z represents an increasing or decreasing trend, respectively. In a two-tailed test, the null hypothesis is no trend, which can be rejected at significance level α if $|Z| > Z_{\alpha/2}$. The $Z_{\alpha/2}$ is the value of the standard normal distribution with an exceedance probability of $\alpha/2$. The significance level α is set to be 0.05 in this study.

4. Results and Discussion

4.1. Multiscale Complexity of Streamflows

Figure 2 displays the composite multiscale sample entropy of the inflow and outflow data of Poyang Lake, respectively. The variations of sample entropy versus time-scale are similar. The sample entropy is increasing when the time-scale is increasing, revealing that the time series have obvious self-similarity and great complexity [25]. Similar results, i.e., the complexity of streamflows is increasing as the time-scale is increasing, have also been found in the multiscale entropy analysis of the streamflows of the Mississippi River in the United States [4], the East River in China [3], and the Yangtze River in China [2]. These previous studies attribute this phenomenon to the existence of the long-range correlation of the streamflow records. Larger sample entropy reflects more randomness and complicated systems, and vice versa. When the time-scale is increasing, the increasing sample entropy indicates an increase in the uncertainty and disorderliness, and a lessened regularity and predictability.

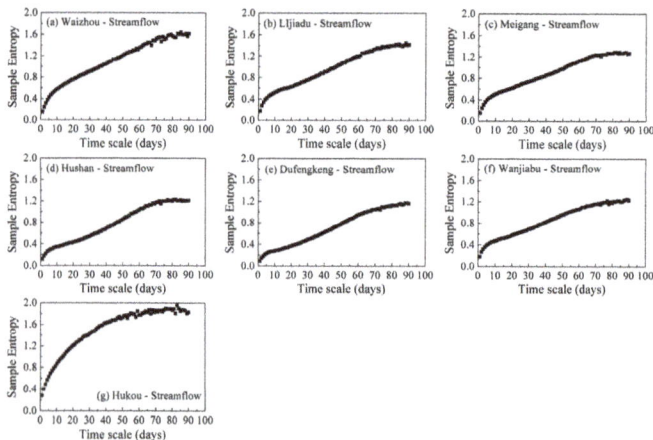

Figure 2. Composite multiscale sample entropy of the streamflows of Poyang Lake.

When comparing Figures 2a–f and 2g, it can be observed that the sample entropy values at different time-scales of the outflows are distributed above those of the inflows. When the sample entropy values of one time series are larger than those of another time series at most time-scales, it reveals that the former is more complex than the latter [26]. Figure 3 compares the average sample entropy of the inflows and outflows of Poyang Lake. The average sample entropy is the mean value of sample entropies at various time-scales. The average sample entropy of the streamflow data at the Hukou station is the largest in Figure 3, indicating that the outflows are more complex than the inflows of Poyang Lake. Due to the spatial and temporal distribution of precipitation, the Yangtze River and Poyang Lake have complex river-lake interactions. The streamflow processes at the Hukou station are determined by the water level in Poyang Lake (relative to the water level in the Yangtze River), which in turn is affected by both the inflow processes and the outflow processes [27]. Poyang Lake acts as a buffer at varying degrees for the Yangtze River streamflow. When the water level of the Yangtze River rises during the flood season, the Poyang Lake helps to absorb some flood and mitigates the peak streamflow of the Yangtze River [13]; hence, the streamflow processes at the Hukou station are more complex than the streamflow processes at the hydrometric stations of the tributaries.

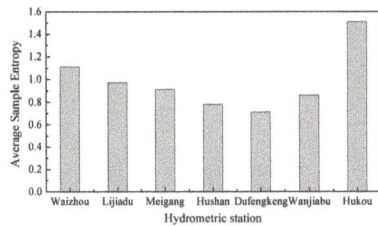

Figure 3. Average sample entropy of the streamflows of Poyang Lake.

4.2. Multiscale Complexity of Water Levels

Figure 4 displays the composite multiscale sample entropy of the water levels of Poyang Lake. The water levels, monitored at the Kangshan, Duchang, Xingzi, and Hukou stations, have similar change patterns in term of sample entropy versus time-scale. The sample entropy increases rapidly until the time-scale reaches about 30 days, after that, the increase becomes small. The sample entropy becomes relatively stable when the time-scale reaches about 30 days. The randomness and irregularity of the water levels are greater with the increased time-scale from a daily to a monthly scale. The water levels have a relatively stable complexity when the time-scale is larger than 30 days, indicating relatively stable regularity and predictability. The water level of Poyang Lake is an integrated response to the inflow, outflow, precipitation, evaporation, leakage, water withdraw, and topography of the lake. Hence, the complexity of the water levels may not be identical to that of the inflows and outflows.

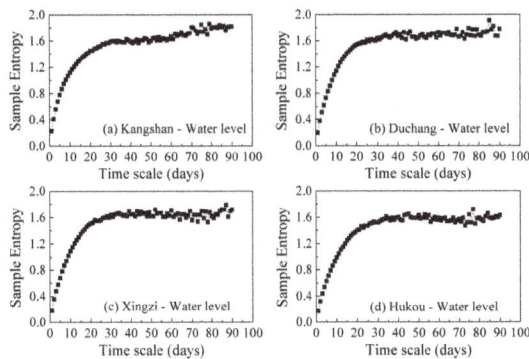

Figure 4. Composite multiscale sample entropy of the water levels of Poyang Lake.

4.3. Temporal Changes in Complexity of Streamflows and Water Levels

Figure 5 shows the statistical Z values of the Mann-Kendall algorithm, which reveals temporal trends in the complexity of the inflows and outflows of Poyang Lake. Figure 6 displays temporal variations in the complexity of the daily streamflows as examples. The sample entropy series of various time-scales at the Waizhou, Lijiadu, Meigang, Hushan, and Wanjiabu stations are dominated by increasing trends, indicating the increased inflow complexity of Poyang Lake. The sample entropy series of various time-scales at the Dufengkeng station are relatively stable. Only a significant decreasing trend is detected at the daily scale. The absolute Z values of the other time-scales are below the 0.05 significance level, indicating no significant trend. The sample entropy series over different time-scales at the Hukou station are also featured by increasing trends, indicating the increased outflow complexity of Poyang Lake.

Figure 7 shows temporal trends in the multiscale complexity of the water levels of Poyang Lake. Figure 8 displays temporal variations in the complexity of the daily water levels as examples. The sample entropy series over different time-scales are characterized by increasing tends, indicating the greater irregularity and randomness of the water levels.

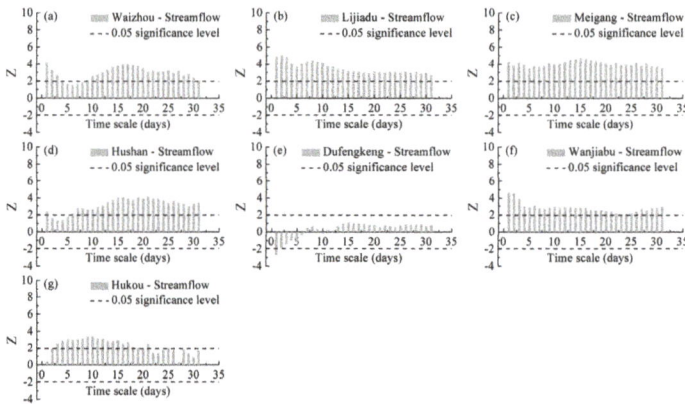

Figure 5. Trends in multiscale complexity of the streamflows of Poyang Lake.

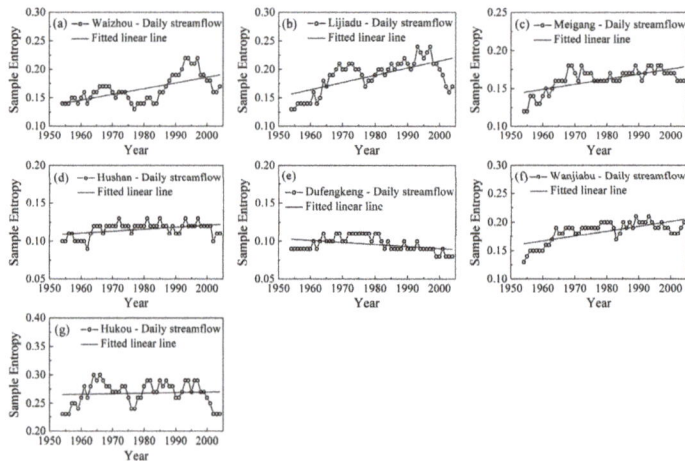

Figure 6. Temporal changes in complexity of the daily streamflows of Poyang Lake.

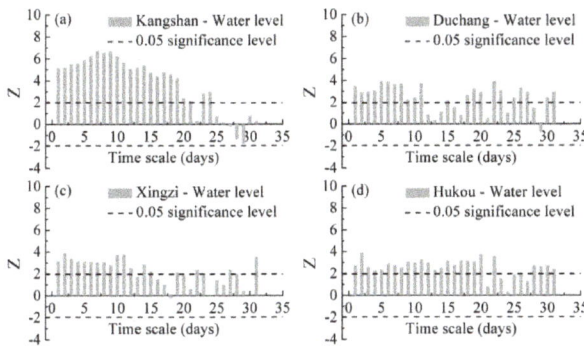

Figure 7. Trends in multiscale complexity of the water levels of Poyang Lake.

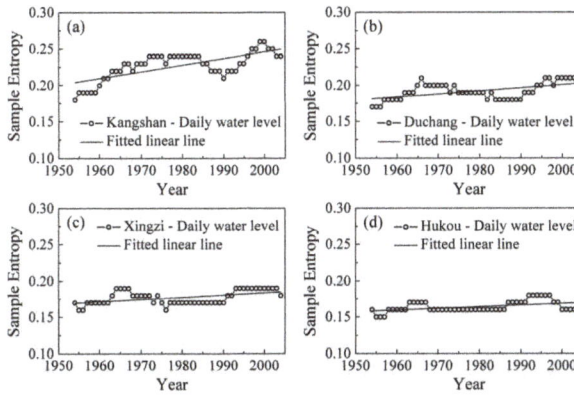

Figure 8. Temporal changes in complexity of the daily water levels of Poyang Lake.

4.4. Discussion on Temporal Changes in Streamflow and Water Level Complexity

Hydrological systems are open, dynamic, complex, giant, and nonlinear compound systems. For the Poyang Lake basin, the outputs of this hydrological system, which are described by the streamflows and the water levels, are highly nonlinear and complex, and are significantly impacted by both climate change and anthropogenic activities. Precipitation is a dominant impact factor of the Poyang Lake hydrological system. In the Jiangxi province, Huang et al. found remarkable differences among the meteorological stations with negative and positive precipitation trends at the annual, seasonal, and monthly scales. Significant increasing trends are mainly found during winter and summer, while significant decreasing trends are mostly observed during autumn [21]. Xiao et al. analyzed the spatial and temporal characteristics of rainfall across the Ganjiang River basin, the largest sub-basin of the Poyang Lake basin, and found significant increasing trends in the annual total rainfall amount [28]. Temperature is another relevant factor which directly affects evaporation. According to a study by Tao et al. of the Poyang Lake basin, the annual mean of the daily minimum temperature has increased significantly, while no significant trend has been detected in the annual mean of daily maximum temperature; thus, resulting in a significant decrease in the diurnal temperature range [29]. Ye et al. detected a significant decreasing trend in the annual reference evapotranspiration in the Poyang Lake basin [30]. Based on prior studies, more attention will be paid in future research to elucidate the complexity features of precipitation, temperature, evaporation, and the mechanisms on how the complexity of those impact factors affect the complexity of the outputs of the Poyang Lake hydrological system.

Including the climate factors, anthropogenic activities are also very important factors affecting the hydrological processes. The anthropogenic activities mainly include dam and reservoir construction, water withdraw, land use, and land cover changes, along with sand extraction from river and lake beds. Mei et al. argued that the average contributions of precipitation variation, human activities in the Poyang Lake catchment, and the Three Gorges Reservoir regulation to the Poyang Lake recession can be quantified as 39.1%, 4.6% and 56.3%, respectively [31]. Zhang et al. found that human-induced and climate-induced influences on streamflows are different in the five Poyang Lake sub-basins. Climate change is the major driving factor for the streamflow increases within the Ganjiang, Xinjiang, and Raohe River basins; however, anthropogenic activities are the principal driving factors for the streamflow increase of the Xiushui River basin and for the streamflow decrease of the Fuhe River basin [32].

Dam and reservoir construction is one of the most important anthropogenic activities, and has attracted worldwide attention due to its significant influence on hydrological processes [33–37]. In the Poyang Lake basin, thousands of reservoirs have been constructed, including 27 large reservoirs ($>10^8$ m^3 in storage capacity) [11]. The reservoirs in the Poyang Lake basin may affect the Poyang Lake hydrological complexity. The reservoirs in the upper Yangtze River basin, especially the Three Gorges Reservoir, may also affect the Poyang Lake hydrological complexity through disturbing the natural river-lake interrelationships. Some interesting conclusions have been obtained in terms of the impact of dam and reservoir construction on hydrological complexity. Huang et al. attributed the loss of streamflow complexity of the upper reaches of the Yangtze River to the underlying surface condition change, which has been influenced by human activities, especially reservoir construction. Huang et al. also argued that the reservoir operation makes the streamflow more regular and self-similar, leading to the streamflow complexity loss [5]. Similar conclusions were obtained in the hydrological complexity analysis of the Colorado River in the United States [38] and the Sao Francisco River in Brazil [39]. Those studies elucidated that dam and reservoir construction induced significant changes in streamflow dynamics, including an increase in regularity and a loss of complexity. However, the study in the East River (China) obtained the opposite conclusion. Zhou et al. argued that reservoir construction greatly increases the complexity of hydrological processes because of reservoir-induced noise of the streamflow [3]. More research is needed to systematically analyze the impact of reservoirs on hydrological complexity.

The enhancement of the streamflow and water level complexity in the Poyang Lake basin is the combined result of climate change and anthropogenic activities. Explaining how climate change and anthropogenic activities affect the hydrological complexity while at the same time distinguishing their individual contributions are our ongoing research objectives.

5. Conclusions

Based on the long term observed daily streamflow and water level data of Poyang Lake, the streamflow and water level complexity over various time-scales and its temporal changes are investigated using the composite multiscale sample entropy and the Mann-Kendall algorithm. The following conclusions can be drawn from the analysis:

(1) The streamflow and water level complexity increases when the time-scale increases. The sample entropy of the streamflows increases when the time-scale increases from the daily to seasonal scale. The sample entropy of the water levels increases when the time-scale increases from the daily to monthly scale.

(2) The complexity of the outflows is greater than that of the inflows. It may be caused by the complex river-lake interrelationships. The outflow processes of Poyang Lake are synthetically impacted by the inflow processes, lake regulation, and the streamflow processes of the Yangtze River.

(3) Significant upward trends can be detected in the sample entropy series, which are calculated using the streamflow and water level data, for most time-scales between the daily to monthly scale. The increased sample entropy indicates the enhanced streamflow and water level complexity, which may be caused by both climate change and anthropogenic activities. The mechanisms of the hydrological complexity changes will be studied in ongoing research.

Acknowledgments: This work is supported by the National Natural Science Foundation Projects of China (grant numbers 41401011, 51309131, 51679118, 41401010, and 41371098); Science and Technology Projects of Water Resources Department of Jiangxi Province (grant number KT201538); CRSRI Open Research Program (grant number CKWV2015237/KY); and Program for Changjiang Scholars and Innovative Research Team in University (grant number IRT13062). The authors acknowledge constructive comments from the editor and anonymous reviewers, which lead to improvement of the paper.

Author Contributions: The authors designed and performed the research together. Feng Huang wrote the draft of the paper. Xunzhou Chunyu, Yuankun Wang, Yao Wu, Bao Qian, Lidan Guo, Dayong Zhao and Ziqiang Xia made some comments and corrections. All authors have read and approved the final manuscript.

Conflicts of Interest: The authors declare no conflict of interest.

References

1. Sang, Y.F.; Wang, D.; Wu, J.C.; Zhu, Q.P.; Wang, L. Wavelet-Based Analysis on the Complexity of Hydrologic Series Data under Multi-Temporal Scales. *Entropy* **2011**, *13*, 195–210. [CrossRef]
2. Zhang, Q.; Zhou, Y.; Singh, V.P.; Chen, X.H. The influence of dam and lakes on the Yangtze River streamflow: long-range correlation and complexity analyses. *Hydrol. Process.* **2012**, *26*, 436–444. [CrossRef]
3. Zhou, Y.; Zhang, Q.; Li, K.; Chen, X.H. Hydrological effects of water reservoirs on hydrological processes in the East River (China) basin: Complexity evaluations based on the multi-scale entropy analysis. *Hydrol. Process.* **2012**, *26*, 3253–3262. [CrossRef]
4. Li, Z.W.; Zhang, Y.K. Multi-scale entropy analysis of mississippi river flow. *Stoch. Environ. Res. Risk Assess.* **2008**, *22*, 507–512. [CrossRef]
5. Huang, F.; Xia, Z.Q.; Zhang, N.; Zhang, Y.D.; Li, J. Flow-Complexity Analysis of the Upper Reaches of the Yangtze River, China. *J. Hydrol. Eng.* **2011**, *16*, 914–919. [CrossRef]
6. Liu, M.; Liu, D.; Liu, L. Complexity research of regional groundwater depth series based on multiscale entropy: a case study of Jiangsanjiang Branch Bureau in China. *Environ. Earth Sci.* **2013**, *70*, 353–361. [CrossRef]
7. Lai, X.J.; Liang, Q.H.; Jiang, J.H.; Huang, Q. Impoundment Effects of the Three-Gorges-Dam on Flow Regimes in Two China's Largest Freshwater Lakes. *Water Resour. Manag.* **2014**, *28*, 5111–5124. [CrossRef]
8. Cao, L.; Fox, A.D. Birds and people both depend on China's wetlands. *Nature* **2009**, *460*, 173. [CrossRef] [PubMed]
9. Zhao, G.J.; Hoermann, G.; Fohrer, N.; Zhang, Z.X.; Zhai, J.Q. Streamflow Trends and Climate Variability Impacts in Poyang Lake Basin, China. *Water Resour. Manag.* **2010**, *24*, 689–706. [CrossRef]
10. Ye, X.C.; Zhang, Q.; Liu, J.; Li, X.H.; Xu, C.Y. Distinguishing the relative impacts of climate change and human activities on variation of streamflow in the Poyang Lake catchment, China. *J. Hydrol.* **2013**, *494*, 83–95. [CrossRef]
11. Lai, X.J.; Huang, Q.; Zhang, Y.H.; Jiang, J.H. Impact of lake inflow and the Yangtze River flow alterations on water levels in Poyang Lake, China. *Lake Reserv. Manag.* **2014**, *30*, 321–330. [CrossRef]
12. Zhang, Z.X.; Chen, X.; Xu, C.Y.; Hong, Y.; Hardy, J.; Sun, Z.H. Examining the influence of river-lake interaction on the drought and water resources in the Poyang Lake basin. *J. Hydrol.* **2015**, *522*, 510–521. [CrossRef]
13. Guo, H.; Hu, Q.; Zhang, Q.; Feng, S. Effects of the Three Gorges Dam on Yangtze River flow and river interaction with Poyang Lake, China: 2003–2008. *J. Hydrol.* **2012**, *416*, 19–27. [CrossRef]
14. Dai, X.; Wan, R.R.; Yang, G.S. Non-stationary water-level fluctuation in China's Poyang Lake and its interactions with Yangtze River. *J. Geogr. Sci.* **2015**, *25*, 274–288. [CrossRef]
15. Chou, C.M. Complexity analysis of rainfall and runoff time series based on sample entropy in different temporal scales. *Stoch. Environ. Res. Risk Assess.* **2014**, *28*, 1401–1408. [CrossRef]

16. Costa, M.; Peng, C.K.; Goldberger, A.L.; Hausdorff, J.M. Multiscale entropy analysis of human gait dynamics. *Physica A* **2003**, *330*, 53–60. [CrossRef]

17. Richman, J.S.; Moorman, J.R. Physiological time-series analysis using approximate entropy and sample entropy. *Am. J. Physiol. Heart Circ. Physiol.* **2000**, *278*, H2039–H2049. [PubMed]

18. Costa, M.; Goldberger, A.L.; Peng, C.K. Multiscale entropy analysis of complex physiologic time series. *Phys. Rev. Lett.* **2002**, *89*, 068102. [CrossRef] [PubMed]

19. Wu, S.D.; Wu, C.W.; Lin, S.G.; Wang, C.C.; Lee, K.Y. Time Series Analysis Using Composite Multiscale Entropy. *Entropy* **2013**, *15*, 1069–1084. [CrossRef]

20. Kundzewicz, Z.W.; Robson, A.J. Change detection in hydrological records—A review of the methodology. *Hydrol. Sci. J.* **2004**, *49*, 7–19. [CrossRef]

21. Huang, J.; Sun, S.L.; Zhang, J.C. Detection of trends in precipitation during 1960–2008 in Jiangxi province, southeast China. *Theor. Appl. Clim.* **2013**, *114*, 237–251. [CrossRef]

22. Zhao, Q.H.; Liu, S.L.; Deng, L.; Dong, S.K.; Yang, J.J.; Wang, C. The effects of dam construction and precipitation variability on hydrologic alteration in the Lancang River Basin of southwest China. *Stoch. Environ. Res. Risk Assess.* **2012**, *26*, 993–1011. [CrossRef]

23. Huang, F.; Xia, Z.Q.; Li, F.; Guo, L.D.; Yang, F.C. Hydrological Changes of the Irtysh River and the Possible Causes. *Water Resour. Manag.* **2012**, *26*, 3195–3208. [CrossRef]

24. Yue, S.; Pilon, P.; Cavadias, G. Power of the Mann-Kendall and Spearman's rho tests for detecting monotonic trends in hydrological series. *J. Hydrol.* **2002**, *259*, 254–271. [CrossRef]

25. Chou, C.M. Applying Multiscale Entropy to the Complexity Analysis of Rainfall-Runoff Relationships. *Entropy* **2012**, *14*, 945–957. [CrossRef]

26. Costa, M.; Goldberger, A.L.; Peng, C.K. Multiscale entropy analysis of biological signals. *Phys. Rev. E* **2005**, *71*, 021906. [CrossRef] [PubMed]

27. Zhang, Q.; Li, L.; Wang, Y.G.; Werner, A.D.; Xin, P.; Jiang, T.; Barry, D.A. Has the Three-Gorges Dam made the Poyang Lake wetlands wetter and drier? *Geophys. Res. Lett.* **2012**, *39*, L20402. [CrossRef]

28. Xiao, Y.; Zhang, X.; Wan, H.; Wang, Y.Q.; Liu, C.; Xia, J. Spatial and temporal characteristics of rainfall across Ganjiang River Basin in China. *Meteorol. Atmos. Phys.* **2016**, *128*, 167–179. [CrossRef]

29. Tao, H.; Fraedrich, K.; Menz, C.; Zhai, J.Q. Trends in extreme temperature indices in the Poyang Lake Basin, China. *Stoch. Environ. Res. Risk Assess.* **2014**, *28*, 1543–1553. [CrossRef]

30. Ye, X.C.; Li, X.H.; Liu, J.; Xu, C.Y.; Zhang, Q. Variation of reference evapotranspiration and its contributing climatic factors in the Poyang Lake catchment, China. *Hydrol. Process.* **2014**, *28*, 6151–6162. [CrossRef]

31. Mei, X.F.; Dai, Z.J.; Du, J.Z.; Chen, J.Y. Linkage between Three Gorges Dam impacts and the dramatic recessions in China's largest freshwater lake, Poyang Lake. *Sci. Rep.* **2015**, *5*, 18197. [CrossRef] [PubMed]

32. Zhang, Q.; Liu, J.Y.; Singh, V.P.; Gu, X.H.; Chen, X.H. Evaluation of impacts of climate change and human activities on streamflow in the Poyang Lake basin, China. *Hydrol. Process.* **2016**, *30*, 2562–2576. [CrossRef]

33. Huang, F.; Chen, Q.H.; Li, F.; Zhang, X.; Chen, Y.Y.; Xia, Z.Q.; Qiu, L.Y. Reservoir-Induced Changes in Flow Fluctuations at Monthly and Hourly Scales: Case Study of the Qingyi River, China. *J. Hydrol. Eng.* **2015**, *20*, 05015008. [CrossRef]

34. Yang, T.; Zhang, Q.; Chen, Y.D.; Tao, X.; Xu, C.Y.; Chen, X. A spatial assessment of hydrologic alteration caused by dam construction in the middle and lower Yellow River, China. *Hydrol. Process.* **2008**, *22*, 3829–3843. [CrossRef]

35. Chen, Q.H.; Zhang, X.; Chen, Y.Y.; Li, Q.F.; Qiu, L.Y.; Liu, M. Downstream effects of a hydropeaking dam on ecohydrological conditions at subdaily to monthly time scales. *Ecol. Eng.* **2015**, *77*, 40–50. [CrossRef]

36. Wang, Y.K.; Wang, D.; Wu, J.C. Assessing the impact of Danjiangkou reservoir on ecohydrological conditions in Hanjiang river, China. *Ecol. Eng.* **2015**, *81*, 41–52. [CrossRef]

37. Wang, Y.K.; Rhoads, B.L.; Wang, D. Assessment of the flow regime alterations in the middle reach of the Yangtze River associated with dam construction: Potential ecological implications. *Hydrol. Process.* **2016**, *30*, 3949–3966. [CrossRef]

38. Serinaldi, F.; Zunino, L.; Rosso, O.A. Complexity-entropy analysis of daily stream flow time series in the continental United States. *Stoch. Environ. Res. Risk Assess.* **2014**, *28*, 1685–1708. [CrossRef]

39. Stosic, T.; Telesca, L.; Ferreira, D.V.D.; Stosic, B. Investigating anthropically induced effects in streamflow dynamics by using permutation entropy and statistical complexity analysis: A case study. *J. Hydrol.* **2016**, *540*, 1136–1145. [CrossRef]

entropy

MDPI

Article

A Connection Entropy Approach to Water Resources Vulnerability Analysis in a Changing Environment

Zhengwei Pan [1,2], Juliang Jin [3,*], Chunhui Li [4], Shaowei Ning [3] and Rongxing Zhou [1]

[1] School of Civil Engineering and Environmental Engineering, Anhui Xinhua University, Hefei 230088, China; pzhwei1023@163.com (Z.P.); zhourx11@163.com (R.Z.)
[2] Institute of Safety and Environmental Assessment, Anhui Xinhua University, Hefei 230088, China
[3] School of Civil Engineering, Hefei University of Technology, Hefei 230009, China; ning@hfut.edu.cn
[4] Ministry of Education Key Lab of Water and Sand Science, School of Environment, Beijing Normal University, Beijing 100875, China; chunhuili@bnu.edu.cn
* Correspondence: jinjl66@126.com; Tel.: +86-0551-62903357

Received: 5 September 2017; Accepted: 1 November 2017; Published: 6 November 2017

Abstract: This paper establishes a water resources vulnerability framework based on sensitivity, natural resilience and artificial adaptation, through the analyses of the four states of the water system and its accompanying transformation processes. Furthermore, it proposes an analysis method for water resources vulnerability based on connection entropy, which extends the concept of contact entropy. An example is given of the water resources vulnerability in Anhui Province of China, which analysis illustrates that, overall, vulnerability levels fluctuated and showed apparent improvement trends from 2001 to 2015. Some suggestions are also provided for the improvement of the level of water resources vulnerability in Anhui Province, considering the viewpoint of the vulnerability index.

Keywords: water resources vulnerability; connection entropy; changing environment; set pair analysis; Anhui Province

1. Introduction

The concept of vulnerability has its roots in geography and natural hazards research. Moreover, the original concept is the degree to which a system is likely to experience harm because of exposure to a hazard [1,2]. Vulnerability is now considered a central concept in a variety of other research contexts including anthropology, sociology, economics, aerography, ecology, management and sustainability science [3]. There are two key research perspectives in water resources vulnerability: a major role key research perspective is vulnerability of single purpose water systems, such as surface water vulnerability, groundwater vulnerability, and vulnerability of the drinking water supplies, etc. Padowski and Gorelick [4] presented a global analysis of urban water supply vulnerability in 70 surface water-supplied cities in a baseline (2010) condition and a future scenario (2040), which considered increased demand from urban population growth and projected agricultural demand under normal climate conditions, but did not account for climate change. Allouchea et al. [5] assessed groundwater vulnerability to contamination from anthropogenic activities and sea water intrusion based on a more robust "global risk index", and the DRASTIC and GALDIT parametric methods were then linked to a novel land use index. Khakhar et al. [6] derived the intrinsic vulnerability of groundwater against contamination using the GIS platform, and applied DRASTIC model for Ahmedabad district in Gujarat (India), that also contributes to validating the existence of higher concentrations of contaminants/indicators with respect to groundwater vulnerability status in the study area. Rushforth and Ruddell [7] spatially mapped and analyzed the Water Footprint of Flagstaff (AZ, USA), using a county-level database of the U.S. hydro-economy, NWED, which can empower city managers to operationalize a city's water footprint information to reduce vulnerability and increase resilience. Martinez et al. [8] assessed the potential effects of climate change and the vulnerability of water sources to support informed decision-making in Mexico City. Padowski and Jawitz [9] presented a quantitative national assessment of urban water availability and vulnerability for 225 U.S. cities with population greater than 100,000. Additionally, urban vulnerability measures developed here were validated using a media text analysis. Sullivan [10] brought forth the water vulnerability index whereby supply-driven vulnerability and demand-driven vulnerability, through the combination of these various dimensions in a mathematical manner.

Another major role key research perspective concerns vulnerability and the factors influencing water systems. Socio-economy and ecology have also been considered, that is the coupled human-environmental system. A major role of the vulnerability concept is the degree to which a system is either susceptible to, or unable to tackle the adverse effects of climate change. This includes climate variability as well as those extremes listed by the Intergovernmental Panel for Climate Change (IPCC) [11]. Furthermore, vulnerability to climate change has been stated as a function of the character, magnitude and rate of climate variation to which a system is exposed, its responsiveness and its adaptive capacity. Safi et al. [12] developed a climate change vulnerability index as a function of physical vulnerability, sensitivity, and adaptive capacity, and discussed its relationships with climate change mitigation policy support. Liersch et al. [13] established a vulnerability framework through the assessment of vulnerability (V) as a function of exposure (E), sensitivity (S) and adaptive capacity (AC), where the impacts on V by E and S can be summarized as external impacts (EI). The assessment result indicates the difference between the current situation and a chosen scenario of change. Yang et al. [14] presented a multifunctional hierarchy indicator system based on sensitivity and adaptability, and established an evaluation model, the Analytic Hierarchy Process Combining Set Pair Analysis (AHPSPA) model. This AHPSPA model was used to assess the water resource vulnerability of Beijing (China). Gain et al. [15] established a generalized methodological framework for vulnerability assessment to support a participatory decision making mechanism in the field of water resources management, with a particular focus on climate change adaptation. This was in addition to facilitating the work of those active in the field of water management in developing countries by moving towards operational solutions. Ionescu et al. [16] presented the contours of a formal framework of vulnerability to climate change developed on the basis of a grammatical probe that stemmed from the everyday meaning of vulnerability to technical consumption in the context of climate change. Al-Saidia et al. [17] developed the Country Vulnerability Index of Water Resources, which is a composite of socioeconomic and natural components. This approach integrates a country-level standpoint while taking into consideration the associated energy and food issues that are capable of reducing water resources vulnerability. Shabbir and Ahmad [18] analyzed the vulnerability status of the water resources system in Rawalpindi and Islamabad (Pakistan) with the help of an analytic hierarchy process, keeping in mind the intricate, integrated, comprehensive, and hierarchical nature of the vulnerability evaluation of water resources. Füssel [3] presented a conceptual framework and a terminology of vulnerability that makes it possible to develop a concise characterization of any vulnerability concept and the key differences between different concepts. Thus, he thereby bridged the gap between various traditions of vulnerability research.

This paper brings forth a synthesis of the two research views discussed above. First, we will identify the factors of the water system by looking at the human socioeconomic system and the coupled human-environmental system. Second, we will analyze the four transformation processes of the water system in a changing environment: sensitive state, damaged state, recovery state, and equilibrium state. The key impact of natural factors has been defined as natural resilience that deals with the change from a damaged state to a recovery state. The major effect of artificial factors is defined as artificial adaptation that aims for the change from a recovery state to an equilibrium state. Research on water resources vulnerability by synthesis sensitivity (S), natural resilience (R) and artificial adaptation (A) will also be performed.

2. Mechanism of Water Resources Vulnerability in a Changing Environment

With the backdrop of natural hazards science, vulnerability research places significant emphasis on the resilience, response, and resilience of the human socioeconomic system in response to disasters. Accordingly, the concept of vulnerability is required to include exposure, sensitivity, adaptation, and resilience when the system is disturbed, destroyed, or affected. This paper presents the concept of water resources vulnerability, which is the nature and state of a water resources system, wherein, the normal structure and function are compromised by a changing environment. Vulnerability also includes, the sensitivity and adaptation to the disturbance and destruction under a changing environment, the ability to bear and cope with such change, and the level of resilience to damage [12–14,19].

In this transformation mechanism, the water system can experience four states: the sensitive state, the damaged state, the recovery state, and the equilibrium state. The water system undergoes change because of the interference and destruction of the changing environment, in addition to the impacts stemming from environmental change factors. In this way, the water system can be said to be in a sensitive state. The degree of change in the sensitive state is the resulting water resources sensitivity [20], expressed by sensitive factors. A water system can be said to be in a damaged state as the result of environmental interference and destruction. In this scenario, the water system is able to self-regulate and repair through the system factors, for instance water resources endowment, and natural ecological environment. This ability to self-regulate and repair is termed natural resilience, which is adaptation in response to experienced environment and its effects, without planning explicitly or consciously focused on addressing changing environment [20].

Natural resilience is the capacity of the water system, but note that it still possesses relative weakness. The system is capable of self-repair when the degree of damage is weak, but while still in the damaged state it cannot meet the social and economic development and ecological environment requirements if the damage exceeds the natural recovery capacity. It is for this reason we are required to reduce water demand with socioeconomic development and the ecological environment through the adjustment of human behavior and the socioeconomic development mode. This is done, with the improvement of, for example, the ecological environment to adapt to the damaged state, which is defined as artificial or social adaptation. Artificial adaptation is an ability of water systems, humans, social economy and ecology to adjust to potential damage, to take advantage of opportunities, or to respond to consequences [21]. The degree of damage to the water system is steadily alleviated by the combined action of natural recovery and artificial adaptation, in addition to being a system of relative equilibrium. This equilibrium state is dynamic, with the system in a changing atmosphere. The best state for the recovery of the damaged water system is prior to the occurrence of environment changes. In general, it is quite difficult to attain the optimal state, so it may be easier to form a new water system in a changing environment. In accordance with the analysis of the mechanism of water resources, vulnerability in changing environment is presented in Figure 1.

2.1. Sensitivity

Water systems are influenced by environmental interference and destruction. These impact factors primarily include water resources utilization, water environment pollution, drought, and floods. Water resources sensitivity is caused and expressed by a sensitive index that looks at water utilization rate (i.e., the ratio of water consumption to water resources availabilities), emission intensity of industrial wastewater (i.e., industrial wastewater discharge to GDP), per capita sewage discharge (i.e., domestic sewage discharge to total population), and the proportion of economic losses of water disasters to GDP (i.e., economic losses of drought and flood disaster to GDP, see Table 1).

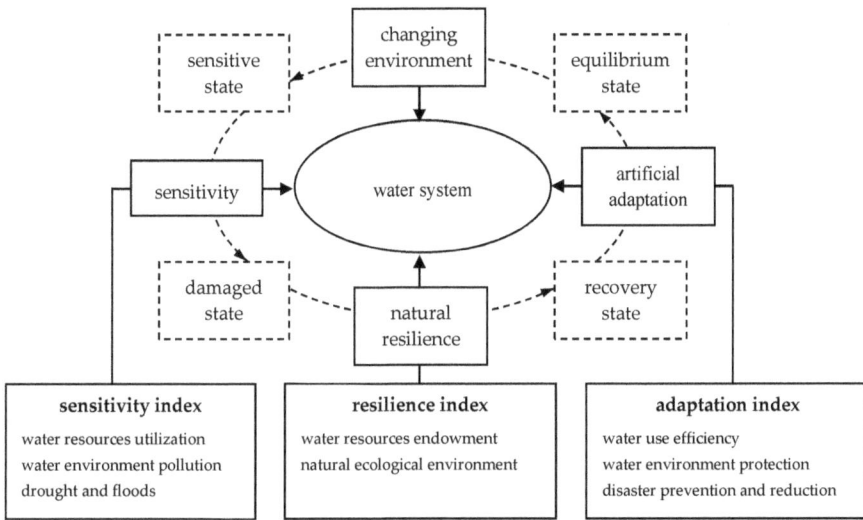

Figure 1. Mechanism diagram of water resources vulnerability in changing environment.

Table 1. Sensitive index of water resources.

Sensitive Factor		Sensitive Index
Water resources utilization	x_1	Water utilization rate (%)
Water environment pollution	x_2	Emission intensity of industrial wastewater (t/10^4Y)
	x_3	Per capita sewage discharge (t/person)
Drought and floods	x_4	Proportion of economic losses of water disasters to GDP (%)

The sensitivity function of the water resources $S(x)$ is proposed to quantitate sensitivity and the sensitive index, which is expressed as follows:

$$S(x) = S(x_1, x_2, x_3, x_4) \tag{1}$$

where $S(x)$ denotes water resources sensitivity, x_1 is the water utilization rate, and x_2 is the emission intensity of industrial wastewater. Furthermore, x_3 depicts the per capita sewage discharge, and x_4 represents the proportion of economic losses of water disasters to GDP.

2.2. Natural Resilience

A water system enters a damaged state because of environmental interference and destruction. The damaged water system is able to self-regulate and repair with the help of water resources endowment, water environmental quality, eco-environmental quality and other factors that aim to encourage the adaption of or repair to the damaged water system. This ability to self-regulate and repair the damaged water system is caused and expressed by resilient index, which looks at the per unit area of available water resources (i.e., the ratio of water resources availabilities to national territorial area), surface water environment quality (i.e., water attainment ratio of monitoring section), and forest cover (i.e., forest coverage, see Table 2).

Table 2. Resilient index of water resources.

Resilient Factor		Resilient Index
Water resources endowment	x_5	Per unit area available amount of water resources ($10^4 \cdot m^3 / km^2$)
Water environmental quality	x_6	Surface water environment quality (%)
Eco-environmental quality	x_7	Forest cover (%)

The natural resilience of the water resources function $R(x)$ is proposed to quantify natural resilience and the resilient index, and is presented as follows:

$$R(x) = R(x_5, x_6, x_7) \tag{2}$$

where $R(x)$ indicates water resources natural resilience, x_5 is the per unit area of available water resources, and x_6 represents surface water environment quality. The term x_7 is the forest cover.

2.3. Artificial Adaptation

The natural resilience of water resources is constrained. The water system remains in a damaged state where the damage exceeds the natural recovery capacity. In this situation, we must take the initiative to adapt to the environmental changes by, for example, increasing water use efficiency, strengthening water environment protection, disaster prevention, and reduction for the purpose of adapting to the damaged state. This artificial adaptation is caused and expressed by the adaptive index, which includes per unit GDP of water consumption (i.e., the ratio of water consumption to GDP), compliance rate of industrial wastewater and treatment rate of urban sewage (i.e., discharge standard of industrial wastewater and urban sewage), and economic benefits of flood control and fight drought (i.e., economic benefits of disaster prevention and mitigation, see Table 3).

Table 3. Adaptive index of water resources.

Adaptive Factor		Adaptive Index
Water use efficiency	x_8	Per unit GDP of water consumption ($m^3 / 10^4 Y$)
Water environment protection	x_9	Compliance rate of industrial wastewater (%)
	x_{10}	Treatment rate of urban sewage (%)
Disaster prevention and reduction	x_{11}	Economic benefits of flood control and fighting drought ($10^8 Y$)

The proposed artificial adaptation function of water resources $A(x)$ to quantify artificial adaptation and the adaptive index, and is presented as follows:

$$A(x) = A(x_8, x_9, x_{10}, x_{11}) \tag{3}$$

where $A(x)$ denotes water resources artificial adaptation, x_8 is the per unit GDP of water consumption, x_9 indicates the compliance rate of industrial wastewater, x_{10} represents treatment rate of urban sewage, and x_{11} depicts the economic benefits of flood control and fight drought.

2.4. Water Resources Vulnerability in a Changing Environment

Water resources vulnerability in a changing environment (V) is the synthetic effect of three elements, namely, water resources sensitivity (S), natural resilience (R) and artificial adaptation (A), and is given as follows:

$$V(x) = f[S(x), R(x), S(x)] \tag{4}$$

where V indicates water resources vulnerability in a changing environment, $S(x)$ denotes sensitivity, $R(x)$ indicates natural resilience, and $A(x)$ presents artificial adaptation.

3. Methodology

Entropy is an concept extensively used in natural and social sciences. It has been developed and perfected in these fields, despite being a simple description of a microcosmic thermodynamic concept with roots in physics research. The evolution of entropy is essentially an extension of "state", which is "state function" in thermodynamics. Entropy is required to possess the attributes of the state function of all, for having used the "mathematic analogy" methods of inference in this type of expansion of the "state". Thereafter, connection entropy is also proposed by the method of "mathematical analogies".

3.1. Development Course

The concept of entropy was primarily introduced in thermodynamics by Clausius in 1854. He initially used the term "equivalence value" to define the concept. In respect of statistical mechanics, the entropy of a system is equal to the logarithm of the number Ω of accessible microstates corresponding to a macroscopical state of this system:

$$S = k \ln \Omega \tag{5}$$

where k denotes Boltzmann's constant and Ω represents the number of microstates having consistency with the given equilibrium macro state.

The use of an intuitive logarithmic measure for information, primarily introduced by Hartley [22], puts forth the suggestion that the self-information of an event increases with the growth of its uncertainty, further implying that the probability of occurrence reduces. In this respect, "S" is termed as a measure of uncertainty.

Schrödinger [23] put forth a proposition regarding a local decrease of entropy for living systems when $(1/D)$ represents the number of states that are prevented from random distribution:

$$-S = k \ln(1/D) \tag{6}$$

where D indicates the number of possible energy states in the system that can be randomly filled with energy. Furthermore, D suggests a measure of disorder, and its reciprocal $1/D$ can be regarded as a direct measure of order.

In 1948, Shannon [24] published his famous paper titled "A Mathematical Theory of Communication", wherein he introduced the entropy of a discrete probability distribution $(p_1, p_2, \ldots , p_n)$, providing an H function of the following form:

$$H = -K \sum_{i=1}^{n} p_i \log p_i \tag{7}$$

where K denotes a positive constant. Moreover, K, merely implying choice of a unit of measurement, plays a focal function in the information theory as measures of information, choice, and uncertainty [24].

In 1972, De Luca and Termini [25] promulgated the concept of entropy again for finite fuzzy sets, similar to Shannon entropy although quite different conceptually, where the range is a set of nonnegative real numbers.

In 1992, Zhao [26] proposed the concepts of identical entropy, discrepancy entropy, contrary entropy, and connection entropy based on the definition of traditional entropy, to measure identity, diversity, opposites, and associative of systems which contains n set pairs.

3.2. Connection Entropy

Set pair analysis (SPA) is a new system analysis approach advanced by the Chinese scholar Zhao [1]. SPA considers both certainty and uncertainty as a system for the purpose of conducting identical-discrepancy-contrary analysis as well as further quantitative mathematical process using a connection number. Because of its realistic approach to deal with uncertainty, SPA has been put to extensive applications in sociology, economics, engineering technology, and management, and has produced numerous research results [14,27–30].

In the case of particular questions, analyses have been conducted on the identity, discrepancy, and opposition of a set pair, the formula of the connection number of set pair "*H*" for given conditions that describe the relationship between certainty and uncertainty. The expression with respect to set pair "*H*" is presented as follows [1]:

$$u = a + bI + cJ \tag{8}$$

where a, b, $c \in [0, 1]$, and $a + b + c = 1$. Furthermore, a denotes the identical degree of set pair H, b indicates the discrepancy degree, c is the contrary degree, and I is the coefficient of discrepancy and $I \in [-1, 1]$; sometimes I may only refer to discrepancy. Finally, J is the coefficient of contrary, and has been ruled to be -1; sometimes it may only be considered as a mark of the opposites.

Here, n connection degrees are attained: $u_1 = a_1 + b_1 I_1 + c_1 J_1$, $u_2 = a_2 + b_2 I_2 + c_2 J_2, \ldots$, and $u_n = a_n + b_n I_n + c_n J_n$ through the analysis of n set pairs.

Identical entropy is defined as follows: $S_s = \sum_{i=1}^{n} a_i \ln a_i$, discrepancy entropy as follows: $S_f = \sum_{i=1}^{n} b_i \ln b_i \times I$, contrary entropy as follows: $S_p = \sum_{i=1}^{n} c_i \ln c_i \times J$, and connection entropy as follows [1]:

$$S = S_s + S_f + S_p = \sum_{i=1}^{n} a_i \ln a_i + \sum_{i=1}^{n} b_i \ln b_i \times I + \sum_{i=1}^{n} c_i \ln c_i \times J \tag{9}$$

Entropy is the "state function" of a system that is considered a measure of randomness or disorder. The variation of thermodynamic entropy was considered as a measure of variation of unavailable energy. Statistical entropy will characterize (in respect of a macro state prepared as per a provided probability law) our uncertainty about the set of all microscopic experiments that can be conceived. Information on Shannon entropy is a measure of uncertainty or information of random events. More specifically, it is a measure of the uncertainty of the test results prior to the randomization test or the amount of information in the event subsequent to the event. Fuzzy entropy is a fuzziness measure of the system.

In the same way, identical entropy is a disorder measure. Moreover, discrepancy entropy is an order measure, and contrary entropy is a chaotic measure of uncertain systems [23]. Discrepancy includes identical and contrary in the connection degree, and therefore, discrepancy entropy is complex entropy, which comprises identical entropy and contrary entropy. The difference can be segregated with the help of the malleability of connection degree. Thus, discrepancy entropy is divided into identical entropy, discrepancy entropy and contrary entropy.

3.3. Improvement of Connection Entropy

It is quite important to improve the correlation entropy, as defined by Zhao [1], because the logarithmic function delivers no meaning, if a, b, c are likely to be 0 in the connection degree. The general expression of improvement of connection entropy is given as follows:

$$S = \sum_{i=1}^{n} a_i \ln(a_i + e) + \sum_{i=1}^{n} b_i \ln(b_i + e) \times I + \sum_{i=1}^{n} c_i \ln(c_i + e) \times J \tag{10}$$

where $S_s = \sum_{i=1}^{n} a_i \ln(a_i + e)$ indicates identical entropy, $S_f = \sum_{i=1}^{n} b_i \ln(b_i + e) \times I$ suggests discrepancy entropy, and $S_p = \sum_{i=1}^{n} c_i \ln(c_i + e) \times J$ denotes contrary entropy.

4. Analysis Method of Water Resources Vulnerability Based on Connection Entropy

4.1. Connection Number of "Identical-Discrepancy-Contrary" Hierarchy Method

As presented in Equation (8), the connection number is established on the basis of the distinction: identical, discrepancy, and contrary, called the identical-discrepancy-contrary connection number or three dimensional connection numbers. In practice, it is not enough to divide the described object into three components and therefore it is important to extend the basic expression of contact number (i.e., Equation (8)) to include more dimensions, which are termed as the malleability of the connection number, extending the connection number, and a multi-dimensional number yield [31]:

$$u = (a_1 + a_2 + \cdots + a_r) + (b_1 I_1 + b_2 I_2 + \cdots + b_s I_s) + (c_1 J_1 + c_2 J_2 + \cdots + c_t J_t) \tag{11}$$

where a_x, b_y, c_z are connection components, a_x, b_y, $c_z \in [0, 1]$, and $\sum_{x=1}^{r} a_x + \sum_{y=1}^{s} b_y + \sum_{z=1}^{t} c_z = 1$; $I_1, I_2, \cdots, I_s$ is the discrepancy coefficients, and $I_1, I_2, \cdots, I_s \in [-1, 1]$, which sometimes perform the functions of a discrepancy mark only. Furthermore, $J_1, J_2, \cdots, J_t$ denote the coefficient of contrary degrees that have been ruled to be minus unity and sometimes only marks the contrary.

On the basis of Equation (11), the multi-connection number and the construction of the identical-discrepancy-contrary hierarchical structure of connection number can be expressed as follows:

$$u = a_1 + a_2 + b_1 I_1 + b_2 I_2 + b_3 I_3 + c_1 J_1 + c_2 J_2 \tag{12}$$

where a_1, and a_2 represent the identical degree and partial differential identical degree, respectively, and their coefficients can be assumed to be unity. Furthermore, $b_1 I_1$, $b_2 I_2$, and $b_3 I_3$ indicate partial similar, middle, and partial opposite discrepancy degrees, correspondingly and their coefficients are $I_1 \in [-1, 0]$, $I_2 \in [-0.5, 0.5]$, and $I_3 \in [0, 1]$. Lastly, $c_1 J_1$ and $c_2 J_2$ indicate partial differential contrary degree and contrary degree, respectively, and J_1, J_2 are their coefficients, which are regulated to be minus unity.

The water resources vulnerability is generally divided into five levels [14,19], and assuming that x_{ij} ($i = 1, 2, \ldots, n; j = 1, 2, \ldots, m$) denotes the index sample and s_{kj} ($k = 0, 1, 2, \ldots, 5$) the threshold value of standard, the connection number of the "identical-discrepancy-contrary" hierarchy method is described as follows [31]:

$$u_{ij} = \begin{cases} \frac{x_{ij}-s_{1j}}{2(s_{0j}-s_{1j})} + 0.5 + \frac{s_{0j}-x_{ij}}{2(s_{0j}-s_{1j})} I_1 + 0 I_2 + 0 I_3 + 0 J_1 + 0 J_2 & x_{ij} \in \text{grade I} \\ 0 + \frac{x_{ij}-s_{2j}}{2(s_{1j}-s_{2j})} + 0.5 I_1 + \frac{s_{1j}-x_{ij}}{2(s_{1j}-s_{2j})} I_2 + 0 I_3 + 0 J_1 + 0 J_2 & x_{ij} \in \text{grade II} \\ 0 + 0 + \frac{x_{ij}-s_{3j}}{2(s_{2j}-s_{3j})} I_1 + 0.5 I_2 + \frac{s_{2j}-x_{ij}}{2(s_{2j}-s_{3j})} I_3 + 0 J_1 + 0 J_2 & x_{ij} \in \text{grade III} \\ 0 + 0 + 0 I_1 + \frac{x_{ij}-s_{4j}}{2(s_{3j}-s_{4j})} I_2 + 0.5 I_3 + \frac{s_{3j}-x_{ij}}{2(s_{3j}-s_{4j})} J_1 + 0 J_2 & x_{ij} \in \text{grade IV} \\ 0 + 0 + 0 I_1 + 0 I_2 + \frac{x_{ij}-s_{5j}}{2(s_{4j}-s_{5j})} I_3 + 0.5 J_1 + \frac{s_{4j}-x_{ij}}{2(s_{4j}-s_{5j})} J_2 & x_{ij} \in \text{grade V} \end{cases} \tag{13}$$

where x_{ij} suggests the ith sample value in the jth index, s_{kj} ($k = 0, 1, 2, \ldots, 5$) indicates the kth standard node value in the same index, and I_1, I_2, I_3, J_1, and J_2 are the same as in Equation (12).

4.2. Connection Entropy of the Vulnerability Index

The index of water resources vulnerability is considered as set $A = \{x_{ij}$ ($i = 1, 2, \ldots, n; j = 1, 2, \ldots, m$)$\}$, and the thresholds of the vulnerability grade are considered as set $B_k = \{s_{kj}$ ($k = 0, 1, 2, \ldots, 5$)$\}$; thereafter, the two sets constitute a set pair $H_k = (A, B_k)$.

The connection numbers of the vulnerability index with Equation (13), and thereafter, connection entropy of the vulnerability index, are described as follows:

$$S_{ij} = a_{1ij} \ln(a_{1ij} + e) + a_{2ij} \ln(a_{2ij} + e) + b_{1ij} \ln(b_{1ij} + e) \times I_1 + b_{2ij} \ln(b_{2ij} + e) \times I_2$$
$$+ b_{3ij} \ln(b_{3ij} + e) \times I_3 + c_{1ij} \ln(c_{1ij} + e) \times J_1 + c_{2ij} \ln(c_{2ij} + e) \times J_2 \tag{14}$$

where $S_{s1} = a_{1ij} \ln(a_{1ij} + e)$ indicates identical entropy, $S_{s2} = a_{2ij} \ln(a_{2ij} + e)$ denotes critical identical entropy, and $S_{f1} = b_{1ij} \ln(b_{1ij} + e) \times I_1$ indicates upper discrepancy entropy, Furthermore, $S_{f2} = b_{2ij} \ln(b_{2ij} + e) \times I_2$ implies medium discrepancy entropy, $S_{f3} = b_{3ij} \ln(b_{3ij} + e) \times I_3$ represents lower discrepancy entropy, $S_{p1} = c_{1ij} \ln(c_{1ij} + e) \times J_1$ is critical contrary entropy, and $S_{p2} = c_{2ij} \ln(c_{2ij} + e) \times J_2$ suggests contrary entropy.

4.3. Synthesis Connection Entropy of Water Resources Vulnerability

In general, in accordance with the significance of the vulnerability index, respective weights for the integrated vulnerability are provided. Consequently, taking advantage of the additively weighted synthesis method (average method), multiplicatively weighted synthesis method (geometric average method) or add-multiplicatively weighted synthesis method [32], the single index connection entropy is multiplied for the purpose of generating the integrated connection entropy S_i. This brings forth the expression:

$$S_i = \prod_{j=1}^{m} w_j \times S_{ij} \tag{15}$$

where $\prod$ indicates the system synthesis method and w_j represents the weight of the jth vulnerability index.

4.4. Decision Criterion of Water Resources Vulnerability

The degree to which water resources vulnerability is subjected to a changing environment can be broken down into five grades (or levels) with 11 indices: low (I), slight (II), moderate (III), high (IV) and extreme (V). The calculation of contact entropy using Equation (14), shows an approximate value of contact entropy S with a range of $(-1.314, 1.314)$. The interval $(-1.314, 1.314)$ is divided into five parts: $[0.877, 1.314), [0.292, 0.877), [-0.292, 0.292), [-0.877, -0.292), (-1.314, -0.877)$, denoting low (I), slight (II), moderate (III), high (IV) and extreme (V) grades of water resources vulnerability, respectively.

5. Case Study

Anhui Province is currently considered as the development frontier in China as it absorbs the economic radiation and industrial transfer from the coastal developed areas. It is also a bridgehead between the development of Western China as well as the rise of Central China. As a consequence, the area possesses unique geographical benefits, functioning as a junction between the East and the West and connects Northern and Southern China. Various obstacles to steady development have emerged in the province, with the water environment becoming a principal factor. The per capita water resources in Anhui Province are 1125 m3, which is approximately half the national average and merely 1/8 the world average, so it goes without saying that there is a shortage of water resources throughout the province. Furthermore, regional differences between annual precipitation and the four seasons are large, giving rise to frequent droughts and flood disasters. Water pollution is another serious concern in the Yangtze River, Huaihe River, and Chaohu Lake in Anhui Province, as parts of them have lost their utility value. Thus, water conflicts have been exacerbated and the vulnerability of the regional water environment has been exposed. In accordance with the above, it is quite apparent that the increasing deterioration of regional water resources vulnerability cannot be overlooked. Therefore, it is necessary to perform an analysis of regional water resources vulnerability in Anhui Province. Here, we study used a connection entropy approach to analyze water resources vulnerability in Anhui

Province from 2001 to 2015, aiming at the provision of a theoretical basis for the improvement of the water situation in the region.

5.1. Standard Interval of the Vulnerability Index in Anhui Province

In accordance with the Section 2, and the research findings of water resources vulnerability [14,30], the standard interval of the vulnerability index in Anhui Province is presented in Table 4.

Table 4. Standard interval of the vulnerability index in Anhui Province.

Index		Standard Interval				
		Low (I)	Slight (II)	Moderate (III)	High (IV)	Extreme (V)
Sensitive	x_1 (%)	0–10	10–25	25–40	40–60	60–100
	x_2 (t/10^4 ¥)	0–10	10–25	25–40	40–55	55–100
	x_3 (t/person)	0–5	5–10	10–40	40–60	60–100
	x_4 (%)	0–1.0	1.0–2.5	2.5–4.0	4.0–4.5	4.5–5.5
Natural resilience	x_5 ($10^4 \cdot m^3/km^2$)	75–100	55–75	35–55	15–35	5–15
	x_6 (%)	90–100	80–90	70–80	50–70	0–50
	x_7 (%)	30–50	20–30	15–20	10–15	5–10
Artificial adaptation	x_8 ($m^3/10^4$ ¥)	5–24	24–140	140–610	610–1060	1060–1600
	x_9 (%)	97.5–100	92.5–97.5	85–92.5	80–85	15–80
	x_{10} (%)	80–100	60–80	40–60	20–40	0–20
	x_{11} (10^8 ¥)	400–600	200–400	100–200	60–100	0–60

5.2. Connection Entropy of Vulnerability Index in Anhui Province

The index of water resources vulnerability in Anhui Province from 2001 to 2015 is considered as set $A = \{x_{ij} \ (i = 1, 2, \ldots, 15; j = 1, 2, \ldots, 11)\}$ in Table 5, where x_{ij} suggests the sample value in 2001 if $i = 1$, the thresholds of vulnerability grade are considered as set $B_k = \{s_{kj} \ (k = 0, 1, 2, \ldots, 5)\}$ in Table 4, and the two sets constitute a set pair $H_k = (A, B_k)$.

Table 5. Index data of the vulnerability index in Anhui Province in 2001–2015.

Year	x_1	x_2	x_3	x_4	x_5	x_6	x_7	x_8	x_9	x_{10}	x_{11}
2001	45.22	17.79	11.47	3.13	34.00	41.00	27.95	603.56	95.24	7.84	128.00
2002	25.36	16.90	12.27	1.48	59.13	51.08	27.95	588.48	95.74	15.80	91.80
2003	16.83	14.88	12.13	5.72	77.65	48.69	27.95	458.79	95.88	20.14	440.67
2004	41.89	13.31	13.04	0.74	35.90	41.70	30.30	435.80	96.91	25.69	70.00
2005	28.92	10.89	14.29	3.49	51.57	44.30	26.06	386.10	97.36	31.28	209.00
2006	41.67	11.40	14.61	1.00	41.62	64.60	26.06	393.80	97.12	32.90	145.30
2007	32.60	9.99	15.24	1.98	51.80	64.70	26.06	305.90	94.77	37.64	420.50
2008	38.09	8.29	15.08	2.41	50.13	69.30	26.06	288.80	96.17	52.08	100.00
2009	39.81	7.30	15.64	1.44	52.56	72.10	26.06	290.04	96.21	72.81	86.00
2010	31.15	5.74	16.66	0.92	67.33	77.70	27.53	238.50	97.67	77.75	260.00
2011	48.94	4.62	25.07	0.61	43.17	65.50	27.53	195.00	97.47	79.04	145.00
2012	41.16	3.90	27.09	0.28	50.26	70.60	27.53	167.60	98.38	86.39	154.80
2013	50.55	3.73	28.16	0.80	41.99	72.40	27.53	155.50	98.51	88.44	187.10
2014	34.95	3.34	29.20	0.14	55.81	77.90	28.65	131.00	98.69	90.19	103.11
2015	31.58	3.25	30.07	0.34	65.54	81.80	28.65	131.20	98.70	91.80	136.89

Sources: Statistics Bureau of Anhui Province, 2002–2016, and Water Resources Bulletin of Anhui Province, 2001–2015.

Thereafter, the connection number and the connection entropy of vulnerability index in Anhui Province can be attained with the application of Equations (13) and (14), respectively. For example, the connection entropy of vulnerability index in 2015 listed in Table 6 is thus obtained.

Table 6. Connection entropy of vulnerability index in 2015.

Index	Connection Entropy	w_j
x_1	$S_{15,1} = 0.0000 + 0.0000 + 0.3083I_1 + 0.5844I_2 + 0.2363I_3 + 0.0000J_1 + 0.0000J_2$	0.120
x_2	$S_{15,2} = 0.3772 + 0.5844 + 0.1717I_1 + 0.0000I_2 + 0.0000I_3 + 0.0000J_1 + 0.0000J_2$	0.100
x_3	$S_{15,3} = 0.0000 + 0.0000 + 0.1754I_1 + 0.5844I_2 + 0.3732I_3 + 0.0000J_1 + 0.0000J_2$	0.100
x_4	$S_{15,4} = 0.3691 + 0.5844 + 0.1791I_1 + 0.0000I_2 + 0.0000I_3 + 0.0000J_1 + 0.0000J_2$	0.080
x_5	$S_{15,5} = 0.0000 + 0.2879 + 0.5844I_1 + 0.2562I_2 + 0.0000I_3 + 0.0000J_1 + 0.0000J_2$	0.150
x_6	$S_{15,6} = 0.0000 + 0.0929 + 0.5844I_1 + 0.4676I_2 + 0.0000I_3 + 0.0000J_1 + 0.0000J_2$	0.090
x_7	$S_{15,7} = 0.0000 + 0.4964 + 0.5844I_1 + 0.0692I_2 + 0.0000I_3 + 0.0000J_1 + 0.0000J_2$	0.060
x_8	$S_{15,8} = 0.0000 + 0.0385 + 0.5844I_1 + 0.5346I_2 + 0.0000I_3 + 0.0000J_1 + 0.0000J_2$	0.090
x_9	$S_{15,9} = 0.2603 + 0.5844 + 0.2838I_1 + 0.0000I_2 + 0.0000I_3 + 0.0000J_1 + 0.0000J_2$	0.075
x_{10}	$S_{15,10} = 0.3254 + 0.5844 + 0.2199I_1 + 0.0000I_2 + 0.0000I_3 + 0.0000J_1 + 0.0000J_2$	0.075
x_{11}	$S_{15,11} = 0.0000 + 0.0000 + 0.1966I_1 + 0.5844I_2 + 0.3502I_3 + 0.0000J_1 + 0.0000J_2$	0.060
$\sum$	$S_{15} = 0.1112 + 0.2776 + 0.3635I_1 + 0.2964I_2 + 0.0867I_3 + 0.0000J_1 + 0.0000J_2$	

5.3. Weight of the Vulnerability Index in Anhui Province

In accordance with the mechanism of water resources vulnerability in a changing environment, the weights of sensitivity, natural resilience and artificial adaptation are 0.4, 0.3, and 0.3, respectively. The weights of the sensitive index are valued at 0.3, 0.25, 0.25, 0.2, the natural resilient index are 0.5, 0.3, 0.2, and those of the artificial adaptive index are 0.3, 0.25, 0.25, 0.2. The weights w_j of the vulnerability index attained for Anhui Province are shown in Table 6.

5.4. Synthesis of Connection Entropy of Water Resources Vulnerability in Anhui Province

By means of a weighted average operator [32] for the synthesis of the connection entropy of water resources vulnerability, the equation for the synthesis connection entropy of water resources vulnerability in Anhui Province is presented as follows:

$$S_i = \sum_{j=1}^{11} w_j \times S_{ij} = a_{1i} + a_{2i} + b_{1i} \times I_1 + b_{2i} \times I_2 + b_{3i} \times I_3 + c_{1i} \times J_1 + c_{2i} \times J_2 \quad (16)$$

where w_j suggests the weight of vulnerability index j.

The connection entropy of the vulnerability index in Anhui Province is synthesized by Equation (16) and the results are presented in Table 7.

Table 7. Synthesize connection entropy of water resources vulnerability in Anhui Province.

Year	Synthesize Connection Entropy	Value	Grade
2001	$S_1 = 0.0000 + 0.0757 + 0.2358I_1 + 0.3786I_2 + 0.3047I_3 + 0.1166J_1 + 0.0336J_2$	−0.109	III
2002	$S_2 = 0.0000 + 0.1302 + 0.3956I_1 + 0.3372I_2 + 0.1765I_3 + 0.0996J_1 + 0.0082J_2$	0.132	III
2003	$S_3 = 0.0144 + 0.2517 + 0.3810I_1 + 0.2092I_2 + 0.1793I_3 + 0.1082J_1 + 0.0012J_2$	0.258	III
2004	$S_4 = 0.0113 + 0.1641 + 0.2431I_1 + 0.2964I_2 + 0.3117I_3 + 0.1139J_1 + 0.0077J_2$	0.019	III
2005	$S_5 = 0.0000 + 0.1185 + 0.3924I_1 + 0.3888I_2 + 0.1694I_3 + 0.0702J_1 + 0.0052J_2$	0.154	III
2006	$S_6 = 0.0002 + 0.1593 + 0.2956I_1 + 0.3815I_2 + 0.2750I_3 + 0.0320J_1 + 0.0000J_2$	0.138	III
2007	$S_7 = 0.0032 + 0.1468 + 0.3988I_1 + 0.4082I_2 + 0.1678I_3 + 0.0170J_1 + 0.0000J_2$	0.249	III
2008	$S_8 = 0.0088 + 0.1120 + 0.3518I_1 + 0.4649I_2 + 0.2066I_3 + 0.0016J_1 + 0.0000J_2$	0.192	III
2009	$S_9 = 0.0142 + 0.1683 + 0.3773I_1 + 0.3927I_2 + 0.1788I_3 + 0.0112J_1 + 0.0000J_2$	0.271	III
2010	$S_{10} = 0.0287 + 0.2735 + 0.4791I_1 + 0.3001I_2 + 0.0585I_3 + 0.0000J_1 + 0.0000J_2$	0.513	II
2011	$S_{11} = 0.0462 + 0.2157 + 0.2949I_1 + 0.3195I_2 + 0.2225I_3 + 0.0395J_1 + 0.0000J_2$	0.259	III
2012	$S_{12} = 0.0928 + 0.2184 + 0.2810I_1 + 0.3596I_2 + 0.1864I_3 + 0.0035J_1 + 0.0000J_2$	0.355	II
2013	$S_{13} = 0.0763 + 0.2184 + 0.2807I_1 + 0.3248I_2 + 0.2020I_3 + 0.0346J_1 + 0.0000J_2$	0.299	II
2014	$S_{14} = 0.1166 + 0.2293 + 0.3222I_1 + 0.3520I_2 + 0.1236I_3 + 0.0000J_1 + 0.0000J_2$	0.445	II
2015	$S_{15} = 0.1112 + 0.2776 + 0.3635I_1 + 0.2964I_2 + 0.0867I_3 + 0.0000J_1 + 0.0000J_2$	0.527	II

The value of the discrepancy entropy coefficients in Equation (16) can be measured with the application of the mean method [1]: $I_1 = 0.5$, $I_2 = 0$ and $I_3 = -0.5$. Moreover, the value of the contrary entropy coefficients, both J_1 and J_2, become equal to −1. Thus, the value of the synthesis of connection entropy of water resources vulnerability in Anhui Province is measured using Equation (16) and the findings are shown in Table 7. The curve for the synthesis of connection entropy of water resources vulnerability in Anhui Province is presented in Figure 2.

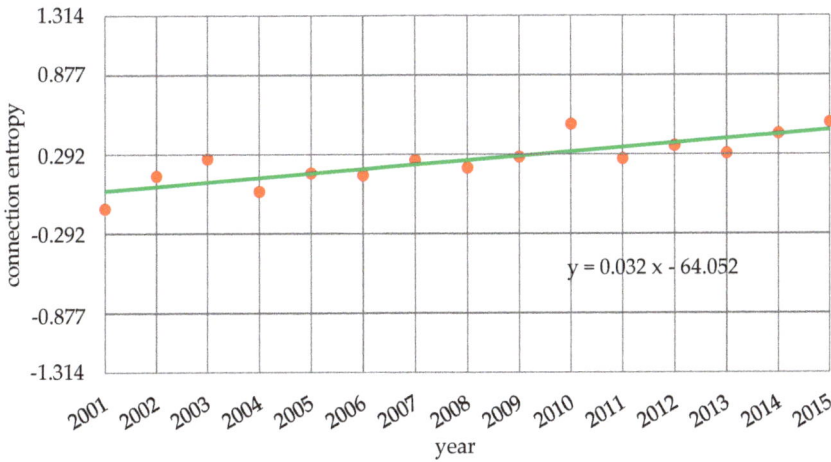

Figure 2. Synthesize connection entropy trend line of water resources vulnerability in Anhui Province.

5.5. Synthesis of the Level of Water Resources Vulnerability in Anhui Province

The synthesis of the level of water resources vulnerability in Anhui Province is attained with the help of a comparison of the value of the synthesis of connection entropy with the decision criterion in Section 4; the findings are presented in Table 7.

5.6. Analysis of Results

The synthesis of the level of water resources vulnerability in Anhui Province appeared to be moderate in 2001–2009, and 2011, and slight in 2010, and 2012–2015 (Table 7). As evident from Figure 2, the following can be observed: the vulnerability level fluctuated and exhibited an apparent improving trend from 2001 to 2015.

The analysis of vulnerability of typical years is as follows: (1) the synthesis of the level of water resources vulnerability in 2003 was highest during 2001–2005 period as it was a wet year. The average annual rainfall was 1460.9 mm, which was 24.4% higher than the multi-year average. Water resources availability was 108.301 billion m^3, which was 51.3% higher than the multi-year average. Thus, the vulnerability index water utilization rate was low, and the per unit area available amount of water resources was high. Good economic benefits are expected because of effective flood control and drought relief measures. (2) The synthesis of the level of water resources vulnerability in 2004 was the lowest during the 2002–2006 period as it was a low flow period. The average annual rainfall was 998.0 mm, 14.9% less than the multi-year average, and water resources availability was 50.065 billion m^3, being 30.1% less than the multi-year average. Thus, the value of the vulnerability index water utilization rate was high, and that of the per unit area available amount of water resources was low, together with the per unit GDP of water consumption being high. (3) The synthesis of the level of water resources vulnerability in 2010 was the highest during the 2008–2012 period as it was a partially wet period. The average annual rainfall was 1308.90 mm, being 11.6% more than the multi-year average, and water resources availability was 93.905 billion m^3, being 31.1% more than the multi-year average. Thus, the vulnerability index water utilization rate appears to be low, and the per unit area available amount of water resources is high, in addition to the reasonable economic benefits of flood control and fight drought.

6. Conclusions and Suggestions

From the finding presented in this paper, the following conclusions may be put forth:

(1) This study considered the mechanism of water resources vulnerability in a changing environment, focusing on the analysis of four key states of a water system: sensitive state, damaged state, recovery state, and equilibrium state. The state of a water system generally changes from damaged to recovery via natural factors, followed by the transition from a state of recovery to one of equilibrium primarily because of artificial factors. The former is defined as natural resilience and the latter as artificial adaptation. This mechanism results offer much-needed insights into water resources vulnerability in a changing environment, but the mechanism of transformation process between four states is the focus of future research.

(2) This paper proposed an analysis method to system uncertainty on the basis of connection entropy, extending the Zhao's [1] original concept. In addition, our analysis of water resources vulnerability in Anhui Province, China showed that the vulnerability level fluctuated, as well as exhibiting apparent improvement trends over a 15-year period. The synthesize grade of water resources vulnerability in Anhui Province appeared to be moderate in 2001–2009, and 2011, and a slight in 2010, and 2012–2015.

(3) The vulnerability level exhibited an apparent improving trend from 2001 to 2015, but still needs to be improved. There is a need to improve the perceptions of the vulnerability index, such as water utilization rate, per capita sewage discharge, surface water environment quality, per unit GDP of water consumption, and the economic benefits of flood control and fight drought.

Based on the results, we suggest the following improvements for water resources vulnerability in Anhui Province: on the one hand, there is an abated water resources sensitivity, whereby an increase in the reuse rate of industrial water can reduce water resources utilization. Moreover, sewage discharge can be reduced by the promotion of water saving policy and knowledge, encouraging water savings, and creating a greater awareness of (as well as building) a water-saving society. On the other hand, there is a need to improve natural resilience and artificial adaptation. Both can be achieved by developing sewage treatment to improve water environment quality, reduce high water consumption, eradicate backward industries, and promote new technology. Such actions will reduce the per unit GDP of water consumption and losses caused by floods and droughts, and improve the economic benefits of flood control and fight drought. Thus, for the purpose of ensuring sustainable development of economic, social and water resources in region, there is a need to improve the level of water resources vulnerability in the Anhui Province.

Acknowledgments: The authors would like to thank the support of the National Key Research and Development Program of China under Grant No. 2016YFC0401303 and No. 2016YFC0401305, the National Natural Science Foundation of China (Grant No. 51579059, No. 51509158, No. 51709071 and No. 51309004).

Author Contributions: Z.P. was responsible for the data analysis and code programming; Z.P. wrote the paper; R.Z. contributed the data; J.J. constructed the research framework and designed the study; C.L. and S.N. reviewed and edited the manuscript. All of the authors have read and approved the final manuscript.

Conflicts of Interest: The authors declare no conflict of interest.

References

1. Zhao, K.Q. *Set Pair Analysis and Preliminary Application*; China Science Technology Press: Hangzhou, China, 2000.
2. Turner, B.L., II; Kasperson, R.E.; Matson, P.A.; McCarthy, J.J.; Corell, R.W.; Christensen, L.; Eckley, N.; Kasperson, J.X.; Luers, A.; Martello, M.L.; et al. A framework for vulnerability analysis in sustainability science. *Proc. Natl. Acad. Sci. USA* **2003**, *100*, 8074–8079. [CrossRef] [PubMed]

3. Füssel, H.M. Vulnerability: A generally applicable conceptual framework for climate change research. *Glob. Environ. Chang.* **2007**, *17*, 155–167. [CrossRef]
4. Padowski, J.C.; Gorelick, S.M. Global analysis of urban surface water supply vulnerability. *Environ. Res. Lett.* **2014**, *9*, 104004. [CrossRef]
5. Allouchea, N.; Maananb, M.; Gontaraa, M.; Rollob, N.; Jmala, I.; Bouria, S. A global risk approach to assessing groundwater vulnerability. *Environ. Model. Softw.* **2017**, *88*, 168–182. [CrossRef]
6. Khakhar, M.; Ruparelia, J.P.; Vyas, A. Assessing groundwater vulnerability using GIS-based DRASTIC model for Ahmedabad district, India. *Environ. Earth Sci.* **2017**, *76*, 440. [CrossRef]
7. Rushforth, R.R.; Ruddell, B.L. The vulnerability and resilience of a city's water footprint: The case of Flagstaff, Arizona, USA. *Water Resour. Res.* **2016**, *52*, 2698–2714. [CrossRef]
8. Martinez, S.; Kralisch, S.; Escolero, O.; Perevochtchikova, M. Vulnerability of Mexico City's water supply sources in the context of climate change. *J. Water Clim. Chang.* **2015**, *6*, 518–533. [CrossRef]
9. Padowski, J.C.; Jawitz, J.W. Water availability and vulnerability of 225 large cities in the United States. *Water Resour. Res.* **2012**, *48*. [CrossRef]
10. Sullivan, C.A. Quantifying water vulnerability: A multi-dimensional approach. *Stoch. Environ. Res. Risk Assess.* **2011**, *25*, 627–640. [CrossRef]
11. McCarthy, J.J.; Canziani, O.F.; Leary, N.A.; Dokken, D.J.; White, K.S. *Climate Change 2001: Impacts, Adaptation, and Vulnerability. Contribution of Working Group II to the Third Assessment Report of the Intergovernmental Panel on Climate Change*; Cambridge University Press: Cambridge, UK, 2001.
12. Safi, A.S.; Smith, W.J.; Liu, Z. Vulnerability to climate change and the desire for mitigation. *J. Environ. Stud. Sci.* **2016**, *6*, 503–514. [CrossRef]
13. Liersch, S.; Cools, J.; Kone, B.; Koch, H.; Diallo, M.; Reinhardt, J.; Fournet, S.; Aich, V.; Hattermann, F.F. Vulnerability of rice production in the Inner Niger Delta to water resources management under climate variability and change. *Environ. Sci. Policy* **2013**, *34*, 18–33. [CrossRef]
14. Yang, X.H.; Sun, B.Y.; Zhang, J.; Li, M.S.; He, J.; Wei, Y.M.; Li, Y.Q. Hierarchy evaluation of water resources vulnerability under climate change in Beijing, China. *Nat. Hazards* **2016**, *84*, 63–76. [CrossRef]
15. Gain, A.K.; Giupponi, C.; Renaud, F.G. Climate Change Adaptation and Vulnerability Assessment of Water Resources Systems in Developing Countries: A Generalized Framework and a Feasibility Study in Bangladesh. *Water* **2012**, *4*, 345–366. [CrossRef]
16. Ionescu, C.; Klein, R.J.T.; Hinkel, J.; Kumar, K.S.K.; Klein, R. Towards a Formal Framework of Vulnerability to Climate Change. *Environ. Model. Assess.* **2009**, *14*, 1–16. [CrossRef]
17. Al-Saidi, M.; Birnbaum, D.; Buriti, R.; Diek, E.; Hasselbring, C.; Jimenez, A.; Woinowski, D. Water Resources Vulnerability Assessment of MENA Countries Considering Energy and Virtual Water. *Procedia Eng.* **2016**, *145*, 900–907. [CrossRef]
18. Shabbir, R.; Ahmad, S.S. Water resource vulnerability assessment in Rawalpindi and Islamabad, Pakistan using Analytic Hierarchy Process (AHP). *J. King Saud Univ. Sci.* **2016**, *28*, 293–299. [CrossRef]
19. Pan, Z.; Jin, J.; Wu, K.; Res, K.D. Earch on the Indexes and Decision Method of Regional Water Environmental System Vulnerability. *Resour. Environ. Yangtze Basin* **2014**, *23*, 518–525. [CrossRef]
20. Field, C.B.; Barros, V.R.; Dokken, D.J.; Mavh, K.J.; Mastrabdrea, M.D.; Bilir, T.E.; Chatterjee, M.; Ebi, K.L.; Estrada, Y.O.; Genova, R.C.; et al. (Eds.) *Climate Change 2014: Impacts, Adaptation, and Vulnerability. Summaries, Frequently Asked Questions, and Cross-Chapter Boxes. A Contribution of Working Group II to the Fifth Assessment Report of the Intergovernmental Panel on Climate Change*; World Meteorological Organization: Geneva, Switzerland, 2014.
21. Hassan, R.; Scholes, R.; Ash, N. (Eds.) Appendix D: Glossary. In *Ecosystems and Human Well-being: Current States and Trends*; Island Press: Washington, DC, USA, 2005; Volume 1, pp. 893–900.
22. Hartley, R.V. Transmission of Information. *Bell Syst. Tech. J.* **1928**, *7*, 535–563. [CrossRef]
23. Schrödinger, E. *What Is Life? The Physical Aspects of Living Cell*; Cambridge University Press: Cambridge, UK, 1944.
24. Shannon, C.E. A mathematical theory of communication. *Bell Syst. Tech. J.* **1948**, *27*, 379–423. [CrossRef]
25. De Luca, A.; Termini, S. A definition of a nonprobabilistic entropy in the setting of fuzzy sets. *Inf. Control.* **1972**, *20*, 301–312. [CrossRef]
26. Zhao, K. Study on set pair analysis and entropy. *J. Zhejiang Univ.* **1992**, *6*, 65–72.

27. Su, M.R.; Yang, Z.F.; Chen, B. Set pair analysis for urban ecosystem health assessment. *Commun. Nonlinear Sci. Numer. Simul.* **2009**, *14*, 1773–1780. [CrossRef]
28. Wang, W.; Jin, J.; Ding, J.; Li, Y. A new approach to water resources system assessment—Set pair analysis method. *Sci. China Ser. E Technol. Sci.* **2009**, *52*, 3017–3023. [CrossRef]
29. Kumar, K.; Garg, H. TOPSIS method based on the connection number of set pair analysis under interval-valued intuitionistic fuzzy set environment. *Comput. Appl. Math.* **2016**, 1–11. [CrossRef]
30. Pan, Z.; Wang, Y.; Jin, J.; Liu, X. Set pair analysis method for coordination evaluation in water resources utilizing conflict. *Phys. Chem. Earth* **2017**. [CrossRef]
31. Pan, Z.; Wu, C.; Jin, J. *Set Pair Analysis Methods for Water Resource System Evaluation and Prediction*; Science Press: Beijing, China, 2016.
32. Jin, J.; Wei, Y. *Generalized Intelligent Evaluation Method for Complex System and Its Application*; Science Press: Beijing, China, 2008.

![entropy logo] *entropy*

MDPI

Article

Testing the Beta-Lognormal Model in Amazonian Rainfall Fields Using the Generalized Space q-Entropy

Hernán D. Salas *, Germán Poveda and Oscar J. Mesa

Facultad de Minas, Departamento de Geociencias y Medio Ambiente, Universidad Nacional de Colombia, Sede Medellín, Carrera 80 # 65-223, Medellín 050041, Colombia; gpoveda@unal.edu.co (G.P.); ojmesa@unal.edu.co (O.J.M.)
* Correspondence: hdsalas@unal.edu.co

Received: 9 October 2017; Accepted: 8 December 2017; Published: 13 December 2017

Abstract: We study spatial scaling and complexity properties of Amazonian radar rainfall fields using the Beta-Lognormal Model (BL-Model) with the aim to characterize and model the process at a broad range of spatial scales. The Generalized Space q-Entropy Function (GSEF), an entropic measure defined as a continuous set of power laws covering a broad range of spatial scales, $S_q(\lambda) \sim \lambda^{\Omega(q)}$, is used as a tool to check the ability of the BL-Model to represent observed 2-D radar rainfall fields. In addition, we evaluate the effect of the amount of zeros, the variability of rainfall intensity, the number of bins used to estimate the probability mass function, and the record length on the GSFE estimation. Our results show that: (i) the BL-Model adequately represents the scaling properties of the q-entropy, S_q, for Amazonian rainfall fields across a range of spatial scales λ from 2 km to 64 km; (ii) the q-entropy in rainfall fields can be characterized by a non-additivity value, q_{sat}, at which rainfall reaches a maximum scaling exponent, Ω_{sat}; (iii) the maximum scaling exponent Ω_{sat} is directly related to the amount of zeros in rainfall fields and is not sensitive to either *the number of bins* to estimate the probability mass function or *the variability of rainfall intensity*; and (iv) for small-samples, the GSEF of rainfall fields may incur in considerable bias. Finally, for synthetic 2-D rainfall fields from the BL-Model, we look for a connection between intermittency using a metric based on generalized Hurst exponents, $M(q_1, q_2)$, and the non-extensive order (q-order) of a system, Θ_q, which relates to the GSEF. Our results do not exhibit evidence of such relationship.

Keywords: hydrology; tropical rainfall; statistical scaling; Tsallis entropy; multiplicative cascades; Beta-Lognormal model

---

1. Introduction

1.1. Statistical Scaling and Multiplicative Random Cascades

Statistical scaling has provided a rich framework to understand and model the spatiotemporal dynamics and the complexity and intermitency of rainfall fields, including (multi-)fractal, multiscaling, and random cascade models [1–19].

The strong variability and intermittence of convective tropical rainfall constitute an adequate setting to study the scaling characteristics of rainfall in a wide range of spatio-temporal scales [19–31]. In particular, Ref. [19] found that 2-D rainfall fields over Amazonia exhibit multiscaling properties in space, which means that the relationship $M_r(\lambda) \sim \lambda^{-\tau(r)}$ exhibits a non-linear behavior, where λ is the spatial scale, r the order of the statistical moment and $\tau(r)$ is the r-th moment scaling exponent. Additionally, they show that both the diurnal cycle and the predominant atmospheric regime of Amazonian rainfall (Easterly or Westerly) exert a strong control on the scaling properties of Amazonian storms, thus shedding light towards understanding the linkages between scaling statistics and physical features of Amazonian rainfall fields.

1.2. Multiplicative Random Cascades and the Beta-LogNormal Model

The Beta-Lognormal Model (hereafter BL-Model) is a discrete 2-D random cascade non-Markovian model [13] based on the observed scaling properties of rainfall, with only two parameters: σ denoting the variability of rainfall intensity, and β representing the rainy area fraction. This model provides a framework to carry out numerical experiments controlling features of rainfall (e.g., σ and β), but also to link diverse statistical and physical characteristics across spatial scales (e.g., Refs. [19,31] and Section 1.5, respectively). In addition, the model provides a tool to investigate the robustness and sensitivity of statistical metrics to poor sampling, data sparsity and intermittency of high-resolution 2-D rainfall fields.

The construction of a spatially distributed discrete random cascade model usually begins with a given mass (or volume) of rainfall over a two-dimensional ($d = 2$) bounded region [6]. The region is successively divided into b equal parts ($b = 2^d$) at each step, and during each iteration the mass obtained at the previous step is distributed into the b subdivisions through multiplication by a set of "cascade generator" W, as shown schematically in Figure 1 (for the case of $d = 2$ and $b = 4$). If the initial area (at level 0) is assigned an average intensity R_0, this gives an initial volume $R_0 L_0^d$, where L_0 is the outer length scale of the study area. Thus, at the first level the volume is subdivided into $b = 4$ subareas denoted by Δ_1^i, $i = 1, 2, ..., 4$. At the second level, each of the previous subareas is further subdivided into $b = 4$ subareas, which are denoted by Δ_2^i, $i = 1, 2, ..., 16$, for a total of $b^2 = 16$ subareas. This subdivision is continued further down the spatial scale, leading at the nth level, to b^n subareas denoted by Δ_n^i, $i = 1, 2, ..., b^n$.

As shown in Figure 1, after the first subdivision, the four subareas (Δ_1^i, $i = 1, 2, ..., b$) are assigned volumes $R_0 L_0^d b^{-1} W_1^i$, for $i = 1, 2, ..., b$. Upon subdivision, the volumes $\mu_n \Delta_1^i$ in subareas at the nth subdivision, Δ_n^i, $i = 1, 2, ..., b^n$, are given by,

$$\mu_n \Delta_1^i = R_0 L_0^d b^{-1} \prod_{j=1}^{n} W_j^i, \tag{1}$$

where, for each cascade's level j, i represents one subarea belonging to the level. The multipliers W in Equation (1) are non-negative random cascade generators, with $E[W] = 1$ to ensure that the mass is conserved on average, from one discretization level to the next one. Over and Gupta [13,32] proposed the so-called BL-Model for the cascade generators W. The BL-Model considers W as a composite generator, $W = BY$, where B is a generator from the "Beta model" and Y is drawn from a Lognormal distribution [33]. Essentially, the Beta model partitions the region into sets with and without rain, while the Lognormal model then assigns a certain amount of rainfall to each rainy area fraction. The Beta model exhibits a discrete probability mass function with just two possible outcomes ($B = 0$ and $B = b^\beta$), given as

$$P(B = 0) = 1 - b^{-\beta} \qquad\qquad P(B = b^\beta) = b^{-\beta}, \tag{2}$$

where b is the branching number and β is a parameter. Since Y belongs to the Lognormal distribution, it can be expressed as $Y = b^{-\frac{\sigma^2 \ln(b)}{2} + \sigma X}$, where X is a standard Normal r.v. and $\sigma^2 (> 0)$ is a parameter equal to the variance of $\log_b Y$, with the condition that $E[Y] = 1$. In such case, it is easy to show that the condition $E[W] = 1$ is also satisfied. The probability distribution function of $W = BY$ can thus be expressed as

$$P(W = 0) = 1 - b^{-\beta}, \tag{3}$$

$$P(W = b^\beta Y = b^{-\frac{\sigma^2 \ln(b)}{2} + \sigma X}) = b^{-\beta}. \tag{4}$$

The parameters of the BL-Model (β and σ^2) can be estimated [13,32] through the so-called Mandelbrot–Kahane–Peyriere (MKP) function [34,35]. The MKP function characterizes the fractal or scale-invariant behavior of the multiplicative cascade process. Over and Gupta [32] theoretically derived an expression for $\chi(r)$ for the BL-Model, in terms of the cascade parameters β, σ^2, b and exponent r, such that,

$$\chi_b(r) = (\beta - 1)(r - 1) + \frac{\sigma^2 \ln(b)}{2}(r^2 - r). \tag{5}$$

Thus, provided that a rainfall field belongs to a discrete random cascade with generators satisfying the BL-Model, the expression given in Equation (5) can be matched with the empirically determined estimators, $\tau(r)/d$, to estimate β and σ^2. The first and second derivatives of $\tau(r) = d\chi(r)$ with respect to r, the latter of which is given by Equation (5), can be used to obtain [20,32],

$$\tau^{(1)}(r) = d\left[\beta - 1 + \frac{\sigma^2 \ln(b)}{2}(2r - 1)\right], \tag{6}$$

$$\tau^{(2)}(r) = d\left[\sigma^2 \ln(b)\right] \tag{7}$$

Both $\tau^{(1)}(r)$ and $\tau^{(2)}(r)$ can be computed by numerically estimating the derivatives of the empirical slopes of the scaling relation between $\tau(r)$ and r, using the log–log plotting of $M(\lambda_n, r)$ versus λ_n as,

$$M(\lambda_n, r) = [\lambda_n]^{\tau(r)}, \tag{8}$$

where $M(\lambda_n, r)$ are the sample moments $M(\lambda_n, r)$ and λ_n is the corresponding scale ratio. Equations (6) and (7) are combined together to express the cascade parameters β and σ^2 in terms of $\tau^{(1)}(r)$ and $\tau^{(2)}(r)$ as follows:

$$\beta = 1 + \frac{\tau^{(1)}(r)}{d} - \frac{\sigma^2 \ln(b)}{2}(2r - 1), \tag{9}$$

$$\sigma^2 = \tau^{(2)}(r)/d\ln(b) \tag{10}$$

Equations (9) and (10) are evaluated for a given value of r. The usual practice is to use $r = 1$, although [6] used $r = 2$ for testing a space-time model of daily rainfall in Australia.

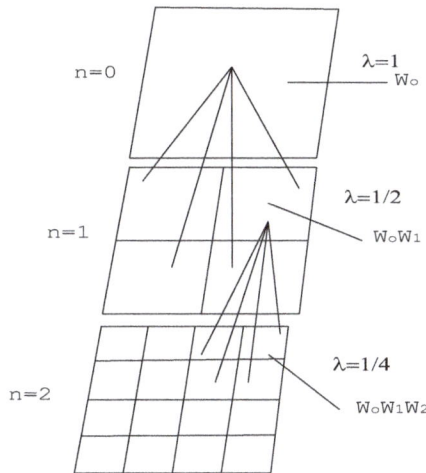

Figure 1. Schematic plot of the random cascade geometry taken from Gupta and Waymire [33].

1.3. q-Entropy

There is a wide family of generalized entropic functions with various degrees of sophistication in the literature [36–40]. In particular, Tsallis [40] proposed the concept of nonadditive entropy S_q (hereafter q-entropy), which has shown to be useful in the study of a broad range of phenomena across diverse disciplines [41,42], related to the well known Rényi entropy [43]. The q-entropy is defined as,

$$S_q = \frac{1 - \sum_{i=1}^{n} p^q(x_i)}{q - 1} \quad \left(\sum_{i=1}^{n} p(x_i) = 1; \ q \in \Re \right). \tag{11}$$

which, in the limit $q \to 1$, recovers the usual Boltzmann–Gibbs–Shannon entropy, S [44], which is additive; in other words, for a system composed of any two (probabilistically) independent subsystems, the entropy S of the sum is the sum of their entropies [45], such that, if A and B are independent,

$$S_{q=1}(A + B) = S(A + B) = S(A) + S(B). \tag{12}$$

It turns out that Tsallis entropy, S_q ($q \neq 1$), violates this property, and is therefore *nonadditive*. Thus, the additivity depends on the functional form of the entropy in terms of probabilities [45]. Therefore, if A and B are independent, then

$$S_q(A + B) = S_q(A) + S_q(B) + (1 - q)S_q(A)S_q(B). \tag{13}$$

More generally, if A and B are not probabilistically independent then,

$$S_q(A + B) = S_q(B) + S_q(A|B) + (1 - q)S_q(B)S_q(A|B). \tag{14}$$

Taking the words of Tsallis [45], the value of q is useful to characterize *the universal classes of nonadditivity*. He argues that it is determined a priori by the microscopic dynamics of the system, which means that the thermostatistical entropy is not universal but depends on the system or, more precisely, on the non-additive universality class to which the system belongs. It is worth mentioning the difference between *additivity* and *extensivity*, which is clearly explained in [45], as follows:

> *"An entropy of a system or of a subsystem is said extensive if, for a large number N of its elements (probabilistically independent or not), the entropy is (asymptotically) proportional to N. Otherwise, it is nonextensive. This means that extensivity depends on both the mathematical form of the entropic functional and the possible correlations existing between the elements of the system. Consequently, for a (sub)system whose elements are either independent or weakly correlated, the additive entropy S is extensive, whereas the nonadditive entropy S_q ($q \neq 1$) is nonextensive. In contrast, however, for a (sub)system whose elements are generically strongly correlated, the additive entropy S can be nonextensive, whereas the nonadditive entropy S_q ($q \neq 1$) can be extensive for a special value of q."*

Additionally, recent studies [28,30,31] have investigated the relationship between the q-entropy and the Generalized Pareto distribution (which is relevant in hydrological analysis). In particular, the maximization of the q-entropy under a prescribed mean leads to a Pareto probability distribution with power-law tail [28,46–48], which belongs to the family of Lévy stable distributions [49], specifically to Type II Generalized Pareto distributions. For $1 < q < 2$, the original distribution takes the form of the Zipf–Mandelbrot type [50–52], which decays as a power law for large values of x, and all moments are divergent when $3/2 < q < 2$ [53].

As such, the q-entropy constitutes a useful tool for the characterization of rainfall, and at the same time motivates interesting discussions about the physical interpretation of entropic non-linear metrics, their connection with stochastic processes (Multiplicative Cascades), and allows revisiting its applicability in geosciences (see Section 5).

1.4. Generalized Space-Time q-Entropy in Rainfall Data

Previous studies [28,30,31] introduced the Generalized Space-Time *q*-Entropy as a new method to study the organization degree and the scaling properties of rainfall, by considering the space-time structure of rainfall as a system conformed by correlated subsystems, which is evident in the hierarchical structure of convective rainfall. The time generalized *q*-entropy is defined as a function of order, *q*, and aggregation interval, *T*, as [28]:

$$S_q(T) = \frac{1 - \sum_{i=1}^{n} p^{\,q}(x_i, T)}{q - 1} \quad \left(\sum_{i=1}^{n} p(x_i) = 1; \ q \in \Re \right),$$ (15)

where *q* is the statistical order and $P(x_i, T)$ is the probability of occurrence of x_i at aggregation interval, *T*. Ever since the original work [28], different characteristics of the Space and Time Generalized *q*-entropy of rainfall were reported later on [30,31], such as:

- $S_q(T)$ decreases monotonically with *q* for all values of *T*.
- For a given value of *q*, estimates are inversely related to *T* for *q* < 0, but directly related for *q* ≥ 0.
- Estimates of $S_q(T)\mid_{q=1}$ recover the standard entropy for different values of *T*.
- Estimates of $S_q(T)$ increase with *T* for values of *q* ≥ 0, up to a certain saturation value (maximum *q*-entropy).
- The function $S(T)$ vs. *T* in log–log space, for different values of *q*, can be considered an (time) entropy analogous of the (space) *structure function* in turbulence [54].
- The scaling exponents, $\Omega(q)$, or the slope of the relation $S(T)$ vs. *T* in log–log space, for different values of *q*, exhibit a non-linear growth with *q*, such that $\Omega_{sat} \approx 0.5$ for *q* ≥ 1. This result allowed extending the conclusions from the standard Shannon entropy to the generalized *q*-entropy.
- The scaling exponents of saturation, Ω_{sat}, are different in time and space for hydrological data, such as time series of rainfall, for which $\Omega_{sat}(q > 1.0) = 0.5$, for time series of streamflows, for which $\Omega_{sat}(q > 1.0) = 0.0$, and for the spatial analysis of radar rainfall fields in Amazonia, for which $\Omega_{sat}(q > 1.0) = 1.0$.

Analogous to Equation (15), the Space Generalized *q*-Entropy was introduced by [31], to study 2-D rainfall fields, using *λ* as the spatial scale, and $P(x_i, \lambda)$ the probability of occurrence of x_i associated with *λ*, such that

$$S_q(\lambda) = \frac{1 - \sum_{i=1}^{n} p^{\,q}(x_i, \lambda)}{q - 1} \quad \left(\sum_{i=1}^{n} p(x_i) = 1; \ q \in \Re \right).$$ (16)

With the aim of linking the present study with the previous ones, some important results are worth mentioning:

- Poveda [28] studied the time scaling properties of tropical rainfall in the Andes of Colombia upon temporal aggregation and introduced the Generalized Time *q*-Entropy Function (GTEF), as a time analogous for *q*-entropy of the *structure function* in turbulence [54]. He showed that the scaling exponents, $\Omega(q)$ of the relation $S_q(T)$ vs. *T* in log–log space, for different values of *q*, exhibit a non-linear growth with *q* up to $\Omega = 0.5$ for *q* ≥ 1, putting forward the conjecture that the time dependent *q*-entropy, $S_q(T) \sim T^{\Omega(q)}$ with $\Omega(q) \simeq 0.5$, for *q* ≥ 1.
- Salas and Poveda [30] revisited results reported in [28], and analyzed the time scaling properties of Shannon's entropy for the same data set in terms of the sensitivity to the record length, and the effect of zeros in rainfall data, and proposed the GTEF to study the scaling properties of river flows. They highlighted two important results: (i) The scaling characteristics of Shannon's entropy differ between rainfall and streamflows owing to the presence of zeros in rainfall series; and (ii) the GTEF exhibits multi-scaling for rainfall and streamflows. For rainfall, the relation $S_q(T)$ vs. *T* in log–log space for different values of *q*, exhibits a non-linear growth with *q*, up to $\Omega = 0.5$ for

$q \geq 1$, in contrast to the scaling properties of river flows which exhibit a non-linear growth with q, up to $\Omega = 0.0$ for $q \geq 1$.

- Poveda and Salas [31] studied diverse topics such as statistical scaling, Shannon entropy and Space-Time Generalized q-Entropy of Mesoscale Convective Systems (MCS) as seen by the Tropical Rainfall Measuring Mission (TRMM) over continental and oceanic regions of tropical South America, and in Amazonian radar rainfall fields. The main result of their study is that both the GTEF and GSEF exhibit linear growth in the range $-1.0 < q < -0.5$, and saturation of the exponent Ω_{sat} for $q \geq 1.0$, but for the spatial analysis (GSEF) the exponent tappers off at $\langle \Omega_{sat} \rangle \sim 1.0$, whereas for the temporal analysis (GTEF) the exponent saturates at $\langle \Omega_{sat} \rangle = 0.5$. In addition, results are similar for time series extracted from radar rainfall fields in Amazonia (radar S-POL) and in-situ rainfall series in the tropical Andes.

1.5. Easterly and Westerly Regimes of Amazonian Rainfall

The WETAMC/LBA campaign (January–February 1999) found that wet-season convection in Amazonia exhibits two general modes, hereinafter, the Westerly and Easterly regimes [55], which are highly correlated to changes in the 850–700 hPa zonal wind direction [56]. According to data from the NCEP-NCAR Reanalysis, as well as from the Fazenda Nossa Senhora radiosonde, the Easterly (negative values) and Westerly (positive) regimes are clearly differentiated in both data sets [19,31]. Furthermore, radar observations from southwestern Amazonia during TRMM-LBA suggest that there was relatively little difference in the daily mean rainfall totals between Easterly and Westerly regimes [57]. In addition, the TRMM-LBA and TRMM satellite observations suggested marked differences in rainfall rate distributions, with the Easterly regime (Altiplano/southern Brazil) associated with a broader rain-rate distribution and greater instantaneous rainfall rates [56]. Precipitation features during both regimes can be summarized considering that during the Easterly regime atmospheric conditions are relatively dry, with increased lightning activity and more intense and deeper convective systems. In contrast, the Westerly regime is characterized by a diminished lightning activity, less deep convection and less intense precipitation rates [56,58,59]. The regime associated with stronger vertical development, more lightning activity, and larger instantaneous rainfall rates must be associated with a more "concentrated" daily latent heat release. In addition, two mechanisms have been proposed to explain the observed changes in the overall convective structure and lightning frequency between the two regimes, these mechanisms are tied to either thermodynamics (changes in CAPE and CIN that modify the energetics of the cloud ensemble), or aerosol loading (e.g., changes in cloud condensation nuclei (CCN) concentration that modify microphysical structure of the cloud ensemble) [56].

The aforementioned results have important practical implications in the spatial and temporal scaling features of rainfall fields [28,30,31] although the connections between entropic and scaling statistics with physical characteristics remain elusive. Furthermore, the effect of the space-time structure of rainfall in the scaling of the q-Entropy as well as its sensitivity to the number of bins and to the variability of rainfall intensity and the sample-size must be investigated in depth. Therefore, in the present study, we aim to investigate how the spatial scaling and complexity of rainfall is reflected in different entropic scaling measures within the frameworks of information theory and non-extensive statistical mechanics. The rationale and objectives of this work are presented next.

1.6. Rationale and Objectives

The objectives of our study are based upon the following considerations:

- The presence of zeros in high resolution rainfall records constitute highly important information to understand, diagnose and forecast the dynamics of rainfall [28,60,61]. Salas and Poveda [30] argued that zeros (inter-storm periods) in time series of tropical convective rainfall are associated with the timescale required by nature to build up the dynamic and thermodynamic conditions of the next storm, as an atmospheric analogous of the time of energy build-up between earthquakes, avalanches and many other relaxational processes in nature [62]. Therefore, the role of zeros and their effect on scaling statistics must be investigated to further understand and model high resolution rainfall.

- The aforementioned previous works [28,30,31] are based on available rainfall data (S-POL radar, TRMM satellite and rain gauges), and it is difficult to understand differences of the *q*-statistics in temporal and spatial scales due to factors such as the intermittency of rainfall, record length, space-time resolution of data sets, and geographic setting.
- The *number of bins* in the probability mass function constitutes a central issue to quantify entropic measures. Previous studies have shown that the scaling exponent of Shannon entropy under aggregation in time it is not sensitive to either *the number of bins* [28] or *record length* [30]. Then, it is necessary to study the sensitivity of the *q*-entropic measures in order to check their robustness to characterizing 2-D tropical rainfall fields.

The objectives of this study are manifold. They involve questions based on the previous studies [28,30,31], and new ones regarding the entropic scaling measures of rainfall. The objectives of our study are thus:

- To test theBL-Model [13] for 2-D Amazonian rainfall fields considering the Easterly and Westerly climatic regimes [55,56], and using the Generalized Space *q*-Entropy [31].
- To examine how the spatial structure of rainfall is reflected in the *q*-entropic scaling measures using the BL-Model and considering the influence of zeros in the GSEF through Montecarlo experiments, aimed at understanding the saturation of the exponent Ω_{sat} reported by [28,30,31].
- To investigate the connection between parameters of the BL-Model [13] and Amazonian rainfall fields considering the identified climatic regimes [55,56].
- To quantify the sensitivity of the *q*-entropic scaling statistics to *the number of bins* and to the *variability of rainfall intensity*, in an attempt to check the robustness of such statistical tools in the multi-scale characterization of rainfall.
- To link two important theoretical frameworks, namely stochastic processes (Multiplicative Cascades) and Information Theory (non-extensive statistical mechanics), to advance our understanding about the scaling properties of tropical rainfall.

The paper is organized as follows. Section 2 describes the study region and data set. Section 3 discusses the methods employed. Section 4 provides an in-depth discussion of results. Section 5 provides a brief discussion about the criteria and conditions to estimate entropy in geophysical data. Finally, Section 6 contains the conclusions.

2. Study Region and Data Sets

General Information

We use a set of 2-D radar rainfall fields gathered in Amazonia during the January–February 1999 Wet Season Atmospheric Meso-scale Campaign/LBA (WETAMC/LBA), which was designed to study the dynamical, microphysical, electrical, and diabatic heating characteristics of tropical convection over southwestern Amazonia [19,59,63,64]. The WETAMC campaign was developed in the state of Rondônia (Brazil). The data set used in the present study consists on radar scans of storm intensities recorded by the S-POL radar (S-band, dual polarimetric) located at 61.9982° W, 11.2213° S. Data consist of 2 km resolution microwave band reflectivity which is directly related to rainfall intensity, over a circle of ~31,000 km². Scans produced by the Colorado State University Radar Meteorology Group were available every 7–10 min at the URL http://radarmet.atmos.colostate.edu/trmm-lba/rainlba.html.

Additionally, information about zonal wind velocity at 700 hPa during the study period was obtained from the NCEP/NCAR Reanalysis [65], over the region inside 61° W to 62.8° W and 10.4° S to 12.1° S, corresponding to the area covered by the S-POL radar. Data were obtained 4 times per day at 0000, 0600, 1200 and 1800 LST. In addition, radiosonde data (62.37° W, 10.75° S) were used from the WETAMC campaign in Ouro Preto d'Oeste at the Fazenda Nossa Senhora site, located inside the S-POL radar coverage region. Radiosonde data were obtained in the URL http://www.master.iag.usp.br/lba/.

3. Methods

A set of experiments were developed to study the sensitivity of the GSEF of 2-D Amazonian rainfall fields attempting to link the spatial structure of rainfall and the emerging scaling exponents of the q-entropic analysis [28,30]. To that aim we use the BL-Model proposed by [13] to generate 2-D rainfall fields as a multiplicative random cascade, by varying the model parameters β and σ. A detailed description of each experiment is presented below.

3.1. Parameters of the BL-Model and Amazonian Precipitation Features

The first experiment is carried out by controlling the percentage of wet (rainy) and dray (non-rainy) areas, and the second one by controlling the average intensity of the rainfall field. A set of 1000 simulations for each parameter were carried out. The mass of rainfall over a two-dimensional ($d = 2$) region was considered the unit, the branching number $b = 4$ for 2-D cascades, and the level of subdivision $n = 6$ (see Figure 1) during all experiments, consistently with the observed scans from S-POL radar which are 64 rows × 64 columns matrices ($b^n = 4^6 = 4096$ values). On the other hand, with the aim to link the numerical results from the BL-Model with the S-POL observations, we compared the samples of the estimated cascade's parameters (Equations (9) and (10)) considering $\beta_{Easterly}$ vs. $\beta_{Westerly}$ and, $\sigma_{Easterly}$ vs. $\sigma_{Westerly}$ using the k-sample tests based on the likelihood ratio [66], which are more robust than traditional methods (e.g., Kolmogorov–Smirnov, Cramer–von Mises, and Anderson–Darling), but also because the climatic regimes prevailing in the study region are statistically different in terms of the continental-scale flow and lightning activity, as well as in the vertical structure of convection and other precipitation features [56].

3.2. Bin-Counting Methods and Entropic Estimators

The correct estimation of diverse informational entropy statistical parameters require to take into account diverse practical considerations. Gong and others [67] argue that there are four practical problems in the estimation of entropy using hydrologic data: (i) the zero effect; (ii) the widely used bin-counting method for estimation of PDFs; (iii) the measure effect; and (iv) the skewness effect. We focus our attention on the second practical issue within the framework of scaling theory using Shannon theoretical entropy inequality [68],

$$S(T) \leq \ln \sqrt{2\pi e V(T)}, \tag{17}$$

where $V(T)$ is the variance of the process at aggregation interval T, with the equality holding just for the for the Gaussian distribution. We use a set of parametric and non-parametric bin-counting methods, such as those introduced by Sturges [69], Dixon and Kronmal [70], Scott [71], Freedman and Diaconis [72], Knuth [73,74], Shimazaki and Shinomoto [75,76], and a recent method for estimating entropy in hydrologic data proposed by Gong et al. [67]. In work, we use common techniques reported in the literature, although there are other methods [77]. Finally, we discuss the effect of *the number of bins* on the scaling of q-entropy in rainfall fields [28,30,31].

3.3. Sample-Size and Entropy Estimators

Information-theory statistics require an adequate sample size for a proper estimation and interpretation. In spite of the existence of a large body of literature dealing with the problem of estimation of distributions for data sparsity and poor sampling [78,79], the problem constitutes a challenge in geosciences. A recent study on the estimation of Shannon entropy in hydrological records under small-samples [80], employed three different estimators: (i) maximum likelihood (ML); (ii) Chao–Shen (CS); and (iii) James–Stein-type shrinkage (JSS). Their results exhibited that the ML estimator had the worst performance of the three methods, with the largest *Mean Squared Errors* (MSE) for all sample sizes. In particular, when sample sizes are small (less than 200 data points), the entropy estimator was dramatically underestimated, although errors turned out to decrease quickly when sample sizes increased. Furthermore, when the sample size was larger than 100 data points, the accuracy of the CS and JSS estimators were basically the same, with MSE nearly equal to zero. It is worth mentioning that the said study [80] did not deal either with the effect of *the number of bins* or the role of zeros in the estimation of entropy.

At the root of the problem of the numerical estimation of entropy, is the sample-size necessary for adequately estimating the underlying probability mass function (pmf) of the process. For example, it is well known that, for the ML estimators, the larger the sample size, the better the estimates will be. In addition, high temporal resolution precipitation data sets (e.g., minutes or hours) are mostly (approximately 90%) constituted by zeros [30], which it is not the case for low-resolution data (e.g., months or years), so an adequate characterization of the tails of the Probability Distribution Function (PDF) requires longer data sets for high-resolution rainfall data than for low-resolution.

With the aim of studying the influence of sample-size in the GSEF, we will compare the q-entropy of the S-POL radar fields (matrices with 64 rows and 64 columns) and synthetic 2-D fields of the BL-Model at different cascade levels, n, to obtain fields with different sizes, ξ, e.g., the cascade level $n = 1$ means a 2×2 field, (2 rows and 2 columns) and, consequently, a cascade level $n = 7$ means a 128×128 field. In other words, a field of the BL-Model with $n = 1$ has a sample-size $\xi_i = 4$ and a field of the BL-Model with $n = 7$ has a sample-size $\xi_i = 16,384$. Using the synthetic fields, we quantify the q-entropy: (i) estimating the pmf for all the values in the synthetic rainfall field (including zeros); (ii) separing values in two subsets $P(x) = \{P(x = 0), P(x > 0)\}$; and (iii) the minimal quantity of non-zero values in the fields, to ensure robustness in the estimation of S_q. Our results are discussed in Section 4.5.

3.4. Intermittency and q-Order

High space-time resolution tropical rainfall is a highly complex and intermittent process. To analyze its intermittent behavior, several techniques have been proposed in the literature such as: spectral scale invariance analysis [31,81–84]; moment-scaling analysis [1,3,18,19,31,32,84], and intermittency exponents [84,85]. In this sense, the Generalized Space q-Entropy Function (GSEF) (Section 1.4) can be thought as a measure of (multi-) fractal behavior of a process [30,31]. In addition, the BL-Model (Section 1.2) is directly related to the Mandelbrot-Kahane-Peyriere (MKP) function [34,35] characterizing the fractal (or scale-invariant) behavior of a multiplicative cascade process. Therefore, an interesting question arises: is there any relationship between the GSEF and intermittency estimators? In order to shed light about such question, the BL-Model is used to estimate the q-order, Θ_q, and the multifractality measure, $M(q_1, q_2)$, defined by Bickel [86], Equation (22).

First, we carry out an experiment for high-resolution-spatial rainfall based on Bickel [86], who showed the relationship between intermittency (multifractality) and the *non-extensive order* (hereafter q-order) for point processes of dimension D ranging from 0.1 to 0.9. The order in a system is defined in terms of its distance from equilibrium, so the higher disordered, the closer to the equilibrium state. The q-order can be defined as,

$$\Theta(q) \equiv 1 - \frac{S_q}{S_q^{max}} \tag{18}$$

where S_q is the q-entropy (Equation (11)) and S_q^{max} is the maximum possible value of S_q for the equilibrium condition, which probabilistically can be denoted as $p_j = 1/N$ for all $j = 1, 2, ..., N$, whose S_q^{max} can be written as,

$$S_q^{max} = \frac{1 - N^{(1-q)}}{q - 1} \tag{19}$$

At this point, it is necessary to emphasize that Equation (19) is a generalization for the *extensive order* defined as a measure of complexity using the Boltzman-Gibbs-Shannon entropy [87]. In addition, the non-extensive order satisfies $0 \leq \Theta(q) \leq 1$ with $\Theta(q) = 0$ if $P_j = 1/N$ $\forall j$ and $\Theta(q) = 1$ if $P_j = \delta_{jl}$ $\forall j$, given any integer l between 1 and N [86,88].

Second, a generalization for $\Theta(q) \equiv \Theta_q$, at multiple spatial or temporal scales, is possible using a similar procedure to Equation (15). Therefore, if the spatial scale is denoted as λ, and the maximum possible value of q-entropy at each scale λ as $S_q^{max}(\lambda)$, the q-order can be rewritten as,

$$\Theta_q(\lambda) \equiv 1 - \frac{S_q(\lambda)}{S_q^{max}(\lambda)}. \tag{20}$$

From previous studies [30,31], it is easy to note that $\Theta_q(\lambda) \neq S_q(\lambda)$ but for the scaling laws $\Theta_q(\lambda) \sim \lambda^{\Omega(q)}$ and $S_q(\lambda) \sim \lambda^{\Omega(q)}$, the scaling exponents $\Omega(q)$ are exactly the same, which means that GSEF, $\Omega(q)$, does not change under such transformation.

Third, considering that a process $x(t)$ exhibiting a multifractal spectrum whose spectrum of generalized Hurst exponents, $H(q)$, is defined as [86],

$$\psi(q, T) = \langle |x(t + T) - x(t)|^q \rangle^{1/q} \propto T^{H(q)}, \tag{21}$$

which holds for some range T with non-extensive parameter q. For $\psi(q = 1, T)$ is the mean (first moment) of the absolute displacement and for $\psi(q = 2, T)$ is the standard deviation of this displacement. Therefore, the intermittency of a nonstationary process $x(t)$ can be quantified by its *multifractality*, $M(q_1, q_2)$, as the difference between two generalized Hurst exponents,

$$M(q_1, q_2) = \begin{cases} -q_1 q_2 \frac{H(q_2) - H(q_1)}{q_2 - q_1} & q_1 \neq q_2 \\ \lim_{q_1 \to q_2} M(q_1, q_2) = -q_2^2 [\partial H(q_2) / \partial q_2] & q_1 = q_2 \end{cases} \tag{22}$$

normalized such that $0 \leq M(q_1, q_2) \leq 1$ for nondegenerate processes, with $M(q_1, q_2) = 0$ for monofractals [89].

From previously mentioned considerations, the generalized Hurst exponents for 2-D rainfall fields can be computed as suggested by Carbone [90] in combination with the Equation (21) as follows,

$$\hat{\psi}_k(q) = \left(\frac{1}{(n_x - k)(n_y - k)} \sum_{i=1}^{n_x - k} \sum_{j=1}^{n_y - k} |x(i + k, j + k) - x(i, j)|^q \right)^{1/q}, \tag{23}$$

where $x \in \Re^2$, n_x and n_y are the number of rows and columns, respectively. The estimates $\hat{\psi}_k(q_1)$ and $\hat{\psi}_k(q_2)$ for $k = k_{min}, 2k_{min}, 4k_{min}, \cdots, k_{max}$, with k_{min} and k_{max} such that $\log \hat{\psi}_k(q_1)$ and $\log \hat{\psi}_k(q_2)$ exhibit linear relationship with $\log k$. Then, $\hat{H}(q_1)$ and $\hat{H}(q_2)$ are the slopes of the least-square regressions of $\log \hat{\psi}_k(q_1)$ vs. $\log k$ and $\log \hat{\psi}_k(q_2)$ vs. $\log k$, respectively. Finally, the *multifractality* is estimated as,

$$\hat{M}(q_1, q_2) = \begin{cases} -q_1 q_2 \frac{\hat{H}(q_2) - \hat{H}(q_1)}{q_2 - q_1} & q_1 \neq q_2 \\ \lim_{q_1 \to q_2} \hat{M}(q_1, q_2) & q_1 = q_2 \end{cases} \tag{24}$$

In this work, we explore the relationship between intermittency (multifractality), $\hat{M}(q_1, q_2)$, and q-order, Θ_q, for synthetic rainfall fields from the BL-Model. The results will be discussed in Section 4.6.

4. Results

4.1. Linking Parameters of the BL-Model with Precipitation Features

Following previous studies [19,31], we classify the available information from the S-POL radar considering the Easterly (negative values) and Westerly (positive) climatic regimes. Then, for both climate regimes, we estimate the parameters of the BL-Model using Equations (9) and (10). Subsequently, we estimate the pmf (histograms) and the Cumulative Distribution Functions (CDFs) of the parameters β and σ^2 (Figures 2 and 3). Results show that the histograms of β and σ^2 are statistically different for both climatic regimes, and, additionally, that the CDFs of the parameter β for the Easterly and Westerly regimes are significantly different according to a k-sample test, based on the likelihood ratio [66] at 95% confidence level, but no so for the parameter σ^2.

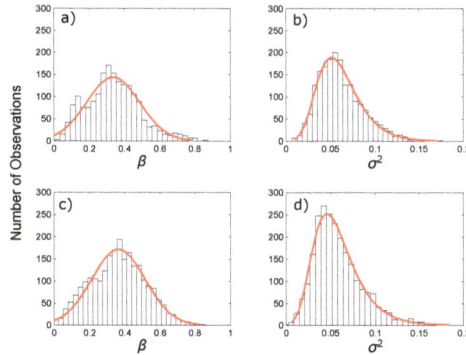

Figure 2. Histograms for Beta-Lognormal Model parameters in rainfall scans of the S-POL radar: (**a,b**) Westerly events; and (**c,d**) Easterly events. (red) the Gaussian function for β and the Generalized Extreme Value function for σ^2.

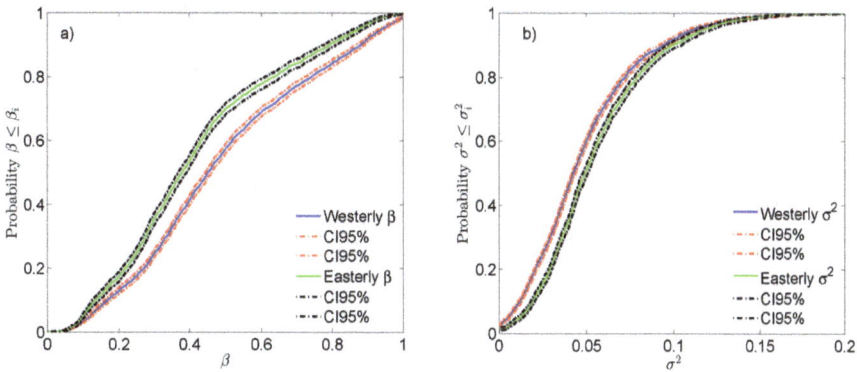

Figure 3. Empirical Cumulative Distribution Functions for cascade parameters: β (**a**); and σ^2 (**b**) using 4227 scans of the S-POL radar considering climatic regimes of Amazonia (1867 scans for Westerly and 2360 for Easterly). The figure shows the 95% confidence bounds using Greenwood's formula.

Secondly, we estimated the GSEF using a set of 1000 synthetic rainfall fields generated with the average values of β and σ^2 from the S-POL scans (see Figure 4). Figure 5b shows the GSEF of rainfall fields generated using the BL-Model, and the average of the observed S-POL rainfall fields. In addition, Figure 6a shows that the scaling exponents of the GSEF, $\Omega(q)$-observed vs. $\Omega(q)$-simulated, exhibit a very good fit. Furthermore, Figure 6b shows that the BL-Model represents adequately the relationship $S_q(\lambda) \sim \lambda^\Omega$, with saturation for $q \geq 2.5$ and $\Omega_{sat} \sim 0.5$. However, observed and simulated rainfall fields exhibit significant differences in the interval $0.5 \leq q \leq 2.5$; the S-POL scans do not exhibit power-laws for $1.0 \leq q \leq 1.5$, albeit in this interval the model shows power-laws with $R^2 \geq 0.7$.

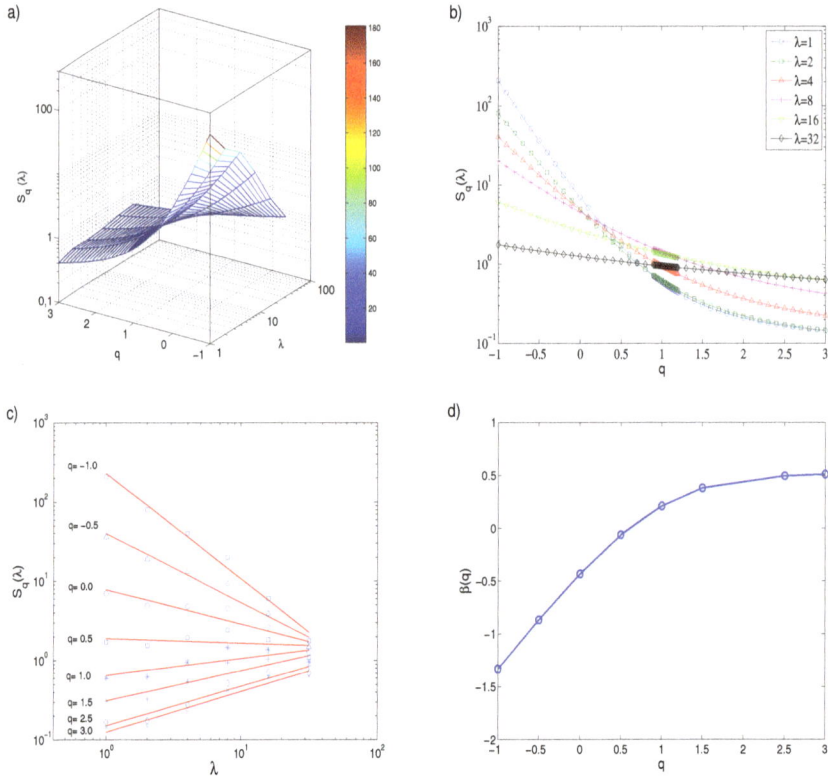

Figure 4. Space Generalized q-Entropy for the S-POL radar scan 01/10/1999 18:23:15 LST: (**a**) 3D plot of the Tsallis' entropy, S_q, for different scale factors, λ, and q-values from -1.0 to 3.0; (**b**) projection of $S(q,\lambda)$ vs. q for different values of λ; (**c**) projection of $S(q,\lambda)$ vs. λ, for different values of q, or spatial structure function for entropy; and (**d**) values of the regression slopes of the spatial structure function for entropy, Ω, as function of q, exhibiting a non-linear growth up to $\langle \Omega \rangle \sim 0.50$ for $q > 2.5$.

Finally, considering that the BL-Model has two parameters, β and σ, it is necessary to link them with precipitation features associated with both climatic regimes in Amazonian rainfall, as follows:

- The cascade parameter, β, (Table 1), for the Easterly events is greater than the Westerly events, indicating more spatially concentrated rainfall fields (more zeros in the Easterly scans). This result is related to diverse precipitation features observed during the Easterly regime, given that the atmospheric conditions are relatively dry, with increased lightning activity and more intense and deeper convective systems [56,58,59].
- The cascade parameter, σ, (Table 1), exhibits smaller (larger) values during the Easterly (Westerly) regime, indicating that the *variability of rainfall intensity* for the Westerly events is higher than for Easterly events. This result is coherent with diverse features observed during the Westerly

regime, which is characterized by less lightning activity, less deep convection and less intense precipitation rates [56,58,59].

Table 1. Scans of S-POL radar according to the two identified Amazonian climate regimes.

Description	Westerly	Easterly
Total number of scans	2607	3884
Average β for all scans	0.421	0.491
Average σ for all scans	0.235	0.221
q-value where the SGEF saturates for all scans	1.50	1.50
Average scaling exponent of saturation, Ω_{sat}, for all scans	1.0	1.0
Scans with more than 200 values non-zero (Denoted as *)	1867	2360
Scans with all values zeros	86	21
Percentage of scans *	71.6%	60.8%
Percentage of scans with less than 200 values non-zero	28.4%	39.2%
Average β for scans *	0.336	0.365
Average σ for scans *	0.248	0.242
q-value where the SGEF saturates for scans *	2.5	2.5
Average scaling exponent of saturation, Ω_{sat}, for scans *	0.38 ± 0.15	0.4 ± 0.15

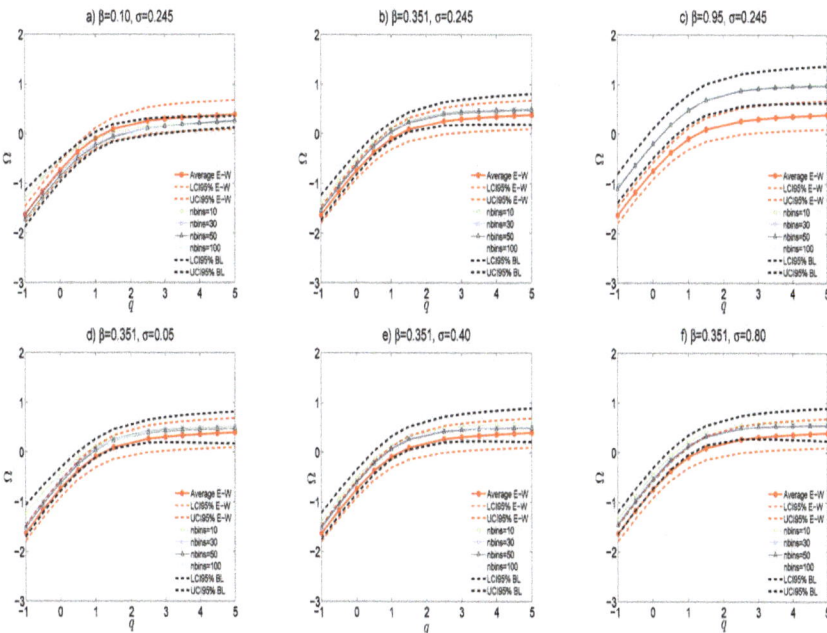

Figure 5. Space Generalized q-Entropy Function for spatially distribute rainfall as a random cascade varying cascade's parameters σ and β for 1000 simulated independently fields including zeros in the histogram. $LCI95\%$ and $UCI95\%$ are the lower and upper confidence intervals for Easterly and Westerly events (E-W) and the BL-Model model (BL): (**a–c**) varying the cascade's parameter β; and (**d–f**) varying the cascade's parameter σ. In all cases, varying the number of bins (nbins = 10, 30, 50 and 100).

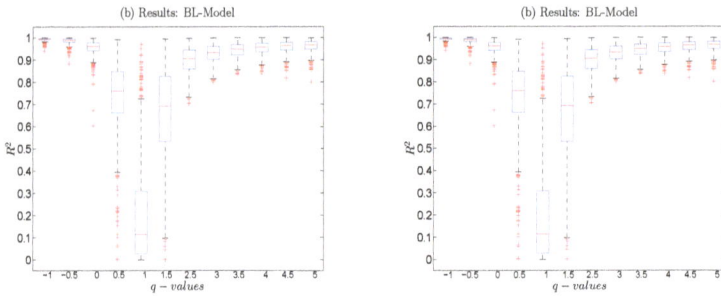

Figure 6. Validation of the BL-Model using Generalized Space q-Entropy: (**a**) comparison $\Omega(q)$-Observed vs. $\Omega(q)$-Simulated, (Solid line) relation 1:1; and (**b**) box plots of the coefficients of determination, R^2, for the power fits $S_q(\lambda) \sim \lambda^\Omega$ from 1000 synthetic fields of BL-Model with $\beta = 0.351$ and $\sigma = 0.245$. $R^2 \geq 0.85$ in the intervals $-1.0 \leq q \leq 0.0$ and $q \geq 2.5$. The histogram for estimate S_q includes zeros.

4.2. The Role of Zeros in the Generalized Space q-Entropy

With the aim of studying the influence of zeros in the estimation of the GSEF, we simulated rainfall fields using the BL-Model with cascade parameters $\beta = 0.10, 0.35$ and 0.95. In addition, 1000 simulations were carried out for each value of β. Modeling a 2-D rainfall field with $\beta = 0$ corresponds to the situation in which the cascade assigns a uniformly distributed unit rainfall intensity over the whole area. In contrast, β close to 1.0 indicates that rainfall is concentrated in a very small area. Our experiments estimate the GSEF for varying values of β, while keeping σ constant and equal to the average value for the Amazonian scans, $\sigma = 0.245$ (see Table 1). Results show that the saturation scaling exponent Ω_{sat} in the GSEF is significantly affected by the fraction of non-rainy cells (Figure 5a–c). A similar result is found for the minimum value of the scaling exponent Ω_{min} in the GSEF, which is significantly increased with the amount of zeros present in the rainfall fields. Figure 7 shows that the scaling exponents Ω_{sat} and Ω_{min} increase with the value of β. This behavior is explained by the loss rate of zeros during the change in spatial resolution, as is explained below.

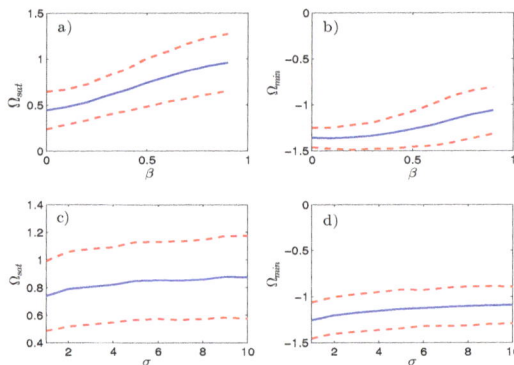

Figure 7. Sensitivity analysis of the saturation Ω_{sat}, and minimum Ω_{min} scaling exponents in the SGEF for 1000 independent rainfall fields generated by random cascade model [13]. Confidence intervals for 95% in dash line and mean value in solid line. (**a**) Varying cascade's parameter β and considering $\sigma = 0.25$ constant; (**b**) varying cascade's parameter β and considering $\sigma = 0.25$ constant; (**c**) varying cascade's parameter σ and considering $\beta = 0.5$ constant; and (**d**) varying cascade's parameter σ and considering $\beta = 0.5$ constant.

4.3. The Role of Rainfall Intensity Variability in Generalized Space q-Entropy

The influence of *the variability in rainfall intensity* on the GSFE was examined with a similar strategy. We generated rainfall fields with cascade parameters σ = 0.05, 0.40 and 0.80 (1000 simulations for each σ value), with a constant parameter β = 0.351 (see Table 1). Results show that the scaling exponent of saturation, Ω_{sat}, in the GSEF increases slowly in comparison with the case where the cascade parameter β is considered variable. This result implies that the increase or decrease in uncertainty across spatial scales is dominated by the dry areas and not by *the variability in rainfall intensity*. In general, the scaling exponents Ω_{sat} and Ω_{min} were not affected by changes in σ (Figure 7c,d). A possible explanation for this result is that the scaling exponents Ω_{sat} and Ω_{min} are directly related to the loss rate of zeros in rainfall fields when data are averaged going from higher resolution (2 km pixel size) to lower resolution scales (32 km pixel size) [30]. Then, if the amount of zeros is constant and the variability of rainfall intensity increases, the loss rate of zeros remains the same regardless of the spatial resolution, which means the scaling exponent Ω_{sat} is not affected by the cascade parameter σ.

4.4. Bin-Counting Methods and the Generalized Space q-Entropy

First, we discuss the numerical estimation of Shannon entropy for an i.i.d. Gaussian random variable under increasing aggregation intervals, T, using the analytic inequality given by Equation (17), and the multiple bin-counting methods mentioned in Section 3.2. From that equation, it is easy to see why entropy increases under aggregation of T. The theorem [91] proves that if $X_1, X_2, X_3, ..., X_n$ are i.i.d. random variables, then the expected value $E(X_j) = \mu$, with finite variance $V(X_j) = \sigma^2$. Defining the sum $S_n = X_1 + X_2 + X_3 + ... + X_n$, then the average is $A_n = \frac{S_n}{n}$, $E(S_n) = n\mu$ and $V(S_n) = n\sigma^2$.

On the other hand, we revisit the classical problem [92] of how and how well diverse information-theoretic quantities, can be estimated given a finite set of i.i.d. r.v., which lies at the heart of the majority of applications of entropy in data analysis. Paninski's paper focuses on the non-parametric estimation of entropy, and compares different estimation methods without delving into the role of the number of bins. For our proposes, we study the sensitivity of Shannon entropy (Equation (25)) to the *number of bins*. According to Shannon [44], discrete data entropy can be estimated as,

$$S(X) = -\sum_{i=1}^{n} p(x_i) \log_a p(x_i) \tag{25}$$

where $p(x_1), p(x_2), ..., p(x_k)$ represents the probability mass function, such that $\sum_{i=1}^{n} p(x_i) = 1$, and $p(x_i) \geq 0, \forall i$. Figure 8 shows that the main differences among the different estimation methods are the following:

- The bin-counting method proposed by Dixon and Kronmal [70] is the nearest to the method presented by Gong et al. (2014) for Gaussian r.v. under aggregation, for a number of aggregation intervals greater than 70.
- The theoretical inequality given by Equation (17) is better captured by Scott's method, although this method shows lower values than the theoretical expression, for aggregation intervals $T \geq 100$.
- The difference between the theoretical inequality (Equation (17)) and Gong et al.'s [67] method is explained because the "Discrete Entropy" and the "Continuous Entropy" (also referred to as "Differential Entropy") are related as:

$$\lim_{\Delta \to 0} [H_\Delta(X_d) + \log(\Delta)] = h(Xc), \tag{26}$$

where $H_\Delta(X_d)$ is the discrete entropy with the bin-width Δ and $h(Xc)$ is the corresponding continuous entropy. Thus, the continuous entropy of a r.v. requires to add $\log(\Delta)$ in the numerical estimation. In our numerical estimation, Δ was selected as the average of the bin-width estimated for the six methods (see the Section 3.2) for each aggregation interval, T, so the behavior of

Gong et al.'s method is approximately the average of the set of methods. At this point, it is worth noting that although Gong et al.'s method is designed for hydrological records, it is not free from sensitivity to *the number of bins*.

- To check the sensitivity of the scaling exponents of the GSEF to *the number of bins*, we developed a numerical experiment using the BL-Model and 1000 independent simulations for each number of bins n = 10, 30, 50 and 100, with parameters β = 0.10, 0.351, and 0.950 and σ = 0.05, 0.40 and 0.80. Figures 5 and 7 show that the GSEF is not statistically affected either by *the number of bins* or by the value of σ when *nbins* > 30 and the sample-size is bigger than 200 data. Consequently, the GSEF is not affect by the bin-counting method.

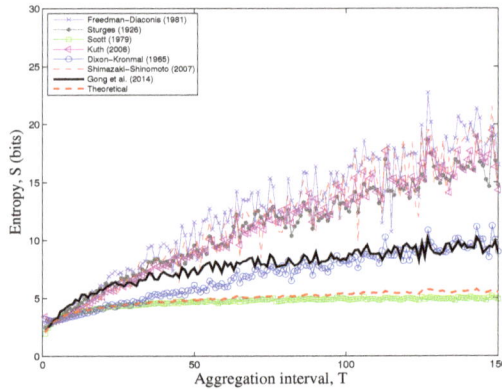

Figure 8. Numerical estimation of Shannon's Entropy using multiple bin-counting methods and the theoretical inequality (Equation (17), for a Gaussian r.v. for different levels of aggregation, T).

4.5. Sample-Size and the Generalized Space q-Entropy

We consider 6491 scans of the S-POL radar, each one with 4096 data (64 rows and 64 columns; pixel-size 2 km), of which on average 82% are zeros. Furthermore, approximately 34% of all scans have less than 200 non-zero values. Thus, using the complete data set, the probability mass function (pmf) of some scans could be concentrated in the first bins, with small informational content to the entropic estimator whereas the highest rain values could appear with a very low probability, contributing to augment the entropy. To quantify the effect of sample-size in estimating q-entropy we performed the following experiments:

First, we generated rainfall fields using the BL-Model with constant values of $\beta = 0.351$, $\sigma = 0.245$, $q = 2.5$ in S_q, and *the number of bins* of the pmf, *nbins* = 50. Then, we calculated by S_q changing the amount of cascade levels $n = 1, 2, ..., 8$ to obtain synthetic rainfall fields with different number of values in the scan (or sample-size) ξ =4, 16, ..., 65,537, and finally we estimated S_q in the following two manners:

- The pmf to calculate S_q was built considering all values in the synthetic rainfall field including zeros.
- The pmf to calculate S_q was built considering all values separated in two subsets: (i) values greater than zero (rain), i.e., $P(x > 0)$; and (ii) values equal to zero (dry), $P(x = 0)$. For the subset (i), the pmf was built and then corrected by the probability of rainfall $(1 - P(x = 0))$ thus, the probability of occurrence of rainfall can be written as $P(x) = \{P(x = 0), P(x > 0)\}$.

Figure 9 shows that, in both cases, S_q decreases with sample size, ξ, but the variance of S_q is slightly greater when the pmf was built using all values including zeros (Figure 9a) than when the pmf was built using two separated subsets (Figure 9b). Figure 10 shows that values of q-entropy S_q differ when zeros are included in the pmf and when $P(x = 0)$ is calculated separately. Figure 10a shows that values of S_q with zeros in the pmf versus S_q without zeros in the pmf are not significantly different when the sample size $\xi \leq 4096$ (i.e., 64×64 matrices or cascade level 6). In contrast, Figure 10b shows that values of S_q with the zeros in pmf versus S_q without the zeros in the pmf are significantly different in fields of sample size $\xi > 4096$ (i.e., 256×256 matrices or cascade level 8).

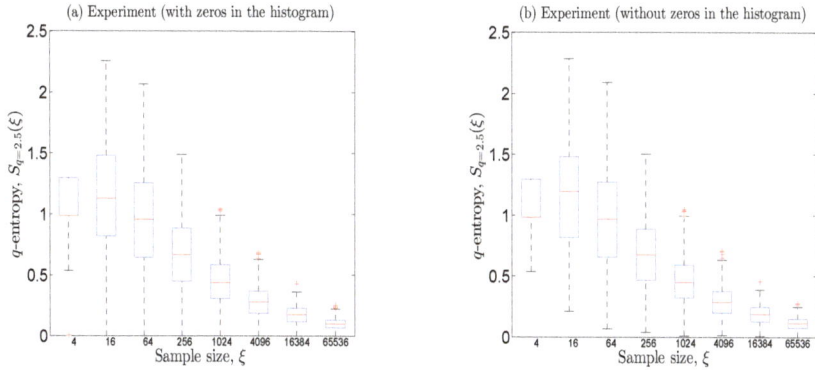

Figure 9. Boxplots for q-Entropy $S_{q=2.5}$ vs. sample sizes, ξ, in 1000 independent random cascade simulations of the Beta-Lognormal model with parameters $\beta = 0.351$ and $\sigma = 0.245$: (**a**) q-Entropy including zeros in the histogram; and (**b**) q-Entropy without including zeros in the histogram. In both cases, (*nbins* $= 50$).

Secondly, we performed a detailed examination of previous results [31] obtained using all the S-POL radar scans, and re-calculated the GSEF varying the sample-size. Results show considerable differences between sample-sizes with *less* and *more* than 200 non-zero values. Figure 11 shows differences between the GSEF for all S-POL scans, and the GSEF considering only scans with more than 200 non-zero values, but including zeros in the pmf. Furthermore, Figure 12 shows that the power laws $S_q(\lambda) \sim \lambda^\Omega$, considering the two climate regimes in Amazonian rainfall, exhibit $R^2 \geq 0.85$ in the intervals $-1.0 \leq q \leq 0.5$ and $q \geq 2.5$. For scans with more than 200 non-zero values, the scaling exponents $\Omega(q)$ of the relation $S_q(\lambda)$ vs. λ exhibit a non-linear growth with q, up to $\Omega \sim 0.5$ for $q \geq 2.5$, while in our previous study [31], the GSEF exhibited a non-linear growth with q, up to $\Omega \sim 1.0$ for $q \geq 1.0$.

These results turned out to be even more interesting with respect to those presented by [31] for the Generalized Time q-Entropy Function (GTEF) in Amazonian rainfall, whose scaling exponents $\Omega(q)$ of the relation $S_q(T)$ vs. T in log–log space, exhibit a non-linear growth with q, up to $\Omega \sim 0.5$ for $q \geq 1.0$. A thorough analysis showed that for the 400 time-series of the S-POL radar used by [31], only the 5% had less than 800 non-zero values, and that the 99% of the time-series had more than 200 non-zero values. Additionally, the scaling exponent of saturation, Ω_{sat}, for the GTEF remains the same, as well as the q-value for saturation. Therefore, our results suggest that the scaling exponent of S_q across a range of scales in space and time reaches the same maximum value $\Omega_{sat} = \Omega \sim 0.5$, but the non-additive q value of saturation differs between space scaling ($q \sim 2.5$) and time scaling ($q \sim 1.0$). According to Tsallis [45], these results reflect the differences between the space and time dynamics of the system, although their connection with the physics of rainfall is an open problem.

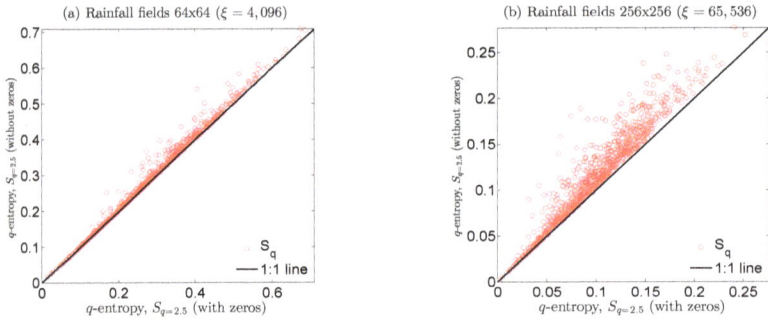

Figure 10. Comparison of q-Entropy $S_{q=2.5}$ in 1000 independent random cascade fields of the Beta-Lognormal model with parameters $\beta = 0.351$ and $\sigma = 0.245$: (a) q-Entropy S_q including zeros in the histogram vs. q-Entropy S_q without zeros in the histogram, cascade's level = 6, i.e., $\xi = 4096$; and (b) q-Entropy S_q including zeros in the histogram vs. q-Entropy S_q without zeros in the histogram, cascade's level = 8, i.e., $\xi = 65,536$. In both cases, (*nbins* = 50).

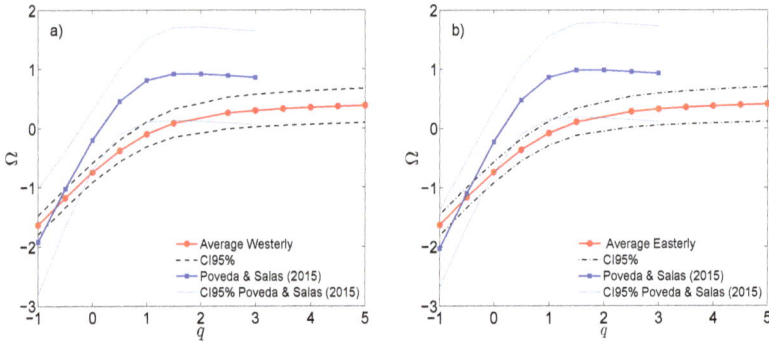

Figure 11. Space Generalized q-Entropy Functions (SGEFs) for the climate regimes of Amazonian rainfall from the S-POL radar: (**a**) Westerly events; and (**b**) Easterly events. (circles) Average SGEF for scans with more than 200 values greater than zero, (dashed lines) 95% confidence intervals (CI); (squares) average SGEF for all the scans available of each climate regime; (solid lines) 95% confidence intervals (CI).

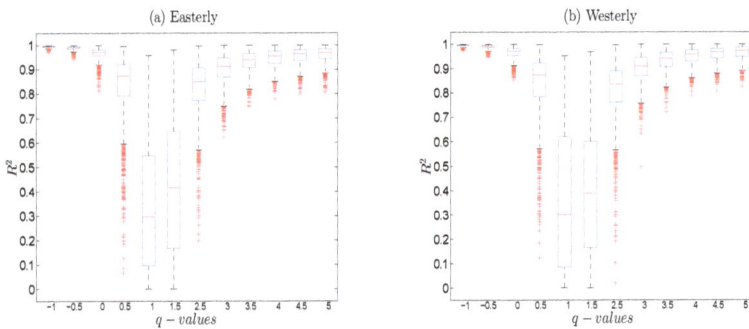

Figure 12. Coefficient of Determination, R^2, for the power fits $S_q(\lambda) \sim \lambda^\Omega$ from scans radars of the climate regimes in Amazonian rainfall. $R^2 \geq 0.85$ in the intervals $-1.0 \leq q \leq 0.5$ and $q \geq 2.5$. The histogram for estimate S_q includes zeros: (**a**) Easterly events; and (**b**) Westerly events.

4.6. Rainfall Intermittency and q-Order

As mentioned in Section 3.4, Bickel [86] showed a positive correlation between multifractality, $\hat{M}(q_1, q_2)$, and q-order, Θ_q, for point process with dimension D ranging from 0.1 to 0.9. In this study, we looked for an analogous relationship for synthetic 2-D rainfall fields from the BL-Model. Our results show that for $\hat{\psi}_k(q_1)$ vs. k and $\hat{\psi}_k(q_2)$ vs. k, both cases exhibit linear relationship in the log–log graph to estimate the generalized Hurst exponents $H(q_1)$ and $H(q_2)$ and subsequently $\hat{M}(q_1, q_2)$ as we explained in Section 3.4. Figure 13 shows a typical regression for a synthetic rainfall field created with the BL-Model with $\beta = 0.351$ and $\sigma = 0.245$, with average intermittency $\langle \hat{M}(q_1, q_2) \rangle = 0.519$. However, there is no evidence of a clear-cut relationship between $\hat{M}(q_1, q_2)$ and Θ_q in our numerical experiments (figures not shown here). Those results can be explained because the point-process model used by Bickel [86] is a Markovian model whose stochastic properties and probability distribution function (PDF) differ from point-process models for rainfall [93,94], which do not explicitly consider statistical scaling properties [95]. In addition, the BL-Model is a non-Markovian rainfall model based on the spatial statistical (multi) scaling properties, whose PDF is well known across spatial scales emerging as power laws, $S_q \sim \lambda^{\Omega(q)}$. The study of the linkages between intermittency and q-entropic statistics are outside the scope of this work that deserves to be explored in detail in future works.

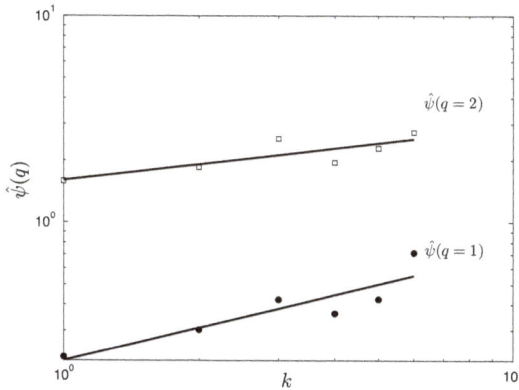

Figure 13. Typical least-square regressions $\hat{\psi}_k(q_1) \sim k^{H(q_1)}$ and $\hat{\psi}_k(q_2) \sim k^{H(q_2)}$ with $H(q = 1) = 0.097$ and $H(q = 2) = 0.412$, for a synthetic 2-D rainfall field from the BL-Model with $\beta = 0.351$, $\sigma = 0.245$ and cascade level $n = 8.0$.

5. Discussion

In spite of the increasing interest in entropic techniques in geosciences, few studies have discussed diverse underlying assumptions regarding the data to guarantee their applicability. In the case studied, high-resolution rainfall records are neither i.i.d. nor with continuous pdf, so the appropriate estimation of entropy needs clarity on the implicit assumptions in data analysis.

First, the i.i.d. condition for rainfall is not satisfied because: (i) the spatial dynamics of mesoscale rainfall has strong spatial correlations [33] (e.g., for Amazonian rainfall see [31]); and (ii) the temporal dynamics of tropical rainfall reflects long-term correlations [28] (see Figure 14). However, by definition, q-entropy, S_q, for $q \neq 1$, considers probabilistically dependent subsystems, with non-negligible global correlations, whereas Shannon entropy (S_q, for $q \rightarrow 1$) considers probabilistically independent subsystems [45,96]. Hence, the non-i.i.d. nature of data is not a restriction in the framework of non-extensive entropy, whereas in the framework of extensive entropy such non-i.i.d nature must be used under specific assumptions (e.g., for weakly correlated sub-systems).

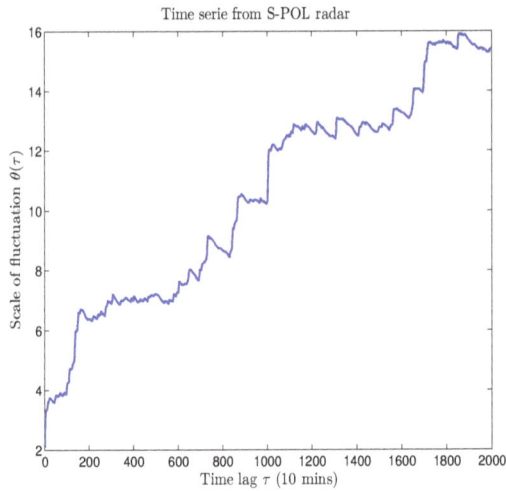

Figure 14. Scale of fluctuation, $\theta(\tau)$, for a time serie of Amazonian rainfall from S-POL radar.

Second, high-resolution Amazonian rainfall records do not satisfy the condition of continuous pdf because zeros constitute more than 80% of data in spatiotemporal scales. Therefore, although continuity in pdf is a fundamental requirement to estimate the additive (Shannon) entropy using the most common estimators [77,97], the condition of pdf's continuity for q-entropy is not clear in the literature. This is a relevant topic for further research.

Finally, an alternative option to deal with the conditions behind entropic estimators consists in finding a transformation that generates i.i.d data exhibiting a continuous pdf. However, that transformation constitutes a great challenge in geosciences, more so having in mind that such a transformation include multi-scale statistical properties.

6. Conclusions

Using 2-D radar rainfall fields from Amazonia, we investigate the spatial scaling and complexity properties of Amazonian rainfall using the Generalized Space q-Entropy Function (GSEF), defined as a set of continuous power laws covering a broad range of spatial scales, $S_q(\lambda) \sim \lambda^{\Omega(q)}$, to test for the validity of the random multiplicative cascade BL-Model in representing 2-D properties of observed rainfall fields. The spatial scaling analysis considered the Westerly and Easterly weather regimes in the Amazon basin. Our results show that for both climate regimes the GSEFs are not statistically different whereas the BL-Model parameters σ and β are statistically different.

We tested the skill of the BL-Model in reproducing the space scaling properties of q-entropy reported in previous works. Our results evidence that the BL-Model appropriately reproduces the relationship $S_q(\lambda) \sim \lambda^{\Omega}$, with saturation for $q \geq 2.5$ and $\Omega_{sat} \sim 0.5$. Furthermore, the power laws, $S_q(\lambda) \sim \lambda^{\Omega}(q)$, observed in S-POL rainfall scans exhibit $R^2 \geq 0.85$ in the intervals $-1.0 \leq q \leq 0.5$ and $q \geq 2.5$, whereas synthetic rainfall fields generated with the BL-Model exhibit power laws in the intervals $-1.0 \leq q \leq 0.5$ and $q \geq 1.5$. This result evidences that the q-entropy allows to successfully characterizing the spatial scaling properties of high resolution Amazonian rainfall, thus confirming the validity of this tool in the study of systems conformed by strongly correlated subsystems, for which the Shannon entropy ($S_{q \to 1}$) is no longer valid. In particular, the spatial scaling structure of Amazonian rainfall can be characterized by a non-additivity value, $q_{sat} \sim 2.5$, at which rainfall reaches its the maximum scaling exponent Ω_{sat}.

Using Montecarlo experiments with the BL-Model, we studied the influence of zeros and rainfall intensity on the estimation of the GSEF, aiming to explain the differences of the saturation exponent Ω_{sat} found between multiple data sets used in previous works [30,31]. Our results evidence that: (i) the scaling exponent of saturation Ω_{sat} is related to the non-rainy area fraction, represented by β; and (ii) the variability in rainfall intensity, represented by σ, does not affect significantly the GSEF. Then, changes in saturation of the scaling exponent Ω_{sat} are related to the intermittence properties of high-resolution rainfall.

In addition, we studied the influence of bin-counting methods and sample-size in the estimation of entropy and q-entropy. We used a set of parametric and non-parametric bin-counting methods showing the difficulties in estimating Shannon entropy with the well-known inequality linking variance and entropy for Gaussian i.i.d. random variables. Furthermore, we explored the sensitivity of the GSEF to *the number of bins* (*nbins*). Our results evidenced that the GSEF is a robust measure provided *nbins* $\geq$ 30. On the other hand, we performed a detailed examination of the results by Poveda and Salas [31] to check the influence of the sample-size, ξ_i, in the estimation of the q-entropy. We studied synthetic 2-D fields of the BL-Model from 2×2 (rows and columns) to 128×128 (rows and columns) quantifying the q-entropy with respect to: (i) all values inside the rainfall fields (including zeros) in the probability mass function (pmf); (ii) pmf considering $P(x) = \{P(x = 0), P(x > 0)\}$; and (iii) the minimum amount of non-zero values inside the rainfall fields. Our results evidenced that for small-samples the generalized space q-entropy function may incur in considerable bias, and our experiments showed that a rainfall field requires at least 200 non-zero values so that the estimation of q-entropy be robust.

Finally, we explored a possible relationship between a measure of multifractality $\hat{M}(q_1, q_2)$ and the q-order Θ_q. Our results suggest that the relationship found by Bickel [86] could be related to the point process used therein. In our case, for the BL-Model based on multiplicative cascades there is not evidence of such links between $\hat{M}(q_1, q_2)$ and Θ_q.

Acknowledgments: We thank the Colorado State University Radar Meteorology Group for radar data and the LBA project for providing access to the Fazenda Nossa Senhora radiosonde. The work of H.D.S. is supported by COLCIENCIAS—Call for National Doctorates 617 (2014). The work of G.P. and O.J.M. is supported by Universidad Nacional de Colombia at Medellin.

Author Contributions: H.D.S., G.P. and O.J.M. conceived and designed the study. H.D.S. prepared the data and performed computations and simulations. H.D.S., G.P. and O.J.M. discussed and analyzed the data and results. H.D.S. and G.P. wrote the manuscript.

Conflicts of Interest: The authors declare no conflict of interest.

References

1. Deidda, R.; Benzi, R.; Siccardi, F. Multifractal modeling of anomalous scaling laws in rainfall. *Water Resour. Res.* **1999**, *35*, 1853–1867.
2. Devineni, N.; Lall, U.; Xi, C.; Ward, P. Scaling of extreme rainfall areas at a planetary scale. *Chaos* **2015**, *25*, 075407.
3. Foufoula-Georgiou, E. On scaling theories of space-time rainfall: Some recent results and open problems. In *Stochastic Methods in Hydrology: Rainfall, Land Forms and Floods*; Barndor-Neilsen, O.E., Gupta, V.K., Pérez-Abreu, V., Waymire, E., Eds.; World Science: Hackensack, NJ, USA, 1998; pp. 25–72.
4. Gupta, V.K.; Waymire, E. Multiscaling properties of spatial rainfall and river flow distributions. *J. Geophys. Res.* **1990**, *95*, 1999–2009.
5. Harris, D.; Foufoula-Georgiou, E.; Droegemeier, K.K.; Levit, J.J. Multiscale statistical properties of a high-resolution precipitation forecast. *J. Hydrometeorol.* **2001**, *2*, 406–418.
6. Jothityangkoon, C.; Sivapalan, M.; Viney, N. Test of a space-time model of daily rainfall in soutwestern Australia based on nonhomogeneous random cascades. *Water Resour. Res.* **2000**, *36*, 267–284.
7. Gentine, P.; Troy, T.J.; Lintner, B.R.; Findell, K.L. Scaling in Surface Hydrology: Progress and Challenges. *J. Contemp. Water Res. Educ.* **2012**, *147*, 28–40.

8. Lovejoy, S. Area-perimeter relation for rain and cloud areas. *Science* **1982**, *216*, 185–187.
9. Lovejoy, S.; Schertzer, D. Multifractal analysis techniques and the rain and cloud fields from 10^3 to 10^6 m. In *Non-Linear Variability in Geophysics: Scaling and Fractals*; Kluwer Academic Publishers: Norwell, MA, USA, 1991; pp. 111–144.
10. Lovejoy, S.; Schertzer, D. Scale invariance and multifractals in the atmosphere. In *Encyclopedia of the Environment*; Pergamon Press: New York, NY, USA, 1993; pp. 527–532.
11. Nordstrom, K.M.; Gupta, V.K. Scaling statistics in a critical, nonlinear physical model of tropical oceanic rainfall. *Nonlinear Process. Geophys.* **2003**, *10*, 1–13.
12. Nykanen, D.K. Linkages between orographic forcing and the scaling properties of convective rainfall in mountainous regions. *J. Hydrometeorol.* **2008**, *9*, 327–347.
13. Over, T.M.; Gupta, V.K. Statistical analysis of mesoscale rainfall: Dependence of a random cascade generator on large-scale forcing. *J. Appl. Meteorol.* **1994**, *33*, 1526–1542.
14. Perica, S.; Foufoula-Georgiou, E. Linkage of scaling and thermodynamic parameters of rainfall: Results from midlatitude mesoscale convective systems. *J. Geophys. Res.* **1996**, *101*, 7431–7448.
15. Perica, S.; Foufoula-Georgiou, E. Model for multiscale disaggregation of spatial rainfall based on coupling meteorological and scaling descriptions. *J. Geophys. Res.* **1996**, *101*, 26347–26361.
16. Singleton, A.; Toumi, R. Super-Clausius-Clapeyron scaling of rainfall in a model squall line. *Q. J. R. Meteorol. Soc.* **2013**, *139*, 334–339.
17. Yano, J.-I.; Fraedrich, K.; Blender, R. Tropical convective variability as $1/f$ noise. *J. Clim.* **2001**, *14*, 3608–3616.
18. Barker, H.W.; Qu, Z.; Bélair, S.; Leroyer, S.; Milbrandt, J.A.; Vaillancourt, P.A. Scaling properties of observed and simulated satellite visible radiances. *J. Geophys. Res. Atmos.* **2017**, *122*, doi:10.1002/2017JD027146.
19. Morales, J.; Poveda, G. Diurnally driven scaling properties of Amazonian rainfall fields: Fourier spectra and order-q statistical moments. *J. Geophys. Res.* **2009**, *114*, D11104, doi:10.1029/2008JD011281.
20. Over, T. M. Modeling Space-Time Rainfall at the Mesoscale Using Random Cascades. Ph.D. Thesis, University of Colorado, Boulder, CO, USA, 1995; 249p.
21. Gorenburg, I.P.; McLaughlin, D.; Entekhabi, D. Scale-recursive assimilation of precipitation data. *Adv. Water Resour.* **2001**, *24*, 941–953.
22. Bocchiola, D. Use of Scale Recursive Estimation for assimilation of precipitation data from TRMM (PR and TMI) and NEXRAD. *Adv. Water Resour.* **2007**, *30*, 2354–2372.
23. Lovejoy, S.; Schertzer, D.; Allaire, V.C. The remarkable wide range spatial scaling of TRMM precipitation. *Atmos. Res.* **2008**, *90*, 10–32.
24. Gebremichael, M.; Over, T.M.; Krajewski, W.F. Comparison of the Scaling Characteristics of Rainfall Derived from Space-Based and Ground-Based Radar Observations. *J. Hydrometeorol.* **2006**, *7*, 1277–1294.
25. Gebremichael, M.; Krajewski, W.F.; Over, T.M.; Takayabu, Y.N.; Arkin, P.; Katayama, M. Scaling of tropical rainfall as observed by TRMM precipitation radar. *Atmos. Res.* **2008**, *88*, 337–354.
26. Hurtado, A.F.; Poveda, G. Linear and global space-time dependence and Taylor hypotheses for rainfall in the tropical Andes. *J. Geophys. Res.* **2009**, *114*, D10105, doi:10.1029/2008JD0110.
27. Varikoden, H.; Samah, A.A.; Babu, C.A. Spatial and temporal characteristics of rain intensity in the peninsular Malaysia using TRMM rain rate. *J. Hydrol.* **2010**, *387*, 312–319.
28. Poveda, G. Mixed memory, (non) Hurst effect, and maximum entropy of rainfall in the Tropical Andes. *Adv. Water Resour.* **2011**, *34*, 243–256.
29. Venugopal, V.; Sukhatme, J.; Madhyastha, K. Scaling Characteristics of Global Tropical Rainfall. In Proceedings of the European Geosciences Union General Assembly, Vienna, Austria, 27 April–2 May 2014.
30. Salas H.D.; Poveda, G. Scaling of entropy and multi-scaling of the time generalized q-entropy in rainfall and streamflows. *Physica A* **2015**, *423*, 11–26.
31. Poveda, G.; Salas, H.D. Statistical scaling, Shannon entropy and generalized space-time q-entropy of rainfall fields in Tropical South America. *Chaos* **2015**, *25*, 075409.
32. Over, T.M.; Gupta, V.K. A space-time theory of mesoscale rainfall using random cascades. *J. Geophys. Res.* **1996**, *101*, 26319–26331.
33. Gupta, V.K.; Waymire, E.C. A statistical analysis of mesoscale rainfall as a random cascade. *J. Appl. Meteorol.* **1993**, *32*, 251–267.

34. Mandelbrot, B.B. Intermittent turbulence in self-similar cascades: Divergence of high moments and dimension of the carrier. *J. Fluid Mech.* **1974**, *62*, 331–358.
35. Kahane, J.P.; Peyriere, J. Sur certains martingales de Benoit Mandelbrot. *Adv. Math.* **1976**, *22*, 131–145.
36. Mittal, D.P. On continuous solutions of a functional equation. *Metrika* **1976**, *22*, 31–40.
37. Rényi, A. On measures of information and entropy. In Proceedings of the Fourth Berkeley Symposium on Mathematics, Statistics and Probability, Berkeley, CA, USA, 20 June–30 July 1960; pp. 547–561.
38. Havrda, J.H.; Charvát, F. Quantification method of classification processes: Concept of structural α-entropy. *Kybernetika* **1967**, *3*, 30–35.
39. Sharma, B.D.; Taneja, I.J. Entropy of type (α, β) and other generalized measures of Information Theory. *Metrika* **1975**, *22*, 205–215.
40. Tsallis, C. Possible Generalization of Boltzmann-Gibbs Statistics. *J. Stat. Phys.* **1988**, *52*, 479–487.
41. Gell-Mann, M.; Tsallis, C. *Nonextensive Entropy—Interdisciplinary Applications*; Oxford University Press: New York, NY, USA, 2004.
42. Singh, V.P. *Introduction to Tsallis Entropy Theory in Water Engineering*; CRC Press: Boca Raton, FL, USA, 2016.
43. Furuichi, S. Information theoretical properties of Tsallis entropies. *J. Math. Phys.* **2006**, *47*, 023302.
44. Shannon, C.E. A mathematical theory of communication. *Bell Syst. Tech. J.* **1948**, *27*, 379–423.
45. Tsallis, C. *Introduction to Nonextensive Statistical Mechanics: Approaching a Complex World*; Springer: New York, NY, USA, 2009; doi:10.1007/978-0-387-85359-8.
46. Bercher, J.-F. Tsallis distribution as a standard maximum entropy solution with 'tail' constraint. *Phys. Lett. A* **2008**, *59*, 5657.
47. Tsallis, C.; Mendes, R.S.; Plastino, A.R. The role of constraints within generalized nonextensive statistics. *Physica A* **1998**, *261*, 534–554.
48. Plastino, A. Why Tsallis' statistics? *Physica A* **2004**, *344*, 608–613.
49. Rathie, P.N.; Da Silva, S. Shannon, Lévy, and Tsallis: A note. *Appl. Math. Sci.* **2008**, *2*, 1359–1363.
50. Zipf, G.K. *Selective Studies and the Principle of Relative Frequency*; Addison Wesley: Cambridge, UK, 1932.
51. Zipf, G.K. *Human Behavior and the Principle of Least Effort*; Addison Wesley: Cambridge, UK, 1949.
52. Mandelbrot, B.B. Structure formelle des textes et communication. *J. Word* **1953**, *10*, 1–27.
53. Abe, S. Geometry of escort distributions. *Phys. Rev. E* **2003**, *68*, 031101.
54. Frisch, U. *Turbulence: The Legacy of A. N. Kolmogorov*; Cambridge University Press: Cambridge, UK, 1995; 296p.
55. Cifelli, R.; Petersen, W.A.; Carey, L.D.; Rutledge, S.A.; da Silva-Dias, M.A.F. Radar observations of the kinematics, microphysical, and precipitation characteristics of two MCSs in TRMM LBA. *J. Geophys. Res.* **2002**, *107*, 8077, doi:10.1029/2000JD000264.
56. Petersen, W.; Nesbitt, S.W.; Blakeslee, R.J.; Cifelli, R.; Hein, P.; Rutledge, S.A. TRMM observations of intraseasonal variability in convective regimes over the Amazon. *J. Clim.* **2002**, *14*, 1278–1294.
57. Carey, L.D.; Cifelli, R.; Petersen, W.A.; Rutledge, S.A. Characteristics of Amazonian rain measured during TRMMLBA. In Proceedings of the 30th International Conference on Radar Meteorology, Munich, Germany, 18–24 July 2001; pp. 682–684.
58. Laurent, H.; Machado, L.A.; Morales, C.A.; Durieux, L. Characteristics of the Amazonian mesoscale convective systems observed from satellite and radar during the WETAM/LBA experiment. *J. Geophys. Res.* **2002**, *107*, 8054, doi:10.1029/2001JD000337.
59. Anagnostou, E.N.; Morales, C.A. Rainfall estimation from TOGA radar observations during LBA field campaign. *J. Geophys. Res.* **2002**, *107*, 8068, doi:10.1029/2001JD000377.
60. Sivakumar, B. Nonlinear dynamics and chaos in hydrologic systems: Latest developments and a look forward. *Stoch. Environ. Res. Risk Assess.* **2009**, *23*, 1027–1036, doi:10.1007/s00477-008-0265-z.
61. Gires, A.; Tchiguirinskaia, I.; Schertzer, D.; Lovejoy, S. Development and analysis of a simple model to represent the zero rainfall in a universal multifractal framework. *Nonlinear Process. Geophys.* **2013**, *20*, 343–356.
62. Peters, O.; Christensen, K. Rain: Relaxations in the sky. *Phys. Rev. E* **2002**, *66*, 036120.
63. Machado, L.A.T.; Laurent, H.; Lima, A.A. Diurnal march of the convection observed during TRMM-WETAMC/LBA. *J. Geophys. Res.* **2002**, *107*, 8064, doi:10.1029/2001JD000338.
64. Silva Dias, M.A.F.; Rutledge, S.; Kabat, P.; Silva Dias, P.L.; Nobre, C.; Fisch, G.; Dolman, A.J.; Zipser, E.; Garstang, M.; Manzi, A.O.; et al. Cloud and rain processes in a biosphere-atmosphere interaction context in the Amazon region. *J. Geophys. Res.* **2002**, *107*, 8072, doi:10.1029/2001JD000335.

65. Kalnay, E.; Kanamitsu, M.; Kistler, R.; Collins, W.; Deaven, D.; Gandin, L.; Iredell, M.; Saha, S.; White, G.; Woollen, J.; et al. The NCEP/NCAR 40-year Reanalysis Project. *Bull. Am. Meteorol. Soc.* **1996**, *77*, 437–471.

66. Zhang, J.; Wu, Y. *k*-Sample tests based on the likelihood ratio. *Comput. Stat. Data Anal.* **2007**, *51*, 4682–4691.

67. Gong, W.; Yang, D.; Gupta, H.V.; Nearing, G. Estimating information entropy for hydrological data: One-dimensional case. *Water Resour. Res.* **2014**, *50*, doi:10.1002/2014WR015874.

68. Cover, T.M.; Thomas, J.A. *Elements of Information Theory*; John Wiley & Sons, Inc.: Hoboken, NJ, USA, 2006.

69. Sturges, H. The choice of a class-interval. *J. Am. Stat. Assoc.* **1926**, *21*, 65–66.

70. Dixon, W.J.; Kronmal, R.A. The choice of origin and scale of graphs. *J. Assoc. Comput. Mach.* **1965**, *12*, 259–261.

71. Scott, D.W. On optimal and data-based histograms. *Biometrika* **1979**, *66*, 605–610.

72. Freedman, D.; Diaconis, P. On the Histogram as a Density Estimator: L_2 Theory, Z. Wahrscheinlichkeitstheorie verw. *Gebiete* **1981**, *57*, 453–476.

73. Knuth, K.H. Optimal data-based binning for histograms. *arXiv* **2006**, arXiv:physics/0605197.

74. Gencaga, D.; Knuth, K.H.; Rossow, W.B. A recipe for the estimation of information flow in a dynamical system. *Entropy* **2015**, *17*, 438–470, doi:10.3390/e17010438.

75. Shimazaki, H.; Shinomoto, S. A method for selecting the bin size of a time histogram. *Neural Comput.* **2007**, *19*, 1503–1527.

76. Shimazaki, H.; Shinomoto, S. Kernel bandwidth optimization in spike rate estimation. *J. Comput. Neurosci.* **2010**, *29*, 171–182.

77. Beirlant, J.; Dudewicz, E.J.; Györfi, L.; Van der Meulen, E.C. Nonparametric entropy estimation: An overview. *Int. J. Math. Stat. Sci.* **1997**, *6*, 17–39.

78. Pires, C.A.L.; Perdigão, R.A.P. Minimum Mutual Information and Non-Gaussianity through the Maximum Entropy Method: Estimation from Finite Samples. *Entropy* **2013**, *15*, 721–752.

79. Trendafilov, N.; Kleinsteuber, M.; Zou, H. Sparse matrices in data analysis. *Comput. Stat.* **2014**, *29*, 403–405.

80. Liu, D.; Wang, D.; Wang, Y.; Wu, J.; Singh, V.P.; Zeng, X.; Wang, L.; Chen, Y.; Chen, X.; Zhang, L.; et al. Entropy of hydrological systems under small samples: Uncertainty and variability. *J. Hydrol.* **2016**, *532*, 163–176, doi:10.1016/j.jhydrol.2015.11.019.

81. Mandelbrot, B.B. *Multifractals and 1/f Noise. Wild Self-Affinity in Physics (1963–1976)*; Springer: New York, NY, USA, 1998; p. 442.

82. Fraedrich, K.; Larnder, C. Scaling regimes of composite rainfall time series. *Tellus* **1993**, *45*, 289–298.

83. Verrier, S.; Mallet, C.; Barthés, L. Multiscaling properties of rain in the time domain, taking into account rain support biases. *J. Geophys. Res.* **2011**, *116*, doi:10.1029/2011JD015719.

84. Mascaro, G.; Deidda, R.; Hellies, M. On the nature of rainfall intermittency as revealed by different metrics and sampling approaches. *Hydrol. Earth Syst. Sci.* **2013**, *17*, 355.

85. Molini, A.; Katul, G.G.; Porporato, A. Revisiting rainfall clustering and intermittency across different climatic regimes. *Water Resour. Res.* **2009**, *45*, doi:10.1029/2008WR007352.

86. Bickel, D.R. Generalized entropy and multifractality of time-series: Relationship between order and intermittency. *Chaos Solitons Fractals* **2002**, *13*, 491–497.

87. Shiner, J.S.; Davison, M. Simple measure of complexity. *Phys. Rev. E* **1999**, *59*, 1459–1464.

88. Badin, G.; Domeisen, D.I.V. Nonlinear stratospheric variability: Multifractal detrended fluctuation analysis and singularity spectra. *Proc. R. Soc. A Math. Phys. Eng. Sci.* **2016**, *472*, 20150864, doi:10.1098/rspa.2015.0864.

89. Bickel, D.R. Simple estimation of intermittency in multifractal stochastic processes: Biomedical applications. *Phys. Lett. A* **1999**, *262*, 251–256.

90. Carbone, A. Algorithm to estimate the Hurst exponent of high-dimensional fractals. *Phys. Rev. E* **2007**, *76*, 056703.

91. Grinstead, C.M.; Snell, J.L. *Introduction to Probability*, 2nd ed.; American Mathematical Society: Providence, RI, USA, 1997.

92. Paninski, L. Estimation of Entropy and Mutual Information. *Neural Comput.* **2003**, *15*, 1191–1253.

93. Rodríguez-Iturbe, I.; Power, B.F.; Valdes, J.B. Rectangular pulses point process models for rainfall: Analysis of empirical data. *J. Geophys. Res. Atmos.* **1987**, *92*, 9645–9656.

94. Cowpertwait, P.; Isham, V.; Onof, C. Point process models of rainfall: Developments for fine-scale structure. *Proc. R. Soc. Lond. A Math. Phys. Eng. Sci.* **2007**, *463*, 2569–2587.

Entropy **2017**, *19*, 685

95. Olsson, J.; Burlando, P. Reproduction of temporal scaling by a rectangular pulses rainfall model. *Hydrol. Process.* **2002**, *16*, 611–630.
96. Boon, J.P.; Tsallis, C. Special issue overview nonextensive statistical mechanics: New trends, new perspectives. *Europhys. News* **2005**, *36*, 185–186.
97. Robinson, P.M. Consisten nonparametric entropy-based testing. *Rev. Econ. Stud.* **1991**, *58*, 437–453.

entropy

MDPI

Article

Randomness Representation of Turbulence in Canopy Flows Using Kolmogorov Complexity Measures [†]

Dragutin Mihailović [1,*], Gordan Mimić [1], Paola Gualtieri [2], Ilija Arsenić [1] and Carlo Gualtieri [2]

[1]   Faculty of Agriculture, University of Novi Sad, Dositej Obradovic Sq. 8, 21000 Novi Sad, Serbia;
      gordan.mimic@polj.uns.ac.rs (G.M.); ilija@polj.uns.ac.rs (I.A.)
[2]   Department of Civil, Architectural and Environmental Engineering, University of Naples Federico II,
      Via Claudio 21, 80125 Naples, Italy; paola.gualtieri@unina.it (P.G.); carlo.gualtieri@unina.it (C.G.)
*   Correspondence: guto@polj.uns.ac.rs; Tel.: +38-121-485-3203
†   This paper is an extended version of our paper published in Proceedings of the 8th International Congress
      on Environmental Modelling and Software (iEMSs), 10–14 July 2016, Toulouse, France.

Received: 11 August 2017; Accepted: 25 September 2017; Published: 27 September 2017

Abstract: Turbulence is often expressed in terms of either irregular or random fluid flows, without quantification. In this paper, a methodology to evaluate the randomness of the turbulence using measures based on the Kolmogorov complexity (KC) is proposed. This methodology is applied to experimental data from a turbulent flow developing in a laboratory channel with canopy of three different densities. The methodology is even compared with the traditional approach based on classical turbulence statistics.

Keywords: turbulent flow; canopy flow; randomness; coherent structures; Shannon entropy; Kolmogorov complexity

1. Introduction

Randomness is narrowly related to the unpredictability of model or observation results and the inability to predict them with sufficient accuracy. Naturally, several questions arise: what in fact constitutes randomness; how can we quantify it; and why do we think about? In addition, randomness and volatility are oft-used agreeably in the same way that precision and accuracy receive the same colloquial treatment. Each means something on its own and merits consideration as such. Here, to the authors' knowledge, we count some papers, from the last two decades, discussing the complexity of the phenomenon from the perspectives of different scientific fields: cosmic radiation [1], statistical mechanics [2,3], chaotic behavior [4] and turbulence [5]. Kreinovich and Kunin [4] suggested the use of the ideas of the Kolmogorov complexity to distinguish singular and non-singular solutions. Lorenz [6] explored how the complexity of wind records (and more generally, of velocity descriptions of thermally-driven flow) relates to the available work and the frictional or viscous dissipation. Weck et al. [7] used the complexity to analyze fluctuating time series of different turbulent plasmas.

Some problems of turbulent flow remain unanswered because a more complete definition of turbulence has not been proposed yet [8]. In particular, randomness, one of the fundamental characteristics of a turbulent flow, is often verbally expressed in terms of the irregularity of the fluid flow, without its quantification. An overview of such definitions is reported by Ichimiya and Nakamura [8]. The only exception is Pope's definition of randomness related to turbulence [9]. Evaluation of the level of randomness in turbulence could be particularly interesting in canopy flows [10–14]. Vegetation in open channel flow considerably affects the flow and turbulence structure, so interactions between the lower layer with the vegetation (i.e., the vegetation layer) and the surface layer that is above the vegetation evolve [13]. In the case of submerged canopy flows, a roughness sublayer (RSL) is formed [15–17]. Within the RSL, three distinct zones exist (Figure 1a): (i) the deep zone (RS1), primarily dominated by small eddies associated with the von Kármán streets; (ii) the RS2 zone, where the superposition of attached eddies and Kelvin–Helmholtz waves cause an inflection point in the mean velocity profile (Figure 1b), which develops when two co-flowing streams have different velocities [11]; in this turbulent mixing region, Kelvin–Helmholtz instability causes coherent turbulent structures that travel downstream in the fluid [15,16,18]; (iii) the RS3 zone, where the rough-wall boundary layer occurs. For a long time, there was an interest in the fluid mechanics for better understanding the physical processes involved in flow-canopy interaction, which includes interaction between the zones described above, e.g., the turbulent boundary layer and the outer laminar region, wall and free-shear turbulent flows exhibiting a coherent structure [8,19]. Based on the Kolmogorov complexity (KC) [20], Lempel and Ziv developed an algorithm for calculating the measure of randomness (LZA) [21].

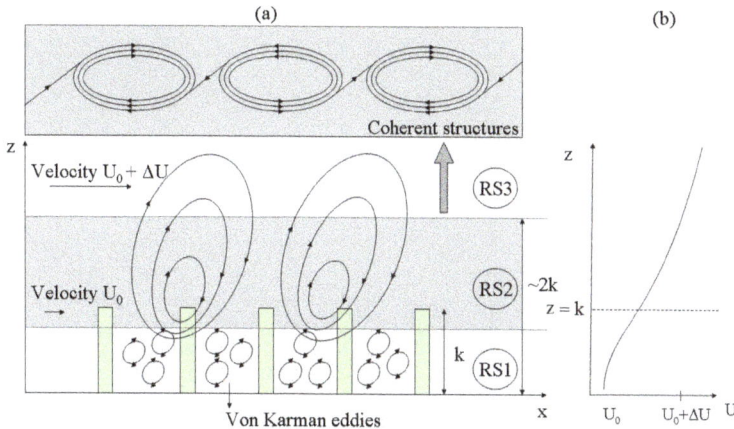

Figure 1. (**a**) A schematic diagram of the eddies structure over and within the canopy: (i) RS1 zone ($z/k < 1$; k is the height of the cylinders used for modeling canopies) where the flow field is primarily dominated by small eddies associated with the von Kármán streets; (ii) RS2 zone straddles the top portion of the canopy, and it is dominated by a mixing layer; and (iii) RS3 zone ($z/k > 2$) is the classical boundary-layer region dominated by eddies with length scales proportional to $(z - d)$, where d is the displacement height; (**b**) the mean velocity profile within the canopy $u(z) = u(k)\sqrt{-\beta_1 e^{-\beta_3 z} + \beta_2 e^{\beta_3 z}}$ is obtained from the solution of the partial differential equation $\partial u^2 / \partial z = [(2C_d \lambda_d k^2)/(\sigma_s P_s)]u^2$ where: $u(k)$ is the velocity at the height k; β_1, β_2 are parameters depending on the morphological and aerodynamic characteristics of the canopy and $\beta_3 = [(2C_d \lambda_d k^2)/(\sigma_s P_s)]$; C_d is the drag coefficient; σ_s is the parameter of proportionality between the turbulent transport coefficient and velocity within the canopy; λ_d is the roughness density; and P_s is the shelter factor [22].

Before we clarify the intention of this paper, we concisely summarize some ideas by Ichimiya and Nakamura [8] in regards to randomness representation in turbulent flows with Kolmogorov complexity in mixing layer. They mentioned that theoretical books do not use phrases such as "the turbulence is the condition whose property is...", but they start from the distinction between mean and irregularly-fluctuating velocities [23,24], and also, discussing different definitions of turbulence, they state that the essential property of turbulence is verbally expressed in terms of irregular, random or chance. Thus, Pope [9] defines randomness in relation to turbulence [25]. This definition, which seems reasonable and practical, cannot be used to measure randomness, and it is difficult to relate it to Kolmogorov's probability axiom. Kolmogorov himself describes the application of probability theory, but does not mention random itself at that time [26].

In the treatment of turbulence based on values of turbulence intensity, it is broadly considered as a measure of randomness. However, this definition, which seems reasonable and practical, cannot be used as a complete measure of randomness [8] in the sense as it is defined in [9]. Let us note that Sato et al. [27] proposed a randomness factor defined as the ratio of the energy of the continuous spectrum to the total energy in the power spectrum distribution of the velocity fluctuations [9]. However, there is a problem that the randomness factor takes the same value in the two locations where the flow conditions are different. The goal of this paper is to examine the degree of turbulence in relation to the KC and measures based on that complexity. To explicate that goal, we use Figure 2. The left side of this figure depicts two time series of the velocity measured at two relative heights (see Figure 1) $z/k = 0.067$ (purple) and $z/k = 2.667$ (red), respectively, while its right side represents the corresponding distributions of frequencies of velocities that occur in certain intervals. Looking at the distributions of velocities (right panel), we can conclude at which relative height the turbulent motion is more random (red distribution) if we used the measure of randomness based on the values of turbulence intensity. In the further text, we will see that if randomness is evaluated using the KC measure, randomness is greater on the other relative height (purple), which could be in some sense confusing.

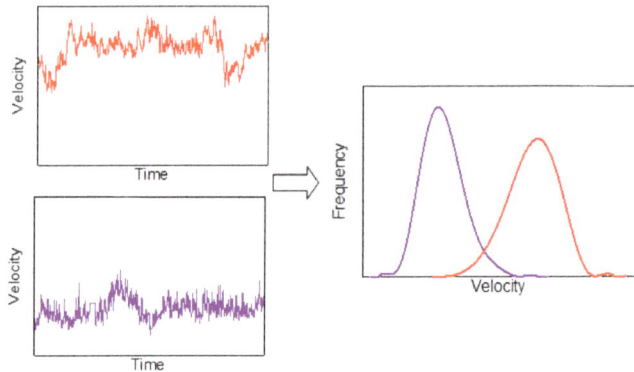

Figure 2. Towards quantification of the randomness of turbulence in canopy flows.

The proposed methodology has been tested with experimental measurements in a canopy flow with a simplified geometry, carried out in a laboratory channel of the Department of Civil, Architectural and Environmental Engineering of the University of Naples Federico II (Naples, Italy), since the focus of this paper is on the application of the KC complexity, not on the base flow. However, it should be pointed out that in the last decade, some authors performed measurements both on more realistic geometries and using more advanced techniques, at least 2C laser Doppler velocimetry (LDV) [28,29].

2. Kolmogorov Complexity and Information Measures Derived from It

2.1. Kolmogorov Complexity

The Kolmogorov complexity $K(x)$ of an object x is the length, in bits, of the smallest program that can be run on a universal Turing machine (U) and that prints object x. The complexity is maximized for random strings [30]. This information measure was developed by Kolmogorov [31], whose comprehensive elaboration can be found in [20]. The complexity $K(x)$ is not a computable function of x. However, a suitable approximation is considered as the size of the ultimate compressed version of x and a lower bound for which a real compressor can achieve [21,32]. The issue of non-computability and the LZA algorithm is comprehensively discussed by Kaspar and Schuster [33]. The calculation of the KC complexity of a time series $\{x_i\}, i = 1, 2, ..., N$ by the LZA algorithm includes the following steps. (1) Encode the time series by constructing a sequence S of the characters 0 and 1 written as $s(i), i = 1, 2, ..., N$ according to the rule $s(i) = 0, x_i < x_*; 1, x_i \geq x_*$, where x_* is the mean value of the time series, selected as the threshold [34]. Other encoding schemes are also available [35,36]. (2) Calculate the complexity counter $c(N)$. The $c(N)$ is defined as the minimum number of distinct patterns contained in a given character sequence [37]. The complexity counter $c(N)$ is a function of the length of the sequence N. The value of $c(N)$ is approaching an ultimate value $b(N)$ as N is approaching infinity, i.e., $c(N) = O(b(N))$ and $b(N) = N/log_2N$. (3) Calculate the normalized information measure $C_k(N)$, which is defined as $C_k(N) = c(N)/b(N) = c(N)log_2N/N$. For a nonlinear time series, $C_k(N)$ varies between zero and one, although it can be larger than one [38]. The LZA algorithm is described in papers [21,33], but in a condensed form. However, here, we will explicate this algorithm through an example that is rather illustrative compared to the examples used in the above mentioned papers. Note that the pattern is a sequence in the coded time series, which is unique and non-repeatable. Recognition of the patterns with the Lempel–Ziv algorithm could be described in the following way: (1) the first digit, no matter if 0 or 1, is always first pattern; (2) define the sequence S, consisting of the digits contained in already recognized patterns; sequence S grows until the whole time series is analyzed; (3) define sequence Q, needed to examine the time series; it is formed by adding new digits until Q is recognized as the new pattern; (4) define the sequence SQ, by adding sequence Q to sequence S; (5) form the sequence $SQ\pi$ by removing the last digit of sequence SQ; (6) now examine if sequence $SQ\pi$ contains sequence Q; (7) if sequence Q is contained in sequence $SQ\pi$, add another digit to sequence Q and repeat the process until the mentioned condition is satisfied; (8) if sequence Q is not contained in sequence $SQ\pi$, then Q is a new pattern. Now, the new pattern is added to the list of known patterns called vocabulary R. Sequence SQ now becomes new sequence S, while Q is emptied and ready for further testing. Now, let us consider a short example with the binary sequence 1011010. The procedure obtained by the LZA can be simply described in this way:

- the first digit is always the first pattern, which implies $\rightarrow R = 1 \cdot$
- $S = 1, Q = 0, SQ = 10, SQ\pi = 1, Q \notin v(SQ\pi) \rightarrow R = 1 \cdot 0 \cdot$
- $S = 10, Q = 1, SQ = 101, SQ\pi = 10, Q \in v(SQ\pi) \rightarrow 1 \cdot 0 \cdot 1$
- $S = 10, Q = 11, SQ = 1011, SQ\pi = 101, Q \notin v(SQ\pi) \rightarrow R = 1 \cdot 0 \cdot 11 \cdot$
- $S = 1011, Q = 0, SQ = 10110, SQ\pi = 1011, Q \in v(SQ\pi) \rightarrow 1 \cdot 0 \cdot 11 \cdot 0$
- $S = 1011, Q = 01, SQ = 101101, SQ\pi = 10110, Q \in v(SQ\pi) \rightarrow 1 \cdot 0 \cdot 11 \cdot 01$
- $S = 1011, Q = 010, SQ = 1011010, SQ\pi = 101101, Q \notin v(SQ\pi) \rightarrow R = 1 \cdot 0 \cdot 11 \cdot 010 \cdot .$

Finally, vocabulary R consists of the patterns 1, 0, 11 and 010, which means that in this particular case, the complexity of the counter is $c(N) = 4$.

2.2. The Kolmogorov Complexity Spectrum and Its Highest Value

The quantification of the complexity of a system is one of the aims of nonlinear time series analysis. Due to artifacts in various forms (spurious experimental results, etc.), it is often not easy to get desirable and reliable information from a series of measurements. Note that the time series, obtained either by measurement or a modeling procedure, is the only source for evaluating the level of complexity of the environmental system (i.e., canopy flow in this paper, through its flow rate). However, the KC, as an information measure, has two drawbacks: (i) it cannot distinguish time series with different amplitude variations and similar random components; and (ii) in the conversion of a time series into a string, its complexity is hidden in the coding rules. Thus, in the procedure of establishing a threshold for a criterion for coding, some information about the structure of the time series can be lost. In time series analysis of the canopy flow, we use two information measures based on the KC complexity: (i) the Kolmogorov complexity spectrum and (ii) the Kolmogorov complexity spectrum's highest value, which are introduced in [39] and briefly described. According to Definition 1 in [39], the time series $\{x_i\}, i = 1, 2, ..., N$ consists of either measured or calculated values. The Kolmogorov complexity spectrum of time series $\{x_i\}$, i.e., the sequence $\{c_i\}, i = 1, 2, ..., N$ is obtained by the LZA algorithm applied N times on time series, where thresholds $\{x_{t,i}\}$ are all elements in $\{x_i\}$. Namely, the original time series samples are converted into a set of 0-1 sequences $\{S_i^{(k)}\}$, $i = 1, 2, ..., N, k = 1, 2, ..., N$, defined by the comparison with a threshold $\{x_{t,k}\}$, i.e., $S_i^{(k)} = \{0, x_i < x_{t,k} \, or1, x_i \geq x_{t,k}\}$. After the LZA algorithm is applied to each element of the series $\{S_i^{(k)}\}$, we get the KC complexity spectrum $\{c_i\}$, $i = 1, 2, ..., N$. This spectrum allows us to explore the range of amplitudes in a time series representing a physical system with highly enhanced stochastic components, i.e., with highest complexity. Note that for a large number of samples in the time series (as in this paper), the computation of the KC spectrum is computationally very consumable. Then, it is recommended to divide the range of the minimal and maximal values in the time series into subintervals, which are further used as thresholds. We used 100 thresholds to obtain each spectrum. The highest value K_{max}^C in this series, i.e., $K_{max}^C = max\{c_i\}$, is called the Kolmogorov complexity spectrum highest value (KCM). The KC complexity carries less information about the complexity of the time series than the KC spectrum, i.e., KCM, does. Namely, the KC gives average information about the complexity of the time series. In contrast, KCM carries the information about the highest complexity among all complexities in the spectrum. Therefore, this measure should be included to better understand the system's randomness and organization.

To illustrate the usefulness of the above measures, we consider two time series of the instantaneous velocity measured at $z/k = 0.067$ (purple) and $z/k = 2.667$ (red), respectively (see Figure 2). Figure 3 shows that the flow dynamics of turbulence varies with the relative height z/k and the Kolmogorov complexity spectrum's highest value (KCM). In fact, at $z/k = 0.067$ (purple curve), the KC is higher than at $z/k = 2.667$ (red curve). Therefore, we can conclude that at $z/k = 0.067$, the turbulence regime is more random, i.e., shows a more pronounced presence of stochastic components than at $z/k = 2.667$. In other words, these spectra can effectively point out the dissimilarities between the turbulent regimes for the two different relative heights.

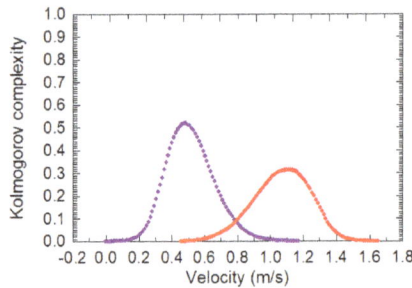

Figure 3. The Kolmogorov complexity spectra for two time series of the velocity measured at $z/k = 0.067$ (purple) and $z/k = 2.667$ (red), respectively. Both lines are fitting curves, which are obtained from the calculated discrete values of the Kolmogorov complexity (KC) spectra.

3. Experimental Details

The experiments were performed in a laboratory channel with a variable bed slope, which was 8 m long and 0.4 m wide. The channel slope is the difference in elevation between two points of the channel bed divided by the distance between them measured along the channel. Vegetation covered the bed of the channel and consisted of rigid cylinder rods of the same height and diameter ($k = 0.015$ m, $d_c = 0.004$ m), set in different aligned arrangements (rectangles or squares), with three different densities λ_d, and Q is the volumetric flow rate (Table 1).

Table 1. Experimental conditions.

Test	D1	D2	D3
λ_d (m$^2 \cdot$m^{-2})	0.024	0.048	0.096
ϑ (°)	0.03	0.02	0.03
Q (l s^{-1})	33	22	22
h_u (cm)	6.35	6.44	6.29
Cylinders per unit area	400	800	1600

Figure 4 depicts the experimental channel where the experiment measurements were carried out and the cylinders, modeling rigid aquatic plants in the literature on vegetated flows [11,13]. In Figure 5, sets of the cylinders for different densities are represented with the vertical cross section in the A-A direction (see figure).

Experimental measurements were carried out in a section set at 12 cm from the hydraulic left of the channel, at the center of a rectangle or a square of cylinders. Vegetation covered the entire length of the channel, and experimental measurements were carried out in a section where the flow attained the uniform condition, i.e., the slope of the free surface was equal to the channel slope ϑ. The friction velocity u_* was calculated as $\sqrt{g\vartheta(h_u - k)}$ where g is the gravitational acceleration, while h_u is the flow depth in uniform conditions. The validity of this method was confirmed comparing its results with the peak values of the Reynolds stress, in another condition of uniform flow on a vegetated bed. Vegetation density λ_d was evaluated as the total canopy frontal area per unit area. The vegetation was always fully submerged with submergence h_u/k of about four.

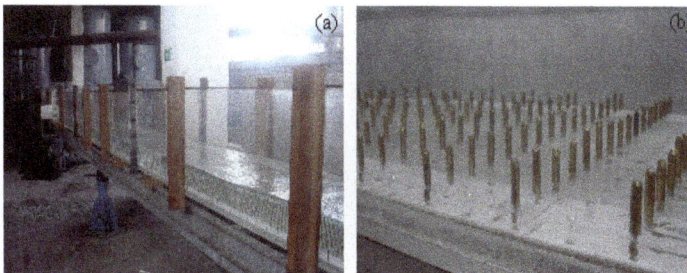

Figure 4. Experimental channel (**a**) and cylinders as the model of rigid vegetation (**b**) [40].

The measurements of the instantaneous velocity in the direction of the main stream were performed with a one-component Polytec LDV system used in back scatter mode. The system was accurately aligned [41]. Due to water characteristics, no artificial seeding of the channel was used. The light source was a laser diode with a maximum power of 65 mW at a wavelength between 810 nm and 840 nm. The sampling frequency for the LDV measurements was 2000 Hz, and in order to obtain a sufficient number of strong bursting events, the acquisition time in each measurement point was equal to 135 s, so the sampling data of N = 270,000 of instantaneous velocity were collected and analyzed for turbulence statistics. The sampling frequency and the sampling time were selected to accurately represent the investigated flow and to result in data files supported by the software package used for computation. For commercial LDV systems, processor information is converted to velocity through software using the equation $V_x = K\lambda f_D$, where: V_x is the component of the instantaneous velocity along the streamwise direction; $K\lambda$ is the calibration factor in nm; $K = 1/2\sin\theta$; θ is the half-angle of the beam intersection angle in the measurement point set in the focus of the lens; λ is the laser wavelength in nm; and f_D is the Doppler frequency in MHz. The uncertainty in the measurement of the Doppler frequency can be assumed negligible [42], and for modern signal processors, the uncertainty of the laser wavelength λ is negligible [43]. The uncertainty in the medium velocity has been evaluated to be about 0.3%. For each measurement point, the analog signal from the processor, by means of an oscilloscope, was accurately checked, to verify the Doppler signal quality [11,44]. LDV provides flow velocity data having high quality, and therefore, it still remains the preferred measuring technique for complex turbulent flows' study [44], confirming the adequacy of fluid velocity measurements based on frequency acquisition and analysis [11,13,45,46]. In Nezu and Sanjou [13] and Nezu and Onitsuka [45], the most accurate measurement device was LDV, and the accuracy of PIV (particle image velocimetry) was evaluated by comparing PIV data with LDA data.

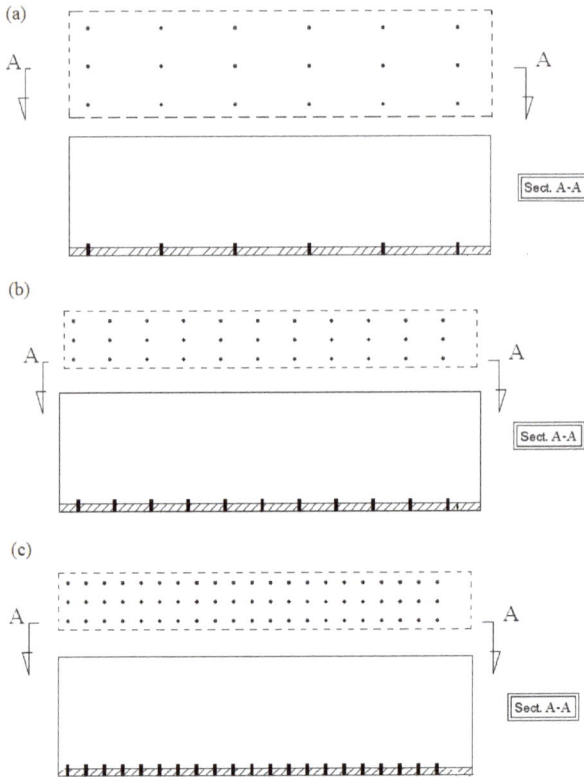

Figure 5. Sets of the cylinders for density equal to: (**a**) D1 = 0.024; (**b**) D2 = 0.048 and (**c**) D3 = 0.096.

4. Results

4.1. Basic Statistical Parameters

Furthermore, in order to deepen the canopy-flow interaction, it could be interesting to evaluate the distributions of skewness and kurtosis of the instantaneous streamwise velocities' distributions, for the different densities of the canopy. In fact, skewness quantifies how symmetrical the distribution of the instantaneous streamwise velocities is, and kurtosis quantifies whether the peak of the data distribution matches the normal distribution. In particular, if the distribution of the instantaneous streamwise velocities is symmetrical, the positive and negative values balance each other, and the skewness is close to zero. If the distribution is not symmetrical, the skewness is positive if the distribution is skewed to the right and negative if skewed to the left. In addition, the skewness values that are generally negative or positive denote the presence, respectively, of lower or higher velocity values from the mean value. Positive kurtosis values indicate a relatively peaked distribution, while negative kurtosis values indicate a relatively flat distribution. Therefore, the distributions of skewness and kurtosis (Table 2) show the differences between the coherent eddy structure within and just above the canopy and the eddy structure in the surface layer above the canopy. In particular, skewness goes from positive inside the canopy to negative above the canopy, where it tends to a Gaussian value. Kurtosis, within the canopy, is positive or negative (for the lowest canopy density) and above the canopy is negative, tending to the Gaussian value. Within the canopy, denser canopies have larger positive skewness and kurtosis. The trend of skewness and kurtosis confirms the scheme of the RSL, proposed for mean velocity and turbulent intensities: in the RS3 zone, for all the densities, skewness and kurtosis tend to the Gaussian value, confirming that the turbulent structure is similar to the one in a boundary layer; in the RS1 zone, larger skewness and kurtosis for denser canopies show the role of the sweeps of high velocity fluid down into the dense canopy space; in the RS2 zone, the turbulent structure is analogous to a mixing layer, where coherent eddies are generated due to the inflection-point instability and govern the vertical transport of momentum [11,13].

Table 2. Basic statistical parameters of the instantaneous streamwise velocities at five relative heights.

	D1			
z/k	Mean	SD	Skew	Kurtosis
2.667	1.49	0.1604	−0.3934	−0.1155
1.8	1.35	0.2035	−0.2483	−0.4652
1	1.19	0.2069	−0.0657	−0.3909
0.4	0.97	0.1878	0.1676	−0.2180
0.067	0.90	0.1715	0.1985	−0.1444
	D2			
z/k	Mean	SD	Skew	Kurtosis
2.667	1.59	0.1412	−0.3644	−0.1181
1.8	0.93	0.1648	−0.1675	−0.4227
1	0.76	0.1677	0.0389	−0.3873
0.4	0.61	0.1490	0.2396	−0.0580
0.067	0.53	0.1313	0.3232	0.0796
	D3			
z/k	Mean	SD	Skew	Kurtosis
2.667	1.65	0.1609	−0.2996	−0.1691
1.8	0.97	0.1871	−0.1501	−0.4304
1	0.73	0.1923	0.1616	−0.4566
0.4	0.58	0.1588	0.3942	0.0678
0.067	0.49	0.1369	0.4273	0.3019

The turbulence statistics applied on measured values of velocity quantify how the flow within and just above the canopy behaves as a perturbed mixing layer. Here, the turbulence intensity is expressed as $\tilde{u} = \sqrt{\sum_{i=1}^{N} u_i^2 / N}$ and normalized by the friction velocity u_*, i.e., $\sigma_u = \tilde{u}/u_*$. The number of the samples in experiments was $N = 270{,}000$. Figure 6 shows the measured profiles of σ_u for all three densities D1, D2 and D3. From this figure, it is seen that for $z/k < 1$, turbulence intensity σ_u is remarkably damped and is more pronounced for higher λ_d. In particular, σ_u is in the range from 1.69 (D1), for a sparse canopy, which is typical for rough-wall layers, to about 1.49 (D3), for a dense canopy, which is typical for mixing layers. Those values are close to the ones reported by Jimenez [10].

4.2. Shannon Entropy as a Measure of the Order or Disorder of the Flow

When stable laminar flows evolve toward the turbulence, they become high-order and complex, exhibiting irregular-like motions with organized dissipative arrangements. In order to precisely specify their fields (velocity and displacement), more parameters are required than for the description of laminar flows, i.e., topological measures that quantify the order or disorder of the flow, as the Shannon entropy, which has been already used in the analysis of geophysical fluids [47,48]. The Shannon entropy SH is defined as $SH = -\sum p_i ln p_i$, where p_i is a discrete probability distribution satisfying the following conditions: $p_i \geq 0; \sum p_i = 1$ and $p_{i \cup j \cup ...} = p_i + p_j + ...$ [49]. In our calculations, p_i is defined as the probability that the velocity amplitude falls within interval $u_i + du$, where du was obtained by dividing the entire interval of velocity amplitudes into one million intervals, due to the fact that the precision of the measurements was up to the sixth decimal place. Figure 7 shows that SH is the highest in the mixing layer ($1 < z/k < 2$), where the turbulence intensity σ_u is the highest. However, it decreases towards the free surface. This trend of SH coincides with the conclusions by Wijesekera and Dillion [47]. The decrease of the SH, going to the rough bed, can be addressed to the occurrence of smaller eddies carrying a smaller amount of energy.

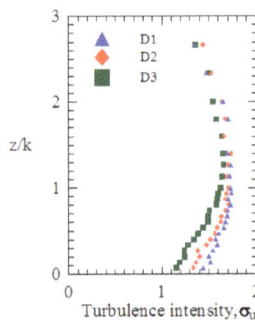

Figure 6. Turbulence intensity $\bar{u}$ normalized by u_* against the relative flow depth ratio of the vertical distance from bed z to the height k of the cylinders used for modeling canopies for all the densities [40].

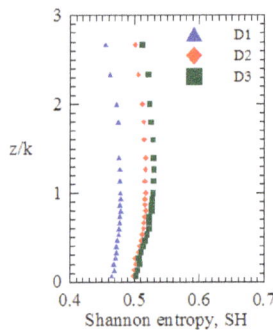

Figure 7. Shannon entropy (SH) versus the relative flow depth ratio of the vertical distance from bed z to the height of the cylinders used for modeling canopies k [40].

To avoid confusion in the following discussion, we make some comments. Namely, the term complexity in physical systems has the connotation of an explicit measure of the probability of the state of the system. It is a mathematical measure that should not be equalized with entropy in statistical mechanics [39]. Thus, Shannon's entropy refers to dissimilarities between amplitudes in a time series, while the Kolmogorov complexity refers to the apparent sequence disorder of amplitudes in a time series. This complexity that we intuitively understand as a measure range between uniformity and total randomness (Figure 3 in [39]) is of interest in this paper. Comparing Figures 7 and 8, they seem overall to have quite symmetrical trends. In Figure 7, the SH entropy weakly increases in the RS1 zone;

it has a constant value in the RS2 zone and then decreases in the RS3 zone. This trend is clearer for sparse bed roughness elements (D1), but anyway, the density of the bed roughness elements seems to affect SH entropy; in fact, lower SH entropy corresponds to sparse density (D1).

4.3. Kolmogorov Complexity

One of the main concerns in turbulence studies is the estimate of the temporal and, if possible, spatial characteristics of the existing eddy in the flow. Therefore, the integral length scale (*L*) is important in characterizing the structure of the turbulence. It is a measure of the longest correlation distance between the flow velocity at two points of the flow field. In Table 3 are given values of integral length scales (*L*) for different vegetation densities normalized by canopy height (*k*) at all relative heights.

Table 3. Integral length scales (*L*) for different vegetation densities normalized by canopy height (*k*) at all relative heights.

Integral Length Scale L/k			
z/k	D1	D2	D3
2.667	3.206	2.710	2.170
2.333	3.310	2.621	2.164
2.000	5.099	2.480	1.919
1.800	3.649	2.465	1.822
1.600	3.403	2.135	1.807
1.400	2.346	2.612	1.537
1.267	2.337	1.843	1.483
1.133	2.014	1.688	1.458
1.000	1.912	1.446	1.341
0.933	1.950	1.407	1.235
0.867	1.650	1.236	1.294
0.800	1.650	1.436	1.245
0.733	1.541	1.262	1.048
0.667	1.481	1.057	1.048
0.600	1.414	1.055	1.044
0.533	1.391	1.065	0.966
0.467	1.242	0.958	0.966
0.400	1.207	0.880	0.861
0.333	1.247	0.777	0.802
0.267	1.064	0.727	0.843
0.200	0.927	0.764	0.751
0.133	1.078	0.642	0.644
0.067	0.776	0.593	0.497

Some characteristics of the non-dimensional distributions of longitudinal integral length scales can be observed. For all the densities, the length scales are largest above the vegetation and are reduced progressively with depth into the vegetation. In the upper part of the vegetation, integral length scales are strongly influenced by the density, increasing as vegetation density decreases [50]. They are larger in sparser vegetation density because the confinement due to vegetation is looser. Near the canopy top, the turbulence length scales are of order of *k*. This is very similar to other reported measurements [51] and shows that the sweeps inferred from the skewness profiles are large turbulent structures, coherent in the streamwise direction over distances comparable with *h*. For all the densities, eddy scales are significantly compressed by the existence of the free surface [13].

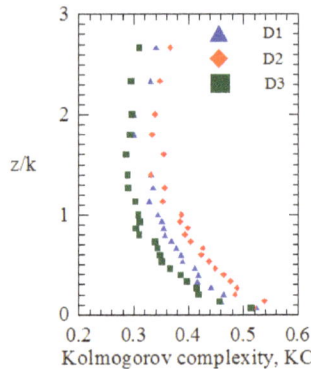

Figure 8. Kolmogorov complexity (KC) versus the relative flow depth ratio of the vertical distance from bed z to the height of the cylinders used for modeling canopies k [40].

We see that neither use of conventional measure for turbulence in canopy flow, nor the Shannon entropy give us quantitative information about the degree of the randomness. These two measures inform either about the intensity of the turbulence in canopy (σ_u) or about dissimilarities between amplitudes in a time series. Figure 9a shows the distributions of velocity frequencies for selected relative flow depths, for D3 canopy density (in further analysis, we use only this density for reasons of simplicity, but without losing the generality of the analysis).

Namely, going from $z/k = 0.067$ towards $z/k = 2.667$, the values of the peak velocity for all curves follow the profile of σ_u versus z/k in Figure 6. Looking at this figure, we can only say, in regard to randomness, that its highest value is expected for velocities around the peak of the distribution. There is no conclusion (neither quantitative, nor descriptive) that we can attain concerning which velocity distribution represents less or high randomness. However, the Kolmogorov complexity spectrum or more precisely its highest value (KCM) carries that information quantitatively through a value expressing the randomness of turbulence in the canopy. From Figure 9b, we see that randomness is higher for lower velocities, but decreases going towards $z/k = 1$. After this relative flow depth, it becomes higher.

Figure 8 shows that the value of KC decreases with height from the rough-wall to the mixing layer ($1 < z/k < 2$). This can be explained by the fact that in the RS1 zone, flow is dominated by smaller eddies (see Figure 1a) contributing to the higher randomness that becomes lower in the mixing layer having a constant value in this zone. This is because eddies in the RS2 zone are larger and coherently organized, without the possibility to introduce more randomness in the flow. Above the mixing layer (RS3 zone), KC slightly increases since eddies become smaller. Following this analysis, it seems that the KC could be related to eddy sizes.

Entropy **2017**, 19, 519

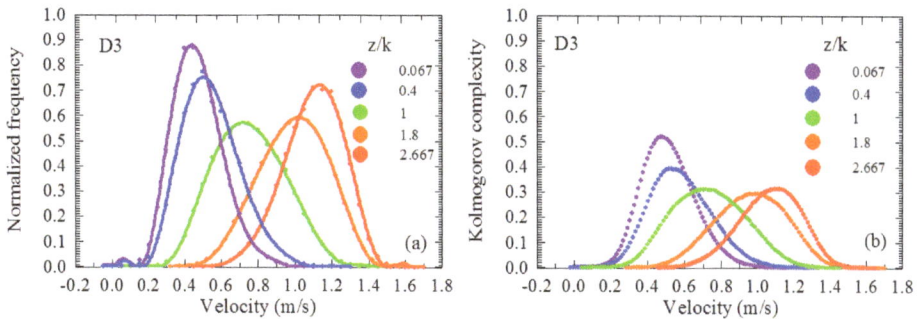

Figure 9. Distributions of velocity frequencies (a) and Kolmogorov complexity spectra (b) for selected relative flow depth, for D3 canopy density.

From Figure 9 is seen an interesting feature that the similarities between velocity spectra and Kolmogorov complexity spectra differ up to the constant. Whether this feature is a consequence either of physical sources, or the mathematical nature of the applied algorithm, or it is just the characteristics of these experiments remains to be deepened. Therefore, for a more reliable conclusion, there is a need for more experiments with different setups for the parameters.

5. Conclusions

In this paper, a methodology to quantify the randomness in turbulent flows with the canopy using experimental data collected in a laboratory channel with a variable bed slope and three different canopy densities is described. An analysis based on the Kolmogorov complexity and measures derived from it (the Kolmogorov complexity spectrum and its highest value) was proposed. First, for all the densities, the turbulence was analyzed with the classical turbulence statistics through the vertical profiles of turbulence intensity. In addition, the Shannon entropy was calculated for all densities, to describe the turbulence in terms of irregular or random flow. Finally, it was shown that the Kolmogorov complexity and measures based on it, for all densities, provides quantification of the degree of turbulence in relation to randomness rather than if it is descriptively expressed in terms of irregular or random. Finally, probably the most important result of this paper is that the value of KC is connected with the turbulent eddies' size, and it could be helpful in explaining the coherent structure in canopy turbulent flow.

Acknowledgments: This paper was realized as a part of Project No. 43007 financed by the Ministry of Education and Science of the Republic of Serbia for the period 2011–2017. Paola Gualtieri and Carlo Gualtieri acknowledge the support by the MIUR PRIN 2010–2011 research project "HYDROCAR".

Author Contributions: Paola Gualtieri and Carlo Gualtieri conceived and designed the experiments; Paola Gualtieri performed the experiments; Gordan Mimić and Ilija Arsenić analyzed the data; Dragutin Mihailović contributed analysis tools; Dragutin Mihailović wrote the paper. All authors have read and approved the final manuscript.

Conflicts of Interest: The authors declare no conflict of interest.

References

1. Gurzadyan, V.G. Kolmogorov Complexity as a Descriptor of Cosmic Microwave Background Maps. *EPL* **1999**, 46, 114–117.
2. Gell-Mann, M.; Lloyd, S. Information Measures, Effective Complexity, and Total Information. *Complexity* **1996**, 2, 44–52.
3. Gell-Mann, M.; Lloyd, S. Effective Complexity. In *Nonextensive Entropy: Interdisciplinary Applications*; Gell-Mann, M., Tsallis, C., Eds.; Oxford University Press: Oxford, UK, 2003.
4. Kreinovich, V.; Kunin, I.A. Kolmogorov Complexity and Chaotic Phenomena. *Int. J. Eng. Sci.* **2003**, 41, 483–493.

5. Abarzhi, S. Review of theoretical modelling approaches of Rayleigh–Taylor instabilities and turbulent mixing. *Philos. Trans. R. Soc. A* **2010**, *368*, 1809–1828.

6. Lorenz, R.D. Maximum Frictional Dissipation and the Information Entropy of Windspeeds. *J. Non-Equilib. Thermodyn.* **2002**, *27*, 229–238.

7. Weck, P.J.; Schaffner, D.A.; Brown, M.R.; Wicks, R.T. Permutation entropy and statistical complexity analysis of turbulence in laboratory plasmas and the solar wind. *Phys. Rev. E* **2015**, *91*, 023101.

8. Ichimiya, M.; Nakamura, I. Randomness representation in turbulent flows with Kolmogorov complexity (in mixing layer). *J. Fluid Sci. Technol.* **2013**, *8*, 407–422.

9. Pope, S.B. *Turbulent Flows*; Cambridge University Press: Cambridge, UK, 2000.

10. Jimenez, J. Turbulent flows over rough walls. *Annu. Rev. Fluid Mech.* **2004**, *36*, 173–196.

11. Poggi, D.; Porporato, A.; Ridolfi, L.; Albertson, J.D.; Katul, G.G. The effect of vegetation density on canopy sub-layer turbulence. *Bound. Lay. Meteorol.* **2004**, *111*, 565–587.

12. Flack, K.A.; Schultz, M.P.; Shapiro, T. Experimental support for Townsend's Reynolds number similarity hypothesis on rough walls. *Phys. Fluids* **2005**, *17*, 035102.

13. Nezu, I.; Sanjou, M. Turbulence structure and coherent motion in vegetated canopy open-channel flows. *J. Hydro-Environ. Res.* **2008**, *2*, 62–90.

14. Cushman-Roisin, B.; Gualtieri, C.; Mihailović, D.T. Environmental Fluid Mechanics: Current issues and future outlook. In *Fluid Mechanics of Environmental Interfaces*, 2nd ed.; Gualtieri, C., Mihailović, D.T., Eds.; CRC Press/Balkema: Leiden, The Netherlands, 2012.

15. Raupach, M.R.; Finnigan, J.J.; Brunet, Y. Coherent eddies and turbulence in vegetation canopies: The mixing-layer analogy. *Bound. Lay. Meteorol.* **1996**, *78*, 351–382.

16. Hussain, A.K.M.F. Coherent Structures—Reality and Myth. *Phys. Fluids* **1983**, *26*, 2816–2850.

17. Nepf, H.M. Flow and transport in regions with aquatic vegetation. *Annu. Rev. Fluid Mech.* **2012**, *44*, 123–142.

18. Rogers, M.M.; Moser, R.D. Direct simulation of a self-similar turbulent mixing layer. *Phys. Fluids* **1994**, *6*, 903–923.

19. Tsuji, Y.; Nakamura, I. The fractal aspect of an isovelocity set and its relationship to bursting phenomena in the turbulent boundary layer. *Phys. Fluids* **1994**, *6*, 3429–3441.

20. Li, M.; Vitanyi, P. *An Introduction to Kolmogorov Complexity and Its Applications*, 2nd ed.; Springer: Berlin/Heidelberg, Germany, 1997.

21. Lempel, A.; Ziv, J. On the complexity of finit sequence. *IEEE Trans. Inf. Theory* **1976**, *22*, 75–81.

22. Sellers, P.; Mintz, Y.; Sud, Y.C.; Dachler, A. A Simple Biosphere Model (SIB) for Use within General Circulation Models. *J. Atmos. Sci.* **1986**, *43*, 505–531.

23. Batchelor, G.K. *The Theory of Homogeneous Turbulence*; Cambridge University Press: Cambridge, UK, 1956.

24. Fricsh, U. *Turbulence: The Legacy of A. N. Kolmogorov*; Cambridge University Press: Cambridge, UK, 1995.

25. Terwijn, S.A. The Mathematical Foundations of Randomness. In *The Challenge of Chance: A Multidisciplinary Approach from Science and the Humanities*; Landsman, K., van Wolde, E., Eds.; Springer: Berlin/Heidelberg, Germany, 2016.

26. Kolmogorov, A.N. *Grundbegriffe der Wahrscheinlichkeitsrechnung*; Springer: Berlin/Heidelberg, Germany, 1933.

27. Sato, H. Laminar-Turbulent Transition in Free Shear Flow. In *Progress in Fluid Mechanics -Turbulent Flow*; Tani, I., Ed.; Maruzen: Tokyo, Japan, 1980. (In Japanese)

28. Pietri, L.; Petroff, A.; Amielh, M.; Anselmet, F. Turbulence characteristics within sparse and dense canopies. *Environ. Fluid Mech.* **2009**, *9*, 297–320.

29. Pietri, L.; Petroff, A.; Amielh, M.; Anselmet, F. Turbulent flows interacting with varying density canopies. *Mech. Ind.* **2009**, *10*, 181–185.

30. Feldman, D.P.; Crutchfeld, J.P. Measures of statistical complexity: Why? *Phys. Lett. A* **1998**, *238*, 244–252.

31. Kolmogorov, A. Logical basis for information theory and probability theory. *IEEE Trans. Inf. Theory* **1968**, *14*, 662–664.

32. Cerra, D.; Datcu, M. Algorithmic relative complexity. *Entropy* **2011**, *13*, 902–914.

33. Kaspar, F.; Schuster, H.G. Easily calculable measure for the complexity of spatiotemporal patterns. *Phys. Rev. A* **1987**, *36*, 842–848.

34. Zhang, X.S.; Roy, R.J.; Jensen, E.W. EEG complexity as a measure of depth of anesthesia for patients. *IEEE Trans. Biomed. Eng.* **2001**, *48*, 1424–1433.

35. Radhakrishnan, N.; Wilson, J.D. An alternate partitioning technique to quantify the regularity of complex time series. *Int. J. Bifurc. Chaos* **2000**, *10*, 1773–1779.
36. Small, M. *Applied Nonlinear Time Series Analysis: Applications in Physics, Physiology and Finance*; World Scientific: Singapore, 2005.
37. Ferenets, R.; Lipping, T.; Anier, A. Comparison of entropy and complexity measures for the assessment of depth of sedation. *IEEE Trans. Biomed. Eng.* **2006**, *53*, 1067–1077.
38. Hu, J.; Gao, J.; Principe, J.C. Analysis of biomedical signals by the Lempel-Ziv complexity: The effect of finite data size. *IEEE Trans. Biomed. Eng.* **2006**, *53*, 2606–2609.
39. Mihailović, D.T.; Mimić, G.; Nikolić-Djorić, E.; Arsenić, I. Novel measures based on the Kolmogorov complexity for use in complex system behavior studies and time series analysis. *Open Phys.* **2015**, *13*, 1–14, doi:10.1515/phys-2015-0001.
40. Mihailović, D.T.; Mimić, G.; Gualtieri, P.; Arsenić, I.; Gualtieri, C. Randomness representation in turbulent flows with bed roughness elements using the spectrum of the Kolmogorov complexity. In Proceedings of the 8th International Congress on Environmental Modelling and Software, Toulouse, France, 10–14 July 2016; pp. 4–11.
41. Annoni, M.; Monno, M.; Scarano, G. Uncertainty components interaction on the water jet velocity measurement by laser velocimetry. In Proceedings of the American WJTA Conference and Expo, Houston, TX, USA, 18–20 August 2009.
42. Annoni, M. Water jet velocity uncertainty in laser Doppler velocity measurements. *Measurement* **2012**, *45*, 1639–1650.
43. ITTC Recommended Procedure and Guidelines. *Uncertainty Analysis Laser Doppler Velocimetry Calibration*; ITTC: Kgs. Lyngby, Denmark, 2008.
44. Ristić, S.; Ilić, J.; Čantrak, D.; Ristić, O.; Janković, N. Estimation of laser Doppler anemometry measuring volume displacement in cylindrical pipe flow. *Therm. Sci.* **2012**, *16*, 1027–1042.
45. Nezu, I.; Onitsuka, K. Turbulent structure in partly vegetated open channel flows with LDA and PIV measurements. *J. Hydraul. Res.* **2011**, *39*, 629–642.
46. Ghannam, K.; Poggi, D.; Porporato, A.; Katul, G. The spatio-temporal statistical structure and ergodic behaviour of scalar turbulence within a rod canopy. *Bound. Lay. Meteorol.* **2015**, *157*, 447–460.
47. Wijesekera, H.W.; Dillion, T.M. Shannon entropy as an indicator of age for turbulent overturns in the ocean thermoclines. *J. Geophys. Res.* **1997**, *102*, 3279–3291.
48. Wesson, H.K.; Katul, G.G.; Siqueira, M. Quantifying organization of atmospheric turbulent eddy motion using nonlinear time series analysis. *Bound. Lay. Meteorol.* **2003**, *106*, 507–525.
49. Shannon, C.E. A mathematical theory of communication. *Bell Syst. Tech. J.* **1948**, *27*, 379–623.
50. Green, S.R.; Grace, J.; Hutchings, N.J. Observations of turbulent air flow in three stands of widely spaced Sitka spruce. *Agric. For. Meteorol.* **1995**, *74*, 205–225.
51. Brunet, Y.; Finnigan, J.J.; Raupach, M.R. A wind tunnel study of air flow in waving wheat: Single-point velocity statistic. *Bound. Lay. Meteorol.* **1994**, *70*, 95–132.

entropy

MDPI

Article

Entropy-Based Investigation on the Precipitation Variability over the Hexi Corridor in China

Liang Cheng, Jun Niu * and Dehai Liao

Center for Agricultural Water Research in China, China Agricultural University, Beijing 100083, China;
cl182234@163.com (L.C.); liaodehai@cau.edu.cn (D.L.)
* Correspondence: junniu.cau@yahoo.com; Tel.: +86-10-6273-7911

Received: 30 September 2017; Accepted: 27 November 2017; Published: 1 December 2017

Abstract: The spatial and temporal variability of precipitation time series were investigated for the Hexi Corridor, in Northwest China, by analyzing the entropy information. The examinations were performed on monthly, seasonal, and annual timescales based on 29 meteorological stations for the period of 1961–2015. The apportionment entropy and intensity entropy were used to analyze the regional precipitation characteristics, including the intra-annual and decadal distribution of monthly and annual precipitation amounts, as well as the number of precipitation days within a year and a decade. The regions with high precipitation variability are found in the western part of the Hexi corridor and with less precipitation, and may have a high possibility of drought occurrence. The variability of the number of precipitation days decreased from the west to the east of the corridor. Higher variability, in terms of both of precipitation amount and intensity during crop-growing season, has been found in the recent decade. In addition, the correlation between entropy-based precipitation variability and the crop yield is also compared, and the crop yield in historical periods is found to be correlated with the precipitation intensity disorder index in the middle reaches of the Hexi corridor.

Keywords: precipitation; variability; marginal entropy; crop yield; Hexi corridor

---

1. Introduction

The Hexi corridor in northwest China (Figure 1), located in 93°21′–104°05′ E and 37°15′–42°48′ N, plays an important role in grain production and regional economic development. The corridor is a long-narrow region in Northwest China with a length of about 1000 km, an average width of about 270 km, and a total area of about 270,000 km^2. The Qiliang Mountains are located in the southern part of the corridor and the Mazong and Longshou Mountains in the north, with the Wushaoling to the east and Dunhuang to the west. The corridor includes three inland water systems, namely the Shule River, the Hei River, and the Shiyang River, from west to east, respectively. The long-term mean annual precipitation is about 100–150 mm. Although it receives relatively limited precipitation, the Hexi corridor has sufficient light and heat resources, with a sunshine duration of about 3000 h/year. Therefore, it is an important crop-producing base, serving both the local food requirements and seeding production for other parts of China. Due to the shortage of irrigation water, the fluctuation of precipitation at different timescales may exert considerable effects on the farmland and oases in this area. The water resources is the most significant factor to affect the regional prosperity and the quality of life.

The crop growth relies heavily on water at different months or seasons for different grow phases. To rationally allocate regional water resources and mitigate the possible effects of drought, we need to analyze the long-term hydro-meteorological variables [1]. Together with the warming effects of climate change, the amount, timing, and distribution of precipitation may be affected, leading to changes in the amount of local available water. Many obvious/abrupt changes have been witnessed, such as the precipitation decline in Sahel [2], greater incidence of extreme precipitation events [3], and an increase in fall precipitation [4]. These observed changes could be partly attributed to the evolutions of large climate signals (e.g., [5]) in the context of climate change.

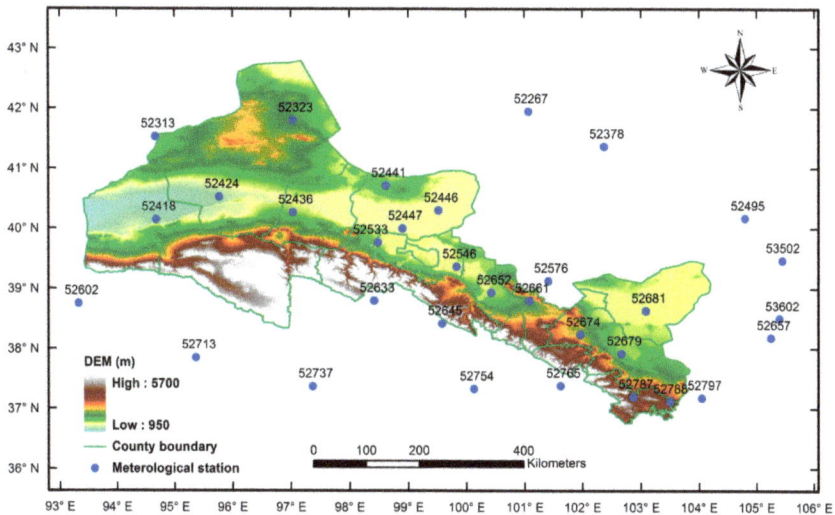

Figure 1. The Hexi corridor and the meteorological stations used in this study.

In order to reveal the underlying disorder features of precipitation variability during different months and different seasons, we use the entropy-based approach to study spatial and temporal precipitation variability over the Hexi corridor. The entropy theory presented by Shannon in the late 1940s [6] and the principle of maximum entropy presented by Jaynes in the late 1950s [7,8] have been applied in various research fields. Among them, there are many valuable hydrological applications of entropy theory (e.g., [9–21]). The aspects we want to address in this study include: (1) the detection of the spatial region and temporal year for high disorder features; (2) the identification of a monthly time series, which dominates the seasonal precipitation in Hexi corridor, and the identification of a seasonal time series, which dominates the annual precipitation in the corridor; (3) the correlations between drought-induced crop reduction and precipitation variability. The methods and data adopted in the present study are first presented in Sections 2 and 3, respectively. The entropy-based precipitation features are analyzed and discussed, and the correlation between crop production and precipitation variability is examined in Section 4. The conclusions are given in Section 5.

2. Materials and Methods

The entropy-based investigation is helpful to determine the least-biased probability distribution of a random variable. A discrete form of entropy $H(x)$ is written as [6,17]:

$$H(x) = -\sum_{k=1}^{K} p(x_k) \, \log_2[p(x_k)] \tag{1}$$

where k is time interval of the K events, x_k is an event corresponding to the interval k, and $p(x_k)$ is the probability of x_k. The application in this study is to measure the spatial and temporal variability/disorder features of precipitation in Hexi corridor in Northwest China. By following the methods in Mishra et al. [1], the several entropy measures, namely, intensity entropy, apportionment entropy, and marginal entropy, are employed for the research purpose. More details about the entropy and its applications can be found in Singh [9].

2.1. Marginal Entropy

The marginal entropy (ME) is the average information of a random variable x with the probability distribution $p(x)$, which is used to quantify uncertainty. The calculation processes are expressed in Equation (1). ME is applicable to different timescales, such as the daily, seasonal, monthly, and annual timescales in the present study.

2.2. Intensity Entropy

We consider the number of precipitation days (n_i) within a month i (i = 1, 2, 3, ..., 12) of a year and the total amount of precipitation days N ($N = \sum n_i$) in the year [1]. The probabilities of precipitation days in each month are expressed as $p_i = n_i/N$. The intensity entropy (IE) for an individual meteorological station for a month is given as:

$$\text{IE} = -\sum p_i(\log_2 p_i) = -\sum (n_i/N)\log_2(n_i/N) \tag{2}$$

2.3. Apportionment Entropy

The apportionment entropy (AE) is used to study the distribution features of precipitation for different months over a year. Assume that the precipitation amount in the ith month is r_i (i = 1, 2, 3, ..., 12), and the total precipitation amounts for twelve months is R ($R = \sum r_i$). The probabilities of precipitation days in each month to the total days in a year are expressed as $p_i = r_i/R$. The AE for an individual meteorological station for a year is calculated as:

$$\text{AE} = -\sum (r_i/R)\log_2(r_i/R) \tag{3}$$

The AE value is in the range of 0 and $\log_2(12)$ [1], in which the precipitation occurs only one out of twelve months and the annual precipitation amount is evenly distributed for twelve month respectively.

2.4. Entropy-Based Variability

The entropy-based variability is characterized by disorder index (DI). The DI is the difference between the maximum possible entropy and the calculated entropy from the individual series. In case the probability of each event is even, the DI reaches the maximum value. Therefore, the DI carries the information of disorder features of the analyzed time series. Accordingly, there is a marginal disorder index (MDI), an apportionment disorder index (ADI), and an intensity disorder index (IDI). In temporal domain, the DI is computed for annual, seasonal, and monthly time series. In spatial and temporal domain, the mean DI (MDI) is given by:

$$\text{MDI} = \frac{1}{N}\sum_{i=1}^{N}\text{DI} \tag{4}$$

where N is the length of entropy time scales. The higher the DI value is, the higher will be the variability [1].

3. Data

The precipitation data used in this study are derived from weather records of national standard meteorological stations for the Hexi corridor in Northwest China, obtained from China Meteorological Data Sharing Service System (http://data.cma.cn). A total of 29 stations, which have continuous precipitation records for the period 1961–2015, were selected for investigating precipitation variability. For the gap filling, the missing data were generally filled using the means of neighboring data, but if the missing period is longer than 1–2 days, the data of the same period from other years were considered occasionally. The geographical information of the 29 meteorological stations in Hexi corridor is listed in Table 1, and the basic statistical properties about annual total precipitation amount (ATP), monthly total precipitation amount (MTP), annual total precipitation days (ATD), and monthly total precipitation days (MTD) for the stations are listed in Table 2. The associated crop yield data were obtained from statistical data of the China Economic and Social Development Statistics Database (http://tongji.cnki.net/kns55/index.aspx), Gansu Water Statistical Yearbook.

Table 1. Geophysical information of 29 meteorological stations.

Station No.	Station ID	Latitude (Decimal Degrees)	Longitude (Decimal Degrees)
1	52267	41.95	101.07
2	52313	41.53	94.67
3	52323	41.80	97.03
4	52378	41.37	102.37
5	52418	40.15	94.68
6	52424	40.53	95.77
7	52436	40.27	97.03
8	52446	40.30	99.52
9	52495	40.17	104.80
10	52533	39.77	98.48
11	52546	39.37	99.83
12	52576	39.13	101.41
13	52602	38.75	93.33
14	52633	38.80	98.42
15	52645	38.42	99.58
16	52652	38.93	100.43
17	52657	38.18	105.25
18	52661	38.80	101.08
19	52674	38.23	101.97
20	52679	37.92	102.67
21	52681	38.63	103.08
22	52737	37.37	97.37
23	52754	37.33	100.13
24	52765	37.38	101.62
25	52787	37.20	102.87
26	52797	37.18	104.05
27	53502	39.47	105.45
28	53513	40.45	107.25
29	53602	38.50	105.40

Table 2. Statistical properties for precipitation time series (1981–2010) of the analyzed 29 stations.

Station No.	Station ID	ATP/MTP (mm)				ATD/MTD (Day)			
		Max.	Min.	Mean	STD	Max.	Min.	Mean	STD
1	52267	77.3/45.5	7.0/0	34.2/2.8	18.2/5.7	32/15	10/0	19.9/1.7	5.4/2.1
2	52313	95.3/40.9	16.9/0	50.5/4.2	19.5/6.9	42/10	22/0	30.0/2.5	5.7/2.1
3	52323	114.9/56.7	30.1/0	63.8/5.3	26.5/9.2	60/13	23/0	33.8/2.8	7.9/2.5
4	52378	105.2/54.6	13.9/0	42.1/3.5	23.9/7.2	26/10	12/0	18.4/1.5	3.6/1.8
5	52418	105.3/81.7	11.6/0	42.6/3.5	21.4/7.2	34/10	14/0	22.4/1.9	4.8/2.0
6	52424	83.6/53.0	19.0/0	50.8/4.2	15.7/6.8	40/10	19/0	25.6/2.1	5.2/2.2
7	52436	123.0/53.3	33.6/0	66.5/5.5	19.6/7.8	49/11	25/0	35.2/2.9	5.9/2.4
8	52446	108.5/58.6	28.4/0	60.4/5.0	21.6/8.6	42/14	19/0	30.8/2.6	6.2/2.7
9	52495	187.5/76.9	42.8/0	104.5/8.7	38.9/13.9	53/13	21/0	36.6/3.0	6.9/3.0
10	52533	157.3/67.8	41.4/0	88.4/7.4	30.3/10.9	60/13	28/0	39.4/3.3	7.5/2.8
11	52546	191.0/69.7	54.9/0	119.0/9.9	30.4/12.6	64/16	35/0	49.1/4.1	7.3/3.3
12	52576	197.6/98.0	76.6/0	134.8/11.2	30.8/16.7	65/16	34/0	48.8/4.1	7.9/3.6
13	52602	35.8/19.5	5.2/0	15.4/1.3	7.2/2.6	22/8	6/0	12.9/1.1	4.3/1.5
14	52633	404.4/152.4	180.7/0	309.7/25.8	57.7/33.5	109/23	74/0	88.3/7.4	10.3/6.3

Table 2. *Cont.*

Station No.	Station ID	ATP/MTP (mm)				ATD/MTD (Day)			
		Max.	Min.	Mean	STD	Max.	Min.	Mean	STD
15	52645	602.3/151.0	274.9/0	426.4/35.5	75.5/41.5	159/28	100/0	125.5/10.5	13.9/7.4
16	52652	216.3/90.3	71.6/0	132.6/11.0	36.6/14.4	73/15	37/0	53.9/4.5	9.4/3.4
17	52657	573.1/162.4	331.0/0	415.1/34.6	61.5/40.1	139/26	87/0	112.0/9.3	12.7/7.1
18	52661	301.2/107.8	97.8/0	202.6/16.9	49.8/20.8	93/17	51/0	68.4/5.7	10.1/3.8
19	52674	294.6/128.0	110.5/0	211.7/17.6	40.4/22.3	95/20	52/0	73.4/6.1	11.9/4.5
20	52679	251.3/111.9	101.6/0	171.0/14.3	40.8/17.8	83/14	46/0	60.9/5.1	7.9/3.6
21	52681	202.0/73.2	52.0/0	113.2/9.4	29.9/13.1	53/14	24/0	39.7/3.3	7.9/3.0
22	52737	354.9/145.1	128.4/0	214.3/17.9	58.1/24.2	87/20	36/0	59.8/5.0	12.2/4.5
23	52754	535.7/162.3	263.7/0	411.9/34.3	61.2/40.6	141/26	89/0	110.7/9.2	11.9/7.5
24	52765	730.7/201.2	404.0/0	531.1/44.3	72.3/44.2	153/26	107/0	130.8/10.9	13.7/6.9
25	52787	543.3/172.2	274.0/0	407.0/33.9	58.6/36.4	183/27	112/0	140.5/11.7	13.7/5.8
26	52797	298.4/95.4	94.8/0	179.8/15.0	54.3/19.1	77/16	42/0	56.1/4.7	9.7/3.9
27	53502	189.1/94.5	56.8/0	101.8/8.5	34.1/13.7	46/13	22/0	33.0/2.8	6.2/2.8
28	53513	267.9/193.6	55.9/0	148.9/12.4	48.7/21.5	52/15	22/0	36.6/3.0	7.4/3.0
29	53602	347.3/103.8	126.2/0	213.7/17.8	55.3/22.3	65/15	37/0	53.2/4.4	7.1/3.3

4. Results and Discussion

4.1. Variability of Annual, Seasonal, and Monthly Precipitation

Figure 2 shows the calculated marginal disorder index (MDI) for annual and seasonal (spring, summer, fall, and winter) precipitation time series over different meteorological stations. Overall, the MDI at annual timescale is lower than those of four seasons. The higher MDI values appear for winter time series, which indicates the higher variability associated with the analyzed stations in winter. Among them, the precipitation recorded by Station 52602 near the western part of the Hexi corridor shows maximum marginal entropy in both annual and four seasonal timescales, followed by Station 52378 (in the northeastern part) and 52446 (in the middle region of the Hexi corridor) in terms of the MDI value in Winter. The spatial distribution of the computed MDI at annual timescale is shown in Figure 3, in comparison with the corresponding mean annual precipitation for the period 1961–2015. It is observed that the annual precipitation in Hexi corridor is increasing from the northwest region to the southeast region, with a range of 50–400 mm/year. Regarding the MDI value, the highest value exits in the west corner of the Hexi corridor, which is in the area of Shule River basin. The MDI gradually decreases from west to the junction area between the Hei and Shiyang River basins, and it further slightly increases in the northeastern part of Shiyang River basin. Therefore, the disorder characteristics associated with the mean annual precipitation vary spatially for the Hexi corridor.

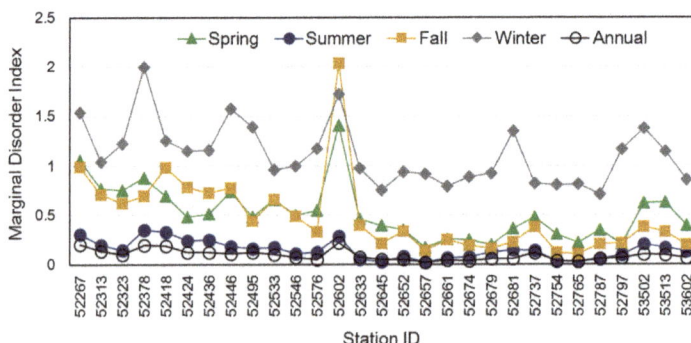

Figure 2. Marginal disorder indices of annual and seasonal precipitation time series over the Hexi region, Northwest China.

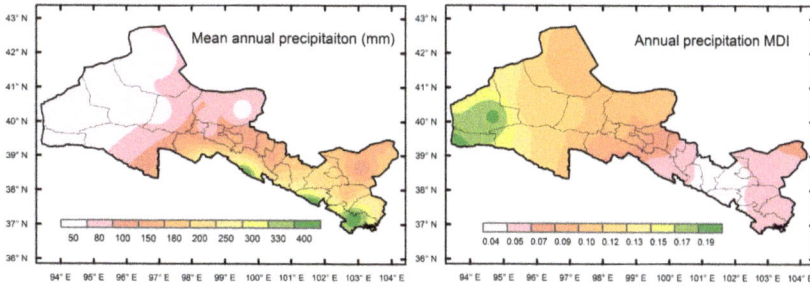

Figure 3. Spatial distribution of mean annual precipitation (**left panel**) and marginal disorder index (MDI) (**right panel**) for annual time series over the Hexi corridor.

The spatial distribution of MDI of seasonal precipitation time series is shown in Figure 4. It is found that the stations with high variability for the four seasons are: Stations 52424, 52436, and 52546 in spring; Station 52418 in summer; Stations 52313, 52418, 52424, and 52446 in fall; and Stations 52602, 52446, 52418, and 52681 in winter. It is interesting that the station 52418 is the only one with relatively higher disorder features in summer over the Hexi corridor. The stations with the higher winter MDI are not concentrated in one region, but distributed in the western, middle, and eastern parts. It is also observed that the winter MDI value is higher than that of other seasons, as shown in Figure 5. The distribution pattern of spring precipitation MDI is similar to that in fall MDI. The MDI value in summer is quite low, which indicates the disorder features are less obvious when the precipitation is relatively high in summer season over the Hexi corridor.

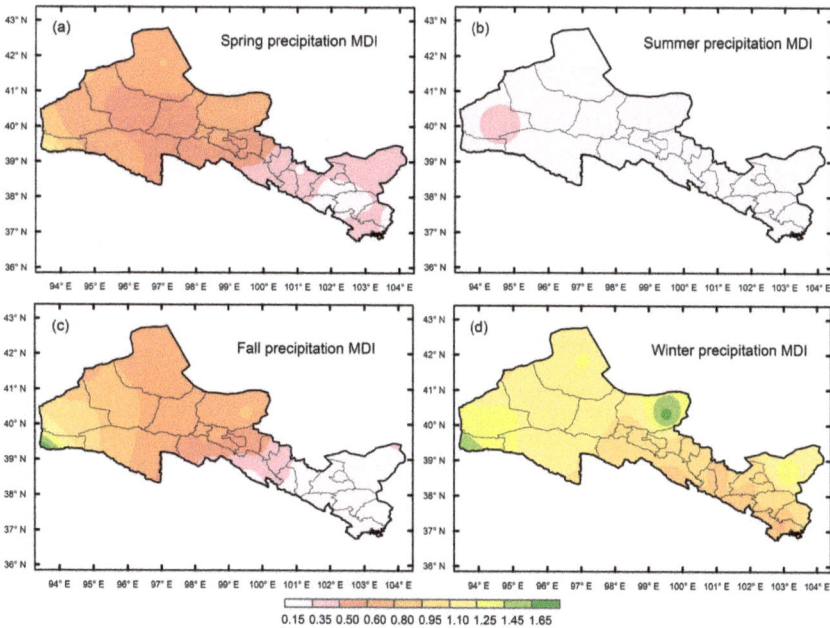

Figure 4. Spatial distribution of marginal disorder index (MDI) of precipitation time series for (**a**) spring; (**b**) summer; (**c**) fall; and (**d**) winter seasons over the Hexi corridor.

To understand which month is responsible for seasonal variability, it is useful to examine the intra-variability of months within a season [1]. Basically, the entropy-based variability of individual month is higher than that of corresponding season, which shows the disordered nature of precipitation. It is observed that the variability of March precipitation time series contributes the most in spring, while the variability in April contributes less. The summer variability is low in terms of the MDI value, and the MDI in July and August exerts the steadiest processes across the different stations in the studied region. Among it, the Stations 52267, 52378, 52418, 52436, and 53513 show relatively higher MDI value, which indicates the precipitation in summer may have more uncertainties. The farmland area located on or near the stations should pay more attention to the precipitation forecasting. The Station 52602, near the west corner of the Hexi corridor, has a distinguished disorder feature through all four seasons, with the most variability appearing in April, August, November, and February, respectively.

4.2. Variability of Precipitation Apportionment and Intensity

The apportionment features of precipitation can be reflected by the variability of precipitation amount of different months within a year. The apportionment disorder index (ADI) serves to quantify the precipitation characteristics, with the higher value indicating the higher unbalanced feature. Figure 6a summarizes the features for the whole studied region for the past 55 years (1961–2015). It is observed that the variability of precipitation amounts within a year are higher during the first 20-year period (1961–1980). These calculation results could be explained in two ways: (1) there are more zero values in precipitation records during 1961–1980 period, which probably demonstrates dire environment in this region compared to that of recent decades; and (2) the observed records may not reliable during the quite earlier years, and those zero values may not be effective. For the period 1980–2015, we can observe that the high variability of precipitation apportionment appears during the last decade in 20th century and after 2010. Meanwhile, Figure 7a reveals the station-based spatial features of ADI for the whole studied period. The stations with high apportionment variability include Stations 52267, 52378, 52418, 52446, and 52602. The variability generally decreases from the northwestern part of the region to the southeastern part area, as shown in Figure 8a.

The intensity entropy is reflected by the number of precipitation days in a month or in a year. The intensity disorder index (IDI) is computed for the involved stations of the studied region. Figure 6b shows the IDI over the past 5 decades. The relatively higher IDI are observed for the first 2 decades, which may be explained by the similar two reasons presented above for ADI results. For the recent 35 years (1980–2015), the years with high precipitation intensity are during 1980s and 2010s. The spatial features over all stations are summarized in Figure 7b. The stations with distinct intensity peak are 52267, 52378, 52418, 52446, and 52602. Overall, the spatial features of IDI generally are in line with the apportionment results in Figure 7a. This can be more easily observed in Figure 8a,b. In addition, Figure 8b reflects precipitation amount contribution within the Hexi corridor.

It is observed that the most contributions are along the south edge of the corridor, and those that are relatively higher are in Shiyang River basin. However, the contributions of precipitation days, shown in Figure 8d, are different from that of precipitation amount. The high contribution area includes both Shiyang River basin and the northern part of the Hei River basin in the middle part of the Hexi corridor.

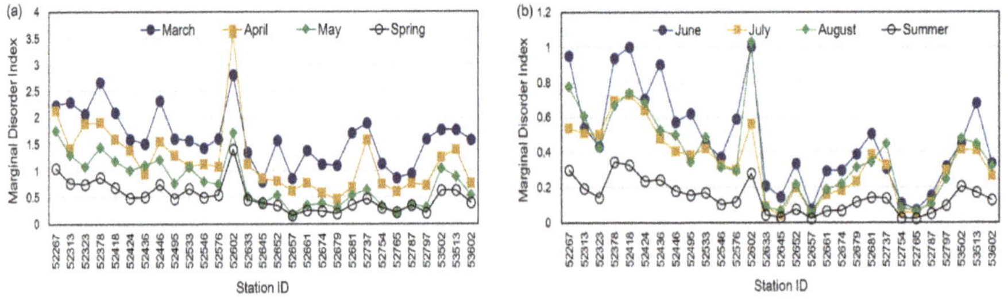

Figure 5. *Cont.*

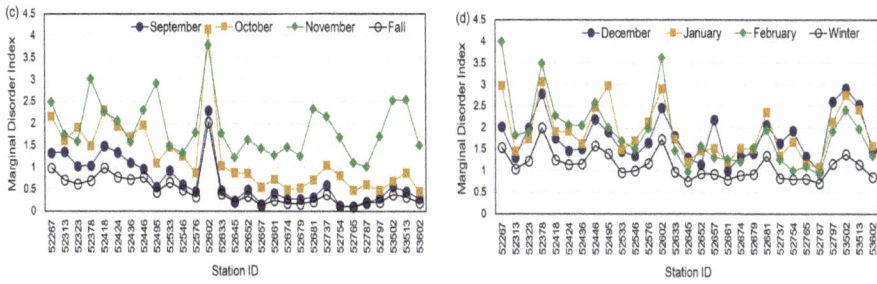

Figure 5. Marginal disorder index (MDI) of precipitation time series over different stations for (**a**) spring; (**b**) summer; (**c**) fall; and (**d**) winter seasons and its respective months.

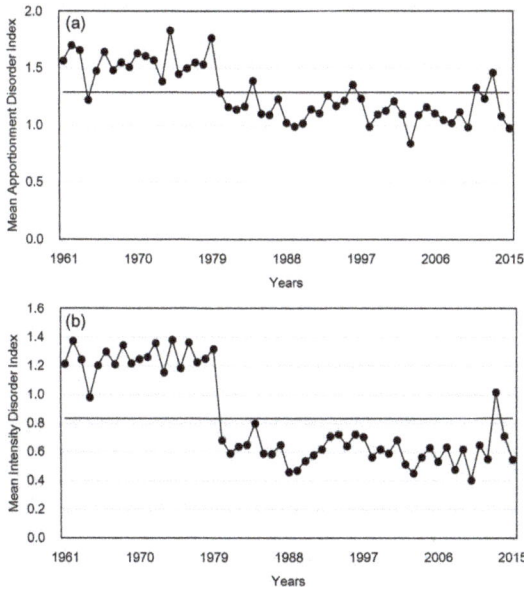

Figure 6. (**a**) Mean apportionment disorder index (MDI) of precipitation time series of all stations for the period of 1961–2015; the central line is the long-term mean value; (**b**) same as (**a**), but for mean intensity disorder index (IDI).

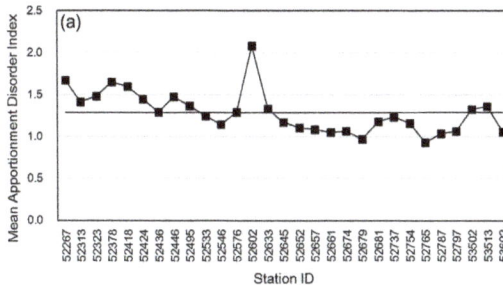

Figure 7. *Cont.*

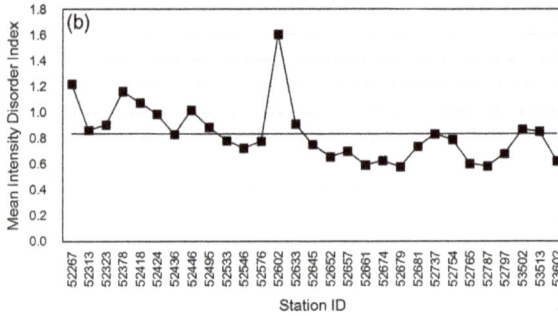

Figure 7. (**a**) Mean apportionment disorder index (MDI) of precipitation time series for different stations for the period of 1961–2015. The central line is the long-term mean value; (**b**) same as (**a**), but for mean intensity disorder index (IDI).

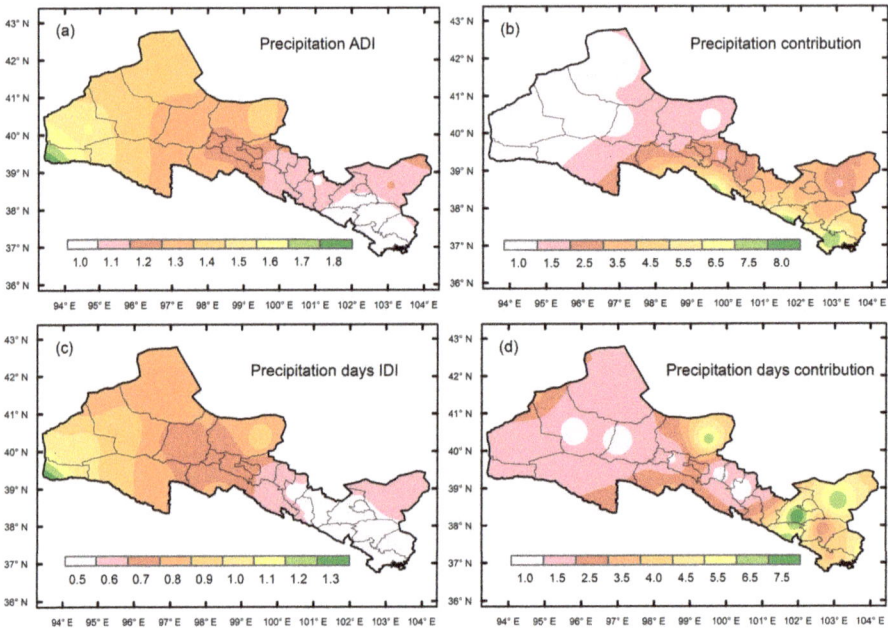

Figure 8. (**a**) The distribution of apportionment disorder index of precipitation time series over the Hexi corridor; (**b**) the contribution of individual station to overall precipitation in the region; (**c**) same as (**a**), but for intensity disorder index; (**d**) same as (**b**), but for a number of precipitation days in the area.

4.3. Decadal Variability of Precipitation Amount and Precipitation Days

To reveal the long-term changing trend of precipitation variability over the Hexi corridor, the decadal apportionment disorder index (DADI) and decadal intensity disorder index (DIDI) were computed for the recent three decades. Figure 9 shows the average entropy value for all stations at annual, seasonal, and monthly time scales for three individual decades. It is observed from Figure 9a that the stations in February, June, and July in the last decade (2001–2010) appear to have more variability in precipitation amount as compared with the two decades (1981–1990 and 1991–2010). In contrast to the variability of precipitation amount, the variability on precipitation days in February in the last decade is less than that during 1981–1990 but slightly higher than the decade of 1991–2010, as shown in Figure 9b. The variability reduction of precipitation days, reflected by DIDI, also occurred in August and October. The variability of precipitation amount and intensity changes for the crop-growing season over the Hexi corridor are very important factors for regional agricultural activities.

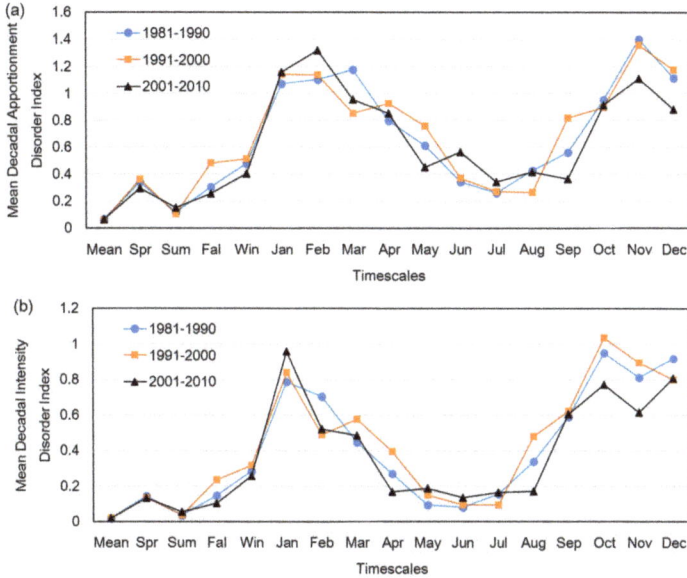

Figure 9. (a) Mean decadal apportionment disorder index of precipitation time series at different timescales; (b) same as (a), but for mean decadal intensity disorder index.

4.4. Implication for Crop Production

Based on the results on the spatial distribution in Figure 8, we found that the entropy-based variability generally decreases from west to east, and the precipitation contributions in terms of both precipitation amount and precipitation days decrease from east to west. Therefore, the middle reaches of the Hexi corridor region has both relatively high variability and precipitation amount/precipitation days. Accordingly, the crop yield in this region may be more sensitive to the evolutions of precipitation.

The crop production rates in four counties in the middle of the Hexi corridor during the period of 2005–2014 were compared with the computed precipitation disorder index and the standard deviations of anuual precipitaion amount/precipitaiton days for the associated stations, as shown in Figure 10. There are mainly four different crop types, namely, corn, wheat, rape, and barley. The crop disorder index (CDI) and standard deviation were computed to represent the variability of crop production. Among them, the standard deviation values were scaled to a comparable magnitude with other entropy index values. It is observed that there is a more consistent trend between the crop disorder index and precipitation intensity disorder index (IDI), compared with the relationship between crop disorder index and apportionment disorder index. This indicates the higher IDI value may introduce more crop variability, which demonstrates the implication of entropy-based variability for regional crop production.

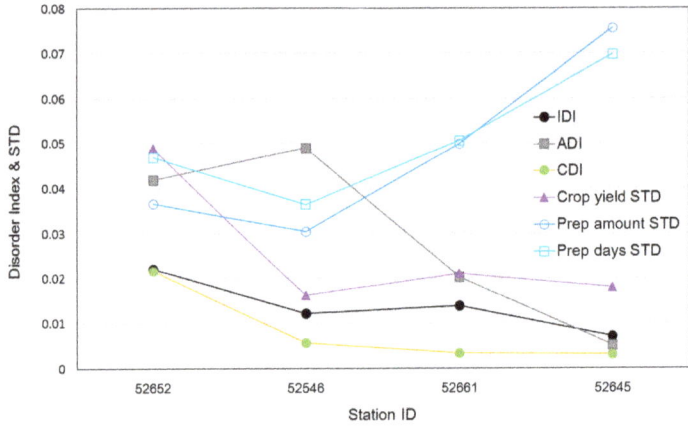

Figure 10. A comparison between crop and precipitation indices for the counties in the middle of the Hexi corridor.

4.5. Co-Variability with Other Meteorological Variables

Screening the co-variability of different meterological or hydrological variations is an important step to revealing the dominant controls in certain geophysical processes, such as for the geochmical hot moments in the study of Arora et al. [21]. To better understand the variability of precipitation and crop production mentioned in Section 4.4, we further checked the entropy-based variabilities of the associated temperature and wind speed. Figure 11 shows the average entropy value for all stations on a monthly scale for three individual decades. More variability appears in May, June, and July for temperature, and in January, November, and December for wind speed. Basically, consistent seasonal distributions are found for past three decades. The relationships between precipitation and temperature on a monthly scale, as shown in Figure 12, displays little precipitation below 0 degrees Celsius, and precipitation reaches its height around 10 degrees Celsius in wet seasons. Furthermore, we compared the crop disorder index CDI in four counties with the computed precipitation disorder index, temperture disorder index (TDI), and wind speed disorder index (WDI) for the associated stations, as shown in Figure 13. The TDI shows relatively steady variability for different stations in the region, and WDI exhibits obvious spatial differences. It is observed that the trend of precipitation intensity disorder index IDI is more consistent with that of CDI, compared with TDI and WDI.

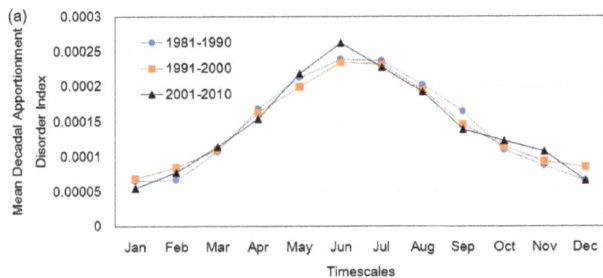

Figure 11. *Cont.*

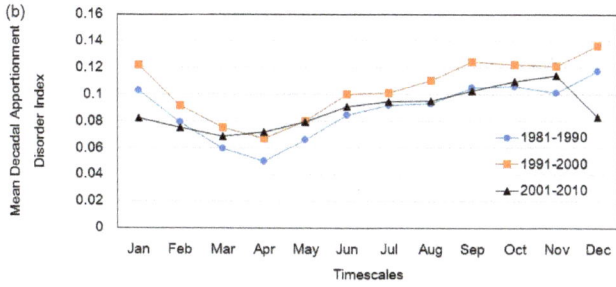

Figure 11. (**a**) Mean decadal apportionment disorder index of monthly temperature time series; (**b**) same as (**a**), but for wind speed.

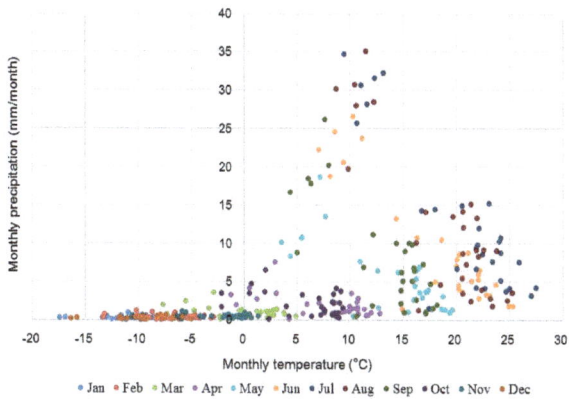

Figure 12. The relationships between temperature and precipitation on a monthly timescale for the studied stations.

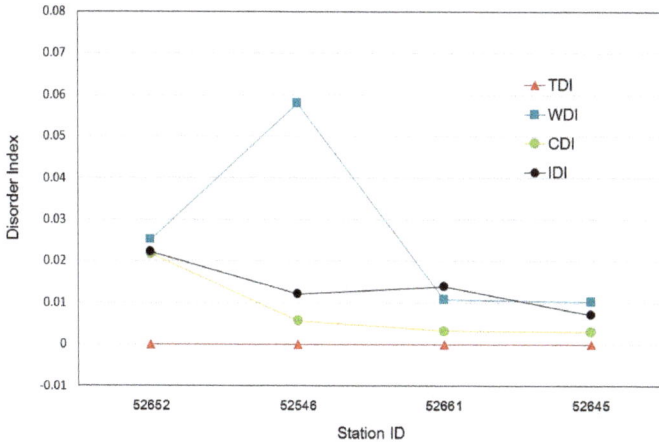

Figure 13. The comparisons between temperature, wind speed, crops, and precipitation disorder index for the counties in the middle of the Hexi corridor.

5. Conclusions

This study undertakes an entropy-based investigation of the spatial and temporal variability of precipitation for the period 1951–2015 over the Hexi corridor in Northwest China. The following conclusions are drawn based on the relevant analyses.

1. The annual variability is mainly contributed to by the winter variability. The precipitation intensity is diversified due to the difference in spatial variability between the monthly precipitation amount and the number of precipitation days. The contributions of different months to seasonal variability are different. A high contribution is detected in March for spring and in December for the winter period.
2. More variability during 2001–2010 is found for the precipitation time series in January, May, June, and July when we consider precipitation days. Regarding the precipitation amount, the variability in February, June, and July has increased in the recent decade.
3. The variability in terms of the precipitation amount and precipitation days decreases from the western part of the Hexi corridor to the eastern part of the region, which indicates more precipitation uncertainty is accommodated in the western part. From the temporal domain, the variability in the recent decade is relatively stronger than the previous decades.
4. The crop yield in historical periods is correlated with the precipitation intensity disorder index for the middle reaches of the Hexi corridor.

Extreme hydrological events may occur for highly disordered regions, which could induce negative effects on agriculture and reduce the crop production. Careful and early water management strategies should be proposed and particularly designed for these areas to mitigate the possible crop losses.

Acknowledgments: This research is supported by the National Training Program of Innovation and Entrepreneurship for Undergraduates (201710019225) and the National Natural Science Foundation of China (51679233).

Author Contributions: The authors designed and performed the research together. Liang Cheng and Dehai Liao collected the data and conducted analysis. Jun Niu and Liang Cheng wrote the draft of the paper. All authors have read and approved the final manuscript.

Conflicts of Interest: The authors declare no conflict of interest.

References

1. Mishra, A.K.; Özger, M.; Singh, V.P. An entropy-based investigation into the variability of precipitation. *J. Hydrol.* **2009**, *370*, 139–154. [CrossRef]
2. Hulme, M.; Doherty, R.; Ngara, T.; New, M.; Lister, D. African climate change: 1900–2100. *Clim. Res.* **2001**, *17*, 145–168. [CrossRef]
3. Easterling, D.R.; Evans, J.L.; Groisman, P.Y.; Karl, T.R.; Kunkel, K.E.; Ambenje, P. Observed variability and trends in extreme climate events: A brief review. *Bull. Am. Meteorol. Soc.* **2000**, *81*, 417–425. [CrossRef]
4. Karl, T.R.; Knight, R.W. Secular trends of precipitation amount, frequency, and intensity in the United States. *Bull. Am. Meteorol. Soc.* **1998**, *79*, 231–241. [CrossRef]
5. Wang, F.; Niu, J. The implication of climate signal for precipitation in the Heihe River basin, Northwest China. *Adv. Meteorol.* **2016**, *2016*, 1078617. [CrossRef]
6. Shannon, C.E. A mathematical theory of communication. *Bell. Syst. Tech. J.* **1948**, *27*, 379–423. [CrossRef]
7. Jaynes, E.T. Information theory and statistical mechanics, I. *Phys. Rev.* **1957**, *106*, 620–630. [CrossRef]
8. Jaynes, E.T. Information and statistical mechanics, II. *Phys. Rev.* **1957**, *108*, 171–190. [CrossRef]
9. Cui, H.; Singh, V.P. On the cumulative distribution function for entropy-based hydrologic modeling. *Trans. ASABE* **2012**, *55*, 429–438. [CrossRef]
10. Cui, H.; Singh, V.P. Computation of suspended sediment discharge in open channels using Tsallis entropy—Based Methods and Empirical Formulas. *J. Hydrol. Eng.* **2013**, *19*, 18–25. [CrossRef]

11. Cui, H.; Singh, V.P. Configurational entropy theory for streamflow forecasting. *J. Hydrol.* **2015**, *521*, 1–7. [CrossRef]
12. Cui, H.; Singh, V.P. Minimum relative entropy theory for streamflow forecasting with frequency as a random variable. *Stoch. Environ. Res. Risk Assess.* **2016**, *30*, 1545–1563. [CrossRef]
13. Singh, V.P. The use of entropy in hydrology and water resources. *Hydrol. Process.* **1997**, *11*, 587–626. [CrossRef]
14. Singh, V.P. *Entropy Theory in Hydrologic Science and Engineering*; McGraw-Hill Education: New York, NY, USA, 2015.
15. Sivakumar, B. Chaos theory in hydrology: Important issues and interpretations. *J. Hydrol.* **2000**, *227*, 1–20. [CrossRef]
16. Sivakumar, B. Chaos Identification and Prediction Methods. In *Chaos in Hydrology*; Springer: Dordrecht, The Netherlands, 2017; pp. 173–197.
17. Dwivedi, D.; Mohanty, B.P. Hot spots and persistence of nitrate in aquifers across scales. *Entropy* **2016**, *18*, 25. [CrossRef]
18. Maruyama, T.; Kawachi, T.; Maeda, S. Entropy-based Assessments of Monthly Rainfall Variability. *J. Rainwater Catchment Syst.* **2002**, *8*, 21–25. [CrossRef]
19. Khan, M.I.; Liu, D.; Fu, Q.; Azmat, M.; Luo, M.; Hu, Y.; Zhang, Y.; Abrar, F.M. Precipitation variability assessment of northeast China: Songhua River basin. *J. Earth Syst. Sci.* **2016**, *125*, 957–968. [CrossRef]
20. Atieh, M.; Rudra, R.; Gharabaghi, B.; Lubitz, D. Investigating the Spatial and Temporal Variability of Precipitation using Entropy Theory. *J. Water Manag. Model.* **2017**. [CrossRef]
21. Arora, B.; Dwivedi, D.; Hubbard, S.S.; Steefel, C.I.; Williams, K.H. Identifying geochemical hot moments and their controls on a contaminated river-floodplain system using wavelet and entropy approaches. *Environ. Model. Softw.* **2016**, *85*, 27–41. [CrossRef]

entropy

MDPI

Article

Comparison of Two Entropy Spectral Analysis Methods for Streamflow Forecasting in Northwest China

Zhenghong Zhou [1], Juanli Ju [1,*], Xiaoling Su [1], Vijay P. Singh [2] and Gengxi Zhang [1]

[1]   College of Water Resources and Architectural Engineering, Northwest A&F University, Yangling 712100, China; zzh199302@nwsuaf.edu.cn (Z.Z.); xiaolingsu@nwsuaf.edu.cn (X.S.); gengxizhang@nwsuaf.edu.cn (G.Z.)

[2]   Department of Biological & Agricultural Engineering and Zachry Department of Civil Engineering, Texas A&M University, 2117 TAMU, College Station, TX 77843, USA; vsingh@tamu.edu

*   Correspondence: jujuanli@nwsuaf.edu.cn; Tel.: +86-29-8708-2902

Received: 27 September 2017; Accepted: 5 November 2017; Published: 7 November 2017

Abstract: Monthly streamflow has elements of stochasticity, seasonality, and periodicity. Spectral analysis and time series analysis can, respectively, be employed to characterize the periodical pattern and the stochastic pattern. Both Burg entropy spectral analysis (BESA) and configurational entropy spectral analysis (CESA) combine spectral analysis and time series analysis. This study compared the predictive performances of BESA and CESA for monthly streamflow forecasting in six basins in Northwest China. Four criteria were selected to evaluate the performances of these two entropy spectral analyses: relative error (RE), root mean square error (RMSE), coefficient of determination (R^2), and Nash–Sutcliffe efficiency coefficient (NSE). It was found that in Northwest China, both BESA and CESA forecasted monthly streamflow well with strong correlation. The forecast accuracy of BESA is higher than CESA. For the streamflow with weak correlation, the conclusion is the opposite.

Keywords: monthly streamflow forecasting; Burg entropy; configurational entropy; entropy spectral analysis time series analysis

---

1. Introduction

Accurate streamflow forecasting is important for developing measures to flood control, river training, navigation, reservoir operation, hydropower generation plan and water resources management. Time series models, such as autoregressive (AR) or autoregressive moving average (ARMA) models, as proposed by Box and Jenkins [1], are generally used for monthly streamflow forecasting [2–4]. These models assume that streamflow time series is stochastic and are linear which limits their application [5]. Monthly streamflow time series not only exhibits stochastic characteristics but also seasonal and periodic patterns. Entropy spectral analysis can extract important information of time series, such as the periodic characteristics [6–11]. Therefore, combining entropy spectral theory with time series analysis provides a new way for streamflow forecasting. Considering frequency f as a random variable, Burg [12] defined entropy, called Burg entropy, and developed an algorithm for the estimation of spectral density function of time series using the principle of maximum entropy (POME). The algorithm is termed Burg entropy spectral analysis (BESA) and has been widely used for spectral analysis of geomagnetic series [13], climate indices [8,14], surface air temperature [15], tide levels [16], precipitation and runoff series [17], and flood stage [18]. BESA is recommended as better than traditional methods for long-term hydrological forecasting [19–23]. Huo et al. [24] applied BESA to simulate and predict groundwater in the west of Shandong province plain of the Yellow River downstream and achieved satisfactory results. Wang and Zhu [25] considered that implicit periodic components of monthly and annual hydrological time series were better identified by BESA. Shen et al. [26] proposed a more rigorous recursion algorithm for maximum entropy spectral estimation method. In addition to meeting forward and backward minimum error of BESA, the algorithm also needed to satisfy a condition that the optimal prediction error was orthogonal to the signal. It was considered that the spectral density resolution of this method was higher than that of BESA. Boshnakov and Lambert-Lacroix [27] proposed an extension of the periodic Levinson-Durbin algorithm which was considered more reliable. However, multi-peak spectral density is difficult to determine under non-stationary conditions. Hence, monthly streamflow that features strong seasonal and periodic characteristics cannot be well simulated [28].

Frieden [29] was the first to use configurational entropy in image reconstruction and Gull and Daniell [30] applied it to radio astronomy. Based on the finite length cepstrum model, Wu [31] deduced an explicit spectral density function estimation formula and solved the complex calculation problem of CESA. Nadeu [32] regarded that spectral estimation precision of CESA was higher than that of BESA for both ARMA and MA, while the corresponding precisions were quite similar for AR. Katsakos et al. [33] found that the precision was higher when the spectral density of white noise series was estimated. Based on the spectral density estimation formula constructed by Wu, Cui, and Singh [28] derived a single variable streamflow forecasting model and found that the forecasting accuracy of CESA was superior to BESA for 19 different rivers in the US. For monthly streamflow forecasting, resolution and reliability of CESA were better than those of BESA.

The objective of this paper therefore was to compare the forecast performances of BESA and CESA for monthly streamflow forecasting in Northwestern China. The paper is organized as follows. First, a brief introduction to streamflow forecasting is given. Second, a maximum entropy spectral analysis prediction model is derived and evaluation methods are discussed. Third, application to streamflow forecasting is discussed. Fourth, results are discussed. Finally, conclusions were given.

2. Derivation and Evaluation of Maximum Entropy Spectral Analysis Prediction Model

2.1. Maximum Entropy Model

Let streamflow time series frequency f be a random variable, and the normalized spectral density $P(f)$ be taken as the probability density function. Thus, the Burg entropy can be defined as

$$H_B(f) = -\int_{-W}^{W} \ln[P(f)]df \tag{1}$$

The configurational entropy is defined in the same form as the Shannon entropy and can be written as

$$H_C(f) = -\int_{-W}^{W} P(f)\ln[P(f)]df \tag{2}$$

where $W = 1/(2\Delta t)$ is the Nyquist fold-over frequency and f is the frequency that varies from $-W$ to W, Δt is the sampling period, $P(f)$ is the normalized spectral density of streamflow series.

2.2. Constraints for Model

For a given streamflow time series, the constraints can be formed from the relationship between the spectral density $P(f)$ and autocorrelation function $\rho(n)$, which can be written as

$$\rho(n) = \int_{-W}^{W} P(f)e^{i2\pi fn\Delta t}df, \quad -N \leq n \leq N \tag{3}$$

where Δt is the discretization or sampling interval, and $i = \sqrt{-1}$. N is normally taken from $1/4$ up to $1/2$ of the series length according to the periodicity of streamflow.

2.3. Determination of Spectral Density

To obtain the least biased spectral density $P(f)$ by entropy maximizing, one needs to maximize the Burg and configurational entropies. Entropy maximizing can be done by using the method of Lagrange multipliers in which the Lagrangian function for the Burg entropy and configurational entropies can be formulated as follows:

$$L_B(f) = -\int_{-W}^{W} \ln[P(f)]df - \sum_{n=-N}^{N} \lambda_n \left[\int_{-W}^{W} P(f)e^{i2\pi fn\Delta t}df - \rho(n) \right] \tag{4}$$

$$L_C(f) = -\int_{-W}^{W} P(f)\ln[P(f)]df - \sum_{n=-N}^{N} \lambda_n \left[\int_{-W}^{W} P(f)e^{i2\pi fn\Delta t}df - \rho(n) \right] \tag{5}$$

where λ_n, $n = 0, 1, 2, \ldots, N$, are the Lagrange multipliers. Taking the partial derivative of Equations (4) and (5) with respect to $P(f)$ and equating the derivative to zero, the least-biased spectral densities $P(f)$ obtained from the maximization of the Burg entropy and configurational entropy, respectively, are

$$P_B(f) = -\frac{1}{\sum\limits_{n=-N}^{N} \lambda_n \exp(-i2\pi fn\Delta t)} \tag{6}$$

$$P_C(f) = \exp\left(-1 - \sum_{n=-N}^{N} \lambda_n e^{i2\pi fn\Delta t}\right) \tag{7}$$

It can be seen from the above two equations that the spectral density derived from the Burg entropy is in the form of inverse of polynomials, while the one from the configurational entropy is in the exponential form, which is easier to manipulate. The form in Equation (6) suggests that BESA is related to a linear prediction process.

2.4. Solution of the BESA Model

The spectral density derived is defined in the same form as the autoregressive model. On the basis of minimum of the forward and backward prediction error, a method of parameter estimation was presented by Burg, which can be written as

$$a_k(i) = \begin{cases} a_{k-1}(i) + k_k a_{k-1}(k-i), & i = 1, 2, \ldots, k-1 \\ k_k, & i = k \end{cases} \tag{8}$$

where $a_k(i)$ is the i-th parameter value of the k-order autoregressive model, and the parameter k_k is estimated by minimizing the forward and backward prediction error.

$$e_k^f(i) = e_{k-1}^f(i) + k_k e_{k-1}^b(i-1)$$
$$e_k^b(i) = e_{k-1}^b(i-1) + k_k e_{k-1}^f(i)$$
$$k_k = \frac{-2\sum\limits_{i=k}^{N-1} e_{k-1}^f(i)e_{k-1}^b(i-1)}{\sum\limits_{i=k}^{N-1}\left|e_{k-1}^f(i)\right|^2 + \sum\limits_{i=k}^{N-1}\left|e_{k-1}^b(i-1)\right|^2} \tag{9}$$

where $e^f{}_0(t) = e^b{}_0(t) = x(t)$, $x(t)$ is the streamflow time series.

For configurational entropy, the Lagrange multipliers and the extension of autocorrelation function can be computed by cepstrum analysis. Then, Wu [31] deduced the explicit solution based on the maximization of configurational entropy. Taking the inverse Fourier transform of the log-magnitude of Equation (7), it becomes

$$\int_{-W}^{W} \{1 + \ln[P(f)]\} \exp(i2\pi k\Delta t) df = \int_{-W}^{W} \left[-\sum_{n=-N}^{N} \lambda_n \exp(i2\pi f n\Delta t) \right] \exp(i2\pi f k\Delta t) df \tag{10}$$

where the second part of the left side of Equation (10) can be denoted as

$$e(k) = \int_{-W}^{W} \ln[P(f)] \exp(i2\pi f k\Delta t) df \tag{11}$$

Doing the integration of both sides of Equation (10), one gets

$$\delta_k + e(k) = -\sum_{n=-N}^{N} \lambda_n \delta_{k-n} \tag{12}$$

where δ_n is the Dirac delta function defined as

$$\delta_n = \begin{cases} 1, & n = 0 \\ 0, & n \neq 0 \end{cases} \tag{13}$$

Equation (12) can be expanded as a set of N linear equations:

$$\begin{aligned}
\lambda_0 &= -e(0) - 1 \\
\lambda_1 &= -e(1) \\
&\vdots \\
\lambda_N &= -e(N)
\end{aligned} \tag{14}$$

Equation (14) shows that the Lagrange multipliers can be determined from the values of cepstrum which entails the spectral density that is obtained from Equation (7). It is the main difference from Burg entropy.

For convenience of solving for the spectral density function, Nadeu [32] developed a simple method for computing cepstrum based on the use of the causal part of autocorrelation, where $\rho(n)$ is used only for $-N \leq n \leq N$. Thus, cepstrum can be estimated by the following recursive relation:

$$e(n) = \begin{cases} 2\left[\rho(n) - \sum_{k=1}^{n-1} \frac{k}{n} e(k)\rho(n-k)\right], & n > 0 \\ 0, & n \leq 0 \end{cases} \tag{15}$$

On the other hand, for the configurational entropy, the autocorrelation is extended with the inverse relationship of Equation (15) using the autocepstrum as

$$\rho(n) = \frac{e(n)}{2} + \sum_{k=1}^{n-1} \frac{k}{n} e(k)\rho(n-k) \tag{16}$$

Therefore, with model order m determined, the autocorrelation function can be estimated as

$$\rho(m) = \sum_{k=1}^{m} a_k \rho(m-k), \quad m \leq N \tag{17}$$

with extension coefficients $a_k = \frac{k}{m} e(k)$, and m is the model order.

Equation (17) extends the autocorrelation function with the configurational entropy maximized. Surprisingly, the autocorrelation extends with a linear combination of past lags, which is the same with the Burg entropy or the AR method. Thus, Equation (17) can be also written as

$$\rho(t) = \sum_{k=1}^{m} a_k \rho(t-k), \ t > T \tag{18}$$

where T is the total time period.

The extended autocorrelation in Equation (18) is a linear combination of the previous values weighted with coefficients a_k. Burg (1975) suggested weighing time series using the extension coefficients as

$$x(t) = \sum_{k=1}^{m} a_k x(t-k), \ t > T \tag{19}$$

Equation (19) represents the forecast using the entropy-based extended autocorrelation. It has been shown by Burg and Krstanovic and Singh [12,19–21] that Equation (19) satisfies the least squares prediction.

2.5. Determination of Model Order

The order of forecasting model m is identified by the Bayesian Information Criterion (BIC) [34]. BIC can reduce the order of the model by penalizing free parameters more strongly compared with AIC (Akaike information criterion).

$$BIC(m) = N \ln \sigma_\varepsilon^2 + m \ln N \tag{20}$$

where N is the length of time series and σ_ε^2 is the variance of residual.

2.6. Procedure for Streamflow Forecasting

The computation procedure for monthly streamflow forecasting is shown in Figure 1. The computation steps are as follows: (1) Streamflow data $x(t)$ are normalized with Equation (21); (2) The parameters in the model (BESA and CESA) are estimated and the cepstrum values are determined for computing the Lagrange multipliers; (3) The forecast order m is identified by the Bayesian Information Criterion (BIC) and monthly streamflow is forecasted; (4) The prediction results of streamflow series are obtained by inverse normalization and exponential transformation.

$$y(t) = zscore[\ln x(t)] \tag{21}$$

where *zscore* is a standardized function and $y(t)$ is a logarithmic sequence minus the mean divided by the standard deviation of the original sequence.

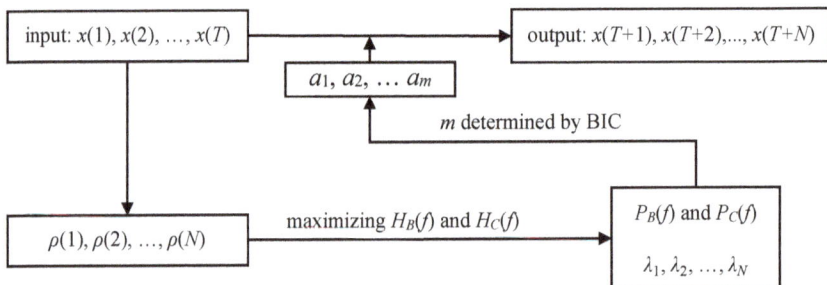

Figure 1. The computation procedure of entropy spectral analysis.

2.7. Evaluation of Model Forecast Performances

Four criteria were selected to evaluate the prediction model performance: relative error (RE), root mean square error (RMSE), coefficient of determination (R^2), and Nash–Sutcliffe efficiency coefficient

(NSE). The relative error provides the average magnitude of differences between observed values and predicted values relative to observed values. RMSE also represents the difference between observed and predicted values, however, it is scale-dependent. The coefficient of determination is defined as the square of the coefficient of correlation. It ranges between 0 and 1, and its higher values indicate better prediction. The Nash–Sutcliffe efficiency coefficient, defined by Nash and Sutcliffe [35], ranges from negative infinity to 1. Higher values of NSE represent more agreement between model predictions and observations, and negative values indicate that the model is worse than the mean value as a predictor.

$$RE = \frac{1}{N} \sum_{i=1}^{N} \left| \frac{Q_f(i) - Q_o(i)}{Q_o(i)} \right| \tag{22}$$

$$RMSE = \sqrt{\frac{\sum_{i=1}^{N} \left(Q_o(i) - Q_f(i)\right)^2}{N - 1}} \tag{23}$$

$$R^2 = \left\{ \frac{\sum_{i=1}^{N} \left(Q_o(i) - \overline{Q_o}\right)\left(Q_f(i) - \overline{Q_f}\right)}{\left[\sum_{i=1}^{N}\left(Q_o(i) - \overline{Q_o}\right)^2\right]^{0.5}\left[\sum_{i=1}^{N}\left(Q_f(i) - \overline{Q_f}\right)^2\right]^{0.5}} \right\}^2 \tag{24}$$

$$NSE = 1 - \frac{\sum_{i=1}^{N} \left|Q_o(i) - Q_f(i)\right|^2}{\sum_{i=1}^{N} \left|Q_o(i) - \overline{Q_o}\right|^2} \tag{25}$$

where N is the number of observed streamflow data, $Q_o(i)$ is the i-th observed streamflow, $Q_f(i)$ is the i-th forecasted streamflow, $\overline{Q_o}$ and $\overline{Q_f}$ are the average values of observed and forecasted streamflow, respectively.

3. Application to Streamflow Forecasting

3.1. Observed Data and Characteristics

The two entropy spectral analysis methods, BESA and CESA, were tested using observed streamflow data from six river sites on the Yellow River, Heihe River, Zamu River, Xiying River, Datong River, and Daxia River. The Yellow River has a large drainage area of 752,443 km², with an average monthly streamflow of 633 m³/s. Datong River and Daxia River are tributaries of the Yellow River. These two rivers have drainage areas of 151,33 km² and 7154 km², with average monthly streamflow of 88 m³/s and 27 m³/s, respectively. Zamu River and Xiying River belong to the Shiyang River watershed, with drainage areas of 851 km² and 1120 km². The Heihe River is the second largest interior river in Northwest China, with a drainage area of 130,000 km². Six hydrological stations selected in this paper are located in the Yellow River, Heihe River and Shiyang River, respectively. Tangnaihai station is located on the mainstream of Yellow River, while Xiangtang and Zheqiao stations are located on the tributary of Yellow River, Zamusi and Jiutiaoling stations are situated on the Shiyang River. Yingluoxia station is located in the Heihe River and it marks the boundary between the upstream and middle reaches. The location and basic information of each station are shown in Figure 2 and Table 1.

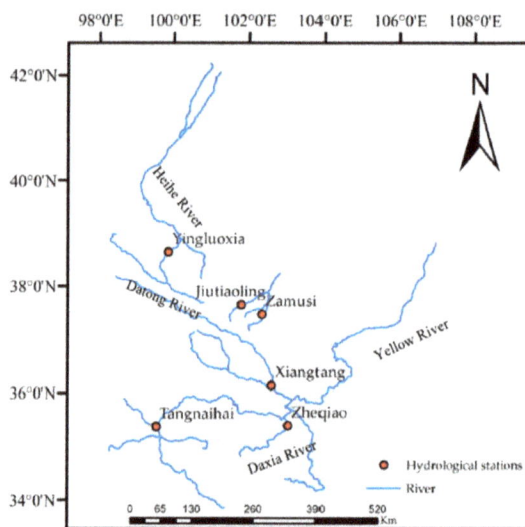

Figure 2. The location of selected stations.

Table 1. Basic information of steamflow data for selected stations.

No.	Station	Longitude	Latitude	River	Basin Area (km²)	Catchment Area (km²)	Record Length	Average (m³/s)	Peak (m³/s)
1	Xiangtang	102°51′E	36°22′N	Datong	15133	15,126	1950–2016	88	506
2	Yingluoxia	100°11′E	38°48′N	Heihe	130,000	10,009	1954–2012	51	214
3	Zamusi	102°34′E	37°42′N	Zamu	851	851	1952–2010	8	58.2
4	Jiutiaoling	102°03′E	37°52′N	Xiying	1120	1077	1972–2010	10	43.7
5	Tangnaihai	100°09′E	35°30′N	Yellow	752,443	121,972	1956–2016	633	3550
6	Zheqiao	103°16′E	35°38′N	Daxia	7154	6843	1963–2016	27	210

Monthly streamflow box-plots with all available data are presented in Figure 3. The bottom (Q1) and top (Q3) of the box are the first and third quartiles of streamflow, and the band inside the box is the median of streamflow. The inter quartile range (IQR) is equal to the difference between first and third quartiles. The limit of whiskers is called the inner fence which is 1.5 IQR from the quartile and the outer fence is 3 IQR from the quartile. Outliers are points that fall outside the limits of whiskers. + represents mild outliers which are between an inner and outer fence. × represents the extreme outliers which are beyond one of the outer fences. As shown in Figure 3, streamflow is concentrated during the flood season (June–September), and it drops down in the non-flood season. Because precipitation is the most important streamflow supply and the precipitation in these basins is concentrated during June–September.

There are many mild and extreme outliers for monthly streamflow data during the flood season at Xiangtang and Zheqiao stations, respectively. This is mainly due to poor vegetation coverage and barren hills in Datong River (Xiangtang station) downstream regions. Meanwhile, rainfall is mainly concentrated from June to September and mainly consists of heavy rain. Daxia River (Zheqiao station) upstream and downstream flow through the rocky mountainous region and loess plateau, separately. Serious soil erosion, heavy rain, mudslides, and landslides are frequent there. Streamflow during the flood season has many positive outliers for every station (Figure 3), and logarithmic processing is able to reduce the skewness of positive outliers in Section 2.7.

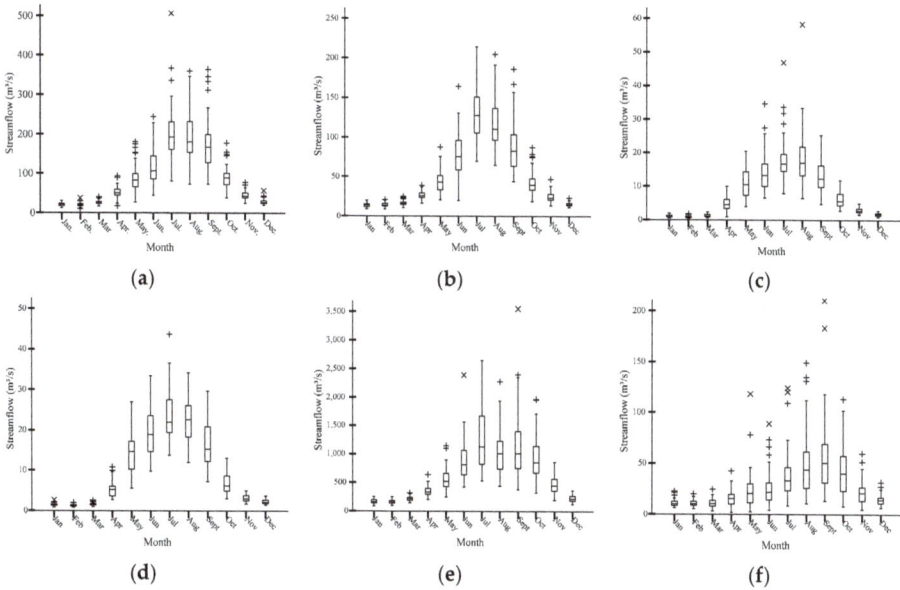

Figure 3. Monthly streamflow for selected stations. (**a**) Xiangtang station; (**b**) Yingluoxia station; (**c**) Zamusi station; (**d**) Jiutiaoling station; (**e**) Tangnaihai station; (**f**) Zheqiao station.

3.2. Comparison the Results of BESA and CESA

It is shown in a previous study that with the increase of the training time, both the accuracy of the training time and the precision of the lead time do not increase. Streamflow is forecasted by the two entropy spectral analysis methods with a five year training time (2003–2007) and a three year lead time (2008–2010) for representative stations. The simulated values and observed values for the two entropy spectrum models during the training period are shown in Figure 4. Both models were capable of simulating preferably streamflow variations at all stations. However, simulation results were better for the short leading time than that of the long leading time. The error between observed and simulated values was increasing with the lead time extension. The simulation values were better in drought seasons than in flood seasons, which mainly reflected the peak position and peak values. The maximum discharge during the flood seasons for six rivers appeared in different months for every year. This may lead to one month in advance or delay for the simulated values than the observed values for both methods. The predicted streamflow in the flood seasons was lower than observed streamflow for some stations. It was mostly at Yingluoxia, Tangnaihai and Zheqiao stations. Compared with CESA, the predicted and observed values were closer in flood seasons for BESA model. Overall, the simulation of streamflow time series at the above stations was superior for BESA to CESA.

The forecasted values and observed values for two entropy spectral models in the lead time are shown in Figure 5. For Xiangtang, Yingluoxia, Zamusi, and Jiutiaoling stations, both models satisfactorily forecasted streamflow. In the first lead year, two models accurately forecasted the time of maximum monthly streamflow at Xiangtang, Yingluoxia, and Jiutiaoling stations. Nevertheless, the predicted maximum streamflow for the last two years appeared one month earlier or later. At Zamusi station, BESA accurately forecasted the bi-modal values of the flood season for the lead time, while CESA did not. BESA forecasted the number of peaks in the following two years, but the peak position appeared one month earlier or later. At Tangnaihai and Zheqiao stations, the difference between the predicted and observed values for BESA model was large. However, CESA still did not forecast the multimodal pattern of partial flood season, while uni-modal year of streamflow in the flood season had better forecast results.

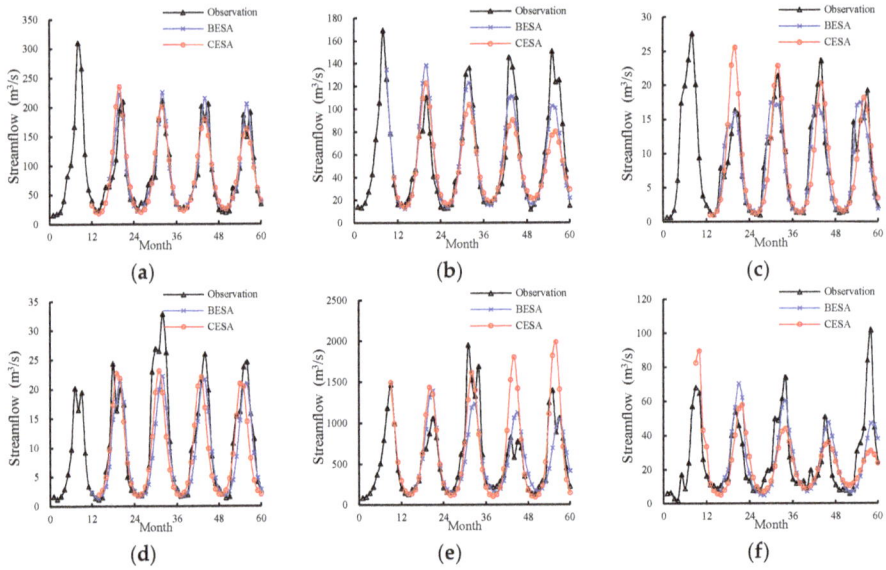

Figure 4. Streamflow forecasted using entropy spectral analysis of representative stations in training time. (**a**) Xiangtang station; (**b**) Yingluoxia station; (**c**) Zamusi station; (**d**) Jiutiaoling station; (**e**) Tangnaihai station; (**f**) Zheqiao station.

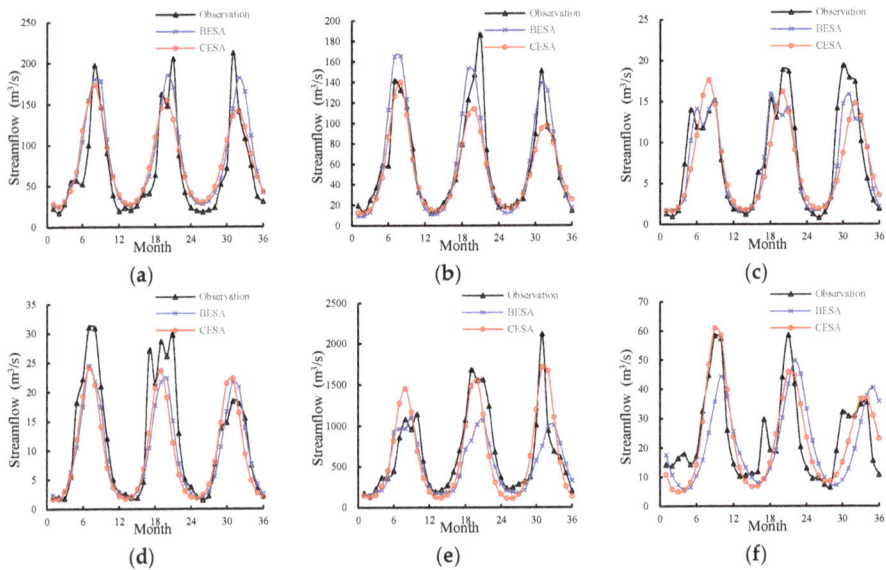

Figure 5. Streamflow forecasted using entropy spectral analysis of representative stations in lead time. (**a**) Xiangtang station; (**b**) Yingluoxia station; (**c**) Zamusi station; (**d**) Jiutiaoling station; (**e**) Tangnaihai station; (**f**) Zheqiao station.

The performance metrics of the models are shown in Table 2. It can be seen that the optimum order of BESA and CESA models range from 8 to 16 and 8 to 13, respectively. The R^2 and NSE values at Xiangtang, Yingluoxia, Zamusi, and Jiutiaoling stations during the training period were relatively high, with the values of over 0.86 and 0.70, respectively. The R^2 and NSE values at Tangnaihai and Zheqiao stations were lower than at the former four stations. The simulation of streamflow time series during the training period for BESA was better than that of CESA for the former four stations, and performance metrics were superior to CESA. For the other stations, the simulation results were equivalent for the two models. Streamflow forecasting by the two entropy spectrum models was good at Xiangtang, Yingluoxia, Zamusi, and Jiutiaoling stations during the verification period. The corresponding R^2 and NSE values were all higher than 0.88 and 0.70, respectively. However, streamflow forecasting at Tangnaihai and Zheqiao stations was relatively poor. Although the R^2 values were more than 0.76, the NSE values were between 0.48 and 0.49. BESA performed better at Xiangtang, Yingluoxia, Zamusi, and Jiutiaoling stations while CESA was more suitable for forecasting streamflow at Tangnaihai and Zheqiao stations.

Table 2. Results of forecasting at representative stations by two entropy methods.

Station	Model	Model Order	Training Time (2003–2007)				Lead Time (2008–2010)			
			RE	RMSE	R^2	NSE	RE	RMSE	R^2	NSE
xiangtang	BESA	16	0.169	19.4	0.953	0.904	0.376	27.3	0.913	0.773
	CESA	11	0.231	24.5	0.920	0.840	0.405	28.1	0.890	0.760
yingluoxia	BESA	8	0.202	17.3	0.915	0.834	0.235	21.2	0.912	0.798
	CESA	10	0.267	22.8	0.874	0.708	0.185	21.1	0.917	0.801
zamusi	BESA	14	0.207	2.7	0.913	0.832	0.268	2.5	0.925	0.838
	CESA	12	0.316	3.7	0.864	0.690	0.382	3.5	0.840	0.684
jiutiaoling	BESA	11	0.229	4.5	0.920	0.768	0.245	4.9	0.915	0.755
	CESA	13	0.346	5.1	0.868	0.697	0.318	5.4	0.886	0.707
tangnaihai	BESA	13	0.312	303.3	0.750	0.548	0.295	354.8	0.759	0.482
	CESA	8	0.335	340.3	0.805	0.447	0.360	273.0	0.861	0.693
zheqiao	BESA	15	0.362	19.0	0.621	0.255	0.291	9.1	0.876	0.618
	CESA	8	0.466	17.8	0.640	0.365	0.369	8.6	0.843	0.659

Comparison of monthly streamflow estimated by BESA and CESA and observed values during the verification period is shown in Figures 6 and 7, respectively. The slope of the trend line was closer to 1, indicating that the bias between predicted and observed values was smaller. The larger R^2 suggested that the correlation between predicted and observed values was better. That is to say, the predicted values were much closer to the observed values. At Xiangtang, Yingluoxia, Zamusi, and Jiutiaoling stations, the trend line slope of BESA was much closer to 1 than that of CESA and the corresponding R^2 was higher. By contrast, the trend line slope of CESA was much closer to 1 than that of BESA at Tangnaihai and Zheqiao stations, and the correlation coefficient was much higher.

Above all, the fitness of BESA for simulating the observed streamflow sequence was better than that of CESA. The forecast accuracy of BESA at Xiangtang, Yingluoxia, Zamusi, and Jiutiaoling stations was better than that of CESA. Nevertheless, it was the opposite at Tangnaihai station. Neither model made better forecasts at Zheqiao station.

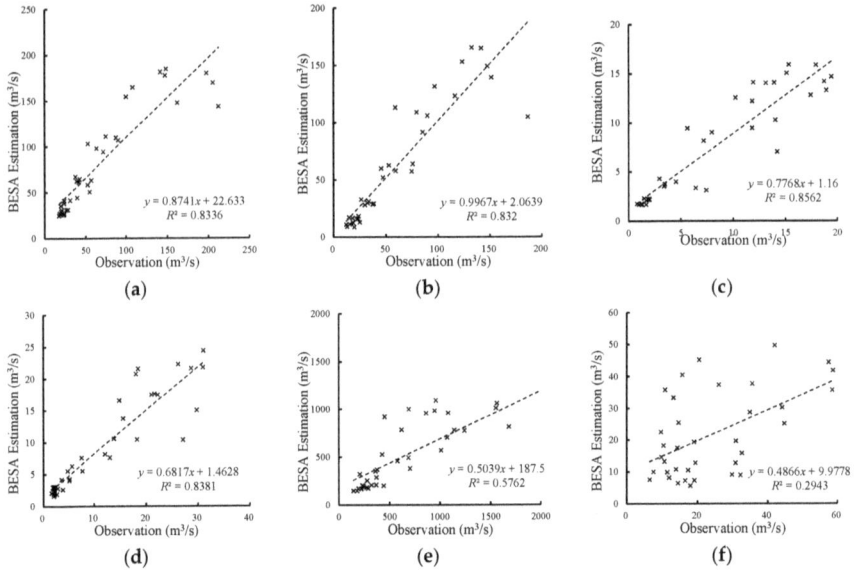

Figure 6. Forecasted values of Burg entropy spectral analysis (BESA) related to observed values in the lead time. (**a**) Xiangtang station; (**b**) Yingluoxia station; (**c**) Zamusi station; (**d**) Jiutiaoling station; (**e**) Tangnaihai station; (**f**) Zheqiao station.

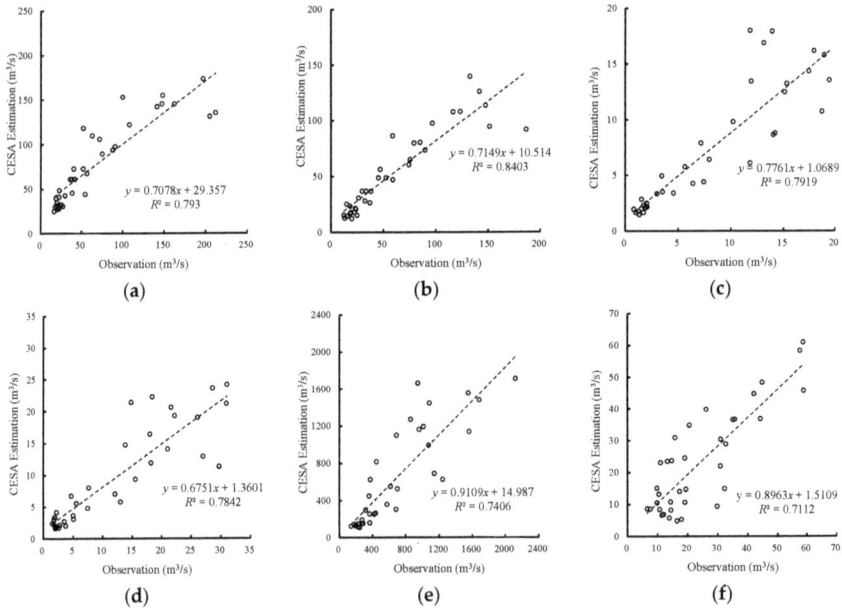

Figure 7. Forecasted values of configurational entropy spectral analysis (CESA) related to observed values in the lead time. (**a**) Xiangtang station; (**b**) Yingluoxia station; (**c**) Zamusi station; (**d**) Jiutiaoling station; (**e**) Tangnaihai station; (**f**) Zheqiao station.

3.3. Comparison with Other Autocorrelation Models

The autoregressive coefficients of BESA and CESA are obtained by maximizing Burg entropy and Configurational entropy, respectively. In order to demonstrate the improved accuracy of the predictions, we performed comparison with two other autocorrelation models. The first one is the AR model, and its coefficients are calculated by Yule-Walker function. The second one is the seasonal autoregressive model (SAR), and it rearranges the streamflow by month to avoid the seasonality. The comparison of the performance metrics is shown in Table 3. It can be seen that the BESA and CESA performed better than the AR and SAR models. This is because the BESA and CESA combine the maximum entropy principle and spectral analysis. The model estimated by maximum entropy principle is unbiased with all available data and no further hypothesis are needed. The spectral analysis can detect the periodical pattern of time series. Thus, BESA and CESA is more accurate and reliable than AR and SAR model.

Table 3. Comparison of the performance metrics by four models.

Station	BESA		CESA		AR		SAR	
	NSE of TT	NSE of LT	NSE of TT	NSE of LT	NSE of TT	NSE of LT	NSE of TT	NSE of LT
xiangtang	0.904	0.773	0.840	0.760	0.686	0.666	0.564	0.631
yingluoxia	0.834	0.798	0.708	0.801	0.612	0.755	0.429	0.624
zamusi	0.832	0.838	0.690	0.684	0.644	0.524	0.479	0.540
jiutiaoling	0.768	0.755	0.697	0.707	0.585	0.511	0.456	0.466
tangnaihai	0.548	0.482	0.447	0.693	0.495	0.34	0.403	0.373
zheqiao	0.255	0.618	0.365	0.659	0.485	0.144	0.364	0.728

Notes: TT represents training time and LT represents lead time. AR: autoregressive; SAR: seasonal autoregressive.

4. Discussion

Despite the fact that the six hydrological stations are in the same area, they differ in factors such as the type of river, control catchment area, vegetation condition, and human activities. Datong River (Xiangtang station), Daxia River (Zheqiao station), and the Yellow River (Yingluoxia station) belong to outflow rivers, whereas Heihe River (Yingluoxia station), Xiying River (Jiutiaoling station), and Zamu River (Zamusi station) belong to interior rivers. As the upstream of Yellow River, the catchment area of Tangnaihai station is the largest, which is about 120,000 km^2. Both Xiangtang and Zheqiao stations are located on the tributary of Yellow River, while Zamusi and Jiutiaoling stations are situated on the Shiyang River. Although four hydrological stations are located on the downstream, the catchment area are all less than 20,000 km^2. Yingluoxia station is located on the upstream Heihe River, with a catchment area of 10,009 km^2. For Xiangtang, Zamusi and Jiutiaoling stations, the upper reaches of piedmont watershed scale have good vegetation coverage and little human activity. By comparison, vegetation coverage is slightly poorer in the middle and lower reaches. Yingluoxia station is located on the Heihe upstream, but its catchment area and vegetation coverage are similar to the upstream of the watershed mentioned above. The impact of human activity is small above Tangnaihai station. In addition, industrial and agricultural water use is rare. Since the 1990s, the grassland has a tendency to gradually degradation. The upstream piedmont of Daxia River (i.e., Zheqiao station-owned) stony mountainous area covered with pasture except for few woods. Its downstream flow through loess plateau with ravines crossbar, poor vegetation and serious soil erosion and concentrated human activities.

Whether outliers, control catchment area, vegetation condition, and human activities, all reflect correlation of streamflow data. The autocorrelation of the six stations is shown in Figure 8. It can be seen that the autocorrelation of Tangnaihai and Zheqiao stations is relatively low (i.e., the autocorrelation coefficient is less than 0.5 at the 12th lag, and other stations are higher than 0.6). Based on the level of autocorrelation, the six stations can be grouped into two categories. The autocorrelation in the first category (Xiangtang, Yingluoxia, Zamusi, and Jiutiaoling stations) is higher than the second category (Tangnaihai and Zheqiao stations). As streamflow forecasting is based on the autocorrelation with the past series, the streamflow series with strong correlation will be more reliable forecasted. Thus, streamflow forecasting effects of the first category are better than the second category. BESA fitness may be better and unbiased because of the difference methods of the two forecast models. For BESA, the autoregressive coefficients are calculated by Levinson–Burg algorithm, which is developed from the AR model. For CESA, the autoregressive coefficients are calculated by cepstrum estimation. Therefore, entropy spectrum analysis methods need to be chosen carefully according to the situation of the study area.

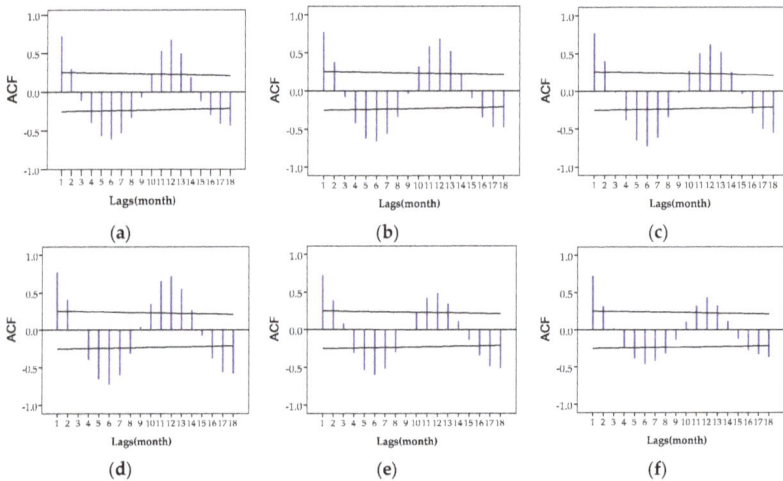

Figure 8. Autocorrelation plot of representative stations. (**a**) Xiangtang station; (**b**) Yingluoxia station; (**c**) Zamusi station; (**d**) Jiutiaoling station; (**e**) Tangnaihai station; (**f**) Zheqiao station.

The six stations selected in this paper are all located in Northwest China. Streamflow is principally composed of precipitation. Precipitation is the main recharge source of streamflow among them. Annual precipitation mainly occurs during June to September. All six rivers originate from alpine regions where April and May are spring flood periods, and flood is mainly formed by snow melt. Although annual precipitation mainly occurs during June to September, rainfall is mostly heavy rain and the month with maximum precipitation is not fixed. Hence, the maximum monthly streamflow is also unset. The input of autoregressive model is only previous monthly streamflow data, which may influence the forecast accuracy of the autoregressive model in the flood season. Therefore, adding precipitation as a predictor, selecting one or more models with high accuracy in the flood season, and using the entropy spectrum model and its combination (such as combined streamflow forecasting based on cross entropy [36]) to forecast can be used as the next research direction.

5. Conclusions

Two entropy spectral analysis methods (Burg entropy and configurational entropy) are mainly developed for streamflow forecasting in Northwest China. The following conclusions are drawn from this study:

1. The autoregressive coefficients obtained by maximizing Burg entropy and configurational entropy leads to more reliable than those by Levinson-Durbin algorithm. So, the streamflow forecasted by BESA and CESA is more accurate than that of the AR and SAR models.
2. For the streamflow with strong correlation, both BESA and CESA forecast monthly streamflow well. The R^2 and NSE were over 0.84 and 0.68, respectively. The forecast accuracy of BESA is higher than that of CESA. For the streamflow with weak correlation, the conclusion is the opposite.

3.   The time of peak flow forecasted by both models (BESA and CESA) may be either earlier or later than observed. The peak flow is generally underestimated by both models. BESA accurately forecasted the bi-modal values of the flood season for the lead time, while CESA had better forecast results for the streamflow data with weak correlation.

4.   In Northwest China, streamflow in the flood periods is principally composed of precipitation. The month with maximum precipitation is not fixed. Hence, the study of streamflow characteristics and spectral pattern associated with the precipitation can be used as the next research direction.

Acknowledgments: We are grateful for the grant support from the National Natural Science Fund in China (Project Nos. 91425302, 51279166 and 51409222), the Fundamental Research Funds for the Central University (Project No. 2452016069), and the Doctoral Program Foundation of Northwest A & F University (Project No. 2452015290). We wish to thank the editor and anonymous reviewers for their valuable comments and constructive suggestions, which were used to improve the quality of the manuscript.

Author Contributions: Z.H.Z. designed the study and performed the experiments. J.L.J. wrote the paper and reviewed it. X.L.S. and V.P.S. reviewed and edited the manuscript. G.X.Z. drew the research area map and gave some feasible suggestions. All authors have read and approved the final manuscript.

Conflicts of Interest: The authors declare no conflict of interest.

References

1.   Box, G.E.; Jenkins, G.M. *Time Series Analysis: Forecasting and Control*; Holden-Day: Oakland, CA, USA, 1976.
2.   Carlson, R.F.; Maccormick, A.J.A.; Watts, D.G. Application of linear random models to four annual streamflow series. *Water Resour. Res.* **1970**, *6*, 1070–1078. [CrossRef]
3.   Hipel, K.W.; McLeod, A.I. *Time Series Modelling of Water Resources and Environmental Systems*; Elsevier: Amsterdam, The Netherlands, 1994; Volume 45.
4.   Haltiner, J.P.; Salas, J.D. Development and testing of a multivariate, seasonal ARMA (1,1) model. *J. Hydrol.* **1988**, *104*, 247–272. [CrossRef]
5.   Elshorbagy, A.; Simonovic, S.; Panu, U. Noise reduction in chaotic hydrologic time series: Facts and doubts. *J. Hydrol.* **2002**, *256*, 147–165. [CrossRef]
6.   Singh, V.P.; Cui, H. Entropy theory for streamflow forecasting. *Environ. Process.* **2015**, *2*, 449–460. [CrossRef]
7.   Fleming, S.W.; Marsh Lavenue, A.; Aly, A.H.; Adams, A. Practical applications of spectral analysis to hydrologic time series. *Hydrol. Process.* **2002**, *16*, 565–574. [CrossRef]
8.   Ghil, M.; Allen, M.; Dettinger, M.; Ide, K.; Kondrashov, D.; Mann, M.; Robertson, A.W.; Saunders, A.; Tian, Y.; Varadi, F. Advanced spectral methods for climatic time series. *Rev. Geophys.* **2002**, *40*. [CrossRef]
9.   Labat, D. Recent advances in wavelet analyses: Part 1. A review of concepts. *J. Hydrol.* **2005**, *314*, 275–288. [CrossRef]
10.   Labat, D.; Ronchail, J.; Guyot, J.L. Recent advances in wavelet analyses: Part 2—Amazon, Parana, Orinoco and Congo discharges time scale variability. *J. Hydrol.* **2005**, *314*, 289–311. [CrossRef]
11.   Marques, C.; Ferreira, J.; Rocha, A.; Castanheira, J.; Melo-Goncalves, P.; Vaz, N.; Dias, J. Singular spectrum analysis and forecasting of hydrological time series. *Phys. Chem. Earth Parts A/B/C* **2006**, *31*, 1172–1179. [CrossRef]
12.   Burg, J.P. Maximum entropy spectral analysis. In Proceedings of the 37th Annual International Meeting, Oklahoma, OK, USA, 31 October 1967.
13.   Currie, R.G. Geomagnetic line spectra-2 to 70 years. *Astrophys. Space Sci.* **1973**, *21*, 425–438. [CrossRef]
14.   Pardo-Igúzquiza, E.; Rodríguez-Tovar, F.J. Maximum entropy spectral analysis of climatic time series revisited: Assessing the statistical significance of estimated spectral peaks. *J. Geophys. Res.* **2006**, *111*. [CrossRef]
15.   Hasanean, H. Fluctuations of surface air temperature in the eastern mediterranean. *Theor. Appl. Climatol.* **2001**, *68*, 75–87. [CrossRef]
16.   Wang, D.; Chen, Y.-F.; Li, G.-F.; Xu, Y.-H. Maximum entropy spectral analysis for annual maximum tide levels time series of the changjiang river estuary. *J. Coast. Res.* **2004**, *43*, 101–108.
17.   Dalezios, N.R.; Tyraskis, P.A. Maximum entropy spectra for regional precipitation analysis and forecasting. *J. Hydrol.* **1989**, *109*, 25–42. [CrossRef]

18. Huang, Z.S. The application of spectrum analysis method in hydrogy. *Hydrology* **1983**, *3*, 101–108.
19. Krstanovic, P.; Singh, V.P. A univariate model for long-term streamflow forecasting. *Stoch. Hydrol. Hydraul.* **1991**, *5*, 173–188. [CrossRef]
20. Krstanovic, P.; Singh, V.P. A real-time flood forecasting model based on maximum-entropy spectral analysis: I. Development. *Water Resour. Manag.* **1993**, *7*, 109–129. [CrossRef]
21. Krstanovic, P.; Singh, V.P. A real-time flood forecasting model based on maximum-entropy spectral analysis: II. Application. *Water Resour. Manag.* **1993**, *7*, 131–151. [CrossRef]
22. Singh, V.P. *Entropy Theory and Its Application in Environmental and Water Engineering*, 1st ed.; Wiley: Hoboken, NJ, USA, 2013.
23. Singh, V.P. *Entropy Theory in Hydrologic Science and Engineering*; McGraw-Hill: New York, NY, USA, 2015.
24. Huo, C. Use of auto-regression model of time series in the simulation and forecast of groundwater dynamic in irrigation areas. *Geotech. Investig. Surv.* **1990**, *1*, 36–38.
25. Wang, D.; Zhu, Y.S. Research on cryptic period of hydrologic time series based on mem1spectral analysis. *Hydrology* **2002**, *2*, 19–23.
26. Shen, H.F.; Li, M.S.; Luo, F. Strict maximum entropy spectral estimation based on recursive algorithm. *Radar Sci. Technol.* **2008**, *4*, 288–291.
27. Boshnakov, G.N.; Lambert-Lacroix, S. A periodic levinson–durbin algorithm for entropy maximization. *Comput. Stat. Data Anal.* **2012**, *56*, 15–24. [CrossRef]
28. Cui, H.; Singh, V.P. Configurational entropy theory for streamflow forecasting. *J. Hydrol.* **2015**, *521*, 1–17. [CrossRef]
29. Frieden, B.R. Restoring with maximum likelihood and maximum entropy. *J. Opt. Soc. Am.* **1972**, *62*, 511–518. [CrossRef] [PubMed]
30. Gull, S.F.; Daniell, G.J. Image reconstruction from incomplete and noisy data. *Nature* **1978**, *272*, 686–690. [CrossRef]
31. Wu, N.L. An explicit solution and data extension in the maximum entropy method. *IEEE Trans. Acoust. Speech Signal Process.* **1983**, *31*, 486–491.
32. Nadeu, C. Finite length cepstrum modelling—A simple spectrum estimation technique. *Signal Process.* **1992**, *26*, 49–59. [CrossRef]
33. Katsakos-Mavromichalis, N.; Tzannes, M.; Tzannes, N. Frequency resolution: A comparative study of four entropy methods. *Kybernetes* **1986**, *15*, 25–32. [CrossRef]
34. Schwarz, G. Estimating the dimension of a model. *Ann. Stat.* **1978**, *6*, 461–464. [CrossRef]
35. Nash, J.E.; Sutcliffe, J.V. River flow forecasting through conceptual models partI—A discussion of principles. *J. Hydrol.* **1970**, *10*, 282–290. [CrossRef]
36. Men, B.; Long, R.; Zhang, J. Combined forecasting of streamflow based on cross entropy. *Entropy* **2016**, *18*, 336. [CrossRef]

entropy

MDPI

Article

Modeling NDVI Using Joint Entropy Method Considering Hydro-Meteorological Driving Factors in the Middle Reaches of Hei River Basin

Gengxi Zhang [1], Xiaoling Su [1,*], Vijay P. Singh [2] and Olusola O. Ayantobo [1,3]

[1] College of Water Resources and Architectural Engineering, Northwest A & F University,
     Yangling 712100, China; gengxizhang@nwsuaf.edu.cn (G.Z.); ayantobo.olusola@nwsuaf.edu.cn or
     ayantobooo@funaab.edu.ng (O.O.A.)
[2] Department of Biological & Agricultural Engineering and Zachry Department of Civil Engineering,
     Texas A&M University, 2117 TAMU, College Station, TX 77843, USA; vsingh@tamu.edu
[3] Department of Water Resources Management and Agricultural-Meteorology, Federal University of
     Agriculture, PMB 2240, Abeokuta 110282, Nigeria
*   Correspondence: xiaolingsu@nwsuaf.edu.cn

Received: 4 August 2017; Accepted: 13 September 2017; Published: 15 September 2017

Abstract: Terrestrial vegetation dynamics are closely influenced by both hydrological process and climate change. This study investigated the relationships between vegetation pattern and hydro-meteorological elements. The joint entropy method was employed to evaluate the dependence between the normalized difference vegetation index (NDVI) and coupled variables in the middle reaches of the Hei River basin. Based on the spatial distribution of mutual information, the whole study area was divided into five sub-regions. In each sub-region, nested statistical models were applied to model the NDVI on the grid and regional scales, respectively. Results showed that the annual average NDVI increased at a rate of 0.005/a over the past 11 years. In the desert regions, the NDVI increased significantly with an increase in precipitation and temperature, and a high accuracy of retrieving NDVI model was obtained by coupling precipitation and temperature, especially in sub-region I. In the oasis regions, groundwater was also an important factor driving vegetation growth, and the rise of the groundwater level contributed to the growth of vegetation. However, the relationship was weaker in artificial oasis regions (sub-region III and sub-region V) due to the influence of human activities such as irrigation. The overall correlation coefficient between the observed NDVI and modeled NDVI was observed to be 0.97. The outcomes of this study are suitable for ecosystem monitoring, especially in the realm of climate change. Further studies are necessary and should consider more factors, such as runoff and irrigation.

Keywords: joint entropy; NDVI; temperature; precipitation; groundwater depth; Hei River basin

---

1. Introduction

Terrestrial vegetation plays a key role in energy, water, and biogeochemical cycles, while variation in vegetation can significantly influence atmospheric and hydrological processes. Vegetation is a sensitive indicator of global change and can also be a natural link between the atmosphere, land surface, soil, and water [1,2]. Changes in vegetation cover are influenced by climate change, human activities, and the atmospheric CO_2 fertilization effect [3–5]. It is necessary to explore the response mechanism of vegetation dynamics to hydrological processes and climate change.

Vegetation emergence and senescence are closely related to characteristics of the lower atmosphere, including the annual cycle of weather pattern shifts, temperature and precipitation [6]. In particular, the amount and duration of precipitation and temperature play a significant role in controlling vegetation development [2,7–9]. During the last decades, the global average land surface air temperature has been increasing systematically in the northern middle and high latitude [10]. Elevated air temperature increased plant growth by lengthening the growing season [11], enhancing photosynthesis, and altering nitrogen availability by accelerating decomposition or mineralization. Precipitation is the dominant controlling climatic factor in water-limited, semi-arid and arid regions such as the northern China [10]. The increase in precipitation contributes to the growth of vegetation. Several studies have reported relationships between vegetation indices and precipitation and temperature [3,12–17]. However, most of these studies focused solely on either precipitation or temperature for vegetation models, thereby resulting in low precision and low correlation coefficients between vegetation indices and climatic variables [17]. For instance, Ichii et al. (2002) found a weaker relationship between vegetation growth and precipitation [18], while Nicholson et al. (1990) reported a weaker relationship between NDVI and precipitation, indicating a nonlinear overall relationship [19]. However, apart from the influence of climatic change, vegetation is also strongly related to the groundwater table level (GTL) in arid and semi-arid regions [2,20–22]. Groundwater is a key impact factor in sustaining the ecological environment, by supplying soil moisture through capillarity for maintaining the root system. The shallower the groundwater depth the more available soil moisture, and vice versa [22]. Many investigators have also reported the relationship between vegetation and groundwater availability [2,21]. For instance, Jin et al. (2016) found a linear correlation between NDVI and groundwater depth in the Qaidam basin [2]. The distribution of GTL is becoming more heterogeneous in the middle reaches of the Hei River as a result of increasing cultivated land area and runoff from Yingluoxia station [21]. The distribution of vegetation is also highly heterogeneous and depends upon multiple factors. In this study, different statistical tools, as well as the joint entropy method, were employed in order to understand the relationships between NDVI and precipitation, temperature, and GTL. This will be necessary for investigating the relevance of coupling variables.

The joint entropy expresses the total amount of uncertainty contained in the union of events [23] either in a linear or nonlinear system [24] and has been employed for various purposes. For instance, Li et al. (2012) employed joint entropy to design appropriate hydrometric networks [25], while Mishra et al. (2012) employed the entropy theory to overcome the nonlinear relationship between precipitation and vegetation, and further identified the downscaling method yielding higher mutual information [23]. Entropy theory has also been applied in ecological studies [5,17,26,27]. For instance, Sohoulande, Djebou and Singh (2015) employed joint entropy to retrieve vegetation growth patterns from climatic variables. However, this study emphasizes the mutual information of NDVI and the influencing factors.

In this paper, the mutual information of NDVI was investigated by coupling cumulative precipitation, the average temperature of growing season, and GTL. To that end, the study area was divided into 5 sub-regions, based on 3 spatial distributions of mutual information. In each sub-region, NDVI was modeled using nested statistical models at the grid and regional scales, respectively.

2. Methodology

2.1. Trend Analysis

Trends for each pixel were represented to reflect the characteristics of vegetation cover over different periods of time [28]. The rate of change of NDVI was calculated as [29].

$$\theta_{slope} = \frac{n \times \sum\limits_{i=1}^{n} i \times NDVI_i - \sum\limits_{i=1}^{n} i \sum\limits_{i=1}^{n} NDVI_i}{n \times \sum\limits_{i=1}^{n} i^2 - (\sum\limits_{i=1}^{n} i)^2} \tag{1}$$

where θ_{slope} is the slope of the trend line of annual NDVI; i represents the years; n is the time span; and $NDVI_i$ is the NDVI value for the i-th year.

θ_{slope} describes the trend in the annual NDVI within the study area. If $\theta_{slope} > 0$, the NDVI increases; or else, the NDVI decreases ($\theta_{slope} < 0$) or remains constant ($\theta_{slope} = 0$).

2.2. Joint Entropy and Mutual Information

Mutual information was employed to explore the relationship between NDVI and influencing factors. Considering a random variable X, for each value of X, x_i represents an event with a corresponding probability of occurrence, p_i. The entropy $H(X)$ can be expressed as:

$$H(X) = - \sum_{i=1}^{n} p_i \log_2(p_i) \tag{2}$$

Where i means the ith event. The logarithm is based on 2, because it is more convenient to use than logarithms based one or 10, however, the base can be taken as other than 2 without alteration [24]. The entropy is thereby measured in bits. The probability density function (PDF) of variable X is obtained using discrete intervals for the values of X. The discrete PDF is defined for n equal-width bins defined for the range of X [17,24,30]. The entropy $H(X)$ is a measurement of information or uncertainty [24]. Similarly, the joint entropy can be computed for a joint probability distribution of two or more variables [24,25]. Specifically, with three variables, the joint PDF is computed based on a three-dimensional contingency analysis as illustrated in Table 1. Considering the three variables precipitation, temperature and NDVI, the process generates a discrete PDF made of n^3 discrete probabilities such that:

$$\sum_{i=1}^{n} \sum_{j=1}^{n} \sum_{k=1}^{n} p(P = P_i, T = T_j, NDVI = NDVI_k) = 1 \tag{3}$$

The joint entropy of the three variables can be defined as:

$$H(P, T, NDVI) = - \sum_{i=1}^{n} \sum_{j=1}^{n} \sum_{k=1}^{n} p(P_i, T_j, NDVI_k) \log_2[p(P_i, T_j, NDVI_k)] \tag{4}$$

where n means the number of events.

Table 1. Illustration of a three-dimension contingency employed to compute the joint PDF. Consider 3 variables X, Y, and Z.

Variables			Discrete Probabilities
$X = \{X_1, X_2\}$	$Y = \{Y_1, Y_2\}$	$Z = \{Z_1, Z_2\}$	
$X = X_1$	$Y = Y_1$	$Z = Z_1$	$p(X = X_1, Y = Y_1, Z = Z_1)$
		$Z = Z_2$	$p(X = X_1, Y = Y_1, Z = Z_2)$
	$Y = Y_2$	$Z = Z_1$	$p(X = X_1, Y = Y_2, Z = Z_1)$
		$Z = Z_2$	$p(X = X_1, Y = Y_2, Z = Z_2)$
$X = X_2$	$Y = Y_1$	$Z = Z_1$	$p(X = X_2, Y = Y_1, Z = Z_1)$
		$Z = Z_2$	$p(X = X_2, Y = Y_1, Z = Z_2)$
	$Y = Y_2$	$Z = Z_1$	$p(X = X_2, Y = Y_2, Z = Z_1)$
		$Z = Z_2$	$p(X = X_2, Y = Y_2, Z = Z_2)$

Note: Each variable is categorized into 2 classes ($n = 2$).

The mutual information represents the amount of information common to both X and Y and provides a general measure of dependence between the random variables. It is superior to the Pearson correlation coefficient, because it captures both linear and nonlinear dependence while the Pearson correlation coefficient is only suitable for linear relationships [25]. It equals the difference between the sum of two marginal entropies and the total entropy:

$$M(X, Y) = H(X) + H(Y) - H(X, Y) \tag{5}$$

The two-dimensional cases of the amount of information transmitted can be extended to three or more dimensional cases [31]. Considering the three variables of precipitation, temperature, and NDVI, precipitation and temperature are considered inputs while NDVI is considered an output. Therefore:

$$M(P, T; NDVI) = H(P, T) + H(NDVI) - H(P, T, NDVI) \tag{6}$$

Moreover, applying the same methodology, the coupling of P and GTL; T and GTL along with NDVI were further explored. The ensemble E of candidate equations can be summarized as:

$$E = \{NDVI_{GTL, P} = f(GTL, P); NDVI_{GTL, T} = f(GTL, T); NDVI_{T, P} = f(T, P)\} \tag{7}$$

3. Study Area and Data

The study area occupied the middle reaches of the Hei River basin (38°30′–39°55′N, 98°55′–100°55′E; Figure 1), which is located in the middle of Hexi corridor of Gansu Province, northwest of China, and belongs to an arid climate zone. The total area is about 10,000 km². The mean annual temperature is around 8 °C, the mean annual precipitation is between 69 and 216 mm and is concentrated between June and September, while the mean annual potential evaporation is between 1453 and 2351 mm [32].

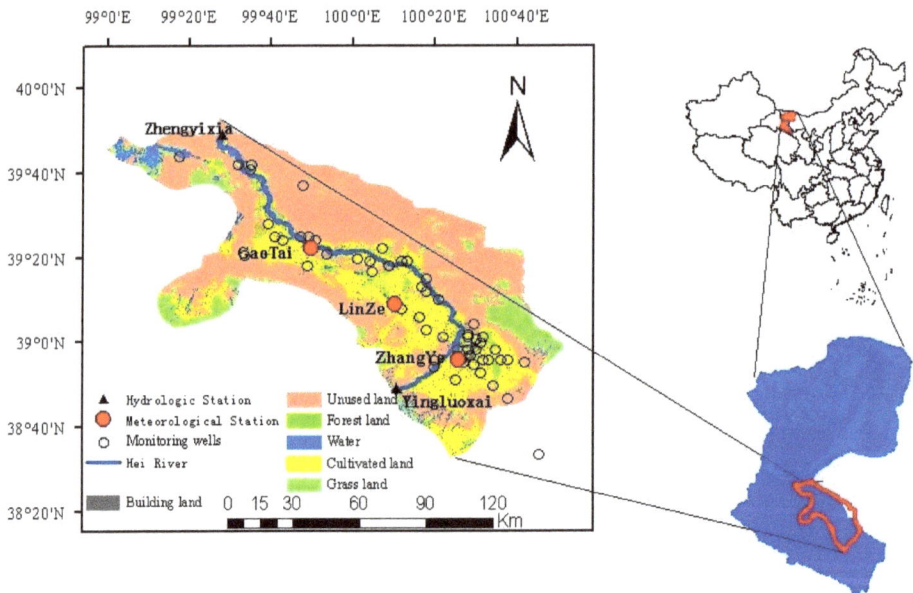

Figure 1. Map of the middle reaches of Hei River basin and the monitoring wells.

The Hei River is the second largest inland river in China with a total length of 821 km and a runoff of 16.0×10^8 m³·year⁻¹. It flows through Yingluoxia and Zhengyixia stations, entering and exiting middle reaches. Runoff has shown an increasing/decreasing trend in the past decade. The annual runoff at Yingluoxai station increased to 16.0×10^8 m³ in the 1990s from around 14.4×10^8 m³ in the 1960s (Figure 2), while the runoff at Zhengyixai station decreased by 3×10^8 m³ during the same time period (Figure 2). This situation accelerated desertification within the northern parts of the basin [33,34]. In order to mitigate deteriorating ecosystems in the Hei River basin, an Ecological Water Diversion Project (EWDP) was initiated by the Chinese government in 2000 to ensure the delivery of water to the lower reaches for ecological water needs [35]. After that, runoff at Zhengyixia station has increased to levels not recorded since the 1960s [33,36]. The implementation of the EWDP alleviated the ecological deterioration in the lower reaches of the Hei River but aggravated the water resources shortage in the middle reaches. In order to meet the increasing demand for water resources, groundwater exploration provided an option. After 2004, the recharge of groundwater also increased due to the increase in precipitation, irrigation and runoff replenishment. The temporal variation in groundwater levels of the observation wells over the period from2001 to 2011 is shown in Figure 3c which shows a rising trend in groundwater level along the Hei River [37]. However, the water level in the region away from the river showed a declining trend.

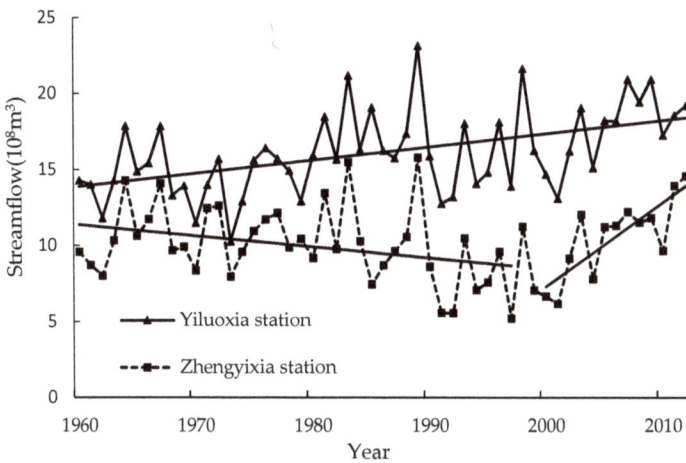

Figure 2. Annual discharge of Hei River at Yinluoxia and Zhengyixia stations.

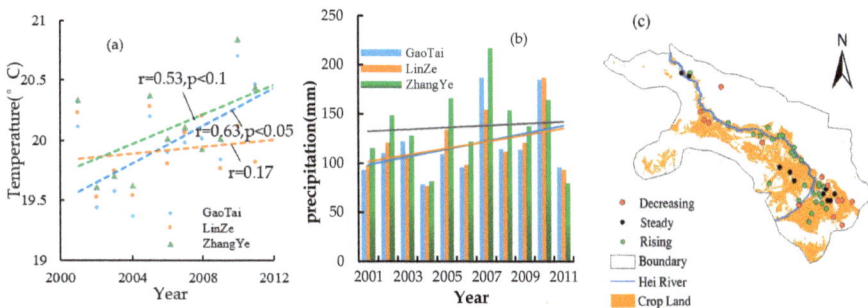

Figure 3. The temporal variation of average temperature of growing season (**a**); cumulative precipitation (**b**) and average ground depths (**c**).

The annual NDVI of each pixel was calculated by the annual maximum value of NDVI in order to eliminate the noise from the cloud and solar altitude. The Moderate-resolution imaging spectroradiometer (MODIS)-based NDVI products(MYD13A2 and MOD13A2)used for this study were acquired from the National Natural Science Foundation of China "Environmental and Ecological

Science Data Center in the West of China (http://westdc.westgis.ac.cn/)" with a spatial resolution of 1 km and a temporal resolution of 1 day, groundwater depths of 53 wells were obtained from the National Natural Science Foundation of China "Hei River Project Data Management Center (http://westdc.westgis.ac.cn/)" while monthly precipitation and temperature datasets were acquired from the China Meteorological Network (http://data.cma.cn/), with a 0.5° spatial resolution, interpolated according to the observed data of 2747 stations in China.

The ordinary kriging method was employed to interpolate the observed values of ground water depths. For consistency regarding the NDVI spatial resolution, the interpolated groundwater table levels, precipitation, and temperature datasets were rescaled to a 1 km spatial resolution.

4. Results and Discussion

4.1. Variation of Hydrometeorological Variables and NDVI

During the period from 2001 to 2011, both the average temperature and cumulative precipitation increased at the three hydrologic stations (Figure 3a,b). The temperature increases significantly at Gaotai station ($p < 0.05$). However, the trend of temperature increase is not obvious at another two stations. The distributions of cumulative precipitation and average temperature are shown in Figure 4a,b. Temperature gradually rises from south to north, while precipitation decreases were also observed. This is mainly determined by topography.

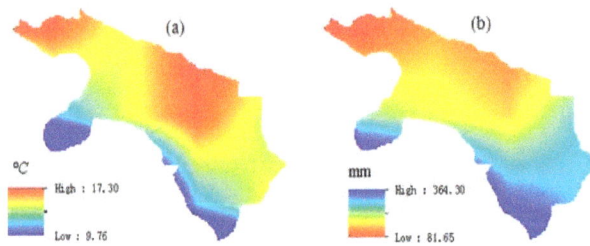

Figure 4. The spatial distribution of (**a**) average temperature of growing season; (**b**) cumulative precipitation.

Figure 5a illustrates the annual maximum NDVI in the middle reaches of the Hei River basin from 2000 to 2011. The slope of the linear regression trend-line was 0.005. The NDVI time series corresponded to the average of all the pixels in the study area and reflected the overall trend. Figure 5b shows the spatial distribution of the linear regression slope (θ_{slope}). In most cases, the area in which NDVI increased occupied about 94% of the total area. It can be seen that NDVI changes obviously. In order to explore the influencing factors on NDVI variation, the whole region was divided into 5 sub-regions based on the joint entropy method, and different regression models were constructed considering hydro-meteorological variables.

4.2. Mutual Information of NDVI with Coupling of Variables

To understand the relationship between NDVI and hydro-meteorological variables, mutual information was employed to measure the dependence between the NDVI time series and each coupled variable. The greater the values of mutual information the higher the correlation. The spatial distribution of mutual information between NDVI and coupled variables are shown in Figure 6a–c, respectively. It can be observed from the figures that the mutual information of NDVI and coupled variables of precipitation and GTL generally decreased from southeast to northwest. The mutual information values of NDVI and the coupled variables of precipitation and temperature were dispersedly distributed, increasing from the center to the edge of regions, while the mutual information of NDVI and the coupled variables of temperature and groundwater table levels were greater in the northwestern and southeastern regions. In order to appropriately model the NDVI using dependent variables, the study region was artificially divided into 5 sub-regions (Figure 6d), based on three spatial distributions of mutual information, topographic and land use maps. In order to simply construct models, as far as possible each sub-region was made to have the same characteristics. For instance, sub-region I and sub-region IV are mainly mountainous and M (P, T; NDVI) is higher than M (T, GTL; NDVI) and M (P, GTL; NDVI) in most areas of these sub-regions. As the same time, the land use types are desert. Likewise, in most areas of sub-region II and sub-region V, M (GTL, P; NDVI) is higher than M (P, T; NDVI) and M (T, GTL; NDVI), and these areas are distributed as artificial oases.

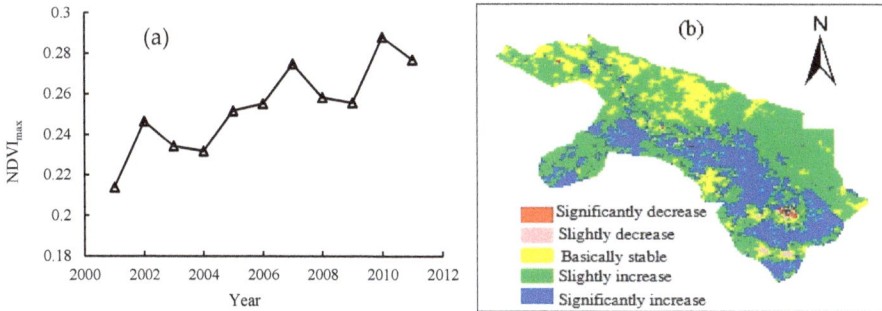

Figure 5. Inter-annual variability of annual NDVI (**a**) and the trends of NDVI (**b**).

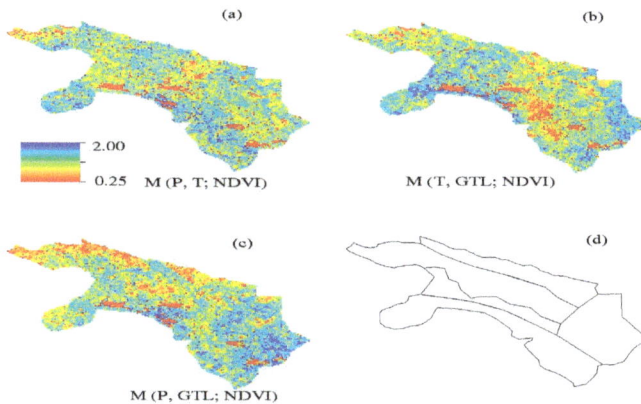

Figure 6. The distribution of mutual information of NDVI and three coupling hydro-meteorological variables (**a**–**c**) and the divided sub-regions (**d**).

4.3. Modelling NDVI

In sub-regions I and IV, precipitation and temperature were used to model the NDVI, since the mutual information of NDVI and the variables of precipitation and temperature were relatively higher. GTL and precipitation were used to model the NDVI in sub-regions II and V, while temperature and GTL were selected to model the NDVI in sub-region III.

Nested statistical models were developed for modeling NDVI in each sub-region. In most cases, the linear fitting of the NDVI time series to precipitation, temperature, and GTL resulted in poor correlation coefficients and high root means square error [17]. The vegetation growth may not be monotonic in relation to the atmospheric variables. For example, the temperature sensitivity of photosynthesis increases up to an optimum and later decreases as temperature gets higher [17]. A two-dimensional fitting was used to model NDVI using each of the hydro-meteorological variables as a factor. In order to select appropriate functional forms for modeling NDVI, four simple functional forms including linear function, quadratic function, exponent function and logarithm function were compared using Akaike information criterion (AIC) criteria. AIC is a measure of the relative quality of statistical models for a given set of data. The comparison of AIC values between different functional forms are as shown in Table 2. The smaller the AIC value, the more appropriate the form of the regression function. It was discovered that a logarithmic transformation of precipitation produced a better correlation with NDVI in some sub-regions (I, II, IV). Considering GTL and temperature, a negative exponential and a quadratic model showed a better relationship with NDVI. However, the logarithm regression model exhibited a better relationship between GTL and NDVI in sub-region III.

Table 2. The comparison of AIC values between different functional forms on 5 sub-regions; AIC in bold indicates most appropriate functional form.

NDVI and Temperature

Function \ Regin	Region I	Region II	Region III	Region IV	Region V
liner	−3.6153		−4.7407	−6.1732	
quadratic	**−3.8733**		−5.4122	**−6.5508**	
exponent	−0.6537		−1.3973	−2.9187	
logarithm	−3.7049		**−5.2447**	−6.5427	

NDVI and Groundwater

Function \ Regin	Region I	Region II	Region III	Region IV	Region V
liner		−1.8812	−3.8162		−2.7076
quadratic		−3.8876	−5.1042		−4.3212
exponent		**−3.9708**	−2.2641		**−4.3329**
logarithm		0.3744	**−5.1401**		0.4571

NDVI and Precipitation

Function \ Regin	Region I	Region II	Region III	Region IV	Region V
liner	−3.5757	−1.7001		−4.9864	−2.8516
quadratic	−4.5808	−4.714		−7.1646	**−4.4654**
exponent	−4.5530	−4.5093		−7.1620	−4.4443
logarithm	**−4.7402**	**−4.7974**		**−7.2847**	−2.5677

Different candidate equations were used to fit NDVI by coupling pair-wise precipitation, temperature, and GTL. The aim of the fitting procedures was to maximize the correlation coefficient and minimize the sum of squared residuals. However, the main importance of the fitting procedure was to maximize the proportion of the area in which the correlation coefficient met with significance testing.

In each sub-region, analysis was performed at the grid scale. In sub-regions I and IV, the coupling of precipitation and temperature resulted in a model of NDVI time series with an average correlation coefficient of 0.68 and 0.58, respectively. Li et al. (2016) also reported that in the desert regions in middle reaches of the Hei River, precipitation is a dominant influencing factor on annual NDVI variation [38]. Zhou et al. (2013) also demonstrated that precipitation and temperature are important factors driving the increase of NDVI in the study area [39]. However, the percentage of the area in which the correlation coefficient met with significance testing was only 42% ($p < 0.05$) and 15% ($p < 0.01$) in sub-region IV. This was mainly because the temperature was relatively high, thereby increasing evapotranspiration and water demand for vegetation growth that precipitation was unable to satisfy. In sub-regions II and V, the best fitting models were obtained by combining pair-wise negative exponentially transformed GTL respectively with the logarithmically and linearly transformed precipitation. In these regions, groundwater depths became an important influencing factor. Zhao et al. (2014) have also reported that in an oasis ecosystem, groundwater plays an important role in the formation of vegetation productivity [40]. In sub-region V, the percentage area in which the correlation coefficient met with the significance testing was only 50% ($p < 0.05$) and 16% ($p < 0.01$). There was a large area of arable land and artificial grassland in sub-region V (Figure 1). In sub-regions III and V, the growth of vegetation was strongly influenced by human activities, e.g., irrigation. Therefore, irrigation was the main influencing factor for vegetation growth. Beyond that, there was a considerable amount of water in sub-region III, thereby influencing prediction accuracy. After various fittings in different sub-regions, the correlation coefficients were combined for the entire study area (Figure 7). Results showed that there was about 53% of the area in which relationships were significant.

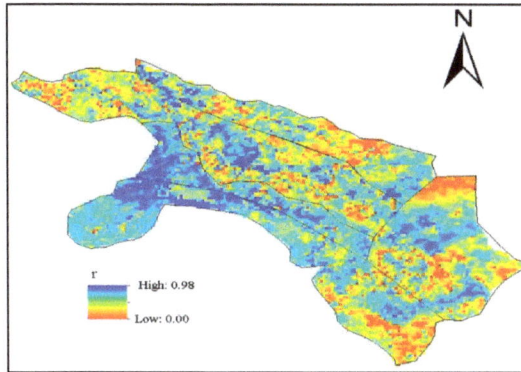

Figure 7. The distribution of correlation coefficients of estimated NDVI by coupling hydro-meteorological variables.

Analysis was also performed on a regional scale. Using the average NDVI time series of different sub-regions as predicted and each of the average hydro-meteorological variables as a predictor, a two-dimensional fitting was employed in the five sub-regions, respectively. This prediction also employed the same formulas used for the prediction on a grid scale (Table 3). Figure 8a–e showed the results of the three-dimensional scatter-plot couplings. Comparisons of these figures demonstrated that sub-regions I and II had a better accuracy of fitting. Figure 8f presents the overall relationship between observed NDVI and modeled NDVI for the whole region, where the correlation coefficient was observed to be 0.97, which indicated a high accuracy. This showed that this method was suitable for NDVI modeling.

Table 3. Fitting models, correlation coefficients and area proportion meeting significance testing in five sub-regions.

Sub-Region No.	Fitting Formula	Average Correlation Coefficient	Area Proportion ($p < 0.05$)	Area Proportion ($p < 0.01$)
I	$NDVI_{T,P} = a \cdot \ln(b \cdot P) + c \cdot T^2 + d \cdot T + e$	0.68	0.75	0.40
II	$NDVI_{GTL,P} = a \cdot e^{-(b \cdot GTL)} + c \cdot \ln(d \cdot P) + e$	0.66	0.72	0.42
III	$NDVI_{GTL,T} = a \cdot \ln(b \cdot GTL) + c \cdot T^2 + d \cdot T + e$	0.47	0.36	0.16
IV	$NDVI_{T,P} = a \cdot \ln(b \cdot P) + c \cdot T^2 + d \cdot T + e$	0.58	0.42	0.15
V	$NDVI_{GTL,P} =$ $a \cdot GTL^2 + b \cdot GTL + c \cdot P^2 + d \cdot P + e$	0.57	0.50	0.16

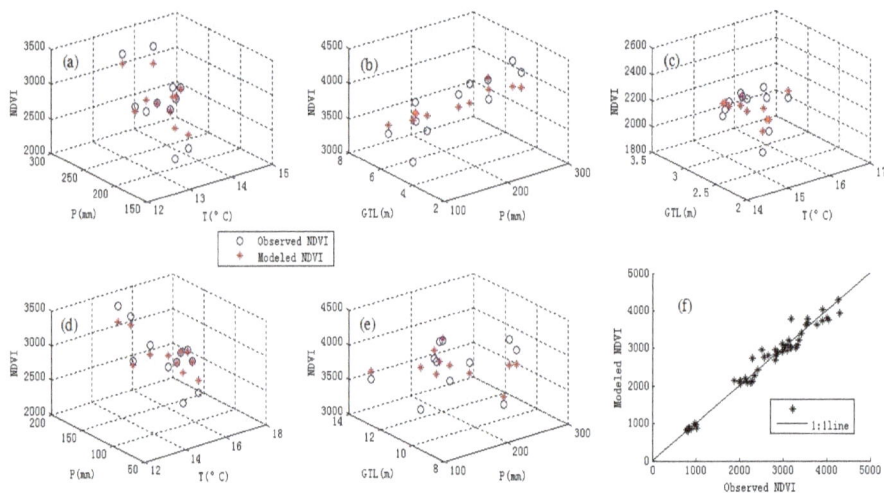

Figure 8. Three-dimensional scatter plots of NDVI (multiplied by 10^4) in five sub-regions (a–e) and the relationship between observed NDVI modeled NDVI in the whole region (f).

5. Conclusions

In this study, the spatio-temporal variation of vegetation cover in the middle reaches of theHei River basin and the main driving factors during the period from 2001 to 2011 were analyzed using MODIS NDVI data and hydro-meteorological data. Considering the nonlinear relationship between hydro-meteorological variables, the entropy theory was employed to calculate the mutual information between the NDVI time series and coupled hydro meteorological variables. Based on the spatial distribution of mutual information, the whole region was further divided into 5 sub-regions, and nested statistical models were employed to simulate NDVI in each sub-region. The main conclusions are as follows:

(1) The average annual NDVI increased at a rate of 0.005/a over the past 11 years in the middle reaches of the Hei River. The percentage area in which NDVI increased occupied 94% of the total area.

(2) In the desert sub-regions (I and IV), temperature and precipitation are the main driving factors for vegetation growth. In sub-region I, NDVI is consistent with the trend of temperature and precipitation (Figure 9a). However, in sub-region IV, the trend of temperature change is not obvious and the change in NDVI is mainly due to the increase in precipitation (Figure 9b). In the oasis regions (sub-region II and sub-region III), groundwater was an important factor for vegetation growth.

(3) In coupling hydro-meteorological variables, a nested statistical model was proposed for modeling NDVI on a regional scale. The overall correlation coefficient between observed NDVI and modeled NDVI was observed to be 0.97. This high simulation accuracy further proves the suitability of this method.

(4) Due to the influence of human activities, the modeling accuracy was not effective within the artificial oasis (sub-region III and sub-region V). For instance, in irrigation areas, vegetation can absorb water from irrigation to sustain growth but in non-irrigation areas the over-exploitation of groundwater caused by the increase in the amount of irrigation breaks the natural ecological balance, affecting the growth of natural vegetation. Additionally, the shortage of the dataset may also be a factor influencing the modeling accuracy. For instance, the scarcity of temperature and precipitation data may cause nearby cells to have similar values of T and P. Therefore, further

Entropy **2017**, *19*, 502

studies are necessary for modeling NDVI that consider additional factors such as runoff and irrigation using long-term and higher resolution datasets.

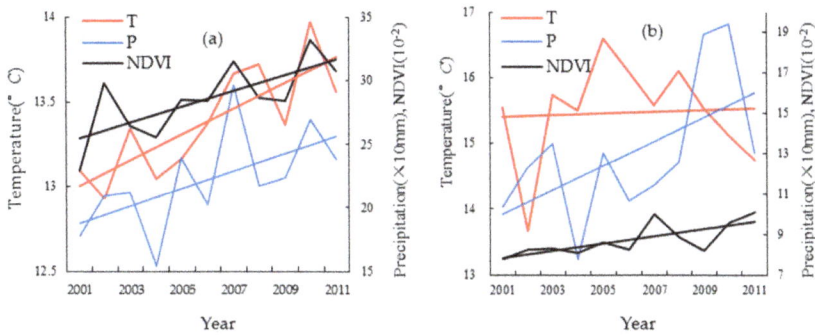

Figure 9. The variation of NDVI, temperature and precipitation on region I (**a**) and region IV (**b**). Note: The black, red and blue lines express the trend of variation in NDVI, temperature and precipitation respectively.

The results can provide some scientific guidance for water resources management and the reasonable exploitation of groundwater, especially in the artificial oasis areas. The outcomes demonstrated the ability to successfully employ the joint entropy method in vegetation dynamics analysis in relation to hydro-meteorological variables.

Acknowledgments: We are grateful for the grant support from the National Natural Science Fund in China (91425302 and 51279166).

Author Contributions: Gengxi Zhang and Xiaoling Su designed the computations and wrote the paper; Vijay P. Singh and Olusola O Ayantobo revised the paper and gave some suggestions; All authors have read and approved the final manuscript.

Conflicts of Interest: The authors declare no conflict of interest.

References

1. Kutiel, P.; Cohen, O.; Shoshany, M.; Shub, M. Vegetation establishment on the southern Israeli coastal sand dunes between the years 1965 and 1999. *Landsc. Urban Plann.* **2004**, *67*, 141–156. [CrossRef]
2. Jin, X.M.; Liu, J.T.; Wang, S.T.; Xia, W. Vegetation dynamics and their response to groundwater and climate variables in Qaidam Basin, China. *Int. J. Remote Sens.* **2016**, *37*, 710–728. [CrossRef]
3. Piao, S.L.; Mohammat, A.; Fang, J.Y.; Cai, Q.; Feng, J.M. NDVI-based increase in growth of temperate grasslands and its responses to climate changes in China. *Glob. Environ. Chang.* **2006**, *16*, 340–348. [CrossRef]
4. Dai, S.P.; Zhang, B.; HaiJun, W.J.; Wang, Y.M. Vegetation cover change and the driving factors over northwest China. *J. Arid Land* **2011**, *3*, 25–33. [CrossRef]
5. Liu, J.H.; Wu, J.J.; Wu, Z.T.; Liu, M. Response of NDVI dynamics to precipitation in the Beijing–Tianjin sandstorm source region. *Int. J. Remote Sens.* **2013**, *34*, 5331–5350. [CrossRef]
6. Reed, B.C.; Brown, J.F.; Darrel, V.Z.; Loveland, T.R.; Merchant, J.W.; Ohlen, D.O. Measuring phenological variability from satellite imagery. *J. Veg. Sci.* **1994**, *5*, 703–714. [CrossRef]
7. Clerici, N.; Weissteiner, C.J.; Gerard, F.G. Exploring the Use of MODIS NDVI-Based Phenology Indicators for Classifying Forest General Habitat Categories. *Remote Sens.* **2012**, *4*, 1781–1803. [CrossRef]
8. Feilhauer, H.; He, K.S.; Rocchini, D. Modeling Species Distribution Using Niche-Based Proxies Derived from Composite Bioclimatic Variables and MODISNDVI. *Remote Sens.* **2012**, *4*, 2057–2075. [CrossRef]
9. Rousvel, S.; Armand, N.; Andre, L.; Tengeleng, S. Comparison between vegetation and rainfall of Bioclimatic Ecoregions in Central Africa. *Atmosphere* **2013**, *4*, 411–427. [CrossRef]

10. Xiao, J.F.; Zhou, Y.; Zhang, L. Contribution of natural and human factors to increases in vegetation productivity in China. *Ecosphere* **2015**, *6*, 1–20. [CrossRef]
11. Nemani, R.; White, M.; Thornton, P.; Nishid, K.; Reddy, S.; Jenkins, J.; Running, S. Recent trends in hydrologic balance have enhanced the terrestrial carbon sink in the United States. *Geophys. Res. Lett.* **2002**, *29*, 1–4. [CrossRef]
12. Brunsell, N.A.; Young, C.B. Land surface response to precipitation events using MODIS and NEXRAD data. *Int. J. Remote Sens.* **2008**, *29*, 1965–1982. [CrossRef]
13. Kawabata, A.; Ichii, K.; Yamaguchi, Y. Global monitoring of interannual changes in vegetation activities using NDVI and its relationships to temperature and precipitation. *Int. J. Remote Sens.* **2001**, *22*, 1377–1382. [CrossRef]
14. Pettorelli, N.; Pelletier, F.; Hardenberg, A.; Festa-Bianchet, M.; Cote, S.D. Early onset of vegetation growth vs. rapid green-up, Impacts on juvenile mountain ungulates. *Ecology* **2007**, *88*, 381–390. [CrossRef] [PubMed]
15. Piao, S.L; Nan, H.J.; Huntingford, C.; Ciais, P.; Friedlingstein, P. Evidence for a weakening relationship between interannual temperature variability and northern vegetation activity. *Nat. Commun.* **2014**, *5*, 5018. [CrossRef] [PubMed]
16. Pielke, R.A.; Avissar, R. Interactions between the atmosphere and terrestrial ecosystems, influence on weather and climate. *Glob. Chang. Biol.* **1998**, *4*, 461–475. [CrossRef]
17. Sohoulande, D.C.; Singh, V.P. Retrieving vegetation growth patterns from soil moisture, precipitation and temperature using maximum entropy. *Ecol. Model.* **2015**, *309*, 10–21. [CrossRef]
18. Ichii, K.; Kawabata, A.; Yamaguchi, Y. Global correlation analysis for NDVI and climatic variables and NDVI trends, 1982–1990. *Int. J. Remote Sens.* **2002**, *23*, 3873–3878. [CrossRef]
19. Nicholson, S; Davenport, M. A comparison of the vegetation response to rainfall in the Sahel and East-Africa, using normalized difference vegetation index from NOAA AVHRR. *Clim. Chang.* **1990**, *17*, 209–241. [CrossRef]
20. Jin, X.M.; Guo, R.H.; Zhang, Q.; Zhou, Y.X.; Zhang, D.R.; Yang, Z. Response of vegetation pattern to different landform and water-table depth in Hailiutu River basin, Northwestern China. *Environ. Earth Sci.* **2014**, *71*, 4889–4898. [CrossRef]
21. Lv, J.J.; Wang, X.S.; Zhou, Y.X.; Qian, K.Z.; Wan, L.; Eamus, D.; Tao, Z.P. Groundwater-dependent distribution of vegetation in Hailiutu River catchment, a semi-arid region in China. *Ecohydrology* **2013**, *6*, 142–149. [CrossRef]
22. Jin, X.M.; Schaepman, M.E.; Clevers, J.; Su, Z.B.; Hu, G.C. Groundwater Depth and Vegetation in the Ejina Area, China. *Arid Land Res. Manag.* **2011**, *25*, 194–199. [CrossRef]
23. Mishra, A.K.; Ines, A.V.M.; Singh, V.P.; Asen, J.W. Extraction of information content from stochastic disaggregation and bias corrected downscaled precipitation variables for crop simulation. *Stoch. Environ. Res. Risk Assess.* **2012**, *27*, 449–457. [CrossRef]
24. Singh, V.P. *Entropy Theory and Its Application in Environmental and Water Engineering*; Wiley: New York, NY, USA, 2013.
25. Li, C.; Singh, V.P.; Mishra, A.K. Entropy theory-based criterion for hydrometric network evaluation and design, Maximum information minimum redundancy. *Water Resour. Res.* **2012**, *48*, 1–15. [CrossRef]
26. Phillips, S.J.; Anderson, R.P.; Schapire, R.E. Maximum entropy modeling of species geographic distributions. *Ecol. Model.* **2006**, *190*, 231–259. [CrossRef]
27. Pueyo, S.; He, F.; Zillio, T. The maximum entropy formalism and the idiosyncratic theory of biodiversity. *Ecol. Lett.* **2007**, *10*, 1017–1028. [CrossRef] [PubMed]
28. Urbani, F.; Alessandro, P.; Frasca, R.; Biondi, M. Maximum entropy modeling of geographic distributions of the flea beetle species endemic in Italy (Coleoptera, Chrysomelidae, Galerucinae, Alticini). *Zoologischer Anzeiger* **2015**, *258*, 99–109. [CrossRef]
29. Song, Y.; Ma, M. Varition of AVHRR NDVI and its relationship with climate in Chinese arid and cold regions. *J. Remote Sens.* **2008**, *12*, 499–505.
30. Sohoulande Djebou, D.C.; Singh, V.P.; Frauenfeld, O.W. Analysis of watershed topography effects on summer precipitation variability in the southwestern United States. *J. Hydrol.* **2014**, *511*, 838–849. [CrossRef]
31. McGill, W.J. Mutivariate information transmisson. *Psychometrica* **1954**, *19*, 97–116. [CrossRef]
32. Nian, Y.; Li, X.; Zhou, J.; Hu, X. Impact of land use change on water resource allocation in the middle reaches of the Heihe River Basin in northwestern China. *J. Arid Land* **2013**, *6*, 273–286. [CrossRef]

33. Wang, Y.; Roderick, M.L.; Shen, Y.; Sun, F. Attribution of satellite-observed vegetation trends in a hyper-arid region of the Heihe River basin, Western China. *Hydrol. Earth Syst. Sci.* **2014**, *18*, 3499–3509. [CrossRef]

34. Guo, Q.; Feng, Q.; Li, J. Environmental changes after ecological water conveyance in the lower reaches of Heihe River, northwest China. *Environ. Geol.* **2009**, *58*, 1387–1396. [CrossRef]

35. Zhang, A.; Zheng, C.; Wang, S.; Yao, Y. Analysis of streamflow variations in the Heihe River Basin, northwest China, Trends, abrupt changes, driving factors and ecological influences. *J. Hydrol. Reg. Stud.* **2015**, *3*, 106–124. [CrossRef]

36. Qin, D.; Zhao, Z.; Han, L.; Qian, Y. Determinnation of groundwater recharge regim and flowpath in the Lower Heihe River basin in an arid area of Northwest China by using environmental tracers, Implications for vegetation degradation in the Ejina Oasis. *Appl. Geochem.* **2012**, *27*, 1133–1145. [CrossRef]

37. Mi, L.; Xiao, H.; Zhang, J.; Yin, Z.; Shen, Y. Evolution of the groundwater system under the impacts of human activities in middle reaches of Heihe River Basin (Northwest China) from 1985 to 2013. *Hydrogeol. J.* **2016**, *24*, 971–986. [CrossRef]

38. Li, F.; Zhao, W. Changes in normalized difference vegetation index of deserts and dunes with precipitation in the middle Heihe River Basin. *Chin. J. Plant Ecol.* **2016**, *40*, 1245–1256. [CrossRef]

39. Zhou, W.; Wang, Q.; Zhang, C.B.; Li, J. Spatiotemporal variation of grassland vegetation NDVI in the middle and upper reaches of the Hei River and its response to climatic factors. *Acta Prataculturae Sin.* **2013**, *22*, 138–147.

40. Zhao, W.; Chang, X. The effect of hydrologic process changes on NDVI in the desert-oasis ecotone of the Hexi Corridor. *Sci. China Earth Sci.* **2014**, *57*, 3107–3117. [CrossRef]

entropy

MDPI

Article

Information Entropy Suggests Stronger Nonlinear Associations between Hydro-Meteorological Variables and ENSO

Tue M. Vu, Ashok K. Mishra * and Goutam Konapala

Glenn Department of Civil Engineering, Clemson University, Clemson, SC 29634, USA;
tuev@g.clemson.edu (T.M.V.); gkonapa@clemson.edu (G.K.)
* Correspondence: ashokm@g.clemson.edu; Tel.: +1-864-656-1209

Received: 13 November 2017; Accepted: 5 January 2018; Published: 9 January 2018

Abstract: Understanding the teleconnections between hydro-meteorological data and the El Niño–Southern Oscillation cycle (ENSO) is an important step towards developing flood early warning systems. In this study, the concept of mutual information (*MI*) was applied using marginal and joint information entropy to quantify the linear and non-linear relationship between annual streamflow, extreme precipitation indices over Mekong river basin, and ENSO. We primarily used Pearson correlation as a linear association metric for comparison with mutual information. The analysis was performed at four hydro-meteorological stations located on the mainstream Mekong river basin. It was observed that the nonlinear correlation information is comparatively higher between the large-scale climate index and local hydro-meteorology data in comparison to the traditional linear correlation information. The spatial analysis was carried out using all the grid points in the river basin, which suggests a spatial dependence structure between precipitation extremes and ENSO. Overall, this study suggests that mutual information approach can further detect more meaningful connections between large-scale climate indices and hydro-meteorological variables at different spatio-temporal scales. Application of nonlinear mutual information metric can be an efficient tool to better understand hydro-climatic variables dynamics resulting in improved climate-informed adaptation strategies.

Keywords: information entropy; mutual information; kernel density estimation; ENSO; nonlinear relation

1. Introduction

For many water resources planning and management studies, reliable preliminary estimates of dependence between two hydroclimatic variables are extremely important. For example, knowledge of dependence between large-scale climate patterns such as El Niño–Southern Oscillation (ENSO) [1], the Pacific Decadal Oscillation (PDO) [2], and the Atlantic Multi-decadal Oscillation (AMO) [3] with local precipitation, temperature, or streamflow has resulted in improved longer lead-time forecasting models [4–6]. Large-scale climate patterns can also predict ecological processes better than local weather [7]. In addition to that, several studies indicated the presence of a significant relationship between large-scale climate phenomena and hydrologic extremes, such as extreme precipitation events [8–10], droughts [11–13], and floods [14,15]. Therefore, the presence of any kind of significant dependence forms the preliminary metric to identify appropriate predictors for forecasting streamflow and other hydroclimatic variables in ungauged river basins [16–18]. As a result, the predictability of these large-scale climate patterns much in advance is extremely important to improve the design of early warning systems of extreme events [19].

A wide range of methods is available for detecting the presence of an association between bivariate data set. Among them, Pearson's correlation coefficient is the most widely used metric to quantify the linear dependence between any two variables [20]. Pearson correlation coefficient is based on the assumption that the considered variables follow a Gaussian distribution. Therefore, using Pearson correlation in case of variables that follow non-Gaussian distributions may be suboptimal.

However, several insights were derived from these linear associations. For example, Zhang et al. [21] quantified the Pearson linear correlation between different sea surface temperature (SST) anomalies and seasonal precipitation for Huai River basin in China. The authors identified some positive/negative correlation with the coefficients ranges from absolute 0.2 to 0.3. Whereas, Wrzesinski [22] characterized and confirmed the strong influences of large-scale climate index NAO to seasonal river flow in Poland by comparing the difference in average runoff during the positive and negative NAO phases. On the same lines, Wrzesinski [23] also found the linkage between NAO and European river streamflow at the 140 gauging station by analyzing the high and low water flows according to the positive and negative NAO phases.

In addition, most of the hydro-climatic series do not follow a Gaussian distribution leading to a possible misinterpretation of the dependency between the variables when using linear measures [24–26]. As a result, the non-parametric rank-based correlation metrics of Kendall's tau and Spearman's rho are applied to quantify the relationship between two given variables [27–29]. However, even though the associations evaluated by these two metrics are independent of any probability distribution assumptions, they are more successful in detecting the monotonous relationships between any two variables. In addition, assuming a monotonous relationship among two hydroclimatic variables might be too restrictive to characterize the existing complex dependence structures between the hydroclimatic variables [30–32].

To overcome the limitations of linear dependence, several studies in hydro-climatology adopted the concepts based on nonlinear statistics to evaluate the strength of association among distinct hydroclimatic variables. For instance, Fleming et al. [33] observed a nonlinear association between northern hemisphere river basin streamflow and teleconnection patterns. In regard to climate extremes, Cannon [34] utilized generalized extreme value distributions to investigate the relationship between ENSO and winter extreme station precipitation in North America. Whereas Lin-Ye et al. [35] investigated the relationship between extreme events of wave storms and large-scale climate covariates using generalized additive and linear models. While Kusumastuti [36] studied the nonlinearity between threshold and rainfall-runoff transformation. More recently, Konapala et al. [37] have investigated the nonlinear relationship between low flows in Texas river basins with large scale climatic patterns.

Recently entropy-based approaches are gaining popularity for different applications, such as, climate variability [38], uncertainty analysis [39], hydrometric network analysis [40–42], selection of predictors [43,44], and drought regionalization [45]. Similarly, the entropy theory based mutual information (*MI*) is widely used for quantifying the non-linear association between multiple variables [40,41]. The *MI* concepts have been utilized to quantify nonlinear dependence structure between several climate variables across the globe [46,47]. Also, several *MI* metrics were developed to assist in predictor selection for hydroclimatic forecasting [48,49]. More importantly, *MI* was able to detect the presence of strong nonlinear associations of streamflow, rainfall, and global mean temperature with several large-scale climatic variables [31,50–54].

Unlike the signals in other research fields, hydroclimatic data being a subset of geophysical research field are non-repeatable, shorter in length and contaminated with significant noise levels [31]. In addition to that, studies have indicated a presence of chaotic component [55]. Therefore, it becomes evident that the selected measure should be relatively robust to noise and chaos and, more importantly, detects signals possibly shorter in length. Thus, the goal of this study is to investigate and compare the performance of nonlinear estimation method based on mutual information (*MI*) with the traditional linear association metric of Pearson correlation. For this purpose, the hydro-meteorological data are considered such as total annual precipitation, annual average streamflow, and extreme precipitation indices specific to Mekong River Basin in Southeast Asia and estimated its association with the large-scale climate index of ENSO.

2. Materials and Methods

2.1. Hydro-Meteorological Data and ENSO

The Mekong River Basin (MRB), originates from Tibetan Plateau in China and flows through the territory of five other countries in the Southeast Asia (Myanmar, Laos PDR, Thailand, Cambodia, and drains to the sea in Vietnam) (Figure 1a). This is currently home to more than 70 million people and it is expected to increase by 100 million in 2050 [56]. The detailed topography map of study region is displayed in Figure 1a with relative high terrain in the upstream of the basin (upper Mekong). In this study, the gridded precipitation data is provided from Asian Precipitation Highly Resolved Observational Data Integration Towards the Evaluation of Water Resources (APHRODITE) with a spatial resolution of 0.25° and daily temporal resolution with the sufficiently long record period from 1951–2007 covering the Asian monsoon domain. The dataset was created primarily with data obtained independently from rain gauge observation network across all the regions of Asia. The daily precipitation values from rain gauges were interpolated using the sphere map technique [57] and the first six components of the fast Fourier transforms were taken to obtain daily data for all land areas in the Asian monsoon domain [58]. This dataset is referred to as "APH" in this paper, for the purpose of brevity. APH has been previously utilized as a reliable gridded rain data set over this area for various hydrological applications [59–63]. The spatial distribution of total precipitation from APH data over MRB is displayed in Figure 1b with high rainfall volumes (around 2000 to 2500 mm) gathering around the eastern part of the river basin near to the Annamite mountain range.

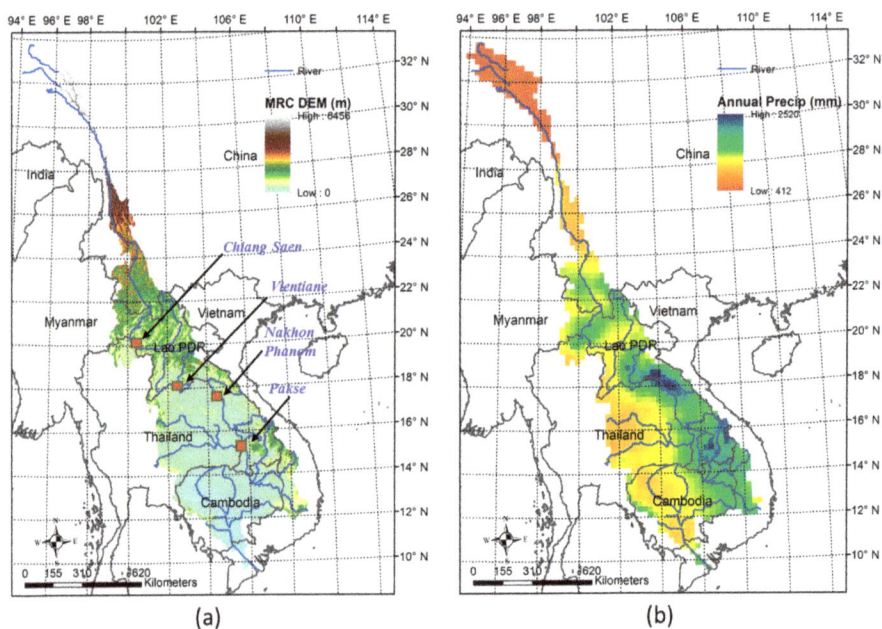

(a)   (b)

Figure 1. (**a**) Mekong River Basin map with digital elevation model terrain and digitized river, location of four hydro-meteorology stations are displayed; (**b**) Annual rainfall distribution over MRB using APHRODITE gridded rainfall 1951–2007.

The southwest monsoon wind originated from the Gulf of Thailand blocked by the Annamite mountain range is the main factor that causes the high seasonal rainfall in summer months (June, July August, September) [61] (Figure 2a). The APH daily gridded rainfall values were extracted to station locations using bilinear interpolation [59,63,64] to four meteorological stations located in MRB—Chiang Saen, Vientiane, Nakhon Phanom, and Pakse (Figure 1a)—for the time period of 1951 to 2007 over the Mekong River Basin. In addition to rainfall data, the corresponding annual streamflow data were also collected at these four hydrological stations from the MRC website for the same period. The monthly mean climatology over 57 years of hydro-meteorological stations is displayed in Figure 2a. Even though different patterns are found over four stations, the hydrographs are quite similar from the upstream station at Chiang Saen to downstream station Pakse, with flooding season from May to November, while the peak discharge months are around August and September. This study considers the seasonal cycle which includes all 12 months of the hydrologic year (starting from the month with the lowest rainfall—i.e., May—and ending in April of the next year; the streamflow month is considered to lag one month which starts from June and ends in May next year).

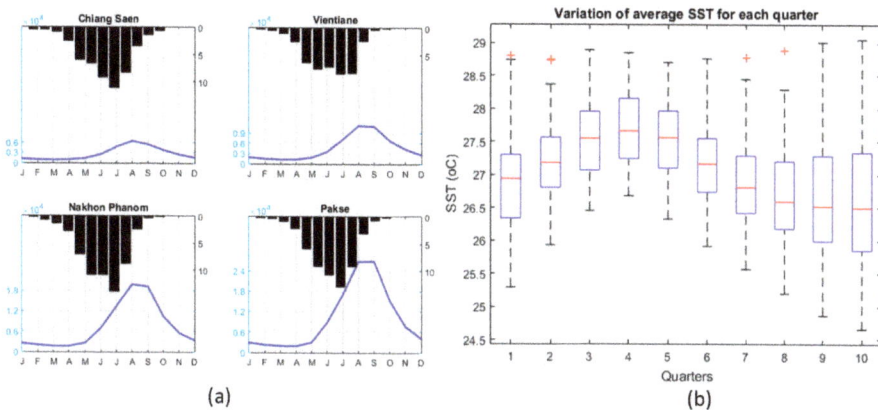

Figure 2. (a) Seasonal cycle of monthly climatology rainfall and streamflow measured at four stations on the mainstream Mekong river; (b) Boxplots of monthly SST Nino 3.4 for different quarters

The study focuses on the total annual rainfall, annual streamflow, as well as extreme precipitation indices [65] for hydrologic years derived from daily precipitation datasets for both wet and dry spell statistics, which are (1) R5d: Max consecutive five-day rainfall, (2) P90p: 90th percentile of the daily precipitation time series for the year, (3) SDII: average daily precipitation on a wet day, and (4) dry spell length computed from daily precipitation for a year. Among four stations, Pakse has the highest streamflow data due to its furthest downstream location. Nakhon Phanom and Pakse have the highest rainfall as seen in Figures 1 and 2.

The assumption has been that the long-term hydro-meteorological data variability can be captured by the fluctuation of the SST over seasonal scales. In this study, the ENSO index has been considered by using the SST over NINO 3.4 region [60,66], which is the area averaged monthly SST over the region bounded by the coordinates 5° N–5° S, 170° W–120° W. This time series could effectively indicate the occurrence of ENSO events [66]. The dataset can be downloaded from the Climate Data Guide website by NCAR [67]. Therefore in this study, the ENSO indices are computed at multiple three-month time windows in terms of quarters [31] consisting of the average SST anomalies over months of JFM, FMA, MAM, etc., for all the years (J: Jan, F: Feb, M: Mar). The SST over Nino 3.4 regions boxplots bound for 57 years for 10 different quarters are displayed in Figure 2b. The red crosses in Figure 2b denote outliers. There are four quarters just before the seasonal cycle, four quarters corresponding to the seasonal cycle, and two quarters after seasonal cycle, computed with the mean monthly SST over Nino 3.4 region. Among all 10 quarters, the fourth quarter (AMJ) has the highest median and whisker values compared to other quarters whilst the last two quarters (9 and 10) have the lowest median values.

2.2. Mutual Information Estimation

The mutual information (*MI*) has been utilized to capture the nonlinear dependence structure between two random variables. When analyzing experimental times series from the non-linear system, the *MI* is especially an important statistics [68]. According to [69], there are three theorems for *MI* between two random variables *X*, *Y*: (1) *MI* is non-negative and is zero if *X* and *Y* are strictly independent; (2) *MI* is infinity if there exists a function "g" such that $X = g(Y)$; (3) *MI* is invariant to separate one to one transformations.

The *MI* can be computed using the relative entropy suggested by Joe [70]. Assuming that a pair of continuous random variables (*X*, *Y*) exist which have a joint probability density function (pdf) p_{XY} and with its marginal pdf accordingly p_X and p_Y. The mutual information or relative entropy can be defined as

$$MI(X, Y) = \int \int p_{XY} \log \frac{p_{XY}}{p_X p_Y} dx dy \tag{1}$$

It is noted that the Equation (1) measures the distance between a joint distribution and the distribution when there is independence [69]. In case of continuous variables, there is no direct way to accurately determine continuous probability distributions. Therefore, several methods have been introduced to approximate the continuous probability distribution functions as discrete distributions. Among them, Khan et al. [71] compared four different methods to estimate the probability distribution function to calculate *MI* using: kernel density estimator (KDE) [68], K-nearest neighbors (KNN) [72], Edgeworth approximation of multivariate differential entropy [73], and adaptive partitioning of the *XY* plane [74]. The authors found that KDE and KNN outperform the other two methods in term of their ability to capture the dependence structure. Khan et al. [31] indicated that KDE is able to capture the underlying nonlinear dependence more consistently compared to KNN and Edgeworth when they are short and noisy assuming such dependence exists. Therefore, this article utilizes the KDE to estimate probability density function using the equation

$$p(x) = \frac{1}{n} \sum_{i}^{n} K(u) \tag{2}$$

in which, $K(u)$ is a multivariate kernel function

$$K(u) = \frac{1}{h^d \sqrt{(2\pi)^d \det(S)}} \exp\left(-\frac{u}{2}\right) \tag{3}$$

$$u = \frac{(x - x_i)^T S^{-1} (x - x_i)}{h^2} \tag{4}$$

From (4), x_i a d-dimensional random vector for the multivariate data set $(x_1, \ldots, x_n)$. *S* is the covariance matrix on the x_i, det(*S*) is a determinant of *S* and *h* is the kernel bandwidth or smoothing variable. The optimal Gaussian bandwidth "*h*" can be computed using Equation (5) [75]

$$h = \left(\frac{4}{d+2}\right)^{\frac{1}{d+4}} n^{-\frac{1}{d+4}} \tag{5}$$

The "*d*" is taken from [76] for the value of $d = 2$, and is similar to [68].

Substituting Equations (3) and (4) to Equation (2) to obtain the approximate probability density function as

$$p(x) = \frac{1}{n} \sum_{i}^{n} \frac{1}{h^d \sqrt{(2\pi)^d \det(S)}} \exp\left[-\frac{(x - x_i)^T S^{-1} (x - x_i)}{2h^2}\right] \tag{6}$$

The detailed procedure using KDE to estimate pdf can be found in [31,68,71]. Here, the discrete formulation of *MI* is shown in Equation (7)

$$MI(X, Y) = \frac{1}{n} \sum_{i=1}^{n} \ln \frac{p_{XY}(x_i, y_i)}{p_X(x_i) p_Y(y_i)} \tag{7}$$

In which $p_{XY}(x_i, y_i)$ is the joint pdf and $p_X(x_i), p_Y(y_i)$ are marginal pdfs at (x_i, y_i). The *MI* values range between independent (*MI* = 0) to completely dependent (*MI* = ∞). In order to make the generalization of the correlation with a range from 0 (independent) to 1 (dependent), Joe [70] proposed a formula to transform the *MI* to nonlinear correlation coefficient (*NLCC*)

$$NLCC = \sqrt{[1 - \exp(-2MI)]} \tag{8}$$

The *NLCC* range is similar to linear correlation coefficients (*LCC*) and has been used in most of the studies by [31,69,70].

Figure 3 modified [77] using their python code to demonstrate the *LCC* and *NLCC* using the scatter plots of two random variables obtained from a different sample of data. The advantage of *NLCC* is based on the fact that *MI* makes no assumption on the distribution of the variables or the nature of the relationship between them and is sensitive to nonlinear and non-monotonic effects [77]. It can be observed that (top row of Figure 3) both the *LCC* and *NLCC* can able to capture the linear relationship between the two random variables with a very close range of values. However, the advantage of *NLCC* compared to *LCC* is due to its ability to recognize the different distributions of the two random variables (bottom row of Figure 3).

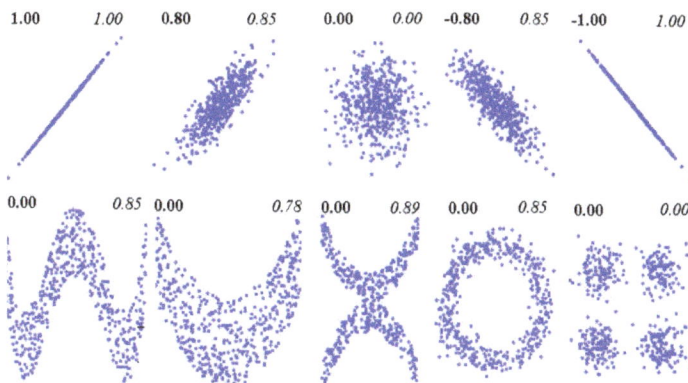

Figure 3. The comparison between linear correlation coefficient (*LCC*) and nonlinear correlation coefficient (*NLCC*) obtained from mutual information (*MI*) computed based on two random variables. For each panel, top left value (bold) shows *LCC* and top right (italic) represents *NLCC*. The top row illustrates the similarity between *LCC* and *NLCC*, whereas, the bottom row exhibits advantages of non-linear relationship based on different distribution. (Figures partially adopted from [77] using their python code. The *NLCC* was computed based on MI values).

Subsequently, the *LCCs* are computed between each of the precipitation indices and compared with *NLCCs* at four hydro-meteorology stations located in MRB (Figure 1a). The four stations are chosen based on their differences in terms of seasonal cycles, total annual rainfall amount and geographical height. Finally, the gridded linear/non-linear CC is constructed for the MRB to showcase the spatial variability of the correlation coefficient.

3. Results

3.1. Linear and Nonlinear Correlation between Annual Precipitation/Streamflow and ENSO Index

We first illustrate the linear dependence as a bivariate normal distribution and nonlinear dependence as a kernel bivariate distribution following the work of Khan et al. [71]. The bivariate normal and kernel density between the annual average streamflow at different hydrologic gauging stations and different quarters of ENSO indices are computed and plotted in Figure 4 for the highest and lowest linear and nonlinear CCs. For kernel density, a Gaussian kernel with an optimal Gaussian bandwidth computed by $h = N^{-1/6}$ with N is the total number of observed points (57 in this case).

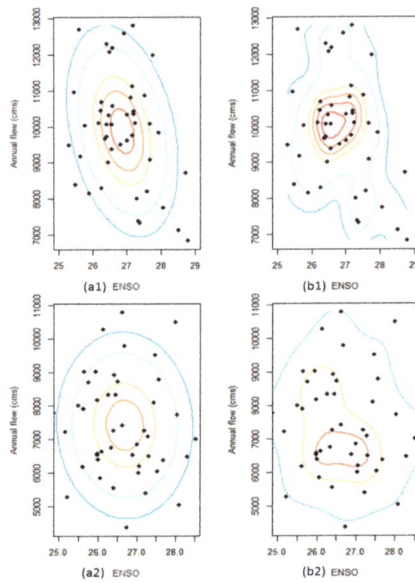

Figure 4. The bivariate normal and kernel density between annual flow and ENSO index for (**a**) normal density; (**b**) kernel density; (1) Quarter 1 at Pakse station (highest linear and nonlinear CCs) (2) Quarter 9 at Nakhon Phanom station (lowest, respectively).

Subsequently, the linear and nonlinear CCs between total annual precipitation and average streamflow were computed at different window lengths of SST Nino 3.4 (quarters) for four different hydro-meteorology stations and display in Figure 5. Based on the formulation Equation (8) *NLCCs* are positive, therefore the absolute value of *LCCs* are computed for comparison plotting in Figure 5 and all other figures henceforth. In order to compute the 90% confidence intervals for the absolute correlation coefficients, the bootstrapping approach [71] was applied using 100 simulations and plotted in Figure 5. It can be seen in Figure 5 that the *NLCCs* have higher values than the *LCCs* for all quarters. It indicates that the KDE captures the more extrabasinal connection between ENSO and precipitation as well as river flows compared to linear modeling [31].

In particular at station scale, Nakhon Phanom has the highest precipitation amount, its *LCCs* and *NLCCs* are also among the highest for the first four quarters Figure 5(c1,c2) which subsequently decrease. The *NLCC* values also have the same patterns as *LCCs* for this station as well as Pakse in Figure 5(d1). Although both stations have highest CCs for the first quarter, however, the lowest CCs values are not found at the same: Quarter 7 (Nakhon Phanom) and Quarter 4 (Pakse). Chiang Saen and Vientiane have variation trends for *LCCs* Figure 5(a1,b1) among all quarters, this is perhaps due to the less rainfall amount observed at these two rain gauges.

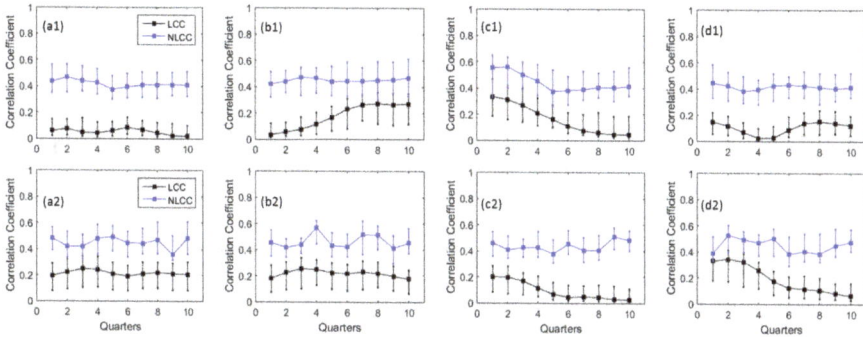

Figure 5. Linear and nonlinear CCs with their 90% confidence intervals between SST Nino 3.4 and (1) total annual precipitation (2) annual average streamflow for (**a**) Chiang Saen; (**b**) Vientiane; (**c**) Nakhon Phanom; and (**d**) Pakse. Note: The correlation coefficients (*CCs*) are calculated between precipitation and ENSO indices in-terms of quarterly temporal windows, such as, JFM, FMA, and MAM and so on. The 90% confidence intervals are computed using 100 simulations by bootstrapping approach.

Among four hydrology gauging stations, Pakse also has highest *LCCs*/*NLCCs* with ENSO at the first four quarters Figure 5(d2). This is because Pakse is the furthest station downstream and it has the highest streamflow measured. Therefore, it shows more dependences by the fluctuations of ENSO indices. The second highest streamflow station is Nakhon Phanom (see Figure 5(c2)) that also indicates similar patterns of *LCCs* and *NLCCs* with higher dependences with the first four quarters of ENSO indices compared with the rest. When the *LCC* values are close to zero (as in the last four quarters of Figure 5(c2)), the differences between *LCCs* and *NLCCs* should be interpreted with caution because of the exponential scaling of *MI* in *NLCC* as shown in Equation (8). The two upstream gauging stations Chiang Saen and Vientiane (Figure 5(a2,b2)) also have similar patterns for *LCCs* and *NLCCs* at different quarters. Compared to total annual precipitation, the annual average streamflow has slightly higher *LCCs* and *NLCCs*, except for Nakhon Phanom station. Overlapping the 90% confidence intervals of *LCCs* and *NLCCs* for streamflow at Chiang Saen, Vientiane, and Pakse (Figure 5(a2,b2,d2)) for the first three quarters indicates that both KDE and linear regressions effectively capture the strong dependence structure.

3.2. Linear and Nonlinear Correlation between Precipitation Extreme Indices and ENSO

In addition to total annual precipitation and streamflow dependences, further investigations on *LCCs* and *NLCCs* are carried out between precipitation indices for extreme values indices (R5d, SDII, P90p, dry spell) with different quarters of ENSO events. This analysis aims to quantify the dependence structure between annual extreme precipitation events and ENSO indices. The wet indices (R5d, P90p, SDII), as well as dry indices (dry spell), are also taken into account for the analyses. The detail dependences based on *LCCs* and *NLCCs* can be found in Figure 6 for four indices and four stations over 10 quarters. Similar to Figure 5, the *LCCs* are displayed as absolute values comparable to *NLCCs*. The 90% confidence intervals for the absolute correlation are computed based on 100 simulations using bootstrapping approach. The detail analyses on Figure 6 reveals several significant dependencies between annual extreme precipitation events and ENSO indices. Chiang Saen exhibits high *LCCs*/*NLCCs* for a maximum five consecutive days of rainfall with the last three quarters of ENSO indices and low *CCs* values for the first three quarters Figure 6(a1). Similar patterns are found for Pakse station in Figure 6(d1) but with the highest *CCs* among Quarters 6, 7, 8 and lowest at the first three quarters. This variation is different from the total annual precipitation in Figure 5(a1,d1). In contrast to R5d, the other two indices: P90p and SDII have similar patterns of *LCCs*/*NLCCs* compared to the total annual precipitation in Figure 5. The last three-quarters of ENSO indices show that it has higher *LCCs/NLCCs* for Vientiane and Pakse Figure 6(b2,b3,d2,d3) compared to other quarters whilst the first three quarters of Nakhon Phanom has the highest values than the last three quarters. The dry spell indices of all four stations show nearly opposite patterns compared to R5d. The overall pattern illustrates that nonlinear *CCs* have higher values (more dependencies) than linear *CCs* for all stations/indices, similar to Figure 5 and [31].

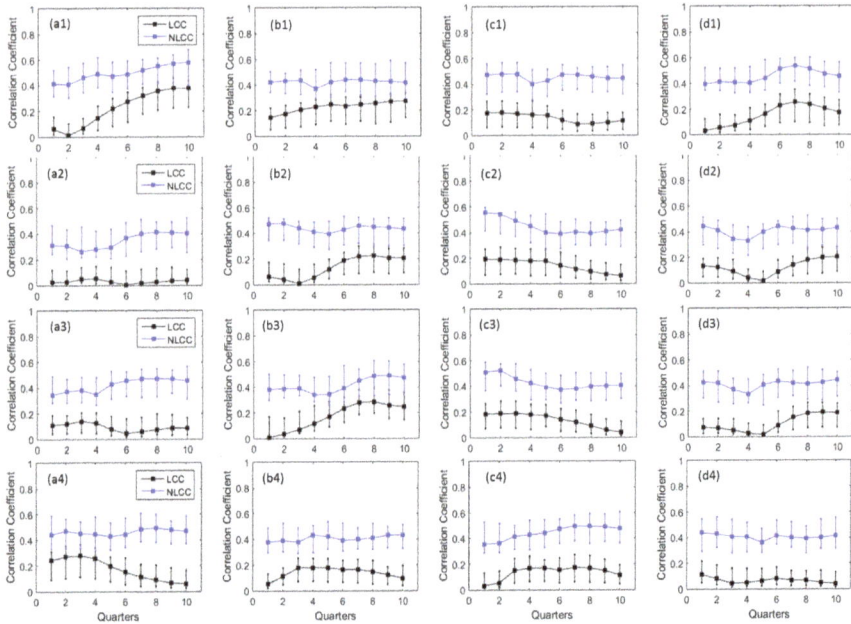

Figure 6. Comparison between linear and nonlinear *CCs* derived between SST Nino 3.4 and (1) R5d; (2) P90p; (3) SDII; (4) dry spell for selected stations located at (**a**) Chiang Saen; (**b**) Vientiane; (**c**) Nakhon Phanom; and (**d**) Pakse. Note: The correlation coefficients (*CCs*) are calculated between precipitation and ENSO indices in-terms of quarterly temporal windows, such as JFM, FMA, MAM, and so on. The 90% confidence intervals are computed using 100 simulations by bootstrapping approach.

The analyses with total annual precipitation, annual average streamflow, and extreme precipitation indices on wet and dry conditions reveal that there exists a nonlinear extrabasinal connection between ENSO and the hydro-meteorological data over Mekong river basin. The analyses based on *LCCs/NLCCs* exhibit the increasing trend in the variation of annual statistics on hydro-meteorology in connection with ENSO by computing nonlinear relationship as compared to linear measures. Therefore, it is expected to give additional support for early prediction (compared to traditional linear measurement) based on the ENSO forecast with the hydro-meteorology connection when *MI*-based approaches are utilized. This approach, somehow, would be helpful in water resources management for drought mitigation, flood control, as well as an irrigation system for agricultural areas.

3.3. Spatial Linear and Nonlinear Correlation Maps

Based on the correlation analysis at four rainfall stations, it was observed that the *NLCCs* have higher values in comparison to *LCCs* at all the selected stations located in MRB. Further analyses of the spatial pattern using the gridded data from APH based on the selected statistics of annual precipitation and extreme indices have been carried out. The linear and nonlinear *CCs* between ENSO Quarter 1 and annual rainfall are displayed in Figure 7 (using the same color bar), and the correlation values range from 0 to 0.6. Quarter 1 was selected arbitrarily as the first quarter, even though, it shows the highest CCs over Nakhon Phanom and Pakse for total annual precipitation but it also displays the lowest CCs for other analyses such as for Vientiane extreme indices (Figure 6b) or Pakse in Figure 6(d1,d3). Figure 7a exhibits the higher *CC* values between total annual precipitation and ENSO Quarter 1 (about 0.4) from Nakhon Phanom to Pakse station, along with Laos, eastern Thailand, and northern Cambodia. This can be clearly observed during the first quarter time frame Figure 5(c1,d1). On the other hand, the *NLCCs* have higher *CC* values (about 0.5 to 0.6) for the same locations compared to other grids. The use of spatial distribution map can be extended to generate the teleconnection patterns in ungauged regions. That, in turn, helps to better inform local stakeholders in building better tools for water resource management.

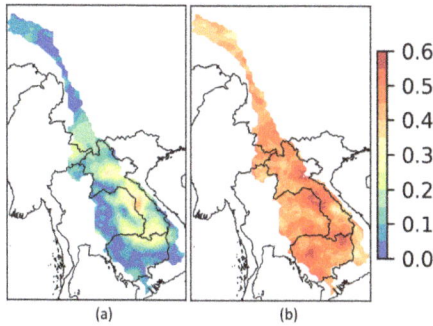

(a)   (b)

Figure 7. (a) Linear and (b) nonlinear CCs between Nino 3.4 at Quarter 1 and annual precipitation.

Similarly, the dependency between selected precipitation indices (R5d, P90p, SDII, and dry spell) and Quarter 1 of ENSO are computed and displayed in Figure 8. The linear *CCs* obtained from precipitation indices (Figure 8) are slightly lower than that of annual precipitation in the previous analysis (Figure 7) for Laos, eastern Thailand, and northern Cambodia. There is slightly higher *LCCs/NLCCs* magnitude observed at southern Cambodia during a dry spell Figure 8(a4,b4). The magnitudes of *NLCCs* (Figure 8) are also slightly lower than *NLCCs* based on the annual precipitation (Figure 7). However, the values obtained from *NLCCs* are still significantly higher than *LCCs* for all extreme indices. The lower Mekong basin (the southern part of river basin spread over Laos, Thailand, and Cambodia to Vietnam) has higher correlations to ENSO indices compared to the northern/upper basin, similar to [78].

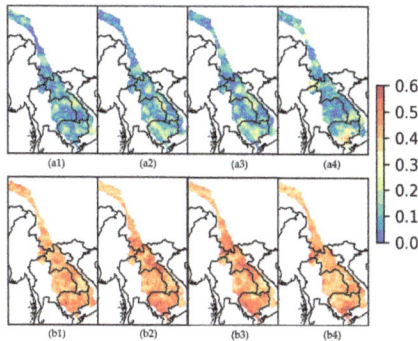

(a1)   (a2)   (a3)   (a4)

(b1)   (b2)   (b3)   (b4)

Figure 8. (a) Linear and (b) nonlinear CCs between Nino 3.4 at quarter 1, and (1) R5d, (2) P90p, (3) SDII, (4) dry spell.

4. Discussion

Although ENSO has a direct influence on rainfall anomalies over the tropical and subtropical regions, only a portion of the variation in the annual flow of rivers located in these regions is associated with ENSO events [31]. This study discusses in detail the possible dependences between different quarters of ENSO indices and hydro-meteorological dataset over Mekong river basin. Other existing studies on MRB highlighted that the linear correlations between ENSO and streamflow over the selected locations in Chiang Saen, Vientiane, and Pakse are around −0.4 to 0.2 [78] and maximum linear correlations between ENSO and precipitation over several stations in Thailand are around −0.18 to 0.22 [79]. The above correlation figures are in line with the study that the maximum LCCs obtained are around 0.35 for both annual precipitation and streamflow as displayed in Figure 5. The results from mutual information derived *NLCCs* therefore suggests an additional approach to look into the dependence structure for multivariate hydro-meteorological data with the large-scale climate patterns. ENSO is a periodic climatic phenomenon with 3–7 years of the cycle and can be predicted several seasons in advances [80]. For instance, Ludescher et al. [81] were able to predict the likelihood of El Niño conditions in 2014 almost a year in advance indicating an improvement in our prediction capacity. In addition, other researchers [82,83] also revealed the intensification of ENSO related precipitation and El Niño frequency in the future due to the warming associated with an increase in greenhouse gas emission. Moreover, Ward et al. [15] indicated that the global flood risk exists during El Niño or La Niña years, or both, in basins spanning 44% of the land surface of the Earth. Therefore, the predictability of ENSO has significant implications and the ability to predict in advance might lead to better local water resources management towards developing more efficient flood and drought early warning systems.

This study limits to using only the kernel density estimation approach to estimate the mutual information values. However, there are several methods that have been used to compute the mutual information, such as K-nearest neighbors (KNN) [72], Edgeworth approximation of multivariate differential entropy [73] and adaptive partitioning of the XY plane [74]. The detailed comparisons among these estimator approaches were carried out in [71] for a different type of data such as linear, quadratic, periodic function, chaotic system. Khan et al. [71] concluded that KDE is the best choice for very short data (50–100 data points). Therefore, KDE approach was utilized to estimate the bivariate mutual information in this study. In addition to mutual information method, there are several approaches used to build the nonlinear relationship between multivariate. Lin-Ye et al. [35] applied the VGAM/VGLM to quantify the nonlinear relationship between storm components and large-scale climate indices (NAO and others) using Global Climate Model data via the regression coefficients that were used to build the location and scale parameters of their statistical model. Zhang et al. [84] investigated the nonlinear relationship by employing the concept of mutual information to evaluate the dependency between the normalized difference vegetation index (NDVI) and meteorological variables for the middle reach of the Hei river basin. Higher dependency between NDVI and coupled precipitation/temperature was observed for the desert area whilst for oasis region, groundwater is an important factor for driving vegetation growth. The authors utilized the mutual information as a method to classify the study region into a smaller area (desert, oasis, artificial oasis).

The *NLCCs* and *LCCs* analysis between ENSO and hydro-meteorological data suggest that there exists a nonlinear correlation for the Mekong River Basin. This finding agrees well with a previous study [28] with a focus on river flow analysis over tropical and sub-tropical river basins. It can be observed that although the linear CCs can be 0, the smallest *NLCCs* can be around 0.3 to 0.4. This needs caution from an artifact of Equation (8) which scales nonlinear CCs exponentially with *MI* [31] when linear CCs are close to 0. It, however, does not affect the other values of CCs higher than 0.

5. Conclusions

This study analyses the possible influence of the large-scale climate index ENSO and hydro-meteorological data in the form of total annual precipitation, annual average streamflow, and extreme precipitation indices using linear and nonlinear correlation over the Mekong River Basin. The nonlinear correlation structure was computed based on mutual information approach using marginal and joint entropy. The kernel density estimation approach was selected among several other techniques to estimate the mutual information values as this approach works best for very short data length of 50 to 100 points. Nonlinear correlations were obtained by transforming the mutual information using Joe formula [70] and thus are comparable with absolute linear correlation coefficients. Bootstrapping approach based on 100 simulations was applied to find the confidence intervals for these absolute correlation coefficients. In-depth analysis was carried out at four hydro-meteorological stations located in the Mekong River Basin. The major conclusions that can be drawn from this study are listed below:

1. Nonlinear correlation is able to reveal the additional dependence structures between hydro-meteorological data over Mekong River Basin and ENSO indices.
2. Both linear and nonlinear correlation exhibits similar varying patterns among different ENSO quarters for most of the stations/indices.
3. The results reveal that higher correlation coefficients can be found using the nonlinear correlation coefficients in comparison to the traditional linear correlation analysis.
4. Spatial correlation structures for *LCCs* and *NLCCs* are also constructed based on extreme precipitation indices and ENSO. The use of spatial maps further complements our analyses based on a single station to other ungagged regions to better inform local stakeholders in building better tools for water resource management.
5. Further analyses are required to reveal the non-linear association between other large-scale climate phenomena (SOI, PDO, NAO, etc.) with local meteorological variables. The mutual information between these indices and local meteorological variables can further help policymakers to improve climate-informed adaptation studies.

Acknowledgments: We appreciate the suggestions provided by the associate editor and three reviewers that helped us to improve quality of our manuscript. Authors would also like to thank the American International Group and the Risk Engineering and Systems Analytics Center, Clemson University for providing financial support.

Author Contributions: Tue Vu conducted experiments analyzed the data/results, Goutam Konapala involved in data collection and preparing Sections 1 and 2. Ashok Mishra supervised the overall procedure and revised the paper. All authors have read and approved the final manuscript.

Conflicts of Interest: The authors declare no conflict of interest.

References

1. Gu, D.; Philander, S.G.H. Interdecadal climate fluctuations that depend on exchanges between the tropics and extratropics. *Science* **1997**, *275*, 805–807. [CrossRef] [PubMed]
2. Mantua, N.J.; Hare, S.R.; Zhang, Y.; Wallace, J.M.; Francis, R.C. A Pacific interdecadal climate oscillation with impacts on salmon production. *Bull. Am. Meteorol. Soc.* **1997**, *78*, 1069–1079. [CrossRef]
3. Enfield, D.B.; Mestas-Nuñez, A.M.; Trimble, P.J. The Atlantic multidecadal oscillation and its relation to rainfall and river flows in the continental US. *Geophys. Res. Lett.* **2001**, *28*, 2077–2080. [CrossRef]
4. Wood, A.W.; Maurer, E.P.; Kumar, A.; Lettenmaier, D.P. Long-range experimental hydrologic forecasting for the eastern United States. *J. Geophys. Res. Atmos.* **2002**, *107*. [CrossRef]
5. Tootle, G.A.; Piechota, T.C.; Singh, A. Coupled oceanic-atmospheric variability and US streamflow. *Water Resour. Res.* **2005**, *41*. [CrossRef]
6. Kalra, A.; Ahmad, S. Using oceanic-atmospheric oscillations for long lead time streamflow forecasting. *Water Resour. Res.* **2009**, *45*. [CrossRef]
7. Hallett, T.B.; Coulson, T.; Pilkington, J.G.; Clutton-Brock, T.H.; Pemberton, J.M.; Grenfell, B.T. Why large-scale climate indices seem to predict ecological processes better than local weather. *Nature* **2004**, *430*, 71–75. [CrossRef] [PubMed]

8. Cayan, D.R.; Redmond, K.T.; Riddle, L.G. ENSO and hydrologic extremes in the western United States. *J. Clim.* **1999**, *12*, 2881–2893. [CrossRef]

9. Jones, C. Occurrence of extreme precipitation events in California and relationships with the Madden-Julian oscillation. *J. Clim.* **2000**, *13*, 3576–3587. [CrossRef]

10. DeFlorio, M.J.; Pierce, D.W.; Cayan, D.R.; Miller, A.J. Western U.S. Extreme Precipitation Events and Their Relation to ENSO and PDO in CCSM4. *J. Clim.* **2013**, *26*, 4231–4243. [CrossRef]

11. Barlow, M.; Nigam, S.; Berbery, E.H. ENSO, Pacific Decadal Variability, and U.S. Summertime Precipitation, Drought, and Stream Flow. *J. Clim.* **2001**, *14*, 2105–2128. [CrossRef]

12. Mo, K.C. Drought onset and recovery over the United States. *J. Geophys. Res.* **2011**, *116*. [CrossRef]

13. Özger, M.; Mishra, A.K.; Singh, V.P. Low frequency drought variability associated with climate indices. *J. Hydrol.* **2009**, *364*, 152–162. [CrossRef]

14. Andrews, E.D.; Antweiler, R.C.; Neiman, P.J.; Ralph, F.M. Influence of ENSO on flood frequency along the California coast. *J. Clim.* **2004**, *17*, 337–348. [CrossRef]

15. Ward, P.J.; Jongman, B.; Kummu, M.; Dettinger, M.D.; Weiland, F.C.S.; Winsemius, H.C. Strong influence of El Niño Southern Oscillation on flood risk around the world. *Proc. Natl. Acad. Sci. USA* **2014**, *111*, 15659–15664. [CrossRef] [PubMed]

16. Sivapalan, M. Prediction in ungauged basins: A grand challenge for theoretical hydrology. *Hydrol. Process.* **2003**, *17*, 3163–3170. [CrossRef]

17. Hrachowitz, M.; Savenije, H.H.G.; Blöschl, G.; McDonnell, J.J.; Sivapalan, M.; Pomeroy, J.W.; Arheimer, B.; Blume, T.; Clark, M.P.; Ehret, U.; et al. A decade of Predictions in Ungauged Basins (PUB)—A review. *Hydrol. Sci. J.* **2013**, *58*, 1198–1255. [CrossRef]

18. Samaniego, L.; Bárdossy, A.; Kumar, R. Streamflow prediction in ungauged catchments using copula-based dissimilarity measures. *Water Resour. Res.* **2010**, *46*. [CrossRef]

19. Chikamoto, Y.; Timmermann, A.; Luo, J.J.; Mochizuki, T.; Kimoto, M.; Watanabe, M.; Ishii, M.; Xie, S.P.; Jin, F.F. Skilful multi-year predictions of tropical trans-basin climate variability. *Nat. Commun.* **2015**, *6*. [CrossRef] [PubMed]

20. Hlinka, J.; Hartman, D.; Vejmelka, M.; Novotná, D.; Paluš, M. Non-linear dependence and teleconnections in climate data: Sources, relevance, nonstationarity. *Clim. Dyn.* **2014**, *42*, 1873–1886. [CrossRef]

21. Zhang, Q.; Wang, Y.; Sing, V.P.; Gua, X.; Kong, D.D.; Xiao, M.Z. Impacts of ENSO and ENSO Modoki+A regimes on seasonal precipitation variations and possible underlying causes in the Huai River basin, China. *J. Hydrol.* **2016**, *533*, 308–319. [CrossRef]

22. Wrzesinski, D. Regional differences in the influence of the North Atlantic Oscillation on seasonal river runoff in Poland. *Quaest. Geogr.* **2011**, *30*, 127–136. [CrossRef]

23. Wrzesinski, D. Typology of spatial patterns seasonality in European rivers flow regime. *Quaest. Geogr.* **2008**, *27A*, 87–98.

24. Lanzante, J.R. Resistant, robust and non-parametric techniques for the analysis of climate data: Theory and examples, including applications to historical radiosonde station data. *Int. J. Clim.* **1996**, *16*, 1197–1226. [CrossRef]

25. Yue, S.; Pilon, P.; Cavadias, G. Power of the Mann–Kendall and Spearman's rho tests for detecting monotonic trends in hydrological series. *J. Hydrol.* **2002**, *259*, 254–271. [CrossRef]

26. Konapala, G.; Mishra, A.K. Three-parameter-based streamflow elasticity model: Application to MOPEX basins in the USA at annual and seasonal scales. *Hydrol. Earth Syst. Sci.* **2016**, *20*, 2545–2556. [CrossRef]

27. Belle, G.; Hughes, J.P. Nonparametric tests for trend in water quality. *Water Resour. Res.* **1984**, *20*, 127–136. [CrossRef]

28. Li, J.; Tan, S. Nonstationary flood frequency analysis for annual flood peak series, adopting climate indices and check dam index as covariates. *Water Resour. Manag.* **2015**, *29*, 5533–5550. [CrossRef]

29. Zhang, Y.; Cabilio, P.; Nadeem, K. Improved Seasonal Mann–Kendall Tests for Trend Analysis in Water Resources Time Series. In *Advances in Time Series Methods and Applications*; Li, W.K., Stanford, D.A., Yu, H., Eds.; Springer: New York, NY, USA, 2016; pp. 215–229, ISBN 978-1-4939-6568-7.

30. Coulibaly, P.; Baldwin, C.K. Nonstationary hydrological time series forecasting using nonlinear dynamic methods. *J. Hydrol.* **2005**, *307*, 164–174. [CrossRef]

31. Khan, S.; Ganguly, A.R.; Bandyopadhyay, S.; Saigal, S.; Erickson, D.J.; Protopopescu, V.; Ostrouchov, G. Nonlinear statistics reveals stronger ties between ENSO and the tropical hydrological cycle. *Geophys. Res. Lett.* **2006**, *33*. [CrossRef]

32. Hao, Z.; Singh, V.P. Review of dependence modeling in hydrology and water resources. *Prog. Phys. Geogr.* **2016**, *40*, 549–578. [CrossRef]

33. Fleming, S.W.; Dahlke, H.E. Parabolic northern-hemisphere river flow teleconnections to El Niño-Southern Oscillation and the Arctic Oscillation. *Environ. Res. Lett.* **2014**, *9*. [CrossRef]

34. Cannon, A.J. Revisiting the nonlinear relationship between ENSO and winter extreme station precipitation in North America. *Int. J. Climatol.* **2015**, *35*, 4001–4014. [CrossRef]

35. Lin-Ye, J.; Garcia-Leon, M.; Gracia, V.; Ortego, M.I.; Lionello, P.; Sanchez-Arcilla, A. Multivariate statistical modeling of future marine storms. *Appl. Ocean Res.* **2017**, *65*, 192–205. [CrossRef]

36. Kusumastuti, D.I.; Struthers, I.; Sivapalan, M.; Reynolds, D.A. Threshold effects in catchment storm response and the occurrence and magnitude of flood events: Implications for flood frequency. *Hydrol. Earth Syst. Sci.* **2007**, *11*, 1515–1528. [CrossRef]

37. Konapala, G.; Veettil, A.V.; Mishra, A.K. Teleconnection between low flows and large-scale climate indices in Texas River basins. *Stoch. Environ. Res. Risk Assess.* **2017**, 1–14. [CrossRef]

38. Mishra, A.K.; Özger, M.; Singh, V.P. An entropy based investigation into the variability of precipitation. *J. Hydrol.* **2009**, *370*, 139–154. [CrossRef]

39. Mishra, A.K.; Özger, M.; Singh, V.P. Association between uncertainty in meteorological variables and water resources planning for Texas. *J. Hydrol. Eng.* **2011**, *16*, 984–999. [CrossRef]

40. Mishra, A.K.; Coulibaly, P. Hydrometric network evaluation for Canadian watersheds. *J. Hydrol.* **2010**, *380*, 420–437. [CrossRef]

41. Mishra, A.K.; Coulibaly, P. Variability in Canadian Seasonal Streamflow Information and its Implication for Hydrometric Network Design. *J. Hydrol. Eng.* **2014**, *19*. [CrossRef]

42. Li, C.; Singh, V.P.; Mishra, A.K. Entropy theory-based criterion for hydrometric network evaluation and design: Maximum information minimum redundancy. *Water Resour. Res.* **2012**, *48*. [CrossRef]

43. Mishra, A.K.; Singh, V.P. Analysis of drought severity-area-frequency curves using a general circulation model and scenario uncertainty. *J. Geophys. Res. Atmos.* **2009**, *114*. [CrossRef]

44. Mishra, A.K.; Özger, M.; Singh, V.P. Trend and persistence of precipitation under climate change scenarios. *Hydrol. Proc.* **2009**, *23*, 2345–2357. [CrossRef]

45. Rajsekhar, D.; Mishra, A.K.; Singh, V.P. Regionalization of drought characteristics using an entropy approach. *J. Hydrol. Eng.* **2013**, *18*, 870–887. [CrossRef]

46. Sharma, A. Seasonal to interannual rainfall probabilistic forecasts for improved water supply management: Part 1—A strategy for system predictor identification. *J. Hydrol.* **2000**, *239*, 232–239. [CrossRef]

47. Harrold, T.I.; Sharma, A.; Sheather, S. Selection of a kernel bandwidth for measuring dependence in hydrologic time series using the mutual information criterion. *Stoch. Environ. Res. Risk Assess.* **2001**, *15*, 310–324. [CrossRef]

48. Song, X.; Zhang, J.; Zhan, C.; Xuan, Y.; Ye, M.; Xu, C. Global sensitivity analysis in hydrological modeling: Review of concepts, methods, theoretical framework, and applications. *J. Hydrol.* **2015**, *523*, 739–757. [CrossRef]

49. Han, M.; Ren, W.; Liu, X. Joint mutual information-based input variable selection for multivariate time series modeling. *Eng. Appl. Artif. Intell.* **2015**, *37*, 250–257. [CrossRef]

50. Knuth, K.H.; Gotera, A.; Curry, C.T.; Huyser, K.A.; Wheeler, K.R.; Rossow, W.B. Revealing relationships among relevant climate variables with information theory. *arXiv* **2013**, arXiv:1311.4632.

51. Khokhlov, V.N.; Glushkov, A.V.; Loboda, N.S. On the nonlinear interaction between global teleconnection patterns. *Q. J. R. Meteorol. Soc.* **2006**, *132*, 447–465. [CrossRef]

52. Hurtado, A.F.; Poveda, G. Linear and global space-time dependence and Taylor hypotheses for rainfall in the tropical Andes. *J. Geophys. Res.* **2009**, *114*. [CrossRef]

53. Naumann, G.; Vargas, W.M. Joint diagnostic of the surface air temperature in southern South America and the Madden–Julian oscillation. *Weather Forecast* **2010**, *25*, 1275–1280. [CrossRef]

54. Yoon, S.; Lee, T. Investigation of hydrological variability in the Korean Peninsula with the ENSO teleconnections. *Proc. IAHS* **2016**, *374*. [CrossRef]

55. Sivakumar, B. *Chaos in Hydrology: Bridging Determinism and Stochasticity*; Springer: Berlin/Heidelberg, Germany, 2016; ISBN 978-90-481-2552-4. [CrossRef]

56. Varis, O.; Kummu, M.; Salmivaara, A. Ten major rivers in monsoon Asia-Pacific: An assessment of vulnerability. *Appl. Geogr.* **2012**, *32*, 441–454. [CrossRef]

57. Wilmott, C.J.; Rowe, C.M.; Philpot, W.D. Small-scale climate maps: A sensitivity analysis of some common assumptions associated with grid-point interpolation and contouring. *Am. Cartogr.* **1985**, *12*, 5–16. [CrossRef]

58. Yatagai, A.; Kamiguchi, K.; Arakawa, O.; Hamada, A.; Yasutomi, N.; Kitoh, A. APHRODITE: Constructing a long-term daily gridded precipitation dataset for Asia based on a dense network of rain gauges. *Bull. Am. Meteorol. Soc.* **2012**, *93*, 1401–1415. [CrossRef]

59. Vu, M.T.; Raghavan, V.S.; Liong, S.Y. SWAT use of gridded observations for simulating runoff—A Vietnam river basin study. *Hydrol. Earth Syst. Sci.* **2012**, *16*, 2801–2811. [CrossRef]

60. Vu, M.T.; Mishra, A.K. Spatial and Temporal Variability of Standardized Precipitation Index over Indochina Peninsula. *Cuad. Investig. Geogr.* **2016**, *42*, 221–232. [CrossRef]

61. Vu, M.T.; Raghavan, S.V.; Liong, S.-Y.; Mishra, A.K. Uncertainties in gridded precipitation observations in characterizing spatio-temporal drought and wetness over Vietnam. *Int. J. Climatol.* **2017**. [CrossRef]

62. Raghavan, V.S.; Vu, M.T.; Liong, S.Y. Ensemble climate projections of mean and extreme rainfall over Vietnam. *Glob. Planet. Chang.* **2017**, *148*, 96–104. [CrossRef]

63. Raghavan, V.S.; Vu, M.T.; Liong, S.Y. Impact of climate change on future stream flow in the Dakbla river. *J. Hydroinform.* **2014**, *16*, 231–244. [CrossRef]

64. Raghavan, V.S.; Liong, S.Y.; Vu, M.T. Assessment of future stream flow over the Sesan catchment of the Lower Mekong Basin in Vietnam. *Hydrol. Proc.* **2012**, *26*, 3661–3668. [CrossRef]

65. Vu, M.T.; Aribarg, T.; Supratid, S.; Raghavan, V.S.; Liong, S.Y. Statistical downscaling rainfall over Bangkok using Artificial Neural Network. *Theor. Appl. Climatol.* **2016**, *126*, 453–467. [CrossRef]

66. Trenberth, K.E. The Definition of El Niño. *Bull. Am. Meteorol. Soc.* **1997**, *78*, 2771–2777. [CrossRef]

67. Trenberth, K.E. The Climate Data Guide: Nino SST Indices (Nino 1+2, 3, 3.4, 4; ONI and TNI). Available online: https://climatedataguide.ucar.edu/climate-data/nino-sst-indices-nino-12-3-34-4-oni-and-tni (accessed on 6 January 2018).

68. Moon, Y.I.; Rajagopalan, B.; Lall, U. Estimation of mutual information using kernel density estimators. *Phys. Rev. E* **1995**, *52*, 2318–2321. [CrossRef]

69. Granger, C.; Lin, J.L. Using the mutual information coefficient to identify lags in nonlinear models. *J. Time Ser. Anal.* **1994**, *15*, 371–384. [CrossRef]

70. Joe, H. Relative entropy measures of multivariate dependence. *J. Am. Stat. Assoc.* **1989**, *84*, 157–164. [CrossRef]

71. Khan, S.; Bandyopadhyay, S.; Ganguly, A.R.; Saigal, S.; Erickson, D.J., III; Protopopescu, V.; Ostrouchov, G. Relative performance of mutual information estimation methods for quantifying the dependence among short and noisy data. *Phys. Rev. E* **2007**, *76*. [CrossRef] [PubMed]

72. Kraskov, A.; Stögbauer, H.; Grassberger, P. Estimating mutual information. *Phys. Rev. E* **2004**, *69*. [CrossRef] [PubMed]

73. Hull, M.M.V. Edgeworth approximation of multivariate differential entropy. *Neural Comput.* **2005**, *17*, 1903–1910. [CrossRef]

74. Cellucci, C.J.; Albano, A.M.; Rapp, P.E. Statistical validation of mutual information calculations: Comparison of alternative numerical algorithms. *Phys. Rev. E* **2005**, *71*. [CrossRef] [PubMed]

75. Silverman, B.W. *Density Estimation for Statistics and Data Analysis*; Chapman and Hall/CRC: London, UK, 1986.

76. Wand, M.P.; Jones, M.C. Multivariate plug-in bandwidth selection. *Comput. Stat.* **1994**, *9*, 97–117.

77. Ince, R.A.A.; Giordano, B.L.; Kayser, C.; Rousselet, G.A.; Gross, J.; Schyns, P.G. A statistical framework for neuroimaging data analysis based on mutual information estimated via a Gaussian Copula. *Hum. Brain Mapp.* **2017**, *38*, 1541–1573. [CrossRef] [PubMed]

78. Räsänen, T.A.; Kummu, M. Spatiotemporal influences of ENSO on precipitation and flood pulse in the Mekong River Basin. *J. Hydrol.* **2013**, *476*, 154–168. [CrossRef]

79. Xu, Z.X.; Takeuchi, K.; Ishidaira, H. Correlation between El Niño–Southern Oscillation (ENSO) and precipitation in Southeast Asia and the Pacific region. *Hydrol. Proc.* **2003**, *18*, 107–123. [CrossRef]

80. Chen, D.; Cane, M.A.; Kaplan, A.; Zebiak, S.E.; Huang, D. Predictability of El Niño over the past 148 years. *Nature* **2004**, *428*, 733–736. [CrossRef] [PubMed]
81. Ludescher, J.; Gozolchiani, A.; Bogacheva, M.I.; Bunde, A.; Havlin, S.; Schellnhuber, H.J. Very early warning of next El Niño. *Proc. Natl. Acad. Sci. USA* **2014**, *111*, 2064–2066. [CrossRef] [PubMed]
82. Power, S.; Delage, F.; Chung, C.; Kociuba, G.; Keay, K. Robust twenty-first-century projections of El Niño and related precipitation variability. *Nature* **2013**, *502*, 541–545. [CrossRef] [PubMed]
83. Cai, W.; Borlace, S.; Lengaigne, M.; Van Rensch, P.; Collins, M.; Vecchi, G.; Timmermann, A.; Santoso, A.; McPhaden, M.J.; Wu, L.; et al. Increasing frequency of extreme El Niño events due to greenhouse warming. *Nat. Clim. Chang.* **2014**, *4*, 111–116. [CrossRef]
84. Zhang, G.; Su, X.; Singh, V.P.; Ayantobo, O.O. Modeling NDVI using Joint Entropy method considering hydro-meteorological driving factors in the middle reaches of Hei river basin. *Entropy* **2017**, *19*, 502. [CrossRef]

entropy

MDPI

Article

Spatio-Temporal Variability of Soil Water Content under Different Crop Covers in Irrigation Districts of Northwest China

Sufen Wang [1],* and Vijay P. Singh [2]

[1] Center for Agricultural Water Research in China, China Agricultural University, Beijing 100083, China
[2] Department of Biological & Agricultural Engineering, and Zachry Department of Civil Engineering, Texas A & M University, College Station, TX 77843-2117, USA; vsingh@tamu.edu
* Correspondence: wangsuf@cau.edu.cn; Tel./Fax: +86-10-6273-8548

Received: 25 April 2017; Accepted: 4 August 2017; Published: 18 August 2017

Abstract: The relationship between soil water content (SWC) and vegetation, topography, and climatic conditions is critical for developing effective agricultural water management practices and improving agricultural water use efficiency in arid areas. The purpose of this study was to determine how crop cover influenced spatial and temporal variation of soil water. During a study, SWC was measured under maize and wheat for two years in northwest China. Statistical methods and entropy analysis were applied to investigate the spatio-temporal variability of SWC and the interaction between SWC and its influencing factors. The SWC variability changed within the field plot, with the standard deviation reaching a maximum value under intermediate mean SWC in different layers under various conditions (climatic conditions, soil conditions, crop type conditions). The spatial-temporal-distribution of the SWC reflects the variability of precipitation and potential evapotranspiration (ET_0) under different crop covers. The mutual entropy values between SWC and precipitation were similar in two years under wheat cover but were different under maize cover. However, the mutual entropy values at different depths were different under different crop covers. The entropy values changed with SWC following an exponential trend. The informational correlation coefficient (R_0) between the SWC and the precipitation was higher than that between SWC and other factors at different soil depths. Precipitation was the dominant factor controlling the SWC variability, and the crop efficient was the second dominant factor. This study highlights the precipitation is a paramount factor for investigating the spatio-temporal variability of soil water content in Northwest China.

Keywords: spatio-temporal variability; soil water content; entropy; arid region

1. Introduction

Soil water content (SWC) varies both spatially and temporally. Characterizing the spatio-temporal variability of SWC is important for water management in agricultural irrigation [1]. The spatio-temporal variability of SWC is related to a number of factors, such as certain soil properties, topography, vegetation, and climate. The interactions among these factors control the spatio-temporal evolution of soil water content at different scales, including gully scale [2], plot scale [3–5], hillslope scale [6–8], catchment scale [9,10] and regional scale [11–13]. The factors that determine SWC depend on the environment of the area of study and the presence or absence of vegetative canopy [14] and the spatio-temporal variability of SWC affect vegetation coverage and density [15]. Investigations on agricultural watersheds have shown that topography, rainfall, and soil texture have mixed effects on SWC at the watershed and regional scales, and that land cover influences runoff, interception and evapotranspiration processes, and in turn, the soil water dynamics [5,16–19]. The spatial-temporal pattern of SWC can be obtained from measurements carried out in different experimental plots [3,4]. Differences in SWC can be small or significant at different soil depths [20,21].

Because of practical limitations, measurements of SWC for deep soil profiles are time consuming and costly [22]. Hence, only the near surface soil moisture has been investigated [3,4,16,23]. Examining the SWC variability in the upper 30 cm of the soil on a Mediterranean abandoned terrace in north-east Spain [24] showed that the spatial variability of SWC followed a bimodal pattern with increasing soil water content. Only a few studies have taken into account the SWC variability at different soil depths [23,25,26] and develop new high throughput approaches to measure soil water content in the field [27]. Soil water contents vary with depth due to the heterogeneity of environmental factors, such as water and energy input, soil texture [28] and genetic control of root architecture [29].

For investigating the spatio-temporal variability of SWC, both statistical and geostatistical methods have been applied using ground-based and remote sensing data [17,19,24,26]. A few studies have applied the entropy theory to investigate the spatio-temporal variability of SWC and dominant factors [30]. Singh developed an entropy theory for describing the one-dimensional movement of soil moisture in unsaturated soils, and tested the theory using experimental observations reported in the hydrologic literature [31]. Gaur and Mohanty used the Shannon entropy to assess the effect of different physical controls on the spatial mean and variability of soil moisture in Oklahoma and Iowa and found that for most soil moisture conditions [30], soil texture as opposed to vegetation and topography is the dominant physical control at both the point and airborne scales. There is little information on the relationship between SWC and climate conditions or vegetation types in crop growing seasons in arid and semi-arid agricultural areas. This relationship can provide a basis for improving agricultural water use efficiency in arid regions.

Maize (being a C4 plant) and wheat (being a C3 plant) are the predominant crops grown in northwest China, where population density and exploitation of water resources are high [32]. How SWC is controlled in the maize and wheat fields is important for agricultural water management in arid regions in northwest China. To address this question, a field scale experiment was conducted to measure soil water content over canopies in northwest China in 2013 and 2014. The objectives of this study therefore were to: (1) determine how environmental variables alter the spatial and temporal variability of SWC; (2) determine the effect of crop growth stage on SWC at different depths; and (3) quantify the environmental control on the variation of SWC using entropy.

2. Materials and Methods

2.1. Experimental Site and Design

A two-year experiment was conducted in two fields at the Shiyanghe Experimental Station for Water-Saving in Agriculture and Ecology of China Agricultural University, located in Wuwei (Gansu Province of Northwest China, 37°52′ N, 102°50′ E, altitude 1581 m) during the crop growing seasons in 2013 and 2014 (Figure 1). The experimental site is located in a typical continental temperate climate zone, where the mean annual temperature is 8 °C, the annual accumulated temperature (>0 °C) is 3550 °C d^{-1}, the annual precipitation is 164.4 mm, the mean annual pan evaporation is about 2000 mm, the aridity index (the ratio of mean annual evaporation to precipitation) is 15–25, the average annual sunshine duration is 3000 h and frost-free days number 150 d. The groundwater table is below 40–50 m [33].

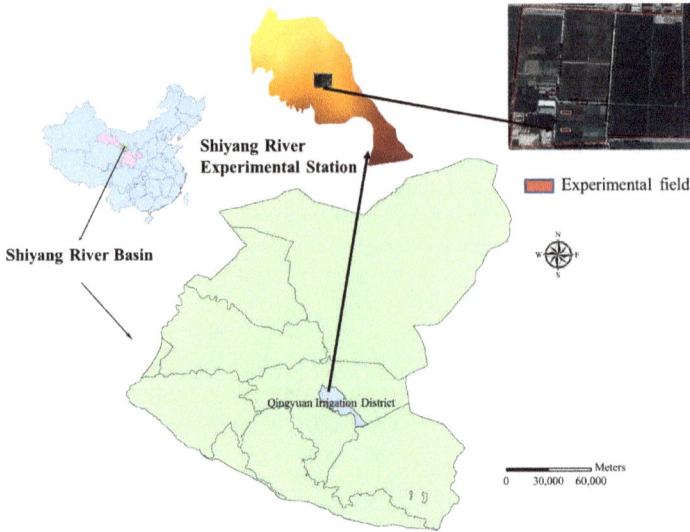

Figure 1. Location of the study area.

The two study sites (field I and field II) have a similar area of about 40 m long and 12 m wide. The SWC (%cm^{3}cm^{-3}) was monitored under two different crops at an interval of 10 cm providing estimates covering 0 to 120 cm depth. A total of 46 sample points for the study site in 2013 (Figure 2) and 24 sample points in 2014 (Figure 3) were monitored.

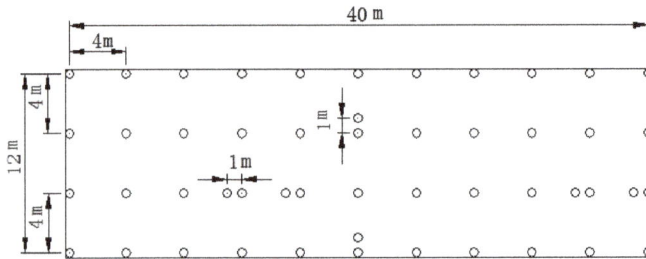

Figure 2. Layout of soil water content measuring probes in the experiment field under maize and wheat covers in 2013.

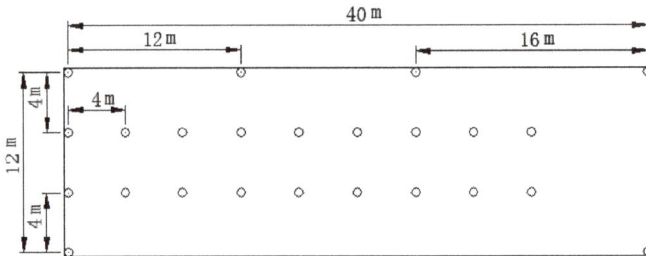

Figure 3. Layout of soil water content measuring probes in the experiment field under maize and wheat covers in 2014.

Spring maize and spring wheat were used in crop rotation, that is, the plot on which wheat was planted in 2013 was planted with maize in 2014. Maize seed was sown through 5.0 cm diameter holes, with a row spacing of 50 cm and a plant spacing of 23.8 cm. Spring wheat was sown in holes of 3.0 cm diameter, with a row spacing of 15 cm and a plant spacing of 10 cm for the two years using a hole sowing machine. Spring wheat was sown with 10–15 seeds in each hole. During each growing season, weeds were removed manually as required. The observation periods for maize were from 14 May to 17 September in 2013 and 28 April to 18 September in 2014; and those for wheat were from 12 May to 27 July in 2013 and 7 April to 23 July in 2014. Precipitation, solar radiation, air temperature, relative humidity and wind speed were measured with a standard automatic weather station at a height of 2.0 m above the ground at the experimental site. A portable Diviner 2000 device (Sentek Pty Ltd., Stepney, Australia) was used to measure SWC at an interval of 3–5 days. Soil samples were collected before and after irrigation events and after rainfall and the probes were calibrated against gravimetric samples for the observation sites. Table 1 lists the physical properties of the soil profile.

Table 1. Physical properties of soil profile at different sites.

Field Site	Soil Depth (cm)	Soil Particle Content (g/g)			Soil Type
		Sand	Silt	Clay	
	10	0.629	0.268	0.103	sandy loam
	20	0.649	0.251	0.0996	sandy loam
	40	0.656	0.241	0.103	sandy loam
I	60	0.443	0.412	0.146	loam
	80	0.653	0.260	0.087	sandy loam
	100	0.931	0.052	0.017	sandy soil
	120	0.616	0.301	0.083	sandy loam
	10	0.620	0.271	0.109	sandy loam
	20	0.663	0.243	0.095	sandy loam
	40	0.531	0.343	0.127	sandy loam
II	60	0.518	0.349	0.134	sandy loam
	80	0.482	0.401	0.117	loam
	100	0.505	0.362	0.133	sandy loam
	120	0.706	0.219	0.074	sandy loam

2.1.1. Statistical Analysis

For each soil depth the coefficient of variation (CV_i) was used for spatial analysis, and temporal standard deviation ($STDV_i$) for temporal analysis. The CV_i and the $STDV_i$ were calculated as:

$$CV_i = \frac{S_i}{\overline{SWC_i}} \tag{1}$$

$$STDV_i = \sqrt{\frac{1}{N} \sum_{i=1}^{n} \left(SWC_i - \overline{SWC_i}\right)} \tag{2}$$

where S_i is mean squared error. $\overline{SWC_i}$ is the average soil water content for the each soil depth, i represents the soil depth. Heterogeneity was considered weak when $CV \leq 10\%$, moderate when $10\% < CV < 100\%$, and strong when $CV \geq 100\%$ [34]. Descriptive statistics of soil water content under different crop covers based on 10 cm-steps are given in Tables 2 and 3. The temporal average versus the standard deviation of SWC in different soil layers under maize and wheat cover in 2013 and 2014 are shown in Figures 4 and 5, respectively.

Table 2. Statistics of soil water content under different land covers in 2013.

Depth (cm)	Land Cover	Field Site	Minimum (%cm³cm⁻³)	Maximum (%cm³cm⁻³)	Mean (%cm³cm⁻³)	Std. Deviation (%cm³cm⁻³)	CV (%)
0–10	maize	I	0.62	32.19	11.86	5.30	44.65
	wheat	II	2.13	32.47	15.17	7.39	58.70
10–20	maize	I	1.53	31.94	15.44	4.71	30.53
	wheat	II	3.29	32.73	17.01	6.46	37.98
20–40	maize	I	3.63	36.39	24.12	4.35	18.10
	wheat	II	7.42	30.00	20.66	4.30	20.81
40–100	maize	I	8.65	41.84	28.52	6.41	22.49
	wheat	II	12.86	34.31	22.33	4.41	19.75
100–120	maize	I	8.19	35.61	20.39	5.75	28.20
	wheat	II	10.48	37.89	23.88	5.81	24.33

Table 3. Statistics of soil water content under different land covers in 2014.

Depth (cm)	Land Cover	Field Site	Minimum (%cm³cm⁻³)	Maximum (%cm³cm⁻³)	Mean (%cm³cm⁻³)	Std. Deviation (%cm³cm⁻³)	CV (%)
0–10	maize	II	4.76	31.59	16.74	5.03	30.05
	wheat	I	6.46	29.09	16.66	5.33	31.99
10–20	maize	II	8.51	31.94	18.69	4.09	21.88
	wheat	I	10.09	30.81	20.74	4.40	21.22
20–40	maize	II	5.14	36.39	32.23	5.15	15.80
	wheat	I	12.76	33.66	22.71	3.46	15.24
40–100	maize	II	4.46	33.16	23.04	5.45	23.66
	wheat	I	13.72	35.89	25.48	4.75	18.64
100–120	maize	II	4.78	38.60	26.79	5.97	22.84
	wheat	I	8.37	30.90	18.33	5.22	28.48

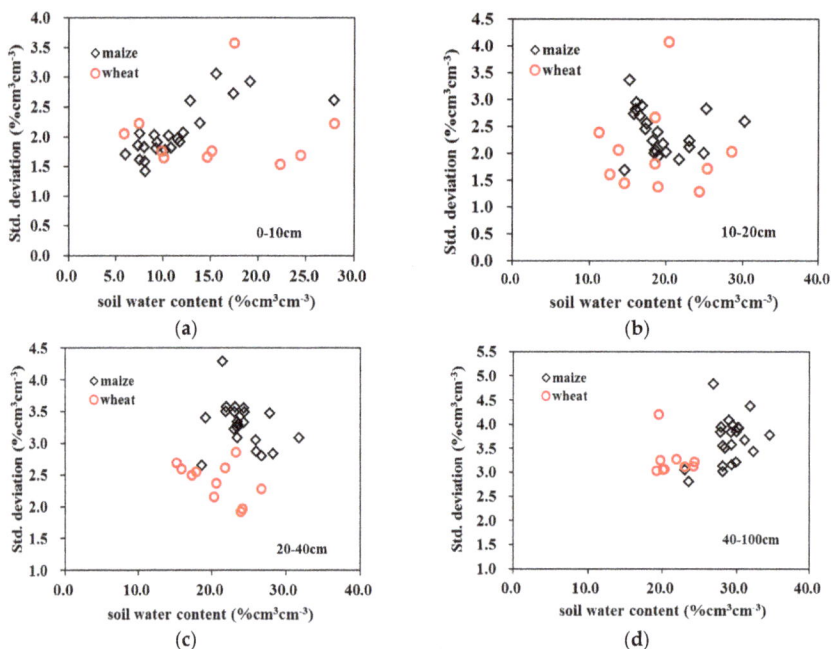

Figure 4. *Cont.*

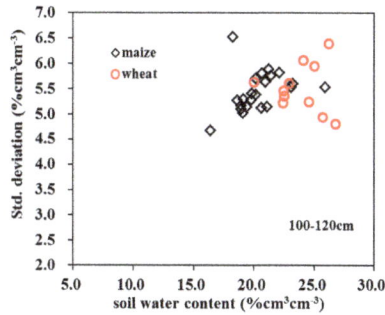

(e)

Figure 4. Temporal standard deviation (*STDV*) in time versus temporally averaged soil water content for different soil depths in 2013. *STDV* values for (**a**) 0–10 cm, (**b**) 10–20 cm, (**c**) 20–40 cm, (**d**) 40–100 cm and (**e**) 100–120 cm soil depths.

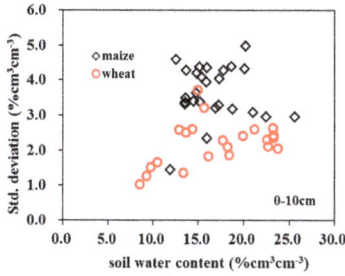

(a)

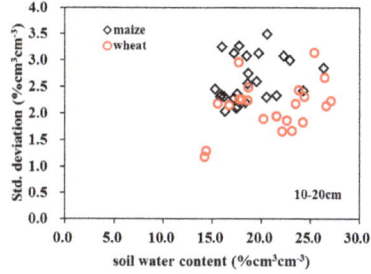

(b)

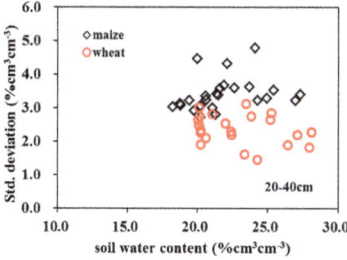

(c)

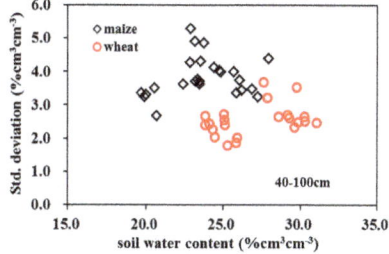

(d)

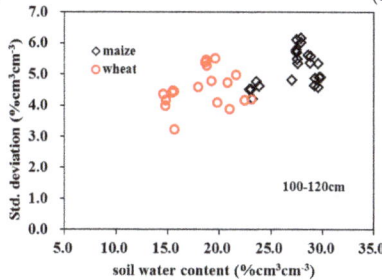

(e)

Figure 5. Temporal standard deviation (*STDV*) in time versus temporally averaged soil water content for different soil depths in 2014. *STDV* values for (**a**) 0–10 cm, (**b**) 10–20 cm, (**c**) 20–40 cm, (**d**) 40–100 cm and (**e**) 100–120 cm soil depths.

2.1.2. Entropy

In this study, the Shannon entropy [35,36] was used to investigate the spatial variability and dominant controls on the variability of SWC during the crop growing period. Entropy has an advantage in that it is capable of incorporating the effect of dependent or independent factors on the SWC spatial distribution, and can be used for short or long data sets. Entropy (*H*) is a statistical quantity representing a measure of information that may be extracted from a system or analogously the uncertainty that the system comprises. For a probability distribution $P = \{p_1, p_2, \cdots, p_n\}$, where $p_1, p_2, \cdots, p_n$ are the probabilities of N outcomes $(x_i, I = 1, 2, \ldots, N)$ of a random variable X or a random experiment, Shannon defined a measure H as a function of probabilities as [35]:

$$H(p_1, p_2, \cdots, p_N) = - \sum_{i=1}^{N} p_i \log p_i \qquad \sum_{i=1}^{N} p_i = 1 \qquad (3)$$

According to the frequency histograms of SWC for each soil depth constructed with empirical frequencies, a probability p_i is assigned to each bin and calculated. We substituted p_i in Equation (3) to find out the marginal entropies.

For two or more independent random variables X and Y, their respective entropy values may be added. However, if there is dependence between the random variables, then the dependence can be accounted for through mutual information $T(X,Y)$, which is the amount of information common to both the random variables. Consider two simultaneous experiments whose outcomes are represented by X and Y. The mutual entropy of X and Y, denoted as $T(X, Y)$, defines the amount of uncertainty reduced in X when Y is known. It equals the difference between the sum of two marginal entropies and the total entropy:

$$T(X,Y) = H(X) + H(Y) - H(X,Y) \qquad (4)$$

Mutual information has an advantage over other measures of information, as it can provide a quantitative measure of a description of the relationship among variables based on their information transmission characteristics. Larger values of T correspond to greater amounts of information transferred. Mutual information is superior to the Pearson correlation coefficient, since it captures both linear and nonlinear dependence, whereas the Pearson correlation coefficient is only suitable for linear relationships, or more generally, for spherical and elliptical dependence structures [37–39]. In reality, the spatial patterns of SWC may be non-stationary and the related processes may be nonlinear [22].

2.1.3. Calculation of Marginal Entropy and Mutual Entropy for Soil Water Content

In this study, the random variables under consideration are the SWC, crop coefficients (K_C), reference crop evapotranspiration (ET_0), and precipitation (P). The highest mutual entropy between soil water content and other factors can be considered to be the most dominant factor that influences the SWC variability.

We arranged SWC values $SWC_{i,d,n}$, where i represents the soil depth, d represents the days (1, 2, ..., *d*), and *n* represents the number of soil water content values (1, 2, ..., *n*), P is daily precipitation, and ET_0 is the reference crop evapotranspiration. In this study, ET_0 was calculated according to the FAO (Food and Agriculture Organization) Penman–Monteith method using daily observed climate data [40,41]. K_C is the daily crop coefficients and was calculated by the single crop coefficient method Equation (5) [40,42–45] as:

$$K_C = \frac{ET_C}{ET_0} \qquad (5)$$

where ET_C is the crop evapotranspiration. ET_C was calculated according to the FAO Penman–Monteith method [46].

The frequency histograms of $SWC_{i,d,n}$ were constructed with empirical frequencies obtained from observed the data with no zero counts [47], and hence the Shannon entropy H of the random variable

$SWC_{i,d}$ from the corresponding observed counts $SWC_{i,d,n}$. The marginal entropy of SWC at different soil depths was computed from Equation (2). The mutual entropy between $SWC_{i,d}$ and ET_0, $SWC_{i,d}$ and K_C, $SWC_{i,d}$ and P were calculated using Equation (3).

2.1.4. Informational Correlation Coefficient

The informational correlation coefficient R_0 is a measure of transferable information between random variables X and Y and measures their mutual dependence and does not assume any type of relationship between them. It is expressed in terms of transinformation as:

$$R_0 = \sqrt{1 - \exp(-2T_0)} \tag{6}$$

where T_0 is the transinformation or mutual information representing the upper limit of transferable information between two variables [38]. In this study, three groups were identified: $SWC_{i,d}$ and ET_0, $SWC_{i,d}$ and K_C, $SWC_{i,d}$ and P. From results one can infer the main dominating factors for the SWC variability.

3. Results and Discussion

3.1. Temporal Variation of Soil Water Content under Different Crops

The two-year sampling in this study can be considered as representative of the usual conditions in Northwest China. The accumulated rainfall during the maize growing period was 54.4 mm in 2013 and 196.6 mm in 2014, respectively, and the accumulated rainfall during the wheat growing period was 29.6 mm in 2013 and 128.8 mm in 2014, respectively. Due to the meteorological forcing (precipitation and ET_0), the SWC observed during the study period showed a succession of increases and decreases in response to the succession of precipitation events (Figures 6a and 7a). This led to a steady increase in SWC at different soil depths from June to mid-July for the maize cover. The mean SWC at different depths increased when precipitation took place. The SWC of layer (20–40 cm) can represent the average soil water content (0–120 cm) under different crop covers. During the growing seasons, the SWC changed only slightly during the early growing period (April to mid-May) under different crop covers (Figures 6 and 7) due to little water demand during this stage. The SWC in the 0–120 cm layer varied greatly over the crop growing period. The temporal variability of SWC changed slightly with the increase of soil depth in the two years. Tables 2 and 3 show that the SWC at different soil depths changed typically with crop and soil conditions.

During the observation growing period of maize (from May to late September) in 2013 (dry year) (Table 2). The minimum SWC varied from $0.62\%cm^3cm^{-3}$ to $8.65\%cm^3cm^{-3}$ and the maximum SWC varied from $30\%cm^3cm^{-3}$ to $41.84\%cm^3cm^{-3}$ for different soil depths. Corresponding to the same soil depths, the ranges were $31.57\%cm^3cm^{-3}$, $30.41\%cm^3cm^{-3}$, $32.76\%cm^3cm^{-3}$, $33.19\%cm^3cm^{-3}$ and $27.42\%cm^3cm^{-3}$, respectively. During the observation period, the range for the 40–100 cm was greater due to the soil texture, where the fraction of sand particles soil in the 40–100 cm layer range from 0.443 to 0.931 (Table 1).

For wheat growing period (from April to late July), The minimum SWC varied from $2.13\%cm^3cm^{-3}$ to $12.86\%cm^3cm^{-3}$ and the maximum SWC varied from $30\%cm^3cm^{-3}$ to $37.79\%cm^3cm^{-3}$ for different soil depths. Corresponding to the same soil depths, the ranges were $30.34\%cm^3cm^{-3}$, $29.44\%cm^3cm^{-3}$, $22.58\%cm^3cm^{-3}$, $21.45\%cm^3cm^{-3}$ and $27.41\%cm^3cm^{-3}$, respectively. The range of SWC under wheat cover was greater for the 0–10 cm soil layer (Table 2). During the growing period of maize in 2014 (wet year), the minimum SWC varied from $4.46\%cm^3cm^{-3}$ to $8.51\%cm^3cm^{-3}$ and the maximum SWC varied from $30\%cm^3cm^{-3}$ to $38.6\%cm^3cm^{-3}$ for different soil depths. Corresponding to the same soil depths, the ranges were $26.83\%cm^3cm^{-3}$, $23.43\%cm^3cm^{-3}$, $31.25\%cm^3cm^{-3}$, $28.7\%cm^3cm^{-3}$ and $33.82\%cm^3cm^{-3}$, respectively. The range of SWC under maize cover was greater for the 100–120 cm soil layer. In this year, maize was arranged in field II. Table 1 shows that the fraction of sand particles in 100–120 cm layer ranged from 0.507 to 0.706. Two years of results of SWC under maize cover showed that soil texture had a direct influence on the SWC variability.

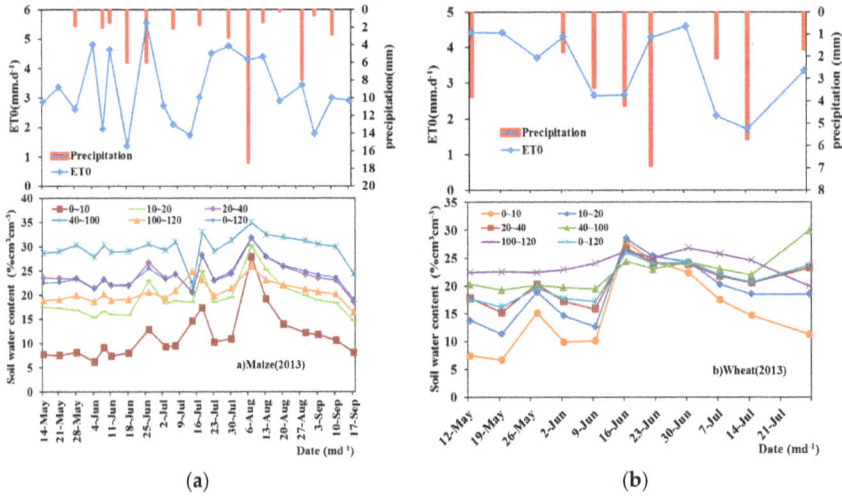

Figure 6. Temporal variation of soil water content under maize cover and wheat cover in 2013. ET$_0$, P and SWC values for (**a**) maize and (**b**) wheat in 2013.

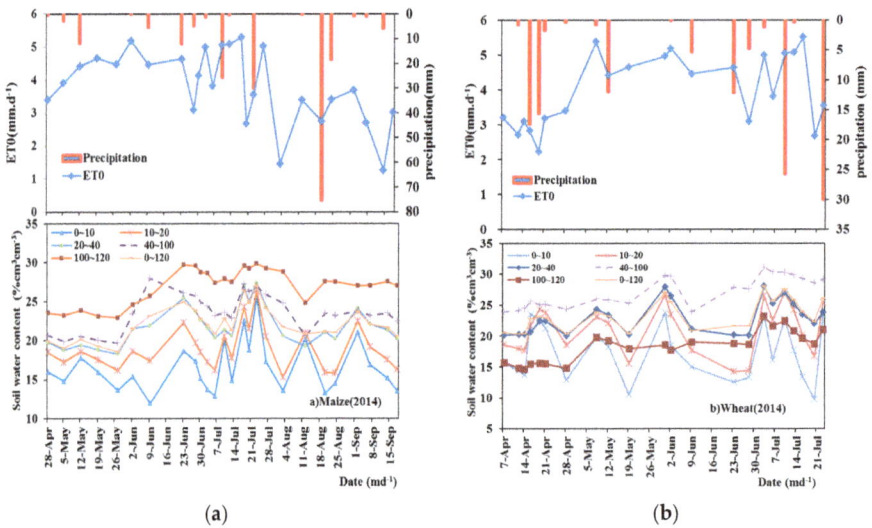

Figure 7. Temporal variation of soil water content under maize cover and wheat cover in 2014. ET$_0$, P and SWC values for (**a**) maize and (**b**) wheat in 2014.

During the observation period under wheat cover, the minimum SWC ranged from 6.46%cm^{3}cm^{-3} to 13.72%cm^{3}cm^{-3} and the maximum SWC ranged from 29.09%cm^{3}cm^{-3} to 35.89%cm^{3}cm^{-3} for different soil depths. Corresponding to the same soil depths, the ranges were 22.63%cm^{3}cm^{-3}, 20.72%cm^{3}cm^{-3}, 20.9%cm^{3}cm^{-3}, 22.17%cm^{3}cm^{-3} and 22.53%cm^{3}cm^{-3}, respectively. The range of SWC under wheat cover was greater for the 0–10 cm soil layer. The result was the same as that of 2013 in spite of the field plot in two years. Table 1 shows that the soil texture of 0–10 cm in field I was similar to that in field II. The fraction of sand particles in the 0–10 cm layer was 0.6291 in field I, and 0.62 in field II.

The temporal variability of SWC in different layers under different crop covers displayed the same trends. It can be seen that the range under wheat cover in different field plots in the growing seasons in the surface layer was higher than in other layers, because the top layer was more sensitive to the environment. The fluctuation of the upper soil layer (0–10 cm) under wheat was greater in the observation period in 2013 and 2014. This result is similar to what found by Odhiambo and Bomke (2007) with no differences in soil water content in depths 20–40 and 40–60 cm among cropping treatments in south coastal British Columbia. However, the range of deep soil layer (≥40 cm) under maize was greater in the two year study period due to different soil textures. The fraction of sand particles above 40 cm layer was from 0.443 to 0.931 in field I, and 0.482 to 0.706 in field II.

Figures 4 and 5 show the temporal average versus the temporal standard deviation of soil water content of maize and wheat in 2013 and in 2014. In contrast with the top layers, a higher $STDV$ was generally found for the deeper layers under two crops. The $STDV$ for deep layers changed from 2.5%cm^3cm^{-3} to 6.5%cm^3cm^{-3} in two years. The temporal standard deviation ($STDV$) of SWC increased with increasing depth. In the 20–40 cm layer, the value of wheat was smaller than that of maize. This implies that roots in this depth had a significant effect. Temporally averaged values in deeper layers (40–100 cm and 100–120 cm) had higher standard deviations. The result is similar to that of [48] that SWC in crop covered deep layer (80–100 cm) exhibited a large degree of change, which was affected by crop root water uptake in the same region.

3.2. Spatial Variation of Soil Water Content under Different Crops

3.2.1. Soil Water Content Analysis Based on Coefficient of Variation

For SWC measurements during crop growth, the coefficient of variation (CV) was used to evaluate the spatial variation in soil water content in the experimental fields. Tables 2 and 3 show the CV of the spatial distribution of SWC in different layers during the crop growing period in 2013 and 2014. Table 2 shows the CV of soil water content under the maize cover varied from 18.10% to 44.65% for different soil depths in 2013. The CV was between 10% and 100%, so that SWC was considered to exhibit moderate heterogeneity. The CV of SWC was the highest in the top layer due to the SWC in the top layer was strongly spatially influenced by precipitation and evapotranspiration, whereas for deeper soil layers, SWC was steady. The CV of SWC under wheat cover varied from 19.75% to 58.70% for different soil depths. The spatial heterogeneity of soil water content under wheat cover was moderate and the CV for the 0–10 cm layer was the highest.

From Table 3, it can be seen that CV of SWC under maize cover changed from 15.80% to 30.05%, for different soil layers. SWC was considered to have moderate heterogeneity due to CV was between 10% and 100%. The extent of heterogeneity was less compared to that in 2013. The CV of SWC was the highest in the top layer, because SWC in the top layer was strongly spatially influenced by precipitation and evapotranspiration during the growing period, whereas for deeper layers, SWC changed more slowly. The CV of SWC under wheat cover varied from 15.24% to 31.99% for different soil layers. It can be seen that the CV range was similar to that under maize. The spatial heterogeneity of SWC under wheat cover was moderate and CV for the 0–10 cm layer was the highest.

Table 2 showed that the CV (from 18.10% to 44.65%) of the spatial distribution of SWC in different layers under maize cover was less than that (from 19.75% to 58.70%) under wheat cover during 2013. In 2014, it can be seen from Table 3 that CV (from 15.80% to 30.05%) of the spatial distribution of SWC in different layers under maize cover was similar to that (from 15.24% to 31.99%) under wheat cover. From Figure 8, it can be seen that CV of spatial SWC decreased with increasing mean SWC under different crop covers, because the evapotranspiration rate decreased with decreasing soil water content

(Figure 5). Results for *CV* of the spatial SWC can be found in other studies [3,23], and the SWC varied with depth due to the heterogeneity of environmental factors [28,48].

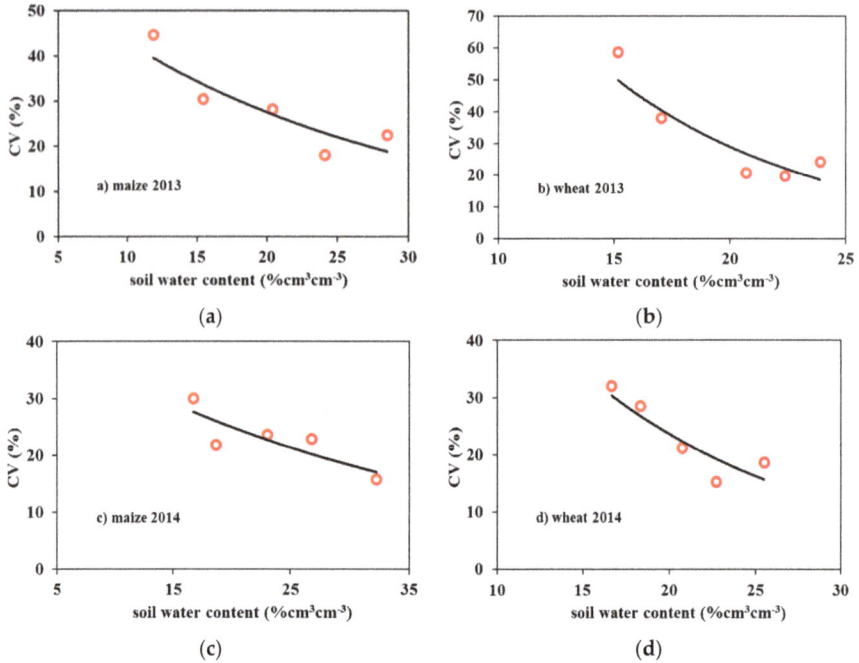

Figure 8. Mean soil water content versus *CV* for maize and wheat in 2013–2014. *CV* values for (**a**) maize, (**b**) wheat cover in 2013, *CV* values for (**c**) maize and (**d**) wheat cover in 2014.

3.2.2. Soil Water Content Analysis Using Entropy

The marginal entropy values of daily soil water content were plotted against the daily mean soil water content for different soil depths under different crop covers in 2013 (Figure 9) and 2014 (Figure 10). Figure 9 shows that entropy was maximum when SWC was the maximum value, and entropy was distributed in the intermediate range under the maize cover in field I in 2013. In the upper soil layer (0–10 cm and 10–20 cm), entropy changed widely during the maize growing period. Entropy was the maximum in the intermediate range under wheat cover for field II in 2013. In the upper soil layer (0–10 cm and 10–20 cm), entropy changed widely during the maize growing period, but the entropy value for the deeper layer changed not so much. The soil texture and crop growth patterns led to the difference under the same weather conditions [49]. In different layers under different crop covers, the marginal entropy of spatial variability and range increased with increasing SWC. In this study, data of marginal entropy and mean SWC were fitted with an exponential model separately for the 0 to 10, 10–20, 20–40, 40–100 and 100–120 cm soil layers, and the results of fitting are shown in Figures 9 and 10.

With all data points included in the analysis, the relationship between marginal entropy and mean SWC apparently was different for different soil layers under the maize cover and the wheat cover in 2013–2014. Entropy changed with soil water content as an exponential model, in which the model parameters indicated the magnitude of the proportional effect on the variability and the dependence of variability on the mean SWC, respectively. Figures 9 and 10 show that the absolute values of both parameters for different soil layers under wheat cover were larger than those of the soil layers under maize cover in 2013–2014. It seems that the soil layer had a relatively greater overall spatial variability under wheat cover and entropy increased relatively faster for the soil layers with increasing SWC.

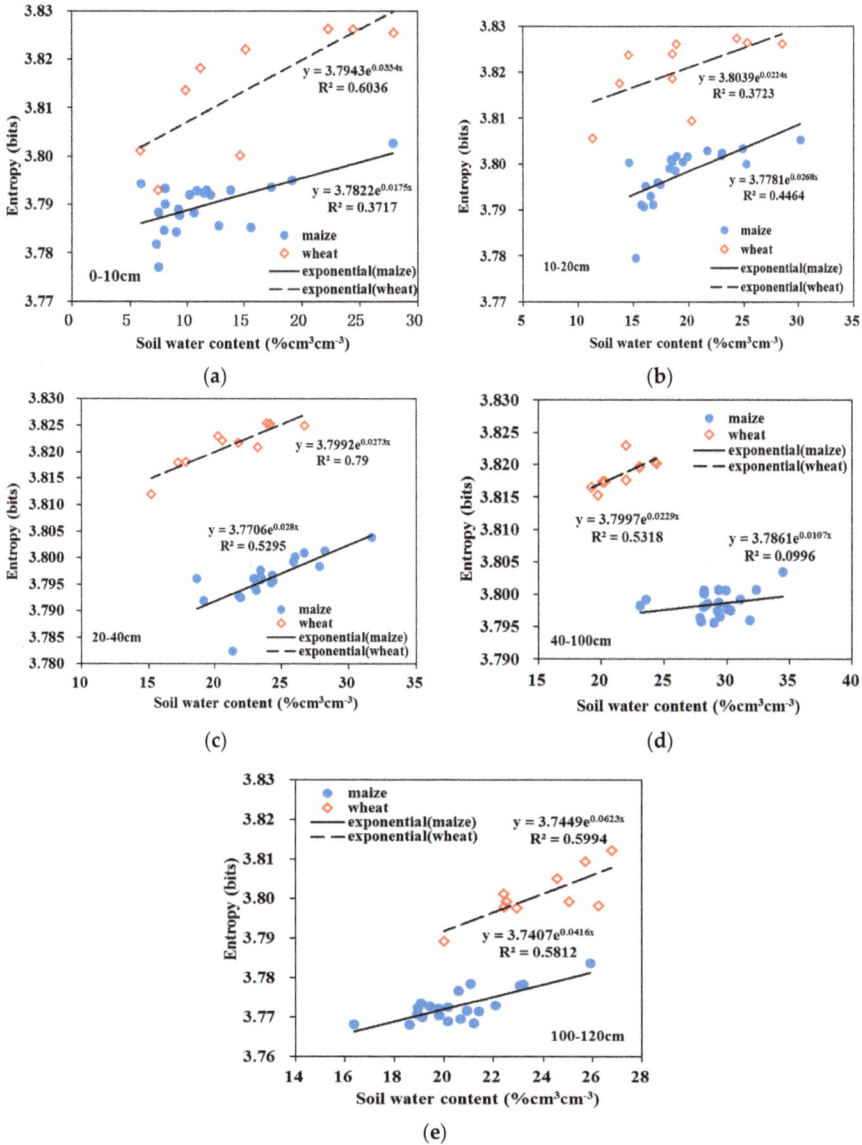

Figure 9. Mean soil water content versus entropy for maize and wheat in 2013. Entropy values for (**a**) 0–10 cm, (**b**) 10–20 cm, (**c**) 20–40 cm, (**d**) 40–100 cm and (**e**) 100–120 cm soil depths.

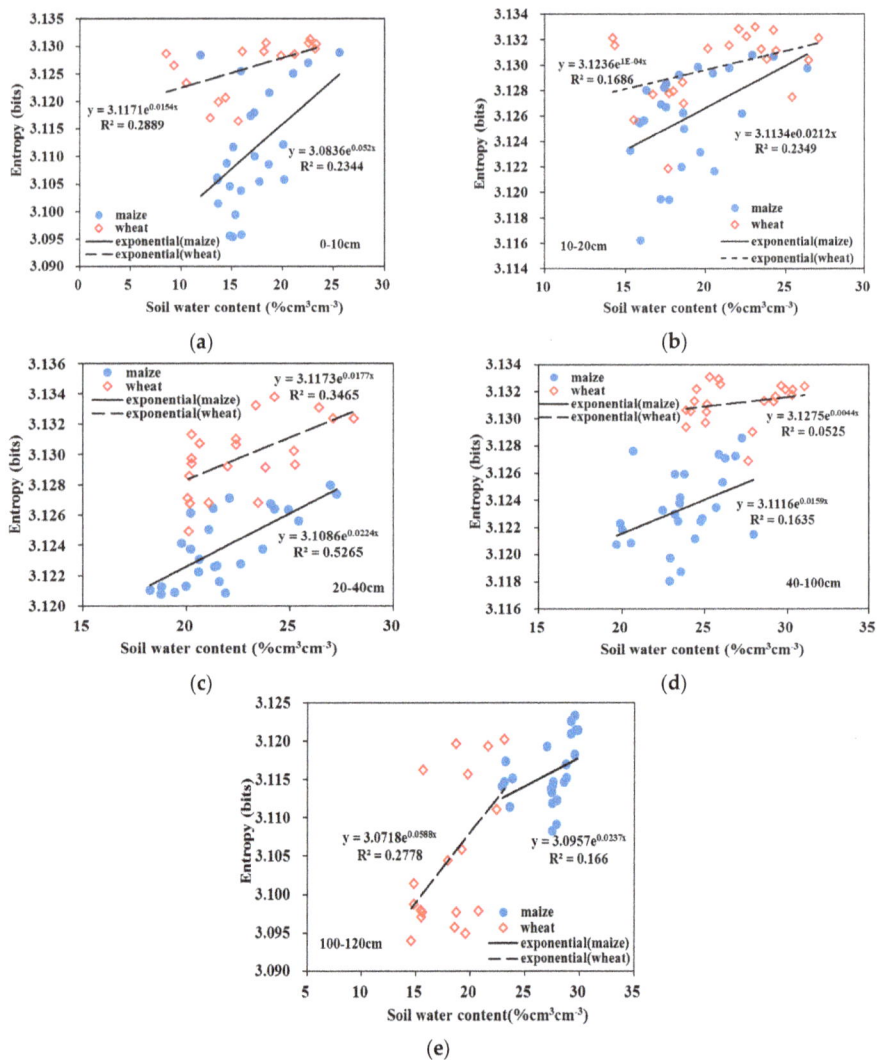

Figure 10. Mean soil water content versus entropy under maize and wheat cover in 2014. Entropy values for (**a**) 0–10 cm, (**b**) 10–20 cm, (**c**) 20–40 cm, (**d**) 40–100 cm and (**e**) 100–120 cm soil depths.

3.3. Calculation of Mutual Entropy for Influencing Factors of Soil Water Content

From Figure 11, it can be seen that a larger value of mutual entropy corresponds to a greater amount of information transferred because the value of mutual entropy represents the amount of information transmitted. In 2013 and 2014, the mutual entropy value of SWC and precipitation was much higher than that of ET_0 and SWC, SWC and K_C under different field plots, which is consistent with previous studies [50]. Crop type was the second contributor that affects the soil water content. Under wheat cover, it was observed that even though the marginal entropy values in 2013 were higher than those in year 2014 (Figures 9 and 10), the mutual entropy values between SWC and precipitation were similar to the value in 2014 (Figure 11). The results reflect that the influence of precipitation to SWC is consistent in spite of the amount. The results reflects the entropy theory dose well in communicating with nature, it helps better understand soil water content and its influence factors in natural systems. The mutual entropy values for different layers were different under wheat cover in different years. For the top soil layers (0–10 cm and 10–20 cm) and observed deepest layer (100–120 cm), the value of mutual entropy between SWC and precipitation (Figure 11b) in dry year (2013) was much higher than that (Figure 11c) in wet year (2014); results were similar to the value between SWC and ET_0.

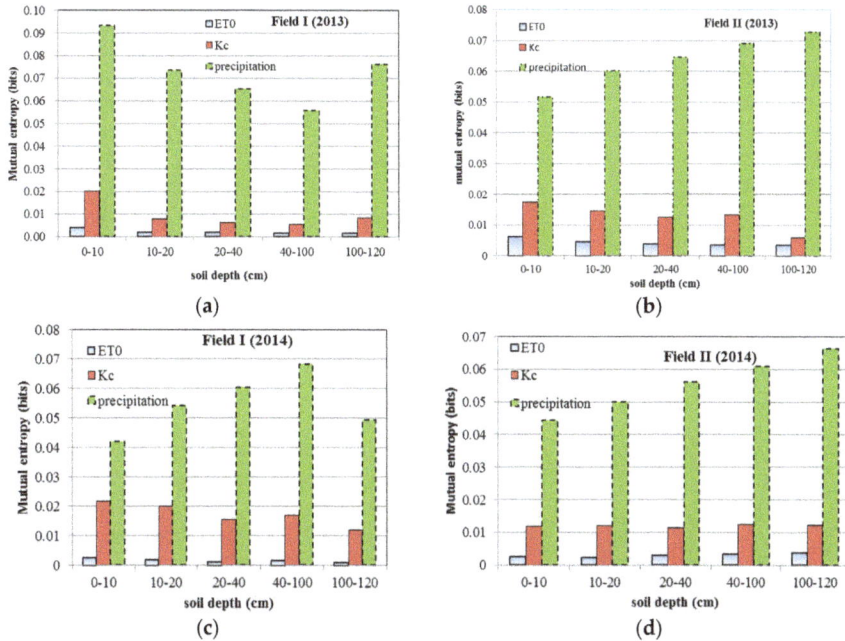

Figure 11. Mutual entropy under different crop covers in 2013–2014. Mutual entropy values for (**a**) Field I and (**b**) Field II in 2013, mutual entropy values for (**c**) Field I and (**d**) Field II in 2014.

Figures 9 and 10 show that the marginal entropy values in 2013 were higher than those in 2014 under maize cover. The mutual entropy values between SWC and precipitation were from 0.055 to 0.091 in 2013 (Figure 11a), which were also bit higher than those (0.045 to 0.067) in 2014 (Figure 11d). The mutual entropy values for different layers were different under wheat cover in the two year period. For the top soil layers (0–10 cm and 10–20 cm), the value of mutual entropy between SWC and precipitation (Figure 11a) in the dry year (2013) was much higher than that (Figure 11d) in the wet year (2014). The value between SWC and ET_0 in 2013 was similar to that in 2014. This result is different from that under wheat cover. The difference resulted from the crop types, wheat (being a C3 plant) extracts more water than maize (being a C4 plant).

Figure 12 shows that informational correlation coefficient (R_0) between SWC and other factors for different soil layers. The value of R_0 between SWC and precipitation was from 0.35 to 0.41 under maize cover (Figure 12a) and from 0.31 to 0.37 under wheat cover (Figure 12b) in 2013. The R_0 value between SWC and precipitation was from 0.29 to 0.36 under maize cover (Figure 12c) and from 0.28 to 0.36 under wheat cover (Figure 12d) in 2014. For the 0–10 cm soil layer, the R_0 value between SWC and precipitation under maize cover was 0.41 and 0.29 in 2013 and 2014, respectively, which was bit higher than the value (0.31 in 2013, 0.28 in 2014) under wheat cover. For the 10–20 cm soil layer, the R_0 value between SWC and precipitation under maize cover was 0.37 and 0.31 in 2013 and 2014, respectively, and the value under wheat cover was 0.34 and 0.32 in 2013 and 2014, respectively. The result was similar to that for the 0–10 cm soil layer.

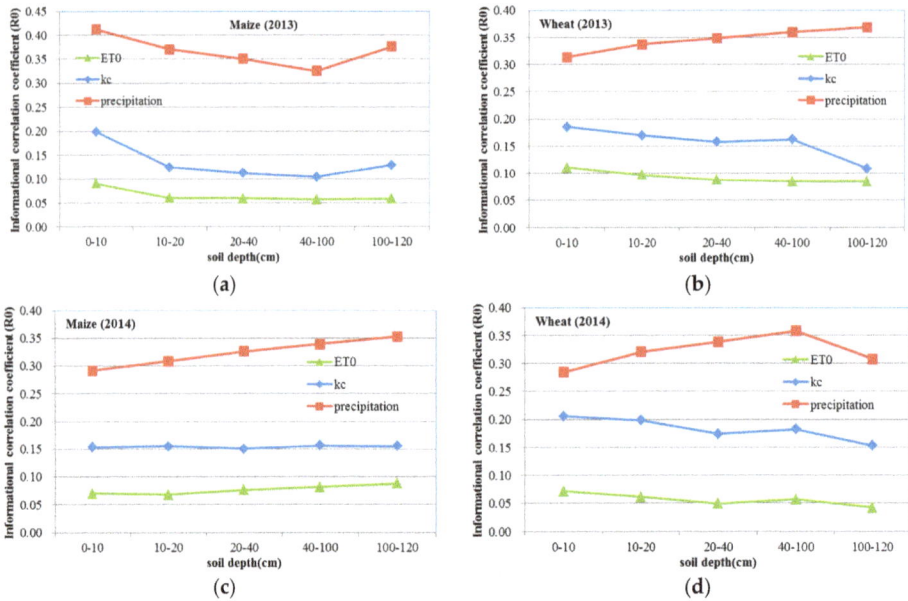

Figure 12. Informational correlation coefficient (R_0) under different crop covers in 2013–2014. R_0 values for (**a**) maize and (**b**) wheat in 2013, R_0 values for (**c**) maize and (**d**) wheat in 2014.

From Figure 12, it can be seen that the transferable information of precipitation for SWC was more than ET_0 and K_C. The value between SWC and K_C was the smallest for different soil layers. The R_0 value was similar under different crop covers in the two years. The influence of K_C on SWC in different layers was steady under different cover crops in the two years. The R_0 value between SWC and K_C under maize cover was from 0.10 to 0.19 in 2013 and from 0.152 to 0.156 in 2014. For wheat cover, the R_0 value between SWC and K_C was from 0.09 to 0.11 and from 0.04 to 0.06 in 2013 and 2014, respectively. In spite of crop type, climate was the dominant factor that affected the SWC. In northwest China, the precipitation is prior to ET_0. So, the irrigation forecast can be according to the surface soil water content and weather conditions for the certain crop cover.

4. Conclusions

In this study, entropy was employed, which was proved as an efficient tool for analyzing the spatio-temporal patterns of SWC and the related environmental conditions. In order to examine the spatial-temporal variability of SWC over the growing season in relation to the environmental variables and to analyze how the influence of these factors vary with the crop growth in different soil depths, both statistical and entropy methods were used to quantify the environmental control on the seasonal variation of SWC based on two-year observations in a field site of Shiyang River basin in Northwest China.

From the distribution of SWC in time, it was found that the temporal variability of SWC in different layers under different cover crops displays the same change trends. The temporal variability of SWC changed slightly with increase of soil depth in the two years. The *CV* of spatial SWC decreased with increasing mean SWC under different covers. The marginal entropy of spatial variability and

Entropy **2017**, *19*, 410

range increased with increasing SWC for different layers under different crop covers. Entropy changed with SWC following an exponential model.

Mutual entropy results showed that precipitation is the dominant factor on SWC regardless of the crop type. The crop coefficient (K_C) is the second dominant factors for SWC in the two years. So the crop type is the second contributor that affects the SWC variability in different soil layers. The results implied that changing crop type is a necessary adaptation to regional water scarcity exacerbated by climate change in arid and semi-arid areas.

The mutual entropy values between SWC and precipitation are similar in different years under wheat cover, but the marginal entropy values for different layers are different. The value of mutual entropy of SWC with precipitation and that with ET_0 for the top soil layer and observed deepest layer are much higher in 2013 (dry year). This work opens a new avenue for further research to use entropy theory for better understanding the variability of soil water contents and its influencing factors.

Acknowledgments: This work was financially supported by the national key Research and Development Plan (2016YFC0400207), the Chinese National Natural Science Fund (91425302, 51439006, 51109211) and the Special Fund for Agro-scientific Research in the Public Interest (201503125). We are grateful for Zhongyi Liu, Xin Liu and Congcong Peng's help with field data collection. We would also like to thank Nandita Gaur and Zhenlei Yang for valuable discussions. We thank the anonymous reviewers for their constructive comments which helped improve this paper.

Author Contributions: Sufen Wang conceived and designed the experiments; Sufen Wang performed the experiments; Sufen Wang analyzed the data; Sufen Wang and Vijay P. Singh contributed reagents/materials/analysis tools; Sufen Wang and Vijay P. Singh wrote the paper. Both authors have read and approved the final manuscript.

Conflicts of Interest: The authors declare no conflict of interest.

References

1. Vereecken, H.; Huisman, J.A.; Bogena, H.; Vanderborght, J.; Vrugt, J.A.; Hopmans, J.W. On the value of soil moisture measurements in vadose zone hydrology: A review. *Water Resour. Res.* **2008**, *44*, W00D06. [CrossRef]
2. Gao, X.; Wu, P.; Zhao, X.; Zhang, B.; Wang, J.; Shi, Y. Estimating the spatial means and variability of root-zone soil moisture in gullies using measurements from nearby uplands. *J. Hydrol.* **2013**, *476*, 28–41. [CrossRef]
3. Brocca, L.; Morbidelli, R.; Melone, F.; Moramarco, T. Soil moisture spatial variability in experimental areas of central Italy. *J. Hydrol.* **2007**, *333*, 356–373. [CrossRef]
4. Brocca, L.; Melone, F.; Moramarco, T.; Morbidelli, R. Soil moisture temporal stability over experimental areas in Central Italy. *Geoderma* **2009**, *148*, 364–374. [CrossRef]
5. Wijewardana, Y.G.N.S.; Galagedara, L.W. Estimation of spatio-temporal variability of soil water content in agricultural fields with ground penetrating radar. *J. Hydrol.* **2010**, *391*, 24–33. [CrossRef]
6. Famiglietti, J.S.; Rudnick, J.W.; Rodelli, M. Variability in surface moisture content along a hillslope transect: Rattlesnake Hill, Texas. *J. Hydrol.* **1998**, *210*, 259–281. [CrossRef]
7. Famiglietti, J.; Devereaux, J.; Laymon, C.A.; Tsegaye, T.; Houser, P.R.; Jackson, T.J.; Graham, S.T.; Rodell, M.; van Oevelen, P.J. Ground-based investigation of soil moisture variability within remote sensing footprints during the Southern Great Plains 1997 (SGP 97) Hydrology Experiment. *Water Resour. Res.* **1999**, *35*, 1839–1851. [CrossRef]
8. Penna, D.; Borga, M.; Norbiato, D.; Dalla, F.G. Hillslope scale soil moisture variability in a steep alpine terrain. *J. Hydrol.* **2009**, *364*, 311–327. [CrossRef]
9. Liang, W.L.; Hung, F.X.; Chan, M.C.; Lu, T.H. Spatial structure of surface soil water content in a natural forested headwater catchment with a subtropical monsoon climate. *J. Hydrol.* **2014**, *516*, 210–221. [CrossRef]
10. Brocca, L.; Tullo, T.; Melone, F.; Moramarco, T.; Morbidelli, R. Catchment scale soil moisture spatial-temporal variability. *J. Hydrol.* **2012**, *422*, 63–75. [CrossRef]
11. Joshi, C.; Mohanty, B.P.; Jacobs, J.M.; Ines, A.V. Spatiotemporal analyses of soil moisture from point to footprint scale in two different hydroclimatic regions. *Water Resour. Res.* **2011**, *47*, W01508. [CrossRef]

12. Rosenbaum, U.; Bogena, H.R.; Herbst, M.; Huisman, J.A.; Peterson, T.J.; Weuthen, A. Seasonal and event dynamics of spatial soil moisture patterns at the small catchment scale. *Water Resour. Res.* **2012**, *48*, W10544. [CrossRef]

13. Joshi, C. Understanding Spatio-Temporal Variability and Associated Physical Controls of Near-Surface Soil Moisture in Different Hydro-Climates. Ph.D. Thesis, Texas A&M University, College Station, TX, USA, 2013.

14. Gómez-Plaza, A.; Mart Nez-Mena, M.; lbaladejo, J.; Castillo, V.M. Factors regulating spatial distribution of soil water content in small semiarid catchments. *J. Hydrol.* **2001**, *253*, 211–226. [CrossRef]

15. Li, H.; Shen, W.; Zou, C.; Jiang, J.; Fu, L.; She, G. Spatio-temporal variability of soil moisture and its effect on vegetation in a desertified aeolian riparian ecotone on the Tibetan Plateau, China. *J. Hydrol.* **2013**, *479*, 215–225. [CrossRef]

16. Baroni, G.; Ortuani, B.; Facchi, A.; Gandolfi, C. The role of vegetation and soil properties on the spatio-temporal variability of the surface soil moisture in a maize-cropped field. *J. Hydrol.* **2013**, *489*, 148–159. [CrossRef]

17. Joshi, C.; Mohanty, B.P. Physical controls of near surface soil moisture across varying spatial scales in an agricultural landscape during SMEX02. *Water Resour. Res.* **2010**, *46*, W12503. [CrossRef]

18. Mohanty, B.P.; Skaggs, T.H. Spatio-temporal evolution and time-stable characteristics of soil moisture within remote sensing footprints with varying soil, slope, and vegetation. *Adv. Water Resour.* **2001**, *24*, 1051–1067. [CrossRef]

19. Zhou, J.; Fu, B.; Gao, G.; Lü, N.; Lü, Y.; Wang, S. Temporal stability of surface soil moisture of different vegetation types in the Loess Plateau of China. *CATENA* **2015**, *128*, 1–15. [CrossRef]

20. Song, X.Y.; Kang, S.Z.; Shen, B. Dynamic Regularity of Soil Moisture in Farmland with Different Planting on Loess Hilly and Gully Region in North-West of China. *J. Soil Water Conserv.* **2003**, *2*, 130–133.

21. Odhiambo, J.J.O.; Bomke, A.A. Cover crop effects on spring soil water content and the implications for cover crop management in south coastal British Columbia. *Agric. Water Manag.* **2007**, *88*, 92–98. [CrossRef]

22. Hu, W.; Si, B.C. Revealing the relative influence of soil and topographic properties on soil water content distribution at the watershed scale in two sites. *J. Hydrol.* **2014**, *516*, 107–118. [CrossRef]

23. Li, T.; Hao, X.M.; Kang, S.Z. Spatiotemporal variability of soil moisture as affected by soil properties during irrigation cycles. *Soil Sci. Soc. Am. J.* **2014**, 598–608. [CrossRef]

24. Molina, A.J.; Latron, J.; Rubio, C.M.; Gallart, F.; Llorens, P. Spatio-temporal variability of soil water content on the local scale in a Mediterranean mountain area (Vallcebre, North Eastern Spain). How different spatio-temporal scales reflect mean soil water content. *J. Hydrol.* **2014**, *516*, 182–192. [CrossRef]

25. Bogena, H.R.; Herbst, M.; Huisman, J.A.; Rosenbaum, U.; Weuthen, A.; Vereecken, H. Potential of wireless sensor networks for measuring soil water content variability. *Vadose Zone J.* **2010**, *9*, 1002–1013. [CrossRef]

26. Hu, W.; Si, B.C. Soil water prediction based on its scale-specific control using multivariate empirical mode decomposition. *Geoderma* **2013**, *193*, 180–188. [CrossRef]

27. Whalley, W.R.; Binley, A.; Watts, C.W.; Shanahan, P.; Dodd, L.C.; Ober, E.S.; Ashton, R.W.; Webster, C.P.; White, R.P.; Hawkesford, M.J. Methods to estimate changes in soil water for phenotyping root activity in the field. *Plant Soil* **2017**, *415*, 407–422. [CrossRef]

28. Wang, Y.Q.; Shao, M.A.; Liu, Z.P. Vertical distribution and influencing factors of soil water content within 21-m profile on the Chinese Loess Plateau. *Geoderma* **2013**, *193*, 300–310. [CrossRef]

29. Dodd, I.C.; Whalley, W.R.; Obe, R.E.S.; Parry, M.A.J. Genetic and management approaches to boost UK wheat yields by ameliorating water deficits. *J. Exp. Bot.* **2011**, *62*, 5241–5248. [CrossRef] [PubMed]

30. Gaur, N.; Mohanty, B.P. Evolution of physical controls for soil moisture in humid and subhumid watersheds. *Water Resour. Res.* **2013**, *9*, 1244–1258. [CrossRef]

31. Singh, V.P. Entropy theory for movement of moisture in soils. *Water Resour. Res.* **2010**, *46*, W03516. [CrossRef]

32. Kang, S.Z.; Su, X.L.; Tong, L.; Zhang, J.H.; Zhang, L.; Davies, W.J. A warning from an ancient oasis: Intensive human activities are leading to potential ecological and social catastrophe. *Int. J. Sustain. Dev. World* **2008**, *15*, 440–447. [CrossRef]

33. Li, S.; Kang, S.Z.; Zhang, L.; Ortega-Farias, S.; Li, F.; Du, T.; Tong, L.; Wang, S.; Ingman, M.; Guo, W. Measuring and modeling maize evapotranspiration under plastic film-mulching condition. *J. Hydrol.* **2013**, *503*, 153–168. [CrossRef]

34. Warrick, A.W.; Nielsen, D.R. *Spatial Variability of Soil Physical Properties in the Field*; Academic Press: New York, NY, USA, 1980; pp. 319–344.

35. Shannon, C.E. A mathematical theory of communication. *Bell Syst. Tech. J.* **1948**, *27*, 623–656. [CrossRef]
36. Shannon, C.E. A mathematical theory of communication. *ACM SIGMOBILE Mob. Comput. Commun. Rev.* **2001**, *5*, 3–55. [CrossRef]
37. Li, C.; Singh, V.P.; Mishra, A.K. Entropy theory-based criterion for hydrometric network evaluation and design: Maximum information minimum redundancy. *Water Resour. Res.* **2012**, *48*, W05521. [CrossRef]
38. Singh, V.P. *Entropy Theory and Its Application in Environmental and Water Engineering*; Wiley: Hoboken, NJ, USA, 2013.
39. Singh, V.P. *Entropy Theory in Hydrologic Science and Engineering*; McGraw-Hill: New York, NY, USA, 2015.
40. Kang, S.Z.; Gu, B.J.; Du, T.S.; Zhang, J.H. Crop coefficient and ratio of transpiration to evapotranspiration of winter wheat and maize in a semi-humid region. *Agric. Water Manag.* **2003**, *59*, 239–254. [CrossRef]
41. Utset, A.; Farre, I.; Martinez-Cob, A.; Cavero, J. Comparing Penman–Monteith and Priestley–Taylor approaches as reference-evapotranspiration inputs for modeling maize water-use under Mediterranean conditions. *Agric. Water Manag.* **2004**, *66*, 205–219. [CrossRef]
42. Ding, R.S.; Kang, S.Z.; Li, F.S.; Zhang, Y.Q.; Tong, L.; Sun, Q.Y. Evaluating eddy covariance method by large-scale weighing lysimeter in a maize field of north-west China. *Agric. Water Manag.* **2010**, *98*, 87–95. [CrossRef]
43. Ding, R.S. Mechanism of Water and Heat Transfer and Evaportranspitation Modeling over Maize Field in an Arid Inland Region. Ph.D. Thesis, China Agricultrual University, Beijing, China, 2012.
44. Ding, R.S.; Kang, S.Z.; Li, F.S.; Zhang, Y.Q.; Tong, L. Evapotranspiration measurement and estimation using modified Priestley–Taylor model in an irrigated maize field with mulching. *Agric. For. Meteorol.* **2013**, *168*, 140–148. [CrossRef]
45. Lian, C.Y.; Ma, Z.M.; Lu, X.D.; Cao, S.Y. Research on main crops water requirement and crop Coefficient in Oasis irrigation Region. *J. Irrig. Drain.* **2012**, *31*, 136–139. (In Chinese)
46. Allen, R.G.; Pereira, L.S.; Raes, D.; Smith, M. *Crop Evapotranspiration: Guidelines for Computing Crop Water Requirements*; Irrigation and Drainage Paper No.56; FAO: Rome, Italy, 1998.
47. Hausser, J.; Strimmer, K. Entropy inference and the James-Stein estimator, with application to nonlinear gene association networks. *J. Mach. Learn. Res.* **2009**, *10*, 1469–1484.
48. Wang, L. Evaluation of Effectiveness of Soil Moisture in Shiyang River Basin Farmland and Its Dynamic Simulation. Ph.D. Thesis, Northwest A&F University, Xianyang, China, 2014.
49. Henning, H.; Olga, T.; Heinz-Josef, K.; Bernward, M. Subsoil properties and cereal growth as affected by a single pass of heavy machinery and two tillage systemson a Luvisol. *J. Plant Nutr. Soil Sci.* **2008**, *171*, 580–590.
50. Ryu, D.; Famiglietti, J.S. Multi-scale spatial correlation and scaling behavior of surface soil moisture. *Geophys. Res. Lett.* **2006**, *33*, L08404. [CrossRef]

![entropy logo] *entropy*

MDPI

Article

Maximum Entropy-Copula Method for Hydrological Risk Analysis under Uncertainty: A Case Study on the Loess Plateau, China

Aijun Guo [1], Jianxia Chang [1,*], Yimin Wang [1], Qiang Huang [1] and Zhihui Guo [2]

[1] State Key Laboratory Base of Eco-Hydraulic Engineering in Arid Area, Xi'an University of Technology, Xi'an 710048, China; aijunguo619@gmail.com (A.G.); wangyimin@xaut.edu.cn (Y.W.); Sy-sj@xaut.edu.cn (Q.H.)

[2] College of Water Conservancy and Hydropower, Hebei University of Engineering, Handan 056038, China; sfagaj@163.com

*   Correspondence: chxiang@xaut.edu.cn; Tel.: +86-029-8231-2906

Received: 25 September 2017; Accepted: 11 November 2017; Published: 15 November 2017

Abstract: Copula functions have been extensively used to describe the joint behaviors of extreme hydrological events and to analyze hydrological risk. Advanced marginal distribution inference, for example, the maximum entropy theory, is particularly beneficial for improving the performance of the copulas. The goal of this paper, therefore, is twofold; first, to develop a coupled maximum entropy-copula method for hydrological risk analysis through deriving the bivariate return periods, risk, reliability and bivariate design events; and second, to reveal the impact of marginal distribution selection uncertainty and sampling uncertainty on bivariate design event identification. Particularly, the uncertainties involved in the second goal have not yet received significant consideration. The designed framework for hydrological risk analysis related to flood and extreme precipitation events is exemplarily applied in two catchments of the Loess plateau, China. Results show that (1) distribution derived by the maximum entropy principle outperforms the conventional distributions for the probabilistic modeling of flood and extreme precipitation events; (2) the bivariate return periods, risk, reliability and bivariate design events are able to be derived using the coupled entropy-copula method; (3) uncertainty analysis highlights the fact that appropriate performance of marginal distribution is closely related to bivariate design event identification. Most importantly, sampling uncertainty causes the confidence regions of bivariate design events with return periods of 30 years to be very large, overlapping with the values of flood and extreme precipitation, which have return periods of 10 and 50 years, respectively. The large confidence regions of bivariate design events greatly challenge its application in practical engineering design.

Keywords: hydrological risk analysis; maximum entropy-copula method; uncertainty; Loess Plateau

1. Introduction

Extreme hydrological events (e.g., floods, rainstorms, droughts) have had disastrous effects on society and the environment in recent years. Specifically, floods, as one of the most frequent and costly natural disasters, have posed a serious threat to the human life and economic development [1–3]. A report issued by UNISDR (2015) highlights the fact that, between 1995 and 2015, floods affected 2.3 billion people, worldwide, accounting for 56% of the people affected by weather-related disasters [4,5]. Flood risk analysis can provide extremely valuable information by estimating the occurrence of floods for flood control and disaster mitigation, hydraulic structure design, reservoir management, and so on [6,7]. It is widely known that, in rain-dominant watersheds, river floods are commonly triggered by extreme precipitation events [8,9]. Therefore, in practice, reducing the flood risk also requires information on extreme precipitation [10–12]. Consequently, the present work focuses on exploring the bivariate risk of annual maximum flood discharge (*AMF*) and associated extreme precipitation (*Pr*) events.

Up until now, copula functions have been used extensively to evaluate the bivariate risk of hydro-meteorological events [13–17]. For instance, Chen et al. (2013) constructed four-dimensional copulas to model the behaviors of drought events; She et al. (2016) applied copula-based severity-duration-frequency curves to evaluate the spatio-temporal variability of dry spells and wet spells. Compared with traditional bivariate hydrologic modeling, the main advantage of copulas is that they allow the joint dependence structure to be modeled, without any restrictions on marginal distributions [18]. Given this, practitioners can flexibly choose marginal and joint probability functions [19,20]. Consequently, the selection of marginal distribution is of crucial importance as it strongly impacts the performance of the copula in modeling bivariate variables [6].

However, distributions that model the univariate hydro-meteorological series are diverse. In terms of hydrologic frequency analysis, the most widely used distributions are parametric ones, such as the general extreme value distribution, normal distribution, lognormal distribution, Pearson type 3 distribution, Log Pearson type 3 distribution, Gamma distribution and so on [6,21–23]. When utilizing these distributions, one obvious drawback is that selecting the appropriate distribution from a variety of candidates is time-consuming [12]. Worse, if the univariate probability distribution is misidentified, results derived from the copulas tend to be underestimated/overestimated [24]. Hence, a widely applicable probability distribution with high accuracy is urgently needed. The maximum entropy principle (MEP), first expounded by [25], offers a methodology for deriving probability distribution functions (PDFs) with a minimum of bias from limited information in a more objective way [7,26]. The MEP proposes a criterion for selecting the most appropriate PDF on the basis of the rationale that the desired PDF possesses maximum uncertainty, subject to a set of constraints [27]. As Zhang and Singh (2012) stated, an entropy-based methodology is able to reach a universal solution, and can better capture the shape of the probability density function, without first knowing the format of the a priori distribution [24]. More moments of observations, beyond just the second moment, can be accounted for in the MEP approach. Additionally, various generalized distributions, such as Pearson type 3 distribution, Gamma distribution, etc., can be derived from the MEP-based distribution using different constraints [28,29]. Attracted by the splendid performance of MEP distribution, therefore, it has been extensively used in the hydrology field [6,12,24,30]. For instance, Mishra et al. (2009) employed the entropy concept to investigate the spatial and temporal variability of precipitation time series for the State of Texas, USA [31]; Rajsekhar et al. (2013) used the entropy concept to identify the homogenous regions based on drought severity and duration [32].

Given the above, the present work takes advantage of the outstanding performance of MEP distribution, and subsequently develops a framework based on a coupled MEP-copula model for bivariate hydrological risk analysis in terms of *AMF* and *Pr*.

Also of note is that uncertainty accompanies the copula-based hydrological risk analysis. As Michailidi and Bacchi (2017) stated, flood risk evaluation without accounting for uncertainty is deceptive [33]. Serinaldi (2013) also stressed that the uncertainty of multivariate design event estimation should be considered carefully for practical application, rather than speculation [34]. However, previous studies have paid considerably less attention to the impact of uncertainty on hydrological risk analysis [34–36]. Therefore, another contribution of this paper is to present a framework aiming to reveal the impact of marginal distribution selection uncertainty and sampling uncertainty on hydrological risk analysis. The two sources of uncertainty are often overlooked in spite of their widely recognized importance; particularly sampling uncertainty, due to its difficult estimation and interpretation [35,36].

The Loess Plateau (LP) is known as the "cradle of Chinese civilization", and is also one of the most serious soil erosion areas worldwide. On the LP, annual average soil erosion reaches to around 2000–2500 t/km², and the area suffering severe soil and water loss covers more than 60% [37].

Sparse vegetation cover, highly intense rainfall events, and the long history (over 5000 years) of human activities are generally considered to be the principal factors causing severe soil loss on the LP [38]. Due to the arid and semi-arid continental monsoon climate, however, most previous studies have primarily focused on low flow and drought conditions [39,40]. Studies investigating the bivariate risk of flood and extreme precipitation events for the LP are still few.

Consequently, the present study primarily aims to advance the coupled MEP-copula model for bivariate risk analysis, and to reveal the impact of the marginal distribution selection uncertainty and sampling uncertainty on hydrological risk analysis. The developed framework is exemplarily applied for two catchments of the LP. The remainder of the paper is constructed as follows. Section 2 describes the study area and data. Section 3 introduces the methods adopted in this study. The results and discussion are presented in Section 4. Section 5 shows the main conclusions drawn from this study.

2. Study Area and Dataset

2.1. Study Area

The Weihe River basin (104–107° E and 33–34° N), located in the southern part of the LP, was selected as our study area (Figure 1). The basin has a typical continental climate, and lies in the semi-humid and semi-arid transitional zone [41]. The Weihe River (hereafter WR) provides the water supply for 9300 km^2 of fertile fields in the Guanzhong Plain, and more than 61% of the Shaanxi Province's population [42]. Additionally, the start-point of the well-known Silk Road Economic Belt, Xi'an City, is situated in this basin. Mean annual precipitation varies between 400 and 600 mm, of which approximately 70% falls between June and September. Floods occur frequently after rainstorms. The largest gauged flood event at the Linjiacun station since 1960 occurred in 1966, and was 4200 m^3/s.

The Weihe River basin is also one of the most serious soil loss areas on the LP. Areas suffering from severe soil loss cover approximately 65% of the total land area of this basin. It is of note that floods accelerate soil and water loss. Accelerated serious soil loss has caused severe sediment deposition in the lower reach of the WR, which poses great challenges for local flood control.

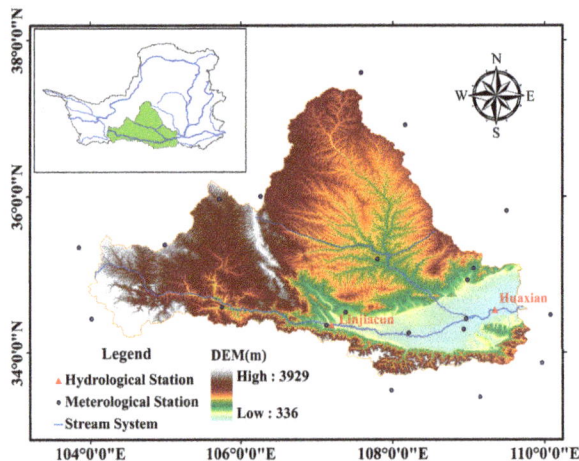

Figure 1. Location of the studied watershed.

2.2. Dataset

The Linjiacun (107°03′ E, 34°22′ N) and Huaxian (109°78′ E, 34°51′ N) stations are important control stations upstream and downstream of the Weihe River basin, respectively. The locations of the two stations are displayed in Figure 1. Annual maximum flood records (1960–2012) from the two stations are utilized. The data quality was strictly controlled by the hydrology bureau of the Yellow River Conservancy Commission before the data was released. Data collected from Linjiacun and Huaxian stations can characterize the water hazard control in the Weihe River basin.

Daily precipitation data (1960–2012) are provided by the China Meteorological Data Sharing Service System (http://cdc.cma.gov.cn). The flood discharge is closely linked to the accumulated rainfall amounts before the occurrence of annual peaks [43]. Given this, the extreme precipitation event (*Pr*) used in present study is defined as:

$$Pr = \sum_{i=l}^{l-n} Rain_i \tag{1}$$

where *Pr* denotes the accumulated rainfall from the 1st to *i*-th day, *l* is the occurrence time of peak discharge *Q*, *n* ($n = 0, 1, 2, 3, 4$) indicates lag time (i.e., time from peak discharge to the beginning of rainfall), $Rain_i$ means the *i*-th day of rainfall. The Pr_1, Pr_2, Pr_3, Pr_4 and Pr_5 represent the accumulated 1-, 2-, 3-, 4- and 5-day consecutive rainfall amounts (i.e., $n = 0$, $n = 1$, $n = 2$, $n = 3$, $n = 4$). The Thiessen polygon method is applied to compute the areal accumulated rainfall.

To select the extreme precipitation events most closely correlated to *AMF*, the Kendall's tau correlation coefficient was computed (Table 1). The Kendall's tau is a rank-based coefficient that is robust to departures from normality. It can be found from Table 1 that Pr_2 and Pr_3 were most closely correlated with *AMF* as gauged at Xianyang and Huaxian stations, respectively.

Table 1. Correlation coefficients between *AMF* and *Pr*. Bold numbers denote the extreme precipitation events most correlated with *AMF*.

Station	Correlation Coefficient				
	(AMF, Pr_1)	(AMF, Pr_2)	(AMF, Pr_3)	(AMF, Pr_4)	(AMF, Pr_5)
Linjiacun	0.1420	**0.4192 ****	0.3654 **	0.3582 **	0.3320 **
Huaxian	0.1586	0.2369 *	**0.3756 ****	0.3175 **	0.3320 **

Note: * and ** indicate that correlation coefficients are significant at the 95% and 99% confidence level, respectively.

3. Methodologies

3.1. Methodological Framework

As mentioned above, the aim of this paper is to disclose the bivariate hydrological risk and to reveal the impact of marginal distribution selection uncertainty and sampling uncertainty on hydrological risk analysis. To achieve this goal, this paper presents the following framework, as shown in Figure 2.

First, appropriate marginal distributions for *AMF* and *Pr* series were ascertained from the MEP distribution, Pearson type III distribution (P3), lognormal distribution (Logn), normal distribution (Norm) and gamma distribution (Gam). These parametric distributions are popular for characterizing the probability distributions of extreme hydrological events due to their better performance [6,44]. Second, we constructed copula models to depict the dependence structure of *AMF* and *Pr* series by joining their marginal distributions. Afterwards, the joint return periods, risk and reliability of *AMF* and *Pr* pairs were estimated for hydrological risk analysis. Last, the bivariate hydrological design events of specific joint return period were selected for hydraulic engineering design. To provide robust information for hydraulic structures design, we examined the impact of the marginal distribution selection uncertainty and sampling uncertainty on bivariate hydrological design event estimation. Here, the 6 candidate marginal distributions were combined with each other to form 36 combinations for modeling the *AMF* and *Pr* series. These combinations were utilized to explore the impact of marginal distribution uncertainty. Moreover, one Monte Carlo-based algorithm was designed to discover the impact of sampling uncertainty.

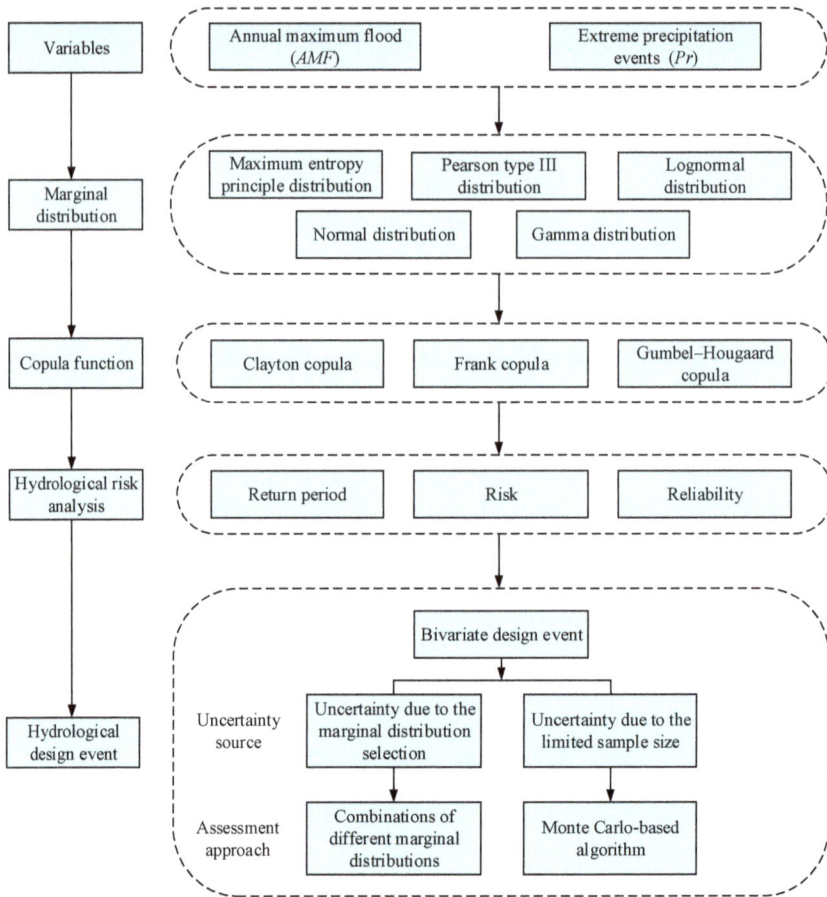

Figure 2. Flow chart of hydrological risk analysis under uncertainty.

3.2. Maximum Entropy Principle Distribution (MEP)

The concept of entropy was first formulated by Shannon (1948) [45]. After that, Jaynes (1957, 1982) developed the maximum entropy principle for deriving a least-biased probability distribution when certain information is given in terms of constraints [25,46]. To date, Shannon entropy has been extensively used in the hydrological field, such as rainfall-runoff modeling, drought analysis, and flow forecasting [29,30,47–49].

The Shannon entropy $H(x)$ of a probability density function (PDF) $f(x)$ for a continuous variable $X = \{x_1, x_2, \cdots, x_n\}$ can be defined as

$$H(x) = -\int_a^b f(x) \ln f(x) dx \tag{2}$$

where a and b denote the lower and upper limits of the variable X, respectively.

In order to attain the least biased probability distribution, the maximum entropy principle is performed by maximizing the entropy given by Equation (2) subject to the following constraints $C_r, r = 0, 1, 2, \cdots, n$

$$C_0 = \int_a^b f(x) dx = 1 \tag{3}$$

$$C_r = \int_a^b g_r(x)f(x)dx = \overline{g_r(x)}, \qquad r = 1, 2, \cdots, n \qquad (4)$$

where $g_r(x), r = 1, 2, \cdots, n$, denotes some known functions of x; n represents the number of constraints; $\overline{g_r(x)}$ means the expectation of $g_r(x)$. Equation (3) states that the probability density function must satisfy the total probability theorem. Here, the $g_r(x)$ can be expressed as the power of x such that

$$g_r(x) = x^n, \qquad r = 1, 2, \cdots, n \qquad (5)$$

In general, different numbers of constraints would obtain different performances [50]. In light of this, the present study employs 3 and 4 constraints to build the MEP-based distributions for *AMF* and *Pr* series, respectively. The corresponding distributions are denoted as MEP-3 and MEP-4, hereafter. The impact of the number of constraints in terms of MEP distribution on bivariate design event identification will be discussed in Section 4.

To achieve the maximization of entropy, the Lagrange multiplier method is one simple and frequently used method [51]. Subject to the constraints expressed in Equations (3) and (4), the Lagrangian function L can be written as

$$L = -\int_a^b f(x)\ln f(x)dx - (\lambda_0 - 1)\left[\int_a^b f(x)dx - C_0\right] - \sum_{r=1}^n \lambda_r\left[\int_a^b f(x)g_r(x)dx - C_r\right] \qquad (6)$$

where $\lambda_r(r = 1, 2, \cdots, n)$ denotes the Lagrange multipliers.

$f(x)$ can be attained through maximizing the function L, and therefore one differentiates L with respect to $f(x)$ being equal to zero:

$$\frac{\partial L}{\partial f} = 0 \Rightarrow -[1 + \ln f(x)] - (\lambda_0 - 1) - \sum_{r=1}^n \lambda_r g_r(x) = 0 \qquad (7)$$

Hence, the resulting maximum entropy-based PDF of a variable X in terms of the given constraints can be written by

$$f(x) = \exp\left[-\lambda_0 - \sum_{r=1}^n \lambda_r g_r(x)\right] \qquad (8)$$

Inserting Equation (8) into Equations (3) and (4), respectively, we can obtain

$$\exp(\lambda_0) = \int_a^b \exp\left[-\sum_{r=1}^n \lambda_r g_r(x)\right]dx, \qquad r = 1, 2, \cdots, n \qquad (9)$$

$$C_r = \int_a^b g_r(x)\exp\left[-\lambda_0 - \sum_{r=1}^n \lambda_r g_r(x)\right]dx, \qquad r = 1, 2, \cdots, n \qquad (10)$$

With the use of Equation (9), the zeroth Lagrange multiplier can be written as

$$\lambda_0 = \ln\int_a^b \exp\left[-\sum_{r=1}^n \lambda_r g_r(x)\right]dx, \qquad r = 1, 2, \cdots, n \qquad (11)$$

Substituting Equation (11) into Equation (8), $f(x)$ can be expressed as

$$f(x) = \exp\left[-\ln\int_a^b \exp\left[-\sum_{r=1}^n \lambda_r g_r(x)\right]dx - \sum_{r=1}^n \lambda_r g_r(x)\right] \qquad (12)$$

Here, the PDF expressed in Equation (12) can preserve the most important statistical moments. The cumulative distribution function (CDF) can be obtained through integration of Equation (12)

$$F(x) = \int_a^x \exp\left[-\ln\int_a^b \exp\left[-\sum_{r=1}^n \lambda_r g_r(x)\right]dx - \sum_{r=1}^n \lambda_r g_r(x)\right]dx \tag{13}$$

Based on the PDF function, the Lagrange multipliers λ_r can be estimated by minimizing the convex function, shown as

$$Z(\lambda) = \ln\left[\int_a^b \exp\left[-\sum_{r=1}^n \lambda_r g_r(x)\right]dx\right] + \sum_{r=1}^n \lambda_r g_r(x) \tag{14}$$

In the present study, the conjugate gradient method is employed to determine the Lagrange multipliers λ_r in Equation (14). The method is superior for solving large-scale nonlinear optimization problems. Its primary advantages are super-linear convergence, simple recurrence formula, and less calculation. Readers interested in the detailed process of determining the Lagrange multipliers λ_r through the conjugate gradient method are referred to the papers of Fan et al. (2016) [6].

3.3. Copula Function

The copula, as introduced by Sklar (1959) is a powerful tool for modeling the dependence structures of individual variables [52]. A d-dimensional copula is defined as a multivariate distribution function $F [0, 1]^d \rightarrow [0, 1]$, linking standard uniform marginal distributions. Formally, the copula can be divided into two components: individual univariate distributions, and a copula function describing dependence structures between variables based on the copula and its parameter(s).

According to Sklar's theorem, one d-dimensional multivariate distribution function F for random variables $X_1, X_2, \ldots, X_d$ with marginal distribution of $F_1, F_2, \ldots, F_d$ can be expressed as

$$F(X_1, \cdots, X_d) = C(F_1(X_1), \cdots, F_d(X_d); \Theta) \tag{15}$$

where Θ is the copula parameter vector and $F_d(X_d) = F(X \le x)$ is the marginal distribution of X_d.

In the field of hydrology, Archimedean copulas are quite popular due to their explicit functional forms. Moreover, they are superior for characterizing a wide range of dependence structures with several desirable properties. In the present study, the Clayton, Frank, and Gumbel-Hougaard copulas, which belong to the Archimedean class of copulas, were employed to evaluate the bivariate hydrological risk. The specific formula of these copulas was first reported by Nelsen (1999) [18].

3.4. Joint Return Periods, Risk and Reliability

To conduct a bivariate frequency analysis, the joint probability behaviors are defined in terms of variables X and Y, with thresholds x and y, respectively:

(1): {$X>x$} OR {$Y>y$}, (2): {$X>x$} AND {$Y>y$}

Accordingly, the joint return periods (also called primary return periods) can be written as [53–55]:

$$T_{OR} = \frac{\mu_T}{1 - C(F_X(x), F_Y(y))} \tag{16}$$

$$T_{AND} = \frac{\mu_T}{1 - F_X(x) - F_Y(y) + C(F_X(x), F_Y(y))} \tag{17}$$

Here, μ_T indicates the average inter-arrival time between two successive events ($\mu_T = 1$ for maximum annual events).

In engineering practice, risk (hereafter, *Risk*) and reliability (hereafter, *Reliability*) are another two important indices for hydraulic structure design [56]. To date, they have received widespread attention within the hydrological community [55–59].

Risk is defined as the probability of occurrence of at least one event exceeding the design event for the project life of n years:

$$Risk = 1 - (1-p)^n \tag{18}$$

where p indicates the annual exceedance probability.

It should be noted that the *Risk* also can be determined from the return period T (T_{OR} and T_{AND} in this study) by substituting $p = \frac{1}{T}$ into Equation (18):

$$Risk = 1 - \left(1 - \frac{1}{T}\right)^n \tag{19}$$

Reliability, which signifies the probability that a dangerous event will not occur within a project life of n years, i.e., that a system will remain in a satisfactory state within its lifetime, is defined as:

$$Reliability = 1 - Risk = (1-p)^n = \left(1 - \frac{1}{T}\right)^n \tag{20}$$

3.5. Bivariate Design Event Derived from Joint Distribution

Practical engineering design applications desire one appropriate multivariate design event or an appropriate subset of multiplets, instead of a large set of potential multiplets for the specific return period [34,60,61]. However, in a multivariate context, there exists a problem of inherent ambiguity, whereby various combinations of random variables X and Y share the same joint probability, and thus produce the same return period. Therefore, Salvadori et al. (2011) proposed a method for solving the ambiguity problem by identifying the most-likely design event [62]. The essence of the method is to identify one design event lying on critical layers $\mathcal{L}_t^F$ for a critical level t with the largest joint probability density. The most-likely design realization δ can be written as:

$$\delta = \operatorname{argmax} w(x,y) = \operatorname{argmax} f(x,y), (x,y) \in \mathcal{L}_t^F \tag{21}$$

Here, f indicates the density of $F(x,y) = C(F_X(x), F_Y(y))$, which can be expressed as

$$f(x,y) = \frac{\partial F(x,y)}{\partial F_X(x) \partial F_Y(y)} = f_X(x) f_Y(y) c(F_X(x), F_Y(y)) \tag{22}$$

where $f_X(x)$ and $f_Y(y)$ represent the probability density function of $F_X(x)$ and $F_Y(y)$, respectively; $c(F_X(x), F_Y(y))$ indicates the probability density function of $C(F_X(x), F_Y(y))$.

$\mathcal{L}_t^F$ is defined as:

$$\mathcal{L}_t^F = \{(x,y) : F(x,y) = t\} \tag{23}$$

where $t \in (0,1)$.

Then, the design event δ can be estimated by determining the largest joint probability density in the logarithmic domain on the critical layers $\mathcal{L}_t^F$, with the corresponding (x^*, y^*) as the design realization with the joint return period T (T_{OR} and T_{AND} in this study).

3.5.1. Uncertainty Due to the Marginal Distribution Selection

To assess the impact of marginal distribution selection uncertainty on the most-likely design event identification, we designed one experiment project by combining the 6 candidate distributions with each other to model the random variables X and Y. The 36 designed combinations are displayed in Figure 3. Then, the 36 fitted combinations $(F_X(x), F_Y(y))$ are carried into Equation (21), and thus the most-likely design events for the different combinations are obtained. Through comparison among these bivariate design events, the impact of marginal distribution uncertainty is expected to be discovered.

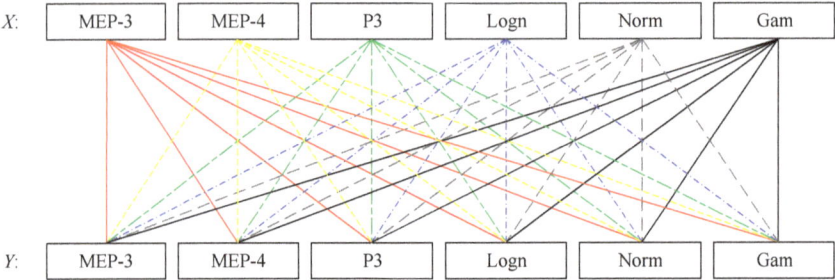

Figure 3. Experimental design for assessing the uncertainty of marginal distribution in bivariate design event estimation.

3.5.2. Sampling Uncertainty

To determine the impact of sampling uncertainty on the most-likely design event estimation, the following Monte Carlo-based procedures were designed:

1. Estimate the parameter Θ of the copula for the observations (i.e., X and Y) as well as the parameters α_x and α_y of the marginal distributions for X and Y, respectively;
2. Simulate B bivariate samples of size n on the basis of the copula parameter, and then apply the marginal backward transformations using the estimated parameters α_x and α_y. The simulated bivariate samples are denoted as $Z^* = (X^*, Y^*) = (x_{ij}, y_{ij})$, $(i = 1, \ldots, n; j = 1, \ldots, B)$. n is equal to the length of the observed sample. Here, B is set equal to 10,000;
3. Estimate the parameters Θ, α_x and α_y for the simulated sample Z_j^* using the same estimation method used for the observations;
4. Identify the most-likely design realization δ for different (x, y) pairs.
5. Estimate the confidence intervals for δ at 95% confidence level by the method of highest density regions (denoted as HDR) propose by Hyndman et al. (1996) [63].

4. Results and Discussion

4.1. Marginal Distribution Selection

To construct the copula model for (AMF, Pr) in the study regions, the first step was to select appropriate marginal distributions. Table 2 lists the relevant parameters for different marginal distributions for the AMF and Pr series in the study regions.

Table 2. Parameters of the fitted distribution for the *AMF* and *Pr* series in the catchments of Linjiacun and Huaxian stations.

Distribution	PDF	Parameter	Linjiacun		Huaxian	
			AMF	Pr_2	AMF	Pr_3
MEP-3	$f(x\|\lambda_1, \lambda_2, \lambda_3)$	λ_1	4.04×10^{-4}	-0.15	3.50×10^{-3}	-0.15
		λ_2	2.64×10^{-7}	4.65×10^{-3}	-1.14×10^{-6}	4.01×10^{-3}
		λ_3	-4.02×10^{-11}	-2.93×10^{-5}	1.19×10^{-10}	-2.25×10^{-8}
MEP-4	$f(x\|\lambda_1, \lambda_2, \lambda_3, \lambda_4)$	λ_1	7.63×10^{-4}	-0.23	-3.20×10^{-4}	0.07
		λ_2	1.57×10^{-6}	9.65×10^{-3}	1.03×10^{-6}	-6.73×10^{-3}
		λ_3	-5.40×10^{-10}	-1.48×10^{-4}	-3.73×10^{-10}	-1.87×10^{-4}
		λ_4	6.01×10^{-14}	9.13×10^{-7}	3.83×10^{-14}	-1.41×10^{-6}
P3	$f(x\|a, b, \alpha)$	a	1.23	13.19	0.99	21.65
		b	88.08	-22.69	866	-26.25
		α	880.82	3.56	1999.35	2.57
Logn	$f(x\|\mu, \sigma)$	μ	6.73	3.01	7.82	3.29
		σ	0.86	0.69	0.56	0.44
Norm	$f(x\|\mu, \sigma)$	μ	1170.32	24.52	2853.66	29.54
		σ	954.42	13.55	1413.64	12.7
Gam	$f(x\|\mu, \sigma)$	μ	1.65	2.82	3.71	5.53
		σ	707.28	8.68	770.06	5.34

Note: PDF of P3: $f(x\|a,b,\alpha) = \frac{1}{b^a\Gamma(a)}(x-\alpha)^{a-1}\exp(-\frac{x-\alpha}{b})$; PDF of Logn: $f(x\|\mu,\sigma) = \frac{1}{x\sigma\sqrt{2\pi}}\exp\left(-\frac{(\ln x-\mu)^2}{2\sigma^2}\right)$;

PDF of Norm: $f(x\|\mu,\sigma) = \frac{1}{\sigma\sqrt{2\pi}}\exp\left(-\frac{(x-\mu)^2}{2\sigma^2}\right)$; PDF of Gam: $f(x\|a,b) = \frac{1}{b^a\Gamma(a)}x^{a-1}\exp(\frac{-x}{b})$, where $\Gamma(\cdot)$ is a complete gamma function.

Figure 4 illustrates the distributions of the *AMF* and *Pr* series in the upper catchments of Linjiacun and Huaxian stations fitted by the MEP-3, MEP-4, P3, Logn, Norm and Gam distributions. The curves of CDF and PDF are exhibited in this figure.

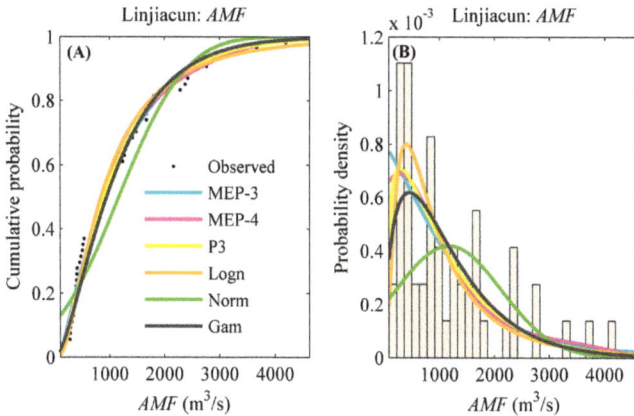

Figure 4. *Cont.*

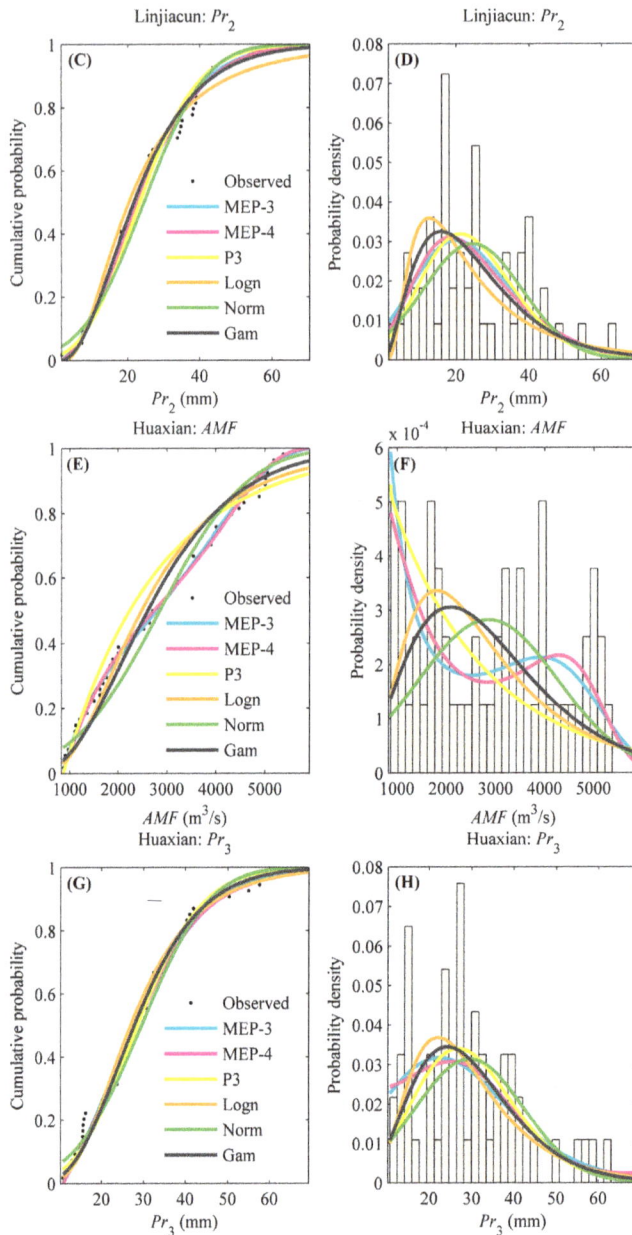

Figure 4. Frequency distributions of the *AMF* and *Pr* series in the upper catchments of Linjiacun and Huaxian stations. (**A,C,E,G**) denote the cumulative probabilities of the *AMF* and *Pr* series, while (**B,D,F,H**) denote the probability density of the *AMF* and *Pr* series.

It can be seen from this figure that the CDFs and PDFs exhibit variations in fitting performance between the theoretical and empirical distributions. In spite of this, it is difficult to select an appropriate fitting distribution by visual assessment. Therefore, the widely used root mean square error (RMSE) was employed to select the most appropriate model from among the candidate distribution models. The marginal distribution characterized by the minimum RMSE value was selected as the preferred model. Moreover, the goodness-of-fit test (the Kolmogorov-Smirnov (K-S) approach) was also performed to provide support in evaluating the validity of these distribution models. The K-S statistic (denoted as S) aims to quantify the largest vertical difference between the empirical and estimated distributions [64]. The p-value of the K-S statistic was obtained by using Miller's approximation. A value of p bigger than 0.05 indicates that the candidate distribution can appropriately fit random variables at the 5% significance level.

The RMSE values and K-S test results are presented in Table 3. It can be seen from Table 3 that the p values were much higher than the significance level 0.05, signifying that these candidate distributions are suitable for fitting the distributions of the *AMF* and *Pr* series. The RMSE values listed in Table 3 indicate that the MEP-4 distribution could be selected as the appropriate distribution for fitting the *AMF* and *Pr* series in the upper catchment of Linjiacun station, while MEP-3 and MEP-4 performed best among the candidates for fitting the *AMF* and *Pr* series, respectively, in the upper catchment of Huaxian station.

Table 3. The goodness-of-fit and RMSE values of the candidate distributions for the *AMF* and *Pr* series in the study regions.

| Station | Series | Functions | K-S Test | | RMSE | Series | Functions | K-S Test | | RMSE |
			S	p				S	p	
Linjiacun	AMF	MEP-3	0.09	0.72	0.0303	Pr_2	MEP-3	0.08	0.89	0.0293
		MEP-4	0.07	0.85	0.0296		MEP-4	0.07	0.95	0.0248
		P3	0.08	0.85	0.0309		P3	0.09	0.80	0.0373
		Logn	0.09	0.80	0.0366		Logn	0.10	0.58	0.0450
		Norm	0.15	0.19	0.0823		Norm	0.12	0.43	0.0473
		Gam	0.11	0.53	0.0405		Gam	0.08	0.87	0.0275
Huaxian	AMF	MEP-3	0.06	0.98	0.0195	Pr_3	MEP-3	0.09	0.73	0.0306
		MEP-4	0.06	0.97	0.0202		MEP-4	0.08	0.78	0.0267
		P3	0.14	0.25	0.0686		P3	0.10	0.65	0.0299
		Logn	0.10	0.59	0.0553		Logn	0.10	0.62	0.0390
		Norm	0.12	0.37	0.0534		Norm	0.09	0.76	0.0382
		Gam	0.09	0.74	0.0513		Gam	0.10	0.63	0.0300

A closer look at the fitting performance of these distributions for the *AMF* series at Linjiacun station, as presented in Figure 4 and Table 3, indicates that the fitting performance among the different distributions were similar, except for the Norm distribution. The RMSE values were 0.0303, 0.0296, 0.0309, 0.0366 and 0.0405 for the MEP-3, MEP-4, P3, Logn and Gam distributions, respectively, and 0.0823 for the Norm distribution. In terms of the Pr_2 series in the upper catchment of Linjiacun station, the Norm distribution had the worst performance of all the candidates.

As for the *AMF* series at the Huaxian station, the MEP-3 distribution outperformed other candidates (RMSE = 0.0195). The RMSE values for MEP-4, P3, Logn, Norm and Gam distributions ranged from 0.0202 to 0.0686. As shown in Figure 4, the histogram of the *AMF* series at Huaxian station has two peaks. MEP distribution can deal with multiple modes, while conventional models fail to fit a distribution with more than one mode. This is the reason that the conventional distributions show poor performance when fitting the *AMF* series at Huaxian station, while for the Pr_3 series in the upper catchment of Huaxian station, a comparison of RMSE values among the candidates indicates that the MEP-4 distribution (RMSE = 0.0267) performed better than other candidates (ranging from 0.0299 to 0.0390).

In light of the above findings, it can be concluded that MEP-related functions provide a better alternative than other conventional distributions for modeling the *AMF* and *Pr* series. Specifically, due to the PDF of the variables consisting of a single mode, the fitting performances of MEP-related distributions were similar to certain conventional distributions. The fitting performance of distributions for the *AMF* and Pr_2 series at Linjiacun station and the Pr_3 series at Huaxian station are able to demonstrate this inference. Moreover, it also can be inferred that the MEP-related distributions outperform the conventional distributions, particularly when modeling the distribution of random variables exhibiting multiple modes in the histogram, such as the *AMF* series at Huaxian station. These findings are similar to those obtained in Zhou et al. (2010) and Liu and Chang (2011) when fitting the distribution of wind speed data [65,66].

Additionally, we should also note that there exist some disadvantages to the MEP distribution; for example, the CDF of MEP-related distributions cannot be expressed in a closed form, parameter estimation is computationally expensive, and the mathematical expression of MEP-related distributions is more complex to develop as a computer program [6,66].

4.2. Copula Function Construction

Once the marginal distribution is chosen, the next step is to estimate the parameters of the copula and select an appropriate copula function from among the Clayton, Frank and Gumbel copulas. The parameters of the copulas are estimated using the maximum pseudo-likelihood method. The constructed MEP-related distributions are used to quantify the marginal probabilities of the *AMF* and *Pr* series. The estimated parameters of these candidate copulas are listed in Table 4.

Table 4. Parameters and goodness-of-fit test values for the candidate copulas.

Station	Copula	Parameter	Cramér–von Mises Test		AICc
			Sn	p-Value	
	Clayton	0.31	0.03	0.23	−10.83
Linjiacun	Frank	4.09	0.04	0.10	−19.54
	Gumbel	**1.59**	0.03	0.15	**−22.38**
	Clayton	0.42	0.03	0.45	−11.67
Huaxian	**Frank**	**3.72**	0.02	0.59	**−17.03**
	Gumbel	1.50	0.03	0.36	−16.22

To select the suitable copulas, the Cramér–von Mises test was used to test goodness-of-fit based on the empirical copula. Table 4 lists the goodness-of-fit statistics S_n corresponding to the Cramér–von Mises criteria, and their associated p-value based on N = 10,000 parametric bootstrap samples. A larger value of S_n indicates greater distance between the estimated and the empirical copulas. p-value > 0.05 means the estimated copula can be accepted at the 5% level. Results displayed in Table 4 illustrate that these candidate copulas are all acceptable for fitting the dependence structures between *AMF* and *Pr* in each region at a 5% significance level.

Further, the corrected Akaike information criterion (AICc) indicator is employed to select the copula with the highest fitting performance (shown in Table 4). The AICc indicator is much stricter than classical AIC, particularly when the size of the hydrological observations is limited [67]. The copula distribution characterized by the minimum AICc value was selected as the preferred model. It can be seen from Table 4 that the Gumbel and Frank copulas should be chosen to model the joint distribution of *AMF* and *Pr* series in the upper catchments of Linjiacun and Huaxian stations, respectively. For simplicity, the bivariate model constructed by integrating the MEP distributions into the copula is denoted as MEP-copula.

4.3. Bivariate Return Period, Risk and Reliability Analysis Based on the MEP-Copula

Exploring the concurrence probabilities of various combinations of *AMF* and *Pr* is of critical importance for practical flood control and disaster mitigation. As expressed as Equations (16) and (17), the bivariate return periods are estimated based on the constructed MEP-copula models. Figure 5 displays the bivariate joint return periods of "AND" and "OR" cases for various (*AMF, Pr*) pairs. It can be seen from Figure 5 that the joint return period level of the "AND" case is concave, while that for the "OR" case is convex. Generally, the joint return period of the "AND" case is lower than that of the "OR" case. For instance, if both the *AMF* and *Pr* observed in the upper catchment of Linjiacun station are in the 20-year return period, the "AND" joint return period of the (*AMF*, *Pr*) pair is 42.23 years, while the "OR" joint return period is 13.10 years. Additionally, in practical terms, water resource managers and policy-makers can identify the return periods for various (*AMF*, *Pr*) pairs of observations or forecasts through Figure 5.

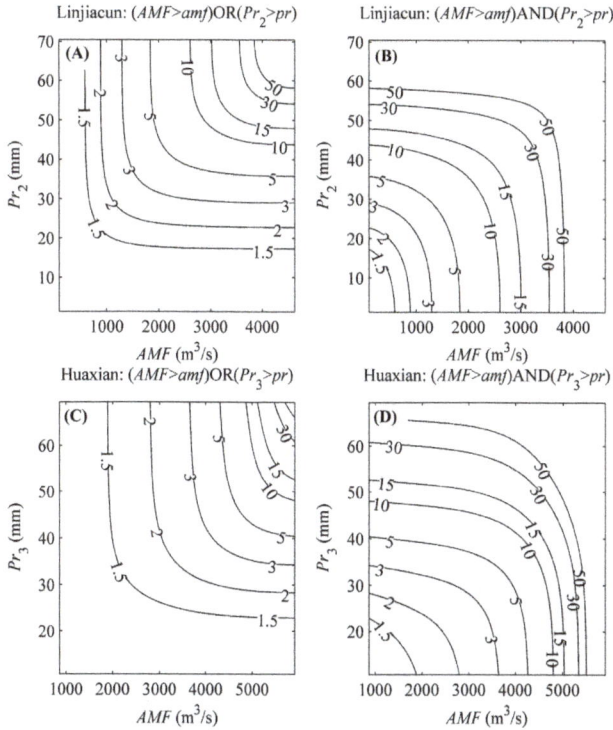

Figure 5. Bivariate return periods for (*AMF*, *Pr*) pairs in the upper catchments of Linjiacun and Huaxian stations. (**A,C**) denote the joint return period for the "OR" case, while (**B,D**) denote the joint return period for the "AND" case.

In engineering design of hydrological infrastructures, risk can be defined as the likelihood of experiencing at least one event exceeding the design event over the design life (denoted as *n*) of the hydraulic structure [6]. Furthermore, reliability over the project life is commonly chosen to describe the probability that the hydraulic structure will remain in a satisfactory state within its project life [60]. In the present study, the "AND" joint return period case is applied to define the bivariate risk and reliability through Equations (18)–(20). Here, the service time of a river levee is assumed to be 10 years, i.e., *n* = 10.

Figure 6 exhibits the bivariate risk and reliability under different designed *AMF* and *Pr* combinations in the two regions. It can be found from Figure 6 that the risk value would decrease as the designed *AMF* or *Pr* increased, which is contrary to the reliability value. In other words, higher values of the designed *AMF* or *Pr* for the hydraulic structure indicate a smaller probability for the occurrence of an undesirable flood event within a project life of *n* years, and a higher probability that the hydraulic structure will remain in a satisfactory state.

The implication for the bivariate risk and reliability of (*AMF*, *Pr*) pairs is to provide decision support for hydraulic structures and practical flood control. Decision-makers can attain the corresponding risk and reliability under various *AMF* and *Pr* scenarios as the design events for the hydraulic structure. For example, if the design event for a river levee in the downstream area of Huaxian station is set to (5000 m^3/s, 39.02 mm), the corresponding risk (reliability) that that event (*AMF* > 5000 m^3/s) AND (*Pr*$_3$ > 39.02 mm) will occur (not occur) during the river levee life of 10 years is 0.3331 (0.6669). Clearly, if decision-makers tend to decrease the risk or increase the reliability, the values of (*AMF*, *Pr*) pair should be improved.

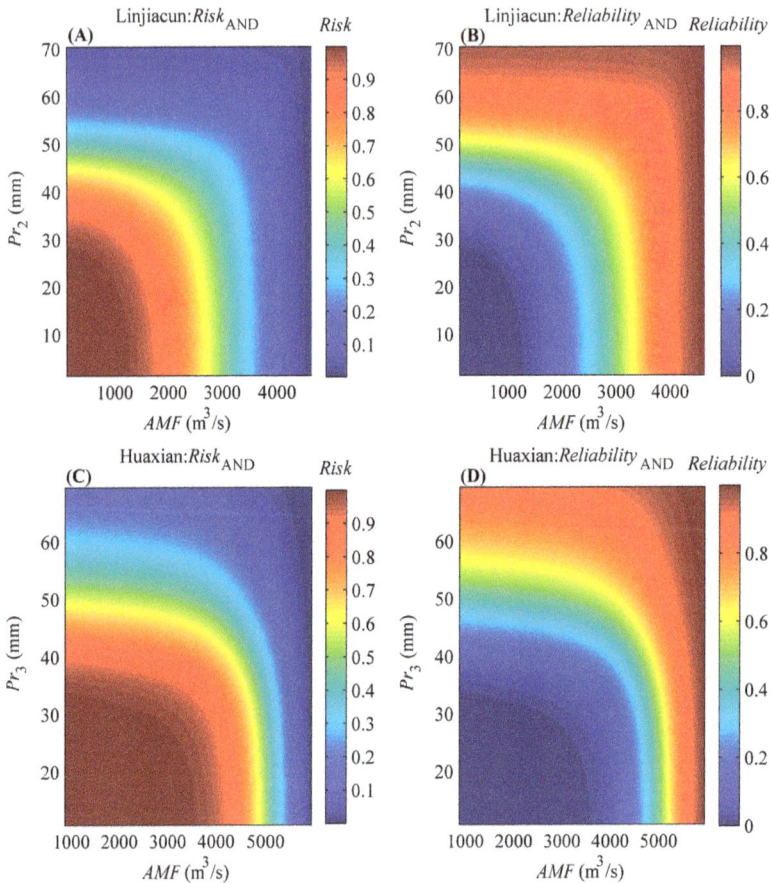

Figure 6. Bivariate risk and reliability under different designed (*AMF*, *Pr*) pairs when the hydraulic project life is given as 10 years. (**A,C**) denote the bivariate risk under different (*AMF*, *Pr*) pairs in the upper catchments of Linjiacun and Huaxian stations, respectively, while (**B,D**) denote the bivariate reliability under different (*AMF*, *Pr*) pairs for the two regions, respectively.

4.4. Bivariate Design Event Identification

In practical hydrological facility design, unique design realization on the isoline is obligatory and crucial. As is widely known, higher values of design event for hydraulic structures lead to poor economy, while lower values have a negative impact on the safety of flood control. However, as displayed in Section 4.3, for the specified return period, risk or reliability, there is a large set of potential

(*AMF*, *Pr*) events. Therefore, appropriate methods for identifying the unique design realization from the return level curve is of great necessity. In the present study, the most-likely design event method is utilized, as described previously in the literature [23,60,62]. The constructed MEP-copula is applied to estimate the most-likely design event. Table 5 lists the most-likely design events determined given $T_{OR} = T_{AND} = 30$ years using Equation (21) in the study regions. It can be seen from Table 5 that the most-likely design events for the upper catchments of Linjiacun and Huaxian stations are (3825.82 m³/s, 56.95 mm) and (5369.10 m³/s, 73.46 mm) given $T_{OR} = 30$ years, respectively, while given $T_{AND} = 30$ years, they are (3052.79 m³/s, 46.76 mm) and (5017.84 m³/s, 41.64 mm), respectively.

Table 5. Most-likely design events under different combinations of marginal distributions.

| Distribution | | Most-Likely Design Event (*AMF*, *Pr*) | | | |
| | | $T_{OR} = 30$ | | $T_{AND} = 30$ | |
AMF	*Pr*	Linjiacun	Huaxian	Linjiacun	Huaxian
MEP-3	MEP-3	(3921.34, 56.26)	(5626.03, 63.42)	(2877.37, 46.53)	(4976.51, 42.52)
	MEP-4	(3920.99, 56.72)	(5369.10, 73.46)	(2864.31, 48.00)	(5017.84, 41.64)
	P3	(3937.32, 57.06)	(5653.22, 55.68)	(2921.17, 46.92)	(4924.88, 41.07)
	Logn	(3930.55, 78.62)	(5678.83, 64.63)	(3065.06, 56.57)	(5028.83, 39.38)
	Norm	(3961.55, 51.29)	(5642.49, 54.98)	(2760.59, 44.54)	(4882.04, 42.83)
	Gam	(3925.95, 60.35)	(5660.16, 59.29)	(2960.32, 48.86)	(4966.30, 40.87)
MEP-4	MEP-3	(3833.86, 57.87)	(5492.61, 65.46)	(3105.70, 45.48)	(4969.43, 42.53)
	MEP-4	(3825.82, 56.95)	(5290.03, 73.46)	(3052.79, 46.76)	(4991.76, 41.49)
	P3	(3849.07, 52.66)	(5521.30, 56.98)	(3034.24, 42.96)	(4928.09, 40.94)
	Logn	(3834.21, 79.18)	(5539.27, 67.13)	(3204.83, 54.23)	(5010.76, 39.49)
	Norm	(3849.58, 51.53)	(5513.93, 56.04)	(2980.19, 43.39)	(4894.56, 42.64)
	Gam	(3830.33, 60.66)	(5526.09, 60.98)	(3126.29, 47.38)	(4960.69, 40.82)
P3	MEP-3	(3828.15, 57.44)	(8613.13, 67.16)	(2772.63, 47.11)	(4897.99, 49.50)
	MEP-4	(3842.96, 57.74)	(7635.10, 73.46)	(2750.37, 48.44)	(5372.34, 46.50)
	P3	(3860.18, 58.30)	(8866.89, 57.89)	(2802.71, 47.44)	(4591.65, 46.52)
	Logn	(3867.76, 81.16)	(8984.54, 68.75)	(2936.73, 57.73)	(5204.69, 46.98)
	Norm	(3868.93, 52.20)	(8820.36, 56.80)	(2654.70, 44.82)	(4400.06, 47.56)
	Gam	(3853.61, 61.71)	(8897.28, 62.14)	(2839.67, 49.46)	(4809.40, 47.38)
Logn	MEP-3	(4609.04, 57.71)	(7749.23, 67.64)	(2843.58, 48.19)	(4499.62, 49.81)
	MEP-4	(4636.79, 57.96)	(6941.66, 73.46)	(2807.61, 49.49)	(4890.92, 46.87)
	P3	(4665.66, 58.58)	(7989.66, 58.14)	(2898.54, 48.52)	(4289.21, 46.69)
	Logn	(4670.78, 81.59)	(8094.99, 69.23)	(3141.70, 60.07)	(4729.38, 47.43)
	Norm	(4681.88, 52.44)	(7948.81, 57.00)	(2618.77, 45.66)	(4155.63, 47.64)
	Gam	(4653.58, 61.99)	(8017.00, 62.46)	(2969.72, 50.66)	(4437.30, 47.64)
Norm	MEP-3	(3108.07, 57.79)	(5793.91, 66.33)	(2524.13, 44.13)	(4699.00, 44.79)
	MEP-4	(3119.85, 58.02)	(5446.16, 73.46)	(2497.94, 45.68)	(4782.48, 43.07)
	P3	(3130.51, 58.61)	(5861.15, 57.45)	(2534.21, 44.25)	(4606.08, 43.07)
	Logn	(3152.78, 82.35)	(5898.04, 67.97)	(2603.77, 47.62)	(4797.38, 41.64)
	Norm	(3124.46, 52.27)	(5846.46, 56.43)	(2481.91, 43.14)	(4535.18, 44.65)
	Gam	(3132.06, 62.18)	(5870.99, 61.58)	(2542.46, 45.41)	(4676.33, 43.10)
Gam	MEP-3	(3586.49, 57.56)	(6654.48, 66.89)	(2633.78, 46.68)	(4677.02, 47.24)
	MEP-4	(3601.58, 57.84)	(6096.07, 73.46)	(2604.74, 48.09)	(4883.48, 44.83)
	P3	(3617.00, 58.41)	(6785.93, 57.75)	(2660.38, 46.99)	(4515.48, 44.90)
	Logn	(3632.96, 81.57)	(6850.96, 68.51)	(2769.53, 56.15)	(4846.30, 44.36)
	Norm	(3620.10, 52.24)	(6760.41, 56.68)	(2540.92, 44.61)	(4401.43, 46.19)
	Gam	(3613.57, 61.86)	(6803.04, 61.96)	(2687.53, 48.87)	(4631.91, 45.35)

4.4.1. Uncertainty Due to Marginal Distribution Selection

In order to explore the impact of the uncertainty of marginal distribution selection on design event estimation, an extended experiment combining different marginal distributions was conducted. These selected marginal distributions all passed the goodness-of-fit test at the 5% significance level.

The copula functions that modeled the joint distributions of (*AMF*, *Pr*) pairs in the upper catchments of the Linjiacun and Huaxian stations were the Gumbel and Frank copulas, respectively. Figure 3 exhibits the analyzed combinations of different marginal distributions. Given $T_{OR} = T_{AND} = 30$ years, and the estimated most-likely design events under different combinations of marginal distributions are shown in Figure 7 and Table 5.

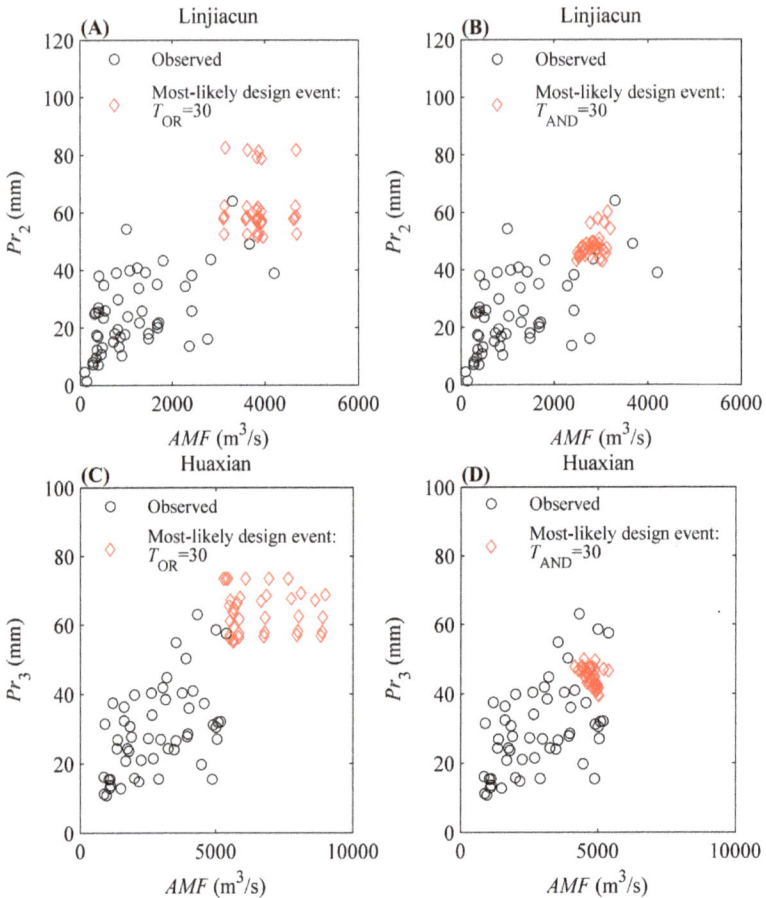

Figure 7. Most-likely design events under different combinations of marginal distributions. (**A,C**) denote the most likely design events under different combinations of marginal distributions when TOR = 30 years in the upper catchments of Linjiacun and Huaxian stations, respectively, while (**B,D**) denote the most likely design events when $T_{AND} = 30$ years in the two regions, respectively.

It can be seen from Figure 7 and Table 5 that there is a remarkable difference among these design events for the "OR" case under different combinations of marginal distributions. Among these estimated (*AMF*, *Pr*) pairs, the *AMF* value ranges from 3108.07 m³/s to 4681.88 m³/s in the upper catchment of Linjiacun station, while the *Pr* value ranges from 51.29 mm to 82.35 mm. The corresponding cumulative probability for the *AMF* series varies between 0.97 and 0.98, while that for the *Pr* series ranges between 0.97 and 0.99.

As for the "AND" case, it can be seen from Figure 7 that the difference among these design events estimated under different combinations of marginal distributions is smaller than that for the "OR" case. Take the upper catchment of Linjiacun station, for example; among these estimated (*AMF*, *Pr*) pairs, the *AMF* value ranges from 2481.91 m³/s to 3204.83 m³/s in the "AND" case, while the *Pr* value ranges from 42.96 mm to 60.07 mm. The corresponding cumulative probability of *AMF* values varies between 0.89 and 0.96, while that for the *Pr* value varies between 0.84 and 0.95.

To further explore the reasons for the difference among design events for the "OR" and "AND" cases, we exemplarily display part of CDF curves for *AMF* and *Pr* series for the upper catchment of Linjiacun station in Figure 8. Figure 8 illustrates that the difference of the fitting performance of different marginal distributions increases with the values of the variables. In other words, the smaller the value of *AMF/Pr* or the cumulative probability of *AMF/Pr* is, the smaller the difference in fitting performance among these distributions is. When $T_{OR} = T_{AND} = 30$ years, the corresponding cumulative probability of *AMF* in the "OR" case ([0.97–0.98]) is larger than that in the "AND" case ([0.89–0.96]), which is the same as that of *Pr*. Therefore, the difference among design events in the "AND" case is smaller than that in the "OR" case. This finding is consistent with that of Dung et al. (2015) [35].

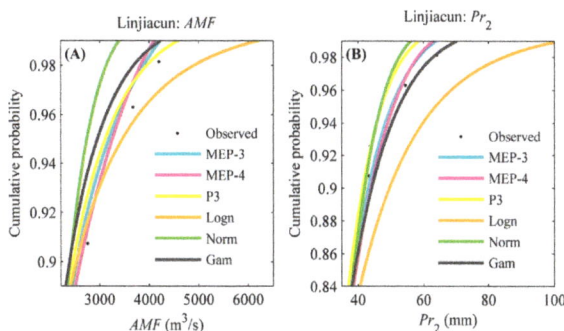

Figure 8. Part of CDF curves for *AMF* (**A**) and *Pr* (**B**) series in the upper catchment of Linjiacun station.

4.4.2. Uncertainty Due to the Limited Size of Hydrological Records

In order to uncover the impact of sampling uncertainty on the most-likely design event estimation, the designed Monte Carlo algorithm in Section 3.5.2 was utilized. Here, the *AMF* and *Pr* series were modeled by the MEP-related distributions. The Gumbel and Frank copulas were applied to describe the joint distribution of (*AMF*, *Pr*) pairs in the upper catchments of Linjiacun and Huaxian stations, respectively.

Moreover, Figure 9 clearly shows the confidence regions of the most-likely design events with $T_{OR} = T_{AND} = 30$ years using the HDR method. The highest density regions (95% and 99%) are exhibited in a two-dimensional plane (*AMF-Pr*) that correspond to a return period of 30 years. It can be seen from Figure 9 that the confidence regions of the most-likely design events for $T_{OR} = T_{AND} = 30$ years are very large in the study regions. The 95% confidence region for the most-likely design events in terms of *AMF* and *Pr* with a return period of 30 years could range between values for *AMF* and *Pr* with return periods of 10 and 50 years, at least. Take the upper catchment of Linjiacun station, for example; in the 95% region for $T_{OR} = 30$ years, the *AMF* could assume values with a univariate return period from 10 to 50 years. Large uncertainties due to sampling uncertainty can also be found in the works of [34,35,68,69], i.e., the return periods of the most-likely design events overlap. These large uncertainties present a significant challenge for reservoir design, flood risk, etc., particularly for Guanzhong plain, as one of the most important Chinese agricultural production regions, and Xi'an city, as one of the four major ancient capitals of civilization, which lie in the floodplain between Linjiacun and Huaxian stations. Uncertainty of copula-based frequency analysis for the study regions should arouse critical concern, particularly when constituting policy for flood control and hazard reduction.

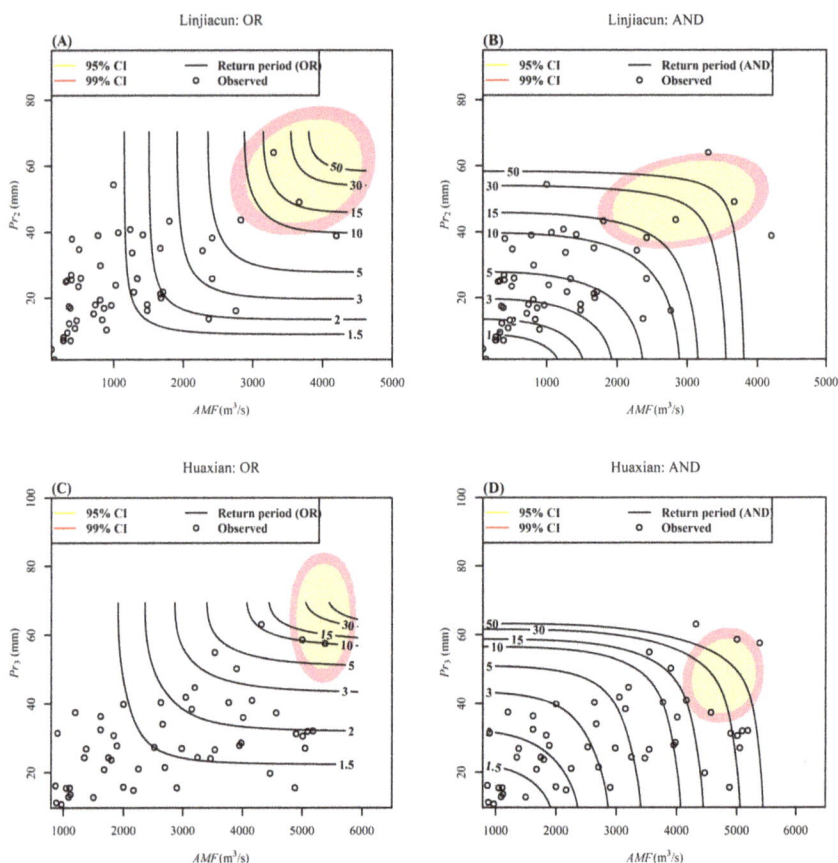

Figure 9. The most-likely design events with the "OR" (**A,C**) and "AND" (**B,D**) joint period of 30 years. Black lines denote the "OR" and "AND" joint periods estimated with the original data series. The shaded area denotes the confidence regions of the most-likely design events with $T_{OR} = T_{AND} = 30$ years at 95% and 99% confidence interval.

5. Summary and Conclusions

In this study, one general framework, aiming to analyze bivariate hydrological risk through a coupled maximum entropy-copula method and to reveal the impact of marginal distribution selection uncertainty and sampling uncertainty on hydrological risk analysis, is proposed.

The framework excels previous studies in applying the maximum entropy principle-based marginal distribution for modeling random variables and accounting for the impact of different uncertainty sources on hydrological risk analysis. The joint return periods, risk reliability, and bivariate design events are derived based on the coupled maximum entropy-copula method. For the purpose of practical engineering design applications, the so-called most-likely design event is identified to characterize the bivariate design event. To reveal the impact of marginal distribution selection uncertainty and sampling uncertainty on the bivariate design event identification, we designed a corresponding experiment project and specific Monte Carlo-based algorithm to achieve the two goals, respectively. Here, to elucidate the impact of marginal distribution uncertainty on the bivariate design event identification, 6 candidate distributions were combined with each other to produce 36 combinations for fitting univariate flood and extreme precipitation series. Then, these combinations concerning the marginal distributions of flood and extreme precipitation events were utilized to derive the bivariate design event. For the second goal, the Monte Carlo-based algorithm was designed to disclose the impact of sampling uncertainty on the bivariate design event identification.

Two sub-catchments of Loess Plateau, which were typical eco-environmentally vulnerable regions, were selected as the study regions. The primary conclusions are drawn as follows:

(1) The maximum entropy principle (MEP)-based distributions outperform the conventional distributions (i.e., P3, Logn, Norm and Gam at least in this study) in quantifying the probability of flood and extreme precipitation events. Results of this study indicate the better performance of MEP distribution, suggesting that it could be an attractive alternative for quantifying the marginal probability of random variables.

(2) The Gumbel and Frank copulas were suitable dependence models for quantifying the joint probabilities of flood and extreme precipitation events in the upper catchments of Linjiacun and Huaxian stations, respectively.

(3) The bivariate return periods, risk and reliability of flood and extreme precipitation events for the two study regions were calculated based on the coupled maximum entropy-copula models, which were expected to provide practical support for the local flood control and disaster mitigation.

(4) The bivariate design realizations were estimated for the study regions. Comprehensive uncertainty analysis revealed that the fitting performance of univariate distribution is closely related to the bivariate design event identification. If the difference of the fitting performance between two marginal distributions is small, values of the bivariate design events are similar, and vice versa. Therefore, advanced univariate distribution is critical for the bivariate design event selection.

Most importantly, the uncertainty related to the limited sample size is considerable, and should arouse critical attention. The bivariate design events of a specific return period exhibit significant variation. In other words, the return periods of the most-likely design events overlap. The 95% confidence regions of bivariate design events for flood and extreme precipitation with a return period of 30 years could reach between the values for flood and extreme precipitation with return periods of 10 and 50 years. The overlap phenomenon poses great challenges for practical engineering design applications, flood control, and so on. To enable a more reliable estimation of the design realization, increasing the information content by expanding the temporal, spatial or causal data is desirable, as proposed by Merz and Blöschl (2008) [70].

Acknowledgments: This work was supported by the National Key Research and Development Program of China (2016YFC0400906), National Natural Science Foundation of China (51679187, 51679189), National Key R & D Program of China (2017YFC0405900), Innovation Fund for doctoral dissertation of Xi'an University of Technology (310-252071606, 310-252071605), and the China Scholarship Council (CSC). Sincere gratitude is extended to the editor and the anonymous reviewers for their professional comments and corrections.

Author Contributions: Jianxia Chang and Aijun Guo designed the experiment; Aijun Guo performed the experiment and wrote the draft of the paper; Jianxia Chang, Yimin Wang and Qiang Huang revised the paper; Zhihui Guo drew figures. All authors have read and approved the final manuscript.

Conflicts of Interest: The authors declare no conflict of interest.

References

1. Nie, C.; Li, H.; Yang, L.; Wu, S.; Liu, Y.; Liao, Y. Spatial and temporal changes in flooding and the affecting factors in China. *Nat. Hazards* **2012**, *61*, 425–439. [CrossRef]
2. Merz, B.; Nguyen, V.D.; Vorogushyn, S. Temporal clustering of floods in Germany: Do flood-rich and flood-poor periods exist? *J. Hydrol.* **2016**, *541B*, 824–883. [CrossRef]
3. Swierczynski, T.; Ionita, M.; Pino González, D. Using archives of past floods to estimate future flood hazards. *EOS Trans.* **2017**, *98*. [CrossRef]
4. UNISDR. The Human Cost of Weather-Related Disasters 1995–2015. 2015. Available online: http://www.unisdr.org/archive/46793 (accessed on 11 November 2017).

5. Svetlana, D.; Radovan, D.; Ján, D. The economic impact of floods and their importance in different regions of the world with emphasis on Europe. *Procedia Econ. Financ.* **2015**, *34*, 649–655. [CrossRef]

6. Fan, Y.R.; Huang, W.W.; Huang, G.H.; Huang, K.; Li, Y.P.; Kong, X.M. Bivariate hydrologic risk analysis based on a coupled entropy-copula method for the Xiangxi river in the three Gorges Reservoir area, China. *Theor. Appl. Climatol.* **2016**, *125*, 381–397. [CrossRef]

7. Li, F.; Zheng, Q. Probabilistic modelling of flood events using the entropy copula. *Adv. Water Res.* **2016**, *97*, 233–240. [CrossRef]

8. Blöschl, G.; Gaál, L.; Hall, J.; Kiss, A.; Komma, J.; Nester, T.; Parajka, J.; Perdigão, R.A.P.; Plavcová, L.; Rogger, M.; et al. Increasing river floods: Fiction or reality? *WIREs Water* **2015**, *2*, 329–344. [CrossRef] [PubMed]

9. Machado, M.J.; Botero, B.A.; López, J.; Francés, F.; Díez-Herrero, A.; Benito, G. Flood frequency analysis of historical flood data under stationary and non-stationary modelling. *Hydrol. Earth Syst. Sci. Discuss.* **2015**, *12*, 525–568. [CrossRef]

10. Kiem, A.S.; Verdon-Kidd, D.C. The importance of understanding drivers of hydroclimatic variability for robust flood risk planning in the coastal zone. *Australas. J. Water Res.* **2013**, *17*, 126–134. [CrossRef]

11. Madsen, H.; Lawrence, D.; Lang, M.; Martinkova, M.; Kjeldsen, T.R. Review of trend analysis and climate change projections of extreme precipitation and floods in Europe. *J. Hydrol.* **2014**, *519*, 3634–3650. [CrossRef]

12. Liu, D.; Wang, D.; Singh, V.P.; Wang, Y.; Wu, J.; Wang, L.; Zou, X.; Chen, Y.; Chen, X. Optimal moment determination in POME-copula based hydrometeorological dependence modelling. *Adv. Water Res.* **2017**, *105*, 39–50. [CrossRef]

13. Zhang, L.; Singh, V.P. Bivariate flood frequency analysis using the copula method. *J. Hydrol. Eng.* **2006**, *11*, 150–164. [CrossRef]

14. Karmakar, S.; Simonovic, S.P. Bivariate flood frequency analysis. Part 2: A copula-based approach with mixed marginal distributions. *J. Flood Risk Manag.* **2009**, *2*, 32–44. [CrossRef]

15. Ozga-Zielinski, B.; Ciupak, M.; Adamowski, J.; Khalil, B.; Malard, J. Snow-melt flood frequency analysis by means of copula based 2D probability distributions for the Narew river in Poland. *J. Hydrol. Reg. Stud.* **2016**, *6*, 26–51. [CrossRef]

16. Chen, L.; Singh, V.P.; Guo, S.; Mishra, A.K.; Guo, J. Drought analysis using copulas. *J. Hydrol. Eng.* **2013**, *18*, 797–808. [CrossRef]

17. She, D.; Mishra, A.K.; Xia, J.; Zhang, L.; Zhang, X. Wet and dry spell analysis using copulas. *Int. J. Climatol.* **2016**, *36*, 476–491. [CrossRef]

18. Nelsen, R.B. *An Introduction to Copulas*, 2nd ed.; Springer: Berlin, Germany, 2006.

19. Khedun, C.P.; Mishra, A.K.; Singh, V.P.; Giardino, J.R. A copula-based precipitation forecasting model: Investigating the interdecadal modulation of ENSO's impacts on monthly precipitation. *Water Resour. Res.* **2014**, *50*, 580–600. [CrossRef]

20. Fan, Y.R.; Huang, G.H.; Baetz, B.W.; Li, Y.P.; Huang, K. Development of copula-based particle filter (CopPF) approach for hydrologic data assimilation under consideration of parameter interdependence. *Water Resour. Res.* **2017**, *53*, 4850–4875. [CrossRef]

21. Laux, P.; Vogl, S.; Qiu, W.; Knoche, H.R.; Kunstmann, H. Copula-based statistical refinement of precipitation in RCM simulations over complex terrain. *Hydrol. Earth Syst. Sci.* **2011**, *15*, 2401–2419. [CrossRef]

22. Bobee, B.; Cavidas, G.; Ashkar, F.; Bernier, J.; Rasmussen, P. Towards a systematic approach to comparing distributions used in flood frequency analysis. *J. Hydrol.* **1993**, *142*, 121–136. [CrossRef]

23. Volpi, E.; Fiori, A. Design event selection in bivariate hydrological frequency analysis. *Hydrol. Sci. J.* **2012**, *57*, 1506–1515. [CrossRef]

24. Zhang, L.; Singh, V.P. Bivariate rainfall and runoff analysis using entropy and copula theories. *Entropy* **2012**, *14*, 1784–1812. [CrossRef]

25. Jaynes, E.T. Information theory and statistical mechanics. *Phys. Rev.* **1957**, *106*, 620. [CrossRef]

26. AghaKouchak, A. Entropy—Copula in hydrology and climatology. *J. Hydrometeorol.* **2014**, *15*, 2176–2189. [CrossRef]

27. Zhao, N.; Lin, W.T. A copula entropy approach to correlation measurement at the country level. *Appl. Math. Comput.* **2011**, *218*, 628–642. [CrossRef]

28. Dong, S.; Wang, N.; Liu, W.; Soares, C.G. Bivariate maximum entropy distribution of significant wave height and peak period. *Ocean Eng.* **2013**, *59*, 86–99. [CrossRef]

29. Chen, L.; Singh, V.P.; Xiong, F. An Entropy-based generalized gamma distribution for flood frequency analysis. *Entropy* **2017**, *19*, 239. [CrossRef]
30. Hao, Z.; Singh, V.P. Integrating entropy and copula theories for hydrologic modeling and analysis. *Entropy* **2015**, *17*, 2253–2280. [CrossRef]
31. Mishra, A.K.; Özger, M.; Singh, V.P. An entropy-based investigation into the variability of precipitation. *J. Hydrol.* **2009**, *370*, 139–154. [CrossRef]
32. Rajsekhar, D.; Mishra, A.K.; Singh, V.P. Regionalization of drought characteristics using an entropy approach. *J. Hydrol. Eng.* **2013**, *18*, 870–887. [CrossRef]
33. Michailidi, E.M.; Bacchi, B. Dealing with uncertainty in the probability of overtopping of a flood mitigation dam. *Hydrol. Earth Syst. Sci.* **2017**, *21*, 2497–2507. [CrossRef]
34. Serinaldi, F. An uncertain journey around the tails of multivariate hydrological distributions. *Water Resour. Res.* **2013**, *49*, 6527–6547. [CrossRef]
35. Dung, N.V.; Merz, B.; Bardossy, A.; Apel, H. Handling uncertainty in bivariate quantile estimation—An application to flood hazard analysis in the Mekong Delta. *J. Hydrol.* **2015**, *527*, 704–717. [CrossRef]
36. Serinaldi, F.; Kilsby, C.G. Stationarity is undead: Uncertainty dominates the distribution of extremes. *Adv. Water Resour.* **2015**, *77*, 17–36. [CrossRef]
37. Zhang, X.; Zhang, L.; Zhao, J.; Rustomji, P.; Hairsine, P. Responses of streamflow to changes in climate and land use/cover in the Loess Plateau, China. *Water Resour. Res.* **2008**, *44*. [CrossRef]
38. Shi, H.; Shao, M. Soil and water loss from the Loess Plateau in China. *J. Arid Environ.* **2000**, *45*, 9–20. [CrossRef]
39. Du, T.; Xiong, L.; Xu, C.Y.; Gippel, C.J.; Guo, S.; Liu, P. Return period and risk analysis of nonstationary low-flow series under climate change. *J. Hydrol.* **2015**, *527*, 234–250. [CrossRef]
40. Ma, M.; Song, S.; Ren, L.; Jiang, S.; Song, J. Multivariate drought characteristics using trivariate Gaussian and student copula. *Hydrol. Process.* **2013**, *27*, 1175–1190. [CrossRef]
41. Peng, H.; Jia, Y.W.; Tague, C.; Slaughter, P. An eco-hydrological model-based assessment of the impacts of soil and water conservation management in the Jinghe river basin, China. *Water* **2015**, *7*, 6301–6320. [CrossRef]
42. Guo, A.; Chang, J.; Huang, Q.; Wang, Y.; Liu, D.; Li, Y.; Tian, T. Hybrid method for assessing the multi-scale periodic characteristics of the precipitation—Runoff relationship: A case study in the Weihe river basin, China. *J. Water Clim. Chang.* **2017**, *8*, 62–77. [CrossRef]
43. Teegavarapu, R.S.V. *Floods in Changing Climate*; Cambridge University Press: New York, NY, USA, 2012.
44. Kamal, V.; Mukherjee, S.; Singh, P.; Sen, R.; Vishwakarma, C.A.; Sajadi, P.; Asthana, H.; Rena, V. Flood frequency analysis of Ganga river at Haridwar and Garhmukteshwar. *Appl. Water Sci.* **2017**, *7*, 1979–1986. [CrossRef]
45. Shannon, C.E. A mathematical theory of communications. *Bell Syst. Tech. J.* **1948**, *27*, 379–423. [CrossRef]
46. Jaynes, E.T. On the rationale of maximum-entropy methods. *Proc. IEEE* **1982**, *70*, 939–952. [CrossRef]
47. Cui, H.; Singh, V.P. Maximum entropy spectral analysis for streamflow forecasting. *Phys. A Stat. Mech. Appl.* **2016**, *442*, 91–99. [CrossRef]
48. Li, C.; Singh, V.P.; Mishra, A.K. Entropy theory-based criterion for hydrometric network evaluation and design: Maximum information minimum redundancy. *Water Resour. Res.* **2012**, *48*. [CrossRef]
49. Mishra, A.K.; Ines, A.V.M.; Singh, V.P.; Hansen, J.W. Extraction of information content from stochastic disaggregation and bias corrected downscaled precipitation variables for crop simulation. *Stoch. Environ. Res. Risk Access.* **2013**, *27*, 449–457. [CrossRef]
50. Singh, V.P. Hydrologic synthesis using entropy theory. *J. Hydrol. Eng.* **2011**, *16*, 421–433. [CrossRef]
51. Singh, V.P.; Rajagopal, A.K. A new method of parameter estimation for hydrologic frequency analysis. *Hydrol. Sci. Technol.* **1987**, *2*, 33–40.
52. Sklar, A. Functions de Repartition à n Dimensions et Luers Marges. *Publ. Inst. Stat. Univ. Paris* **1959**, *8*, 229–231.
53. Salvadori, G.; De Michele, C.; Kottegoda, N.T.; Rosso, R. *Extremes in Nature: An Approach Using Copulas*; Springer: New York, NY, USA, 2007.
54. Vandenberghe, S.; Verhoest, N.E.C.; Onof, C.; De Baets, B. A comparative copula-based bivariate frequency analysis of observed and simulated storm events: A case study on Bartlett-Lewis modeled rainfall. *Water Resour. Res.* **2011**, *47*. [CrossRef]

55. Fan, Y.R.; Huang, W.; Huang, G.H.; Li, Y.P.; Huang, K. Hydrologic risk analysis in the Yangtze river basin through coupling Gaussian mixtures into copulas. *Adv. Water Resour.* **2016**, *88*, 170–185. [CrossRef]

56. Salas, J.D.; Obeysekera, J. Revisiting the concepts of return period and risk for nonstationary hydrologic extreme events. *J. Hydrol. Eng.* **2013**, *19*, 554–568. [CrossRef]

57. Mood, A.; Graybill, F.; Boes, D.C. *Introduction to the Theory of Statistics*, 3rd ed.; McGraw-Hill: New York, NY, USA, 1974.

58. Tung, Y.K. *Risk/Reliability-Based Hydraulic Engineering Design in Hydraulic Design Handbook*; Mays, L., Ed.; McGraw-Hill: New York, NY, USA, 1999.

59. Read, L.K.; Vogel, R.M. Reliability, return periods, and risk under nonstationarity. *Water Resour. Res.* **2015**, *51*, 6381–6398. [CrossRef]

60. Corbella, S.; Stretch, D.D. Predicting coastal erosion trends using non-stationary statistics and process-based models. *Coast. Eng.* **2012**, *70*, 40–49. [CrossRef]

61. Brunner, M.I.; Seibert, J.; Favre, A.C. Bivariate return periods and their importance for flood peak and volume estimation. *WIRs Water* **2016**, *3*, 819–833. [CrossRef]

62. Salvadori, G.; De Michele, C.; Durante, F. On the return period and design in a multivariate framework. *Hydrol. Earth Syst. Sci.* **2011**, *15*, 3293–3305. [CrossRef]

63. Hyndman, R.J.; Bashtannyk, D.M.; Grunwald, G.K. Estimating and visualizing conditional densities. *J. Comput. Graph. Stat.* **1996**, *5*, 315–336.

64. Massey, J.F. The Kolmogorov-Smirnov test for goodness of fit. *J. Am. Stat. Assoc.* **1951**, *46*, 68–78. [CrossRef]

65. Zhou, J.; Erdem, E.; Li, G.; Shi, J. Comprehensive evaluation of wind speed distribution models: A case study for North Dakota sites. *Energy Convers. Manag.* **2010**, *51*, 1449–1458. [CrossRef]

66. Liu, F.J.; Chang, T.P. Validity analysis of maximum entropy distribution based on different moment constraints for wind energy assessment. *Energy* **2011**, *36*, 1820–1826. [CrossRef]

67. Burnham, K.P.; Anderson, D.R. Multimodel inference: Understanding AIC and BIC in model selection. *Sociol. Methods Res.* **2004**, *33*, 261–304. [CrossRef]

68. Serinaldi, F. Can we tell more than we can know? The limits of bivariate drought analyses in the United States. *Stoch. Environ. Res. Risk A* **2016**, *30*, 1691–1704. [CrossRef]

69. Zhang, Q.; Xiao, M.; Singh, V.P. Uncertainty evaluation of copula analysis of hydrological droughts in the East River basin, China. *Glob. Planet Chang.* **2015**, *129*, 1–9. [CrossRef]

70. Merz, R.; Blöschl, G. Flood frequency hydrology: 1. Temporal, spatial, and causal expansion of information. *Water Resour. Res.* **2008**, *44*. [CrossRef]

![entropy logo] *entropy*

MDPI

Article

Application of Entropy Ensemble Filter in Neural Network Forecasts of Tropical Pacific Sea Surface Temperatures

Hossein Foroozand [1], Valentina Radić [2] and Steven V. Weijs [1],*

[1]   Department of Civil Engineering, University of British Columbia, Vancouver, BC V6T 1Z4, Canada;
      hosseinforoozand@civil.ubc.ca
[2]   Department of Earth, Ocean and Atmospheric Sciences, University of British Columbia, Vancouver,
      BC V6T 1Z4, Canada; vradic@eoas.ubc.ca
*   Correspondence: steven.weijs@civil.ubc.ca; Tel.: +1-(604)-822-6301

Received: 1 February 2018; Accepted: 15 March 2018; Published: 20 March 2018

Abstract: Recently, the Entropy Ensemble Filter (EEF) method was proposed to mitigate the computational cost of the Bootstrap AGGregatING (bagging) method. This method uses the most informative training data sets in the model ensemble rather than all ensemble members created by the conventional bagging. In this study, we evaluate, for the first time, the application of the EEF method in Neural Network (NN) modeling of El Nino-southern oscillation. Specifically, we forecast the first five principal components (PCs) of sea surface temperature monthly anomaly fields over tropical Pacific, at different lead times (from 3 to 15 months, with a three-month increment) for the period 1979–2017. We apply the EEF method in a multiple-linear regression (MLR) model and two NN models, one using Bayesian regularization and one Levenberg-Marquardt algorithm for training, and evaluate their performance and computational efficiency relative to the same models with conventional bagging. All models perform equally well at the lead time of 3 and 6 months, while at higher lead times, the MLR model's skill deteriorates faster than the nonlinear models. The neural network models with both bagging methods produce equally successful forecasts with the same computational efficiency. It remains to be shown whether this finding is sensitive to the dataset size.

Keywords: entropy ensemble filter; ensemble model simulation criterion; EEF method; bootstrap aggregating; bagging; bootstrap neural networks; El Niño; ENSO; neural network forecast; sea surface temperature; tropical Pacific

---

1. Introduction

Most data-mining algorithms require proper training procedures [1–14] to learn from data. The Bootstrap AGGregatING (bagging) method is a commonly used tool in the machine learning methods to increase predictive accuracy. Despite its common application, the bagging method is considered to be computationally expensive, particularly when used to create new training data sets out of large volumes of observations [15–17]. To improve the computational efficiency, Wan et al. [15] proposed a hybrid artificial neural network (HANN), while Kasiviswanathan et al. [17] combined the bagging method with the first order uncertainty analysis (FOUA). The combined method reduced the computational time of simulation for uncertainty analysis with limited statistical parameters such as mean and variance of the neural network weight vectors and biases. Wang et al. [16] showed that sub-bagging (SUBsample AGGregatING) gives similar accuracy but is computationally more efficient than bagging. This advantage is highlighted for Gaussian process regression (GPR) since its computational time increases in cubic order with the increase of data. Yu and Chen [18] compared different machine learning techniques and found that the fully Bayesian regularized artificial neural network (BANN) methods are much more time consuming than support vector machine (SVM) and maximum likelihood estimation (MLE)-based Gaussian process (GP) model. Table 1 provides a brief overview of studies that applied the bagging method in a range of different machine learning algorithms.

The Entropy Ensemble Filter (EEF) method, as a modified bagging procedure, has been proposed recently by Foroozand and Weijs [19]. The EEF method uses the most informative training data sets in the ensemble rather than all ensemble members created by the conventional bagging method. The EEF method achieved a reduction of the computational time of simulation by around 50% on average for synthetic data simulation, showing a potential for its application in computationally demanding environmental prediction problems.

In this paper, the first application of the EEF method on real-world data simulation is presented. We test its application in forecasting the tropical Pacific sea surface temperatures (SST) anomalies based on the initially proposed neural-network model of Wu et al. [20]. We chose this particular application due to the numerous studies of the El Nino-southern oscillation (ENSO) phenomenon and its use for water resources management. The ENSO is the strongest climate fluctuation on time scales ranging from a few months to several years and is characterized by inter-annual variations of the tropical Pacific sea surface temperatures, with warm episodes called El Niño, and cold episodes, La Niña. As ENSO affects not only the tropical climate but also the extra-tropical climate [21,22], the successful prediction of ENSO is of great importance. One example is the Pacific Northwest of North America, where water management operations depend on the accuracy of seasonal ENSO forecasts. For the Columbia River hydropower system, the use of ENSO information, in combination with adapted operating policies, could lead to an increase of $153 million in expected annual revenue [23]. Successful long-term forecasts of ENSO indices themselves could increase forecast lead-times, potentially further increasing benefits from hydropower operations. Vu et al. [24] recently argued that information entropy suggests stronger nonlinear links between local hydro-meteorological variables and ENSO, which could further strengthen its predictive power. Also, recent drought in coastal British Columbia, Canada, has increased the need for reliable seasonal forecasts to aid water managers in, for example, anticipating drinking water supply issues.

Since the early 1980s, much effort has been allocated to forecasting the tropical Pacific SST anomalies with the use of dynamical, statistical and hybrid models [21,25,26]. Because of ENSO's nonlinear features [20,21,27–30], many studies applied nonlinear statistical models such as a neural network (NN) model. Detailed comparisons between linear and nonlinear models in ENSO forecasts have been conducted in [2,20,30]. Wu et al. [20] developed a multi-layer perceptron (MLP) NN approach, where sea level pressure (SLP) field and SST anomalies over Tropical Pacific were used to predict the five leading SST principal components at lead times from 3 to 15 months. The performance of the MLP model, when compared to the multiple-linear regression (MLR) models, showed higher correlation skills and lower root mean square errors over most Nino domains. In this study, we incorporate the EEF method in both MLP and MLR models and evaluate their performance and computational efficiency relative to the original models. In addition to the original MLP model that uses Bayesian neural network (BNN), henceforth labeled as BNN model, we also test the MLP model that applies a cross-validation with Levenberg-Marquardt optimization algorithm [31–33], henceforth labeled as NN model. The main difference between BNN and NN models is their procedure to prevent overfitting. The NN model splits the provided data into training and validation and uses the early stop training procedure to prevent overfitting, while the BNN model uses all of the provided data points for training and uses weight penalty function (complexity penalization) to prevent overfitting (see [2,31,32,34] for details).

This paper is structured as follows: in Section 2 we give a brief explanation of the EEF method and model structures, followed by a description of data, predictors, and predictands in Section 3. In Section 4 we present and discuss the results of the three models (MLR, BNN, NN) run with the conventional bagging method in comparison to the runs with the EEF method. Finally, a conclusion and outlook are presented in Section 5.

Table 1. Examples of studies on machine learning algorithms with bagging methods and summary of their discussion on computational efficiency.

Authors (Year)	Machine Learning Method *	Computational Efficiency of the Bagging Method
Wan et al. (2016) [15]	HANN and BBNN	The bootstrapped NN training process is extremely time-consuming. The HANN approach is nearly 200 times faster than the BBNN approach with 10 hours runtime.
Liang et al. (2016) [35]	BNN and BMH	The bootstrap sample cannot be very large for the reason of computational efficiency.
Zhu et al. (2016) [27]	BNN	The proposed improvement in accuracy comes at the cost of time-consumption during the network training.
Gianola et al. (2014) [36]	GBLUP	Bagging is computationally intensive when one searches for an optimum value of BLUP-ridge regression because of the simultaneous bootstrapping.
Faridi et al. (2013) [37]	ANN	Each individual network is trained on a bootstrap re-sampling replication of the original training data.
Wang et al. (2011) [16]	ANN and GPR	Subagging gives similar accuracy but requires less computation than bagging. This advantage is especially remarkable for GPR since its computation increases in cubic order with the increase of data.
Mukherjee and Zhang (2008) [38]	BBNN	Dividing the batch duration into fewer intervals will reduce the computation effort in network training and batch optimisation. However, this may reduce the achievable control performance . . .
Yu and Chen (2005) [18]	BNN, SVM, and MLE-GP	Fully Bayesian methods are much more time consuming than SVM and MLE-GP.
Rowley et al. (1998) [39]	ANN	To improve the speed of the system different methods have been discussed, but this work is preliminary and is not intended to be an exhaustive exploration of methods to optimize the execution time.

* ANN (artificial neural network), BNN (Bayesian neural network), HANN (hybrid artificial neural network), BBNN (bootstrap-based neural network), BMH (bootstrap Metropolis-Hastings), GPR (Gaussian process regression), GBLUP (genomic best linear unbiased prediction), MLE-GP (maximum likelihood estimation-based Gaussian process), SVM (support vector machine) and V-SVM (virtual support vector machine).

2. Methods

2.1. Entropy Ensemble Filter

The EEF method is a modified bagging procedure to improve efficiency in ensemble model simulation (see [19] for details). The main novelty and advantages of the EEF method are rooted in using the self-information of a random variable, defined by Shannon's information theory [40,41] for selection of most informative ensemble models which are created by conventional bagging method [42]. Foroozand and Weijs [19] proposed that an ensemble of artificial neural network models or any other machine learning technique can use the most informative ensemble members for training purpose rather than all bootstrapped ensemble members. The results showed a significant reduction in computational time without negatively affecting the performance of simulation. Shannon information theory quantifies information content of a dataset based on calculating the smallest possible number of bits, on average, to convey outcomes of a random variable, e.g., per symbol in a message [40,43–47]. The Shannon entropy H, in units of bits (per symbol), of ensemble member M in a bagging dataset, is given by:

$$H_M(X) = -\sum_{k=1}^{K} p_{x_k} \log_2 p_{x_k},$$ (1)

where p_{x_k} is the probability of occurrence outcome k of random variable X within ensemble member M. This equation calculates the Shannon entropy in the units of "bits" because logarithm's base is 2. H gives the information content of each ensemble member in a discretized space, where the bootstrapped members are processed using K bins of equal bin-size arranged between the signal's minimum and maximum values. These bin sizes are user-defined and chosen to strike a balance between having enough data points per bin and keeping enough detail in representing the data distribution of the time series. In this case, we chose K = 10.

Figure 1 illustrates the flowchart of the EEF procedure as applied in this study. The EEF method will assess and rank the ensemble members, initially generated by the bagging procedure, to filter and select the most informative ones for the training of the NN model. As it is expected in machine learning, the overall computational time depends roughly linearly on the number of retained ensemble members which potentially leads to significant time savings.

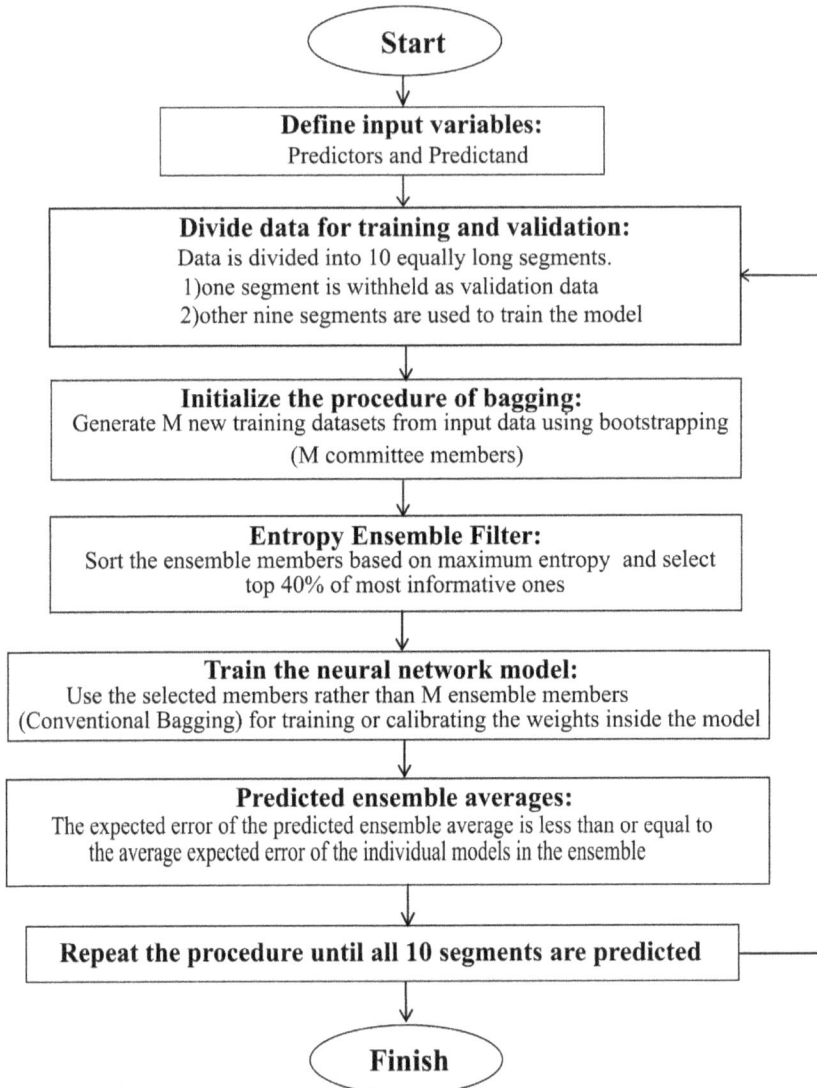

Figure 1. The flowchart of Entropy Ensemble Filter (EEF) method applied in the study.

2.2. Models

Following Wu et al. [20], we adopt the same structure for the NN models used with both the conventional bagging and the EEF method. For details on the model's training and cross-validation procedures to prevent overfitting, we refer to that paper [20]. The original model, a standard feed-forward multilayer perceptron neural network model with Bayesian regularization (BNN model) is run in MATLAB using the 'trainbr.m' function of the Neural Network Toolbox [34]. We introduce an additional NN model that applies the Levenberg-Marquardt optimization algorithm instead of Bayesian regularization and is run using the 'trainlm.m' MATLAB function (NN model) [32,48]. Finally, we run a multiple linear regression (MLR) model using the same 12 PCs as predictors. Following the recommendation of Wu et al. [20], in both neural network models, we optimize the number of hidden neurons (m) which is varying from 1 to 8 (the network architectures of 12-m-1) based on the correlation skill for each predictand out of five SST PCs. For all models we apply the following three schemes in order to produce the model ensemble runs:

(1) MLR, NN, BNN scheme (following the original method in Wu et al [20]): Using the conventional bagging method, training data from the nine segments is randomly drawn multiple times in order to train the BNN and NN model separately. In each 'bagging' draw we include the model run into the ensemble only if the model's correlation skill is higher than that of the MLR model, and its mean-square-error (MSE) less than that of the MLR model; if otherwise, the model run is rejected. This entire procedure is repeated until 30 ensemble members for each NN model are accepted. The ensemble average is then used as the final forecast result. With 10 segments for cross-validation and each segment producing 30 models, a total of 300 models are used for forecasting over the whole record.

(2) MLR_E, NN_E, BNN_E scheme: Using the EEF bagging method (Figure 1), training data from the nine segments is randomly drawn 12 times, producing 12 model ensemble for each data segment. We chose the model ensemble size to be 40% of the original one above, i.e., 12 out of 30, following the recommendations for EEF method application [19]. No selection criteria involving the comparison with the MLR model is applied here. A total of 120 models are used for forecasting over the whole record.

(3) MLR_{rand}, NN_{rand}, BNN_{rand} scheme: This scheme is the same as (2) except the conventional bagging scheme is used instead of the EEF method. This scheme mainly serves as a control run, i.e., for direct comparison of its performance with the scheme (2), both yielding the same total amount of 120 models over the whole period.

The above procedures are repeated until all five SST PCs at all lead times (3, 6, 9, 12 and 15 months) are predicted. For each lead time, we therefore have the ensemble-mean forecast from each of the nine models (MLR, MLR_E, MLR_{rand}, NN, NN_E, NN_{rand}, BNN, BNN_E, BNN_{rand}).

3. Data, Predictors, and Predictands

Monthly SST and SLP data in this study are from European Re-Analysis Interim (ERA-Interim) which is a global atmospheric reanalysis dataset [49] downloaded from the ECMWF website (https://www.ecmwf.int/en/forecasts/datasets/reanalysis-datasets/era-interim) for the tropical Pacific region (124° E–90° W, 20° N–20° S) at 0.75° × 0.75° for the period January 1979 to August 2017. Anomalies in both variables are calculated by subtracting their monthly climatology based on the 1979–2017 period. Following Wu et al. [20], we define the predictand as one of the five leading principal components (PCs) of the SST anomalies over the whole spatial domain, i.e., each of the five SST PCs is predicted separately. The corresponding spatial patterns of the eigenvectors (also called empirical orthogonal functions, EOFs) of the predictands, together explaining 80% of the total variance of the SST anomalies, are displayed in Figure 2. Note that our eigenvectors and SST PCs are somewhat different from those in Wu et al. [20] since their study used different reanalysis data and different time period (1948–2004). For the predictors, after applying a 3-month running mean to the gridded anomaly data of SLP and SST, a separate principal component analysis (PCA) is performed over the whole spatial domain with 7 SLP PCs and 9 SST PCs retained. These retained PCs are then separately normalized by

dividing them by the standard deviation of their first PC. To set up the predictors' structure, the 7 SLP
PCs supplied at time leads of 0, 3, 6, and 9 months and the 9 SST PCs at time leads of 0 months are
stacked together, altogether yielding 37 PC time series (4 × 7 SLP PCs and 9 SST PCs). Finally, another
PCA is performed on the 37 PC time series to yield 12 final PCs that are used as the predictors in the
models. As in Wu et al. [20], the lead time is defined as the time from the center of the period of the
latest predictors to the center of the predicted period. The data record was partitioned into 10 equal
segments (Figure 1), each 44 months long; one segment is withheld to provide independent data for
testing the model forecasts, while the other nine segments are used to train the model. By repeating
this procedure until all 10 segments are used for testing, we provide the forecast over the whole period.
Following the recommendation in [20] for model evaluation criteria, we then calculate the correlation
and root mean square error (RMSE) between the predicted SST anomalies and the corresponding
target data (ERA-Interim) over the whole record. This is consistent with the Gaussianity assumptions
underlying PCA and routinely employed in forecasting SST anomalies. Mean squared error (MSE) has
been used inside of both neural network optimization procedures.

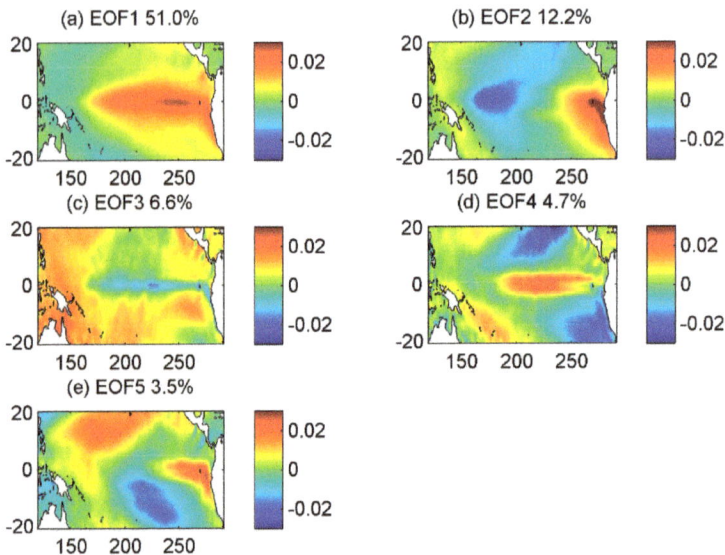

Figure 2. Spatial patterns (eigenvectors) of the first five PCA modes for the SST anomaly field.
The percentage variance explained by each mode is given in the panel titles.

4. Results and Discussion

We first inter-compare the performance of the nine modelling schemes across all five SST PCs and
all five lead times by looking at the correlation between modelled and observed predictands (Figure 3).
Note that, relative to Wu et al. [20] all our correlations are higher. It is important to note, though, that
different time periods were used for the forecast, due to data availability. The pattern across PCs is
the following: for the first PC, which carries most of the variance and spatially best resembles the
ENSO pattern, all models perform equally well at all lead times with correlation coefficient greater
than 0.9 for the lead times of 3–9 months. As we move to higher PCs in our experiments, with the
exception of PC4, the neural network models out-perform the MLR model, especially at the lead times
12 and 15 months. The better performance of neural network models is particularly striking for the
PC2 with lead times 12 and 15 months where the correlation of MLR model substantially drops from
the value greater than 0.8 at the 3-month lead time to the value less than 0.2 at the 15-month lead time.

This result indicates the importance of using nonlinear models for higher lead times, corroborating the findings in Wu et al. [20]. The BNN and NN models perform similarly well for PC1, PC2 and PC4, while the BNN model scores higher for PC3 and lower for PC5. The variation in skill between different methods seems to increase with increasing lead time and with higher PC modes. In general, the expected decrease of skill at higher lead times is visible throughout all PCs, except for PC4 where the predictability peaks at 9-months lead time.

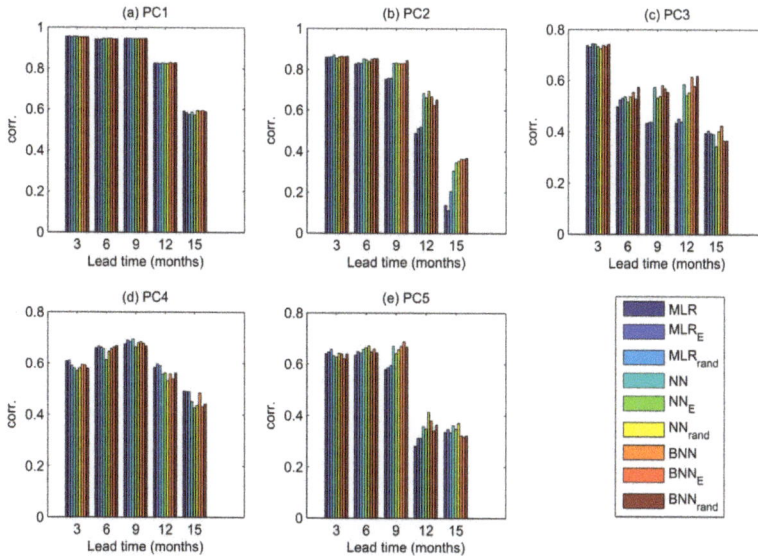

Figure 3. Correlation skill of predictions of the five leading principal components of the SST fields at lead times from 3 to 15 months for all 9 models.

To assess the overall model performance, i.e., combining the skill across all five PCs, we derive the weighted mean correlation across all the PCs, for each lead time, assigning the weights to each PC mode according to the amount of variance explained by the mode (Figure 4a). The following patterns emerge: (1) all models perform equally well for the lead time 3 and 6 months; (2) MLR model's skill drops at 9 months lead time more substantially than the skill of other models; (3) at the lead times of 9 and 15 months, the NN models outperform MLR models by roughly 0.05 difference in mean correlation (the difference between these models' correlations is statistically significant, Steiger's Z = 4.174 [50], $p < 0.01$ and Steiger's Z = 1.73, $p < 0.05$ at the lead times of 9 and 15 months respectively); (4) the NN models all produce correlations very close to each other (within 0.01 difference in correlation); (5) at the 15-month lead time the BNN model is the best performing model; and (6) overall, EEF method achieved a significant reduction in computational time and performed well especially in the forecast at the first 3 lead times. However, its underperformance at 12 and 15 lead time can be regarded as a compromise on computational time-saving.

Next, we look into the computational time for the models to produce the forecast for each PC mode (Figure 4b). The assessment is performed on the computer with 8 parallel quad-core processors (Intel® Core™ i7-4790 CPU @ 3.60GHz × 8). As expected, the BNN simulation is the most computationally expensive with 30 hours runtime in total. Considering the similar performance among all the models at the lead time of 3 and 6 months, the faster algorithms (e.g., MLR model, NN model with EEF method or with random selection) have the computational advantage over the BNN model. It appears that the original Wu et al. [20] model, which relies on the selection criteria, i.e., the inclusion of ensemble members that perform better than their MLR equivalents, does not have an advantage over the modeling scheme without this selection criteria. Finally, the EEF method in BNN and NN models performs equally well and is equally computationally efficient as the application of random reduced ensemble selection, i.e., the conventional bagging. As expected, computation time is mainly driven by ensemble size, as training the models is the most computationally expensive step of the forecasting procedure.

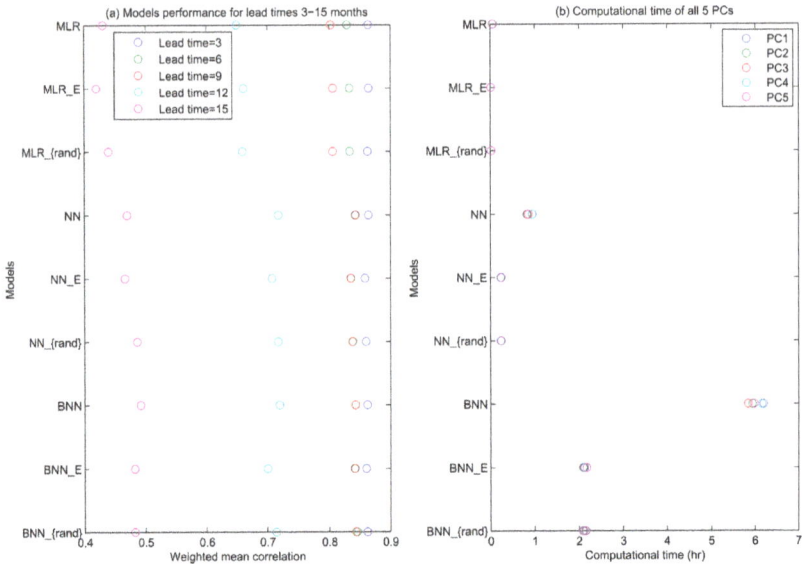

Figure 4. Weighted mean correlation and computational time for all models.

Regional Forecast

To estimate the regional forecast skills over the whole tropical Pacific, SST anomaly fields were reconstructed from five predicted PCs multiplied by their corresponding EOF spatial patterns. We spatially averaged the reconstructed SST anomalies over the Niño 4 (160° E–150° W, 5° S–5° N), Niño 3.4 (170° W–120° W, 5° S–5° N), Niño 3 (150° W–90° W, 5° S–5° N), and Niño 1+2 (90° W–80° W, 10° S–0°) regions, and then computed the correlation skills and root mean squared error (Figure 5). We focus on the difference among the model performance with the EEF method (MLR$_E$, NN$_E$, and BNN$_E$). Overall, their performances are at the same level at lead times of 3–12 months. The BNN model provides a better correlation skill than the other two models at 15 months lead time for all regions and especially for Niño 1+2 domain. We also look into the correlation skills spatially across each domain, i.e., correlation between reconstructed modeled and observed SST anomalies for each grid cell in the domain (Figure 6). At 15 months lead time, the large part of the domain is best simulated with the BNN model. There are significant variations between BNN and NN forecast skill at different lead times in central- northern equatorial region of the tropical Pacific. We also compared the time series of modeled vs observed SST anomalies averaged within the Niño 3 and Niño 1+2 regions at lead times of 3–15 months for the three models with the EEF method (see Appendix A). As expected, all models produced successful forecasts for the major El Niño and La Niña episodes at 3 months lead time. The BNN model outperformed other models at higher lead times, especially in Niño 1+2 region.

The spatial distribution of prediction skill was tested and compared for the various models at different lead times. Note that the performance of the BNN model relative to the NN model changes within the same region at different lead times (Figure 6). The same is true for the linear versus the nonlinear model. For example, for a lead time of 12 months, the nonlinear model outperforms the linear model in the Western Pacific, but not near Middle America, while the results for 15 month lead time are opposite.

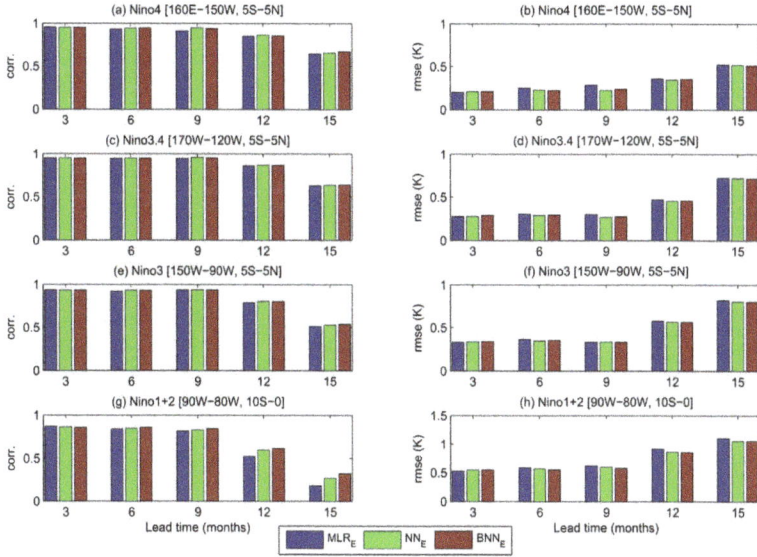

Figure 5. Correlation skills (**left column**) and RMSE scores (**right column**) of SST anomaly forecasts at lead times of 3–15 months for the Niño 4, Niño 3.4, Niño 3 and Niño 1+2 regions.

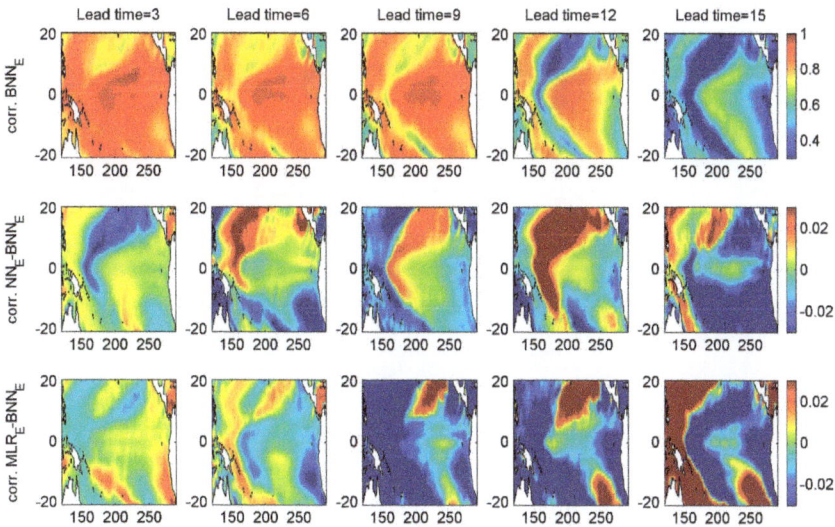

Figure 6. Forecast performance (correlation) per pixel of the forecast reconstructed from 5 leading principal components at lead times of 3–15 months for the period 1979–2017. Top row: BNN_E model, middle and bottom rows: improvement of performance of NN_E and MLR_E over BNN_E.

5. Conclusions

In this study, we performed sea-surface temperature (SST) forecasts over the tropical Pacific using both linear and nonlinear (neural network) models with different training and ensemble generation schemes. In addition to the conventional bagging scheme (randomly generated samples for model training), we applied the ensemble entropy filter (EEF) method. This method reduces the original model ensemble size, and thus computation time, while trying to maintain prediction quality by prioritizing the retention of the most informative training data sets in the ensemble. We incorporated the EEF method in a multiple-linear regression (MLR) model and two neural network models (NN and BNN) and evaluated their performance and computational efficiency relative to the same models when conventional bagging is used. The predictands were the principal components (PCs) of the first five modes of SST monthly anomaly fields over the tropical Pacific for the period 1979–2017. The models' skills were tested for five different lead times: from 3 to 15 months, with 3 months increment.

We show that all models perform equally well at the lead time of 3 and 6 months, while a significant drop in MLR model's skill occurs at 9 month lead time and progressively deteriorates at 12 and 15 months lead time. At the higher lead times, the NN models outperform MLR models, while at the 15 months lead time the BNN model is the best performing model. Models with the EEF method perform equally well as the same models with the conventional bagging method with larger and equal ensemble sizes. Although the EEF method does not improve the correlation skill in the nonlinear forecast, it does not deteriorate it either. Considering that the EEF method selects only a portion of the data to be used in the forecasting, the improvement of computational efficiency with a minimum reduction in model skill makes this method attractive for the application on big datasets. For this particular case, however, the conventional bagging draws random ensembles that closely resemble the optimal ensembles from the EEF method. Thus, the neural network model with both bagging methods produced equally successful forecasts with the same computational efficiency. It remains to be shown, however, whether this finding is sensitive to the number of observations in the dataset.

A limitation of this study is that we deal with deterministic predictions only. Using an ensemble of models in prediction would, in theory, lend itself to generate probabilistic forecasts if an appropriate post-processing scheme can be developed. This would then allow for information-theoretical analysis of prediction skill. Also, a more detailed analysis of variations in predictability could be undertaken, investigating links with slower modes of large scale circulation patterns, such as the Pacific Decadal Oscillation (PDO). This is best investigated in a practical prediction context.

Future work will focus on further exploring the possibilities for forecasting streamflows in the Pacific Northwest, where seasonal forecasts can provide significant benefits for hydropower production, prediction of drinking water shortages and water management in general. Reducing computation time would enable smaller individual organizations to use seasonal forecasts tailored to their specific river basins and water resources management problems, rather than using agency issued ENSO index forecasts that may not always exploit the maximum information in the teleconnection to their most important variables. Also, the use of the ensemble prediction techniques to produce uncertainty estimates with the forecasts is a promising area of research that can inform risk-based decision making for water resources.

Acknowledgments: This research was supported by funding from H. Foroozand's NSERC scholarship and S.V. Weijs's NSERC discovery grant.

Author Contributions: Hossein Foroozand, Valentina Radic, and Steven V. Weijs designed the method and experiments; Hossein Foroozand, Valentina Radic, and Steven V. Weijs performed the experiments; Hossein Foroozand, Valentina Radic, and Steven V. Weijs wrote the paper.

Conflicts of Interest: The authors declare no conflict of interest.

Appendix A. Results of Regional Forecast

In this appendix, the results are shown for predictions of various regional averages that are often used SST indices. The figures show forecasts for various lead times, reconstructed for the 1979–2017 period. Figure A1 shows the Niño 3 SST anomalies, while Figure A2 shows the Niño 1+2 SST anomalies.

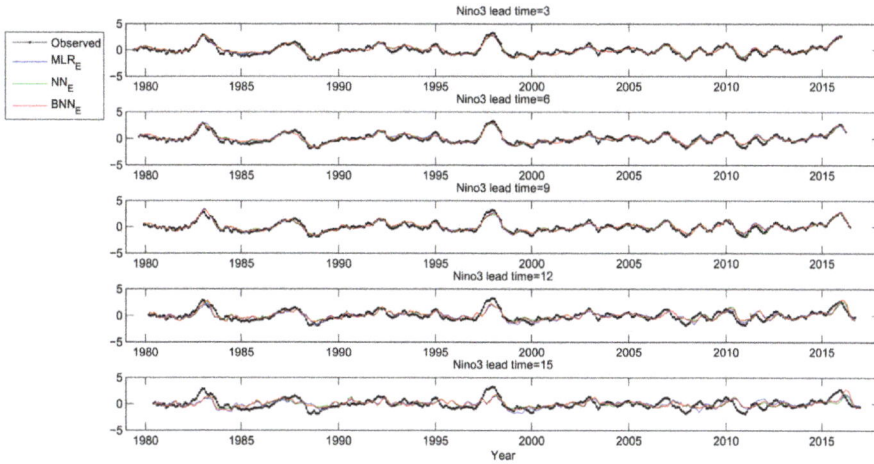

Figure A1. Time observed series of the Niño 3 SST anomalies as well as those predicted by different models at lead times of 3–15 months for the period 1979–2017.

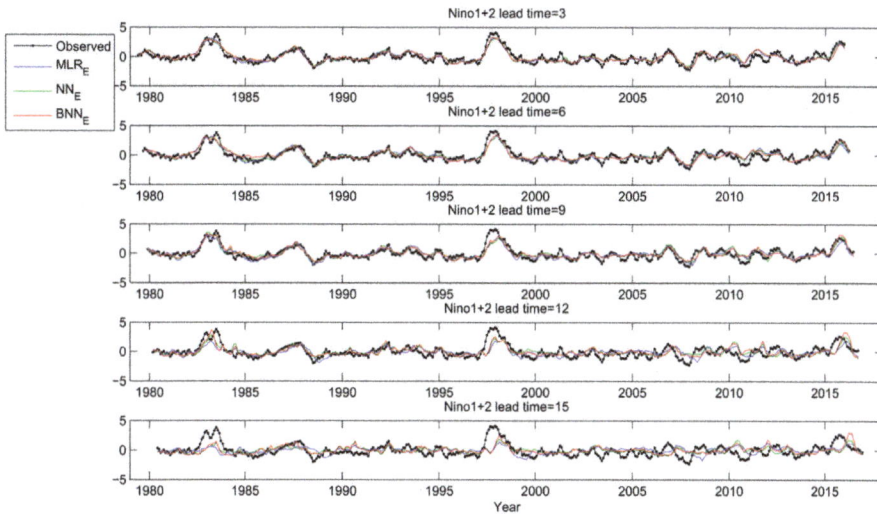

Figure A2. Time observed series of the Niño 1+2 SST anomalies as well as those predicted by different models at lead times of 3–15 months for the period 1979–2017.

References

1. Chau, K. Use of meta-heuristic techniques in rainfall-runoff modelling. *Water* **2017**, *9*, 186. [CrossRef]
2. Hsieh, W.W. *Machine Learning Methods in the Environmental Sciences: Neural Networks and Kernels*; Cambridge University Press: Cambridge, UK, 2009.
3. Lazebnik, S.; Raginsky, M. Supervised learning of quantizer codebooks by information loss minimization. *IEEE Trans. Pattern Anal. Mach. Intell.* **2009**, *31*, 1294. [CrossRef] [PubMed]
4. Zaky, M.A.; Machado, J.A.T. On the formulation and numerical simulation of distributed-order fractional optimal control problems. *Commun. Nonlinear Sci. Numer. Simul.* **2017**, *52*, 177. [CrossRef]
5. Ghahramani, A.; Karvigh, S.A.; Becerik-Gerber, B. HVAC system energy optimization using an adaptive hybrid metaheuristic. *Energy Build.* **2017**, *152*, 149. [CrossRef]
6. Foroozand, H.; Afzali, S.H. A comparative study of honey-bee mating optimization algorithm and support vector regression system approach for river discharge prediction case study: Kashkan river basin. In Proceedings of the International Conference on Civil Engineering Architecture and Urban Infrastructure, Tabriz, Iran, 29–30 July 2015.
7. Niazkar, M.; Afzali, S.H. Parameter estimation of an improved nonlinear muskingum model using a new hybrid method. *Hydrol. Res.* **2017**, *48*, 1253. [CrossRef]
8. Sahraei, S.; Alizadeh, M.R.; Talebbeydokhti, N.; Dehghani, M. Bed material load estimation in channels using machine learning and meta-heuristic methods. *J. Hydroinformatics* **2018**, *20*, 100. [CrossRef]
9. Nikoo, M.R.; Kerachian, R.; Alizadeh, M. A fuzzy KNN-based model for significant wave height prediction in large lakes. *Oceanologia* **2017**. [CrossRef]
10. Sivakumar, B.; Jayawardena, A.W.; Fernando, T.M.K.G. River flow forecasting: Use of phase-space reconstruction and artificial neural networks approaches. *J. Hydrol.* **2002**, *265*, 225. [CrossRef]
11. Moosavian, N.; Lence, B.J. Nondominated sorting differential evolution algorithms for multiobjective optimization of water distribution systems. *J. Water Resour. Plan. Manag.* **2017**, *143*, 04016082. [CrossRef]
12. Moosavian, N.; Jaefarzadeh, M.R. Hydraulic analysis of water distribution network using shuffled complex evolution. *J. Fluid.* **2014**, *2014*, 979706. [CrossRef]
13. Chen, X.Y.; Chau, K.W. A hybrid double feedforward neural network for suspended sediment load estimation. *Water Resour. Manag.* **2016**, *30*, 2179. [CrossRef]
14. Olyaie, E.; Banejad, H.; Chau, K.W.; Melesse, A.M. A comparison of various artificial intelligence approaches performance for estimating suspended sediment load of river systems: A case study in United States. *Environ. Monit. Assess.* **2015**, *187*, 189. [CrossRef] [PubMed]
15. Wan, C.; Song, Y.; Xu, Z.; Yang, G.; Nielsen, A.H. Probabilistic wind power forecasting with hybrid artificial neural networks. *Electr. Power Compon. Syst.* **2016**, *44*, 1656. [CrossRef]
16. Wang, K.; Chen, T.; Lau, R. Bagging for robust Non-Linear Multivariate Calibration of Spectroscopy. *Chemom. Intell. Lab. Syst.* **2011**, *105*, 1. [CrossRef]
17. Kasiviswanathan, K.S.; Sudheer, K.P. Quantification of the predictive uncertainty of artificial neural network based river flow forecast models. *Stoch. Environ. Res. Risk Assess.* **2013**, *27*, 137. [CrossRef]
18. Yu, J.; Chen, X.W. Bayesian neural network approaches to ovarian cancer identification from high-resolution mass spectrometry data. *Bioinformatics* **2005**, *21*, 487. [CrossRef] [PubMed]
19. Foroozand, H.; Weijs, S.V. Entropy ensemble filter: A modified bootstrap aggregating (Bagging) procedure to improve efficiency in ensemble model simulation. *Entropy* **2017**, *19*, 520. [CrossRef]
20. Wu, A.; Hsieh, W.W.; Tang, B. Neural network forecasts of the tropical Pacific sea surface temperatures. *Neural Netw.* **2006**, *19*, 145. [CrossRef] [PubMed]
21. Aguilar-Martinez, S.; Hsieh, W.W. Forecasts of tropical Pacific sea surface temperatures by neural networks and support vector regression. *Int. J. Oceanogr.* **2009**, *2009*, 167239. [CrossRef]
22. Wallace, J.M.; Rasmusson, E.M.; Mitchell, T.P.; Kousky, V.E.; Sarachik, E.S.; von Storch, H. On the structure and evolution of ENSO-related climate variability in the tropical Pacific: Lessons from TOGA. *J. Geophys. Res. Oceans* **1998**, *103*, 14241. [CrossRef]
23. Hamlet, A.F.; Huppert, D.; Lettenmaier, D.P. Economic value of long-lead streamflow forecasts for columbia river hydropower. *J. Water Resour. Plan. Manag.* **2002**, *128*, 91. [CrossRef]
24. Vu, T.M.; Mishra, A.K.; Konapala, G. Information entropy suggests stronger nonlinear associations between hydro-meteorological variables and ENSO. *Entropy* **2018**, *20*, 38. [CrossRef]

25. Goddard, L.; Mason, S.J.; Zebiak, S.E.; Ropelewski, C.F.; Basher, R.; Cane, M.A. Current approaches to seasonal to interannual climate predictions. *Int. J. Climatol.* **2001**, *21*, 1111. [CrossRef]

26. Barnston, A.G.; Tippett, M.K.; L'Heureux, M.L.; Li, S.; DeWitt, D.G. Skill of real-time seasonal ENSO model predictions during 2002–11: Is our capability increasing? *Bull. Am. Meteorol. Soc.* **2011**, *93*, 631. [CrossRef]

27. Zhu, L.; Jin, J.; Cannon, A.J.; Hsieh, W.W. Bayesian neural networks based bootstrap aggregating for tropical cyclone tracks prediction in south China sea. In Proceedings of the 23rd International Conference ICONIP, Kyoto, Japan, 16–21 October 2016.

28. Tangang, F.T.; Hsieh, W.W.; Tang, B. Forecasting the equatorial Pacific sea surface temperatures by neural network models. *Clim. Dyn.* **1997**, *13*, 135. [CrossRef]

29. Cannon, A.J.; Hsieh, W.W. Robust nonlinear canonical correlation analysis: Application to seasonal climate forecasting. *Nonlinear Process. Geophys.* **2008**, *15*, 221. [CrossRef]

30. Tang, B.; Hsieh, W.W.; Monahan, A.H.; Tangang, F.T. Skill comparisons between neural networks and canonical correlation analysis in predicting the equatorial Pacific sea surface temperatures. *J. Clim.* **2000**, *13*, 287. [CrossRef]

31. Levenberg, K. A method for the solution of certain non-linear problems in least squares. *Q. Appl. Math.* **1944**, *2*, 164. [CrossRef]

32. Marquardt, D. An algorithm for least-squares estimation of nonlinear parameters. *J. Soc. Ind. Appl. Math.* **1963**, *11*, 431. [CrossRef]

33. Taormina, R.; Chau, K.W.; Sivakumar, B. Neural network river forecasting through baseflow separation and binary-coded swarm optimization. *J. Hydrol.* **2015**, *529*, 1788. [CrossRef]

34. MacKay, D.J.C. Bayesian Interpolation. *Neural Comput.* **1992**, *4*, 415. [CrossRef]

35. Liang, F.; Kim, J.; Song, Q. A bootstrap metropolis-hastings algorithm for Bayesian analysis of big data. *Technometrics* **2016**, *58*, 304. [CrossRef] [PubMed]

36. Gianola, D.; Weigel, K.A.; Krämer, N.; Stella, A.; Schön, C.C. Enhancing genome-enabled prediction by bagging genomic BLUP. *PLoS ONE* **2014**, *9*, 91693. [CrossRef] [PubMed]

37. Faridi, A.; Golian, A.; Mousavi, A.H.; France, J. Bootstrapped neural network models for analyzing the responses of broiler chicks to dietary protein and branched chain amino acids. *Can. J. Anim. Sci.* **2013**, *94*, 79. [CrossRef]

38. Mukherjee, A.; Zhang, J. A reliable multi-objective control strategy for batch processes based on bootstrap aggregated neural network models. *J. Process Control* **2008**, *18*, 720. [CrossRef]

39. Rowley, H.A.; Baluja, S.; Kanade, T. Neural network-based face detection. *IEEE Trans. Pattern Anal. Mach. Intell.* **1998**, *20*, 23. [CrossRef]

40. Shannon, C.E. A mathematical theory of communication. *Bell Syst. Tech. J.* **1948**, *27*, 379. [CrossRef]

41. Singh, V.P.; Byrd, A.; Cui, H. Flow duration curve using entropy theory. *J. Hydrol. Eng.* **2014**, *19*, 1340. [CrossRef]

42. Breiman, L. Bagging predictors. *Mach. Learn.* **1996**, *24*, 123. [CrossRef]

43. Cover, T.M.; Thomas, J.A. *Elements of Information Theory*; Wiley-Interscience: Hoboken, NJ, USA, 2006.

44. Shannon, C.E. Communication in the Presence of Noise. *Proc. IRE* **1949**, *37*, 10. [CrossRef]

45. Weijs, S.V.; van de Giesen, N. An information-theoretical perspective on weighted ensemble forecasts. *J. Hydrol.* **2013**, *498*, 177. [CrossRef]

46. Weijs, S.V.; van de Giesen, N.; Parlange, M.B. HydroZIP: How hydrological knowledge can be used to improve compression of hydrological data. *Entropy* **2013**, *15*, 1289–1310. [CrossRef]

47. Cui, H.; Singh, V.P. Maximum entropy spectral analysis for streamflow forecasting. *Phys. Stat. Mech. Its. Appl.* **2016**, *442*, 91. [CrossRef]

48. Hagan, M.T.; Menhaj, M.B. Training feedforward networks with the Marquardt algorithm. *IEEE Trans. Neural Netw.* **1994**, *5*, 989. [CrossRef] [PubMed]

49. Dee, D.P.; Uppala, S.M.; Simmons, A.J.; Berrisford, P.; Poli, P.; Kobayashi, S.; Andrae, U.; Balmaseda, M.A.; Balsamo, G.; Bauer, P.; et al. The ERA-interim reanalysis: Configuration and performance of the data assimilation system. *Q. J. R. Meteorol. Soc.* **2011**, *137*, 553. [CrossRef]

50. Steiger, J.H. Tests for comparing elements of a correlation matrix. *Psychol. Bull.* **1980**, *87*, 245. [CrossRef]

MDPI

Article

Cross Mean Annual Runoff Pseudo-Elasticity of Entropy for Quaternary Catchments of the Upper Vaal Catchment in South Africa

Masengo Ilunga

Department of Civil and Chemical Engineering, College of Science, Engineering and Technology, Florida Campus, University of South Africa, Florida 1710, South Africa; ilungm@unisa.ac.za; Tel.: +27-11-471-2791

Received: 2 November 2017; Accepted: 21 December 2017; Published: 13 April 2018

Abstract: This study focuses preliminarily on the intra-tertiary catchment (TC) assessment of cross MAR pseudo-elasticity of entropy, which determines the impact of changes in MAR for a quaternary catchment (QC) on the entropy of another (other) QC(s). The TCs of the Upper Vaal catchment were used preliminarily for this assessment and surface water resources (WR) of South Africa of 1990 (WR90), of 2005 (WR2005) and of 2012 (WR2012) data sets were used. The TCs are grouped into three secondary catchments, i.e., downstream of Vaal Dam, upstrream of Vaal dam and Wilge. It is revealed that, there are linkages in terms of mean annual runoff (MAR) between QCs; which could be complements (negative cross elasticity) or substitutes (positive cross elasticity). It is shown that cross MAR pseudo-elasticity can be translated into correlation strength between QC pairs; i.e., high cross elasticity (low catchment resilience) and low cross elasticity (high catchment resilience). Implicitly, catchment resilience is shown to be associated with the risk of vulnerability (or sustainability level) of water resources, in terms of MAR, which is generally low (or high). Besides, for each TC, the dominance (of complements or substitutes) and the global highest cross MAR elasticity are determined. The overall average cross MAR elasticity of QCs for each TC was shown to be in the zone of tolerable entropy, hence the zone of functioning resilience. This could assure that water resources remained fairly sustainable in TCs that form the secondary catchments of the Upper Vaal. Cross MAR pseudo-elasticity concept could be further extended to an intra-secondary catchment assessment.

Keywords: entropy; cross elasticity; mean annual runoff; water resources; resilience; quaternary catchment; complement; substitute

1. Introduction

Elasticity, a concept borrowed from economic sciences has been used in hydrology and water resources to determine a relative change in a variable, e.g., rainfall, with respect to the change in runoff generated [1–3]. When rainfall is considered to be the most influential variable for generating runoff (e.g., bivariate rainfall-runoff relationships), rainfall elasticity of streamflow is generally positive, i.e., an increase in rainfall is translated into an increase in runoff [1]. This is contrary to the pure economic law on the price elasticity of demand, which is always negative [4,5], however market forces may dictate the sign (positive or negative) of price elasticity [5,6]. Nonetheless other parameters have been shown to influence positively or negatively rainfall elasticity of streamflow, e.g., temperature/evapotranspiration [7–9], land use, water use [10]. The notion of elasticity has been linked recently to entropy concept to assess catchment resilience, i.e., mean annual runoff (MAR) pseudo elasticity [11]. The notion of pseudo-elasticity of entropy was derived for linear regression models and measured the relative change in entropy with respect to the relative change in MAR for tertiary catchments (TC), which are comprised of quaternary catchments (QCs). For the specific case of TCs of the Upper Vaal, MAR pseudo-elasticity of entropy was shown to be relatively positive when considering regression models. In hydrology and water resources, the computation of elasticity is usually related to the same catchment that undergoes change/transformation, i.e., what will be the relative change in runoff/streamflow corresponding to 1% change in rainfall for a given catchment? Likewise, for MAR pseudo-elasticity of entropy, a certain % change in entropy will be associated with 1% change in MAR, for a given catchment. Pseudo-elasticity and elasticity concepts in water related studies could be referred to as self/own-elasticities since changes in the different variables have been formulated with respect to the same catchment. Self/own-elasticity is commonly used to mean elasticity [12–14]. Moreover, in hydrology and water resources, elasticity/pseudo-elasticity concept for a given catchment is often defined without the influence of changes in variables from other catchments. The current paper takes this opportunity by assessing how changes in MAR for a given catchment will likely impact on changes in entropy associated with MAR for another (other) catchment(s). Hence this was done by introducing explicitly the notion of cross elasticity of entropy associated with hydrological change (i.e., change in MAR) occurring in two different catchments belonging to the same hydrological region. In this study, the cross MAR pseudo-elasticity of entropy is defined as the ratio between the change in entropy of a given QC and the change in MAR of another catchment. Cross elasticity is assessed preliminarily at the smaller scale of catchment, i.e., QC. Currently, there is almost no literature in hydrology and water resources that deals explicitly with cross elasticity concept. Cross elasticity concept originated from economy, i.e., cross price-elasticity of demand and shows how the change in price of a given good is likely linked to the demand of another good [12–16]; as opposed to (self/own)-elasticity, which considers the change in price and demand for the same good. Cross price-elasticity of demand yields complementary goods when price and demand vary in different directions and to substitutable goods when the signs are the same [13,15]. Complementary goods and substitutable goods are usually referred to as complements and substitutes respectively [15,17]. Cross elasticity concept was later applied to other fields; e.g., transportation [6,18], marketing [13], electricity [19], etc. The following could also support the introduction of cross elasticity concept in hydrology and water resources:

- In the same hydrological zone (TC) and beyond, different QCs are not considered in isolation (hence are interdependent) and are usually subject to activities such as domestic, industrial, social and economic [20], which are likely to impact on MAR [11]. For instance exaggerated inefficient water use from one part of the catchment may impact over time on other parts of the catchment, e.g., uneven distribution of water availability coupled with unbalanced water demand/supply; change in MAR, etc.
- Water strategy exists to enable proper management and efficient use of water resources in a given region [21]. Thus an integrated water management resource approach is always required to balance water demand and supply. This is possible only if water is well managed at the smaller scale; i.e., quaternary level, within the TC and that water management takes into consideration of how changes in MAR for one QC is likely to affect the changes in another (other) catchment(s). Water resources could be explored from specific catchments of the region without compromising the sustainability of water resources in the region.
- In reality, water transfers can be made intermittently among different catchments, in view of balancing water demand and supply. In a complementary way, water resources could be jointly sustainably used among two or more catchments. These could be seen as "complementary" catchments. Alternatively, water resources from specific catchments could be used to support several activities (domestic, industrial, social and economic) in other catchments. Catchments supplying water could be seen as "substitutes" of those receiving water. A practical example is the Vaal catchment of South Africa (with the Vaal dam) that supplies water from its catchments to several other catchments in Gauteng and beyond. Hence it is a fact that the Upper Vaal catchment impacts on the water balance of several catchments within South Africa and beyond [20].

Inspired by an economic aspect, e.g., [5], this study suggested that when water resources from two catchments can be used jointly sustainably to support their activities, the catchments will be referred to as complementary catchments (complements). One catchment could enhance the insufficient yield of another catchment, but not the opposite; hence the former will be seen as "complement" of the latter. When water resources from one catchment are used preferably over another (other) catchment(s) and vice-versa, these catchments will be referred to as substitutable catchments (substitutes). Water resources from one catchment could used to support another catchment, but not the opposite; the former will be referred to as "substitute" of the latter.

In the light of the above, this study investigates preliminarily, the change in uncertainty of MAR for a given QC with respect to the change in MAR of another (other) QC(s), hence cross MAR elasticity of entropy. Climatic conditions and human activities may cause changes in hydrological variables such as streamflow (runoff) [20,22–26], which has been associated with a degree of uncertainty expressed as Shannon entropy, e.g., [23,25,27,28]. As mentioned earlier, natural climatic conditions [7–9,11,29–31] and/or human activities [10] may influence positively or negatively streamflow elasticity. Nonetheless, naturalized streamflows, thus naturalized MAR plays an important role in South Africa for water resources planning, development and management; hydraulic structure control, natural variability associated with climatic conditions; and ecological water requirements [32]. Hence surface water resources of South Africa of 1990 (WR90), of 2005 (WR2005), and of 2012 (WR2012) usually present the naturalized MAR data of the different catchments, that include the Upper Vaal catchment. In their derivation, naturalized flows are observed flows adjusted with the net effect of upstream land use changes [33]. It is also acknowledged that a problem of unknown origin may occur during streamflow adjustments at a given gauge [33]. Meteorological data with limited human influence could be focused on climatic controls [31]. MAR is assumed to be relevant to a catchment under virgin conditions, i.e., prior to any human activities that have influenced the hydrology of the catchment [33]. Therefore naturalized MAR could assume adjustments of observed flows with quantifiably manageable land use changes. For that, cross MAR pseudo-elasticity could be assumed dependent on natural factors, e.g., climatic variations (temperature, rainfall, evapotranspiration), natural vegetation, natural wetlands and aquifers, soil type, topography, etc. Data at QC level is very important since a QC is usually considered as the smallest unit for hydrological studies, at tertiary, secondary and primary level. The Upper Vaal catchment is made of three secondary catchments, which comprise three TCs each. TCs are finally subdivided into QCs.

The implications of cross elasticity (in comparison with self-elasticity) on catchment resilience were assessed in so far as water resources are concerned. Cross elasticity of entropy was applied to the Upper Vaal catchment and the spatial assessment of cross MAR pseudo-elasticity at QC level, which is fundamental for tertiary and secondary catchments, was also investigated. It is noted that the assessment of spatial *distribution* (*variation*) of rainfall elasticity of streamflow [2,3,7–9,29–31], and entropy associated with streamflow [22,23] and with precipitation [34–37] have been documented. The use of maps for the spatial distribution analysis of the catchment under investigation, is common. This study is built on MAR pseudo-elasticity concept that was introduced recently [11]. It is a preliminary study that investigates cross MAR pseudo-elasticity concept within TCs. It is natural to consider QCs, which constitute a TC. It does not deal yet in detail with cross MAR pseudo-elasticity among TCs, within secondary catchments. Hence a separate study should be conducted at secondary level to determine the impact of change in entropy of a TC with respect to the change in MAR at another TC. Moreover, the individual as well as the combined effect of natural factors on cross MAR pseudo elasticity could be further investigated.

The paper is organized as follows: the first section gives an overview on cross elasticity, which focuses on cross price-elasticity concept from economics, as compared with elasticity (self-/own-elasticity). Main characteristics of the concept are given and its applications to other fields are outlined. The second section explores the main characteristics of cross-elasticity and extends the concept to hydrology and water resources, by focusing preliminarily on the change in MAR and its uncertainty (entropy), hence cross MAR pseudo-elasticity of entropy. MAR is a very important hydrological variable. The third section gives the methods for assessing cross MAR elasticity of entropy and its implication for catchment resilience/sustainability of water resources. The fourth section applies the methods to the Upper Vaal catchment; hereby presents the results and discussion on cross MAR pseudo-elasticity, by outlining the implications on hydrological resilience. The last section summarizes the findings from the previous section and presents recommendations and outlines further research.

In what follows, the prefix pseudo will be sometimes omitted from elasticity. Self/own-elasticity will mean elasticity as compared to cross elasticity. Hence cross MAR pseudo-elasticity will often mean cross MAR elasticity. MAR will mean naturalized mean annual runoff. Complementary catchments and substitutable catchments will be referred to as complements and substitutes respectively. Uncertainty and entropy could mean the same. Upper Vaal, Upper Vaal catchment and Upper Vaal region will be used interchangeably.

2. Overview on Cross Elasticity Concept

From an economic point of view, price-elasticity has been defined as the change in the demand of a good (brand) over the change in price of the same good [14,15]. Elasticity is referred some times to as self- or own-elasticity [13,14], since these changes are assessed with respect to the same good (brand). When changes are assessed with regard to two different goods, the term cross price-elasticity is used; i.e., how the demand of a good is linked to the change in price of another good [5]. The importance of cross price-elasticity has been shown to give power to the consumer for switching to other brand(s), when one good becomes unaffordable or rare, for instance [13,15].

Some of the important characteristics of cross price-elasticity are as follows: firstly a positive cross elasticity between two brands shows that these brands are substitutable; secondly brands are complementary when their cross price-elasticities vary in opposite direction (i.e., negative) [5,14]; and lastly the two brands do not compete at all when their cross elasticities are null [15]. These characteristics among brands are referred to as substitutability, complementarity and independence respectively [15]. The context of the market determines the extent to which substitutable, complementary goods could both appear together in the perpetual space [15]. It was shown also that a good (palm oil) could be a substitute for another good (soy oil) and not the opposite [5]. In specific situations, complementarity can be related to the purchase of two goods on a single buying opportunity, while substitutability can be related to switching between goods of interpurchase times [15]. The higher the cross elasticity between two brands, the more correlated the goods should be [13–15]. Results on cross-elasticity are usually summarized into a matrix called self and cross elasticity matrix; which was shown to be asymmetric generally with the upper left diagonal of the matrix being the line of asymmetry, for which the diagonal entries are self-elasticities [14,38]. The strength of a brand with respect to the interbrand price competition is measured by the amount of asymmetry [17]. In exceptional cases, constraints could be imposed to have a symmetric cross elasticity matrix [17,39].

These characteristics are so appealing that they were adapted to the case of water resources sustainability; in particular catchment resilience assessed using entropy associated with MAR. For instance, water managers could decide to switch to a water surplus catchment to ensure good water demand/supply balance in the region (substitutability). Hence water transfers among different catchments could be possible for water sustainability of the hydrological region. Water managers could also decide to use jointly water resources from different catchments (complementarity). Water transferred from one catchment to another catchment and possibly vice-versa, may not be necessarily the same (asymmetry); hence the impact of change in MAR of a catchment on entropy of another catchment and vice-versa may not necessarily be the same. MAR is one of the important hydrological variables for catchment resilience or water resources sustainability since the yield of a catchment depends on MAR [11]. Hence for the purpose of the current study, cross MAR pseudo-elasticity was used to assess how changes in uncertainty for a given catchment were possibly linked to changes in MAR for another catchment, within the same TC. In line with Ilunga (2017), the link between cross elasticity concept and catchment resilience in terms of water resources sustainability, was determined by using the concept of linear zoning of catchment resilience. Similar to selected studies on entropy of streamflow [22,23], entropy of rainfall [35–37] and in particular rainfall elasticity of streamflow [2,3,7–9,29–31], spatial distribution (variation) assessment of cross MAR pseudo-elasticity across the Upper Vaal catchment was carried out by using a map representation. This is a preliminarily study that introduces cross MAR pseudo-elasticity of entropy at QCs level within a TC and could be further extended to TCs within secondary catchments.

3. Cross MAR Pseudo-Elasticity in Hydrology and Water Resources

The literature of cross elasticity concept is almost inexistent in hydrology and water resources. As outlined earlier, MAR pseudo-elasticity of entropy concept has been introduced recently [11] and on which cross-MAR elasticity is built. A bivariate model (i.e., linear model) was used to derive MAR pseudo-elasticity of entropy for the different TCs of the Upper Vaal catchment. In particular, MAR pseudo-elasticity was interpreted as the regression coefficient of the linear regression. MAR pseudo-elasticity of entropy (ε_i) was defined as the ratio between relative changes in uncertainty associated with MAR and changes in MAR for different QCs of a given TC.

It is useful to give important equations that are critical in the determination of cross MAR elasticity as adapted from Ilunga [11].

Equation (1) below is derived from the Shannon entropy index of the variable z_i, and shows the uncertainty associated with z_i. Entropy can be written as $H(z_i)$. The variable z_i can be MAR for a specific ith QC in a given TC. The i values ($i = 1, 2, 3, \ldots, k$) represent the i QCs in the TC of the Upper Vaal catchment . Each QC contributes to the total MAR of their TC.

$$H(z_i) = -\frac{z_i}{Z} \log \frac{z_i}{Z} \tag{1}$$

In Equation (1), the base of the logarithm is the unit of $H(z_i)$. It is in bits if the base is 2, in Napiers if the base is e, and in decibels (dB) if the base is 10. If Z is MAR of the TC in either WR90, 2005 or WR2012 data set, therefore $Z = \sum_{i=1}^{k} z_i$.

For a specific ith QC in a given TC, the relative change/variation in MAR and relative change in entropy are given by Equations (2) and (3) respectively as shown below when using W90 and W2005 data sets:

$$\delta_i = \delta(z_i) = [\frac{z_{i(WR2005)} - z_{i(WR90)}}{z_{i(WR90)}}] \times 100 \tag{2}$$

$$\delta H_i = \delta H(z_i) = [\frac{H(z_{iWR2005}) - H(z_{iWR90})}{H(z_{iWR90})}] \times 100 \tag{3}$$

where $\delta(z_i)$, $\delta H(z_i)$ are the relative change/variation in MAR and relative change/variation in entropy for a specific i-th QC respectively. The variable z_i is MAR as defined earlier.

Similar Equations to (2) and (3) can be established for WR2005/WR2012.

For hydrological changes occurring between 2 data sets, the MAR pseudo-elasticity for the i-th QC of a given TC, is given by Equation (4):

$$\varepsilon_i = \frac{\delta H_i}{\delta_i} \tag{4}$$

where δH_i is the relative change in entropy associated with MAR; δ_i is the relative change in MAR for the i-th QC in the TC.

Equation (4) can be referred to as self/own-MAR pseudo elasticity of entropy for the i-th QC since changes in the equation are defined with reference to the same i-th QC. Changes are expressed in relative terms. Hence cross MAR pseudo-elasticity (ε_{ij}) can be defined as the ratio between the change in entropy of i-th QC and the change in MAR of j-th QC. Hence cross MAR pseudo-elasticity (ε_{ij}) is given by Equation (5):

$$\varepsilon_{ij} = \frac{\delta H_i}{\delta_j} \tag{5}$$

Equation (5) determines the % change in entropy of i-th QC associated with a 1% change in MAR of the j-th QC.

Conversely, cross MAR pseudo-elasticity (ε_{ji}) is the ratio between the change in entropy of j-th QC and the change in MAR of i-th QC. This is translated into Equation (6) as given below:

$$\varepsilon_{ji} = \frac{\delta H_j}{\delta_i} \tag{6}$$

The terms in Equation (6) can be defined in a similar way as in Equation (4).

In hydrology and water resources studies, rainfall elasticity of streamflow [1] and MAR pseudo elasticity of entropy for bivariate relationships [11] were generally positive. This is not in line with the ideal economic law of the price-elasticity of demand, which is negative [4,5]. However, other hydro-climatic variables (i.e., temperature and evapotranspiration) [9] and human activities (i.e., land use, water use, etc.) [10] could impact positively or negatively rainfall elasticity. Pseudo-elasticity of entropy varied in opposite direction with mean annual evaporation, and in the same direction with mean annual precipitation for TCs of the Upper Vaal [11]. A positive MAR pseudo-elasticity of entropy means that an increase/decrease in entropy is associated with an increase/decrease in MAR. A negative MAR pseudo-elasticity of entropy implies the variations in entropy and in MAR are of opposite signs. Similar to economic concept, two catchments will be said "complements" in terms of water resource sustainability, when the cross-MAR elasticities are negative. The sustainability of water resources is characterized by the level of catchment resilience defined with respect to the uncertainty associated with MAR [11]. Complementarity refers to the increase/decrease in entropy for one catchment, linked to the decrease/increase in MAR for another catchment and vice-versa. Hence water resources could be used jointly from the 2 catchments to balance water demand/supply. When cross MAR elasticities are positive, QCs could be regarded as substitutes, in terms of water resource sustainability (catchment resilience). In this case, water transfer between different QCs could also be promoted to balance water demand/supply. Substitutability is referred to the increase/decrease in entropy for one catchment linked to the increase/decrease in MAR for the other catchment and vice-versa. There could be a situation where one catchment could be a substitute/complement for another catchment, but not vice-versa [5]. Complementarity and substitutability for different QCs should always be comprehended within the zone of acceptable uncertainty (entropy), i.e., the zone of functioning resilience associated with MAR. The zone of acceptable uncertainty that defines catchment resilience was determined by the interval (10%, 25%) increase/decrease for pseudo elasticity; when considering a change in 10% MAR [11]. That is, MAR pseudo-elasticity of entropy should vary between -25% and 25%. This interval was confirmed valid when a statistical analysis was conducted on the significance of MAR pseudo-elasticity, within the context of linear regression. Hence, the cross MAR elasticity of entropy associated with MAR could be assumed acceptable, if it could fall into the above-mentioned interval.

The following could also enhance the introduction of cross MAR elasticity concept:

At a lower scale, the different QCs in a TC are not considered in isolation or may have a certain level of interdependence, as far as water resources are concerned. Streamflows from one catchment may have an impact in one way or the other on the runoff of another catchment. This interdependence may cause a change in runoff in the different catchments and change in uncertainty associated with MAR. For example surrounding catchments of the Vaal dam catchments may contribute to the inflow into Vaal dam. In return, water stored in the Vaal dam will be supplied and contribute to several activities in the QCs of the TCs and beyond.

At a relatively larger scale, water transfers can be made among different water management areas, in view of balancing water demand and supply between two or several catchments, i.e., in the Upper Vaal, water resources should be well managed and co-ordinated with other interdependent catchments, in a way to have an integrated water resource system [20]. Hence, there is a need for water resources to be properly managed strategically at the smaller scale; i.e., quaternary level, within the TC as well as at a larger scale such as secondary and primary catchments. Water management should take into consideration of how the change in MAR of a given catchment is likely to affect the change in uncertainty associated with MAR in another catchment, within the same hydrological zone and beyond.

4. Methods and Data Availability

4.1. Data Used

The Upper Vaal catchment is one of the three parts of the Vaal catchment (i.e., a primary catchment) of South Africa. The two other parts of the Vaal catchment are the Lower Vaal and Middle Vaal. The Upper Vaal falls partly into the Gauteng province, which is considered as the economic hub of the country. The Upper Vaal catchment is situated approximately between 26.5° and 30.5° E longitudes and between 25.2° and 26.2° S latitudes (Figure 1).

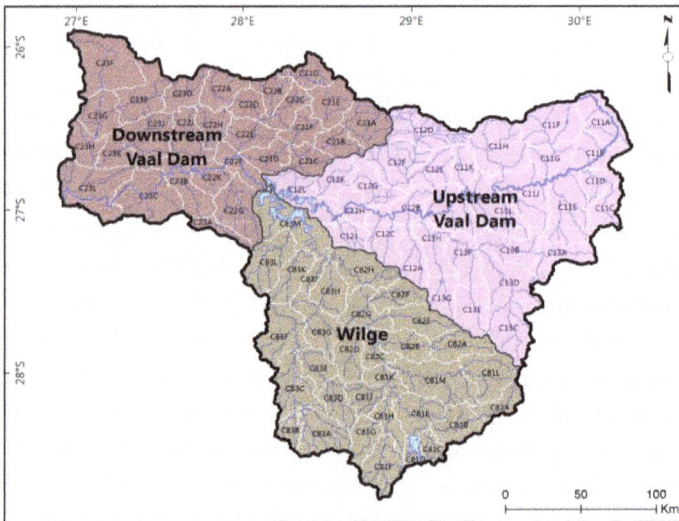

Figure 1. Quaternary, tertiary and secondary catchments of the Upper Vaal catchment [11].

The Upper Vaal is subdivided into three secondary catchments: upstream of Vaal dam (C1), downstream of Vaal dam (C2) and Wilge (C8). Upstream of Vaal dam comprises the following TCs: C11, C12 and C13. Downstream of Vaal dam is formed of the following TCs: C21, C22 and C23. Wilge is subdivided into the following TCs: C81, C82 and C83. Each TC is subdivided into several QCs. A QC is considered to be the basic areal unit for WR90, WR2005 and WR2012 data sets of South Africa as published by Water Research Commission [40]. According to the nomenclature used by the Water Research Commission of South Africa (WRC), a given QC, e.g., C11A means that C is the primary catchment (i.e., Vaal catchment), 1 is the secondary catchment, 1 refers to the TC and the letter at the end is specific to the QC. Upper case letters are usually used in this nomenclature. A good summary of data used for the Upper Vaal was given previously [11]. Nonetheless, of importance is the number of QCs for each TC (Table 1). Details pertaining to catchment areas of QCs, MAR, mean annual precipitation (MAP), mean annual evaporation (MAE), are detailed in surface water resources of 2012 (WR2012) reports, which are the latest publications for water resources of South Africa [40]. MAP and MAE are around 700 mm and 1400 mm. The climate in the Upper Vaal is temperate and fairly uniform, with strongly seasonal rainfall, occurring during the summer period. Savannah grassland, with sparse bushveld is the type of vegetation; the foothills of the Maluti constitute the main topographic feature; the soils are arable and dolomitic. The surface water resources of 2012 (WR2012) reports contained all three data sets; i.e., surface water resources of 1990 (WR90), of 2005 (WR2005) and of 2012 (WR2012). WR90 included MAR data from 1920 to 1989; WR2005 included MAR data from 1920 to 2004 and WR2012 contained MAR data from 1920 to 2012 [40]. The reader should use this reference for further details. The methods laid down in the next section will be performed for both hydrological changes between data sets WR90 and WR2005, and between WR2005 and WR2012. These changes were also written as WR90/WR2005 and WR2005/2012 respectively. MAR is assumed for a catchment under virgin conditions, i.e., without any human influence that have affected catchment hydrology [33].

Table 1. The number of QCs in each tertiary catchment (TC), and their MAR, in the Upper Vaal region, as extracted from [11] and originally published by the Water Research Commission of South Africa [40].

Tertiary Catchment (TC)	Number of QC	MAR (WR90) × 10⁶ m³	MAR (WR2005) × 10⁶ m³	MAR (WR2012) × 10⁶ m³
C11	12	548.2	546.28	527.34
C12	11	296.7	231.32	211.96
C13	8	291.8	322.49	343.05
C21	7	141.5	89.88	98.98
C22	10	131.5	150.02	157.51
C23	11	238.7	185.68	219.00
C81	12	450.6	515.02	529.08
C82	8	198	214.82	151.72
C83	12	283.8	236.76	252.56

In the above table, QC is quaternary catchment, MAR is mean annual runoff, WR90 is surface water resources of 1990 data sets, WR2005 is surface water resources of 2005 data sets, WR2012 is surface water resources of 2012 data sets.

Figure 2 shows the land use/cover of Upper Vaal catchment, with a total area of 55,565 km². It is comprised of natural, cultivation, degraded, urban built-up, water bodies and plantations. The natural part is the most dominant part and constitutes more than the half of the total area. It is followed by the water bodies and the urban built-up area which constitute, each, approximately 3% of the total area. The urban built-up and mining areas (coal and gold) are mostly confined into the downstream of Vaal, particularly, in the northern and the north-west parts of the Upper Vaal catchment. Due to the presence of the large numbers of financial and business service institutions and head offices in Gauteng, the financial sector (with a number of business institutions in Gauteng) plays an important role in the Upper Vaal catchment.

In the next section, for the reason given earlier, the method is mainly based on cross MAR pseudo-elasticity for QCs, within TCs.

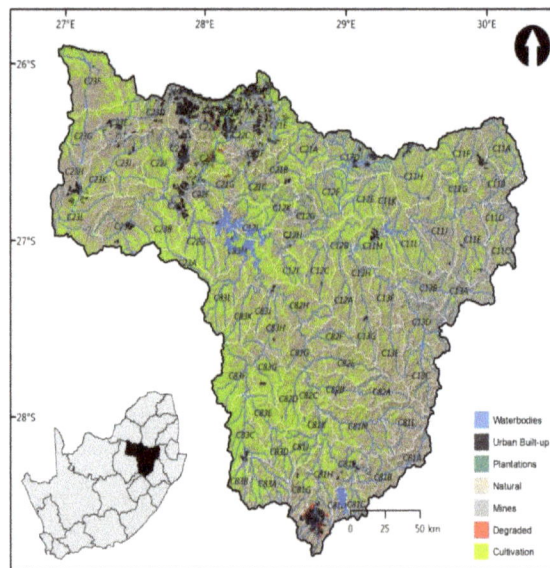

Figure 2. Land use/cover in the Upper Vaal of 2009.

4.2. Methods

4.2.1. Complementarity and Substitutability of QCs

For different QCs in a given TC, MAR pseudo-elasticities and cross MAR pseudo-elasticities were computed according to Equations (1)–(6). The matrix of cross MAR elasticities could be set up and was assessed for symmetry/asymmetry. The matrix is a square matrix $k \times k$, where k is the number of QCs in a given TC. For a specific QC pair, matrix symmetry helped to assess whether or not the change in MAR of one QC had the same impact on the change in entropy of the other QC, and vice-versa. This yielded the determination of complementarity and substitutability of QC pairs. The level of complementarity and substitutability of QC pairs was determined from the signs and magnitudes of cross MAR elasticities of entropy, as explained previously. Hence in this manner, complementary QCs (complements) and substitutable QCs (substitutes) could be determined. In line with Ilunga (2017), three zones of catchment resilience were essentially determined for cross MAR pseudo-elasticity, based on the linear resilience approach zoning:

- Zone of high uncertainty: this is a zone of high entropy and assumes that water resources are not sustainable. It corresponds to higher values of cross elasticities, which are the result of relatively small changes in MAR associated with relatively higher values of entropy. This zone was considered for cross elasticity values above 25%, i.e., 2.5 in unit terms and was characterized with low resilience. Substitutability of QCs becomes impractical due to the higher magnitudes of relative changes in entropy as compared to the changes in MAR and these changes are of the same sign. Natural factors such as climatic conditions, topography, soil type, natural vegetation, etc. could dominate.

- Zone of functioning resilience: this is a zone of tolerable uncertainty and assumes that water resources could be managed sustainably. It corresponds to the values of cross elasticities, which are derived from small changes in MAR associated with reasonable changes in entropy. This zone was considered for cross elasticity values between −25% and 25%, i.e., −2.5 and −2.5 units. Substitutability and complementarity of QCs for water resources planning and management could be sustainable.

- Zone of uniformity: this is a zone, where a catchment is prompt to water resources vulnerability due to relatively low values of entropy and water resources cannot be sustainable if further changes in MAR occur. The zone corresponded to values of cross elasticity below −25%, i.e., −2.5 in unit terms. Complementarity of QCs could be assumed impractical due to higher relative changes in entropy as compared to the change in MAR and these changes could be of opposite directions. Once again natural conditions could also prevail over land use/cover changes. The first and the last zones are characterized by low resilience.

In general, the interval (−25%, 25%) was used for both elasticities [11] and cross elasticities to assess catchment resilience/water resources sustainability.

4.2.2. Cross MAR Pseudo-Elasticities under Constant Change in MAR/Entropy

For any *j*-th column of the matrix, cross MAR pseudo-elasticity values were computed from the changes in entropy of the k QCs and a change in MAR of the *j*-th QC. Each value shows the extent to which a constant change in MAR for the *j*-th QC affects the entropy of remaining QCs. The average of these values is given by Equation (7) below:

$$\varepsilon_j = (1/k)\sum_{i=1}^{k} \varepsilon_{ij} \ (j = 1,2,\dots,k) \tag{7}$$

Likewise, for any *i*-th row of the matrix, the average cross MAR elasticities computed from a change in entropy of the *i*-th QC and the changes in MAR of the k QCs, is given by Equation (8) below:

$$\varepsilon_i = (1/k)\sum_{j=1}^{k} \varepsilon_{ij} \ (i = 1,2,\dots,k) \tag{8}$$

ε_j shows on average how a constant change in MAR of the *j*-th QC may impact on entropy of k QCs, whereas ε_i shows on average how changes in MAR of k QCs may impact on the *i*-th QC, such that the change in its entropy remains constant. If ε_i (ε_j) is positive, then the positive cross elasticities dominate over the negative cross-elasticities. Othewise ε_i (ε_j) is negative, hence the negative cross elasticities dominate over the positive cross elasticities. Equations (7) and (8) were useful in drawing the map of the spatial distribution of cross MAR elasticities of the different QCs, as explained in Section 4.2.6.

The average cross MAR pseudo-elasticity (ε_p) for any *p*-th TC in a secondary catchment as presented in Equation (9) below, was computed from either Equations (7) or (8):

$$\varepsilon_p = \frac{\sum \varepsilon_i}{k} \ (p = 1,\dots,3) \tag{9}$$

where k is the number of QCs in a given TC.

Similarly, the overall average of cross MAR pseudo-elasticity for a given secondary catchment was roughly computed from Equation (9) by using an average of three TCs. As mentioned in the previous section, the number of TCs in each secondary catchment is 3. Normally, a detailed study should be conducted at secondary catchment level to determine the impact of the change in entropy of one TC with respect to the change in MAR at another TC.

4.2.3. Strengths of Correlation between QCs

Similar to economics [15], cross MAR elasticity could show the correlation between QCs, in terms of water resources sustainability/vulnerability. The higher the cross MAR elasticity between two catchments, the more correlated the two catchments should be. That is, a strong correlation could correspond to low catchment resilience (high elastic instability of QCs) or zone of high uncertainty. Conversely, an acceptable catchment resilience could be associated with low cross MAR pseudo-elasticity. This could imply that resilient catchments could be correlated (linked) through the zone of acceptable uncertainty (elastic stability) for water resources sustainability. Therefore, in terms of MAR, high and weak cross elasticity of entropy will imply promptness to vulnerable and sustainable water resources respectively. Using the linear zoning of resilience for the different QCs as described earlier, the following conditions for the correlation strength between i-th and j-th QCs were applied:

- Strong correlation between QCs i and j (low catchment resilience or vulnerability): $\varepsilon_{ij}, \varepsilon_{ji} < -2.5$ or $\varepsilon_{ij}, \varepsilon_{ji} > 2.5$.
- Low correlation between QCs i and j (high catchment resilience or sustainability): $\varepsilon_{ij}, \varepsilon_{ji} \geq -2.5$ and $\varepsilon_{ij}, \varepsilon_{ji} \leq 2.5$. (This includes the case of complete independence between QCs, i.e., where $\varepsilon_{ij}, \varepsilon_{ji} = 0$).
- In % form, the cross elasticity thresholds -2.5 and 2.5 are -25% and 25%, respectively.

4.2.4. Dominance between Complements and Substitutes

The number of occurrences of complements and substitutes in the different QCs pairs enabled to derive their frequencies, which were determined from the $k \times k$ entries of the matrix. Frequency in % showed the dominance between complements and substitutes, in each TC. This indicates whether or not water resources could be used jointly from different QCs or predominantly from specific catchments, alternatively to allow transfers among QCs.

4.2.5. Catchment Vulnerability Risk

As a flip side of resilience, catchment vulnerability could be seen as a risk for water resources not being sustainable, in terms of MAR. The focus is on strong cross MAR elasticities and their magnitudes within a TC, which may cause catchments having relatively low resilience. Catchment vulnerability was understood mainly in the context of the risk for a catchment not sustaining the yield, i.e., MAR. It shows the proportion of occurrences that a catchment could operate within the zone of elastic instability, i.e., either zone of high uncertainty or zone of uniformity [11]. In this way, the dominance of complementary catchments or substitutable catchments was determined.

Mathematically in terms of risk concept, catchment vulnerability (r), could be given by Equation (10):

$$r = \frac{N}{k^2} \tag{10}$$

where N is the number of occurrences of strong cross MAR elasticities of entropy; k is the number of QCs in a TC.

Hence the level of assurance for a catchment not being vulnerable (R) will be defined by Equation (11)

$$R = 1 - r = \frac{k^2 - N}{k^2} \tag{11}$$

Lastly the highest cross elasticity per TC helped to assess the level of resilience. This is the highest variation in entropy versus change in MAR, which could be positive or negative. When highest cross MAR elasticities do not fall in the acceptable range of resilience, negative and positive values corresponded to the zone of uniformity (i.e., very low entropy) and zone of high entropy respectively. In other words, as explained previously cross MAR elasticity values were checked against the interval $(-25\%, 25\%)$ for a change of 10% in MAR to decide whether the zone of functioning resilience/zone of acceptable entropy applied. Outside this range, water resources in QCs could be prompt to vulnerability unless found otherwise.

4.2.6. Spatial Distribution of Cross MAR Pseudo-Elasiticities

Similar to precipitation elasticity of streamflow [2,3,7–9,29–31], the analysis of the spatial distribution (variation) of cross MAR pseudo-elasticities was carried out by using a map representation of the Upper Vaal catchment depicting the different QCs. Threshold values of elasticities (e.g., 1.0 to 2.0 in absolute values) derived from statistical analysis were used for spatial distribution analysis of streamflow elasticity [1] and at the most 1.0 to 2.5 [11]. For example, a value above 2.0 was adopted and used for elastic conditions of streamflow, hence below this value, streamflows were considered inelastic [2,29]. The intent of spatial distribution, in this study was simply to assess the differences in the cross MAR elasticities across Upper Vaal catchment. In the current study, the spatial analysis was confined mainly to the thresholds values (−2.5 and 2.5) for the zone of acceptable resilience, as defined earlier. The thresholds for the linear zoning approach of catchment resilience were verified statistically for the Upper Vaal by conducting a significance test [11] and were adapted to this study. Two cases were mainly considered for spatial distribution analysis: the average cross MAR pseudo-elasticity under constant change in MAR and cross MAR pseudo-elasticity under constant change in entropy at a given QC. Hence Equations (7) and (8) were used to determine the values of cross MAR elasticities that were drawn on the map of Upper Vaal catchment.

5. Results and Discussion

The assessment of cross MAR elasiticities, was based on the methods outlined in Section 4. In the tables presented later in the current section, the computations revealed that most QCs in the different TCs were dominated by positive MAR elasticity values. Complementarity and substitutability for the different QCs of the TC are summarized as shown, i.e., from Tables 2–19. Hence resilient QCs (in terms of water resources) displayed generally weak cross elasticity of entropy values, as opposed to unstably elastic QCs. It was observed that some QC pairs displayed both positive and negative cross MAR elasticity of entropy. Hence in a QC pair, one catchment could be regarded as complement and the other as substitute, when dealing with their changes in MAR respectively. The analysis of cross MAR pseudo-elasticity and interpretation of results was focused on specific QC pairs and could be naturally extended to the remaining QC pairs. Nonetheless the spatial distribution gave further insights in the differences of cross MAR elasticity across Upper Vaal. Catchment vulnerability risk was also assessed to have implicitly the level of assurance for sustainable water resources. The analysis of cross MAR pseudo-elasticity in hydrology and water resources has been carried out preliminarily at QC level, as shown in this section. For each TC, the number of QCs determined the dimension of the cross MAR elasticity matrix and two matrices were presented in a tabular form (e.g., for C11, Tables 2 and 3). The matrices are representative of hydrological changes between WR90 and WR2005, and between WR2005 and WR2012 data sets. The last row in the tables depicts the average cross elasticity at constant change in MAR of each QC and the last column corresponds to the average cross elasticity at constant change in entropy of each QC.

Table 2. Cross MAR elasticities for tertiary catchment C11, between 1990 and 2005.

	C11A	C11B	C11C	C11D	C11E	C11F	C11G	C11H	C11J	C11K	C11L	C11M	ε_i
C11A	0.56	0.36	0.29	0.36	0.40	0.99	−4.31	−0.30	−0.25	0.22	−0.86	−0.23	−0.23
C11B	1.01	0.64	0.51	0.64	0.71	1.77	−7.74	−0.53	−0.45	0.39	−1.55	−0.41	−0.42
C11C	1.25	0.80	0.63	0.79	0.88	2.19	−9.56	−0.66	−0.55	0.48	−1.91	−0.50	−0.51
C11D	1.09	0.69	0.55	0.69	0.76	1.90	−8.29	−0.57	−0.48	0.41	−1.66	−0.43	−0.44
C11E	0.74	0.47	0.37	0.47	0.52	1.29	−5.64	−0.39	−0.33	0.28	−1.13	−0.30	−0.30
C11F	0.31	0.19	0.15	0.19	0.21	0.53	−2.33	−0.16	−0.13	0.12	−0.47	−0.12	−0.13
C11G	−0.12	−0.08	−0.06	−0.08	−0.08	−0.21	0.93	0.06	0.05	−0.05	0.18	0.05	0.05
C11H	−0.86	−0.55	−0.43	−0.54	−0.60	−1.50	6.55	0.45	0.38	−0.33	1.31	0.34	0.35
C11J	−1.22	−0.78	−0.62	−0.77	−0.85	−2.13	9.31	0.64	0.54	−0.47	1.86	0.49	0.50
C11K	1.89	1.20	0.96	1.19	1.32	3.31	−14.44	−0.99	−0.84	0.72	−2.89	−0.76	−0.78
C11L	−0.38	−0.24	−0.19	−0.24	−0.27	−0.67	2.93	0.20	0.17	−0.15	0.59	0.15	0.16
C11M	−1.51	−0.96	−0.77	−0.96	−1.06	−2.65	11.57	0.80	0.67	−0.58	2.31	0.61	0.62
ε_j	0.23	0.15	0.12	0.15	0.16	0.40	−1.75	−0.12	−0.10	0.09	−0.35	−0.09	−0.09

Table 3. Cross MAR elasticities for tertiary catchment C11, between 2005 and 2012.

	C11A	C11B	C11C	C11D	C11E	C11F	C11G	C11H	C11J	C11K	C11L	C11M	ε_i
C11A	0.92	1.03	0.82	0.95	1.24	1.11	1.54	−0.43	−0.24	0.20	−0.67	−0.34	0.51
C11B	0.97	1.08	0.86	1.01	1.31	1.17	1.62	−0.45	−0.26	0.21	−0.71	−0.36	0.54
C11C	1.09	1.21	0.97	1.13	1.47	1.32	1.82	−0.51	−0.29	0.23	−0.80	−0.41	0.60
C11D	1.08	1.21	0.97	1.12	1.47	1.31	1.82	−0.51	−0.29	0.23	−0.80	−0.41	0.60
C11E	0.70	0.78	0.63	0.72	0.94	0.85	1.18	−0.33	−0.19	0.15	−0.52	−0.26	0.39
C11F	0.80	0.89	0.72	0.83	1.09	0.97	1.34	−0.37	−0.21	0.17	−0.59	−0.30	0.44
C11G	0.84	0.93	0.75	0.86	1.14	1.01	1.40	−0.39	−0.22	0.18	−0.61	−0.31	0.46
C11H	−0.75	−0.84	−0.67	−0.78	−1.02	−0.91	−1.25	0.34	0.20	−0.16	0.55	0.28	−0.42
C11J	−1.94	−2.16	−1.73	−2.00	−2.64	−2.34	−3.23	0.89	0.51	−0.41	1.42	0.72	−1.08
C11K	3.67	4.10	3.28	3.79	5.00	4.43	6.11	−1.69	−0.96	0.79	−2.70	−1.37	2.04
C11L	−0.46	−0.51	−0.41	−0.47	−0.62	−0.55	−0.76	0.21	0.12	−0.09	0.34	0.17	−0.25
C11M	−1.39	−1.55	−1.25	−1.44	−1.90	−1.68	−2.32	0.64	0.36	−0.28	1.04	0.52	−0.77
ε_j	0.46	0.51	0.41	0.48	0.62	0.56	0.77	−0.22	−0.12	0.10	−0.34	−0.17	0.26

Table 4. Cross MAR elasticities for tertiary catchment C12, between 1990 and 2005.

	C12A	C12B	C12C	C12D	C12E	C12F	C12G	C12H	C12J	C12K	C12L	ε_i
C12A	0.20	0.72	0.18	−0.38	0.17	0.18	0.17	0.16	0.16	0.21	0.22	0.18
C12B	−0.36	−1.28	−0.32	0.67	−0.30	−0.32	−0.30	−0.29	−0.28	−0.37	−0.39	−0.32
C12C	0.31	1.11	0.27	−0.58	0.26	0.28	0.26	0.25	0.25	0.32	0.34	0.28
C12D	−0.49	−1.73	−0.43	0.91	−0.40	−0.43	−0.41	−0.40	−0.38	−0.50	−0.53	−0.43
C12E	0.38	1.33	0.33	−0.70	0.31	0.34	0.32	0.31	0.30	0.38	0.41	0.34
C12F	0.25	0.89	0.22	−0.47	0.21	0.22	0.21	0.20	0.20	0.26	0.27	0.22
C12G	0.34	1.20	0.30	−0.63	0.28	0.30	0.28	0.28	0.27	0.35	0.37	0.30
C12H	0.44	1.55	0.38	−0.82	0.36	0.39	0.37	0.36	0.34	0.45	0.47	0.39
C12J	0.48	1.70	0.42	−0.89	0.39	0.43	0.40	0.39	0.38	0.49	0.52	0.43
C12K	0.19	0.65	0.16	−0.34	0.15	0.16	0.15	0.15	0.14	0.19	0.20	0.16
C12L	0.12	0.42	0.10	−0.22	0.10	0.11	0.10	0.10	0.09	0.12	0.13	0.11
ε_j	0.17	0.60	0.15	−0.31	0.14	0.15	0.14	0.14	0.13	0.17	0.18	0.15

Table 5. Cross MAR elasticities for tertiary catchment C12, between 2005 and 2012.

	C12A	C12B	C12C	C12D	C12E	C12F	C12G	C12H	C12J	C12K	C12L	ε_i
C12A	0.50	0.66	0.46	−4.47	1.61	1.56	1.58	1.63	1.61	1.56	1.53	0.75
C12B	0.28	0.37	0.26	−2.52	0.91	0.88	0.89	0.92	0.91	0.88	0.86	0.42
C12C	0.53	0.69	0.49	−4.72	1.69	1.65	1.67	1.71	1.69	1.65	1.61	0.79
C12D	−0.10	−0.13	−0.09	0.88	−0.31	−0.31	−0.31	−0.32	−0.31	−0.31	−0.30	−0.15
C12E	−0.01	−0.02	−0.01	0.12	−0.04	−0.04	−0.04	−0.04	−0.04	−0.04	−0.04	−0.02
C12F	−0.01	−0.01	−0.01	0.06	−0.02	−0.02	−0.02	−0.02	−0.02	−0.02	−0.02	−0.01
C12G	−0.01	−0.01	−0.01	0.10	−0.04	−0.04	−0.04	−0.04	−0.04	−0.04	−0.03	−0.02
C12H	−0.02	−0.03	−0.02	0.18	−0.06	−0.06	−0.06	−0.06	−0.06	−0.06	−0.06	−0.03
C12J	−0.02	−0.02	−0.02	0.16	−0.06	−0.06	−0.06	−0.06	−0.06	−0.06	−0.05	−0.03
C12K	0.00	0.00	0.00	0.03	−0.01	−0.01	−0.01	−0.01	−0.01	−0.01	−0.01	0.00
C12L	0.00	0.00	0.00	0.02	−0.01	−0.01	−0.01	−0.01	−0.01	−0.01	−0.01	0.00
ε_j	0.10	0.14	0.10	−0.92	0.33	0.32	0.33	0.34	0.33	0.32	0.32	0.15

Table 6. Cross MAR elasticities for tertiary catchment C13, between 1990 and 2005.

	C13A	C13B	C13C	C13D	C13E	C13F	C13G	C13H	ε_i
C13A	−0.27	−0.15	−0.20	−0.16	−0.17	−0.08	−0.17	−0.13	−0.17
C13B	0.04	0.02	0.03	0.02	0.02	0.01	0.02	0.02	0.02
C13C	−0.12	−0.07	−0.09	−0.07	−0.07	−0.04	−0.07	−0.06	−0.07
C13D	−0.02	−0.01	−0.02	−0.01	−0.01	−0.01	−0.01	−0.01	−0.01
C13E	−0.03	−0.02	−0.02	−0.02	−0.02	−0.01	−0.02	−0.01	−0.02
C13F	0.81	0.46	0.59	0.50	0.50	0.24	0.51	0.38	0.50
C13G	−0.06	−0.03	−0.04	−0.03	−0.03	−0.02	−0.04	−0.03	−0.03
C13H	0.25	0.14	0.19	0.16	0.16	0.08	0.16	0.12	0.16
ε_j	0.08	0.04	0.06	0.05	0.05	0.02	0.05	0.04	0.05

Table 7. Cross MAR elasticities for tertiary catchment C13, between 2005 and 2012.

	C13A	C13B	C13C	C13D	C13E	C13F	C13G	C13H	ε_i
C13A	−0.10	−0.07	−0.09	−0.08	−0.08	−0.07	−0.07	−0.06	−0.08
C13B	0.07	0.05	0.06	0.05	0.06	0.05	0.05	0.04	0.05
C13C	−0.06	−0.04	−0.05	−0.05	−0.05	−0.04	−0.04	−0.03	−0.05
C13D	0.02	0.01	0.02	0.01	0.01	0.01	0.01	0.01	0.01
C13E	−0.01	0.00	−0.01	0.00	−0.01	0.00	0.00	0.00	0.00
C13F	0.03	0.02	0.03	0.02	0.03	0.02	0.02	0.02	0.02
C13G	0.10	0.07	0.09	0.08	0.08	0.08	0.07	0.06	0.08
C13H	0.27	0.19	0.24	0.21	0.22	0.20	0.19	0.16	0.21
ε_j	0.04	0.03	0.04	0.03	0.03	0.03	0.03	0.02	0.03

Table 8. Cross MAR elasticities for tertiary catchment C21, between 1990 and 2005.

	C21A	C21B	C21C	C21D	C21E	C21F	C21G	ε_i
C21A	0.22	0.49	0.47	0.58	0.57	0.56	0.71	0.51
C21B	−0.11	−0.23	−0.22	−0.28	−0.27	−0.26	−0.33	−0.24
C21C	−0.10	−0.22	−0.21	−0.26	−0.25	−0.25	−0.31	−0.23
C21D	−0.16	−0.35	−0.34	−0.42	−0.41	−0.40	−0.51	−0.37
C21E	−0.13	−0.28	−0.28	−0.34	−0.33	−0.32	−0.41	−0.30
C21F	−0.15	−0.32	−0.31	−0.39	−0.38	−0.37	−0.47	−0.34
C21G	−0.22	−0.49	−0.47	−0.58	−0.57	−0.56	−0.70	−0.51
ε_j	−0.09	−0.20	−0.19	−0.24	−0.23	−0.23	−0.29	−0.21

Table 9. Cross MAR elasticities for tertiary catchment C21, between 2005 and 2012.

	C21A	C21B	C21C	C21D	C21E	C21F	C21G	ε_i
C21A	−0.06	−0.04	−0.04	−0.05	−0.05	−0.04	−0.12	−0.06
C21B	0.16	0.10	0.10	0.14	0.14	0.11	0.31	0.15
C21C	0.18	0.12	0.12	0.16	0.16	0.13	0.36	0.17
C21D	−0.01	0.00	0.00	0.00	0.00	0.00	−0.01	−0.01
C21E	−0.02	−0.01	−0.01	−0.02	−0.02	−0.02	−0.04	−0.02
C21F	0.13	0.08	0.08	0.11	0.11	0.09	0.25	0.12
C21G	−0.32	−0.21	−0.21	−0.28	−0.29	−0.22	−0.64	−0.31
ε_j	0.01	0.01	0.01	0.01	0.01	0.01	0.02	0.01

Table 10. Cross MAR elasticities for tertiary catchment C22, between 1990 and 2005.

	C22A	C22B	C22C	C22D	C22E	C22F	C22G	C22H	C22J	C22K	ε_i
C22A	0.28	0.31	0.31	0.29	−2.84	−1.00	−1.06	−1.00	−0.99	−0.83	−0.65
C22B	0.30	0.33	0.33	0.31	−3.02	−1.06	−1.13	−1.07	−1.05	−0.88	−0.70
C22C	0.27	0.30	0.30	0.28	−2.77	−0.97	−1.04	−0.98	−0.97	−0.81	−0.64
C22D	0.33	0.36	0.37	0.34	−3.39	−1.19	−1.27	−1.20	−1.18	−0.99	−0.78
C22E	−0.18	−0.20	−0.20	−0.19	1.83	0.64	0.69	0.65	0.64	0.54	0.42
C22F	−0.31	−0.34	−0.34	−0.32	3.14	1.10	1.18	1.11	1.09	0.92	0.72
C22G	−0.27	−0.29	−0.30	−0.27	2.69	0.94	1.01	0.95	0.94	0.79	0.62
C22H	−0.31	−0.34	−0.35	−0.32	3.17	1.11	1.19	1.12	1.11	0.93	0.73
C22J	−0.29	−0.32	−0.33	−0.30	2.96	1.04	1.11	1.04	1.03	0.87	0.68
C22K	−0.35	−0.38	−0.39	−0.36	3.54	1.24	1.33	1.25	1.23	1.03	0.82
ε_j	−0.05	−0.06	−0.06	−0.05	0.53	0.19	0.20	0.19	0.19	0.16	0.12

Table 11. Cross MAR elasticities for tertiary catchment C22, between 2005 and 2012.

	C22A	C22B	C22C	C22D	C22E	C22F	C22G	C22H	C22J	C22K	ε_i
C22A	−0.05	−0.05	−0.03	−0.03	−0.06	−0.02	−0.02	−0.02	−0.02	−0.02	−0.03
C22B	−0.09	−0.10	−0.05	−0.06	−0.10	−0.04	−0.04	−0.04	−0.04	−0.04	−0.06
C22C	0.33	0.34	0.18	0.21	0.37	0.15	0.12	0.15	0.15	0.15	0.21
C22D	0.21	0.22	0.11	0.13	0.23	0.09	0.08	0.09	0.09	0.09	0.14
C22E	−0.14	−0.14	−0.07	−0.09	−0.15	−0.06	−0.05	−0.06	−0.06	−0.06	−0.09
C22F	0.64	0.67	0.34	0.41	0.72	0.29	0.24	0.29	0.28	0.29	0.42
C22G	−2.18	−2.30	−1.18	−1.41	−2.46	−1.00	−0.84	−0.98	−0.97	−0.98	−1.43
C22H	0.67	0.70	0.36	0.43	0.76	0.31	0.26	0.30	0.30	0.30	0.44
C22J	0.64	0.67	0.34	0.41	0.72	0.29	0.24	0.29	0.28	0.29	0.42
C22K	0.67	0.70	0.36	0.43	0.75	0.31	0.26	0.30	0.30	0.30	0.44
ε_j	0.07	0.07	0.04	0.04	0.08	0.03	0.03	0.03	0.03	0.03	0.04

Table 12. Cross MAR elasticities for tertiary catchment C23, between 1990 and 2005.

	C23A	C23B	C23C	C23D	C23E	C23F	C23G	C23H	C23J	C23K	C23L	ε_i
C23A	0.58	0.58	0.61	0.93	0.93	−0.27	0.89	0.78	0.80	0.80	0.81	0.68
C23B	0.49	0.50	0.52	0.79	0.79	−0.23	0.76	0.66	0.68	0.68	0.69	0.58
C23C	0.39	0.40	0.41	0.63	0.63	−0.18	0.60	0.53	0.54	0.54	0.55	0.46
C23D	0.12	0.13	0.13	0.20	0.20	−0.06	0.19	0.17	0.17	0.17	0.17	0.14
C23E	0.12	0.12	0.13	0.19	0.19	−0.06	0.18	0.16	0.17	0.17	0.17	0.14
C23F	−0.96	−0.97	−1.02	−1.54	−1.54	0.45	−1.48	−1.30	−1.33	−1.33	−1.35	−1.12
C23G	0.15	0.16	0.16	0.25	0.25	−0.07	0.24	0.21	0.21	0.21	0.22	0.18
C23H	0.26	0.26	0.27	0.42	0.42	−0.12	0.40	0.35	0.36	0.36	0.36	0.30
C23J	0.18	0.18	0.19	0.29	0.29	−0.09	0.28	0.25	0.25	0.25	0.26	0.21
C23K	0.23	0.23	0.24	0.37	0.37	−0.11	0.35	0.31	0.32	0.32	0.32	0.27
C23L	0.15	0.15	0.16	0.24	0.24	−0.07	0.23	0.20	0.21	0.21	0.21	0.18
ε_j	0.16	0.16	0.17	0.25	0.25	−0.07	0.24	0.21	0.22	0.22	0.22	0.18

Table 13. Cross MAR elasticities for tertiary catchment C23, between 2005 and 2012.

	C23A	C23B	C23C	C23D	C23E	C23F	C23G	C23H	C23J	C23K	C23L	ε_i
C23A	0.40	0.40	0.49	−1.73	−1.86	−6.73	−1.73	0.72	1.27	1.98	2.03	−0.51
C23B	0.31	0.32	0.38	−1.36	−1.46	−5.29	−1.36	0.57	1.00	1.56	1.60	−0.41
C23C	0.21	0.21	0.26	−0.91	−0.99	−3.56	−0.91	0.38	0.67	1.05	1.07	−0.27
C23D	−0.24	−0.24	−0.29	1.04	1.12	4.06	1.04	−0.44	−0.76	−1.20	−1.23	0.31
C23E	−0.23	−0.23	−0.28	0.98	1.06	3.83	0.98	−0.41	−0.72	−1.13	−1.15	0.29
C23F	−0.05	−0.05	−0.06	0.20	0.22	0.79	0.20	−0.08	−0.15	−0.23	−0.24	0.06
C23G	−0.24	−0.24	−0.30	1.05	1.13	4.10	1.05	−0.44	−0.77	−1.21	−1.24	0.31
C23H	0.18	0.18	0.22	−0.77	−0.83	−3.00	−0.77	0.32	0.57	0.88	0.91	−0.23
C23J	0.05	0.05	0.06	−0.22	−0.24	−0.86	−0.22	0.09	0.16	0.25	0.26	−0.07
C23K	0.01	0.01	0.01	−0.05	−0.05	−0.18	−0.05	0.02	0.03	0.05	0.05	−0.01
C23L	0.01	0.01	0.01	−0.03	−0.03	−0.10	−0.03	0.01	0.02	0.03	0.03	−0.01
ε_j	0.04	0.04	0.05	−0.16	−0.17	−0.63	−0.16	0.07	0.12	0.19	0.19	−0.05

Table 14. Cross MAR elasticities for tertiary catchment C81, between 1990 and 2005.

	C81A	C81B	C81C	C81D	C81E	C81F	C81G	C81H	C81J	C81K	C81L	C81M	ε_i
C81A	4.06	2.51	16.42	5.02	1.38	2.02	−1.35	−0.99	−5.98	0.95	2.26	1.29	2.30
C81B	4.20	2.59	16.97	5.19	1.42	2.09	−1.40	−1.03	−6.18	0.98	2.34	1.34	2.38
C81C	1.62	1.00	6.52	2.00	0.55	0.80	−0.54	−0.39	−2.38	0.38	0.90	0.51	0.91
C81D	1.86	1.14	7.50	2.29	0.63	0.92	−0.62	−0.45	−2.73	0.43	1.03	0.59	1.05
C81E	4.12	2.54	16.64	5.09	1.39	2.05	−1.37	−1.01	−6.06	0.96	2.29	1.31	2.33
C81F	6.64	4.09	26.80	8.19	2.25	3.30	−2.20	−1.62	−9.76	1.55	3.69	2.11	3.75
C81G	−0.92	−0.57	−3.72	−1.14	−0.31	−0.46	0.31	0.22	1.35	−0.21	−0.51	−0.29	−0.52
C81H	−2.42	−1.49	−9.77	−2.99	−0.82	−1.20	0.80	0.59	3.56	−0.56	−1.35	−0.77	−1.37
C81J	0.49	0.30	1.98	0.61	0.17	0.24	−0.16	−0.12	−0.72	0.11	0.27	0.16	0.28
C81K	4.23	2.61	17.10	5.23	1.43	2.11	−1.41	−1.03	−6.23	0.99	2.36	1.35	2.39
C81L	4.93	3.04	19.91	6.09	1.67	2.45	−1.64	−1.20	−7.25	1.15	2.74	1.57	2.79
C81M	5.22	3.22	21.08	6.44	1.77	2.60	−1.73	−1.27	−7.68	1.22	2.91	1.66	2.95
ε_j	2.84	1.75	11.45	3.50	0.96	1.41	−0.94	−0.69	−4.17	0.66	1.58	0.90	1.60

Table 15. Cross MAR elasticities for tertiary catchment C81, between 2005 and 2012.

	C81A	C81B	C81C	C81D	C81E	C81F	C81G	C81H	C81J	C81K	C81L	C81M	ε_i
C81A	−0.30	−0.40	0.06	−0.01	−0.29	−0.46	−0.22	−0.21	−0.41	−0.13	−0.17	−0.16	−0.23
C81B	−0.45	−0.62	0.09	−0.02	−0.44	−0.70	−0.34	−0.32	−0.63	−0.20	−0.26	−0.24	−0.34
C81C	−4.68	−6.37	0.90	−0.23	−4.52	−7.24	−3.52	−3.29	−6.53	−2.05	−2.68	−2.47	−3.56
C81D	12.05	16.39	−2.33	0.59	11.64	18.63	9.07	8.47	16.80	5.27	6.89	6.36	9.15
C81E	−0.34	−0.46	0.07	−0.02	−0.33	−0.52	−0.25	−0.24	−0.47	−0.15	−0.19	−0.18	−0.26
C81F	−0.34	−0.47	0.07	−0.02	−0.33	−0.53	−0.26	−0.24	−0.48	−0.15	−0.20	−0.18	−0.26
C81G	−0.19	−0.26	0.04	−0.01	−0.18	−0.30	−0.14	−0.13	−0.27	−0.08	−0.11	−0.10	−0.15
C81H	−0.13	−0.18	0.03	−0.01	−0.13	−0.20	−0.10	−0.09	−0.18	−0.06	−0.08	−0.07	−0.10
C81J	−0.63	−0.86	0.12	−0.03	−0.61	−0.98	−0.48	−0.45	−0.88	−0.28	−0.36	−0.33	−0.48
C81K	0.46	0.63	−0.09	0.02	0.45	0.72	0.35	0.33	0.65	0.20	0.27	0.25	0.35
C81L	0.07	0.09	−0.01	0.00	0.06	0.10	0.05	0.05	0.09	0.03	0.04	0.04	0.05
C81M	0.14	0.19	−0.03	0.01	0.13	0.21	0.10	0.10	0.19	0.06	0.08	0.07	0.11
ε_j	0.47	0.64	−0.09	0.02	0.45	0.73	0.35	0.33	0.66	0.21	0.27	0.25	0.36

Table 16. Cross MAR elasticities for tertiary catchment C82 between 1990 and 2005.

	C82A	C82B	C82C	C82D	C82E	C82F	C82G	C82H	ε_i
C82A	0.42	−0.14	−0.22	0.32	0.05	0.51	0.60	3.33	0.61
C82B	−1.46	0.50	0.75	−1.11	−0.17	−1.77	−2.09	−11.59	−2.12
C82C	−1.19	0.40	0.62	−0.91	−0.14	−1.45	−1.70	−9.45	−1.73
C82D	0.68	−0.23	−0.35	0.52	0.08	0.83	0.98	5.42	0.99
C82E	0.34	−0.12	−0.18	0.26	0.04	0.42	0.49	2.74	0.50
C82F	0.41	−0.14	−0.21	0.31	0.05	0.50	0.59	3.28	0.60
C82G	0.30	−0.10	−0.16	0.23	0.04	0.37	0.43	2.40	0.44
C82H	0.01	0.00	0.00	0.01	0.00	0.01	0.01	0.07	0.01
ε_j	−0.06	0.02	0.03	−0.05	−0.01	−0.07	−0.09	−0.47	−0.09

Table 17. Cross MAR elasticities for tertiary catchment C82 between 2005 and 2012.

	C82A	C82B	C82C	C82D	C82E	C82F	C82G	C82H	ε_i
C82A	−0.17	−2.72	−0.10	−0.12	−0.12	1.24	−0.07	−0.11	−0.27
C82B	−0.24	−2.81	−0.14	−0.17	−0.18	1.81	−0.10	−0.17	−0.25
C82C	0.32	−1.08	0.19	0.22	0.24	−2.42	0.13	0.22	−0.27
C82D	0.11	−1.24	0.07	0.08	0.09	−0.86	0.05	0.08	−0.20
C82E	0.04	−2.76	0.02	0.03	0.03	−0.29	0.02	0.03	−0.36
C82F	−0.93	−4.44	−0.55	−0.65	−0.70	6.99	−0.37	−0.64	−0.16
C82G	0.93	0.62	0.55	0.64	0.69	−6.92	0.37	0.64	−0.31
C82H	0.11	1.62	0.07	0.08	0.08	−0.84	0.04	0.08	0.16
ε_j	0.02	−1.60	0.01	0.01	0.02	−0.16	0.01	0.01	−0.21

Table 18. Cross MAR elasticities for tertiary catchment C83 between 1990 and 2005.

	C83A	C83B	C83C	C83D	C83E	C83F	C83G	C83H	C83J	C83K	C83L	C83M	ε_i
C83A	−5.53	1.84	0.98	−0.88	1.62	−1.44	−0.83	−0.83	16.04	−1.02	−1.02	−1.00	0.66
C83B	−7.84	2.60	1.40	−1.24	2.30	−2.05	−1.18	−1.17	22.75	−1.45	−1.44	−1.42	0.94
C83C	−9.58	3.18	1.71	−1.52	2.81	−2.51	−1.44	−1.44	27.82	−1.77	−1.76	−1.74	1.15
C83D	0.26	−0.09	−0.05	0.04	−0.08	0.0	0.04	0.04	−0.75	0.05	0.05	0.05	−0.03
C83E	−8.19	2.72	1.46	−1.30	2.40	−2.14	−1.23	−1.23	23.77	−1.51	−1.51	−1.49	0.98
C83F	−3.30	1.10	0.59	−0.52	0.97	−0.86	−0.50	−0.49	9.57	−0.61	−0.61	−0.60	0.39
C83G	−0.50	0.17	0.09	−0.08	0.15	−0.13	−0.08	−0.07	1.45	−0.09	−0.09	−0.09	0.06
C83H	−0.07	0.02	0.01	−0.01	0.02	−0.02	−0.01	−0.01	0.20	−0.01	−0.01	−0.01	0.01
C83J	−5.53	1.84	0.99	−0.88	1.62	−1.45	−0.83	−0.83	16.06	−1.02	−1.02	−1.00	0.66
C83K	−1.52	0.51	0.27	−0.24	0.45	−0.40	−0.23	−0.23	4.42	−0.28	−0.28	−0.28	0.18
C83L	−2.49	0.83	0.44	−0.39	0.73	−0.65	−0.37	−0.37	7.23	−0.46	−0.46	−0.45	0.30
C83M	−3.21	1.07	0.57	−0.51	0.94	−0.84	−0.48	−0.48	9.33	−0.59	−0.59	−0.58	0.38
ε_j	−3.96	1.32	0.71	−0.63	1.16	−1.03	−0.59	−0.59	11.49	−0.73	−0.73	−0.72	0.47

Table 19. Cross MAR elasticities for tertiary catchment C83 between 2005 and 2012.

	C83A	C83B	C83C	C83D	C83E	C83F	C83G	C83H	C83J	C83K	C83L	C83M	ε_i
C83A	0.97	−0.74	−0.60	−1.15	−0.51	−0.43	−1.23	−1.65	−1.37	−1.30	−1.44	−1.32	−0.90
C83B	−0.28	0.22	0.18	0.34	0.15	0.13	0.36	0.48	0.40	0.38	0.42	0.39	0.26
C83C	−0.34	0.26	0.21	0.40	0.18	0.15	0.43	0.58	0.48	0.46	0.50	0.46	0.31
C83D	0.02	−0.01	−0.01	−0.02	−0.01	−0.01	−0.02	−0.03	−0.03	−0.03	−0.03	−0.03	−0.02
C83E	−0.60	0.46	0.37	0.71	0.32	0.26	0.76	1.02	0.84	0.80	0.89	0.81	0.55
C83F	−0.70	0.53	0.43	0.83	0.37	0.31	0.89	1.19	0.99	0.94	1.04	0.95	0.65
C83G	0.05	−0.04	−0.03	−0.06	−0.03	−0.02	−0.07	−0.09	−0.07	−0.07	−0.08	−0.07	−0.05
C83H	0.18	−0.13	−0.11	−0.21	−0.09	−0.08	−0.22	−0.30	−0.25	−0.24	−0.26	−0.24	−0.16
C83J	0.11	−0.08	−0.07	−0.13	−0.06	−0.05	−0.14	−0.19	−0.16	−0.15	−0.16	−0.15	−0.10
C83K	0.08	−0.06	−0.05	−0.09	−0.04	−0.03	−0.10	−0.13	−0.11	−0.10	−0.11	−0.10	−0.07
C83L	0.10	−0.08	−0.06	−0.12	−0.05	−0.05	−0.13	−0.18	−0.14	−0.14	−0.15	−0.14	−0.10
C83M	0.07	−0.05	−0.04	−0.08	−0.03	−0.03	−0.08	−0.11	−0.09	−0.09	−0.10	−0.09	−0.06
ε_j	−0.03	0.02	0.02	0.03	0.02	0.01	0.04	0.05	0.04	0.04	0.04	0.04	0.03

5.1. Complementarity and Substitutability of QCs

5.1.1. Tertiary Catchment C11

The results of cross MAR elasticities are displayed in Tables 2 and 3. As in economics related studies [14,36], the two matrices were found to be asymmetric as shown in these tables. The asymmetry shows the degree of complementarity or substitutability in terms of water resource utilization within the same QC pair. For example, the cross elasticity values for the pair (C11A, C11B) were 0.3 and 1. It was observed from Tables 2 and 3 that all the entries (self-elasticities) in the upper left diagonal were positive and acceptable for the zone of functioning resilience. In the above-mentioned tables, it was revealed that the cross MAR elasticities were either complements or substitutes and were not necessarily always in the zone of functioning resilience.

For instance, between 1990 and 2005, and between 2005 and 2012, the QC pair (C11A, C11B) was in the zone of functioning resilience. Since the cross elasticities were positive, they could be considered as substitutes with different strengths, i.e., between 1990 and 2005, a decrease of 10% in MAR for catchment C11A, corresponded to a 10% decrease in entropy for C11B, whereas a decrease of 10% in MAR for C11B was associated with 3.6% decrease in entropy for C11A. It could mean that water resources for these two QCs could still be used interchangeably in a sustainable manner, at different periods. However, the level of catchment resilience for C11A was relatively slightly higher than that for C11B. Between 2005 and 2012, a decrease of 10% in MAR for C11A (C11B) implied almost the same decrease in entropy for C11B (C11A).

Between 1990 and 2005, for the pair of complements (C11F, C11K), it was observed that a decrease of 10% in MAR for catchment C11K, corresponded to a decrease of 1.2% in entropy for C11F whereas a decrease of 10% in MAR for C11F was associated with 33% decrease in entropy for C11K. Although these two QCs could be substitutes, it could mean that the change in MAR for C11K was associated with the zone of functioning resilience for C11F, while the opposite implied promptness to vulnerability for C11K. This could be an alarming situation for water managers, of the implications of a decrease in MAR for C11F. This situation would coincide with the zone of uniformity; where water resources were regarded as unsustainable [11]. Between 2005 and 2012; 10% increase in MAR for C11F was associated with 44% increase in entropy for C11K. In the same perspective, between 2005 and 2012, it was observed that a 10% increase in MAR (for QCs, from C11A to C11G), corresponded to an increase between 33 and 61% in entropy for C11K. This situation would correspond to water resources vulnerability for C11K, i.e., with C11K operating in the zone of high entropy [11]. Overall, C11K was subjected to higher increase and decrease in uncertainty, when slight changes in MAR occurred in other catchments. A small positive constant change in MAR for C11G yielded to strong positive cross elasticities of 66%, 93%, 29% and 116% for C11H, C11J, C11L and C11M respectively. It could be suggested that these QCs were in the zone of high uncertainty or low resilience. The cause of these higher cross MAR pseudo-elasticities could be further investigated in the natural factors, e.g., climatic variations (temperature, rainfall, evapotranspiration), natural vegetation, natural wetlands and aquifers, soil type, topography, etc. The effect of unquantifiable land use changes should be taken into consideration. On the other hand, the cause of relatively low cross MAR elasticities could be investigated in the combined effect among natural conditions. This combined effect could be complex with the increasing number of factors. They could influence positively or negatively cross MAR pseudo-elasticities. Previous studies on hydrological elasticity [2,9,11,29–31], showed such an influence. A similar analysis could be extended to the rest of substitutes presented in the tables.

The results in Tables 2 and 3 revealed also that C11A and C11L were substitutable QCs between 2005 and 2012 and displayed an acceptable level of resilience for water resources. However, they were complementary QCs with different strengths, i.e., between 1990 and 2005, a decrease of 10% in MAR for catchment C11A, corresponded to an increase of 3.8% in entropy for C11L, whereas an increase of 10% in MAR for C11L was associated with 8.6% decrease in entropy for C11A. These results could mean that the relatively high uniform distribution of MAR for C11L (i.e., entropy increase) could be an indication of sustaining water resources. Water resources from C11L could be used as a complement of the yield from C11A to balance water demand/supply. Between 2005 and 2012, an increase of 10% in MAR for catchment C11A, corresponded to a decrease of 4.6% in entropy for C11L, whereas a decrease of 10% in MAR for C11L was associated with 6.7% increase in entropy for C11A. It could mean that a relatively high yield could be available for C11A, and hence used to assure the sustainability of water resources. Water managers could use more water from C11A than from C11L to balance water demand/supply. Water resources from C11A could be used to complement MAR for C11L.

Between 2005 and 2012, all cross MAR elasticity values were relatively weak (low) and implied that all complements could maintain a good level of resilience for water resources. This was not the case between 1990 and 2005; where there were instances of vulnerability of water resources for selected QC pairs. Hence complements displayed elastic instability, between 1990 and 2005, whereas they were elastically stable between 2005 and 2012.

5.1.2. Tertiary Catchment C12

According to the results displayed in Tables 4 and 5, there were complementary and substitutable QC pairs in C12. In addition, there were QC pairs which displayed both positive and negative cross MAR elasticities, hence having a mixture of complement and substitute in the same QC pairs. Between 1990 and 2005, all QCs had relatively low cross MAR elasticities, hence displayed an acceptable level of resilience, except for few QCs pairs between 2005 and 2012. All substitutes had relatively low cross elasticities, hence displayed an acceptable level of resilience. The implications in terms of water sustainability are similar to substitutes as discussed for TC C11.

Between 2005 and 2012, for the pair (C12A, C12D), 10% increase in MAR for C12D, led to 45% decrease in entropy for C12A, while 10% decrease in MAR for C12A, led to 1% increase in entropy for C12D. The increase in MAR for C12D might cause water resources for C12A to be elastically unstable. Hence a change in MAR from catchments other than C12D could be considered to maintain the level of acceptable resilience.

It was observed in selected QC pairs that one catchment was a substitute while the other was complement and vice-versa. Just an illustration; e.g., between 1990 and 2005, and between 2005 and 2012, the QC pair (C12E, C12A) was in the zone of functioning resilience. Interestingly, between 1990 and 2005, both C12A and C12E were substitutes; however between 2005 and 2012, C12E was a complement of C12A and C12A was a substitute of C12E.

Between 2005 and 2012, a decrease of 10% in MAR for catchment C12A, corresponded to an increase of 0.1% in entropy for C12E, whereas a decrease of 10% in MAR for C12E was associated with 16% decrease in entropy for C12A. It is believed that an increase in entropy for C12E could help have fairly a good distribution of MAR in the pair, for water resource sustainability. Hence water resources for C12E could be used as a complement to balance water demand/supply from C12A. When MAR for C12E decreases, water resources from C12A could be used as a substitute for C12E, without compromising the zone of acceptable uncertainty.

5.1.3. Tertiary Catchment C13

From 1990 to 2012, the results in Tables 6 and 7 revealed that the complementary QC pairs displayed very low cross elasticities, i.e., close to 0. Hence, a variation of MAR for a given QC was insignificant to cause an opposite variation in uncertainty for another QC. These weak cross elasticities could mean that the water resources from QCs could be used jointly to maintain high resilience.

Similar to complements, very low cross MAR elasticities were observed for substitutable QC pairs. As shown in Tables 6 and 7, variations of MAR for specific QCs caused in the same direction, insignificant variations of uncertainty for other QCs. Water resources from those specific QCs could be used as substitutes of other QCs and still maintaining a high level of resilience. The results in these tables showed that there was a mixture of complement and substitute in selected QC pairs. The implications of this mixture on water sustainability can be interpreted as done earlier.

5.1.4. Tertiary Catchment C21

As displayed in Tables 8 and 9, the values of cross MAR elasticities were relatively low; hence catchments could operate in the zone of tolerable entropy. Complementary, substitutable QC pairs as well as a mixture of complement and substitute in specific QC pairs were observed.

5.1.5. Tertiary Catchment C22

The results of cross MAR elasticities were summarized in Tables 10 and 11. Weak cross elasticities were observed, from 2005 and 2012, however cases of high elastic instability of water resources could be observed from 1990 to 2005 due to high changes in entropy associated with a constant change in MAR for C22E. Water management could focus on constant changes in MAR for QCs other than C22E to avoid elastic instability, in terms of water resources.

5.1.6. Tertiary Catchment C23

From 1990 to 2005, low cross elasticities were observed as displayed in Tables 12 and 13. However, from 2005 and 2012, cases of low catchment resilience could be observed due to high changes in entropy associated with a constant change in MAR for C23F; except relatively low changes in uncertainty were observed for C23J, C23K and C23L. In situations where a change in MAR for C23F remained constant, water management could focus more on C23J, C23K and C23L for water resource sustainability. Alternatively, a constant change in MAR for QCs other than C23F should be considered to assure elastic stability in catchment yields.

5.1.7. Tertiary Catchment C81

As depicted in Tables 14 and 15, several cases of very low resilience for QCs were observed between 1990 and 2005. C81M and C81C had the highest cross MAR elasticity; that is 210% change in entropy for C81M was associated with 10% change in MAR for C81C. During the same period, TC C81 scored cases of vulnerability of water resources for C81A, C81B, C81C, C81F and C81L, due to relatively high self-elasticities. QC pairs in C81 which do not fall in the range of acceptable resilience would need particular attention as far as water resources management is concerned. Between 2005 and 2012, cases of high elastic instability were observed only for a constant value in entropy for C81D associated with different changes in MAR for other QCs.

5.1.8. Tertiary Catchment C82

Low cross MAR elasticity values were observed, from 1990 to 2005, and from 2005 and 2012, as shown in Tables 16 and 17 respectively. However, between 1990 to 2005 a constant decrease in MAR for C82H yielded high values of cross elasticity (high elastic instability). Between 2005 and 2012, a decrease in MAR for C82B, was linked to a higher increase in uncertainty for C82A, C82B, C82E and C82F. Likewise, an increase in MAR for C82F was associated with a drastic decrease in uncertainty for C82G, which could yield water resource vulnerability, within the zone of uniformity.

5.1.9. Tertiary Catchment C83

Between 2005 and 2012, the results in Tables 18 and 19, displayed weak cross elasticities. However, between 1990 and 2005, a constant decrease in MAR for C83A, yielded elastic instability, with C83B, C83C, C83E, C83F, C83J and C83M, due to high entropy in these QCs. Similarly, a change in MAR for C83J, yielded vulnerable water resources with all QCs, except C83D, C83G and C83H. It was also observed that high self-elasticity values were recorded for C83A, C83B and C83J. Hence a decrease in MAR for these QCS could be associated with unsustainable water resources in the same QCs.

5.2. Dominance of Complements or Substitutes

The number of complements and substitutes per TC were computed and their corresponding frequencies within the TC were summarized in Figure 3a,b. It could be observed that, in most TCs, substitutable QCs dominated over complementary QCs.

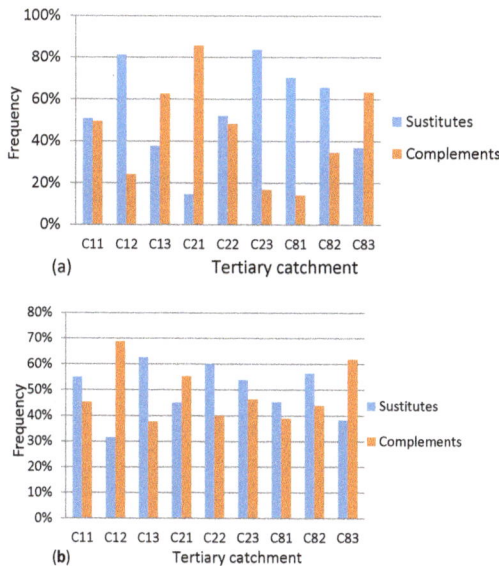

Figure 3. (a) Cross mean annual elasticity dominance between 1990 and 2005. (b) Cross mean annual runoff elasticity dominance between 2005 and 2012.

5.3. Catchment Vulnerability (Risk)

The level of assurance for QCs to sustain water resources QCs in the Upper Vaal and the risk for not sustaining water resources were computed and presented in Figure 4a,b, respectively. It was observed that the risk associated with water resources vulnerability varied between 0 and 16%. Hence between 1990 and 2012, sustainability level (i.e., the level of assurance) to sustain water resources was relatively high and varied between 84 and 100%. Nonetheless, between 1990 and 2005, in most cases QCs displayed slightly higher level of assurance, as compared to the period between 2005 and 2012. In reality, when sustainability notion of water resources includes several variables other than MAR, it could be very likely that the values of the risk of vulnerability would increase from the above-mentioned values, hence the level of assurance to sustain water resources would decrease further.

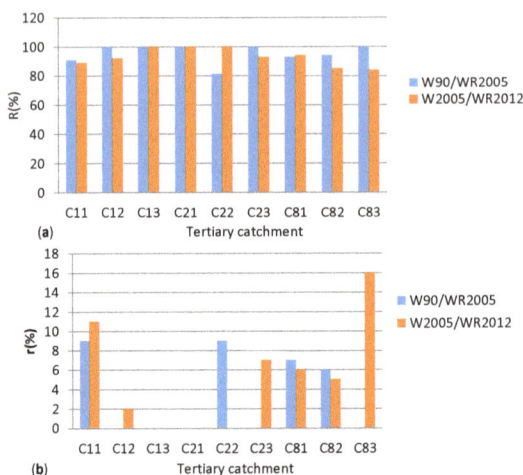

Figure 4. (**a**) Level of assurance of water resources sustainability (R) in tertiary catchments of Upper Vaal. (**b**) Vulnerability risk of water resources (r) in tertiary catchments of Upper Vaal.

Table 20 depicts the highest cross MAR elasticity for each TC, between 1990 and 2005 and, between 2005 and 2012. Between 1990 and 2005, the following QC pairs C11K & C11G, C22K & C22E, C81M & C81C and C83C & C83J scored their highest cross elasticity values outside the zone of functioning resilience; hence these QCs could be associated with the lowest resilience, in terms of water resources. Likewise, between 2005 and 2012, the QC pairs C11G & C11K, C12C & C11D, C23A & C23F, C81D & C81F and C82G & C82GF were associated with low resilience. For example 10% change in MAR for C11G was linked to 144% reduction and 61% increase in entropy for C11K, between 1990 and 2005 and between 2005 and 2012 respectively. On the other hand 10% change in MAR for C83J was linked to 278% increase in entropy for C83C, between 1990 and 2005. This reduction and increase in entropy implied that C11G and C11K were in the zone of elastic instability associated with a relatively smaller change in MAR as compared to the change in entropy. Situations of higher cross MAR pseudo-elasticities could be due to climatic influences on changes in MAR. Water management and planning should take care of this situation.

Table 20. Highest cross mean annual runoff pseudo-elasticity of QCs in TCs of the Upper Vaal region.

Tertiary Catchment	Highest Cross MAR Elasticity (%)	
	W90/WR2005 (Quaternary Catchment Pair)	W2005/WR2012 (Quaternary Catchment Pair)
C11	−144.4 (C11K & C11G)	61 (C11G & C11K)
C12	−17.2 (C12D & C12B)	−47.1 (C12C & C11D)
C13	8 (C13F & C13A)	2.7 (C13F & C11A)
C21	7.1 (C21A & C11G)	3.6 (C21C & C11G)
C22	35 (C22K & C22E)	−23 (C22G & C22B)
C23	−15 (C23F & C23D)	−67 (C23A & C23F)
C81	211 (C81M & C81C)	186 (C81D & C81F)
C82	−116 (C82B & C82H)	−70 (C82G & C82GF)
C83	278(C83C & C83J)	−17(C83A & C83H)

The results in Table 20 have been plotted in Figure 5. This figure shows that selected QCs in C11, C22, C81, C82 and C83 their highest cross MAR elasticity outside the range the zone of acceptable entropy/resilience. The global highest cross MAR elasticity of entropy value was located in the tertiary catchment C83, followed by C81 and C82. These TCs could be prompt to vulnerability when their highest cross elasticity values were considered.

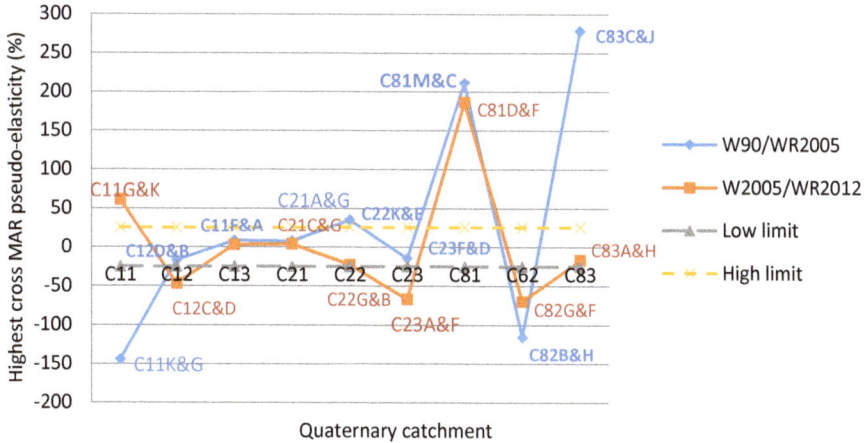

Figure 5. Highest cross mean annual runoff pseudo-elasticity for tertiary catchments of the Upper Vaal region.

From the results shown in Table 20, the highest cross MAR elasticity values were derived roughly for secondary catchments C1, C2 and C3. For each secondary catchment, the QC pair was selected based on the overall highest value among three TCs. In this manner, the QC pairs were C11K & C11G, C22K & C22E and C83C & C83J respectively for C1, C2 and C3, between 1990 and 2005. Between 1990 and 2005, the QC pairs corresponding to the secondary catchments were C12C & C11D, C23A & C23F, C81D & C81F. Since these QCs scored their highest cross-elasticity values; they could be associated with a high risk of vulnerability, with regard to water resources.

5.4. Cross MAR Pseudo Elasticities under Constant Change in MAR/Entropy

5.4.1. Constant Change in MAR

The last row of each table (i.e., from Tables 2–19) showed that, on average a constant change in MAR for any QC was associated with a relatively low change in entropy for the remaining QCs, hence yielded low cross MAR elasticity values. This could lead to QCs operating in the zone of functioning resilience. However, higher values of cross MAR elasticity (low catchment resilience) were scored by a change in MAR for C81A, C81C, C81D, C81J, C83A and C83J, between 1990 and 2005. It was observed generally that on average, positive cross MAR elasticities overshadowed negative cross elasticities. For a change in MAR for a given QC, the remaining QCs could be used as substitutes of the given QC in most cases, without compromising the sustainability of water resources. Nonetheless, between 2005 and 2012, exceptions were observed for changes in MAR for C11H, C11J and C11L; hence the remaining QCs could be used as complements of these 3 QCs respectively. Between 1990 and 2005, only C11G made an exception. Similarly, between 1990 and 2005, such an exception was observed for C12D, C21B, C21C, C21D, C21E, C21F, C21G, C22A, C22B, C22C, C22D, C23E, C81G, C81H, C81J, C82B, C82F and C83A; and between 2005 and 2012 for C12D, C23D, C23E, C23F, C23G, C81C, C82A, C82D, C82E, C82F, C82G and C82H.

5.4.2. Constant Change in Entropy

The last column of each table (i.e., from Tables 2–19) showed that on average, changes in MAR of QCs could lead to a constant change in entropy for any QC, within the zone of functioning resilience. However, between 2005 and 2012, cases of high cross MAR elasticities were observed for C81C and C81D; likewise, for C81F, C81L and C81M, between 1990 and 2005. On average, for a constant change in entropy for a given QC, it was observed that positive cross MAR elasticities overweighed negative cross elasticities and vice-versa. Hence, for a constant change in entropy of a given QC, the remaining QCs could be used as substitutes or complements of the given QC, without compromising the sustainability of water resources, within the TC.

5.5. Spatial Distributions of Cross MAR Pseudo-Elasticity of Entropy

The spatial distribution of entropy MAR cross elasticities computed earlier in the different tables, were presented in Figure 6a,b and Figure 7a,b. For each of the following hydrological changes WR90/WR2005 and WR2005/WR2012, two cases were considered for spatial distribution, i.e., average cross MAR pseudo-elasticities under constant change in MAR and under constant entropy for a given QC. In this manner, it was easier to plot the value of cross-elasticity for each QC.

Figure 6a,b show the spatial distributions of average cross elasticities at constant entropy for the different QCs, between WR90 and WR2005; and between WR2005 and WR2012 data sets. In Figure 6a, it was observed that the majority of QCs were situated in the zone of functioning resilience since the cross elasticity values were in the interval (−2.5, 2.5). The results in this figure revealed that there were 64% of substitutable QCs as compared to 33% of complementary QCs and 3% of vulnerable QCs in the Upper Vaal. Substitutes were distributed across the majority of QCs and in all TCs. The Upper Vaal catchment could be a good location for water transfers between a given QC and its surrounding QCs of the same TC. Complements were situated mostly in the North West (urban built-up areas of the downstream of Vaal dam) and in the North East of upstream of Vaal dam. Besides water transfers, this could mean that in these locations of Upper Vaal, water could also be jointly used to balance water demand/supply in a specific QC, due to the change in MAR of the surrounding QCs, within their TC. Nonetheless, QCs associated with relatively higher cross-elasticities (>2.5) were located in the Wilge region. This region has been found to have the lowest water requirements currently and as projected in 2025 and this projection showed small differences in water requirements between 2000 and 2025 [20]. Further small increases of MAR as compared to entropy could only make more vulnerable water resources as far as a joint water use or water transfers between a specific QC and the remaining QCs are concerned.

From Figure 6b, it was observed that 54% of complementary QCs, 44% of substitutable QCs were in the zone of acceptable resilience, while 2% QCs were in the zone of low resilience. Hence the complements dominated over substitutes, but both were distributed fairly in all TCs. This could imply that while water transfers between QCs could be promoted in the Upper Vaal, there could be a huge potential for joint water use in over 50% of QCs, to maintain a good water balance in the different TCs. Once again, it was observed that vulnerable QCs were situated in Wilge and their number was insignificant as compared to resilient QCs. However, the level of vulnerability could be very high in vulnerable QCs, as shown through the computation of cross MAR pseudo-elasticities.

Figure 7a revealed that 57% of complementary QCs, 36% of substitutable QCs were in the zone of functionning resilience. Substitutes and complements were fairly distributed across the Upper Vaal. Although the majority of QCs were situated in the zone of functioning resilience, there were few QCs in the zone of low resilience (i.e., cross-elasticity <−2.5 or >2.5). These QCs constituted 7% and were mostly concentrated in the Wilge sub-region. For a TC of this sub-region, water resources planners should monitor relatively small changes (negative or positive) in MAR for a given QC as compared to the changes in entropy of surrounding QCs. Changes in climatic conditions (e.g., precipitation, evapotranspiration) that could cause such changes in entropy, should be assessed. A decrease in MAR and in yield could be experienced in South Africa, which is located in the South Hemisphere [41], where evaporation losses are higher than rainfall [20,42].

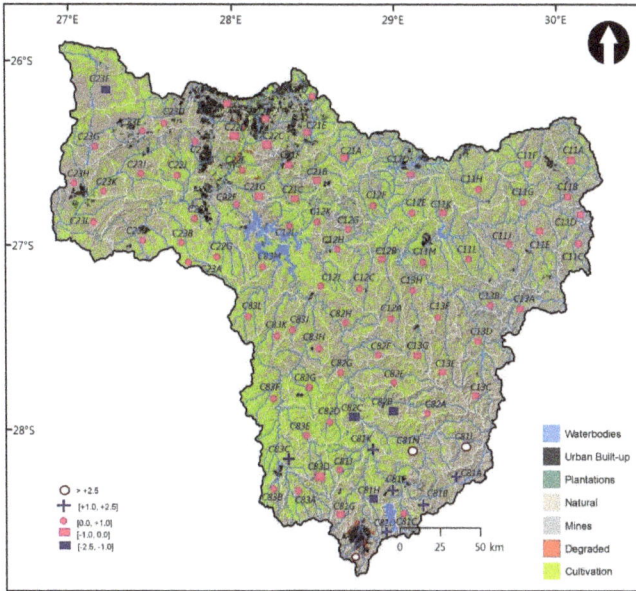

(a)

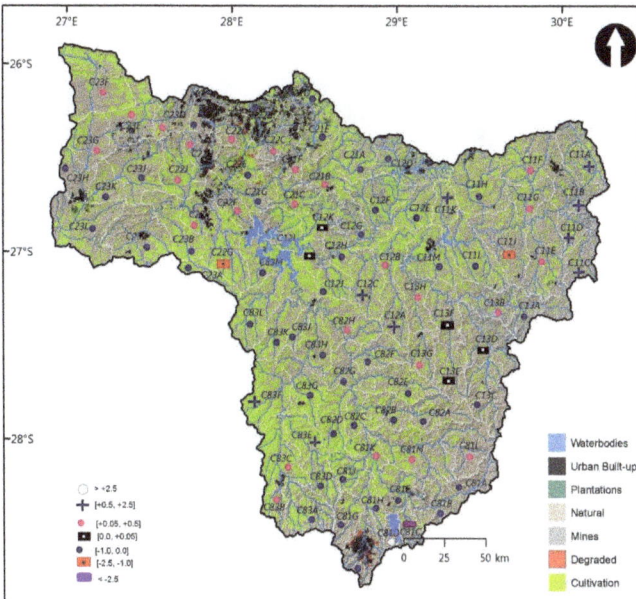

(b)

Figure 6. (**a**) Spatial distribution of average cross MAR pseudo-elasticities under constant entropy of QCs for WR90/WR2005. (**b**) Spatial distribution of average cross MAR pseudo-elasticities under constant entropy of QCs for WR2005/WR2012.

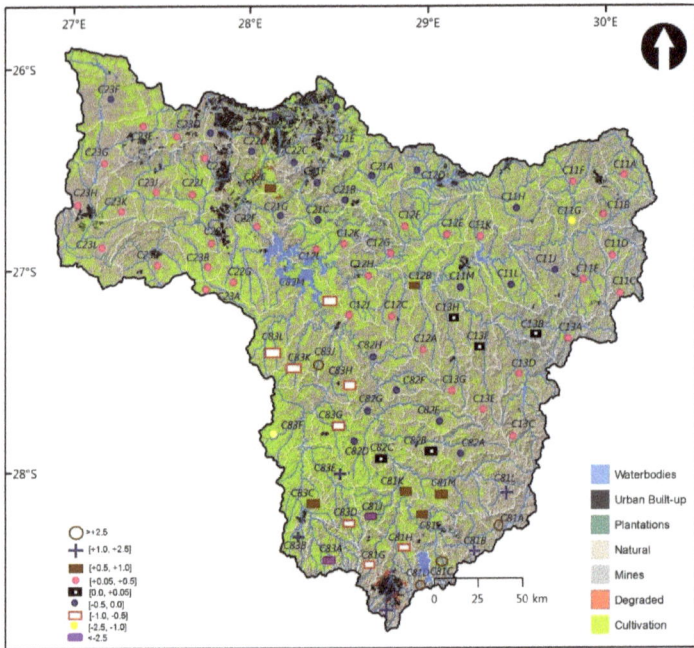

(a)

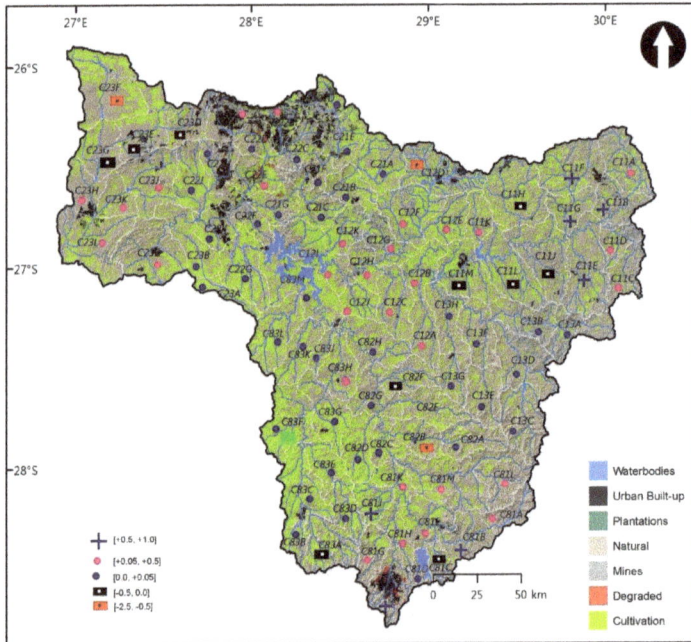

(b)

Figure 7. (**a**) Spatial distribution of average cross MAR pseudo-elasticities under constant MAR of QCs for WR90/WR2005. (**b**) Spatial distribution of average cross MAR pseudo-elasticities under constant MAR of QCs for WR2005/WR2012.

Figure 7b revealed that the substitutable QCs were distributed in the majority of the QCs and in all TCs, across Upper Vaal while complementary QCs were located in the North West of the downstream of dam Vaal, towards the centre of upstream of Vaal dam and the Wilge. All QCs were in the zone of acceptable resilience. Substitutes represented 86% as compared to 14% of complements. This showed a strong dominance of substitutes. Hence water transfer could prevail as compared to joint uses of water within the Upper Vaal. The spatial differences in cross MAR pseudo-elasticities for the Upper Vaal did not give any information related to natural factors such as soil, natural vegetation, topology, aquifers, etc.

5.6. Overall Cross MAR Elasticity of Entropy for Each Tertiary and Each Secondary Catchment in the Upper Vaal

The average cross elasticity of each TC was found to be positive and was relatively small (<1), as shown in Table 21. This showed that the different TCs were elastically stable (i.e., relatively inelastic) as far as changes in MAR were concerned. Hence water transfers, joint use of water from different QCs, etc. could result in the zone of acceptable uncertainty. Overall, the increase in MAR for the Upper Vaal catchment was associated with an increase in entropy, such that the catchment was in the zone of functioning resilience.

Table 21. Overall average cross MAR pseudo elasticity for tertiary catchments of Upper Vaal.

Tertiary Catchment	Elasticity [-]	
	WR90/2005	WR2005/WR2012
C11	−0.094	0.26
C12	0.15	0.16
C13	0.046	0.032
C21	−0.211	0.08
C22	0.122	0.045
C23	0.183	−0.048
C81	1.604	0.36
C82	−0.087	−0.209
C83	0.47	0.027

As presented in Table 22, the cross MAR pseudo-elasticity for each secondary catchment was roughly determined by averaging the values of cross elasticity for its TCs. The positive cross MAR pseudo-elasticities in the secondary catchments of Upper Vaal supported that substitutable QCs remained dominant within the zone of acceptable resilience for water resources. As said previously, a complete study should be conducted for cross MAR elasticity with TC pairs within the secondary catchments.

Table 22. Cross MAR pseudo elasticity for secondary catchments of Upper Vaal.

Secondary Catchment	Elasticity [-]	
	WR90/2005	WR2005/WR2012
Upstream of Vaal dam (C1)	0.03	0.15
Downstream of Vaal dam (C2)	0.03	0.077
Wilge (C8)	0.66	0.059

In this study, the influence of climatic parameters on cross MAR pseudo-elasticity was not investigated, but could be assumed to be the cause of changes in the naturalized MAR data sets of the Upper Vaal catchment. Besides, the influence of other natural conditions associated with the soil type, natural vegetation, natural water storages, and geology could also be informative in the determination of cross elasticity. The combined effect of climatic conditions, together with human activities could impact positively or negatively on cross MAR pseudo-elasticity. When considering the influence of human activities, historical data could be used to assess the change in cross MAR pseudo-elasticity.

6. Conclusions

For the first time, cross elasticity concept has been extended to hydrology and water resources using entropy concept; i.e., cross MAR pseudo-elasticity of entropy. The level of linkage between changes in MAR and changes in entropy of another (other) catchment(s) was established in the Upper Vaal. The analysis of cross MAR pseudo-elasticity of entropy was carried out on surface water resource data sets; i.e., WR90, WR2005 and WR2012. These data were in their naturalized form when extracted from WRC database and this could assume there was little influence of the human factor. Climatic controls could be associated with these data [36] as far as cross MAR elasticity was concerned. Thresholds of a linear resilience zoning [11] were extended to the assessment of cross elasticity values. The notion of cross MAR pseudo-elasticity was shown as a fair tool to assess the complementarity and substitutability for the different QCs of the TCs in the Upper Vaal region. That is, the use of water resources for some QCs could be done jointly, while in some cases, water transfer between QCs could be an option to sustain the yield (or water resources) from selected QCs, in a way to balance demand/supply in the region. However, the Wilge sub-region of Upper Vaal displayed promptness to water resources vulnerability in fewer QCs. Cross MAR elasticity characterized the correlation strength between QC pairs (weak versus high cross elasticity) to determine the resilience of catchments. Resilient QCs (in terms of water resources) displayed relatively weak cross elasticity values as opposed to vulnerable QCs, which were characterized by high cross elasticity values. In particular, instances of cross elasticities, close to 0 were observed; showing that a variation of MAR for a given QC was insignificant to cause an opposite variation in entropy for another QC. Although the resilience was acceptable in most cases, it was revealed that in some instances some QCs displayed very strong correlations. For instance, the global highest cross MAR elasticity of entropy was located in the TCs C83, followed by C81 and could go above 100% in magnitude for specific QCs. This implied that decision-makers could pay attention to water resources planning and management, and control of water infrastructures. A spatial analysis conducted on cross MAR pseudo-elasticity revealed that, in general, there was a fair distribution of complementary and substitutable QCs in the TCs. In most cases, the latter were dominant as compared to the former. Hence, this supported that water transfers between the different QCs and beyond could prevail in the Upper Vaal as it is the current situation [20]. However, this analysis did not give more information on the cross MAR elasticity differences related to the natural characteristics of the catchments. In general, it found out that the level of assurance to sustain water resources was relatively acceptable; hence low vulnerability risk of water resources, generally less than 20%. Overall, cross MAR elasticity for QCs of the Upper Vaal region yielded the zone of tolerable entropy.

However, it should be noted that entropy through cross MAR elasticity does not show how complementary or substitutable QCs should be managed and operated, a different time scales. The concept of cross MAR pseudo-elasticity was built on MAR pseudo-elasticity, therefore it considered only one parameter; i.e., MAR. Therefore, the inclusion of variables other than runoff, such as temperature, evaporation, precipitation should be investigated. This is a preliminary study dealing specifically with the intra-tertiary catchment assessment of cross MAR elasticity for QCs. Further work could be done on cross MAR elasticity between TCs within secondary catchments and on the assessment of its implications on the Upper Vaal catchment resilience and beyond. The impacts of climatic parameters on cross MAR pseudo-elasticity and other natural conditions associated with soil type, topography, natural vegetation, natural water storages and wetlands, could be investigated. Similar to previous studies on hydrological elasticity [2,7–9,11,29–31], the impact of these factors on cross elasticities investigated individually (bivariate relationships) could be different when these impacts are considered simultaneously in a multivariate configuration. The effect of unquantifiable land use changes should be taken into consideration. The combined effect of natural conditions, together with human activities such as land use, water use, etc. could impact positively or negatively on cross MAR pseudo-elasticity and would need to be also researched when historical data are considered. The concept should be investigated when considering runoff data of different time scales, i.e., monthly, seasonal, and annual, etc. The inclusion of parameters other than MAR in the determination of catchment vulnerability risk should be investigated. The linkage between cross MAR elasticity and the existing cross entropy concept could also be researched.

Acknowledgments: The author thanks the Water Research Commission of South Africa for making available data used in this study, on its website.

Conflicts of Interest: The author declare no conflict of interest.

References

1. Sankarasubramanian, A.; Vogel, R.M.; Limbrunner, J.F. Climate elasticity of streamflow in the United States. *Water Resour. Res.* **2001**, *37*, 1771–1781. [CrossRef]
2. Chiew, F.H.S. Estimation of rainfall elasticity of streamflow in Australia. *Hydrol. Sci. J.* **2006**, *51*, 613–625. [CrossRef]
3. Andréassian, V.; Coron, L.; Lerat, J.; Le Moine, N. Climate elasticity of streamflow revisited—An elasticity index based on long-term hydrometeorological records. *Hydrol. Earth Syst. Sci.* **2016**, *20*, 4503–4524. [CrossRef]
4. Mulugeta, D.; Greenfield, J.; Bolen, T.; Conley, L.; Health, C. Price- and cross-price elasticity estimation using SAS. In Proceedings of the SAS Global Forum 2013, San Francisco, CA, USA, 28 April–1 May 2013; pp. 1–17.
5. Santeramo, F.G. Cross-Price Elasticities for Oils and Fats in the US and the EU. 2017. Available online: http://www.theicct.org/sites/default/files/publications/Cross-price-elasticities-for-oils-fats-US-EU_ICCT_consultant-report_06032017.pdf (accessed on 21 October 2017).
6. Nowak, W.P.; Savage, I. The cross elasticity between gasoline prices and transit use: Evidence from Chicago. *Transp. Policy* **2013**, *29*, 38–45. [CrossRef]
7. Fu, G.; Charles, S.P.; Chiew, F.H.S. A two-parameter climate elasticity of streamflow index to assess climate change effects on annual streamflow. *Water Resour. Res.* **2007**, *43*, W11419. [CrossRef]
8. Vano, J.A.; Das, T.; Lettenmaier, D.P. Hydrologic sensitivities of Colorado River runoff to changes in precipitation and Temperature. *J. Hydrometeorol.* **2012**, *13*, 932–949. [CrossRef]
9. Meng, F.; Su, F.; Yang, D.; Tong, K.; Hao, Z. Impacts of recent climate change on the hydrology in the source region of the Yellow River basin. *J. Hydrol. Reg. Stud.* **2016**, *6*, 66–81. [CrossRef]
10. Allaire, M.C.; Vogel, R.M.; Kroll, C.N. The hydro-morphology of an urbanizing watershed using multivariate elasticity. *Adv. Water Resour.* **2015**, *86*, 147–154. [CrossRef]
11. Ilunga, M. Assessing catchment resilience using entropy associated with mean annual runoff for the Upper Vaal catchment in South Africa. *Entropy* **2017**, *19*, 147. [CrossRef]
12. Taragin, C.; Sandfort, M. The Antitrust Package. 2015. Available online: https://cran.r-project.org/web/packages/antitrust/vignettes/manual.pdf (accessed on 19 April 2017).
13. Buklin, R.E.; Russel, G.J.; Srinivasan, V. A relationship between market share elasticities and brand switching probabilities. *J. Mark. Res.* **1998**, *35*, 94–113. [CrossRef]
14. Song, I.; Chintagunta, P.K. Measuring cross-category price effects with aggregate store data. *Manag. Sci.* **2006**, *52*, 1594–1609. [CrossRef]
15. Cooper, L.G. Competitive maps: The structure underlying asymmetric cross elasticities. *Manag. Sci.* **1988**, *34*, 707–723. [CrossRef]
16. Kushkuley, A.; Wu, S.-M. A note on Lerner Index, cross-elasticity and revenue optimization Invariants. *arXiv* **2014**, arXiv:1402.1995.
17. Russel, G.J. A model of latent symmetry in cross price elasticities. *Mark. Lett.* **1992**, *3*, 157–169. [CrossRef]
18. Sanders, S.; Weisman, D.L.; Li, D. Child safety seats on commercial airliners: A demonstration of cross-price elasticities. *J. Econ. Educ.* **2008**, *39*, 135–144. [CrossRef]
19. Tambe, V.J.; Joshi, S.K. Estimating price elasticity of electricity for the major consumer categories of Gujarat State. *J. Electr. Eng.* **2015**, *1*, 367–374.

20. Department of Water Affairs. *Upper Vaal Management Area: Overview of Water Resources Availability and Utilisation*; Report No. PWA 08/000/00/0203; Department of Water Affairs: Pretoria, South Africa, 2003.

21. National Water Resource Strategy. Department of Water Affairs (Pretoria: DWA), South Africa. 2013. Available online: http://www.dwa.gov.za/nwrs/NWRS2013.aspx (accessed on 29 September 2016).

22. Wrzesinski, D. Use of Entropy in the Assessment of uncertainty of river runoff Regime in Poland. *Acta Geophys.* **2016**, *64*, 1825–1839. [CrossRef]

23. Wrzesinski, D. Uncertainty of flow regime characteristics of rivers in Europe. *Quaest. Geogr.* **2013**, *32*, 44–53. [CrossRef]

24. Zhou, F.; Xu, Y.P.; Chen, Y.; Xu, C.Y.; Gao, Y.Q.; Du, J.K. Hydrological response to urbanisation at different spatio-temporal scales simulated by coupling of CLUE-S and the SWAT model in the Yangtze River Delta region. *J. Hydrol.* **2013**, *485*, 113–125. [CrossRef]

25. Fan, J.; Huang, Q.; Chang, J.; Sun, D.; Cui, S. Detecting Abrupt change of streamflow at Lintong Station of Wei River. *Math. Probl. Eng.* **2013**, *2013*, 976591. [CrossRef]

26. Pan, S.; Liu, D.; Wang, Z.; Zhao, Q.; Zou, H.; Hou, Y.; Liu, P.; Xiong, L. Runoff responses to climate and land use/cover changes under future scenarios. *Water* **2017**, *9*, 475. [CrossRef]

27. Ilunga, M.; Singh, V.P. Measuring spatial variability of land use associated with hydrological impact in urbanized quaternary catchment using entropy. *Water SA* **2015**, *41*, 41–54. [CrossRef]

28. Zhang, L.; Singh, V.P. Bivariate rainfall and runoff analysis using entropy and Copula theories. *Entropy* **2012**, *14*, 1784–1812. [CrossRef]

29. Chiew, F.H.S.; Peel, M.C.; McMahon, T.A.; Siriwardena, L.W. Precipitation elasticity of streamflow in catchments across the world. In Proceedings of the Fifth FRIEND World Conference, Havana, Cuba, 27 November–1 December 2006; IAHS: Wallingford, UK, 2006; Volume 308, pp. 256–262.

30. Zhou, X.; Zhang, Y.; Yang, Y. Comparison of two approaches for estimating precipitation elasticity of Streamflow in China's Main River Basins. *Adv. Meteorol.* **2015**, *2015*, 1–8. [CrossRef]

31. Konapala, G.; Ashok, K.; Mishra, A.K. Three-parameter-based streamflow elasticity model: Application to MOPEX basins in the USA at annual and seasonal scales. *Hydrol. Earth Syst. Sci.* **2016**, *20*, 2545–2556. [CrossRef]

32. Middleton, B.J.; Bailey, A.K. *Water Resources of South Africa*, 2nd ed.; 2005 Study (WR2005), WRC Report No. TT 512/11; Water Research Commission: Pretoria, South Africa, 2011.

33. Midgley, D.C.; Pitman, W.V.; Middleton, B.J. *Surface Water Resources of South Africa 1990*, 1st ed.; WRC Report Nos. 298/1.1/94, User's Manual; Water Research Commission: Pretoria, South Africa, 1994; Volumes I–IV.

34. Atieh, M.; Rudra, R.; Gharabaghi, B.; Lubitz, D. Investigating the spatial and temporal variability of Precipitation using Entropy Theory. *J. Water Manag. Model.* **2017**, *25*, 1–8. [CrossRef]

35. Wei, C.; Yeh, H.-C.; Chen, Y.-C. Spatiotemporal scaling effect on rainfall network design using entropy. *Entropy* **2014**, *16*, 4626–4647. [CrossRef]

36. Liu, Q.; Yang, Z.; Cui, B. Spatial and temporal variability of annual precipitation during 1961–2006 in Yellow River Basin, China. *J. Hydrol.* **2008**, *361*, 330–338. [CrossRef]

37. Mishra, A.K.; Mehmet, O.; Singh, V.P. An entropy based investigation into the variability of precipitation. *J. Hydrol.* **2009**, *370*, 139–154. [CrossRef]

38. Price, D.W.; Mittelhammer, R.C. A matrix of demand elasticities for fresh fruit. *West. J. Agric. Econ.* **1979**, *151*, 69–86.

39. Houck, J.P. A Look at flexibilities and elasticities. *J. Farm Econ.* **1966**, *48*, 225–232. [CrossRef]

40. Water Resources of South Africa. 2012 Study (WR2012). Available online: http://waterresourceswr2012.co.za/ (accessed on 29 September 2016).

41. McMahon, T.A.; PeeL, M.C.; Pegram, G.G.S.; Smith, I.N. A Simple Methodology for Estimating Mean and Variability of Annual Runoff and Reservoir Yield under Present and Future Climates. *J. Hydrometeorol.* **2011**, *11*, 135–146. [CrossRef]

42. Jury, M.R. Economic Impacts of Climate Variability in South Africa and Development of Resource Prediction Models. *J. Meteorol.* **2001**, *41*, 46–55. [CrossRef]

entropy

MDPI

Article

Entropy Production in Stochastics

Demetris Koutsoyiannis

Department of Water Resources, School of Civil Engineering, National Technical University of Athens, 15780 Athina, Greece; dk@itia.ntua.gr

Received: 14 September 2017; Accepted: 23 October 2017; Published: 30 October 2017

Abstract: While the modern definition of entropy is genuinely probabilistic, in entropy production the classical thermodynamic definition, as in heat transfer, is typically used. Here we explore the concept of entropy production within stochastics and, particularly, two forms of entropy production in logarithmic time, unconditionally (EPLT) or conditionally on the past and present having been observed (CEPLT). We study the theoretical properties of both forms, in general and in application to a broad set of stochastic processes. A main question investigated, related to model identification and fitting from data, is how to estimate the entropy production from a time series. It turns out that there is a link of the EPLT with the climacogram, and of the CEPLT with two additional tools introduced here, namely the differenced climacogram and the climacospectrum. In particular, EPLT and CEPLT are related to slopes of log-log plots of these tools, with the asymptotic slopes at the tails being most important as they justify the emergence of scaling laws of second-order characteristics of stochastic processes. As a real-world application, we use an extraordinary long time series of turbulent velocity and show how a parsimonious stochastic model can be identified and fitted using the tools developed.

Keywords: entropy production; conditional entropy production; stochastic processes; scaling; climacogram; turbulence

1. Introduction

Entropy was first recognized as a probabilistic concept in 1887 by Boltzmann [1], who established a relationship of entropy with probabilities of statistical mechanical system states, thus explaining the Second Law of Thermodynamics as the tendency of the system to run toward more probable states. In 1948 Shannon [2] used an essentially similar, albeit more general, definition of entropy as a probabilistic concept, a measure of information or, equivalently, uncertainty. In 1957 Jaynes [3] introduced the principle of maximum entropy thus equipping the entropy concept with a powerful tool for logical inference.

A decade later, probabilistic entropy and the principle of maximum entropy were used in geophysical sciences and particularly hydrology, initially for parameter estimation of models [4] and probability distributions [5]. Detailed reviews on the use of entropy in applications in hydrology, and water and environmental engineering have been provided by Singh ([6,7] and more recently [8–11]). Most applications of probabilistic entropy are static, disregarding time. In studying systems out of the equilibrium, time should necessarily be involved (e.g., [12]) and the paths of the systems should be inferred. The more recently developed, within non-equilibrium thermodynamics, extremum principles of entropy production attempt to predict the most likely path of system evolution [13–15]. Central among the latter are Prigogine's minimum entropy production principle [16] and Ziegler's maximum entropy production principle [17,18].

Entropy production in thermodynamic systems is typically defined in terms of the derivative of entropy with respect to time, while the entropy flux in open systems is also considered. Niven and Ozawa [19] provide a general definition of entropy production along with a brief review of applications of extremum principles in geophysics and particularly hydrology. It can be seen that in these applications, the entropy concept is typically used with its classical thermodynamic definition and in deterministic terms, without reference to its modern probabilistic definition.

Deterministic descriptions of the evolution of uncertain physical and geophysical systems are generally inefficient [20,21], while a much more powerful alternative is offered by stochastics. The term 'stochastics' describes the mathematics of random variables and stochastic processes, and comprises probability theory, statistics and stochastic processes. While the entropy concept is well defined within stochastics, the entropy production does not have a scientifically mature definition. By analogy to classical thermodynamic versions, we may define entropy production in stochastics as the derivative of (probabilistic) entropy with respect to time. As here we deal with a single stochastic process and not with interaction of many processes, there are no entropy fluxes and, thus, entropy production should merely rely on the time derivative of entropy. In an earlier study [22], the derivative with respect to the logarithm of time was introduced and termed 'entropy production in logarithmic time' (EPLT). Differentiation with respect to the logarithm of time has some attractive elements, such as:

- It is dimensionless as $d(\ln t) = dt/t$ and thus, given that entropy is a dimensionless quantity per se (see next section), the EPLT remains a dimensionless quantity.
- It is consistent with the notion of the 'arrow of time'. Usually entropy per se is related to the time asymmetry and the Second Law is regarded as the origin of the arrow. However, it may be simpler to think that it is the action of observation that creates time asymmetry. In this respect, we can set $t = 0$ for the observation time, i.e., the present, while $t < 0$ denotes the observable past and $t > 0$ denotes the unknown future. The fact that the past is observable, but the future not, generates the asymmetry. It is clarified that when the future becomes observable, it is no longer future; rather it has become present or past. Once the future ($t > 0$) is treated separately from the past, it is legitimate to differentiate with respect to the logarithm of time.
- It makes extremization of entropy production easier, particularly for asymptotic times $t \rightarrow 0$ and $t \rightarrow \infty$, as it avoids infinite or zero values of entropy production [22]; indeed, the asymptotic values of EPLT are always bounded (see Section 2.4).
- It provides the means to study or even explain the scaling behaviour often observed in geophysical processes (see Sections 2.2 and 2.4).

Here we advance the study of the EPLT concept, with particular emphasis on the conditional entropy production (CEPLT), when the past and present have been observed (Section 2.2). The theoretical properties of EPLT and CEPLT are studied in a general setting and in application to a broad set of specific types of stochastic processes (Sections 3.1 and 3.2). A main question studied is how the entropy production can be estimated from time series, i.e., realizations of stochastic processes. It turns out that there is a strong link of the EPLT with the climacogram, with the slope of a log-log plot of the latter representing the EPLT. Further, we introduce two new tools, additional to the climacogram and based on it, the differenced climacogram and the climacospectrum, where the latter has properties similar to those of the power spectrum. The slope of a log-log plot of either of the two additional tools can be used as an estimator of CEPLT. Of particular importance are the asymptotic slopes for time scale tending to zero or infinity (Section 3.3). To illustrate the theoretical and empirical properties of EPLT and CEPLT using real-world data and to test the applicability of the specific models proposed, a long time series of very fine resolution is needed in order to allow viewing the real-world behaviour at the far-left and the far-right tails. Fine resolution measurements of turbulence can provide an ideal test series and indeed a time series of length 36×10^6 of turbulent velocities is utilized for illustration and testing (Section 3.4).

To avoid bothering the reader with derivations and theoretical details, most of them have been put in a series of annexed sections organized as Appendix A.1. In addition, the details of the specific stochastic models and their behaviour have been grouped in Appendix A.2. While the two appendices are made in a stand-alone form separate from the body of the article, they are perhaps the most essential part of the study.

2. Methods

2.1. General Context

The evolution of a system state x over time t is usually represented as a trajectory $x(t)$, but such a description is often inefficient. Instead, the evolution can be represented as a stochastic process $\underline{x}(t)$, which is a collection of (infinitely many) random variables $\underline{x}$ indexed by t. A realization (sample) $x(t)$ of $\underline{x}(t)$ is a trajectory; if it is known at certain points t_i, $i = 1, 2, \ldots$, the realization is called a time series.

A random variable $\underline{x}$ is an abstract mathematical entity associated with a probability density (or mass) function $f(x)$; the realizations x of $\underline{x}$ belong to a set of possible numerical values. Notice the different notation of random variables and hence stochastic processes (underlined, according to the Dutch notation [23]) from regular ones.

Most natural processes evolve in continuous time but they are observed in discrete time, typically by averaging. Accordingly, the stochastic processes devised to represent the natural processes should evolve in continuous time and be converted into discrete time, as illustrated in Figure 1.

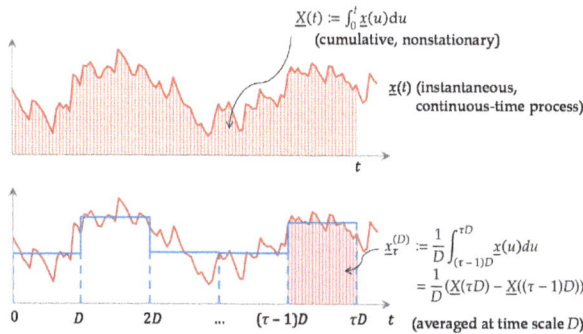

$$\underline{X}(t) := \int_0^t \underline{x}(u)du$$
(cumulative, nonstationary)

$\underline{x}(t)$ (instantaneous, continuous-time process)

$$\underline{x}_\tau^{(D)} := \frac{1}{D}\int_{(\tau-1)D}^{\tau D} \underline{x}(u)du$$
$$= \frac{1}{D}(\underline{X}(\tau D) - \underline{X}((\tau-1)D))$$

(averaged at time scale D)

Figure 1. Explanatory sketch for a stochastic process in continuous time and in discrete time. Note that the graphs display a realization of the process (it is impossible to display the process as such) while the notation is for the process per se (adapted from [24]).

While a stochastic process denotes, by conception, change (process = a series of changes), there should be some properties that are unchanged in time. This implies the concept of stationarity [25], which is central in stochastics. According to Kolmogorov's definition [26], a stochastic process $\underline{x}(t)$ is stationary if the distributions of $(x(t_1), x(t_2), \ldots, x(t_n))$ and $(x(t_1 + \tau), x(t_2 + \tau), \ldots, x(t_n + \tau))$ coincide for any n, t_1, t_2, $\ldots$, t_n, τ. By negation, in a nonstationary process the probability density $f(x(t_1), x(t_2), \ldots, x(t_n))$ for some (or all) τ is not equal to $f(x(t_1 + \tau), x(t_2 + \tau), \ldots, x(t_n + \tau))$, which means that the mathematical expression of the latter should explicitly contain the time shift τ, or else that it should be a deterministic function of τ.

Stationarity is closely related to ergodicity, which in turn is a prerequisite to make inference from data, that is, induction. Within the stochastics domain ([20,27] (p. 427)) ergodicity is defined in the following manner: A stochastic process $\underline{x}(t)$ is ergodic if the time average of any function $g(\underline{x}(t))$, as time tends to infinity, equals the true (ensemble) expectation $E[g(\underline{x}(t))]$. This allows the estimation (i.e., approximate calculation) of the true but unknown quantity $E[g(\underline{x}(t))]$ (e.g., the true average $E[\underline{x}(t)]$) from the available data, i.e., from the sample mean). Without stationarity there is no ergodicity and without ergodicity inference from data is impossible. More details about stationarity and ergodicity, and their importance, are provided by Koutsoyiannis and Montanari [25], along with highlights of the misconceptions and abuses of these concepts which abound in the literature.

For the remaining part of this article, unless otherwise stated, the processes are assumed to be stationary and ergodic, noting that nonstationary processes should be converted to stationary before their study. For example, the cumulative process $\underline{X}(t)$ in Figure 1 is nonstationary, but by differentiating it in time we obtain the stationary process $\underline{x}(t)$.

Without loss of generality we assume that the mean of the process $\underline{x}(t)$ is zero ($E[\underline{x}(t)] = 0$) and we denote its variance as:

$$\gamma_0 := \mathrm{Var}[\underline{x}(t)] \tag{1}$$

Let:

$$\Gamma(t) := \mathrm{Var}[\underline{X}(t)], \ \gamma(k) := \mathrm{Var}[(1/k)(\underline{X}(t+k) - \underline{X}(t))] = \Gamma(k)/k^2 \tag{2}$$

be the variances of the cumulative and the averaged process, respectively (note that $\Gamma(0) = 0$ and $\gamma(0) = \gamma_0$). While $\Gamma(t)$, the variance of the nonstationary process $\underline{X}(t)$, is a function of t, $\gamma(k)$ is the variance of the stationary process $\underline{x}_\tau^{(k)}$, i.e., $\gamma(k) = \mathrm{Var}[\underline{x}_\tau^{(k)}]$ and thus it is not a function of time τ. We call $\Gamma(t)$ and $\gamma(k)$, as functions of time t and time scale k, the *cumulative climacogram* and the *climacogram* of the process, respectively.

The climacogram is the second central moment of the process, as a function of time scale, and thus it is a second-order characteristic of the process. Other customary second order characteristics of a stationary process are the autocovariance function $c(h)$, where h denotes time lag, the power spectrum $s(w)$, where w denotes frequency, and the structure function $v(h)$ (also known, for a stationary process, as the semivariogram or variogram). The latter has a direct analogy based on the climacogram, the climacostructure function $\xi(k)$. Definitions and notation for these second order characteristics are contained in Table 1 for the continuous time domain and in Table 2 for the discrete time case. All of these functions are transformations of one another, as shown in Table 3.

The climacogram, like the autocovariance function, is a positive definite function (see proof in Appendix A.1.2) but of the time scale k, rather than the time lag h. It is not as popular as the other tools but it has several good properties due to its simplicity, close relationship to entropy (see below), and more stable behaviour, which is an advantage in model identification and fitting from data. In particular, when estimated from data, the climacogram behaves better than all other tools, which involve high bias and statistical variation [24,28].

Table 1. Definition of main characteristics and notations for a stochastic process in continuous time; see Figure 1 for clarification.

Name of Quantity or Characteristic	Symbol and Definition	Remarks	Equation
Stochastic process of interest	$\underline{x}(t)$	Assumed stationary	
Time, continuous	t	Dimensional quantity	
Cumulative process	$\underline{X}(t) := \int_0^t \underline{x}(\xi)\,d\xi$	Nonstationary	(3)
Variance, instantaneous	$\gamma_0 := \mathrm{Var}[\underline{x}(t)]$	Constant (not a function of t)	(4)
Cumulative climacogram	$\Gamma(t) := \mathrm{Var}[\underline{X}(t)]$	A function of t, $\Gamma(0) = 0$	(5)
Climacogram	$\gamma(k) := \mathrm{Var}[(1/k)(\underline{X}(t+k) - \underline{X}(t))]$ $= \mathrm{Var}[\underline{X}(k)/k] = \Gamma(k)/k^2$	Not a function of t, $\gamma(0) = \gamma_0$	(6)
Time scale, continuous	k	Units of time	
Autocovariance function	$c(h) := \mathrm{Cov}[\underline{x}(t), \underline{x}(t+h)]$	$c(0) = \gamma_0$	(7)
Time lag, continuous	h	Units of time	
Structure function (or semivariogram or variogram)	$v(h) := \frac{1}{2}\mathrm{Var}[\underline{x}(t) - \underline{x}(t+h)]$		(8)
Climacostructure function	$\xi(k) := \gamma_0 - \gamma(k)$		(9)
Power spectrum (or spectral density)	$s(w) := 4\int_0^\infty c(h)\cos(2\pi wh)\,dh$		(10)
Frequency, continuous	$w = 1/k$	Units of inverse time	(11)

Table 2. Definition of main characteristics and notations for a stochastic process in discrete time and their relationship with those in continuous time; see Figure 1 for clarification.

Name of Quantity or Characteristic	Symbol and Definition	Remarks	Equation
Stochastic process, discrete time	$\underline{x}_\tau^{(D)} := \frac{1}{D}\int_{(\tau-1)D}^{\tau D}\underline{x}(u)du = \frac{1}{D}\left(\underline{X}(\tau D) - \underline{X}((\tau-1)D)\right)$		(12)
Time unit = discretization time step	D	Time step or length of time window of averaging	
Time, discrete	$\tau := t/D$	Dimensionless quantity, integer	(13)
Characteristic variance	$\mathrm{Var}[\underline{x}_\tau^{(D)}] = \gamma(D)$		(14)
Climacogram	$\gamma_\kappa^{(D)} := \gamma(\kappa D) = \frac{\Gamma(\kappa D)}{(\kappa D)^2}$	$\gamma_1^{(D)} = \gamma(D)$	(15)
Time scale, discrete	$\kappa := k/D$	Dimensionless quantity	(16)
Autocovariance function	$c_\eta^{(D)} := \mathrm{Cov}[\underline{x}_\tau^{(D)}, \underline{x}_{\tau+\eta}^{(D)}]$	$c_0^{(D)} = \gamma(D)$	
Time lag, discrete	$\eta := h/D$	Dimensionless quantity	(17)
Structure function	$v_\eta^{(D)} := \gamma(D) - c_\eta^{(D)}$		(18)
Power spectrum	$s_d^{(D)}(\omega) := \frac{1}{D}\sum_{j=-\infty}^{\infty}s\left(\frac{\omega+j}{D}\right)\mathrm{sinc}^2(\pi(\omega+j))$		(19)
Frequency, discrete	$\omega := wD = 1/\kappa$	Dimensionless quantity	(20)

Note: In time-related quantities, Latin letters denote dimensional quantities and Greek letters dimensionless ones. The Latin i, j, l may also be used as integers to denote quantities τ, η, κ, depending on the context.

Table 3. Relationships between several characteristics of a process in continuous and discrete time.

Related Characteristics	Direct Relationship	Inverse Relationship	Equation
$\gamma(k) \leftrightarrow c(h)$	$\gamma(k) = 2\int_0^1(1-\chi)c(\chi k)d\chi$	$c(h) = \frac{1}{2}\frac{d^2(h^2\gamma(h))}{dh^2}$	(21)
$s(w) \leftrightarrow c(h)$	$s(w) := 4\int_0^\infty c(h)\cos(2\pi wh)dh$	$c(h) = \int_0^\infty s(w)\cos(2\pi wh)dw$	(22)
$\gamma(k) \leftrightarrow s(w)$	$\gamma(k) = \int_0^\infty s(w)\,\mathrm{sinc}^2(\pi wk)dw$	$s(w) := 2\int_0^\infty \frac{d^2(h^2\gamma(h))}{dh^2}\cos(2\pi wh)dh$	(23)
$v(h) \leftrightarrow c(h)$	$v(h) = \gamma_0 - c(h)$	$c(h) = v(\infty) - v(h)\quad (v(\infty) = \gamma_0)$	(24)
$\xi(k) \leftrightarrow \gamma(k)$	$\xi(k) := \gamma_0 - \gamma(k)$	$\gamma(k) = \xi(\infty) - \xi(k)\quad (\xi(\infty) = \gamma_0)$	(25)
$\xi(k) \leftrightarrow v(h)$	$\xi(k) = 2\int_0^1(1-\chi)v(\chi k)d\chi$	$v(h) = \frac{1}{2}\frac{d^2(h^2\xi(h))}{dh^2}$	(26)
$\gamma_\kappa^{(D)} \equiv \gamma(\kappa D)$ $\leftrightarrow c_\eta^{(D)}$	$\gamma_\kappa^{(D)} = \frac{1}{\kappa}\left(c_0^{(D)} + 2\sum_{\eta=1}^{\kappa-1}(1-\frac{\eta}{\kappa})c_\eta^{(D)}\right)$ Alternatively, $\gamma_\kappa^{(D)} = \frac{\Gamma(\kappa D)}{(\kappa D)^2}$ where, in recursive mode, $\Gamma(\kappa D) = 2\Gamma((\kappa-1)D) - \Gamma((\kappa-2)D) + 2c_{\kappa-1}^{(D)}D^2$ with $\Gamma(0) = 0$, $\Gamma(D) = c_0^{(D)}D^2$	$c_\eta^{(D)} = \frac{1}{D^2}\left(\frac{\Gamma(|\eta+1|D)+\Gamma((|\eta|-1|D)}{2} - \Gamma(|\eta|D)\right)$	(27)
$c_\eta^{(D)} \leftrightarrow s_d^{(D)}(\omega)$	$s_d^{(D)}(\omega) = 2c_0^{(D)} + 4\sum_{\eta=1}^\infty c_\eta^{(D)}\cos(2\pi\eta\omega)$	$c_\eta^{(D)} = \int_0^{1/2}s_d^{(D)}(\omega)\cos(2\pi\omega\eta)d\omega$	(28)
$v_\eta^{(D)} \leftrightarrow c_\eta^{(D)}$	$v_\eta^{(D)} = \gamma(D) - c_\eta^{(D)}$	$c_\eta^{(D)}\gamma(D) - v_\eta^{(D)}$	(29)

The climacogram involves bias too, but this can be determined analytically and included in the estimation. Specifically, assuming that we have n observations of the averaged process $\underline{x}_i^{(k)}$, at scale $k = \kappa D$, so that $T = nk$ is the observation period (rounded off to an integer multiple of k), the standard statistical estimator of variance has expected value:

$$E\left[\hat{\underline{\gamma}}(k)\right] = \theta(k, T)\gamma(k) \tag{30}$$

where θ is the bias correction coefficient, given by [22,24]:

$$\theta(k,T) = \frac{1 - \gamma(T)/\gamma(k)}{1 - k/T} = \frac{1 - (k/T)^2 \Gamma(T)/\Gamma(k)}{1 - k/T} \tag{31}$$

It becomes clear from the above equations, which are valid for any process, that direct estimation of the variance $\gamma(k)$ for any time scale k, let alone that of the instantaneous process γ_0, is not possible merely from the data. We need to know the ratio $\gamma(T)/\gamma(k)$ and thus we should assume a stochastic model which evidently influences the estimation of $\gamma(k)$. Once the model is assumed and its parameters estimated based on the data, we can expand our calculations to estimate the variance for any time scale, including that of the instantaneous scale γ_0.

2.2. Entropy and Entropy Production

Historically entropy was introduced in thermodynamics but its rigorous definition was given within probability theory. Thermodynamic and probabilistic entropy are regarded by many as different concepts having in common only their name. However, here we embrace the view that they are essentially the same thing, as has articulated elsewhere [29,30]. According to the latter interpretation, the mathematical description of thermodynamic systems could be produced by the probabilistic definition of entropy. As an example indicating how powerful this interpretation could be, the law of phase change transition of water (Clausius-Clapeyron equation) has been produced in [30], with impressive agreement with reality, by maximizing, in a formal probabilistic frame, the entropy, i.e., the total uncertainty about the state of a single molecule of water.

In this frame, entropy (often called Boltzmann-Gibbs-Shannon entropy to give credit to its pioneers and distinguish it from other forms of generalized entropies) is a *dimensionless* measure of uncertainty defined as follows:

- For a discrete random variable $\underline{x}$ with probability mass function $P_j := P\{\underline{x} = x_j\}, j = 1, \dots, w$:

$$\Phi[\underline{x}] := E[-\ln P\underline{x}] = -\sum_{j=1}^{w} P_j \ln P_j \tag{32}$$

- For a *continuous random variable* $\underline{z}$ with probability density function $f(z)$:

$$\Phi[\underline{x}] := E\left[-\ln \frac{f(\underline{x})}{m(\underline{x})}\right] = -\int_{-\infty}^{\infty} \ln \frac{f(x)}{m(x)} f(x) dx \tag{33}$$

where $m(x)$ is the density of a background measure (usually Lebesgue, i.e., $m(x) = 1$, with dimensions $[x^{-1}]$).

Entropy acquires its importance from the *principle of maximum entropy* [3], which postulates that the entropy of a random variable should be at maximum, under some conditions, formulated as constraints, which incorporate the information that is given about this variable. Its physical counterpart, the tendency of entropy to become maximal (Second Law of thermodynamics) is the driving force of natural change.

The definition of the entropy of a stochastic process directly follows that of a random variable. Thus, the entropy of the cumulative process $\underline{X}(t)$ with probability density function $f(X; t)$ is a dimensionless quantity defined as:

$$\Phi[\underline{X}(t)] := E\left[-\ln \frac{f(\underline{X}; t)}{m(\underline{X})}\right] = -\int_{-\infty}^{\infty} \ln \frac{f(X; t)}{m(X)} f(X; t) dX \tag{34}$$

In a stochastic process, which by definition involves time, it is natural to consider the change of entropy in time. Intuitively, for the stochastic process $\underline{X}(t)$ we can define entropy production as the time derivative, i.e., $\Phi'[\underline{X}(t)] := d\Phi[\underline{X}(t)]/dt$. This has dimensions of inverse time. Koutsoyiannis [22] defined the entropy production in logarithmic time (EPLT) as the derivative of entropy in logarithmic time:

$$\phi(t) \equiv \phi[\underline{X}(t)] := \Phi'[\underline{X}(t)]\, t \equiv d\Phi[\underline{X}(t)]/d(\ln t) \tag{35}$$

and showed that it has some advantages over $\Phi'[\underline{x}(t)]$ as described above.

Assuming that the background density is constant, $m(X) \equiv m$, and using simple constraints related to the preservation of the mean and variance, the principle of maximum entropy yields that the probability distribution of $\underline{X}$ will be Gaussian. For a Gaussian process the entropy depends on its variance $\Gamma(t)$ and is:

$$\Phi[\underline{X}(t)] = (1/2)\ln(2\pi e\, \Gamma(t)/m^2) \tag{36}$$

Hence:

$$\phi(t) = \Gamma'(t)\, t/2\Gamma(t) = 1 + \gamma'(t)\, t/2\gamma(t) \tag{37}$$

Because of the simplicity of the Gaussian case, we are basing further analyses on that assumption, noting that for large times, due to the Central Limit Theorem, the cumulative process $\underline{X}(t)$ will tend to Gaussian, even if for the smallest scales it deviates from Gaussian.

A question arises whether $\phi(t)$ is a totally abstract concept useful for theoretical analyses only or it can be related to data, and visualized or estimated from them. The reply is fortunate if, in addition, we consider another concept, the log-log derivative (LLD) of a function $f(x)$, formally expressed by:

$$f^{\#}(x) := \frac{d(\ln f(x))}{d(\ln x)} = \frac{x f'(x)}{f(x)} \tag{38}$$

This is visualized as the (local) slope on the popular log-log plot (plot of $\log f(x)$ vs. $\log x$). It is then seen that (37) can be written as:

$$\phi(t) = \tfrac{1}{2}\,\Gamma^{\#}(t) = 1 + \tfrac{1}{2}\,\gamma^{\#}(t) \tag{39}$$

In other words, EPLT is visualized and estimated by the slope of a log-log plot of the climacogram.

The quantity $\phi(t)$ represents the entropy production when nothing is known (observed) about the process realization. More interesting is the conditional EPLT (CEPLT, $\phi_C(t)$) when the past ($t < 0$) and the present ($t = 0$) are observed. In this case, instead of the unconditional variance $\Gamma(t)$ we should use a variance $\Gamma_C(t)$ conditional on the past and present. It has been shown [22,24] that for a Markov process (see definition of the Markov process in Section 3.1 and its details in Appendix A.2.1), the conditional variance is:

$$\Gamma_C(t) \approx 2\Gamma(t) - \frac{\Gamma(2t)}{2}, \qquad \gamma_C(k) \approx 2(\gamma(k) - \gamma(2k)) \tag{40}$$

We note that the difference $\gamma(k) - \gamma(2k)$ is always nonnegative (this is shown in Appendix A.1.3, Equation (A21)) and thus it can indeed represent a (conditional) variance.

For a non-Markov process the following adaptation is necessary:

$$\gamma_C(k) = \varepsilon(\gamma(k) - \gamma(2k)) \tag{41}$$

where ε is a constant, ensuring that as $k \to \infty$, $\gamma_C(k)/\gamma(k) \to 1$; the rationale of this limit is that the information about the past and present will tend not to affect the far distant future. As shown in the Appendix A.1.4, its value is:

$$\varepsilon = \frac{1}{1 - 2\gamma^{\#}(\infty)} \tag{42}$$

It is easily verified that for a Markov process in which $\gamma^{\#}(\infty) = -1$ (see Appendix A.2.1), $\varepsilon = 2$, while for a Hurst-Kolmogorov process (original or filtered; see details in [24], in next session and in Appendix A.2) for which $\gamma^{\#}(\infty) = 2H - 2$, $\varepsilon = 1/(1 - 2^{2H-2})$, where H is the Hurst parameter.

Due to its mathematical form, we will refer to $\gamma_C(k)$ as the differenced climacogram. The inverse transformation, i.e., that giving the climacogram $\gamma(k)$ once the differenced climacogram $\gamma_C(k)$ is known, can be easily deduced by iterative evaluations also observing that $\gamma(\infty) = \gamma_C(\infty) = 0$ (this is a necessary condition in order for a stochastic process to be ergodic [27] (p. 429)) and takes both the following forms:

$$\gamma(k) = \frac{1}{\varepsilon} \sum_{i=0}^{\infty} \gamma_C\left(2^i k\right) = \gamma(0) - \frac{1}{\varepsilon} \sum_{i=1}^{\infty} \gamma_C\left(2^{-i} k\right) \tag{43}$$

whereas for numerical evaluations it should be kept in mind that for large time scales k, $\gamma(k) \approx \gamma_C(k)$. Note that (43) entails:

$$\gamma(0) = \frac{1}{\varepsilon} \sum_{i=-\infty}^{\infty} \gamma_C\left(2^i k\right) \tag{44}$$

and this is valid for any k. The importance of the differenced climacogram stems from the fact that it is directly related to the CEPLT, through its obvious relationship (similar to (39)):

$$\varphi_C(k) = 1 + 1/2\gamma_C^{\#}(k) \tag{45}$$

As the differenced climacogram can be easily visualized from data, because it only needs calculation of variances, it directly provides a means for estimation and visualization of the CEPLT in terms of the slope in a log-log plot of the differenced climacogram.

In Appendix A.1.5 it is shown that the asymptotic properties of $\gamma_C(k)$ are:

$$\gamma_C(0) = \gamma_C(\infty) = 0, \quad \gamma_C^{\#}(0) = \xi^{\#}(0), \quad \gamma_C^{\#}(\infty) = \gamma^{\#}(\infty) \tag{46}$$

In other words, we need both the climacogram $\gamma(k)$ and the climacostructure function $\xi(k)$ to express the asymptotic behaviour of $\gamma_C(k)$. Once this is known, the asymptotic behaviour of CEPLT is also known from (45).

As shown in Appendix A.1.6, the differenced climacogram for persistent processes (see their definition in next subsection) is not integrable in $(0, \infty)$. However, by slightly changing it multiplying with k, we can obtain an additional tool which is integrable and has some additional good properties. Specifically, we introduce the following tool, which, as we will see next, resembles the power spectrum and thus we refer to it as the climacospectrum:

$$\zeta(k) := \frac{k\big(\gamma(k) - \gamma(2k)\big)}{\ln 2} \tag{47}$$

The climacospectrum is also written in an alternative manner in terms of frequency $w = 1/k$:

$$\tilde{\zeta}(w) := \zeta(1/w) = \frac{\gamma(1/w) - \gamma(2/w)}{(\ln 2)w} \tag{48}$$

The inverse transformation, i.e., that giving the climacogram $\gamma(k)$ once the climacospectrum $\zeta(k)$ is known, can be easily deduced with the same method as before, and takes the form:

$$\gamma(k) = \ln 2 \sum_{i=0}^{\infty} \frac{\zeta(2^i k)}{2^i k} = \gamma(0) - \ln 2 \sum_{i=1}^{\infty} \frac{\zeta(2^{-i} k)}{2^{-i} k} \tag{49}$$

A very good approximation in which the sum is replaced by an integral, is:

$$\gamma(k) \approx \frac{(\varepsilon \ln 2)k + 2}{2k + 2} I(k), \quad I(k) := \int_k^{\infty} \frac{\zeta(x)}{x^2} dx = \int_0^{1/k} \tilde{\zeta}(w) dw \tag{50}$$

This holds for all k and its details are given in Appendix A.1.7. For small k, it is easy to see that the multiplier of the integral $I(k)$ is 1. It follows that for small time scales k or large frequencies $w = 1/k$ the climacospectrum, expressed in terms of frequency, has an area $I(k)$ on the left of w equal to the variance $\gamma(1/w)$. As shown in Appendix A.1.6, the entire area under the curve $\tilde{\zeta}(w)$ (i.e., $I(0)$) is precisely equal to the variance $\gamma(0)$ of the instantaneous process, a property also shared by the power spectrum $s(w)$. This is not the only connection with the power spectrum. The climacospectrum has also the same asymptotic behaviour with the power spectrum, i.e.,

$$\tilde{\zeta}^{\#}(0) = -\zeta^{\#}(\infty) = s^{\#}(0), \quad \tilde{\zeta}^{\#}(\infty) = -\zeta^{\#}(0) = s^{\#}(\infty) \tag{51}$$

(see Appendix A.1.9). This property holds almost always, with the exception of the cases where $\zeta^{\#}(0)$ is a specific integer ($\zeta^{\#}(\infty) = -1$ or $\zeta^{\#}(0) = 3$, as explained in Appendix A.1.9).

Again the connection of the climacospectrum with the CEPLT emerges through log-log derivatives. Specifically, combining (41), (45) and (47) we find that $\zeta^{\#}(k) = 1 + \gamma_C^{\#}(k) = 1 + 2\varphi_C(k) - 2$; hence:

$$\varphi_C(k) = \tfrac{1}{2}\left(1 + \zeta^{\#}(k)\right) = \tfrac{1}{2}\left(1 - \tilde{\zeta}^{\#}(1/k)\right) \tag{52}$$

By virtue of (51) and with the exceptions mentioned, the asymptotic CEPLT can also be inferred from the slopes of the power spectrum and, as the climacospectrum slopes do not differ substantially from those of the power spectrum even for intermediate time scales, the power spectrum slopes could also be used for a qualitative estimation if the CEPLT.

Obviously, however, the climacospectrum is more advantageous than the power spectrum in this respect, because the connection with conditional entropy production is more precise and without exceptions at all. In addition, intuitively the variance, on which the definition of the climacospectrum is based, is more closely related to uncertainty, and hence entropy of the process, than the power spectrum and the autocovariance. Other advantages are the easy calculation only using the concept of variance, without any need to perform laborious transformations (such as Fourier) and the fact that, like the climacogram, it is not affected by discretization (while autocovariance and power spectrum are). The empirical climacogram and climacospectrum are easily determined from data using nothing more than the standard statistical estimator of variance and they have a smooth shape, much smoother than those of the empirical autocovariance and power spectrum, thus enabling better model identification and fitting.

When the objective is the model fitting, we should be aware that any of the statistical tools entail estimation bias, which should be accounted for in the fitting. Again the climacogram and the climacospectrum are preferable than other tools as the bias is explicitly calculated from the assumed model (Equations (30) and (31)). In particular, the bias of the climacospectrum is usually very small (see the relevant graphs in Appendix A.2) because of its definition as a difference of two variances, in which the biases tend to cancel out. In comparison with another version of climacogram-based spectrum [24], the climacospectrum introduced here has advantages. Namely, these are its simpler expression, as it is a linear transformation (difference) of the climacogram, and the absence of the variance $\gamma_0 \equiv \gamma(0)$ from the definition, which makes possible the calculation of the empirical climacospectrum prior to specifying the model (note that γ_0 is not known before a model is specified and its parameters are estimated).

2.3. Scaling

In the previous subsection we have shown that the EPLT and the CEPLT are related to LLDs or slopes of log-log plots of second order tools such as climacogram, climacospectrum, power spectrum, etc. With a few exceptions, these slopes are nonzero asymptotically, hence entailing asymptotic scaling or asymptotic power laws with the LLDs being the scaling exponents. It is intuitive to expect that an emerging asymptotic scaling law would provide a good approximation of the true law for a range of scales. If the scaling law was appropriate for the entire range of scales, then we would have a simple scaling law. Such simple scaling sounds attractive from a mathematical point of view, but, for reasons that will be explained in the next subsection, it turns out to be impossible to make it appropriate for physical processes.

It is thus physically more realistic to expect two different types of asymptotic scaling laws, one in each of the ends of the continuum of scales. We call these two behaviours local scaling or fractal behaviour, when $k \to 0$ or $w \to \infty$, and global scaling, persistence or Hurst-Kolmogorov behaviour, when $k \to \infty$ or $w \to 0$. The respective scaling exponents are (see Appendix A.1.8):

- For local scaling: $\gamma_C^\#(0) = \xi^\#(0) = v^\#(0) = \zeta^\#(0) - 1 = 2\varphi_C(0) - 2 = -s^\#(\infty) - 1$
- For global scaling: $\gamma_C^\#(\infty) = \gamma^\#(\infty) = c^\#(\infty) = \zeta^\#(\infty) - 1 = 2\varphi_C(\infty) - 2 = -s^\#(0) - 1$

Here, the emergence of scaling has been related to maximum entropy considerations, and this may provide the theoretical background in modelling complex natural processes by such scaling laws. However, as shown in Appendix A.1.8, scaling laws are a mathematical necessity and could be constructed for virtually any continuous function defined in $(0, \infty)$. In other words, there is no magic in power laws, except that they are, logically and mathematically, a necessity. (No assumption of criticality, self-organization, fractal or multi-fractal generating mechanisms, is necessary to justify their emergence). Here we have focused on second order characteristics of a stochastic process. If we also considered higher order (raw) moments, it is natural to expect that some other (different) power laws would emerge (or could be constructed using the procedure described in Appendix A.1.8), and then one would speak about multifractals everywhere, etc.

2.4. Bounds of Scaling and Entropy Production

It is not well known that the asymptotic scaling exponents cannot take on any arbitrary values but they should be constrained in rather narrow ranges. Instead, misconceptions and incorrect applications resulting in inconsistent values abound in the literature. One of the most typical examples is the asymptotic scaling of the power spectrum for $w \to 0$, for which often steep slopes are reported for low frequencies (down to 0), based on data analyses. Usually these reports are accompanied with a claim that the process with $-s^\#(0) > 1$ is nonstationary, going further to characterize the power spectrum as a tool to identify whether a process is stationary or nonstationary (with values $-s^\#(0)$ below or above 1 suggesting stationarity and nonstationarity, respectively). As thoroughly studied elsewhere [31], this entire line of thought is theoretically inconsistent and flawed, while such results often reported are usually artefacts due to insufficient data or inadequate estimation algorithms. In fact, it has been shown [31] that a process with $-s^\#(0) > 1$ is nonergodic. Inference from data is only possible when the process is ergodic and thus, claiming that $-s^\#(0) > 1$ based on data is self-contradictory. Furthermore, claiming nonstationarity using the power spectrum as a function of frequency is also self-contradictory as in a nonstationary process both the autocovariance and the spectral density, i.e., the Fourier transform of the autocovariance, are functions of two variables, one being related to "absolute" time (see e.g., [32]).

An example of the conditions leading to misinterpreting a stationary process as nonstationary is discussed in Section 3.3 and Appendix A.2.2. We must clarify though that steep slopes $(-s^\#(w) > 1)$ are mathematically and physically possible for medium and large w (see, e.g., Figure A7) and indeed they are quite frequent in geophysical and other processes. In this respect, the estimation from data of steep slopes is possible and not problematic, if we are conscious that such slopes are for intermediate or large frequencies (actually, the estimation algorithms are bandpass filtered with the bounds of the filter determined by the resolution and length of the series). The problem arises when such slopes are interpreted as asymptotic ones and particularly when they are used for inference resulting in self-contradicting conclusions.

Because of the equality of slopes of power spectrum and climacospectrum, the ergodicity limitation holds also for the slope of the climacospectrum, i.e., $\zeta^\#(\infty) = -\tilde{\zeta}^\#(0) < 1$. On the other hand, too steep negative asymptotic slopes of the climacospectrum are also impossible. Indeed, $\zeta^\#(k) = -\tilde{\zeta}^\#(1/k) < -1$ would entail (because of (52)) $\varphi_C(k) < 0$ and (because of (37) and (45)),

$\Gamma'_C(k) < 0$. This means that the variance of the cumulative process would be a decreasing function of time, which is absurd. This holds both for the global case ($k \to \infty$, in which the conditional variance $\Gamma_C(\infty)$ equals the unconditional $\Gamma(\infty)$) and the local case ($k \to 0$, for the conditional variance $\Gamma_C(0)$).

However, in the local case there is a more severe limitation imposed by physical reasoning. As demonstrated in Appendix A.1.10, the case $\zeta^\#(0) = -s^\#(\infty) < 1$ would entail infinite variance. Infinite variance would require infinite energy to emerge, which is physically inconsistent. Therefore, the physical lower limit for $\zeta^\#(0) = -s^\#(\infty)$ is 1. Interestingly, the inconsistency if this constraint is violated stems from the behaviour at high frequencies (see Appendix A.1.10), even though the power spectrum at high frequencies in such cases tends to zero rather diverging to infinity. Therefore, the related "catastrophe" is of "ultraviolet" type, while the infinite value of the power spectrum at low frequencies, which was highlighted in several texts as "infrared catastrophe" [33] is not actually a problem at all, nor does it impose any limitation on the scaling law additional to the limitation for ergodicity discussed above.

A final—and quite severe—limitation is an upper bound of the local scaling exponent, which is 3 for $\zeta^\#(0) = -s^\#(\infty)$. Proof is provided in Appendix A.1.11. The problem if this limitation is violated is that the resulting autocovariance function is not positive definite or, equivalently, that the resulting power spectrum is not always (for any frequency w) positive but takes on negative values for some w. Likewise, the Fourier transform of the climacogram takes on negative values for some w.

Because of the relationship of the entropy production with the scaling exponents, the bounds on the spectrum slopes translate directly into bounds of the CEPLT, i.e., $1 \leq \varphi_C(0) \leq 2$ and $0 \leq \varphi_C(\infty) \leq 1$. These define the "green square" of admissible values of φ_C in Figure 2, which is also depicted in terms of admissible values of slopes $\zeta^\#$ and $s^\#$ (noting that, as discussed above, $s^\#$ can, by exception, take on values out of the square when $\varphi_C(0) = 2$ or $\varphi_C(\infty) = 0$. The reasons why a process out of the square would be impossible or inconsistent, as discussed above, are also marked in Figure 2.

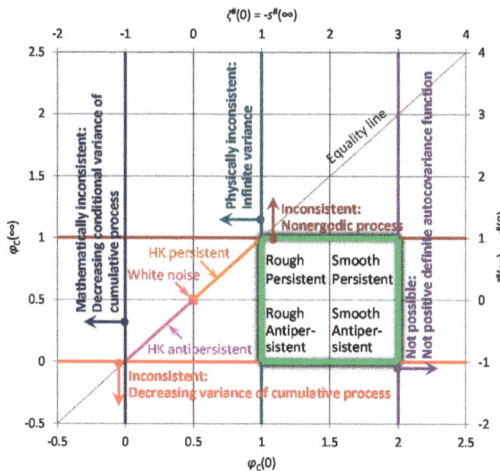

Figure 2. Bounds of asymptotic values of CEPLT, $\varphi_C(0)$ and $\varphi_C(\infty)$, and corresponding bounds of the log-log slopes of power spectrum and climacospectrum The "green square" represents the admissible region (note that $s^\#$ can, by exception, take on values out of the square when $\varphi_C(0) = 2$ or $\varphi_C(\infty) = 0$). The reasons why a process out of the square would be impossible or inconsistent are also marked. The lines $\varphi_C(0) = 3/2$ and $\varphi_C(\infty) = 1/2$ represent neutrality and support the classification of stochastic processes into the indicated four categories (smaller squares within the "green square").

Large values of $\varphi_C(0)$ indicate a smooth process and small ones a rough process. Also, large values of $\varphi_C(\infty)$ indicate a persistent process and small ones an antipersistent process. The centre of the square, with coordinates $\varphi_C(0) = 3/2$, $\varphi_C(\infty) = 1/2$ represents a neutral process. As will be seen in next section, neutrality corresponds to a Markov process, which is neither rough nor smooth, and neither antipersistent nor persistent. Even though sometimes it is said to reflect short-term persistence, we must have in mind that a discretized Markov process at time scale k tends to be uncorrelated in time as k increases and therefore we may not have it in mind as persistent. In other words, here persistence is used as synonymous to long-term persistence, while high autocorrelations at small lags or scales, which die off at large lags or scales, are better described with the term smoothness. Both smoothness and persistence reflect high entropy production, locally and globally, respectively.

A useful observation in Figure 2 is that the entire "green square" lies below the equality line, which means that the same scaling exponent is not possible for both local and global behaviour, or else, it is impossible to have a physically realistic simple scaling process. There is one exception, the upper-left corner of the "green square", which corresponds to the so-called "pink noise" or "$1/f$ noise" and will be discussed further in Section 3.3.

Since we have pointed out several inconsistent or contradictory results that are reported in the literature, it may be useful to mention another case, which is frequently met in geostatistics. The so-called *intrinsic* models are common in geostatistics but susceptible to inconsistent use, even though the literature presents them as nonstationary to avoid some of the emerging problems. Such models are defined in terms of their structure function (variogram), $v(h)$, which tends to infinity as lag h tends to infinity (e.g., [34]). An example is $v_I(h) = a\,h^b$ with $b > 0$, in which we note that its formulation with respect to a single scalar argument, h, does not reveal the nonstationarity. Thus, it may be treated as if it was stationary. In that case, from (24) we have $\gamma_0 = v_I(\infty) = \infty$ and thus the intrinsic process violates the constraint for physical consistency. Moreover, since $v_I(h) = \gamma_0 - c(h) < \infty$ for $h < \infty$, it turns out that $c(h) = \infty$ for any h and, hence, (21) entails that $\gamma(k) = \infty$ for any k. In other words, the infinity of $v_I(\infty)$ is transferred to the entire autocorrelation function and the entire climacogram. In particular, the property $\gamma(k) = \infty$ means that the process is non-ergodic (in ergodic processes $\gamma(k)$ should tend to zero as k tends to infinity; see [27] (p. 429)). Thus, in addition to being physically inconsistent such a treatment of the process is mathematically inconsistent and logically self-contradictory. A consistent way of treatment is to identify the intrinsic process with the (nonstationary) cumulative process $\underline{X}(t)$ [35], derive the stationary process $\underline{x}(t)$ and treat the latter further regularly. In that case the variogram is no longer related to the structure function of $\underline{x}(t)$ but to its climacogram; namely, it is identical to the cumulative climacogram $v_I(h) = \Gamma(h) = \gamma(h)\,h^2$.

Another peculiarity in geostatisical analyses is the so-called "nugget effect", which is also problematic or enigmatic [36]. Namely, this is a discontinuity of the structure function at the origin. An example is $v_E(h) = a\,(h + c)^b$, $h > 0$, with $c, b > 0$, while $v_E(0) = 0$. Investigation shows that the "nugget effect" does not necessarily create inconsistency. It is obviously associated with an infinite derivative of the structure function at the origin, i.e., $v'(0) = \infty$. However, the LLD, $v^{\#}(h) = v'(h)h/v(h)$ can be finite at the limit as $h \to 0$ (because of the multiplication by h). Therefore, the resulting $\varphi_C(0) = 1 + v^{\#}(0)/2$ does not necessarily lie out of the "green square".

Figure 3 illustrates both the "intrinsic" case and the "nugget effect" and provides hints how to avoid both, adopting much better modelling alternatives. The above example of $v_E(h)$ was used with parameters indicated in the figure caption. It is easy to find that a FHK-C model which is both persistent ($H = 0.75 > 1/2$) and rough ($M = 0.2 < 1/2$) can replace the $v_E(h)$ and be used further.

Conversely, we can imagine that in the model identification and fitting of $v_E(h)$, persistence was regarded as an "intrinsic" characteristic and roughness was interpreted as "nugget effect". Note that because $M < 1/2$, $v'(h) = -c'(h) = \infty$. Thus, a rough process seems like if it exhibited the "nuggest effect". However, if instead of the standard plot of the variogram, a logarithmic plot was made (also shown in Figure 3 with respect to the upper horizontal axis), then the "nugget" would disappear.

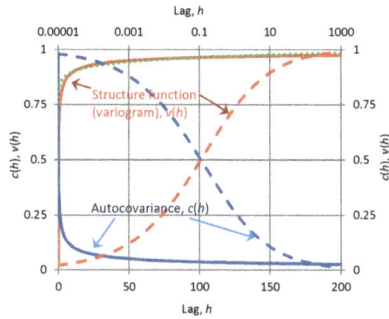

Figure 3. A power type variogram model with "nugget effect" (green dotted line), with equation $v_E(h) = 0.83\ (h + 0.05)^{0.03}$ (corresponding to a "nugget" value $v(0) = 0.83\ (0.05)^{0.03} = 0.76$) and its replacement with a FHK-C model with $H = 0.75$ and $M = 0.2$.

3. Results

3.1. Revisiting Earlier Results

In an earlier study (already mentioned, [22]) on subjects related to the present one, it was suggested that extremization of EPLT in a continuous time representation could determine the entire dependence structure of the process of interest based on simple constraints. The specific premises for EPLT extremization were:

(a) Lebesgue background measure;
(b) constrained mean μ and variance $\gamma(1)$ at a specified (observation) time scale;
(c) constrained lag-one autocorrelation ρ at the specified time scale;
(d) an inequality constraint $\phi_C(t) \geq \phi(t)$ to ensure physical realism as, naturally, by observing the present and past state of a process, the future entropy is reduced, whereas as $t \to \infty$ conditional and unconditional entropies should tend to be equal, which, however, cannot happen if the entropy production is consistently lower in the conditional than in the unconditional case;
(e) extremization of entropy production at asymptotic time scales, i.e., $t \to 0$ and $t \to \infty$.

These premises after systematic analyses resulted in two processes extremizing entropy production:

• A Markov process:

$$c(h) = \lambda e^{-h/\alpha}, \ \gamma(k) = \frac{2\lambda}{k/\alpha}\left(1 - \frac{1 - e^{-k/\alpha}}{k/\alpha}\right) \tag{53}$$

maximizes entropy production for small times ($t \to 0$) but minimizes it for large times ($t \to \infty$).

• A *Hurst-Kolmogorov* (HK) process:

$$\gamma(k) = \lambda(\alpha/k)^{2-2H} \tag{54}$$

maximizes entropy production for large times ($t \to \infty$) but minimizes it for small times ($t \to 0$).

In these definitions α and λ are scale parameters with dimensions of $[t]$ and $[x^2]$, respectively. The parameter H, known as the Hurst parameter, determines the global properties of the process with the notable property:

$$H = \varphi(\infty) = \varphi_C(\infty) \tag{55}$$

Here we revisit these results as well as the premises, in light of the theoretical analyses of the previous section and incorporating empirical experience based on various data sets. These allow the following notes:

1. The minimal entropy production of the HK process at small times may not be plausible for complex processes; rather we should generally expect that complex processes maximize entropy production at both small and large times (asymptotically as $t \to 0$ and ∞).

2. The HK process is characterized by infinite variance ($\gamma(0) = \infty$), which, as discussed in the previous section, makes it physically inconsistent.

3. The premise $\phi_C(t) \geq \phi(t)$ should certainly hold asymptotically (for $t \to 0$ and $t \to \infty$) but not necessarily at intermediate times. A specific example is the process shown in Figure 4h which includes a periodic component where the curve of $\phi_C(t)$ intersects that of $\phi(t)$ (the latter is not shown in the figure), even though asymptotically the inequality holds. It can thus be conjectured that if a process incorporates a deterministic component (e.g., periodic) while it is treated stochastically without separating the deterministic component, the inequality for EPLT and CEPLT could be violated at intermediate times.

4. While the Markov process represents the maximal EPLT as $t \to 0$ ($\phi(0) = 1$), this property is shared by many other processes (see next subsection). Furthermore, in terms of CEPLT, while the Markov process corresponds to $\phi_C(0) = 3/2$, there are processes which have higher CEPLT than the Markov, up to $\phi_C(0) = 2$. These are smoother than the Markov process and have structure function with $v^\#(0) = \zeta^\#(0) > 1$, where the value 1 corresponds to the Markov process.

Based on these remarks, the Markov process remains a valid candidate of physically consistent processes maximizing entropy production, if only the local asymptotic behaviour is of interest. However, additional constraints to the above premises (a)–(e) (from [22]) are needed to confirm its appropriateness, both on small scales (to enable determination of $\phi_C(0)$) and large scales (to enable determination of $\phi_C(\infty) = \phi(\infty)$).

On the contrary, the HK process should not be regarded as a candidate of general validity, even though it is valid for large scales. Indeed, it is still quite useful as it results in high entropy production at large scales, a property that cannot hold for processes with exponential decrease of autocovariance such as the Markov process. We can thus regard the HK process as a useful mathematical construct which does not appear in nature but can be filtered appropriately to make a physically consistent process. This is the same with white noise (WN), which again is not physically consistent as it involves infinite variance in continuous time (see the specific location and the relationship of the WN and HK processes in the diagram of Figure 2).

It is well known that if a WN process $y(t)$ is the input to a moving average filter $x(t) = \frac{1}{2T} \int_{t-T}^{t+T} y(a)\,da$, then it produces a physically consistent process $x(t)$ with finite variance and autocovariance linearly varying in the interval $[0, 2T]$ ([27] (p. 325)). Also, if $y(t)$ is the input to a linear system corresponding to the linear differential equation $x'(t) + \alpha x(t) = y(t)$, then the output $x(t)$ is a Markov process ([27] (p. 326)), physically consistent.

Likewise if in these two cases the input $y(t)$ is an HK process, then it is easy to see that the filtered output is a physically realistic process with finite variance $\gamma(0)$, practically unaffected climacogram $\gamma(k)$ at large scales, with $\gamma^\#(\infty) = 2H - 2$ (as in the original HK process) but highly modified climacogram at small scales, thus having a valid structure (and climacostructure) function with $v^\#(0) = \zeta^\#(0) = 2H$.

3.2. Specific Processes

Following the theoretical analyses of Section 2 and the remarks of the Section 3.1, we describe here a number of stochastic processes specified through any of their second order functions. All of these processes respect the limitations discussed in Section 2.4 and thus they are consistent mathematically and physically. Some of them correspond to high entropy production, but we also discuss processes with low entropy production. The complete details of these processes, along with graphs illustrating their properties and results of their application, are contained in Appendix A.2.

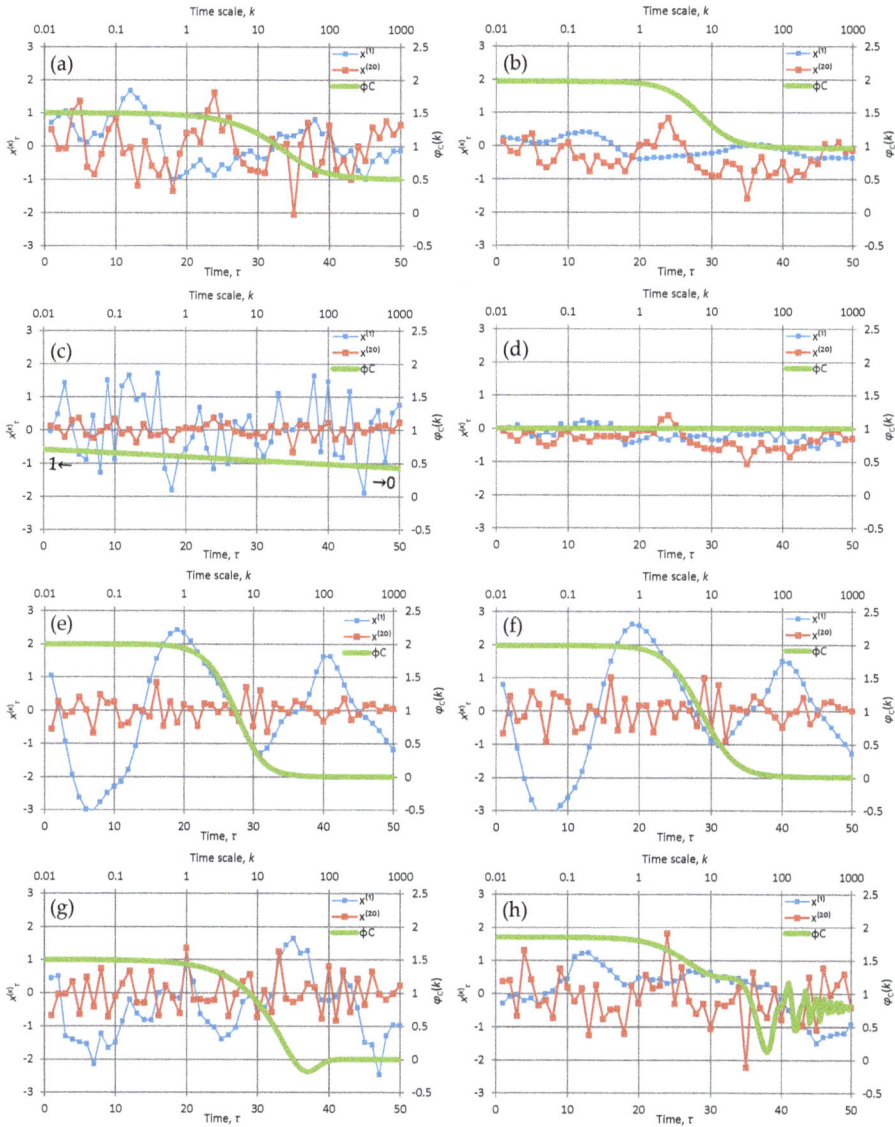

Figure 4. The first fifty terms at time scales $k = 1$ and 20 of time series produced by various models, along with "stamps" of the models (green lines plotted with respect to the secondary axes) represented by the CEPLT, $\phi_C(k)$. The different models are (**a**) Markov; (**b**) FHK, with CEPLT close to the absolute maximum ($H = M = 0.97$); (**c**) FHK, with CEPLT close to the absolute minimum ($H = M = 0.05$); (**d**) FHK, with CEPLT close to the absolute maximum for large scales ($H = 0.99$) and close to the absolute minimum for small scales ($M = 0.01$); (**e**) GP with $n = 1$ (blackbody spectrum) with CEPLT equal to the absolute minimum (0) for large scales and to the absolute maximum (2) for small scales; (**f**) Approximation of (**e**) by FHK ($H = 0.01$, $M = 0.99$, also using the same initial sequence of random numbers as in (**e**) to visualize similarity); (**g**) AE with $n = 2$; (**h**) average of a FHK with $H = M = 0.8$ and a harmonic oscillation with $T = 5a = 50$. In all cases the discretization time scale is $D = 1$, the characteristic time scale $a = 10$, and the characteristic variance scale λ is chosen so that for time scale D, $\gamma(D) = 1$. The mean is 0 in all cases and the marginal distribution is normal. The FHK is implemented using the Cauchy-type climacogram.

We start with a general Filtered Hurst-Kolmogorov (FHK) process (which in some cases, e.g., [24], has also been called the Hybrid Hurst-Kolmogorov—HHK—process). In Section 3.1 we briefly discussed two types of linear filters to the HK process, whose asymptotic properties are determined based on a single parameter, H. Here we generalize this result also making the asymptotic behaviour on the left independent from that on the right (cf. [37]) through introducing another parameter denoted as M (in honour of Mandelbrot, following [31]) which is called the smoothness (or fractal) parameter. We avoid specifying the linear filter needed to convert the HK process into the types of the FHK process given as this is not necessary (and in some cases it would be too involved). Rather we specify these types in terms of a convenient expression of the climacogram. The first case is defined through a generalized Cauchy-type climacogram (FHK-C) (see also [24]):

$$\gamma(k) = \lambda\left(1 + (k/\alpha)^{2M}\right)^{\frac{H-1}{M}} \tag{56}$$

Both M and H are dimensionless parameters, and M and $1 - H$ vary in the interval $(0, 1]$ with $M < 1/2$ or $> 1/2$ indicating a rough or a smooth process, respectively, and with $H < 1/2$ or $> 1/2$ indicating an antipersistent or a persistent process, respectively.

The second case is defined through a Dagum-type (FHK-D) climacogram:

$$\gamma(k) = \lambda\left(1 - \left(1 + (k/\alpha)^{2(H-1)}\right)^{\frac{M}{H-1}}\right) \tag{57}$$

and, despite its different expression, its behaviour, as well as the meaning and ranges of parameters are the same as in the Cauchy-type climacogram. For $M = 1 - H$ both result in precisely the same special case:

$$\gamma(k) = \frac{\lambda}{1 + (k/\alpha)^{2(1-H)}} \tag{58}$$

The third case is obtained by summing a Cauchy-type climacogram with $M = 1$ and a Dagum-type climacogram with $H = 0$. The climacogram of the thus produced model (FHK-CD) is:

$$\gamma(k) = \lambda_1(1 + (k/\alpha_1)^2)^{H-1} + \lambda_2(1 - (1 + (k/\alpha_2)^{-2})^{-M}) \tag{59}$$

This is a convenient form, in which the first additive term determines merely the persistence of the process and the second one the smoothness of the process. In addition, it is more flexible and richer than its constituents, as it contains two couples of scale parameters; however, if parsimony is sought, then it can take the same number of parameters as each of the constituents by setting $\alpha_1 = \alpha_2$ and $\lambda_1 = \lambda_2$ (note that, for dimensional consistency, one λ and one α are minimal parameter requirements).

Even in its most parsimonious form, the FHK in any of the above three variants can cover the entire admissible range, i.e., the entire "green square" in Figure 2. The different patterns in time series generated by different M and H (specifically for the Cauchy-type climacogram) are illustrated in some plots of Figure 4 also in comparison with other models, first of which is the most customary Markov model (Figure 4a). The Markov model is good as a benchmark for comparisons because, as already discussed in Section 2, it is fully neutral (neither rough nor smooth as $\phi_C(0) = 3/2$, and neither antipersistent nor persistent as $\phi_C(\infty) = 1/2$).

Each of the panels shows the first fifty terms of time series produced by each of the model implementations at time scales $k = 1$ and 20. In addition, each panel contains a "stamp" of the specific model represented by the plot of CEPLT, $\phi_C(k)$. Additional plots for all second order tools of each of the models are given in Appendix A.2. The time series for these models were generated quite easily using the generic model proposed in [24]; their length is 1024 and this is also the length of the series of coefficients a used for the generating symmetric moving average (SMA) scheme. All calculations are of algebraic type (based on the equations of Section 2 and Appendix A.2) and were performed in Excel spreadsheets without difficulties.

In Figure 4b the CEPLT is close to the absolute maximum both for small and large scales ($H = M = 0.97$ so as to obtain $\phi_C(0) = 1.97 \approx 2$ and $\phi_C(\infty) = 0.97 \approx 1$); notable is the very smooth shape at scale 1 and the large departures from the mean (which is 0) at scale 20. On the contrary, in Figure 4c the CEPLT is close to the absolute minimum for all scales ($H = M = 0.05$, so as to obtain $\phi_C(0) = 1.05 \approx 1$ and $\phi_C(\infty) = 0.05 \approx 0$—for better visualization it was preferred not to use values of H and $M < 0.05$). Furthermore, in Figure 4d the CEPLT is close to the absolute maximum for large scales ($H = \phi_C(\infty) = 0.99 \approx 1$) and close to the absolute minimum for small scales ($M = 0.01$ resulting in $\phi_C(0) = 1.01 \approx 1$). Finally, in Figure 4f the conditions are opposite to those in 4d, i.e., the CEPLT is equal to the absolute minimum for large scales ($H = \phi_C(\infty) = 0.01 \approx 0$) and to the absolute maximum (2) for small scales ($M = 0.99$ resulting in $\phi_C(0) = 1.99 \approx 2$).

The particular case of Figure 4d is close to what is usually called "pink noise" or "$1/f$ noise", as the power spectrum has almost constant slope -1 for the entire frequency domain (which is the same in the climacospectrum). This means that using the FHK model we can theoretically represent and practically produce even "pink noise" in a consistent stationary setting without linking it to a nonstationary process [38,39], which actually involves several theoretical inconsistencies [31]. Indeed, as can be seen more thoroughly in the detailed graphs of Figure A5 in Appendix A.2.2, which are for the same application with that of Figure 4d, the small change of slope of from 0.99 to 1.01 is not actually visible, especially in view of the very rough shape of the empirical periodogram, which certainly cannot support differentiation between 0.99 and 1. The FHK model can be used also in other ways to produce "pink noise", that is, by selecting a very large (small) parameter α so as to expel from our field of vision the asymptotic behaviour on large (small) scales. And we can imagine that in several cases of empirical explorations using observations of natural processes, the observation resolution and length, compared to characteristic scale(s) of the process, are such as to hide the asymptotic behaviour of the process.

We can use this as a trick to obtain virtually constant power spectrum slopes much steeper than -1. This is illustrated Figure A7 of Appendix A.2.2 where the FHK was used with $H = M = 0.75$ and $\alpha = 100$. These yield theoretical slopes $\tilde{\zeta}^\#(\infty) = s^\#(\infty) = -(1 + 2M) = -2.5$ and $\tilde{\zeta}^\#(0) = s^\#(0) = -(2H - 1) = -0.75$. However, the large α does not allow viewing the asymptotic behaviour at low frequencies or large scales and the slope -2.5 dominates everywhere. Actually the empirical periodogram estimates an even steeper constant slope, $s^\# = -2.6$. This should not mislead us to conclude that the process is nonstationary because the slope is steeper than -1 (as happens in many studies in the literature). Likewise, in the case that the time series represented observations of a natural process, such a result must not be misinterpreted as evidence that natural processes can lead to slopes constant for all scales and steeper than -1. Clearly here, the process is stationary, the slope -2.5 refers to large frequencies or small scales but, because of the large characteristic scale and the limited observation period (1024 in this example), we cannot see what happens at larger scales. If we saw, then the slope would be -0.75, in accord with the theory.

As discussed, the Markov process is neutral but, by modifying its power spectrum, we can obtain processes which are smooth or antipersistent. Two types of modifications are studied in Appendix A.2, introduced in terms of their power spectrum, i.e.,

$$s(w) = \frac{4^{n+1}(n!)^2}{(2n)!} \frac{\lambda\alpha}{\left(1 + (2\pi\alpha w)^2\right)^{n+1}} \tag{60}$$

and:

$$s(w) = \frac{4^{n+1}(n!)^2}{(2n)!} \frac{\lambda\alpha(2\pi\alpha w)^{2n}}{\left(1 + (2\pi\alpha w)^2\right)^{n+1}} \tag{61}$$

where $n = 0, 1, 2 \ldots$, and the value $n = 0$ results precisely in a Markov process. For $n > 0$, (60) corresponds to a smooth process with $\phi_C(0) = 2$ and (61) corresponds to an antipersistent process with $\phi_C(\infty) = 0$. The resulting autocovariance function retains the characteristic exponential form of the Markov process but is multiplied by a polynomial function of lag (see Appendix A.2). Therefore, we call these two processes the smooth exponential (SE) and the antipersistent exponential (AE), respectively. In each of the cases, a most extreme value of CEPLT is achieved (maximal for small scales in SE and minimal for large scales in AE). Figure 4g, shows an example of AE for $n = 2$; the hole in the function $\phi_C(k)$ for moderate k is a characteristic of this process.

An interesting process that is simultaneously extremely smooth ($\phi_C(0) = 2$) and extremely antipersistent ($\phi_C(\infty) = 0$) is defined by the following power spectrum:

$$s(w) = \frac{c_n \lambda \alpha (\alpha w)^{2n+1}}{e^{2\pi \alpha w} - 1} \tag{62}$$

where c_n is a normalizing constant making the area of the power spectrum equal to $\gamma_0 = \lambda$. This process, which we call generalized Planck (GP) is obtained by a generalization of Planck's law of black-body radiation, whereas it is identical to this law if $n = 1$. The detailed equations of this process are also given in Appendix A.2. A characteristic plot of time series generated from this process is shown in Figure 4e, and is almost indistinguishable from Figure 4f which, as already discussed, was produced as an approximation by FHK.

A final case examined in the Appendix A.2 is a harmonic oscillation, which is very easily modelled as a deterministic process but here it is treated as a stochastic process. Of course stochastic treatment is not advisable in this case, but often such periodic behaviours appear as components of stochastic processes. Obviously the second-order characteristics of such processes are affected by periodic components and therefore we need to know which equations should be superimposed in those of the pure stochastic process (see also [40,41]). As an example of such a case, Figure 4h shows the behaviour of a process defined as the average of a FHK with $H = M = 0.8$ and a harmonic oscillation. It is difficult to identify the presence of the oscillation from visual inspection of the time series, but the detailed graph of any of the second order characteristics, plotted in Appendix A.2., captures the periodic behaviour.

3.3. Comparison of Asymptotic Properties

Most of the models examined in the previous subsection can only reproduce specific values of asymptotic properties. An exception is the FHK model which is the most powerful as it can perform in the entire admissible domain of asymptotic properties. A visual comparison of all models examined, it terms of the asymptotic values of the different second order characteristics is made in Figure 5, while Figure 6 provides an envelope for all models, with classification of the ranges in terms of smoothness and persistence.

Further comparative information about which model can perform for the most extremal cases is provided in Table 4. These figures and the table can be useful for the model selection based on data.

Table 4. Models that can perform in each of the specified extremal or neutral cases of smoothness and persistence.

Persistence	Smoothness		
	Maximal Roughness $\varphi_C(0) = 1$	Neutral Smoothness $\varphi_C(0) = 3/2$	Maximal Smoothness $\varphi_C(0) = 2$
Maximal Persistence, $\varphi_C(\infty) = 1$	~FHK	~FHK	~FHK
Neutral Persistence $\varphi_C(\infty) = 1/2$	~FHK	Markov, FHK	SE, GP, FHK
Maximal Antipersistence $\varphi_C(\infty) = 0$	~FHK	AE, FHK	GP, FHK

Note: the symbol '~' indicates that the model can perform in close approximation but not precisely.

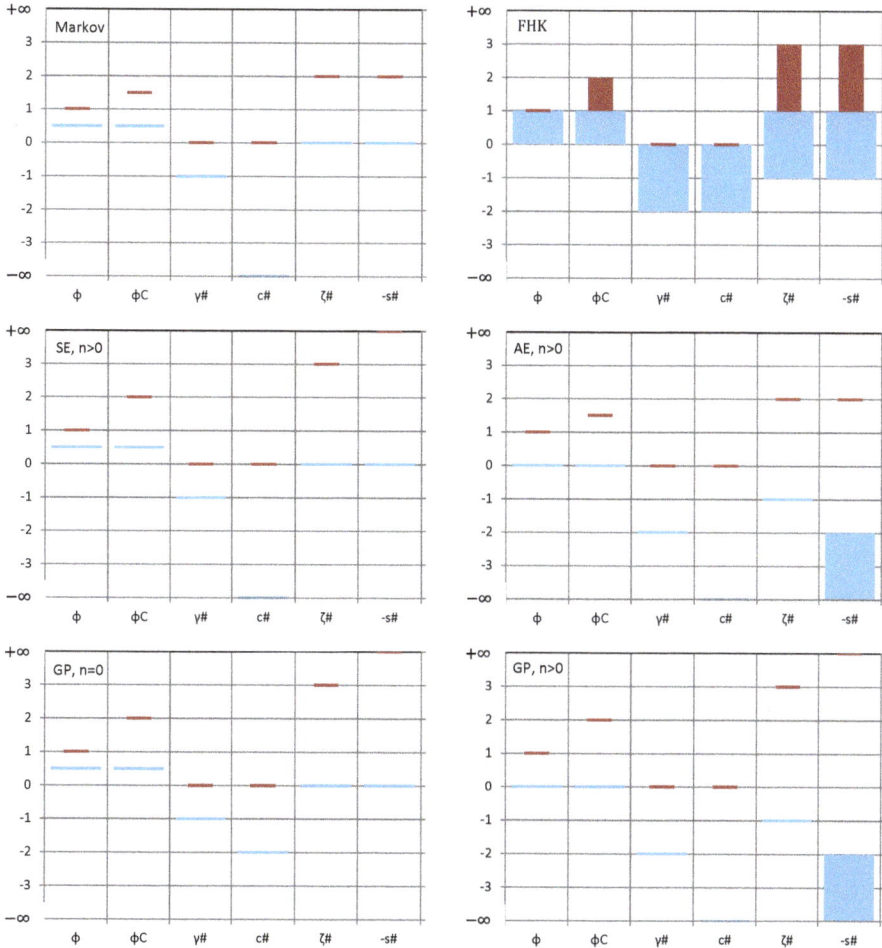

Figure 5. Ranges of asymptotic values for each of the most important second order functions of the examined processes. Wide light blue lines/boxes represent global properties ($k, h \to \infty$, $w \to 0$) while narrow dark red lines/boxes represent local properties ($k, h \to 0, w \to \infty$). Markov: Markov process; FHK: filtered Hurst-Kolmogorov process in any of the three forms; SE: smooth process with generalized exponential-type autocovariance; AE: antipersistent process with generalized exponential-type autocovariance; GP: generalized Planck. Note that the Markov process is identical with the SE and PE for $n = 0$.

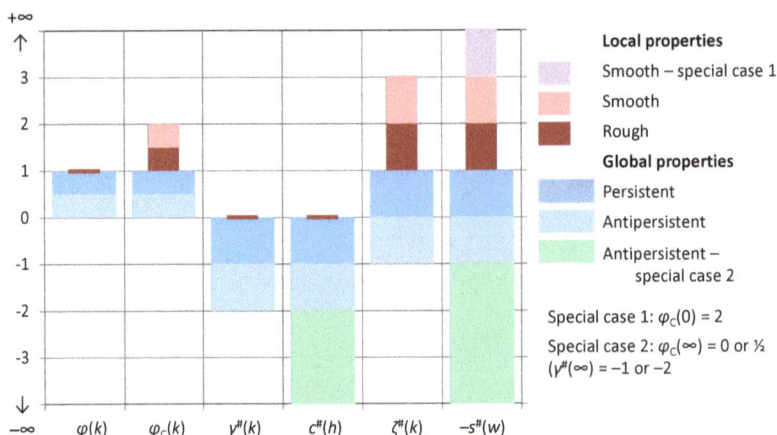

Figure 6. Envelopes of ranges of asymptotic values for each of the most important second order functions for all examined processes.

3.4. Real World Case Study

Empirical testing of the theoretical analyses presented here needs fine resolution of measurements and long time series, in order to reliably determine the asymptotic properties from data. Geophysical time series usually do not satisfy either of these two conditions. However, laboratory measurements of turbulent velocity can meet the conditions. Indeed, in laboratory experiments at sampling intervals of μs, very large samples can be formed which can enable viewing the asymptotic behaviour. Here we use grid data of nearly isotropic turbulence from the Corrsin Wind Tunnel at a high-Reynolds-number [42], which were made available on the Internet by the Johns Hopkins University. This dataset consists of 40 time series with $n = 36 \times 10^6$ data points of wind velocity along the flow direction and an equal number of time series of cross-stream velocity, all measured at a sampling time interval $D = 25$ μs by X-wire probes placed downstream of the grid. Here we use part of the data, namely the series of wind velocity along the flow direction at the first of the probes (first column of files in http://pages.jh.edu/~cmenevel/datasets/Activegrid/M20/H1/). More data and analyses have been used by Dimitriadis and Koutsoyiannis [43] at a somewhat different context.

Characteristic plots of the turbulent velocity time series are shown in the Figure 7a for various time scales. The large persistence of turbulence is evident from the large variability at the largest time scales. Impressively, an FHK-CD model with five parameters describes well the data at the entire range of available time scales which spans almost seven orders of magnitude.

The fitted model parameters are: $H = 5/6 = 0.83$, $M = 1/3 = 0.33$, $a_1 = a_2 = 2170\, D = 54$ ms, $\lambda_1 = 1.13$ m^2/s^2, $\lambda_2 = 2.39$ m^2/s^2. These confirm the intense persistence and a small roughness, which is reflected in a climacospectrum slope of $-5/3$ for large frequencies (the neutral would be -2), in accord with Kolmogorov's theory.

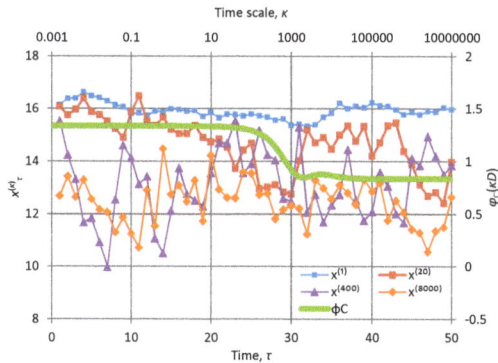

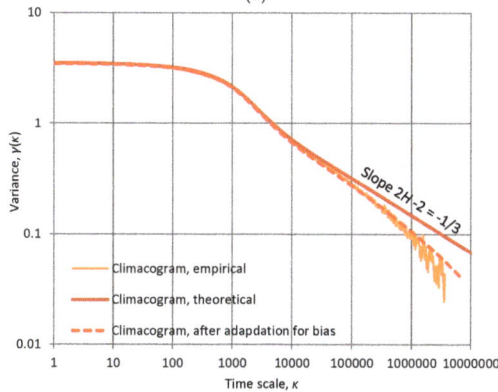

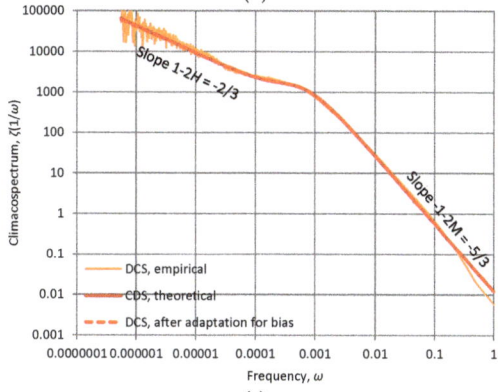

Figure 7. Characteristic graphs of the turbulent velocity time series: (**a**) plot of the first fifty terms of the time series at time scales $\kappa = 1, 20, 400$ and 8000, along with the "stamp" $\phi_C(k)$ of the fitted model (green line plotted with respect to the secondary axes); (**b**) empirical and fitted theoretical climacogram; (**c**) empirical and fitted theoretical climacospectrum.

4. Discussion and Conclusions

The question whether the probabilistic (information) entropy incorporates, as a special case, thermodynamic entropy or the two are different concepts, is still debated [44,45]. If we adopt the former proposition of the dilemma, as we have done here, the requirement emerges that entropy production should be formulated in a stochastic context, while current attempts mostly use the classical thermodynamic entropy definition as in heat transfer. This is not an easy task and certainly requires exploration of a broad spectrum of alternatives for a definition of the concept of entropy production. Exploration of extremum principles for entropy production in a stochastic framework is also crucial. The two versions proposed here, entropy production in logarithmic time, unconditionally (EPLT) or conditionally on the past and present having been observed (CEPLT), may have some usefulness and potential as explanatory and modelling tools. The findings of this study include the following:

- Definition of entropy production with respect to logarithmic time has some attractive characteristics and is consistent with the notion of the time arrow.
- The asymptotic values of entropy production for time zero and infinity seem to be crucial for its extremization and define important features of a process.
- In particular, the asymptotic value of CEPLT for zero time defines the roughness (fractality) of the stochastic process while the common asymptotic value of EPLT and CEPLT for infinite time defines the (long-term) persistence of the process and equals the Hurst parameter.
- Both ELPT and CEPLT are finite and constrained within specific theoretically justified bounds. The scaling laws of all second order characteristics of stochastic processes are a result of these finite values and are also subject to bounds (with few exceptions fully described). The proofs of existence of these bounds offer the means to fight common misunderstanding, misuse, false reporting and inconsistent estimation of characteristics of geophysical processes, as well as to develop consistent estimation algorithms.
- The climacogram and its derivative tools introduced here, namely, the differenced climacogram and the climacospectrum are very powerful and support the estimation of entropy production, in either of its forms, from data series.
- The broad family of models for stochastic processes investigated here covers the entire admissible range of asymptotic behaviours and offers the means for modelling any type of processes, either using a single model or a combination of more than one model.
- The real-world application presented with the extraordinary long time series of turbulent velocity shows how a parsimonious stochastic model can be identified and fitted using the tools developed.

Measurements of diverse turbulent phenomena can potentially provide equally long time series with fine resolution, shaping a broader empirical basis for future work on identifying natural behaviours, specifically on larger spatial and temporal scales, and further testing the adequacy of the models. Characteristics of order higher than second, along with the effects of non-Gaussianity on entropy production, are also interesting to examine in light of the developments for second-order ones. A recent study [43] has already provided interesting results with respect to high-order moments. Connections of microscale turbulence and macroscale atmospheric phenomena are also a broad field for future studies. Further applications, also extending similar recent works in hydrology, geophysics and ecosystems [30,40,41], would enrich our knowledge on natural behaviours.

Acknowledgments: I thank Huijuan Cui, Bellie Sivakumar and Vijay P. Singh for the invitation to write an article for the Special Issue, and the two anonymous reviewers for the encouraging and constructive comments. I gratefully acknowledge discussions about several issues on stochastics covered here with members of the Itia research team, colleagues and friends, and particularly with Panayiotis Dimitriadis, Federico Lombardo, Spencer Stevens, Any Iliopoulou, Hristos Tyralis, Andreas Efstratiadis, Katerina Tzouka, Georgia Papacharalambous, Panagiotis Kossieris, Yannis Tsoukalas, Andreas Langousis, Yannis Markonis, Christoforos Pappas, Simon Papalexiou, John Halley, Elena Volpi, Yannis Dialynas, Ilias Deligiannis, Harry Lins, Christian Onof, Alin Carsteanu, Amilcare Porporato and Alberto Montanari, and earlier with Tim Cohn and Vit Klemes who are no longer with us.

Conflicts of Interest: The author declares no conflict of interest.

Appendix A

Appendix A.1. Derivations and Theoretical Details

Appendix A.1.1. Properties of the LLD

Here we examine some asymptotic properties of the log-log derivative of general validity, which are useful in other derivations herein. First we examine the product of two functions, i.e.,

$$g(t) := f_1(t)f_2(t) \tag{A1}$$

By applying the definition of LLD (Equation (38)) and using standard calculus, it is easy to show that:

$$g^{\#}(t) = f_1^{\#}(t) + f_2^{\#}(t) \tag{A2}$$

which we can conveniently denote as:

$$(f_1(t)f_2(t))^{\#} = f_1^{\#}(t) + f_2^{\#}(t) \tag{A3}$$

Likewise, it is easily shown that:

$$(f_1(t)/f_2(t))^{\#} = f_1^{\#}(t) - f_2^{\#}(t) \tag{A4}$$

By setting in (A3) $f_1(t) = \lambda$ (constant), $f_2(t) = f(t)$ and observing that $f_1^{\#}(t) = 0$, it is easily verified that:

$$(\lambda f(t))^{\#} = f^{\#}(t) \tag{A5}$$

By setting in (A3) $f_1(t) = f_2(t) = f(t)$ it is easily verified that:

$$\left(f(t)^2\right)^{\#} = 2f^{\#}(t) \tag{A6}$$

and likewise, extending for any exponent λ:

$$\left(f(t)^{\lambda}\right)^{\#} = \lambda f^{\#}(t) \tag{A7}$$

We now examine the sum of two functions, i.e.,

$$g(t) := f_1(t) + f_2(t) \tag{A8}$$

By applying the definition of LLD and using standard calculus, it is easy to show that:

$$g^{\#}(t) := \frac{f_1(t)f_1^{\#}(t) + f_2(t)f_2^{\#}(t)}{f_1(t) + f_2(t)} \tag{A9}$$

In other words:

$$(f_1(t) + f_2(t))^{\#} = \frac{f_1(t)f_1^{\#}(t) + f_2(t)f_2^{\#}(t)}{f_1(t) + f_2(t)} \tag{A10}$$

Finally, we examine the composition of two functions, i.e.,

$$(f \circ g)(t) = f(g(t)) \tag{A11}$$

Again by applying the definition of LLD and using standard calculus, we easily find:

$$((f \circ g)(t))^{\#} = (f(g(t)))^{\#} = f^{\#}(g(t)) \, g^{\#}(t) \tag{A12}$$

Appendix A.1.2. Proof of the Positive Definiteness of the Climacogram

We will demonstrate that $\gamma(k)$ is a positive definite function based on the well-known fact that the autocovariance function $c(h)$ is positive definite. It suffices to show that the Fourier transform of $\gamma(k)$, i.e.,

$$s_\gamma(w) = 4 \int_0^\infty \gamma(k) \cos(2\pi wk) dk \tag{A13}$$

is positive. Expressing $\gamma(k)$ in terms of $c(k)$ from (21) we have:

$$s_\gamma(w) = 8 \int_0^\infty \int_0^1 (1-\chi) c(\chi k) d\chi \, \cos(2\pi wk) dk = 8 \int_0^1 (1-\chi) \int_0^\infty c(\chi k) \, \cos(2\pi wk) dk \, d\chi \tag{A14}$$

Setting $\chi k = \psi$, $\chi dk = d\psi$, we find:

$$s_\gamma(w) = 8 \int_0^1 (1-\chi) \int_0^\infty c(\psi) \, \cos(2\pi w\psi/\chi)(1/\chi) d\psi \, d\chi \tag{A15}$$

and invoking the definition of the power spectrum as the Fourier transform of $c(h)$ (Equation (10)), we get:

$$s_\gamma(w) = 2 \int_0^1 \frac{1-\chi}{\chi} s(w/\chi) d\chi \tag{A16}$$

Both quantities $s(w/\chi)$ and $(1-\chi)/\chi$ are positive for $0 < \chi < 1$ and thus $s_\gamma(w) \geq 0$.

Appendix A.1.3. Proof that the Differenced Climacogram Is Nonnegative

First we will prove a more general inequality that will be used for the proof, i.e.,

$$(h+r)\sqrt{\gamma(h+r)} \leq h\sqrt{\gamma(h)} + r\sqrt{\gamma(r)}, \; \forall h, r > 0 \tag{A17}$$

With reference to the cumulative process, as seen in Figure 1, we have:

$$\begin{aligned} \Gamma(h+r) &= \mathrm{Var}[\underline{X}(h+r)] = \mathrm{Var}[(\underline{X}(h+r) - \underline{X}(h)) + \underline{X}(h)] \\ &= \mathrm{Var}[\underline{X}(h+r) - \underline{X}(h)] + \mathrm{Var}[\underline{X}(h)] + 2\mathrm{Cov}[(\underline{X}(h+r) - \underline{X}(h)), \underline{X}(h)] \\ &\leq \mathrm{Var}[\underline{X}(h+r) - \underline{X}(h)] + \mathrm{Var}[\underline{X}(h)] + 2\sqrt{\mathrm{Var}[\underline{X}(h+r) - \underline{X}(h)] \, \mathrm{Var}[\underline{X}(h)]} \end{aligned} \tag{A18}$$

and since $\mathrm{Var}[\underline{X}(h+r) - \underline{X}(h)] = \mathrm{Var}[\underline{X}(r)]$, we find:

$$\Gamma(h+r) \leq \Gamma(r) + \Gamma(h) + 2\sqrt{\Gamma(r)\Gamma(h)} = \left(\sqrt{\Gamma(r)} + \sqrt{\Gamma(h)}\right)^2 \tag{A19}$$

or, taking the square roots as the quantities are positive:

$$\sqrt{\Gamma(h+r)} \leq \sqrt{\Gamma(h)} + \sqrt{\Gamma(r)} \tag{A20}$$

By substituting $\Gamma(h) = h^2 \, \gamma(h)$ we get (A17). As a corollary, by setting $r = h$ we derive:

$$\gamma(2h) \leq \gamma(h) \tag{A21}$$

which ensures that the differenced climacogram is nonnegative.

Appendix A.1.4. Determination of the Characteristic Constant ε of CEPLT

From (41) and the condition put as desideratum (i.e., that, as $k \to \infty$, $\gamma_C(k)/\gamma(k) \to 1$), it follows that:

$$\frac{1}{\varepsilon} = 1 - \lim_{t \to \infty} \frac{\gamma(2t)}{\gamma(t)} \tag{A22}$$

Now assuming that $b = \gamma^\#(\infty)$, it follows that the limit:

$$\lim_{t \to \infty} t^{-b}\gamma(t) = c \tag{A23}$$

is finite. Likewise:

$$\lim_{t \to \infty} (2t)^{-b}\gamma(2t) = c \tag{A24}$$

By dividing the last two equations we find:

$$\lim_{t \to \infty} \frac{\gamma(2t)}{\gamma(t)} \frac{(2t)^{-b}}{(t)^{-b}} = 1 \tag{A25}$$

and finally:

$$\lim_{t \to \infty} \frac{\gamma(2t)}{\gamma(t)} = 2^b \tag{A26}$$

which proves (42).

Appendix A.1.5. Asymptotic Properties of CEPLT

The fact that $\gamma_C(0) = \gamma_C(\infty) = 0$ follows directly from the definition. For the LLDs, starting from the definition of the differenced climacogram we proceed as follows:

$$\gamma_C(k) = \varepsilon(\gamma(k) - \gamma(2k)) = \varepsilon\gamma(k)\left(1 - \frac{\gamma(2k)}{\gamma(k)}\right) \tag{A27}$$

Hence, by virtue of (A3):

$$\gamma_C^\#(k) = \gamma^\#(k) + \left(1 - \frac{\gamma(2k)}{\gamma(k)}\right)^\# \tag{A28}$$

According to the derivations in Appendix A.1.4, the quantity in parentheses has a finite limit as $k \to \infty$, which specifically is $1 - 2^{\gamma^\#(\infty)} = 1/\varepsilon$. Thus, its LLD tends to zero as $k \to \infty$ (see explanation in the last paragraph in Appendix A.1.8) and hence $\gamma_C^\#(\infty) = \gamma^\#(\infty)$. Likewise, $\gamma^\#(0) = 0$ because $\gamma(0) = \gamma_0$ (finite). However the term in parenthesis is nonzero for $k \to 0$, so to determine the asymptotic behaviour of $\gamma_C^\#(k)$ in that case we follow a different path.

Specifically, we can express $\gamma_C(k)$ in an alternative manner in terms of the climacostructure function. Using the definition of the latter, we can write:

$$\gamma_C(k) = \varepsilon(\check{\varsigma}(2k) - \check{\varsigma}(k)) = \varepsilon\check{\varsigma}(k)\left(\frac{\check{\varsigma}(2k)}{\check{\varsigma}(k)} - 1\right) \tag{A29}$$

Hence:

$$\gamma_C^\#(k) = \check{\varsigma}^\#(k) + \left(\frac{\check{\varsigma}(2k)}{\check{\varsigma}(k)} - 1\right)^\# \tag{A30}$$

Proceeding as above we derive:

$$\lim_{k \to 0} \frac{\check{\varsigma}(2k)}{\check{\varsigma}(k)} = 2^{\check{\varsigma}^\#(0)} \tag{A31}$$

and hence the quantity in parentheses in (A30) has a finite limit as $k \to 0$, which specifically is $2^{\check{\varsigma}^\#(0)} - 1$. Thus, its log-log derivative tends to zero as $k \to 0$ and hence $\gamma_C^\#(0) = \check{\varsigma}^\#(0)$.

We further note that from the definition of the climacostructure function and using Equation (A10) we get:

$$\varsigma^{\#}(k) = \frac{-\gamma^{\#}(k)\gamma(k)}{\gamma_0 - \gamma(k)} \tag{A32}$$

As $k \to \infty$, the numerator is zero, while the denominator is γ_0, so that $\varsigma^{\#}(\infty) = 0$. The concluding remark is that, because $\gamma^{\#}(0) = \varsigma^{\#}(\infty) = 0$, none of the two tools, climacogram and climacostructure function suffices alone to express both the local and global asymptotic properties of $\gamma_C(k)$. On the other hand, the combination of the two suffices.

Appendix A.1.6. Area of Climacospectrum

From the definition of climacospectrum (48), the entire area under it is:

$$A := \int_0^{\infty} \tilde{\zeta}(w)dw = \frac{1}{\ln 2} \int_0^{\infty} \frac{\gamma(1/w) - \gamma(2/w)}{w}dw \tag{A33}$$

By changing variable, setting $k = 1/w$, $dk = -(1/w^2)\,dw$, we get:

$$A = \frac{1}{\ln 2} \int_0^{\infty} \frac{\gamma(k) - \gamma(2k)}{k}dk \tag{A34}$$

We will evaluate A by taking the limit as $x \to 0$ of:

$$A(x) = A_1(x) - A_2(x), \quad A_1(x) := \frac{1}{\ln 2} \int_x^{\infty} \frac{\gamma(k)}{k}dk, \quad A_2(x) := \frac{1}{\ln 2} \int_x^{\infty} \frac{\gamma(2k)}{k}dk \tag{A35}$$

The latter can be written as:

$$A_2(x) = \frac{1}{\ln 2} \int_{2x}^{\infty} \frac{\gamma(k)}{k}dk = A_1(2x) \tag{A36}$$

so that:

$$A(x) = \frac{1}{\ln 2} \int_x^{2x} \frac{\gamma(k)}{k}dk \tag{A37}$$

For small x, because $\gamma(k)$ is finite for any k and we are interested about the limit of $A(k)$ as $x \to 0$, at the limit we can replace $\gamma(k)$ with $\gamma(0)$ and get:

$$\lim_{x \to 0} A(x) = \frac{\gamma(0)}{\ln 2} \int_x^{2x} \frac{1}{k}dk = \frac{\gamma(0)}{\ln 2}(\ln 2x - \ln x) = \frac{\gamma(0)\ln 2}{\ln 2} = \gamma(0) \tag{A38}$$

For completeness we note that, as can be easily seen, the integral:

$$B := \int_0^{\infty} (\gamma(k) - \gamma(2k))dk \tag{A39}$$

evaluates to:

$$B = \frac{1}{2} \int_0^{\infty} \gamma(k)dk \tag{A40}$$

However, this is finite only for antipersistent processes, while it diverges to $+\infty$ for persistent processes.

Appendix A.1.7. Details and Derivation of the Inverse Transformation of Climacospectrum

Combining both forms of (49) we find:

$$\gamma(0) = \ln 2 \sum_{i=-\infty}^{\infty} \frac{\zeta(2^i k)}{2^i k} \tag{A41}$$

or by setting $k = 2^x$:

$$\gamma(0) = \ln 2 \sum_{i=-\infty}^{\infty} \frac{\zeta(2^{x+i})}{2^{x+i}} \tag{A42}$$

and this is valid for any x. This means that the function $\zeta(x)$ cannot take any arbitrary form but its form should be such that (A42) is satisfied for any x. At the same time, as shown in Appendix A.1.6:

$$\gamma(0) = \int_0^\infty \tilde{\zeta}(w)dw = \int_0^\infty \frac{\zeta(k)}{k^2} dk = \ln 2 \int_{-\infty}^{\infty} \frac{\zeta(2^x)}{2^x} dx \tag{A43}$$

where for the very last term we used the transformation $k = 2^x$, $dk = (\ln 2) \, 2^x \, dx$. Combining (A42) and (A43) we find that, for any k:

$$\sum_{i=-\infty}^{\infty} \frac{\zeta(2^i k)}{2^i k} = \int_{-\infty}^{\infty} \frac{\zeta(2^x)}{2^x} dx = \frac{\gamma(0)}{\ln 2} \tag{A44}$$

Note, however, that the equality of the sum with the integral holds only if both limits for summation and integration are infinite. It can be used also as an approximation for small k, by setting the lower integration limit to k; thus the following approximations hold:

$$\gamma(k) = \begin{cases} I(k) & k \to 0 \\ \varepsilon \ln 2 \frac{\zeta(k)}{k} = \frac{\varepsilon \ln 2}{2} I(k) & k \to \infty \end{cases} \tag{A45}$$

where $I(k)$ has been defined in Equation (50). Combining these two cases we heuristically derive the left part of Equation (50), which was shown by an extensive numerical investigation to be almost indistinguishable from the exact solution (49). The advantage of (50) over (49) is that, to calculate a sufficient approximation by (49), we need to go to very large time scales ($2^i k$), which may not be feasible when working with data series, while (50) can perform well in reasonably smaller time scales. In numerical applications, the evaluation of the integral $I(t)$ goes sequentially from large to small time scales. To start the calculation, for the largest time scale $k = q$ we exploit the second case of (A45) and get $I(q) = 2 \zeta(q)/q$. It should be kept in mind that both $I(k)$ and $\zeta(k)/k$ are decreasing functions of k for large scale. Even if we assumed that $\zeta(k)/k$ is constant, rather than decreasing, for a range of scales close to q, we would derive from the second case of (A45) that $I(q - iD) = (I(q)/q) (2 + 1/(q/D - 1) + \dots + 1/(q/D - i))$, which clearly shows that $I(q)$ is a decreasing function of k.

Appendix A.1.8. Remarks on Asymptotic Scaling

We maintain that every nonzero continuous function $f(x)$ defined in $(0, \infty)$, whose limits as $x \to 0$ and $x \to \infty$ exist, is associated with two asymptotic power laws, or else two asymptotic scaling behaviours, one on each of the two ends, 0 (local scaling behaviour) and ∞ (global scaling behaviour). To prove this, first we examine the asymptotic behaviour of $f(x)$ as $x \to \infty$ and we try to identify the first asymptotic law. We perform the following steps:

1. We find $\beta_1 := \lim_{x \to \infty} f(x)$. If $\beta_1 \neq \pm\infty$ then we replace $f(x)$ with $1/(f(x) - \beta_1)$, otherwise we keep it as it is. Clearly then, $|\lim_{x \to \infty} f(x)| = \infty$.

2.   We find $\beta_2 := \lim_{c \to \infty}\left(\lim_{x \to \infty} x^{-c}f(x)\right)$. If $\beta_2 \neq 0$ then we replace $f(x)$ with $\ln|f(x)|$ (which preserves the property $\lim_{x \to \infty} f(x) = \infty$); if necessary, we make iterations so that eventually $\lim_{x \to \infty} f(x) = \infty$ and $\lim_{c \to \infty}\left(\lim_{x \to \infty} x^{-c}f(x)\right) = 0$.

3.   Given the properties in point 2, there exists a unique b, $0 < b < \infty$ satisfying:

$$\left|\lim_{x \to \infty} x^{-b}f(x)\right| < \infty \tag{A46}$$

so that for any $b' \neq b$:

$$\left|\lim_{x \to \infty} x^{-b'}f(x)\right| = \begin{cases} 0, & \forall b' < b \\ \infty, & \forall b' > b \end{cases} \tag{A47}$$

4.   The constant b defines an asymptotic power law with exponent b (cf. Hausdorff dimension; the case $b = 0$ signifies an improper scaling).

To identify the second asymptotic law as $x \to 0$, we define $\widetilde{f}(x) := f(1/x)$ and we proceed in the same manner to construct a function $\widetilde{f}(x)$ satisfying the conditions in point 2 above. Proceeding likewise, we determine the unique a for which $\left|\lim_{x \to \infty} x^{-a}\widetilde{f}(x)\right| < \infty$. This determines an asymptotic power law with exponent a for $f(x)$ as $x \to 0$.

We remark that the two power laws refer to the same function $f(x)$ but, due to the replacement steps of the procedure, may eventually correspond to different functions, say, $f_a(x)$ and $f_b(x)$ for the asymptotic behaviours as $x \to 0$ (local behaviour) and $x \to \infty$ (global behaviour), respectively. However, it is easy to construct a single function that combines both, making a final replacement, e.g., $f(x) = f_a(x)f_b(x)$, but many different $f(x)$ can actually be constructed. Thus, as well as any object has a dimension, any continuous function entails asymptotic power laws; generally not one but two, which in the special case of simple scaling can be identical.

Now, coming to the asymptotic values of $f^{\#}(x)$ for $x \to 0$ and ∞, symbolically $f^{\#}(0)$ and $f^{\#}(\infty)$, these are:

$$f^{\#}(\infty) = b, \; f^{\#}(0) = a \tag{A48}$$

To prove the former case, we proceed as follows:

$$\lim_{x \to \infty} x^{-b}f(x) = \lim_{x \to \infty} \frac{f(x)}{x^b} = \lim_{x \to \infty} \frac{f'(x)}{b\,x^{b-1}} = \lim_{x \to \infty}\left(\frac{xf'(x)}{b\,f(x)}\frac{f(x)}{x^b}\right) = \lim_{x \to \infty}\left(\frac{f^{\#}(x)}{b}\frac{f(x)}{x^b}\right) = \lim_{x \to \infty}\frac{f^{\#}(x)}{b}\lim_{x \to \infty}\frac{f(x)}{x^b} \tag{A49}$$

This implies that $\lim_{x \to \infty} f^{\#}(x)/b = 1$ and thus $f^{\#}(\infty) = b$. The latter case $(f^{\#}(0) = a)$ can be proved in the same manner.

A useful property is that if $\lim_{x \to \infty} f(x)$ is a finite value $(<\infty)$, then $b = f^{\#}(\infty) = 0$, because for any $b' > 0$, $\lim_{x \to \infty} x^{-b'}f(x) = \lim_{x \to \infty} f(x)/x^{b'} = 0$. Likewise, if $\lim f(x)$ is a finite value, then $a = f^{\#}(0) = 0$. Thus, to have asymptotic (for $x \to 0$ and ∞) values of $f^{\#}$ different for 0 and ∞, the asymptotic values of f should be either 0 or ∞. These properties are useful and are utilized in other proofs herein.

Appendix A.1.9. Asymptotic Scaling of the Different Second Order Tools

We assume that the asymptotic behaviour of the climacogram-based second order tools is given as:

$$\gamma_C^{\#}(0) = \xi^{\#}(0) = \zeta^{\#}(0) - 1 = 2\varphi_C(0) - 2 = a > 0$$
$$\gamma_C^{\#}(\infty) = \gamma^{\#}(\infty) = \zeta^{\#}(\infty) - 1 = 2\varphi_C(\infty) - 2 = b < 0 \tag{A50}$$

and we will determine that of the autocovariance-based tools. Note that the equality of the different function limits in (A50) has been proved in Appendix A.1.5.

We will fist examine the asymptotic properties of the autocovariance function. Initially we note that, because $\gamma(0) = c(0) = \gamma_0$ (finite), it follows that $\gamma^\#(0) = c^\#(0) = 0$. Now, since $b = \gamma^\#(\infty) < 0$, it follows that:

$$\lim_{k \to \infty} k^{-b}\gamma(k) = c \tag{A51}$$

where c is finite. We use the relationship between $\gamma(k)$ and $c(h)$ (Equation (21)) to find:

$$\lim_{h \to \infty} h^{-b}c(h) = \lim_{h \to \infty} \frac{c(h)}{h^b} = \lim_{h \to \infty} \frac{1}{2} \frac{d^2\left(h^2\gamma(h)\right)/dh^2}{h^b} = \lim_{h \to \infty} \left(\frac{\gamma(h)}{h^b} + 2\frac{\gamma'(h)}{h^{b-1}} + \frac{1}{2}\frac{\gamma''(h)}{h^{b-2}}\right) \tag{A52}$$

Because both $\gamma(h)$ and $h^b \to 0$ as $h \to \infty$, applying l'Hôpital's rule we find:

$$\lim_{h \to \infty} \frac{\gamma(h)}{h^b} = \lim_{h \to \infty} \frac{\gamma'(h)}{bh^{b-1}} = \lim_{h \to \infty} \frac{\gamma''(h)}{b(b-1)h^{b-2}} \tag{A53}$$

Since $b < 0$, $\gamma'(h)$, $\gamma''(h)$, h^{b-1} and h^{b-2} will also $\to 0$ as $h \to \infty$. Thus:

$$\lim_{h \to \infty} h^{-b}c(h) = \lim_{h \to \infty} \left(\frac{\gamma(h)}{h^b} + 2b\frac{\gamma(h)}{h^b} + \frac{b(b-1)}{2}\frac{\gamma(h)}{h^b}\right) = \tfrac{1}{2}(b+1)(b+2)\lim_{h \to \infty}\frac{\gamma(h)}{h^b} = \tfrac{1}{2}(b+1)(b+2)c \tag{A54}$$

Unless $b = -1$ or $b = -2$, the limit $\lim_{h \to \infty} h^{-b}c(h)$ is 0, finite or ∞, if and only if $\lim_{h \to \infty} h^{-b}\gamma(h)$ is 0, finite or ∞, respectively. Therefore, if $b \neq -1, -2$, then $c^\#(\infty) = \gamma^\#(\infty) = b$, otherwise $\lim_{h \to \infty} h^{-b}c(h) = 0$, which does not guarantee (nor does it exclude) that $c^\#(\infty) = \gamma^\#(\infty) = b$.

Next we will examine the asymptotic properties of the structure function. Initially we note that, because $\xi(\infty) = v(\infty) = \gamma_0$ (finite), it follows that $\xi^\#(\infty) = v^\#(\infty) = 0$. Now, since $a = \xi^\#(0) > 0$, it follows that:

$$\lim_{k \to 0} k^{-a}\xi(k) = d \tag{A55}$$

where d is finite. We use the relationship between $\xi(k)$ and $v(h)$ (Equation (26)) to find:

$$\lim_{h \to 0} h^{-a}v(h) = \lim_{h \to 0} \frac{v(h)}{h^a} = \lim_{h \to 0} \frac{1}{2}\frac{d^2\left(h^2\xi(h)\right)/dh^2}{h^a} = \lim_{h \to 0}\left(\frac{\xi(h)}{h^a} + 2\frac{\xi'(h)}{h^{a-1}} + \frac{1}{2}\frac{\xi''(h)}{h^{a-2}}\right) \tag{A56}$$

Because both $\xi(h)$ and $h^a \to 0$ as $h \to 0$, applying l'Hôpital's rule we find:

$$\lim_{h \to 0} \frac{\xi(h)}{h^a} = \lim_{h \to 0} \frac{\xi'(h)}{ah^{a-1}} \tag{A57}$$

If $a > 1$, then both $\gamma'(h)$ and h^{a-1} will also $\to 0$ as $h \to 0$. In this case, again applying l'Hôpital's rule:

$$\lim_{h \to 0} \frac{\xi'(h)}{ah^{a-1}} = \lim_{h \to 0} \frac{\xi''(h)}{a(a-1)h^{a-2}} \tag{A58}$$

If $a < 1$, then $\gamma'(h)$ and h^{a-1} will both $\to \infty$ as $h \to 0$. Thus, again applying l'Hôpital's rule, we will conclude with (A58) again. Thus, in both cases:

$$\lim_{h \to 0} h^{-a}v(h) = \lim_{h \to 0}\left(\frac{\xi(h)}{h^a} + 2a\frac{\xi(h)}{h^a} + \frac{a(a-1)}{2}\frac{\xi(h)}{h^a}\right) = \tfrac{1}{2}(a+1)(a+2)\lim_{h \to 0}\frac{\xi(h)}{h^a} = \tfrac{1}{2}(a+1)(a+2)d \tag{A59}$$

Since $a > 0$ the rightmost expression cannot be zero unless $d = 0$. Therefore, the limit $\lim_{h \to 0} h^{-a}v(h)$ is 0, finite or ∞, if and only if $\lim_{h \to 0} h^{-a}\xi(h)$ is 0, finite or ∞, respectively. Therefore, $v^\#(0) = \xi^\#(0) = a$. If $a = 1$ precisely, then $h^{a-1} = 1$ (it does not tend to 0 or ∞), while $\lim_{h \to 0} h\xi''(h) = 0$. Thus, from (A56) we get $\lim_{h \to 0} h^{-1}v(h) = 3d$ (finite), which again means that $v^\#(0) = \xi^\#(0) = a$. Also, if $a = 2$ precisely, then $h^{a-2} = 1$ and from (A56) we get $\lim\limits_{h\to 0} h^{-1}v(h) = 6d$ (finite). In conclusion, in all cases it holds that $v^{\#}(\infty) = \xi^{\#}(\infty) = 0$ and $v^{\#}(0) = \xi^{\#}(0) = a$, which means that $v(k)$ has the same asymptotic behaviour with $\xi(k)$ in both ends. We can thus rewrite (A50) as:

$$\gamma_C^{\#}(0) = \xi^{\#}(0) = v^{\#}(0) = \zeta^{\#}(0) - 1 = 2\varphi_C(0) - 2 = a > 0$$
$$\gamma_C^{\#}(\infty) = \gamma^{\#}(\infty) = c^{\#}(\infty) = \zeta^{\#}(\infty) - 1 = 2\varphi_C(\infty) - 2 = b < 0$$
$$\text{Possible exceptions for } c^{\#}(\infty): \ b = \gamma_C^{\#}(\infty) = \gamma^{\#}(\infty) = \{-1, -2\},$$
$$\zeta^{\#}(\infty) = \{0, -1\}, \ \varphi_C(\infty) = \{1/2, 0\}$$

(A60)

Now coming to the asymptotic behaviour of the power spectrum, according to Gneiting and Schlather [37] (who also cite [46–48]), the power spectrum is "typically" associated with the asymptotic relationship $s(w) \sim w^{-a-1}$ as $w \to \infty$ and $s(w) \sim w^{-b-1}$ as $w \to 0$, where a (>0) and b (<0) are the LLDs of the structure function, $v^{\#}(0)$ and autocorrelation function $c^{\#}(\infty)$, respectively, same to the above discourse. By inspection, it is seen that the slopes $-(a+1)$ and $-(b+1)$ are precisely the log-log slopes of the climacospectrum expressed in terms of frequency, $\tilde{\zeta}(w)$, or equivalently $-\zeta^{\#}(0)$ and $-\zeta^{\#}(\infty)$, respectively. It can be conjectured that, since this is the "typical" behaviour, the non-typical cases correspond to integral values of a and b.

In plain language, the above results can be expressed as follows:

1. The asymptotic behaviour of the autocovariance function is the same with that of the climacogram.
2. The asymptotic behaviour of the structure function is the same with that of the climacostructure function.
3. The asymptotic behaviour of the power spectrum is the same with that of the climacospectrum.

Exceptions to the rule 1 may occur when $b = -1$ or $b = -2$ and this was indeed verified by thorough analytical work using the series of models examined in Appendix A.2. Exception to rule 2 can also appear and indeed they were found when $b = -2$ or $a = 2$. The entire list of exceptions found in this analytical work follows, also mentioning the specific models in which they appeared:

1. $b = \gamma^{\#}(\infty) = -1$ and $c^{\#}(\infty) = -\infty$: Markov, Smooth Exponential models
2. $b = \gamma^{\#}(\infty) = -1$ and $c^{\#}(\infty) = -2$: Generalized Planck model for $n = 0$
3. $b = \gamma^{\#}(\infty) = -2$ and $c^{\#}(\infty) = -\infty$, also $-s^{\#}(0) = -2n$ while $\zeta^{\#}(\infty) = -1$: Antipersistent Exponential model for $n > 0$
4. $b = \gamma^{\#}(\infty) = -2$ and $c^{\#}(\infty) = -2(n+1)$, also $-s^{\#}(0) = -2n$ while $\zeta^{\#}(\infty) = -1$: Generalized Planck model for $n > 0$
5. $a = \zeta^{\#}(0) = 2$ and $-s^{\#}(\infty) = \infty$ while $\zeta^{\#}(0) = 3$: Generalized Planck model
6. $a = \zeta^{\#}(0) = 2$ and $-s^{\#}(\infty) = 2(n+1)$ while $\zeta^{\#}(0) = 3$: Smooth Exponential for $n > 0$

Appendix A.1.10. Proof of Physical Inconsistency of too Mild Slopes at High Frequencies in Power Spectrum

Let us assume that for small scales $k < \epsilon$ or high frequencies $w > 1/\epsilon$, with ϵ however small, the log-log derivative is $s^{\#}(w) \approx \beta$, or else $s(w) \sim w^{\beta}$ where β is a constant satisfying $\beta > -1$. As the area of the power spectrum equals the variance of the continuous time process, we will have:

$$\gamma(0) = \int_0^{\infty} s(w)dw = \int_0^{1/\epsilon} s(w)dw + \int_{1/\epsilon}^{\infty} s(w)dw$$

(A61)

Because $s(w) \sim w^{\beta}$, there exist $\alpha > 0$ so that $s(w) \geq \alpha \, w^{\beta}$ for $w > 1/\epsilon$. Thus the, rightmost of the above integrals can be evaluated as:

$$\int_{1/\epsilon}^{\infty} s(w)dw \geq \int_{1/\epsilon}^{\infty} \alpha w^{\beta} \, dw = \left. \frac{\alpha w^{\beta+1}}{\beta+1} \right|_{1/\epsilon}^{\infty}$$

(A62)

Clearly if $\beta > -1$, $w^{\beta+1} \to \infty$ as $w \to \infty$ and thus $\gamma(0) = \infty$. The same result could be obtained if the climacospectrum $\tilde{\zeta}(w)$ was used instead of the power spectrum. Because infinite variance means infinite energy, which is physically inconsistent, it turns out that the assumption $\beta > -1$ is physically absurd.

Appendix A.1.11. Proof of the Inconsistency of too Steep Slopes at High Frequencies in Power Spectrum

We prove that $\varphi_C(0) \leq 2$ or, equivalently, $\zeta^\#(0) \leq 2$, by contradiction, i.e., assuming $\zeta^\#(0) = 2M$ with $M > 1$. In the latter case for small k we may approximate $\zeta(k)$ as:

$$\zeta(k) \approx \lambda \frac{k^{2M}}{1 + k^{2M}} \tag{A63}$$

which indeed yields $\zeta^\#(0) = 2M$. The climacogram for small k is thus:

$$\gamma(k) = \frac{\lambda}{1 + k^{2M}} \tag{A64}$$

The positive definiteness of $\gamma(k)$ entails that any matrix A with elements $a_{i,j} = \gamma(k_i - k_j)$ should be positive definite. Choosing three points, $k_i = k_0$, $k_0 + k$ and $k_0 + 2k$, we form the 3×3 matrix:

$$A = \begin{bmatrix} \gamma(0) & \gamma(k) & \gamma(2k) \\ \gamma(k) & \gamma(0) & \gamma(k) \\ \gamma(2k) & \gamma(k) & \gamma(0) \end{bmatrix} \tag{A65}$$

whose determinant can be easily evaluated. Specifically, after algebraic manipulations we obtain:

$$B(k, M) := \mathrm{Det}(A)/\lambda^3 = 1 - \frac{2}{(1 + k^{2M})^2} - \frac{1}{\left(1 + (2k)^{2M}\right)^2} + \frac{2}{(1 + k^{2M})^2 \left(1 + (2k)^{2M}\right)} \tag{A66}$$

This can also be written as:

$$B(k, M) = \frac{2^{2M} k^{4M} G(k,M)}{(1 + k^{2M})^2 \left(1 + (2k)^{2M}\right)^2},$$
$$G(k, M) := 4 - 2^{2M} + 2(1 + 2^{2M}) k^{2M} + 2^{2M} (k^{2M})^2 \tag{A67}$$

We observe that $G(k, M)$ is a second-order polynomial in terms of $y := k^{2M}$ with discriminant $D(M) = 1 + 2^{2M+1}(2^{2M} - 1)$, which is positive for any $M > 0$ and thus the equation $G(k, M) = 0$ has two real solutions:

$$y_{1,2} = 2^{-2M}\left(-1 - 2^{2M} \mp \sqrt{1 + 2^{2M+1}(2^{2M} - 1)}\right) \tag{A68}$$

Clearly, $y_1 < 0$ for any M, while $y_2 < 0$ only when $M < 1$ (for $M = 1$, $y_2 = 0$ precisely). Therefore, if $M > 1$, then for $0 < k^{2M} < y_2$, $G(k, M) < 0$ and hence $B(k, M) < 0$, which proves that A is not positive definite as its principal determinant is negative.

Solved for k, the interval where $B(k, M) < 0$ is:

$$0 < k < \frac{1}{2}\left(-1 - 2^{2M} + \sqrt{1 + 2^{2M+1}(2^{2M} - 1)}\right)^{1/2M} \tag{A69}$$

This means that, if $M > 1$, $B(k, M)$ is negative for arbitrarily small k (in the neighbourhood of 0), which justifies the approximation we have used for $\zeta(k)$. Therefore the assumption $M > 1$ should be rejected.

As an illustration by a different approach, Figure A1 shows for the FHK-C model, a single value of the power spectrum, namely $s(\alpha)$, where α is the time scale parameter of the model, as a function of the parameter M. As seen in the plot, for $M > 1$, $s(\alpha) < 0$ which confirms absence of positive definiteness, now of the autocorrelation function. (We note that for $M > 1$ but very close to 1, the plot shows $s(\alpha)$ slightly positive but in fact it is negative again for some $w > \alpha$—not shown in the figure).

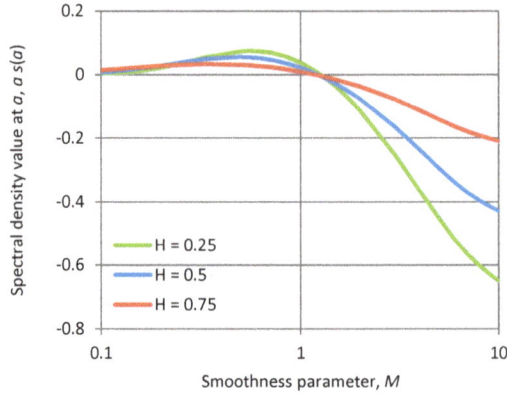

Figure A1. Plot of the spectral density value at $w = \alpha$, standardized by α, for the FHK-C model with smoothness parameter M varying in [0.1, 10] and for the indicated values of the Hurst parameter H.

Appendix A.2. Specific Stochastic Models and Their Behaviour

In Appendix A.2 we provide detailed tables with mathematical expressions for all the models used in the paper and detailed graphs with applications of these models. All models include two scale parameters, α (scale parameter of time with dimensions of $[t]$) and λ (scale parameter of variance, with dimensions of $[x^2]$). These are absolutely necessary for dimensional consistency. Some models contain additional dimensionless shape parameters, such as the smoothness parameter M, the persistence (Hurst) parameter H, or other as specified in each of the models. The theoretical characteristics depicted in the graphs are determined from the expressions given in the tables. The time series also depicted in the graphs, along with and their empirical characteristics, were produced by the SMA model as described in the text; their length is 1024 in all cases. For facilitating comparisons, in all applications the parameter α is the same, $\alpha = 10$ (except if stated otherwise in the caption), while λ is chosen so that $\gamma(1) = 1$.

Appendix A.2.1. Markov Process

The Markov (also known as Ornstein–Uhlenbeck) process is the most convenient as all its second-order characteristics have simple expressions (see Table A1) and, simultaneously, the most parsimonious as it includes no other parameter additional to the two necessary scale parameters α and λ. For these reasons it is the most common in applications. On the other hand, its neutrality in terms of smoothness and persistence, and more specifically the low entropy production for large time scales, does not make it a good candidate to model natural behaviours.

The application depicted in Figure A2 corresponds to an unusually high lag-one autocorrelation ($\rho = 0.94$). Note that the model does not allow control of autocorrelation once the parameter α is specified.

Table A1. Mathematical expressions of the second order characteristics of the Markov process and their asymptotic properties.

Characteristic	Mathematical Expression	Local Asymptotic Properties	Global Asymptotic Properties	Equation
Climacogram	$\gamma(k) = \frac{2\lambda}{(k/a)^2}\left(-1 + k/a + e^{-k/a}\right)$	$\gamma^{\#}(0) = 0$	$\gamma^{\#}(\infty) = -1$	(A70)
Climacospectrum	$\zeta(k) := \frac{k(\gamma(k)-\gamma(2k))}{\ln 2}$ $= \frac{\lambda(-3+2k/a+4e^{-k/a}-e^{-2k/a})}{2(\ln 2)k/a}$	$\zeta^{\#}(0) = 2$	$\zeta^{\#}(\infty) = 0$	(A71)
Entropy production	$\varphi(k) = \frac{2\lambda(1-e^{-k/a})}{\gamma(k)(k/a)}$ $= \frac{(k/a)(1-e^{-k/a})}{2(-1+k/a+e^{-k/a})}$ $\varphi_C(k) = \frac{\varphi(k)\gamma(k)-\varphi(2k)\gamma(2k)}{\gamma(k)-\gamma(2k)}$	$\varphi(0) = 1$ $\varphi_C(0) = 3/2$	$\varphi(\infty) = \varphi_C(\infty) = 1/2$	(A72)
Autocovariance function	$c(h) = \lambda e^{-h/a}$	$c^{\#}(0) = 0$	$c^{\#}(\infty) = -\infty$	(A73)
Power spectrum	$s(w) = \frac{4\lambda a}{1+(2\pi a w)^2}$	$-s^{\#}(\infty) = 2$	$-s^{\#}(0) = 0$	(A74)

Appendix A.2.2. FHK Process with Cauchy-Type Climacogram (FHK-C)

General description of the FHK process with Cauchy-type climacogram (FHK-C) is given in Section 3.2 and its detailed mathematical expressions and asymptotic processes are given in Table A2.

Table A2. Mathematical expressions of the second order characteristics of the FHK process with Cauchy-type climacogram (FHK-C) and their asymptotic properties.

Characteristic	Mathematical Expression	Local Asymptotic Properties	Global Asymptotic Properties	Equation
Climacogram	$\gamma(k) = \lambda(1+(k/a)^{2M})^{\frac{H-1}{M}}$	$\gamma^{\#}(0) = 0$	$\gamma^{\#}(\infty) = 2H-2$	(A75)
Climacospectrum	$\zeta(k) := \frac{k(\gamma(k)-\gamma(2k))}{\ln 2}$	$\zeta^{\#}(0) = 1+2M$	$\zeta^{\#}(\infty) = 2H-1$	(A76)
Entropy production	$\varphi(k) = H + \frac{1-H}{1+(k/a)^{2M}}$ $\varphi_C(k) = \frac{\varphi(k)\gamma(k)-\varphi(2k)\gamma(2k)}{\gamma(k)-\gamma(2k)}$	$\varphi(0) = 1$ $\varphi_C(0) = 1+M$	$\varphi(\infty) = \varphi_C(\infty) = H$	(A77)
Autocovariance function	$c(h) = \gamma(h) \times$ $\frac{1+(3H+2MH-2M-1)(h/a)^{2M}+H(2H-1)(h/a)^{4M}}{1+2(h/a)^{2M}+(h/a)^{4M}}$	$c^{\#}(0) = 0$	$c^{\#}(\infty) = 2H-2$	(A78)
Power spectrum	Analytical expression not possible except in special cases; numerical solution easily derived from (10)	$-s^{\#}(\infty) = 1+2M$	$-s^{\#}(0) = 2H-1$	(A79)

This model is very rich as it covers the entire admissible range, by appropriate choice of the smoothness parameter ($0 < M \leq 1$) and persistence parameter ($0 \leq H < 1$). It represents smooth processes (if $M > \frac{1}{2}$), rough processes (if $M < \frac{1}{2}$), persistent processes (if $H > \frac{1}{2}$) and antipersistent processes (if $H < \frac{1}{2}$). For $M = H = \frac{1}{2}$ it becomes a neutral process, practically identical to Markov.

The FHK process can even produce near pink ($1/f$) noise if $M = 1 - H \approx 0$, which is associated with maximal entropy production for large scales and minimal for small scales. In this case the quantity $(k/a)^{2M}$ is close to 1 and the following approximation holds:

$$\gamma(k) = \frac{\lambda}{1+(k/a)^{2(1-H)}} \approx \frac{\lambda(1+(H-1)\ln(k/a))}{2} \tag{A80}$$

This can be verified by observing that for x close to 1 the quantities $\ln x$ and $(x^2 - 1)/(x^2 + 1)$ are close to each other, differing only by a factor of $O[x - 1]^3$.

For $M = H \approx 0$, we have minimal entropy production for all scales. In this case:

$$\gamma(k) = \lambda\left(1+(k/a)^{2H}\right)^{\frac{H-1}{H}} \approx \lambda(k/a)^{H-1} \tag{A81}$$

This can be verified by observing that for x close to 1 the logarithms of the two quantities differ only by a factor of $H\,O[x-1]^2$. We note that both (A80) and (A81) violate the condition for physical consistency (finite variance).

A set of applications of the FHK-C are depicted in detail in Figures A3–A7, which are also discussed in Section 3.2. It is observed that the SMA generation scheme performs almost perfectly even in the most extreme cases and that the climacogram-based characteristics perform better than the autocorrelation-based ones.

Appendix A.2.3. FHK Process with Dagum-Type Climacogram (FHK-D)

The FHK process with Dagum-type climacogram (FHK-D) has expressions for its second order characteristics different from those of the Cauchy-type climacogram (see Table A3). However, the behaviour of the two models, as well as the meaning and ranges of parameters, are the same. For $M = 1 - H$ the two models become precisely identical. The model is particularly useful in combination with the FHK-C model as shown below.

Table A3. Mathematical expressions of the second order characteristics of the FHK process with Dagum-type climacogram (FHK-D) and their asymptotic properties.

Characteristic	Mathematical Expression	Local Asymptotic Properties	Global Asymptotic Properties	Equation
Climacogram	$\gamma(k) = \lambda\left(1 - \left(1 + (k/\alpha)^{2(H-1)}\right)^{\frac{M}{H-1}}\right)$	$\gamma^{\#}(0) = 0$	$\gamma^{\#}(\infty) = 2H - 2$	(A82)
Climacospectrum	$\zeta(k) := \frac{k(\gamma(k) - \gamma(2k))}{\ln 2}$	$\zeta^{\#}(0) = 1 + 2M$	$\zeta^{\#}(\infty) = 2H - 1$	(A83)
Entropy production	$\varphi(k) = 1 - \frac{\lambda - \gamma(k)}{\gamma(k)} \frac{M}{(1+(k/\alpha)^{2-2H})}$ $\varphi_C(k) = \frac{\varphi(k)\gamma(k) - \varphi(2k)\gamma(2k)}{\gamma(k) - \gamma(2k)}$	$\varphi(0) = 1$ $\varphi_C(0) = 1 + M$	$\varphi(\infty) = \varphi_C(\infty) = H$	(A84)
Autocovariance function	$c(h) = \lambda - (\lambda - \gamma(h)) \times$ $\frac{1+(2+M+2MH)(h/\alpha)^{2H-2}+(1+M)(1+2M)(h/\alpha)^{4H-4}}{1+2(h/\alpha)^{2H-2}+(h/\alpha)^{4H-4}}$	$c^{\#}(0) = 0$	$c^{\#}(\infty) = 2H - 2$	(A85)
Power spectrum	Analytical expression not possible except in special cases; numerical solution easily derived from (10)	$-s^{\#}(\infty) = 1 + 2M$	$-s^{\#}(0) = 2H - 1$	(A86)

Appendix A.2.4. FHK Process with Climacogram Equal to the Sum of a Cauchy and a Dagum Climacograms (FHK-CD)

By summing a Cauchy-type climacogram with $M = 1$ and a Dagum-type climacogram with $H = 0$ we get the FHK-CD model whose second-order characteristics are shown in Table A4. This is a convenient model, in which the first additive term determines merely the persistence of the process and the second one the smoothness of the process. It is more convenient that its constituents as most of mathematical expressions are simpler; in particular, it results in an explicit relationship of the power spectrum, which however is not very simple. In addition, it is more flexible and richer than its constituents, as it contains two couples of scale parameters; however, if parsimony is sought, then it can take the same number of parameters as each of the constituents by setting $\alpha_1 = \alpha_2$ and $\lambda_1 = \lambda_2$ (it is reminded that one λ and one α are absolutely minimal requirements of parameters).

An application of the model based on real measurements of turbulent velocities is contained in Section 3.4.

Another option, of the same type, not studied in detail here, would be to sum a Cauchy-type climacogram with $M = \frac{1}{2}$ and a Dagum-type climacogram with $H = \frac{1}{2}$. The climacogram of the thus formed model is:

$$\gamma(k) = \lambda_1(1 + k/\alpha_1)^{2H-2} + \lambda_2(1 - (1 + \alpha_2/k)^{-2M}) \tag{A87}$$

The model (A87), however, can only produce rough persistent processes with $0 < \varphi_C(0) \le 3/2$ and $1/2 \le \varphi_C(\infty) < 1$. Indeed, it is easily shown that in this case $\varphi_C(0) = \min(1 + M, 3/2)$ and

$\varphi(\infty) = \varphi_C(\infty) = \max(H, 1/2)$. The turbulent velocity case, studied in Section 3.4, belongs to this domain and thus the model (A87) could also be used as an alternative.

Table A4. Mathematical expressions of the second order characteristics of the FHK process formed as the sum of a Cauchy and a Dagum process (FHK-CD) dealing separately with persistence and smoothness, respectively, and their asymptotic properties.

Characteristic	Mathematical Expression	Local Asymptotic Properties	Global Asymptotic Properties	Equation
Climacogram	$\gamma(k) = \gamma_1(k) + \gamma_2(k)$ $\gamma_1(k) = \lambda_1(1 + (k/\alpha_1)^2)^{H-1}$ $\gamma_2(k) = \lambda_2(1 - (1 + (k/\alpha_2)^{-2})^{-M})$	$\gamma^\#(0) = 0$	$\gamma^\#(\infty) = 2H - 2$	(A88)
Climacospectrum	$\zeta(k) := \frac{k(\gamma(k) - \gamma(2k))}{\ln 2}$	$\zeta^\#(0) = 1 + 2M$	$\zeta^\#(\infty) = 2H - 1$	(A89)
Entropy production	$\varphi(k) = \frac{\varphi_1(k)\gamma_1(k) + \varphi_2(k)\gamma_2(k)}{\gamma(k)}$ $\varphi_1(k) = \frac{1 + H(k/\alpha_1)^2}{1 + (k/\alpha_1)^2}$ $\varphi_2(k) = 1 - \frac{M}{(1 + (k/\alpha_2)^2)((1 + (k/\alpha_2)^{-2})^M - 1)}$ $\varphi_C(k) = \frac{\varphi(k)\gamma(k) - \varphi(2k)\gamma(2k)}{\gamma(k) - \gamma(2k)}$	$\varphi(0) = 1$ $\varphi_C(0) = 1 + M$	$\varphi(\infty) = \varphi_C(\infty) = H$	(A90)
Autocovariance function	$c(h) = c_1(h) + c_2(h)$ $c_1(h) = \gamma_1(h) \times$ $\frac{1 + (5H-3)(h/\alpha_1)^2 + H(2H-1)(h/\alpha_1)^4}{1 + 2(h/\alpha_1)^2 + (h/\alpha_1)^4}$ $c_2(h) = \lambda_2 - (\lambda_2 - \gamma_2(h)) \times$ $\frac{(1+M)(1+2M) + (2+M)(h/\alpha)^2 + (h/\alpha)^4}{1 + 2(h/\alpha)^2 + (h/\alpha)^4}$	$c^\#(0) = 0$	$c^\#(\infty) = 2H - 2$	(A91)
Power spectrum	Analytical expression too complex, derived from (10)	$-s^\#(\infty) = 1 + 2M$	$-s^\#(0) = 2H - 1$	(A92)

Appendix A.2.5. Smooth Process with Generalized Exponential-Type Autocovariance (SE)

By generalizing the power spectrum of a Markov process, as seen in Table A5, we can obtain a smooth processes (the smooth exponential process—SE). Generalization is made by introducing an integral parameter $n = 0, 1, 2, \ldots$, where the value $n = 0$ results precisely in a Markov process. A notable characteristic of the process is that it corresponds to the absolute maximum of entropy production for small scales.

Appendix A.2.6. Antipersistent Process with Generalized Exponential-Type Autocovariance (AE)

This is again a generalization of the Markov process by introducing an integral parameter $n = 0$, $1, 2, \ldots$, where the value $n = 0$ results precisely in a Markov process. Its characteristics are shown in Table A6. The process is antipersistent and its most notable characteristic is that it corresponds to the absolute minimum of entropy production for large scales. An application of the process is shown in Figure A8 for $n = 2$; the hole in the function $\phi_C(k)$ for moderate k is a characteristic of this process.

Appendix A.2.7. Generalized Planck Model (GP)

The generalized Planck (GP) model is obtained by a generalization of Planck's law of black-body radiation, introducing an integral parameter $n = 0, 1, 2, \ldots$, where the value $n = 1$ corresponds to Planck's law precisely. The process, whose detailed equations are given in Table A7, is simultaneously extremely smooth ($\phi_C(0) = 2$) and extremely antipersistent ($\phi_C(\infty) = 0$). An application for $n = 1$ is depicted in Figure A9.

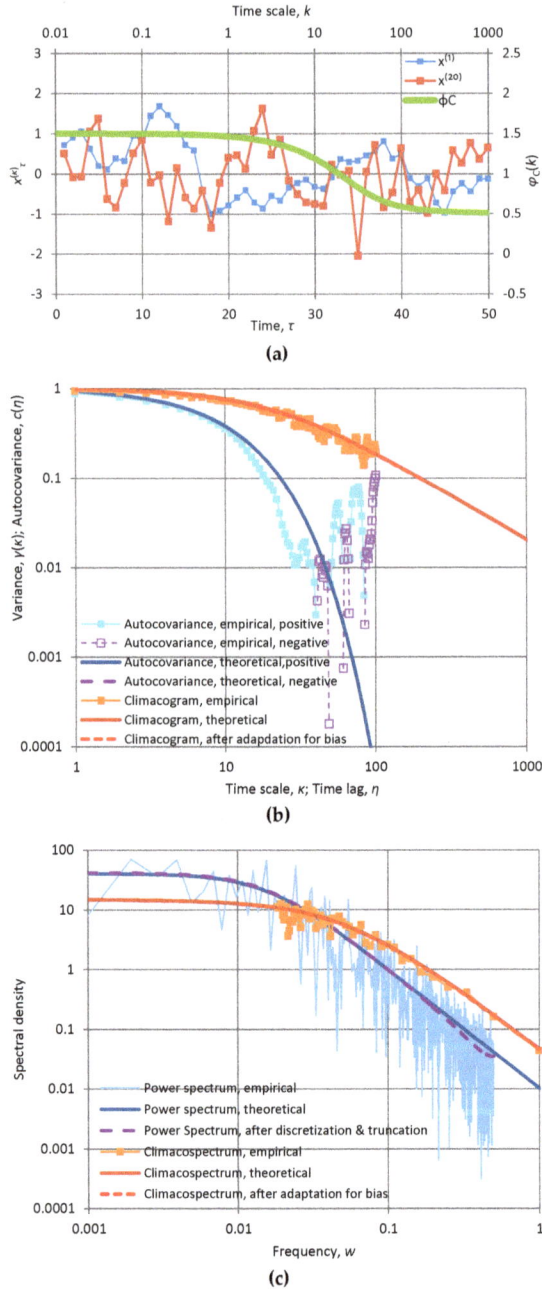

Figure A2. Characteristic graphs of a time series of length 1024 generated by the Markov model with $\alpha = 10$ and $\lambda = 1.03$ (so that $\gamma(1) = 1$): (**a**) plot of the first fifty terms of the at time scales $\kappa = 1$ and 20, along with the "stamp" $\phi_C(k)$ of the fitted model (green line plotted with respect to the secondary axes); (**b**) empirical and theoretical climacogram and autocovariance function; (**c**) empirical and fitted theoretical climacospectrum and power spectrum.

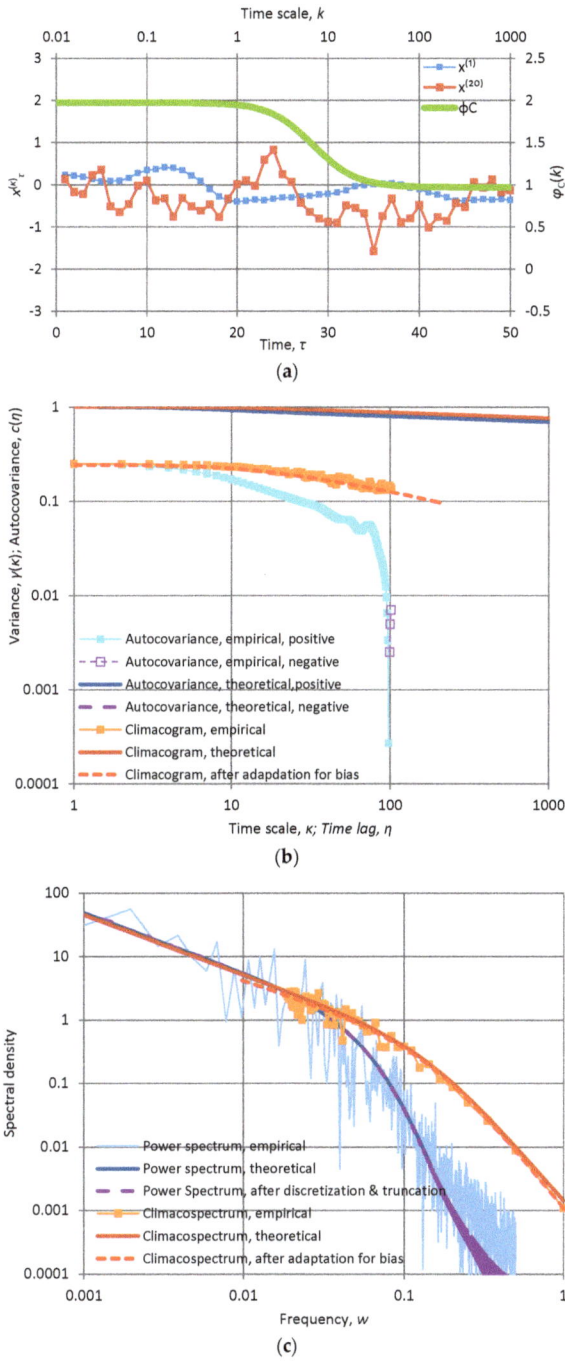

Figure A3. Characteristic graphs as in Figure A2 but for a time series generated by the FHK-C model with $\gamma(1) = 1$, $\alpha = 10$ and $H = M = 0.97$.

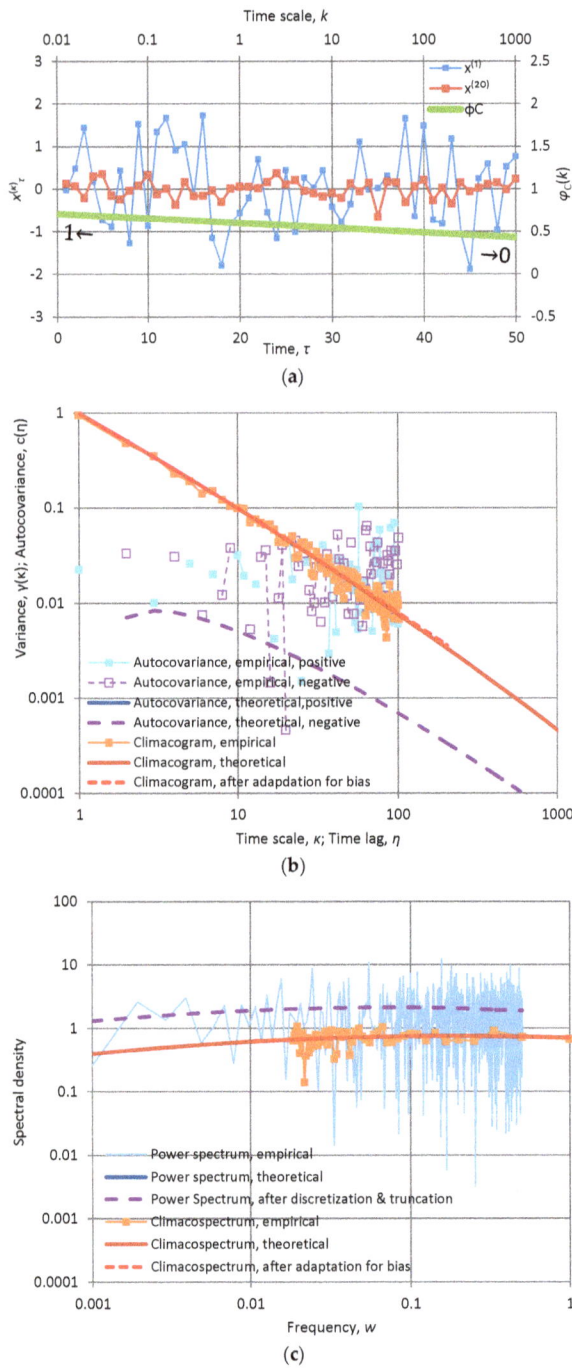

Figure A4. Characteristic graphs as in Figure A2 but for a time series generated by the FHK-C model with $\gamma(1) = 1$, $\alpha = 10$, and $H = M = 0.05$.

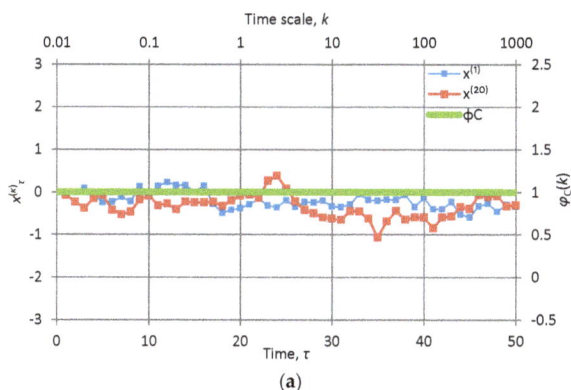

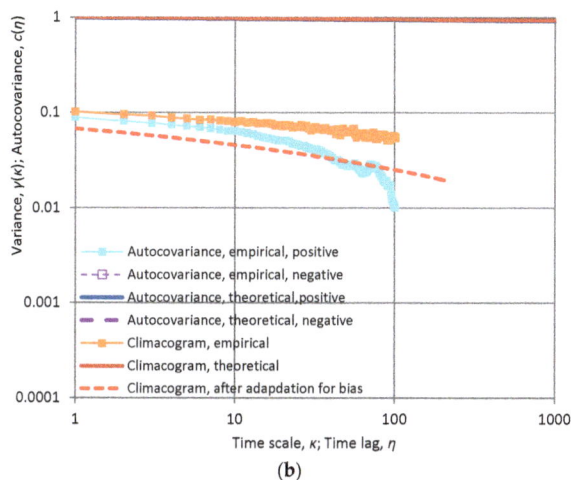

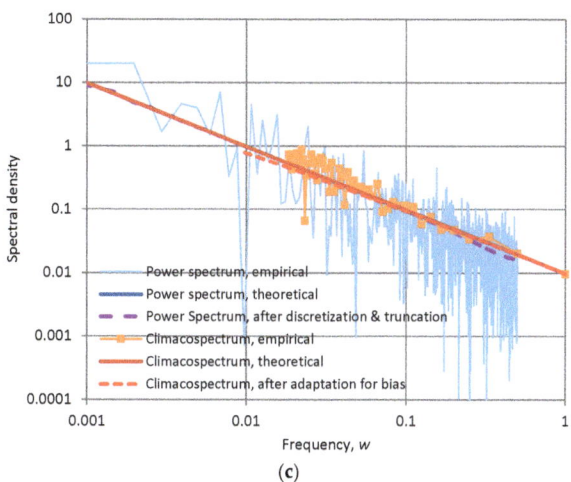

Figure A5. Characteristic graphs as in Figure A2 but for a time series generated by the FHK-C model with $\gamma(1) = 1$, $\alpha = 10$, $H = 0.99$ and $M = 0.01$ (close to "pink noise").

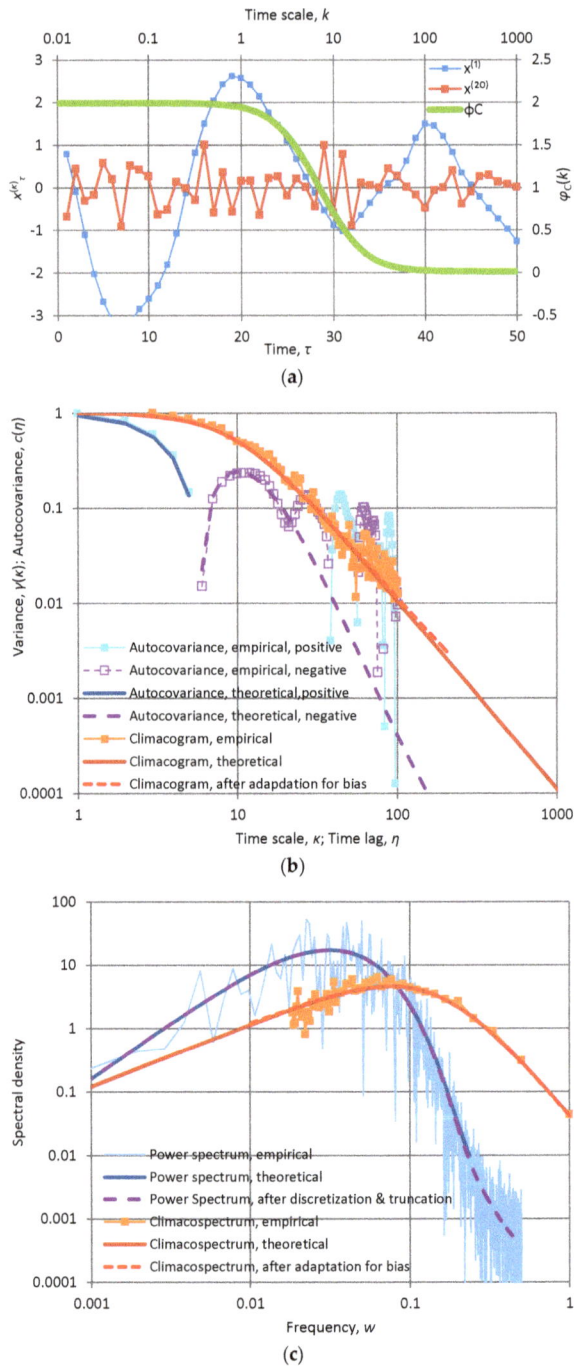

Figure A6. Characteristic graphs as in Figure A2 but for a time series generated by the FHK-C model with $\gamma(1) = 1$, $\alpha = 10$, $H = 0.01$ and $M = 0.99$.

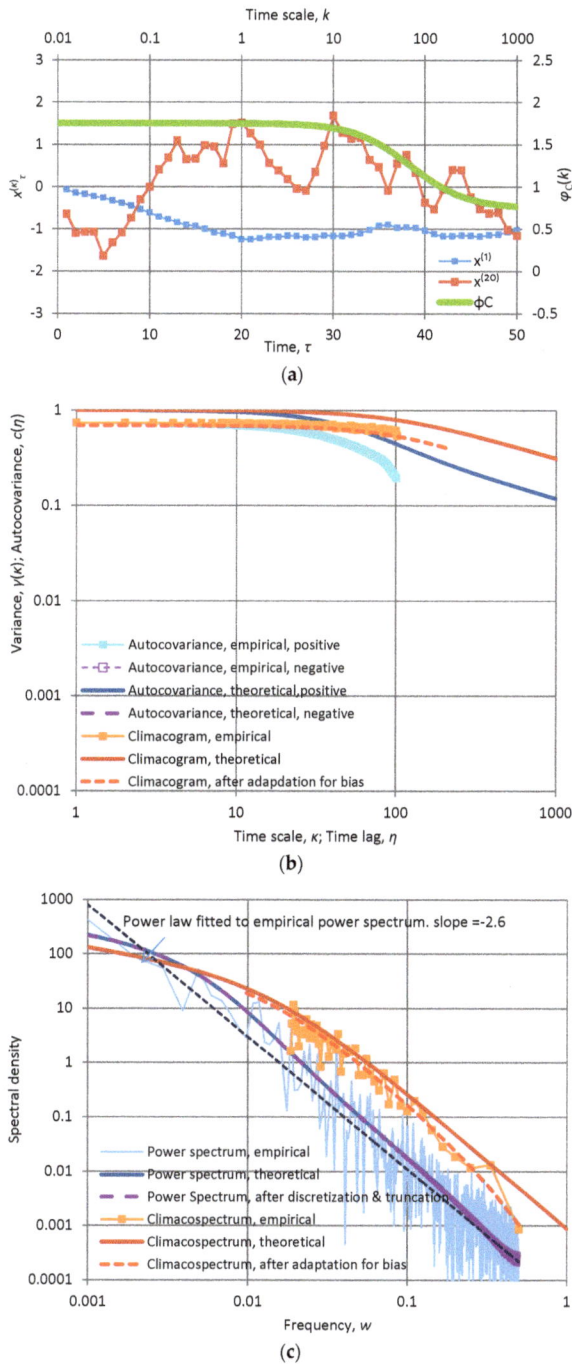

Figure A7. Characteristic graphs as in Figure A2 but for a time series generated by the FHK-C model with $\gamma(1) = 1$, $\alpha = 100$ and $H = M = 0.75$.

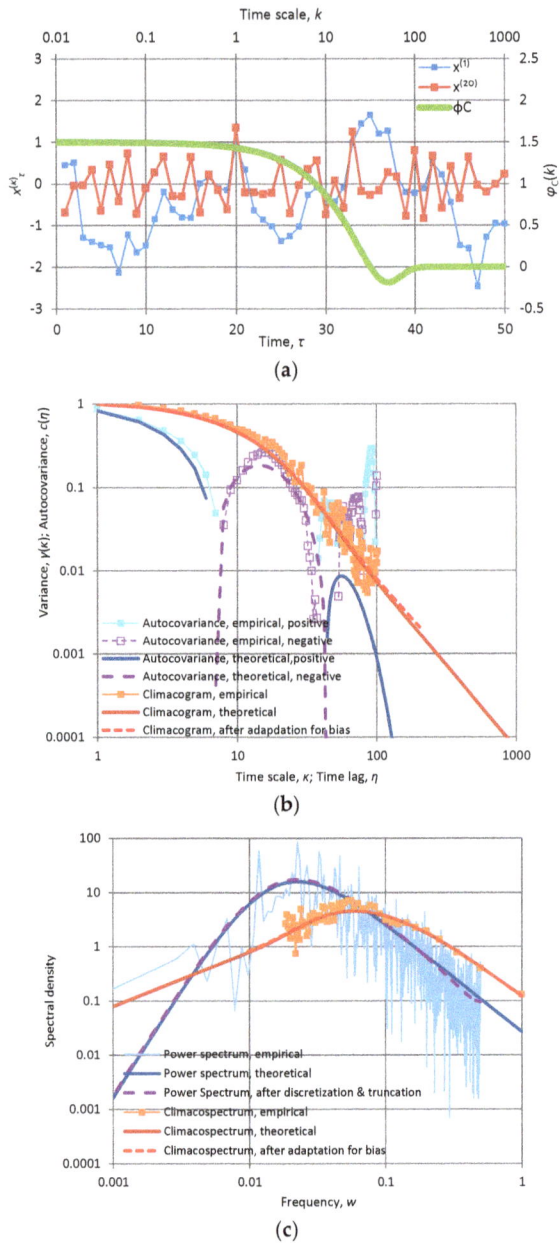

Figure A8. Characteristic graphs as in Figure A2 but for a time series generated by the AE model with $\gamma(1) = 1$, $\alpha = 10$ and $n = 2$.

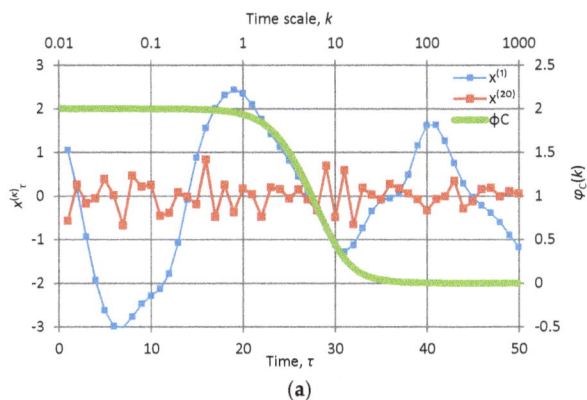

(a)

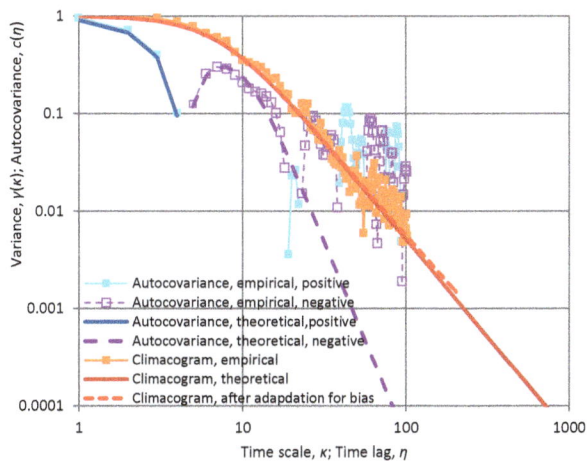

(b)

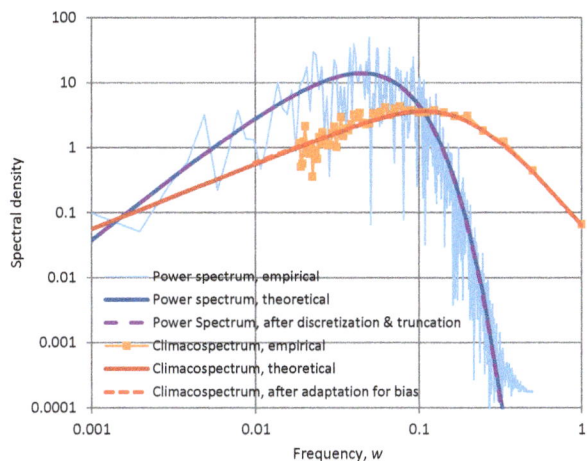

(c)

Figure A9. Characteristic graphs as in Figure A2 but for a time series generated by the GP model with $\gamma(1) = 1$, $\alpha = 10$ and $n = 1$.

Table A5. Mathematical expressions of the second order characteristics of the SE process and their asymptotic properties.

Characteristic	Mathematical Expression	Local Asymptotic Properties	Global Asymptotic Properties	Equation
Climacogram	$\gamma(k) = \frac{2\lambda}{(k/\alpha)^2} \times \left(-2n - 1 + a_n^{PE} k/\alpha + e^{-k/\alpha} P_n^{PE}(k/\alpha)\right)$ where $n = 0, 1, 2, \ldots, P_n^{PE}(x)$ is a polynomial and a_n^{PE} a coefficient, both determined using (23); the first five are: $a_0^{PE} = 1, P_0^{PE}(x) = 1$ $a_1^{PE} = 2, P_1^{PE}(x) = 3 + x$ $a_2^{PE} = \frac{8}{3}, P_2^{PE}(x) = 5 + \frac{7}{3}x + \frac{1}{3}x^2$ $a_3^{PE} = \frac{16}{5}, P_3^{PE}(x) = 7 + \frac{57}{15}x + \frac{12}{15}x^2 + \frac{1}{15}x^3$ $a_4^{PE} = \frac{128}{35}, P_4^{PE}(x) = 9 + \frac{561}{105}x + \frac{141}{105}x^2 + \frac{18}{105}x^3 + \frac{1}{105}x^4$	$\gamma^\#(0) = 0$	$\gamma^\#(\infty) = -1$	(A93)
Climacospectrum	$\zeta(k) := \frac{k\gamma(k) - \gamma(2k)}{\ln 2}$	$\zeta^\#(0) = 2, \; n = 0$ $\zeta^\#(0) = 3, \; n > 0$	$\zeta^\#(\infty) = 0$	(A94)
Entropy production	$\varphi(k) = \frac{2\lambda}{\gamma(k)(k/\alpha)}\left(a_n^{PE} - e^{-k/\alpha} R_n^{PE}(k/\alpha)\right)$ where a_n^{PE} are coefficients as above and $R_n^{PE}(x)$ are polynomials determined using (39); the first five are: $R_0^{PE}(x) = 1$ $R_1^{PE}(x) = 2 + x$ $R_2^{PE}(x) = \frac{8}{3} + \frac{5}{3}x + \frac{1}{3}x^2$ $R_3^{PE}(x) = \frac{48}{15} + \frac{33}{15}x + \frac{9}{15}x^2 + \frac{1}{15}x^3$ $R_4^{PE}(x) = \frac{384}{105} + \frac{279}{105}x + \frac{87}{105}x^2 + \frac{14}{105}x^3 + \frac{1}{105}x^4$ $\varphi_C(k) = \frac{\varphi(k)\gamma(k)-\varphi(2k)\gamma(2k)}{\gamma(k)-\gamma(2k)}$	$\varphi(0) = 1$ $\varphi_C(0) = 3/2, \; n = 0$ $\varphi_C(0) = 2, \; n > 0$	$\varphi(\infty) = \varphi_C(\infty) = 1/2$	(A95)
Autocovariance function	$c(h) = \lambda e^{-h/\alpha} Q_n^{PE}(h/\alpha)$ where $Q_n^A(x)$ is a polynomial determined by the power spectrum (see below) using (22); the first five are: $Q_0^{PE}(x) = 1$ $Q_1^A(x) = 1 + x$ $Q_2^A(x) = 1 + x + \frac{1}{3}x^2$ $Q_3^A(x) = 1 + x + \frac{6}{15}x^2 + \frac{1}{15}x^3$ $Q_4^A(x) = 1 + x + \frac{45}{105}x^2 + \frac{10}{105}x^3 + \frac{1}{105}x^4$	$c^\#(0) = 0$	$c^\#(\infty) = -\infty$	(A96)
Power spectrum	$s(w) = \frac{4^{n+1}(n!)^2}{(2n)!} \frac{\lambda\alpha}{\left(1 + (2\pi\alpha w)^2\right)^{n+1}}$	$-s^\#(\infty) = 2(n+1)$	$-s^\#(0) = 0$	(A97)

Entropy **2017**, *19*, 581

Table A6. Mathematical expressions of the second order characteristics of the AE process and their asymptotic properties.

Characteristic	Mathematical Expression	Local Asymptotic Properties	Global Asymptotic Properties	Equation
Climacogram	$\gamma(k) = \frac{2\lambda}{(k/\alpha)^2}\left(\frac{1}{2n-1} - e^{-k/\alpha}\, P_n^{AE}(k/\alpha)\right)$ where $n = 0, 1, 2 \ldots$, $P_0^{AE}(x) = -1 - xe^x$ while for $n > 0$ $P_n^{AE}(x)$ is a polynomial determined using (23); the first four are: $P_1^{AE}(x) = 1 + x$ $P_2^{AE}(x) = \frac{1}{3} + \frac{1}{3}x - \frac{1}{3}x^2$ $P_3^{AE}(x) = \frac{3}{15} + \frac{3}{15}x - \frac{6}{15}x^2 + \frac{1}{15}x^3$ $P_4^{AE}(x) = \frac{15}{105} + \frac{15}{105}x - \frac{45}{105}x^2 + \frac{14}{105}x^3 - \frac{1}{105}x^4$	$\gamma^{\#}(0) = 0$	$\gamma^{\#}(\infty) = -1,\ n = 0$ $\gamma^{\#}(\infty) = -2,\ n > 0$	(A98)
Climacospectrum	$\zeta(k) := \frac{k(\gamma(k) - \gamma(2k))}{\ln 2}$	$\zeta^{\#}(0) = 2$	$\zeta^{\#}(\infty) = 0,\ n = 0$ $\zeta^{\#}(\infty) = -1,\ n > 0$	(A99)
Entropy production	$\varphi(k) = \frac{2\lambda e^{-k/\alpha}}{\gamma(k)}\, R_n^{AE}(k/\alpha)$ where $R_0^{AE}(x) = \frac{-1-e^x}{h}$, while for $n > 0$ $R_n^{AE}(x)$ are polynomials determined using (39); the first four are: $R_1^{PE}(x) = 1$ $R_2^{PE}(x) = 1 - \frac{1}{3}x$ $R_3^{PE}(x) = 1 - \frac{9}{15}x + \frac{1}{15}x^2$ $R_4^{PE}(x) = 1 - \frac{87}{105}x + \frac{18}{105}x^2 - \frac{1}{105}x^3$ $\varphi_C(k) = \frac{\varphi(k)\gamma(k) - \varphi(2k)\gamma(2k)}{\gamma(k) - \gamma(2k)}$	$\varphi(0) = 1$ $\varphi_C(0) = 3/2$	$\varphi(\infty) = \varphi_C(\infty) = 1/2,\ n = 0$ $\varphi(\infty) = \varphi_C(\infty) = 0,\ n > 0$	(A100)
Autocovariance function	$c(h) = \lambda e^{-h/\alpha}\, Q_n^{A}(h/\alpha)$ where $Q_n^{A}(x)$ is a polynomial determined by the power spectrum (see below) using (22); the first five are: $Q_0^{A}(x) = 1$ $Q_1^{A}(x) = 1 - x$ $Q_2^{A}(x) = 1 - \frac{5}{3}x + \frac{1}{3}x^2$ $Q_3^{A}(x) = 1 - \frac{33}{15}x + \frac{12}{15}x^2 - \frac{1}{15}x^3$ $Q_4^{A}(x) = 1 - \frac{279}{105}x + \frac{141}{105}x^2 - \frac{22}{105}x^3 - \frac{1}{105}x^4$	$c^{\#}(0) = 0$	$c^{\#}(\infty) = -\infty$	(A101)
Power spectrum	$s(w) = \frac{4^{n+1}(n!)^2}{(2n)!}\, \frac{\lambda\alpha(2\pi\alpha w)^{2n}}{\left(1+(2\pi\alpha w)^2\right)^{n+1}}$	$-s^{\#}(\infty) = 2$	$-s^{\#}(0) = -2n$	(A102)

Table A7. Mathematical expressions of the second order characteristics of the Generalized Planck (GP) process and their asymptotic properties.

Characteristic	Mathematical Expression	Local Asymptotic Properties	Global Asymptotic Properties	Equation
Climacogram	$\gamma(k) = \frac{\gamma}{(\pi k/\alpha)} P_n^{GP}(\pi k/\alpha)$ where $n = 0, 1, 2, \ldots$, $P_n^{GP}(x)$ is a function determined using (23) the first three are: $P_0^{GP}(x) = 3\left(\ln\frac{e^{2x}-1}{2x} - x\right)$ $P_1^{GP}(x) = \frac{5}{2}\left(1 - \frac{3}{x^2} + \frac{3}{\sinh x}\right)$ $P_2^{GP}(x) = \frac{21}{40}\left(1 - \frac{45}{x^4} - \frac{30-2\cosh 2x}{\sinh^4 x}\right)$	$\gamma^\#(0) = 0$	$\gamma^\#(\infty) = -1$ $n = 0$ $\gamma^\#(\infty) = -2$ $n > 0$	(A103)
Climacospectrum	$\zeta(k) := \frac{k(\gamma(k)-\gamma(2k))}{\ln 2}$	$\zeta^\#(0) = 3$	$\zeta^\#(\infty) = 0$ $n = 0$ $\zeta^\#(\infty) = -1$ $n > 0$	(A104)
Entropy production	$\phi(k)$ is determined from (39); the first three terms are: $\phi_0(k) = \frac{1-(\pi k/\alpha)\coth(\pi k/\alpha)}{2\pi k/\alpha - 2\ln(e^{2\pi k/\alpha}-1)/(2\pi k/\alpha)}$ $\phi_1(k) = \frac{3-3(\pi k/\alpha)^3\coth(\pi k/\alpha)}{3(\pi k/\alpha)^2+((\pi k/\alpha)^2-3)\sinh^2(\pi k/\alpha)}$ $\phi_2(k) = \frac{180-15(\pi k/\alpha)^5(11\cosh(\pi k/\alpha)+\cosh(3\pi k/\alpha))}{30(\pi k/\alpha)^4(2+\cosh(2\pi k/\alpha))\sinh(\pi k/\alpha)-(2(\pi k/\alpha)^4+90)\sinh^5(\pi k/\alpha)}$ $\phi_C(k) = \frac{\phi(k)\gamma(k)-\phi(2k)\gamma(2k)}{\gamma(k)-\gamma(2k)}$	$\phi(0) = 1$ $\phi_C(0) = 2$	$\phi(\infty) = \phi_C(\infty) = 1/2, n = 0$ $\phi(\infty) = \phi_C(\infty) = 0, n > 0$	(A105)
Autocovariance function	$c(h) = \frac{\lambda}{(\pi k/\alpha)^2} Q_n^{GP}(\pi k/\alpha)$ where $Q_n^A(x)$ is a function determined from (22); the first three are: $Q_0^{GP}(x) = 3\left(1 - \frac{x^2}{\sinh^2 x}\right)$ $Q_1^{GP}(x) = -\frac{45}{x^2} + \frac{15x^2(2+\cosh 2x)}{\sinh^4 x}$ $Q_2^{GP}(x) = \frac{945}{2x^4} - \frac{x^2(2079+1648\cosh 2x + 63\cosh 4x)}{8\sinh^6 x}$	$c^\#(0) = 0$	$c^\#(\infty) = -2(n+1)$	(A106)
Power spectrum	$s(w) = \frac{c_n\lambda\alpha(\alpha w)^{2n+1}}{e^{2\pi\alpha w}-1}$ where c_n is a normalizing constant making the area of the power spectrum $\gamma_0 = \lambda$; the first three are $c_0 = 24$, $c_1 = 240$, $c_2 = 504$	$-s^\#(\infty) = \infty$	$-s^\#(0) = -2n$	(A107)

Appendix A.2.8. Harmonic Oscillation Treated as a Stochastic Process

A harmonic oscillation, here expressed as

$$x(t) = \sqrt{2\lambda} \, \cos\left(2\pi(t+b)/T\right), \ 0 \leq b \leq T \tag{A108}$$

is very easily modelled as a deterministic process but, when it is superimposed to a stochastic process, the resulting process is a stochastic process, too. Obviously the second-order characteristics of this composite process are affected by periodic components and therefore we need to know which equations should be superimposed to those of the stochastic component (see also [40,41]). These equations are given in Table A8. Remarkably, the climacogram retains information about the phase $(2\pi b/T)$ of the process (through the spikes appearing at $k = (m + 1/2)T$, $m \in N_0$, which depend on the phase). Note that this information is completely lost in the autocovariance function and the power spectrum.

An application is depicted in Figure A10. The example process is the average of a FHK with $H = M = 0.8$ and a harmonic oscillation. It is difficult to identify the presence of the oscillation from visual inspection of the time series, but the detailed graph of any of the second-order characteristics, plotted in Figure A10, captures the periodic behaviour. Note that for the application, the generation was done by the SMA model for the entire (composite) process and not by separating the generation of the stochastic and the deterministic component. This is an indication of the flexibility and generality of the generation algorithm.

Table A8. Mathematical expressions of the second order characteristics of a harmonic oscillation treated as a stochastic process and their asymptotic properties.

Characteristic	Mathematical Expression	Local Asymptotic Properties	Global Asymptotic Properties	Equation
Climacogram	$\gamma(k) = \begin{cases} \frac{\lambda T^2}{\pi^2 k^2} \sin^2\left(\frac{\pi k}{T}\right), & k \neq \left(m+\frac{1}{2}\right)T \\ \frac{2\lambda T^2}{\pi^2 k^2} \sin^2\left(\frac{2\pi b}{T}\right), & k = \left(m+\frac{1}{2}\right)T \end{cases}$ where T is the period, $m \in N_0$ and the harmonic oscillation is described by $x(t) = \sqrt{2\lambda} \, \cos(2\pi(t+b)/T), 0 \leq b \leq T$	$\gamma^\#(0) = 0$	$\gamma^\#(\infty) = -2$	(A109)
Climacospectrum $\zeta(k) := \frac{k(\gamma(k) - \gamma(2k))}{\ln 2}$		$\zeta^\#(0) = 3$	$\zeta^\#(k)$ does not converge as $k \to \infty$	(A110)
Entropy production	$\varphi(k) = (\pi k/T) \cot(\pi k/T)$ $\varphi_C(k) = (2\pi k/T)\cot(\pi k/T)$ (for $k \neq \left(m+\frac{1}{2}\right)T)$	$\varphi(0) = 1$ $\varphi_C(0) = 2$	$\varphi(k)$ and $\varphi_C(k)$ do not converge as $k \to \infty$	(A111)
Autocovariance function	$c(h) = \lambda \cos(2\pi h/T)$	$c^\#(0) = 0$	$c^\#(k)$ does not converge as $k \to \infty$	(A112)
Power spectrum	$s(w) = \lambda T \delta(Tw - 1)$	(not applicable)		(A113)

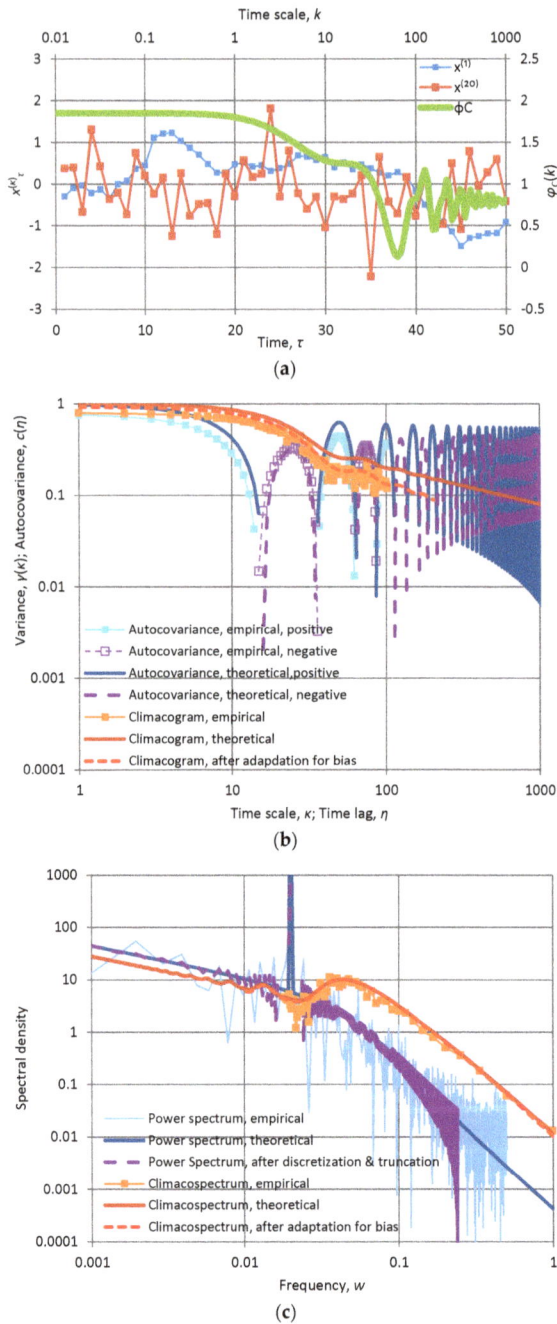

Figure A10. Characteristic graphs as in Figure A2 but for a time series generated by the FHK-C model with $\lambda = 1$, $\alpha = 10$, and $H = M = 0.8$ in which a harmonic oscillation with $T = 5\alpha = 50$ is superimposed.

References

1. Boltzmann, L. Über die Beziehung zwischen dem zweiten Hauptsatze der mechanischen Wärmetheorie und der Wahrscheinlichkeitsrechnung respektive den Sätzen über das Wärmegleichgewicht. *Wien. Ber.* **1877**, *76*, 373–435. (In German)
2. Shannon, C.E. The mathematical theory of communication. *Bell Syst. Tech. J.* **1948**, *27*, 379–423. [CrossRef]
3. Jaynes, E.T. Information theory and statistical mechanics. *Phys. Rev.* **1957**, *106*, 620–630. [CrossRef]
4. Sonuga, J.O. Principle of maximum entropy in hydrologic frequency analysis. *J. Hydrol.* **1972**, *17*, 177–191. [CrossRef]
5. Jackson, D.R.; Aron, G. Parameter estimation in hydrology: The state of the art. *J. Am. Water Resour. Assoc.* **1971**, *7*, 457–472. [CrossRef]
6. Singh, V.P.; Fiorentino, M. A Historical Perspective of Entropy Applications in Water Resources. In *Entropy and Energy Dissipation in Water Resources*; Singh, V.P., Fiorentino, M., Eds.; Springer: Dordrecht, The Netherlands, 1992.
7. Singh, V.P. The use of entropy in hydrology and water resources. *Hydrol. Process.* **1997**, *11*, 587–626. [CrossRef]
8. Singh, V.P. Hydrologic synthesis using entropy theory: review. *J. Hydrol. Eng.* **2011**, *16*, 421–433. [CrossRef]
9. Singh, V.P.; Cui, H. Entropy Theory for Groundwater Modeling. Available online: http://www.aggs.in/Issues/JGWR-2015101.pdf (accessed on 24 October 2017).
10. Singh, V.P.; Cui, H. Entropy theory for streamflow forecasting. *Environ. Process.* **2015**, *2*, 449–460. [CrossRef]
11. Singh, V.P. Entropy theory. In *Handbook of Applied Hydrology*, 2nd ed.; Singh, V.P., Ed.; McGraw Hill: New York, NY, USA, 2017.
12. Porporato, A.; Rigby, J.R.; Daly, E. Irreversibility and fluctuation theorem in stationary time series. *Phys. Rev. Lett.* **2007**, *98*, 94101. [CrossRef] [PubMed]
13. Kondepudi, D.; Prigogine, I. *Modern Thermodynamics*; Wiley: Chichester, UK, 1998.
14. Pujol, T.; Llebot, J.E. Extremal principle of entropy production in the climate system. *Q. J. R. Meteorol. Soc.* **1999**, *125*, 79–90. [CrossRef]
15. Ozawa, H.; Ohmura, A.; Lorenz, R.D.; Pujol, T. The second law of thermodynamics and the global climate system: A review of the maximum entropy production principle. *Rev. Geophys.* **2003**, *41*. [CrossRef]
16. Prigogine, I. Moderation et transformations irreversibles des systemes ouverts. *Bull. Classe Sci. Acad. R. Belg.* **1945**, *31*, 600–606.
17. Ziegler, H. *Progress in Solid Mechanics*; Sneddon, I.N., Hill, R., Eds.; North-Holland: Amsterdam, The Netherlands, 1963; Volume 4.
18. Martyushev, L.M.; Seleznev, V.D. Maximum entropy production principle in physics, chemistry and biology. *Phys. Rep.* **2006**, *426*. [CrossRef]
19. Niven, R. Entropy production extremum principles. In *Handbook of Applied Hydrology*, 2nd ed.; Singh, V.P., Ed.; McGraw Hill: New York, NY, USA, 2017.
20. Koutsoyiannis, D. HESS Opinions: A random walk on water. *Hydrol. Earth Syst. Sci.* **2010**, *14*, 585–601. [CrossRef]
21. Montanari, A.; Koutsoyiannis, D. A blueprint for process-based modeling of uncertain hydrological systems. *Water Resour. Res.* **2012**, *48*, W09555. [CrossRef]
22. Koutsoyiannis, D. Hurst-Kolmogorov dynamics as a result of extremal entropy production. *Phys. A Stat. Mech. Appl.* **2011**, *390*, 1424–1432. [CrossRef]
23. Hemelrijk, J. Underlining random variables. *Stat. Neerl.* **1966**, *20*. [CrossRef]
24. Koutsoyiannis, D. Generic and parsimonious stochastic modelling for hydrology and beyond. *Hydrol. Sci. J.* **2016**, *61*, 225–244. [CrossRef]
25. Koutsoyiannis, D.; Montanari, A. Negligent killing of scientific concepts: The stationarity case. *Hydrol. Sci. J.* **2015**, *60*, 1174–1183. [CrossRef]
26. Kolmogorov, A.N. A simplified proof of the Birkhoff-Khinchin ergodic theorem. *Uspekhi Mat. Nauk* **1938**, *5*, 52–56.
27. Papoulis, A. *Probability, Random Variables and Stochastic Processes*, 3rd ed.; McGraw-Hill: New York, NY, USA, 1991.

28. Dimitriadis, P.; Koutsoyiannis, D. Climacogram versus autocovariance and power spectrum in stochastic modelling for Markovian and Hurst–Kolmogorov processes. *Stoch. Environ. Res. Risk Assess.* **2015**, *29*, 1649–1669. [CrossRef]

29. Koutsoyiannis, D. Physics of uncertainty, the Gibbs paradox and indistinguishable particles. *Stud. Hist. Philos. Mod. Phys.* **2013**, *44*, 480–489. [CrossRef]

30. Koutsoyiannis, D. Entropy: from thermodynamics to hydrology. *Entropy* **2014**, *16*, 1287–1314. [CrossRef]

31. Koutsoyiannis, D.; Dimitriadis, P.; Lombardo, F.; Stevens, S. From fractals to stochastics: Seeking theoretical consistency in analysis of geophysical data. In *Advances in Nonlinear Geosciences*; Tsonis, A., Ed.; Springer: New York, NY, USA, 2018.

32. Dechant, A.; Lutz, E. Wiener-Khinchin theorem for nonstationary scale-invariant processes. *Phys. Rev. Lett.* **2015**, *115*, 80603. [CrossRef] [PubMed]

33. Mandelbrot, B. *Gaussian Self-Affinity and Fractals: Globality, the Earth, 1/f Noise, and R/S (Vol. 8)*; Springer: New York, NY, USA, 2002.

34. Kitanidis, P.K. Geostatistics. In *Handbook of Hydrology*; Maidment, D.R., Ed.; McGraw-Hill: New York, NY, USA, 1993.

35. Gaetan, C.; Guyon, X. *Spatial Statistics and Modeling*; Springer: New York, NY, USA, 2010.

36. Clark, I. Statistics or geostatistics? Sampling error or nugget effect? *J. South. Afr. Inst. Min. Metall.* **2010**, *110*, 307–312.

37. Gneiting, T.; Schlather, M. Stochastic models that separate fractal dimension and the Hurst effect. *SIAM Rev.* **2004**, *46*, 269–282. [CrossRef]

38. Keshner, M.S. 1/f noise. *Proc. IEEE* **1982**, *70*, 212–218. [CrossRef]

39. Wornell, G.W. Wavelet-based representations for the 1/f family of fractal processes. *Proc. IEEE* **1993**, *81*, 1428–1450. [CrossRef]

40. Markonis, Y.; Koutsoyiannis, D. Climatic variability over time scales spanning nine orders of magnitude: Connecting Milankovitch cycles with Hurst–Kolmogorov dynamics. *Surv. Geophys.* **2013**, *34*, 181–207. [CrossRef]

41. Pappas, C.; Mahecha, M.D.; Frank, D.C.; Babst, F.; Koutsoyiannis, D. Ecosystem functioning is enveloped by hydrometeorological variability. *Nat. Ecol. Evol.* **2017**, *1*, 1263–1270. [CrossRef] [PubMed]

42. Kang, H.S.; Chester, S.; Meneveau, C. Decaying turbulence in an active-grid-generated flow and comparisons with large-eddy simulation. *J. Fluid Mech.* **2003**, *480*, 129–160. [CrossRef]

43. Dimitriadis, P.; Koutsoyiannis, D. Stochastic synthesis approximating any process dependence and distribution. *Stoch. Environ. Res. Risk Assess.* **2017**. submitted.

44. Jaynes, E.T. *Probability Theory: The Logic of Science*; Cambridge University Press: Cambridge, UK, 2003; p. 728.

45. Swendsen, R.H. How physicists disagree on the meaning of entropy. *Am. J. Phys.* **2011**, *79*, 342–348. [CrossRef]

46. Stein, M.L. *Interpolation of Spatial Data*; Springer: New York, NY, USA, 1999.

47. Beran, J. *Statistics for Long-Memory Processes*; Chapman and Hall: New York, NY, USA, 1994.

48. Taqqu, M.S. Fractional Brownian motion and long-range dependence. In *Theory and Applications of Long-Range Dependence*; Doukhan, P., Oppenheim, G., Taqqu, M.S., Eds.; Birkhäuser: Boston, MA, USA, 2003.

MDPI

St. Alban-Anlage 66

4052 Basel

Switzerland

Tel. +41 61 683 77 34

Fax +41 61 302 89 18

www.mdpi.com

Entropy Editorial Office

E-mail: entropy@mdpi.com

www.mdpi.com/journal/entropy

www.ingramcontent.com/pod-product-compliance
Lightning Source LLC
Chambersburg PA
CBHW051701210326
41597CB00032B/5328